T0186671

ESTUARINE ECOLOGY

Mngazana, a mangrove estuary (Photo: A.E.F.Heydorn)

Estuarine Ecology

with particular reference to southern Africa

Edited by

J.H. Day D.F.C., Ph.D., D.Sc., F.L.S., F.R.S. (S.A.)

A.A.Balkema / Rotterdam / 1981

Copyright is retained by the writers of the individual chapters.
A.A.Balkema P.O.Box 1675 NL 3000 BR Rotterdam Netherlands
ISBN 90 6191 205 9
Printed in the Netherlands

Contents

LIST OF PHOTOGRAPHS

Preface

World-wide alarm at the effects of pollution and the decline of marine fisheries has focussed attention on estuaries. Estuarine ecology is of recent development and the reactions within the ecosystem are complex. Can we learn how to dispose of sewage without damaging the ecosystem? Can we use these highly productive waters effectively for aquaculture and as nurseries for the sea? The need to answer these questions is urgent.

Marine biologists in Europe and America became interested in estuaries in the 1930's and the number of publications increased rapidly after the war. A review of many of these early studies and references to many of the important papers will be found in Day (1951). Remane (1934) studying the fauna of the Danish waddens, Hartley & Spooner working on the Tamar, Bassindale (1938, 1943) working on the Mersey and the Severn, related faunistic changes to decreased salinities. Hedgpeth (1947) working on the Laguna Madre found similar changes in hypersaline waters. Studies of osmoregulation and other adaptations to salinity variations followed, based on the early work of Krogh in Denmark. However, it soon became evident that salinity variation is not the only important factor. Nicol (1935) had already shown that the nature of the substrate is at least as important as the salinity. Later Holme (1949) found that clean, coarse sand at the mouth of the Exe has a poor fauna while silty sand within the estuary has a much richer fauna. Apparently some factor associated with silt is important although Percival (1929) had shown that very soft, slurry mud carries a poor macrofauna.

As early as 1911, Petersen & Jensen had shown that the main source of food in the Danish wadden is the detritus derived from the decay of *Zostera* and other salt marsh plants. MacGinitie (1935) working on Elkhorn Slough in California drew attention to the predominance of detritus feeders among benthic animals and suggested that bacteria are an important source of food.

It has long been known that unlike coastal seas, most estuaries have poor plankton populations. This appeared surprising for estuaries are richer in nutrients than the sea. It is only in deep, slow-flowing estuaries such as Chesapeake Bay and the almost tideless Norwegian fjords where the turbidity is minimal that large plankton populations develop. In retrospect it is obvious that the current system which determines the distribution of salinity, the suspension and distribution of sediments, and the rate at which estuarine water is transported to the sea, is of basic importance in the ecology of estuaries.

Mathematical formulations of estuarine circulation were developed by Stommel, Pritchard and their associates in the early 1950's. The origin of estuarine sediments had been investigated by Dutch workers including van Veen (1936) and Crommelin (1940, 1948). Postma (1954) and later van Straaten (1960), published a series of papers explaining the distribution of these sediments. Meanwhile Day (1951) drew attention to the deterioration of an estuarine fauna when the mouth of the estuary is closed by littoral drift. It remained for Brunn & Gerritsen (1960) to determine the mathematical relationship between littoral drift, the velocity of the currents through tidal inlets and the cross-sectional area of the mouth.

These are but a few of the papers on various aspects of estuarine research. The rate of research was accelerating and an exchange of views across the board was needed. The First International Conference on Estuaries was held in 1964 on Jekyll Island in Georgia. The island had been a pirate's nest but at the time of the conference the sale of liquor was strictly controlled. But not too strictly for Joel Hedgpeth in his concluding remarks found that the spirit of the meeting was high. Tongues were well-oiled and international discussions were stimulating. The proceedings of the conference edited by George Lauff appeared in 1967 under the title *Estuaries*. It revealed the complex reactions between the many scientific disciplines needed in estuarine research. It also stressed the inadequacy of our knowledge of estuarine microbiology and the many biochemical reactions facilitated by these organisms.

Since then physical, chemical and sedimentological

research has developed rapidly and there has been a more quantitative approach to biological work. Systems ecology is developing with the aim of predicting the effects of human activities on the ecosystem. Many papers on these lines are discussed in the chapters that follow. I am indebted to the authors and publishers of several books and journals for permission to reproduce many diagrams and tables in this book. Detailed references to these will be found at the ends of the relevant chapters.

Estuarine research in southern Africa commenced in 1947 with the study of Knysna, Hermanus Lagoon and the St Lucia system. These pioneer studies soon revealed that there are several types of estuaries in southern Africa some of which differ from those in Europe and the United States. While pollution is minimal, many estuaries are badly silted and others are closed by sandbars in the dry months and their biological characteristics change. Such estuaries provide a complementary field of study to those in Europe.

Work has spread from the University of Cape Town to many institutions. The National Institute of Water Research (NIWR) is investigating the chemistry of rivers, estuaries and the sea as part of the international programme for monitoring pollution. The National Research Institute for Oceanology (NRIO) has studied the development of sandbars in those inlets where harbours are needed and is now turning to pollution research in the Cape. Biologists at the coastal universities of Cape Town, Port Elizabeth and Natal as well as the Oceanographic Research Institute at Durban (ORI), the Natal Parks Board and the Cape Department of Nature and Environmental Conservation are studying the biomass and productivity of estuarine plants and animals. Rhodes University is doing similar work on rivers and coastal lakes. The ultimate aim of course is to lay the foundations for systems ecology. Much of this work is funded by the CSIR through the South African National Committee for Oceanographic Research (SANCOR).

In 1976, SANCOR decided to review what had been learnt of the estuaries of southern Africa. The review was to be based on international researches so as to reveal the significance of observations in southern Africa and the gaps in our knowledge. The aim was to provide a firm basis for further research and, even more important, to provide a guide for management. This book is the result. It has been written for senior students and research workers in estuarine ecology. Chapters 1 to 5 deal with environmental conditions; chapters 6 to 13 deal with different biological groups; chapters 14 and 15 summarise what is known of a series of estuaries and the coastal lakes of southern Africa. Chapter 16 is an attempt to analyse the interrelations within estuarine ecosystems and the constraints which limit their biota. Chapter 17 is quite different. It deals with the management of estuaries based on the evidence presented in earlier chapters. It is written in non-technical terms for busy legislators and administrators who do not have the time to read the whole book.

Funds were provided by the CSIR and SANCOR commissioned me to act as editor. As many specialists were busy with their researches I have written several chapters which others would have been more competent to undertake. But the book is really a team effort. Original data and unpublished reports were kindly supplied by the many institutions mentioned earlier. Chapter 2 on coastal hydrodynamics and the stability of inlets is really the work of Mr Jan Rossouw, Dr Peter Swart and Mr Kenneth Russell of NRIO but they have modestly left my name as the author. Other officers of NRIO have also been most helpful. Dr Burg Flemming provided constructive criticisms for the section on sediments, and both Dr Michael Orren and Dr Garth Eagle have vetted the chemical section. The Bolus Herbarium identified many of the plants and Dr Eugene Moll of the Botany Department of the University of Cape Town has helped with chapter 6. Many workers provided the notes on which the summaries of 43 estuaries in chapter 14 are based. Staff and students of the University of Cape Town took part in the early estuarine surveys; later Dr Dan Baird of the University of Port Elizabeth, Dr Burke Hill of Rhodes University and Mrs Ann Palmer of the same institution provided notes on the estuaries in the Eastern Cape. George Begg of ORI not only allowed me a preview of his book *The Estuaries of Natal,* but also took me on a conducted tour of estuaries near Durban and corrected many mistakes in my earlier drafts. I should like to take this opportunity of thanking them all. My special thanks are due to Dr George Branch who has made helpful comments on most of the chapters and Dr Allan Heydorn who has been a tower of strength in many ways.

Most of all I wish to thank the authors and co-authors of the various chapters. The scope of the book would have been very limited without their contributions. Professor John Grindley, Professor Brian Allanson and Dr Burke Hill have also helped in many other ways.

So many people have helped to prepare the manuscript that it is impossible to name them all. I am particularly grateful to Mrs M.I. Cousins for typing many drafts and Mrs Sandra Hardman and Mr Rod Bally for secretarial assistance. Miss Ann Westoby prepared many of the illustrations and as the captions show, the photographs were taken by many friends.

Finally I wish to thank the CSIR and the University of Cape Town for financial assistance during the three and a half years it has taken to edit the book and for generous subventions towards the cost of publication.

J.H. Day

University of Cape Town
30 November 1979

Chapter 1

The nature, origin and classification of estuaries

J.H. Day

Department of Zoology, University of Cape Town

DEFINITION OF AN ESTUARY

To most people an estuary is simply the saline mouth of a river where it meets the sea. The word is derived from the Latin *aestuarium* or tidal inlet of the sea. This omits any reference to a river and permits the inclusion of deep bays and wholly marine lagoons. Webster's dictionary is more specific: 'An estuary is a water passage where the tide meets a river current; an arm of the sea at the lower end of a river'. Again there is a stress on tidal conditions and no reference to salinity. Thus the Amazon which is tidal for 480 km but lacks any saline intrusion would be regarded as an estuary while rivers which flow into tideless seas such as the Black Sea would not. In the First International Conference on Estuaries held in Georgia in 1964, a confusing array of definitions was proposed and these will be found in the proceedings of the conference edited by George Lauff (1967) under the title *Estuaries*. Physicists proposed definitions based on current systems and salinity variations; geologists stressed the origin of estuaries and the distribution of sediments while biologists were concerned with the nature of the fauna and flora. McHugh (1967) even discussed the estuarine fisheries of the Bering Straits and other areas of reduced salinity such as the Bay of Bengal. The word 'estuary' seemed to be losing any meaning but nonetheless there was wide agreement that variable salinity is an essential feature of all estuaries and the definition proposed by Pritchard (1967) was gratefully accepted. According to Pritchard:

'An estuary is a semi-enclosed coastal body of water which has a free connection with the open sea and within which sea water is measurably diluted with fresh water derived from land drainage'. This definition thus excludes saline lakes whose waters contain salts with a different chemical composition from that of the sea. It also excludes marine inlets and lagoons without fresh water inflow such as Langebaan Lagoon in South Africa and inlets on arid coasts whose salinity is the same as the sea. According to Pritchard, the Baltic whose salinity is effectively stable over wide areas, is not an estuary but this is not accepted by many Scandinavian workers. A point of practical importance is that the definition excludes the tidal reaches of rivers beyond the limit of saline intrusion. This point not only marks a change in the fauna and flora but also a change from brak water to fresh water suitable for drinking and irrigation. In a number of countries the legal definition of an estuary includes all tidal waters and difficulties arise in regard to administrative control, conservation and the rights of farmers.

There are, however, two difficulties with Pritchard's definition. The phrase '. . . a free connection with the open sea . . .' would exclude those estuaries which are temporarily cut off from the sea during the dry season. Secondly, the phrase '. . . diluted with fresh water . . .' would exclude estuaries which, far from being diluted sea water, become hypersaline when evaporation exceeds fresh water inflow. In southern Africa, Western and Southern Australia and other arid countries, there are many 'blind' estuaries which are closed by

1

Merimbula, New South Wales; a coastal plain estuary (Photo: J.H. Day)

sandbars for longer or shorter periods. All of them burst open to the sea during floods but thereafter some of them become hypersaline. St Lucia in Natal and the Laguna Madre in Texas are well-known examples. For these reasons the amended definition given by Day, 1980 is:

An estuary is a partially enclosed coastal body of water which is either permanently or periodically open to the sea and within which there is a measurable variation of salinity due to the mixture of sea water with fresh water derived from land drainage.

THE ORIGIN OF ESTUARIES

Although several geological events have contributed to the formation of estuaries, most of them are drowned river valleys due to eustatic changes of sea level. During the major glaciations of the Pleistocene, over one-third of the land surface was covered with ice one to two kilometres thick and the sea surface on different coasts was 100 to 150 m below present sea level, corresponding approximately with the edge of the continental shelf. According to different authorities, this was between 16 000 and 30 000 years ago. Since then the climate has fluctuated but become measurably warmer and as the ice melted the sea level rose about one metre per century. Raised beaches and wave-cut terraces at different levels both above and below present sea level indicate that the climatic changes were irregular and the consequent rise or fall in sea level was arrested at different times. There was a major arrest about 3 000 to 5 000 years ago (Russell 1967), and Steers (1967) suggests that the minor changes in level which occurred subsequently were mainly isostatic. In high latitudes, the Pleistocene glaciers eroded gorges that were rectangular in section and deposited the spoil as terminal moraines where they met the sea. These deep gorges now form the fjords of Norway, Canada and Western New Zealand. Many of them are 50 to 500 m deep although the sill formed by the moraine at the entrance of the fjord may be fairly shallow. In lower latitudes, the Pleistocene rivers eroded V-shaped valleys to the edge of the continental shelf. With the rise in sea level the lower ends of these valleys were completely submerged. Some remain as shelf valleys leading to submarine canyons at the edge of the continental shelf while others have been filled with sediment. The upper valleys were only partly submerged and these drowned valleys now form the majority of estuaries. Many have been filled with sediment high above present sea level but this has been eroded and redeposited many times as the sea level changed. Most estuaries are only a few metres deep but the bed rock of the original channel may be buried more than 70 m below the sediment.

The drowning of river valleys to form coastal plain estuaries may be simply due to a rise in sea level in the post-glacial period. Krige (1927) states that the South African coast is: 'typically a young to mature coast-

Doubtful Sound; a New Zealand fjord (Photo: NZ Oceanographic Institute)

line of submergence modified by later emergence phenomena . . .' and speaking of the rivers he says: 'Although not recognisable from the map as such, practically all South African estuaries are true drowned valleys more or less silted up'. Alternatively the drowning of some river valleys may be due to warping of the continental margin with a downward tilt of the coastal plain. King & King (1959) suggest that this occurred in Natal. A few estuaries including San Francisco Bay, have been formed by block faulting on the continental margin with subsidence of the graben between nearby fault zones. Bar-built estuaries have been formed from bays in shallow seas by littoral drift. Here an off-shore line of breakers builds up a bar across the mouth of the bay and sandspits grow out from the promontories on either side. Once such a bar reaches intertidal levels, winds blow the drying sand into dunes and these are later stabilized by vegetation. Storm waves breach the barrier and a chain of barrier islands is formed. The gaps act as tidal channels for the enclosed waters which form a sound parallel to the original coast. Rivers flowing into the sound reduce its salinity and a bar-built estuary may sometimes have several rivers flowing into it and several exits to the sea. A whole series of estuarine sounds has developed along the Atlantic coast of the United States and along much of the coastline of the Gulf of Mexico. A similar complex series of estuaries forms the Wadden Zee from Holland to the mouth of the Elbe and there are others on the coast of Nigeria, the Bay of Bengal and South Australia. Gorsline (1967) describes estuaries east of the Mississippi delta as having two rows of lagoons running parallel to the seashore with a sand ridge partially

separating the inner series from the outer. Jennings & Bird (1967) describe similar formations at the mouth of the Murray River in South Australia and in the Gippsland Lakes in Victoria.

According to van Heerden (1976), several geological events contributed to the formation of Lake St Lucia on the Zululand coast. Until the break-up of Gondwanaland, southern Africa was completely landlocked. There is evidence of coastal erosion in the early Cretaceous. Thus exposures of Lower Cretaceous sediments include fluvial and beach deposits while the Upper Cretaceous rocks include shales and sandy limestones with a rich ammonite fauna laid down in fairly deep water. All this is evidence of a progressive marine transgression due either to a rise of sea level or to a down-warping of the continental margin. The Middle and Upper Tertiary beds include 'sublittoral beach and coastal dune deposits (laid down) during marine regression' (van Heerden 1976). For further details refer to the description of the St Lucia System in chapter 14.

Orme (1973) has given an interesting account of the formation of the coastal sand dunes along the Zululand coast and the estuarine lagoons which have developed behind them. Many of them are now silting up rapidly. As Gorsline (1967) remarks, estuaries are transient features in the geological time scale and the length of life of an estuary is determined by the rate of sedimentation. Many estuaries are probably no more than 3 000 years old and within recent years erosion in the drainage basin and sedimentation in the estuary has been accelerated by bad agricultural practices and the canalisation of swamps.

West and East Kleinemond; two temporarily closed estuaries (Photo: A.E.F. Heydorn)

THE CLASSIFICATION OF ESTUARIES

There is an infinite variety of estuaries and it is only for convenience that they are grouped into a few classes. The geological history, the gradient of the coastal plain and the hardness of the rocks determine the width and the depth of the original estuary basin but its present shape is related to the supply of sediment and the strength of the distributive forces. Most of the estuaries in southern Africa are short and the mouth is usually constricted by a sandspit behind which the estuary broadens to form a lagoon. Often the whole estuary is shallow but in a few cases, where the supply of sediment is limited, the middle and upper reaches are deep. The maximum recorded depth is 35 m. The factors that affect the form and nature of estuaries on different stretches of the coasts of southern Africa are considered in chapter 14. Thereafter there are maps and brief descriptions of 43 estuaries between southern Mozambique and Namibia.

The estuaries of Europe and North America are typically wedge-shaped and broaden towards the mouth which is beset with shoals. In calm seas these form a projecting delta as at the mouth of the Mississippi. Elsewhere high-energy waves and the resulting long-shore currents plane off such projecting banks while littoral drift leads to the formation of sandspits and a bar across the mouth. This is very common in South Africa and Australia. The size and permanence of the mouth will be discussed in the next chapter. During periods of low river flow or after a violent storm, the mouth may close completely and a blind estuary is formed. There are many examples along the coasts of southern Africa: the Mhlanga near Durban, the Diep River lagoon at Hermanus and Milnerton Lagoon at Cape Town are well-known. The ratio of river discharge to tidal inflow of sea water determines the current patterns within the estuary. This will be discussed in chapter 3.

The velocity of river flow is one of the several factors which influence the penetration of sea water into the mouth of a river and thus affect its length. In some fast-flowing rivers, the discharge of fresh water is so great that no sea water penetration occurs and the mixing of fresh and salt water takes place in the sea itself. The mighty Amazon is a well-known example and variations of salinity extend for many kilometres from the mouth. Thus the water within the confines of the river banks is not estuarine. Such a system is

4

best referred to as a river mouth in spite of the fact that tidal rise and fall in water level extends more than 300 km along the river.

In rivers where the rate of flow varies from the dry to the rainy season or the freezing and melting of ice, the river may for some period flow fresh to the sea and at other times become estuarine. The shallow exit channel of the Orange River between Namibia and South Africa is seldom saline and is referred to here as a river mouth. So too, is the mouth of the Tugela River in Natal although the last kilometre may become an estuary for two or three months. The same must hold for several other large shallow rivers in different parts of the world. Conversely, estuaries with a normal salinity gradient for most of the year, may run fresh to the sea during a heavy flood. The basic definition of an estuary demands that it must have a measurable variation of salinity within its banks. If this is of brief duration, the exit channel is here classed as a river mouth; if it is only fresh during an occasional flood, it is classed as an estuary. The benthic macrophytes and the benthic macrofauna, both of which tolerate brief periods of salt or fresh water but die if such conditions persist, provide useful evidence of estuarine or fresh water conditions. River mouths are not included in the classification of estuaries which are outlined below and discussed in more detail in chapter 3. Descriptions of the Orange and Tugela river mouths are given in chapter 14.

Geologists, physicists, chemists and biologists have classified estuaries according to their particular interests. Several such classifications will be found in Lauff (1967) and Dyer (1973) but as the current system affects so many other features, it is generally accepted as the basis of classification.

1. Normal estuaries

Most estuaries are normal or 'positive' in the sense that there is an increase in salinity from the head where the river enters, towards the sea. Further, there is a net flow seaward over a full tidal cycle. Normal estuaries may be subdivided according to the degree of stratification or mixing of salt and fresh water.

1a. *Saltwedge estuaries.* These are normal estuaries with a wedge of sea water on the bottom and a layer of fresh water flowing out at the surface but no mixing between the two. Such a condition is rare if not entirely theoretical, but possibly occurs in some tranquil fjords.

1b. *Highly stratified estuaries.* These are normal estuaries with a layer of sea water flowing in along the bottom, a layer of fresh water flowing out at the surface and between the two is a layer of mixed water separated by marked haloclines. Most fjords belong to this class.

1c. *Partially mixed estuaries.* These are normal estuaries in which the vertical salinity gradient shows varying degrees of mixing or stratification between the outward-flowing surface layer and the inward-flowing bottom layer. Many estuaries belong to this class including the Thames and the Mersey in England, the Seine, Scheldt and Elbe in Europe, and the Hudson and the Chesapeake in the United States.

1d. *Vertically homogeneous estuaries.* These are normal estuaries with the salinity decreasing from the mouth towards the head but without a vertical gradient in salinity at any point. There may, however, be differences in salinity across the width of the estuary with the net current flowing landward along one side and seaward along the other. The lack of a vertical salinity gradient is due to turbulent mixing which often occurs in the strong tidal currents near the mouth of a shallow estuary. Many South African estuaries are homogeneous near the mouth but become partially mixed or even highly stratified in the calm upper reaches. The Bashee, Swartkops, Knysna and Breede estuaries show these features.

2. Hypersaline estuaries

These have a reversed or 'negative' salinity gradient with the salinity increasing from sea water values at the mouth to hypersaline values in the upper reaches where the water level is below mean sea level, so that the net flow is landward. Such conditions occur during severe droughts. The Laguna Madre in Texas is a classical example. St Lucia in Zululand becomes hypersaline periodically and Milnerton Lagoon near Cape Town becomes hypersaline every summer.

3. Closed or blind estuaries

These are estuaries which are temporarily closed by a sandbar across the sea mouth. At such times there is no tidal range and thus no tidal currents. Fresh water enters from the river and the circulation is dependent on the residual river current and the stress of the wind on the water surface. According to the ratio between evaporation and seepage through the bar on the one hand, and fresh water inflow plus precipitation on the other, the salinity will vary. The estuary may become hypersaline, it may retain its normal value when the mouth is closed or it may become hyposaline. Examples of all three types will be quoted in chapters 3, and 14.

LAGOONS

Rather surprisingly the word 'lagoon' has recently been used in a special sense. Begg (1978), p 11, states that 'Natal's estuaries can *initially* be separated into two major types, these being estuaries and lagoons'. He then goes on to discuss the differences between true estuaries, embayments and river mouths. In Schedule 1 (p 14), he lists the estuaries, lagoons, embayments and river mouths of Natal. His discussion need not be quoted for it is lengthy and it is clear that he uses the term 'lagoon' to mean estuaries that are temporarily closed by sandbars. Such bodies of water have been classed here as 'closed or blind estuaries'.

Noble & Hemens (1978), p 47, define estuaries as: 'partially enclosed bodies of water receiving fresh water inflow, in which mixing between fresh and sea-water occurs, which are connected to the sea and at least at some time under tidal influence'. This is essentially similar to the definition of an estuary given earlier although the importance of measurable variations in salinity is not explicit. In distinction to an estuary, Noble & Hemens (p 50) define lagoons as 'shallow brakish standing water bodies connected intermittently with the sea'. From their subsequent remarks it is clear that like Begg, they use the term lagoon to mean what I have referred to as a closed or blind estuary.

It would seem a pity, indeed incorrect to limit the meaning of the word 'lagoon' is this way. The meaning given in Webster's dictionary is: 'a shallow sound, channel or pond near or communicating with a larger body of water'. There is no reference to estuaries or indeed to salt or fresh water. Thus a lagoon could be an arm of the sea partly enclosed by the land or by a coral reef; it could be part of an estuary or a semi-enclosed part of a large freshwater lake. In fact the word is used in all these senses in common speech. Essentially, a lagoon is an expanse of sheltered, tranquil water, and the word is used in this sense in this book. The Laguna Madre in Texas is a hypersaline lagoon; Langebaan Lagoon in the Cape is a sheltered arm of the sea with a normal marine salinity; Knysna lagoon is an expanded part of a normal estuary and Hermanus Lagoon is a temporarily closed estuary. It is most convenient to have a term which is not rigidly defined.

It may be thought that this discussion of what is meant by an estuary, a lagoon or a river mouth is purely of theoretical interest. When it is remembered that these words are used in legislation and that different regulations may apply to estuaries and other bodies of water, the practical importance of definitions becomes apparent.

REFERENCES

BEGG, G.M. 1978. *The estuaries of Natal.* Vol 41, Natal Town and Regional Planning Commission, Pietermaritzburg.

DAY, J.H. 1980. What is an estuary? *S.Afr. J. Sci.* **76**:198.

DYER, K.R. 1973. *Estuaries: a physical introduction.* Wiley, London.

GORSLINE, D.S. 1967. Contrasts on coastal bay sediments on the Gulf and Pacific coasts. *In:* G. Lauff (ed), *Estuaries.* Am. Ass. Adv. Sci., Washington.

JENNINGS, J.N. & E.C.F. BIRD 1967. Regional morphological characteristics of some Australian estuaries. *In:* G. Lauff (ed), *Estuaries.* Am. Ass. Adv. Sci., Washington.

KING, L.C. & L.A. KING 1959. A reappraisal of the Natal monocline. *S.Afr. Geol. J.* **41**: 15-30.

KRIGE, A.V. 1927. An examination of the Tertiary and Quarternary changes of sea level in South Africa with special stress on evidence in favour of recent world-wide sinking of ocean level. *Ann. Univ. Stellenbosch* **5**: 1-18.

LAUFF, G. (ed) 1967. *Estuaries.* Am. Ass. Adv. Sci., Washington.

McHUGH, J.L. 1967. Estuarine nekton. *In:* G. Lauff (ed), *Estuaries.* Am. Ass. Adv. Sci., Washington.

NOBLE, R.G. & J.R. HEMENS 1978. Inland water ecosystems in South Africa: a review of research needs. *S.Afr. natl. Sci. Programmes.* Rep **34**, CSIR, Pretoria.

ORME, A.R. 1973. Barrier and lagoon systems along the Zululand coast, South Africa. *In:* D.R. Coates (ed), *Coastal geomorphology.* Geomorph. State Univ. NY, Binghampton, NY.

PRITCHARD, D.W. 1967. What is an estuary, physical viewpoint. *In:* G. Lauff (ed), *Estuaries.* Am. Ass. Adv. Sci., Washington.

RUSSELL, R.J. 1967. Origins of estuaries. *In:* G. Lauff (ed), *Estuaries.* Am. Ass. Adv. Sci., Washington.

STEERS, J.A. 1967. Geomorphology and coastal processes. *In:* G. Lauff (ed), *Estuaries.* Am. Ass. Adv. Sci., Washington.

VAN HEERDEN, J.L. 1976. The geology of Lake St Lucia and some aspects of its sedimentation. *In:* A.E.F. Heydorn (ed), *St Lucia Scientific Advisory Council Workshop – Charters Creek, February 1976.* Natal Parks Bd, Pietermaritzburg.

Chapter 2

Coastal hydrodynamics, sediment transport and inlet stability
J.H. Day

Department of Zoology, University of Cape Town

INTRODUCTION

This account is based on contributions by Mr Jan Rossouw, Dr Harry Swart and Mr Kenneth Russell of the South African National Research Institute for Oceanology (NRIO). It is not intended for the specialist but should provide an ecologist with some understanding of waves, tides, coastal currents and the characteristics of sediments. These factors determine the size and location of sandspits and shoals which form the outer bar or ebb tide delta. The inner bar or flood tide delta will be considered in chapter 4 after estuarine currents have been discussed in more detail.

The editor has modified the original contributions and is responsible for any errors or omissions.

The contributors specifically acknowledge the use of material from the following sources:

Waves – *Shore protection manual* (US Army Corps of Engineers, 1973)

Tides – *Admiralty manual of tides* (Doodson & Warburg, 1941)

Currents and coastal sediment movement – *Estuarine and coastline hydrodynamics* (Ippen, 1966); *Shore protection manual* (US Army Corps of Engineers, 1973)

Inlet stability – *Stability of coastal inlets* (Brunn & Gerritsen, 1960); *Tidal inlets and littoral drift* (Brunn, 1966); *Offset coastal inlets* (Hayes & Goldsmith, 1970).

WAVES

The stress of the wind on any water surface causes sur-

face drift and also generates waves. They are oscillatory waves and may be defined by their height, length and period (Fig. 2.1). The wave height H is the vertical distance from the top of the crest to the bottom of the trough; the wave length L is the horizontal distance between successive crests and the wave period T is the time taken for successive crests to pass a fixed point. The wave steepness is the ratio H/L and the velocity or celerity C of the waves is related to the period and the wave length so that $C = L/T$. As waves propagate in water the progressive movement of water particles is very slight; it is the form and wave energy that move forward.

As gusts of wind blow over the surface of the sea, different types of waves are generated. The height, length and period of any wave in the deep sea is determined by the velocity of the wind, its duration and its fetch or the distance the wind blows over the sea surface. In general, the longer the fetch of the wind, the higher the wind velocity and the longer it blows, the larger is the wave. If the winds of a local storm blow towards the shore, the waves will reach the coast in nearly the same form in which they were generated. Under such conditions the waves are steep, ie, the wave length is 10 to 20 times the wave height. Such waves are often termed seas. As waves travel away from the generating area they decay and if the storm centre is far away, steep waves are eliminated and long, low-amplitude waves remain which may travel hundreds or even thousands of kilometres. Such waves have

DIRECTION OF PROPAGATION

WAVE LENGTH

WAVE HEIGHT

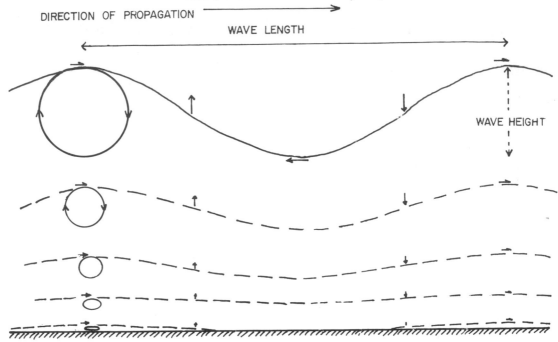

Figure 2.1
Diagram of an oscillatory wave showing orbits of water particles (after Perkins, 1974).

lengths of 30 to 500 times their heights and are termed swells although there is no basic difference between seas and swells.

The South African coasts are subjected to high energy waves. The highest waves are generated by storm centres in the Roaring Forties and thus occur along the extreme southern and south-western coasts where individual wave heights in excess of 12 m are recorded every year. Along the east and west coasts, the maximum wave heights are significantly lower and seldom exceed 8 m. The dominant wave period along the entire South African coastline lies between 65 and 15 seconds and does not seem to vary much from place to place.

In deep water the velocity of wave propagation is entirely a function of the wave period. Where not affected by the wind, the wave form is approximately sinusoidal (Fig. 2.1). As the wave progresses, each particle of water passes through an almost circular orbit, rising and moving forward in the wave crest and sinking and moving backward in the trough. The energy of the wave and thus the diameter of the orbits decreases with depth so that there is little motion at a depth d greater than half the wave length L. Where $d = L$ the orbit of each water particle is reduced to 1/534,5 of that at the surface. For example, if the wave height is 10 m and the wave length is 100 m, then, at a depth of 100 m, the orbit will be less than 2 cm in diameter.

As a wave enters water shallower than half its wave length the velocity or celerity of the wave is reduced and the orbital motion of the water particles changes with depth. At the surface the orbits remain circular but towards the bottom they become more and more elliptical until, at the very bottom, the water particles are moving backward and forward over the sea bed.

In shallow water all the characteristics of a wave change except the wave period. As the wave moves into progressively shallower water, the velocity of wave propagation is continuously reduced whereas the horizontal velocities of the water particles increase until eventually the water particles near the surface move faster than the speed at which the wave travels. The wave thus becomes unstable and eventually breaks. The water depth in which the wave will break and the type of breaker formed, are mainly functions of the slope of the sea bed and the wave steepness. For fairly flat beach slopes and steep waves (seas), the ratio of breaker height to water depth may be as low as 0,5 whereas for steep beach slopes and low amplitude waves (swells) the ratio may be as high as 1,3. In the first case spilling breakers will be formed and in the second case plunging breakers. Spilling breakers cause shore accretion and plunging breakers cause shore erosion.

When a wave advances normal to a straight, evenly shelving shore and eventually reaches the critical depth for its steepness, the whole crestline breaks simultaneously and the surf rolls forward parallel to the beach (Fig. 2.2A). Wave refraction will occur on the sides of a headland (Fig. 2.2C) or where a wave advances obliquely towards the shore (Fig. 2.2B). Due to the reduction in wave celerity with decreasing water depth, the part of the crestline which reaches

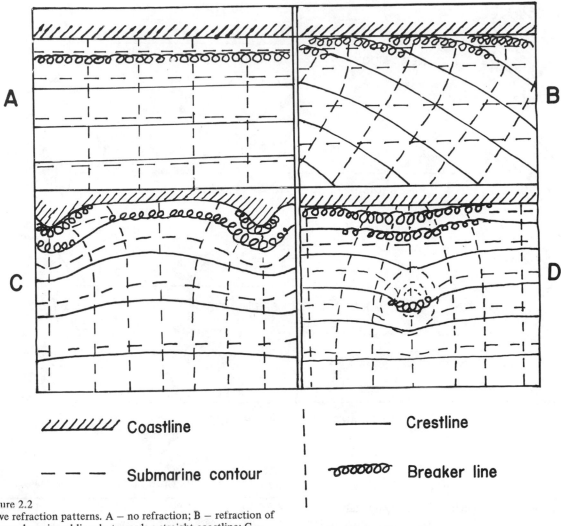

/////////// Coastline

——— Crestline

— — — Submarine contour

ᴑᴑᴑᴑᴑᴑ Breaker line

Figure 2.2
Wave refraction patterns. A – no refraction; B – refraction of
waves advancing obliquely towards a straight coastline; C –
refraction towards headlands; D – refraction around the sides
of an offshore shoal (modified from King, 1962).

ORTHOGONAL

the shallows first will slow down, while the rest of the
wave continues to advance rapidly. Thus the crestline
will swing around in such a way that the angle
between the crestline and the shore is reduced.

Refraction will also occur on either side of a shoal
(Fig. 2.2D) and as a result, the orthogonals beyond
the shoal will converge and the wave energy will be
concentrated there and be reduced on either side. This
may have tragic consequences. High energy swells
derived from storms in the Roaring Forties enter False
Bay from the south-west. At the entrance, Shipley
(1964) has shown that they are refracted around either
side of a large submerged reef known as Rocky Bank.
The two convergent trains of waves come to a focus
on the north-eastern shore of the bay near Steenbras
River mouth. Anglers on rocks which are normally well
above the reach of the waves may suddenly be swept

away by trains of waves up to 4 m high. Many fatalities
have occurred.

TIDES AND TIDAL RANGE

Tides are the result of the gravitational attraction of
the moon and the sun on the earth and its mobile
waters and are observed as a regular rise and fall of the
water surface. The original theory of tides was developed by Sir Isaac Newton following his work on
gravitation. Although his theory has since been modified as explained by Doodson & Warburg (1941) and
by King (1962), it does provide an understanding of
tidal range and the variation from neap to spring
tides. The moon, being much closer to the earth than
the sun has slightly more than twice the effect on the
tides and may be considered first.

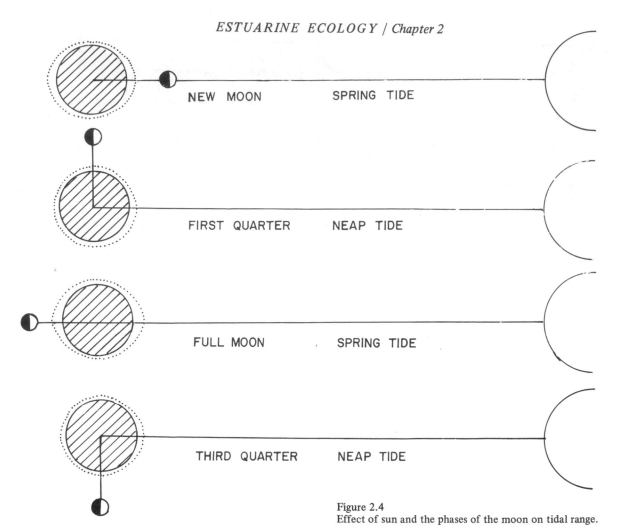

Figure 2.4
Effect of sun and the phases of the moon on tidal range.

If the earth were a perfect sphere completely covered with water (Fig. 2.3), the tractive force of the moon would cause the water facing the moon to bulge in that direction and at the opposite side of the world there would be a slightly lower bulge. Thus the water surface would have two bulges or two high tides and two troughs or two low tides at right angles. As the earth rotates each day, the two tidal cycles follow the moon, causing semi-diurnal tides. The moon encircles the earth every 27,3 days while the earth rotates on its own axis every 24 hours so that the moon passes over each locality about 50 minutes later each successive day and the semi-diurnal tides are separated by about 13 hours 25 minutes.

The tidal cycle caused by the sun has a period of 24 hours and is thus termed a diurnal tide. When the sun and the moon act in the same direction, either on the same side of the earth (new moon) or on diametri-cally opposite sites (full moon), the tidal range should be maximal and spring tides should occur. However, due to inertial effects, there is a lag and maximal spring tides occur two days after full moon and two days after new moon. At the same locality maximal spring tides occur at approximately the same time of

Figure 2.3
Moon's influence on oceans (after Doodson & Warburg, 1941).

day. Spring and neap tides alternate every 14 days, the spring tide range increasing gradually to a maximum and then decreasing. At the equinoces in spring and autumn, the earth, the moon and the sun approach most closely to a linear alignment and the spring tides-range is maximal. Towards the summer and winter solstices, the declination of the moon increases and the spring tide range is reduced.

As noted, Newton's theory postulated the earth as covered with water so that the two tidal waves were progressive and encircled the earth. Such a condition is approached in the southern oceans but elsewhere

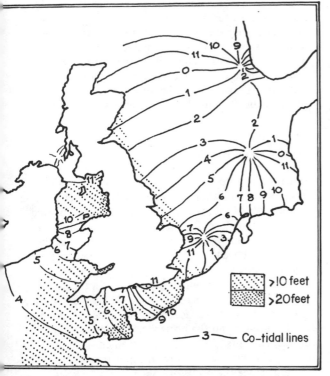

Figure 2.5
The tides around Britain. Explanation in text (after King, 1962).

amphidromic tidal system results. Cotidal lines radiate from the nodal point and the tidal range increases along the length of each cotidal line. Thus coasts near the nodal point have a relatively small tidal range while those distant from the nodal point have a larger range. As shown in Fig. 2.5, there are three resonating basins with three amphidromic systems in the North Sea and others in those tidal basins whose natural period of oscillation is nearly equal to one or other of the tidal components.

As noted, the tidal forces may be resolved into two components: a diurnal component of one tidal cycle per day and a semi-diurnal component of two tidal cycles per day. One or other of these components may predominate in each amphidromic system. On many coasts there are semi-diurnal tides; on the northern shores of the Gulf of Mexico, diurnal tides predominate and in some areas the two components are more nearly equal, and as they are usually out of phase, they interfere to produce mixed tides as on the Pacific coast of North America. In the South Atlantic and south Indian Oceans, semi-diurnal tides predominate as seen on the coasts of southern Africa.

On the shores of oceanic islands the spring tide range is usually less than one metre but on open continental coasts the range at springs is often of the order of 2 m although it may vary from 1 to 6 m. On South African shores the maximal range is less than 2 m but it increases to 5 m on the coasts of northern Mocambique. This increase in tidal range on gently shelving shores and in narrowing channels is well-illustrated on the British shores of the English Channel. On the coast of Cornwall where the channel is widest the range at springs is about 3 m, but in the Straits of Dover it increases to 5 m and as the channel widens again at the entrance to the North Sea the range falls to 1,2 m.

The average tidal levels for southern African ports are given in Table 2.1. In general the levels are computed from at least one complete year's observations and are adjusted for long period variations to give values which are the average over the whole of the cycle. The values of Lowest Astronomical Tide (LAT) and Highest Astronomical Tide (HAT) are noted over

the oceans are separated by continents which are so shaped as to form a number of natural basins whose waters resonate with the tidal forces. In each ocean basin or sea with an entrance sufficient to transmit tidal waves, a seiche movement develops. This may be likened to the oscillatory movement produced by tilting a basin of water backwards and forwards in a regular manner related to the length and depth of the basin. The elevation of the water surface will be maximal at the ends of the basin and minimal across the middle where a nodal line is formed. In a natural tidal basin, Corioli's force transforms the oscillatory movement of the tidal waves into a circular movement around a fixed nodal point of no tidal motion and an

Table 2.1
Tidal levels at ports in southern Africa

	Latitude	MLWS	MLWN	ML	MHWN	MHWS	HAT	Years of observation
Walvis Bay		0,20	0,60	0,91	1,22	1,62		1960
Lüderitz	− 0,21	0,06	0,44	0,74	1,03	1,41		1967-75
Port Nolloth	− 0,21	0,09	0,55	0,87	1,20	1,66	2,03	1966-75
St Helenã Bay		0,14	0,60	0,90	1,20	1,65		1968
Saldanha Bay		−0,06	0,44	0,69	0,94	1,44		
Cape Town	− 0,01	0,27	0,71	0,98	1,26	1,69	2,02	1968-75
Simon's Bay	− 0,21	0,07	0,53	0,81	1,09	1,55	1,86	1958-75
Hermanus		0,09	0,53	0,81	1,09	1,53		1964
Mossel Bay	− 0,15	0,11	0,70	0,99	1,27	1,86	2,28	1968-71
Knysna	− 0,17	0,08	0,62	0,88	1,15	1,68	2,01	1966-75
Port Elizabeth		0,23	0,78	1,03	1,29	1,84		1967
East London		0,23	0,80	1,05	1,30	1,86		1967
Durban	− 0,09	0,18	0,79	1,04	1,29	1,90	2,22	1971-75
Richards Bay		0,19	0,83	1,09	1,35	1,99		

The above levels are referred to CHART DATUM

a span of years and are defined below.

(a) HAT (Highest Astronomical Tide). LAT (Lowest Astronomical Tide). These are the highest and lowest levels which can be predicted to occur under average meteorological conditions and under any combination of astronomical conditions; these levels will not be reached every year. HAT and LAT are not the extreme levels which can be reached, as storm surges may cause considerably higher and lower levels to occur.

(b) MHWS (Mean High Water Springs). MLWS (Mean Low Water Springs). The height of mean high water springs is the average, throughout a year when the maximum declination of the moon is 23°, of the heights of two successive high waters during those periods of 24 hours (approximately once a fortnight) when the range of the tide is greatest. The height of mean low water springs is the average height obtained by the two successive low waters during the same periods.

(c) MHWN (Mean High Water Neaps). MLWN (Mean Low Water Neaps). The height of mean high water neaps is the average, throughout a year as defined in (b) above, of the heights of two successive high waters during those periods (approximately once a fortnight) when the range of the tide is least. The height of mean low water neaps is the average height obtained from the two successive low waters during the same periods.

Note: The values of MHWS, etc, vary from year to year in a cycle of approximately 18,6 years. The tidal levels given in Table 2.1 are computed average values for the whole cycle.

(d) ML (Mean Level) as given in the table, is the mean of the heights of MHWS, MHWN, MLWS and MLWN.

(e) Years of observations. Tidal predictions of South African ports are based on at least one complete year's observations, with the exception of Richard's Bay where they are based on 40 days' observations.

Sea level is also affected by the barometric pressure; it is lowered during the passage of a high pressure system and raised during the passage of low pressure system. The advance of a low is associated with strong winds and all bodies of water, whether seas, lakes or estuaries are affected by persistent winds which raise the water level on windward shores and lower the level on lee shores. If high spring tide coincides with a strong onshore wind the coast may be flooded as happened in Britain and Holland in 1953. An offshore wind coinciding with low spring tide exposes banks which are normally covered. Where the astronomic tide is less than a metre as in the Gulf of Mexico and on the south-western coast of Australia, meteorological effects on sea level are of major importance.

The depth of the coastal sea is also important. Where the depth of the coastal sea is very great as it is around oceanic islands and steep continental margins, the tidal phenomena may be completely described by the sum of the astronomic tidal components, given the amplitude and phase of each. On such coasts the tidal

wave remains sinusoidal and the time interval from low to high water is equal to that from high to low. However, when a tidal wave advances over shallow water it is distorted by friction and possible changes in depth. The crest below which the water depth is greater, tends to overtake the trough below which the water is shallower. Thus the duration of the flood tide is reduced while that of the ebb is increased.

COASTAL CURRENTS

Since this chapter is primarily concerned with the forces which transport sediments and affect conditions at the inlets or mouths of estuaries, little will be said about ocean currents. In near-shore waters, the main currents include those due to tides, the stress of the wind and the waves. All such currents are affected by Corioli's force.

In the deep sea where the passage of a tidal wave causes the water particles to move slowly through open circular orbits, the vertical components correspond to the tidal range and the horizontal components result in tidal currents. The forward movement is termed the flood current and the return movement is the ebb current. Since the orbits are open and not closed circles, there is also a small, net horizontal movement termed mass transport. Since the tidal range in the open ocean is of the order of a metre and the tidal period is over six hours, all the movements are very slow but in shallowing coastal areas where the tidal range increases and the orbits of the water particles become elliptical with the major axis horizontal, the velocity of tidal currents increases. The flood current moves shoreward with the rising tide and reaches maximum velocity at high tide. Towards midtide the velocity decreases to zero and then reverses direction and the seaward velocity increases to a maximum at low tide. Thus the currents are in phase with the tides, their velocities increase with tidal range and maximum velocities occur at spring tides.

On open continental coasts which shelve fairly steeply as do those around South Africa, the tidal range is small and the tidal currents are still slow, but in narrow seas with gently shelving shores the tidal range increases and tidal currents become important. The direction of any current flow is affected by Corioli's force which acts to the right of the current flow in the northern hemisphere and to the left in the southern hemisphere. In channels and broad estuaries this causes rotating currents to develop. For example, when the flood tide is flowing eastwards in the English Channel, the current on British coasts is deflected to the south and when the ebb current is flowing westwards it is deflected to the north. The deflection is less powerful than the main current so that the whole ebb and flow pattern will be elliptical. The shelving sides of a channel add to the effect, particularly at midtide. At high water

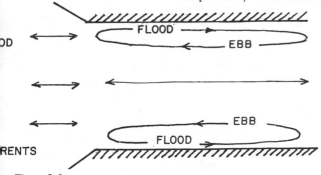

Figure 2.6
Tidal current patterns in a channel with shelving sides.

there is no modification of the tidal current, but towards midtide when the water level on the shore is falling rapidly, an offshore current is flowing and the tidal current is decreasing to nil. At low water the tidal current is flowing in the reverse direction and the onshore current is nil. Similarly, when the water level on the shore is rising an onshore current is flowing and the main tidal current falls to nil at midtide. The combined effect is an elliptical current along each side of the channel with a clockwise direction along northerly shores (see Fig. 2.5). The ebb and flood currents erode separate channels in the bottom sediments and these become shallower in the direction of flow.

The stress of the wind on the water causes surface drift which is particularly important in tideless bodies of water. In Lake St Lucia, the level on a windward shore may be increased by 20 cm and there is a similar decrease in level on lee shores. As the water surface is tilted in this way, compensatory subsurface currents develop which maintain the vertical circulation.

Currents are also caused by wind-generated waves. As an unbroken wave advances towards a shelving shore and passes the critical depth of half its own length, its height increases and the crest becomes steeper. There is an onshore acceleration under the crest and a slower and more prolonged offshore movement under the trough. Mass transport due to the net forward movement of water particles also increases, so that there are important differences between long flat waves and short steep waves. King (1962) quotes observations that long flat waves induce a slight shoreward movement at the surface, a slow seaward movement through most of its depth but near the bottom there is a much stronger shoreward movement which transports sediments towards the breaker zone. In contrast to this, there is a net landward movement about mid-depth under short, steep waves and a net seaward movement at lower levels. Turbulence due to the passage of the wave crest suspends the finer particles of the bottom sediments and they are transported into deeper water.

Only a small part of the wave energy is dissipated by bed friction outside the breaker zone and practically all the energy is dissipated within the breaker zone by the breaking of the wave.

If a wave approaches normal to the shoreline and breaks, it generates onshore currents and offshore currents but there are no longshore currents. If, however, the incident wave approaches the shoreline at an oblique angle, or if the bottom contours are not parallel with the shoreline, the wave energy will have components in both the longshore and onshore-offshore directions. The onshore-offshore component leads to a gradient in the mean water level relative to the still water level from the break point of the waves to the beach. This increase in water level on the beach, known as the wave set-up, varies from about 14 % of the wave height in the case of a plunging breaker to about 18 % of the height of a spilling breaker. The wave set-up generates currents, either an undertow normal to the shoreline or a large-scale three dimensional circulation of the water due to variations of the wave set-up along the shore. This variation in the longshore component of the energy drives a longshore current which is primarily confined to the breaker zone. The water eventually escapes from the breaker zone as rip currents which often develop where a promontary has blocked the longshore flow.

The most important variable in determining the velocity of the longshore current is the angle between the wave crestline and the shoreline. However, the volume of flow and the longshore transport rate depend mostly on the breaker height. Since the waves break in water depths approximately proportional to the wave heights, the width of the breaker zone increases with wave height and so does its cross-sectional area. If the cross-section is approximately triangular, an increase in wave height increases its area and thus the volume of flow as the square of the wave height. It is thus obvious that the height of the waves has important effects on the width and volume of flow of longshore currents in the breaker zone.

SEDIMENT CHARACTERISTICS AND TRANSPORT

The entrainment, transportation and subsequent deposition of a sediment depend not only on the characteristics of the flow involved but also on the properties of the sediment itself. These can be divided into the properties of the particles and those of the sediment as a whole.

The settling velocity of a particle directly characterises its reaction to flow, and ranks next to size in importance. Frequency distribution of properties, eg size and settling velocity, are necessary for the description of sediments. Flocculation is of importance in the behaviour of very fine sediments and, in many cases, may be the major factor in determining the settling velocity and the specific weight of a deposit. Specific weight is determined by the sediment properties and by the environment and manner in which it is laid down. Thus specific weight is not a true sediment property; however, it is included in this section because it appears to depend more on sediment properties than on other factors.

Sediments include organic and mineral particles of all sizes and composition, but those in estuaries are seldom coarser than pebbles and among these, shell fragments or shingle are common. Sand is abundant at the mouths of most estuaries and varies from very fine to coarse according to the current velocity, while silt and particles with diameters approaching colloidal dimensions are mainly deposited in the upper reaches. The finer fractions include not only mineral particles, but also the siliceous shells of diatoms, the calcareous shells of foraminifera and many organic aggregates.

Practically all mineral sediments have their origin in rock material, and all constituents of the parent material can usually be found in the sediments. However, as the materials become finer due to weathering and abrasion, the less stable minerals tend to weather faster and are carried away as fine particles or in solution, leaving the more stable components. The highest degree of sorting of minerals is to be expected in the fine fractions of sediments. Coarse material, eg boulders, may be a part of the parent rock and contain all the constituents of the original material. Although quartz, because of its great stability, is by far the most common mineral found in sediments moved by water and wind, numerous other minerals are also present. For example, specific gravity analyses of sand taken from the bed of the Missouri River at Omaha in the USA, indicated that only about half the sand is quartz, the remainder being shale, carbonate, chert, feldspar and heavy minerals.

Because the sizes and shapes of grains making up a sediment vary over wide ranges, it is convenient to group sediments into different size classes or grades. Such classifications are essentially arbitrary and many grading systems are to be found in engineering and geological literature. The Wentworth scale of units is the most commonly used because the grain sizes are grouped in a geometric series with a ratio of 2, and because the sizes correspond closely to the mesh openings of sieves in common use. The Wentworth scale

characterises grain sizes in ϕ units, where $\phi = -\log_2 D$, with D the grain size in mm. The method is described by Morgans (1956). A shortened version of the scale is given in Table 2.2 below.

Natural sediment particles are of irregular shape so that any single length or diameter that is to characterize the size of a group of grains must be chosen arbitrarily according to some convenient method of measurement. The three such diameters recommended by the subcommittee on sediment terminology of the American Geophysical Union (Lane, 1947) are defined as follows:

1. Sieve diameter is the length of the side of a square sieve opening through which the given particle will just pass.
2. Sedimentation diameter is the diameter of a sphere of the same specific weight and the same terminal settling velocity as the given particle in the same sedimentation fluid.
3. Nominal diameter is the diameter of a sphere of the same volume as the given particle.

The size of sand grains is commonly measured by sieving, although recently, the sedimentation diameter also has been obtained by determining the settling velocity. The size of silt and clay particles is generally expressed as a sedimentation diameter. This is actually a fictitious size that enables the settling velocity to be calculated. For this reason it is of greater physical significance than the other two diameters. The nominal diameter has little significance in sediment transportation, but it is useful when discussing the nature of sedimentary deposits.

The size distributions in a sample are generally based on weight, rather than the number of particles. Quantitatively, the size distribution may be characterised by a diameter that is in some way typical of the sample, and by the way that coarser and finer particles are distributed. A size distribution is described as well sorted if all particles have sizes that are close to the typical size. If the particle sizes are distributed evenly over a wide range of sizes, the sample is said to be well-graded. A well-graded sample is poorly sorted; a well-sorted sample is poorly graded.

The median (M_d) and the mean (M) define typical sizes of a sample of littoral materials. The median size, M_d in millimetres, is the most common measure of sand size in engineering and ecological reports. It may be defined as:

$$M_d = D_{50}$$

where D_{50} is the size in millimetres that divides the sample so that half by weight has particles coarser than the D_{50} size. An equivalent definition holds for the median of the phi size distribution, using the symbol $M_{d\phi}$ instead of M_d.

Several formulae have been proposed to compute an approximate mean (M) from the cumulative size distribution of the sample. These formulae are averages of two, three or more symmetrically selected percen-

Table 2.2
Shortened version of the Wentworth scale

Class	Size range in mm	ϕ – units
Boulders and cobbles	64 and over	−6 and less
Very coarse to very fine gravel	64 to 2	−6 to −1
Very coarse to coarse sand	2 to 0,5	−1 to +1
Medium sand	0,5 to 0,25	+1 to +2
Fine to very fine sand	0,25 to 0,063	+2 to +4
Coarse to medium silt	0,063 to 0,016	+4 to +6
Fine to very fine silt	0,016 to 0,004	+6 to +8
Coarse to medium clay	0,004 to 0,001	+8 to +10
Fine to very fine clay	0,001 and less	+10 and more

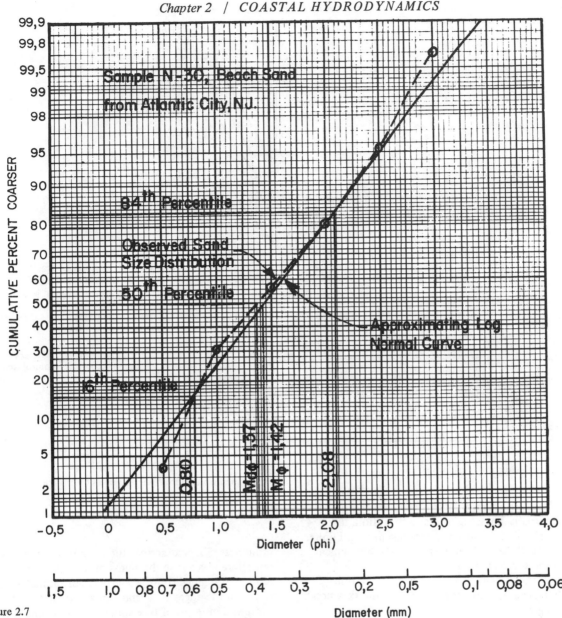

Figure 2.7
The relation of the median and mean particle diameters in
natural beach sands (from US Army Corps of Engineers, 1973).

tiles of the phi frequency distribution such as:

$$M_\phi = \frac{\phi_{16} + \phi_{84}}{2} \quad \text{or} \quad M_\phi = \frac{\phi_{16} + \phi_{50} + \phi_{84}}{3}$$

where ϕ is the particle size in phi units from the distribution curve at the percentiles equivalent to the sub-
scripts 16, 50 and 84 (see Fig. 2.7). ϕ_x is the size in
phi units that is exceeded by x percent (by dry weight)
of the total sample.

To a good approximation, the median M_d is inter-
changeable with the mean M for most beach sediments.
For example, in one CERC study of 465 sand samples
from three New Jersey beaches, the mean computed
by the method of moments averaged only 0,01 mm
smaller than the median for sands whose average

median was 0,3 mm (1,74 phi).

Since the actual size distributions are such that the
log of the size is approximately normally distributed,
the approximate distribution can be described in phi
units by the two parameters that describe a normal
distribution. These are the mean and the standard
deviation. In addition to these two parameters, skewness and kurtosis describe how far the actual size distribution of the sample departs from this theoretical
lognormal distribution. The standard eviation may be
approximated by:

$$0_\phi = \frac{\phi_{84} - \phi_{16}}{2}$$

where ϕ_{84} is the sediment size in phi units that is finer

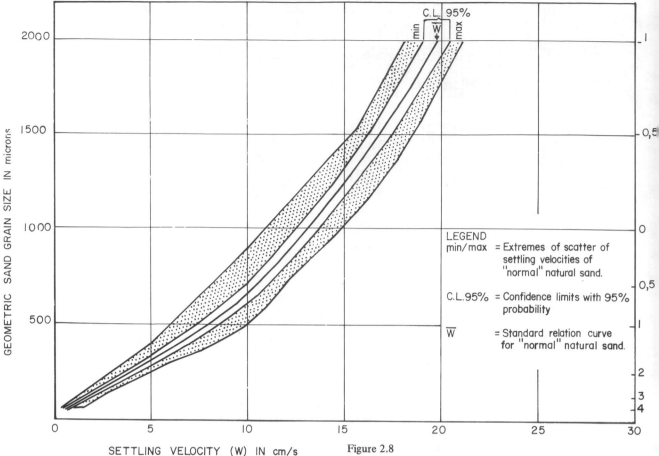

Figure 2.8
Envelope of observed settling velocities of 'normal' natural sands and the 'standard relation curve' between grain size and settling velocity (after Fromme, 1977).

than 84 % of the sample. If the sediment size in the sample actually has a lognormal distribution, then 0_ϕ is the standard deviation of the sediment in phi units. For a perfectly sorted sediment, $0_\phi = 0$. For typical well-sorted sediments, $0_\phi = 0,5$.

The degree by which the phi size distribution departs from symmetry is measured by the skewness SK_ϕ or a_ϕ as:

$$a_\phi = \frac{M_\phi - M_{d\phi}}{0_\phi}$$

where M_ϕ is the mean, $M_{d\phi}$ is the median, and 0_ϕ is the standard deviation in phi units. For a perfectly symmetrical distribution, the mean equals the median and the skewness is zero.

Currently, the median grain size is the most commonly reported sand grain size characteristic, probably because there are only limited data to show the usefulness of other size distribution parameters in coastal engineering design. In ecology, the median grain size is also used and the percentage of fines or sub-sieve particles (<0,063 mm) is often quoted. Other sediment parameters will be discussed in chapter 4.

The size of a sediment particle alone is not sufficient to describe it. The characteristics that seem most important for sediment transportation are shape and roundness. Shape describes the form of a particle without references to the sharpness of its edges, while roundness depends on the sharpness or radius of curvature of the edges. The shape of a grain has been expressed in terms of true sphericity, which has been defined as the ratio of the surface area of a sphere with the same volume as the grain, to the surface area of the particle.

The fall velocity of particles may be expressed in terms of the sieve diameter and a shape factor S.F., where S.F. is defined by S.F. = $c/\sqrt{ab}$, in which a, b and c are respectively the lengths of the longest, intermediate and shortest mutually perpendicular axes of the particle.

Fall velocity of natural sand grains

Although the fall velocity of sand grains is determined by grain size, shape, specific weight and water temperature, a comprehensive investigation (Fromme, 1977) into the relationship between fall velocity and grain size, as found from all available publications on the subject, indicated that a unique relationship exists for natural sand (S.F. = 0,6 to 0,7). The resultant relationship is shown in Fig. 2.8.

Silt and humus with particle sizes of 0,002 mm or less, tend to remain suspended in river water but flocculate and are eventually deposited in the saline water of estuaries. The mechanism will be discussed in chapter 4.

Sediment movement

Water flowing over a sandy bed exerts forces on the grains that tend to move or entrain them. The forces that resist the entraining action differ according to the grain size and grain size distribution of the sediment. For coarse sediments, such as sands and gravels, the forces resisting motion are caused mainly by the weight of the particles. Finer sediments that contain appreciable fractions of silt, clay or both, tend to be cohesive and their resistance to entrainment is mainly due to cohesion rather than the weight of the individual grains. Also, in a fine sediment, groups of grains are entrained as units, whereas coarse, non-cohesive sediments are moved as individual grains. Cohesion of the fine sediments is a complicated phenomenon that appears to vary with mineral composition and environment. The roots of aquatic plants and the carpet of microscopic algae also increase the cohesion of surface sediments.

The forces acting on the grains of a non-cohesive sediment over which a fluid is flowing are the gravity forces of weight and buoyancy, hydrodynamic lift normal to the sediment bed and drag parallel to the bed. As the intensity of flow increases from rest, the hydrodynamic lift and the drag increase in magnitude. A condition is eventually reached at which particles in the bed are dislodged and eventually moved. This movement is not an instantaneous one for all particles of a given size in the top layer. In fact, at any given hydraulic condition, some move and some do not. If this motion is one of rolling, sliding and jumping (saltating), and takes place close to the bed, the mode of transport is called *bed load (or contact load)*. If the entire motion of the solid particles is such that they are surrounded by fluid, they are said to *move in suspension*. Owing to the weight of the particles, there is a tendency for settling, which, however, is counterbalanced by turbulent velocity components. Thus the hydraulic conditions determine when a given size fraction will be in suspension. Furthermore, sediment particles that are part of the suspended load at one time may be part of the bed load at another time, and vice versa. There is not only an active interchange between the suspended load and bedload, but also between the bedload and the bed itself.

When the sediment is in suspension the coarser fractions are transported near the bed while the finer fractions are more uniformly distributed over the water depth. As the velocity of the current increases,

there is a delay between the instant it reaches the critical velocity for particle movement and the moment when the particle is fully suspended. This delay is termed the *scour lag*. When the flow velocity in a stream carrying sediment in suspension decreases, the suspended sediment will start settling again. There will, however, be a delay before the particle actually settles on the bed. This delay is called the *settling lag*. Both the scour lag and settling lag have important consequences on the sorting of sediments. In Fig. 2.9 an example is given of a suggested relationship between current velocity, particle size, erosion, transportation and sedimentation.

As may be seen, there exists for each grain size a certain velocity below which it will experience sedimentation, while above a certain velocity, called the critical scour velocity, it will be eroded. These two phenomena are only slightly related.

COASTAL SEDIMENT MOVEMENT

Littoral transport is the movement of sedimentary material by waves and currents in the littoral zone. The extent of the littoral zone, according to engineers, is from the shoreline to the limit of the most seaward breakers. Ecologists have used 'littoral zone' in the same sense but more recently there has been a tendency to define the littoral zone as the shore between tide marks. Since there is obviously ambiguity, this area is best referred to as the intertidal zone, although difficulties will remain in tideless seas. Below the intertidal zone the deeper areas which are affected by waves of normal height may be termed the turbulent zone. The term littoral zone will be used here in the engineering sense.

Littoral transport includes onshore-offshore transport and longshore transport. Littoral transport occurs in two modes: bed load transport and suspended load transport and, as noted, it is hard to distinguish where bed load transport ends and suspended load transport begins. It is more useful to identify two zones of transport based on the type of fluid motion initiating sediment movement. These are the offshore region or zone and the surf zone, shown in Fig. 2.10.

The offshore region or zone

In the offshore zone, sediment transport is initiated by the passage of a wave crest over sand ripples on the sea bed and in the surf zone transport is initiated by a passing breaker. Net sediment transport is due to the wave-induced currents in both zones.

In the offshore zone, the waves reach a depth where the elliptical orbits of the water particles near the bottom affect the sediment. At first only low density material, such as seaweed, moves and this oscillates

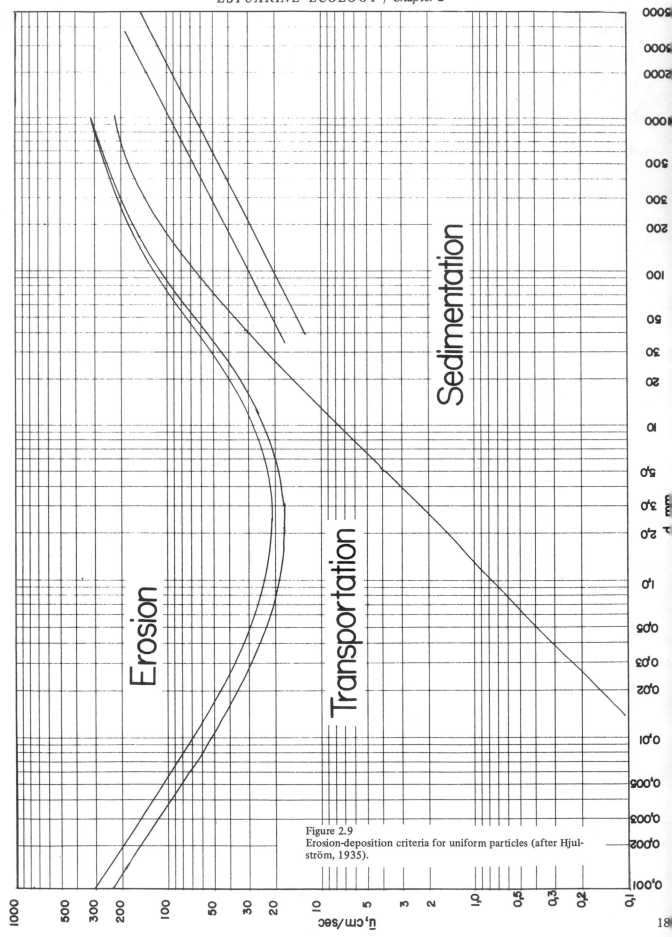

Figure 2.9
Erosion-deposition criteria for uniform particles (after Hjulström, 1935).

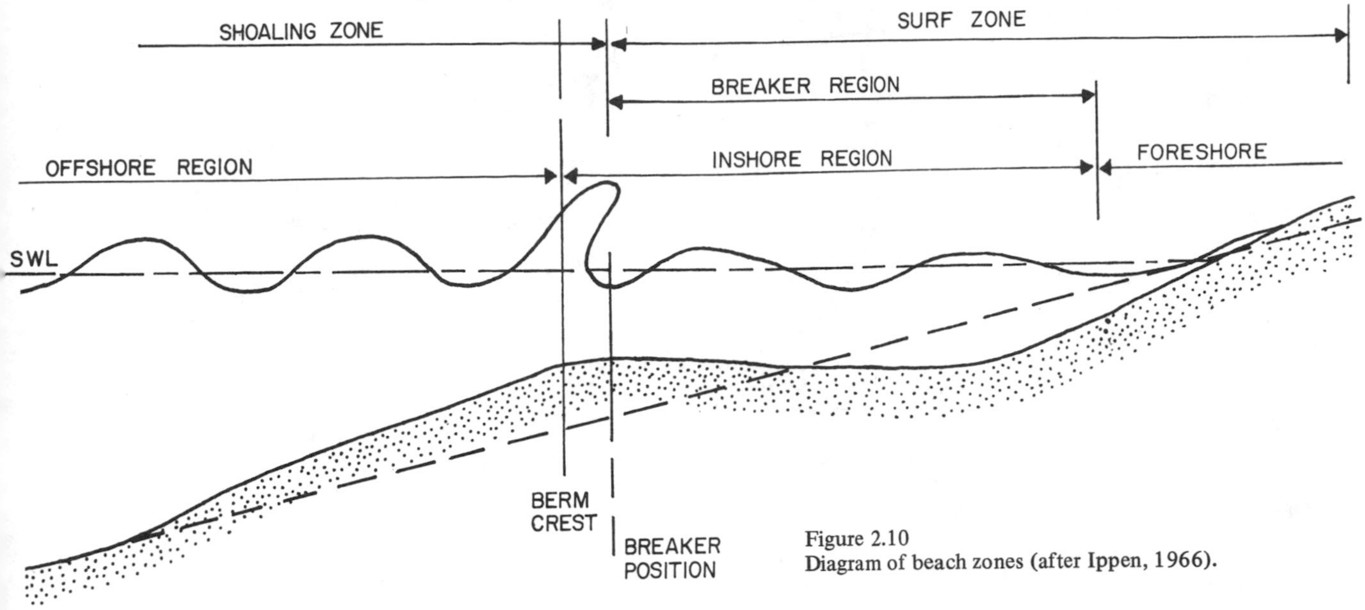

Figure 2.10
Diagram of beach zones (after Ippen, 1966).

back and forth under the waves often forming ripple-like ridges parallel with the wave crests. In shallower depths the fluid motion over the bottom increases until it exerts enough shear to move sand particles and the sand itself forms ripples (Fig. 2.11). These sand ripples are typically uniform and periodic with the sand grains moving from one side of the crest to the other with the passage of each wave. As the water depth decreases to about seven times the wave height, the oscillatory bottom current develops a high shoreward component during the brief passage of the wave crest, and a lower seaward component during the longer duration of the trough. As the shoreward velocity decreases and the bottom current reverses direction over a sand ripple, a cloud of sand erupts from the landward face of the ripple. The sand cloud drifts seaward under the trough and some of the particles are deposited. As the next wave crest approaches, the remainder of the sand cloud reverses its direction of motion and moves landward. The cumulative effect of this cyclic process seems to be a gradual propagation of the ripple form in the shoreward direction. There is a degree of sorting going on, however, since the lighter particles remain in suspension and are continually drifting seaward. As the fluid velocity increases, the ripples increase in height and forward velocity. Beyond a critical fluid velocity, however, both the ripple height and wave length decrease and if the fluid velocity is sufficiently high the sand ripples vanish completely.

The direction of the oscillatory sand movement is always normal to the wave crestline so that, if the waves approach the coast at an angle, the net sand motion will have a longshore component called longshore transport. For the nearshore bottom profile to remain in equilibrium there must be no net erosion or accretion. Any net change will be determined by the

net residual currents near the bottom which transport the sediment set in motion by the waves.

The surf zone

Perhaps the most striking characteristic of the sea bed under the surf, is the system of large-scale bars and troughs formed by the action of breaking waves. It is the location of the seaward-most bar which defines the outer limit of the inshore region. The inner limit is defined by the intersection of the beach surface and mean low water. Beyond this the swash of the waves extends up to cover the foreshore.

Within the surf zone there is a high degree of turbulence and suspended sediment motion is predominant. The deep troughs between the bars form natural channels for longshore currents. Because of the high turbulence there is a constant removal of fines from this area.

The foreshore is generally defined as the fairly steep beach face extending from mean low water to the limit of uprush from the breaking waves. Sediment transport in this zone may be bed load or suspended load depending on the breaker type. In spilling or surging breakers, which are most common on very flat slopes, the suspended sediment load will be low and bed load transport will predominate. The uprush from plunging breakers may be loaded with suspended material.

When waves approach the beach obliquely, the bed load in the uprush and downwash on the foreshore follows a zigzag pattern producing a net longshore sediment movement known as beach-drift. This same fluid motion produces longshore currents onshore of the breaker. These are often called 'feeder' currents as they feed the rip currents. Several rip currents may be

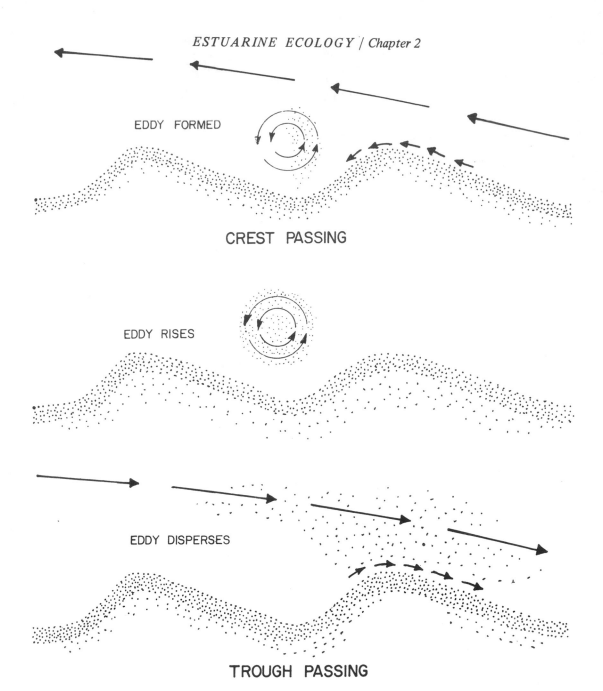

EDDY FORMED

CREST PASSING

EDDY RISES

EDDY DISPERSES

TROUGH PASSING

Figure 2.11
Mechanism of sand movement over bed ripples (after Ippen, 1966).

formed along a surf-beaten shore between headlands and as they transport sediment, the bottom contours become irregular.

The direction of onshore-offshore sediment movement is primarily determined by the beach slope and the steepness of the waves. Offshore movement occurs mostly on steep beaches during storms which generate high, steep waves; onshore sediment movement occurs more frequently on flat beaches during calm periods when the waves are low and flat.

Wave and wind effects on the coasts of southern Africa

As noted earlier, wave heights in excess of 12 m are recorded annually on the coasts of the south-western and southern Cape. Further north the waves decrease so that on both the eastern and western coasts of South Africa the wave height seldom exceeds 8 m. Thus in comparison to European and to the Atlantic and Gulf coasts of North America where wave heights in excess of 3 m seldom occur except during tornadoes, the wave energy on the shores of South Africa is very great. From personal observation it seems greater than on the coast of south-western Australia. Moreover, the prevailing winds on the South African coasts often blow parallel to the shore, the main directions being

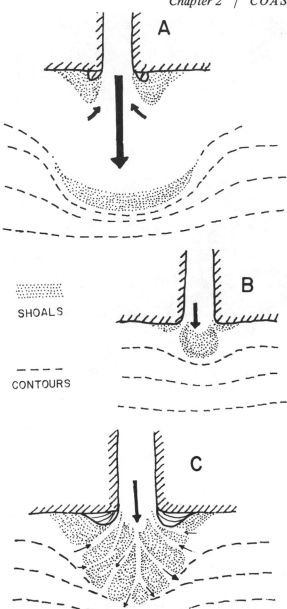

SHOALS

CONTOURS

Figure 2.12
Pattern of ebb tidal deltas with variations of tidal forces, distributive forces and sediment volume (longshore currents omitted).

SE and NW along westerly shores, and SE and SW along southerly shores and NE and SW along easterly shores (Schulze, 1965). As a result, longshore currents and the littoral drift of sediment is very marked. The contours of a Natal beach have been reported to change by the height of 2 m during a single storm. Under such conditions, it is not surprising that the coastline of southern Africa is very even, with few promontories projecting normal to the coastline and few sheltered bays. There are many eroding rocky shores and many long, sandy, surf beaches with little to interrupt the heavy longshore drift of sand. Since fine sediments

are transported seaward well beyond the surf zone, the main sediment that remains is medium to coarse sand.

INLET STABILITY

As stated in the introduction, the type of inlet, whether it be the mouth of an estuary or the entrance to a land-locked bay, depends on several factors. These are:
1. Currents in the inlet caused by tides and river discharge;
2. Currents and waves in the coastal area;
3. Longshore transport of marine sediments;
4. The sediment load discharged by the estuary.
The ratio between the supply of sediment and the distributive forces is of basic importance for it determines whether the sediment is distributed in the sea by waves and currents as fast as it is supplied, or whether it accumulates until the inlet closes for part of the year. The tidal currents in an inlet form two types of channels and bars called the outer and inner bars or, in the terminology adopted by Hayes (1977), ebb tidal deltas and flood tidal deltas. The ebb current, flowing out of an inlet transports sediment derived from the river or from the inner bar and where the current velocity decreases in the sea, the sediment is deposited. Only the shoals and channels of the outer bar will be discussed in this chapter for those within the inlet depend on the estuarine currents which will be discussed later.

Inlet currents

As long as the inlet or mouth of a bay or estuary is wide and deep in proportion to the intertidal volume of the enclosed water, tidal effects will predominate. If, however, the cross-sectional area of the inlet becomes small, the propagation of the tidal wave will be reduced and may become negligible. In such a case the flow through the inlet will be hydraulic, ie the water surface will have a slope. Such a small inlet will probably close. Normally the volume of sea water that enters an estuary basin with the flood tide is substantially augmented by river discharge; the scouring action of the combined outflow during the ebb keeps the inlet open and may enlarge its cross-sectional area.

Presuming that there are no longshore currents, the sandspits and shoals which form the ebb tide delta would be symmetrically arranged at the entrance of the inlet. The channel formed by the stronger ebb current would be directed straight offshore and the sediment would be deposited where the current slackened sufficiently. The weaker flood currents which develop as the tide rises and the water spreads over the shoals, tend to form shallower channels on either side of the central ebb channel with ramps or levees between the two (Hayes, 1977). If the force of the ebb current is strong in comparison to the distributive forces, the sediment is deposited further offshore in the form of a crescentric bar (Fig. 2.12A). If the ebb current

Effect of longshore drift on estuary mouth in West Australia (Photo: J.H. Day)

weakens the material is deposited closer to the estuary mouth and its form becomes more shield-shaped (Fig. 2.12B). If the volume of sediment transported out of the estuary increases while the distributive forces remain constant, the ebb flow delta becomes larger and more complex with branching ebb channels and intervening flood channels, eventually forming 'bird's foot' distributaries (Fig. 2.12C) as at the mouth of the Mississippi.

The elevation of the shoals and sandspits will be controlled by the force of the waves which tend to plane off the surface of the shoals and distribute the material elsewhere. Even waves approaching normal to the coastline will be refracted by the sides of the shoals and a component of longshore transport will develop. Prograding deltas develop best where wave energy is low, as at the mouth of the Mississippi and the Zambesi.

Longshore transport and offset tidal inlets

The direction of longshore transport depends on the direction of the dominant wave attack and consequently on the direction of the prevailing wind relative to the coastline. As the sediment particles are transported away or deposited, the beach is eroded or built up. A mean regime is established on every shore although there may be marked seasonal changes, particularly on coasts subject to typhoons.

Longshore currents may, as stated earlier, be diverted by the flood tide entering an inlet, and the sediments they transport will be deposited within the estuary as part of the flood tide delta. The operation

is reversed by the ebb tide, but some of the sediment remains. Longshore transport is thus of prime importance in a study of the dynamics of inlets. A rough estimate of the magnitude of longshore transport is given by a formula derived by Galvin (1970) from prototype observations on straight coasts where no additional currents caused by tides or other factors were present.

$$S = k_o(H_b)^2 m^3 day^{-1}$$

where

S = an estimate of longshore sediment transport in $m^3 day^{-1}$

H_b = height (in metres) of the breaking wave inducing sediment transport

k_o = a factor of proportionality which varies with the grain size.

Values for k_o for a few typical grain sizes are given in Table 2.3 below:

Table 2.3. Relation between median grain size and k_o

D_{50} (mm)	k_o (m/day)
0,1	7 000
0,2	5 700
0,4	3 700
0,6	2 500
0,8	1 700
1,0	650

It is also possible to get some indication of the amount of longshore transport from the position of the inlet relative to the main axis of the estuary and the shape and position of the sandspits, shoals and channels forming the ebb tide delta.

The sediment on the outer bar, brought into suspension by the turbulence of the waves, is distributed

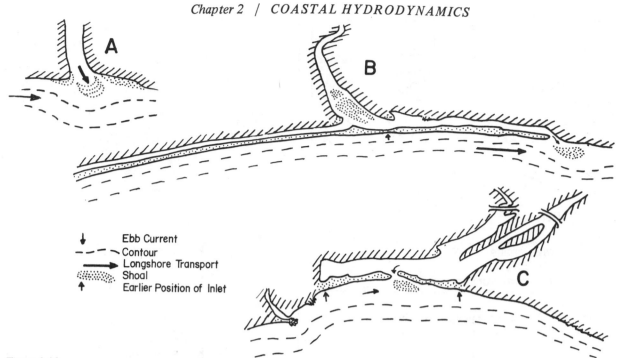

Ebb Current
Contour
Longshore Transport
Shoal
Earlier Position of Inlet

Figure 2.13
Effects of longshore transport and ebb current variations on
positions of inlet, shoals and sandspits.

by longshore currents in the downdrift direction. Thus
the updrift side of the outer bar forms more slowly.
At the same time, longshore transport deposits material
on the near side sandspit and the ebb current erodes
the opposite bank. The combined effect is that the
inlet or mouth of the estuary with its sandspits and
bars migrates along the coast in the direction of the
longshore transport. Figure 2.13B illustrates an exit
channel running parallel to the coastline due to heavy
longshore transport and Figure 2.13A illustrates a
moderate effect. The exit channel of the Matigulu
estuary in Natal has been deflected from the main axis
of the estuary for a length of 4 km and the older parts
of the sandspit which separates the exit channel from
the sea have been elevated by wind-blown sand to
form a dune ridge which has been colonised by coastal
vegetation. Cloete & Oliff (1976) report that after
heavy rains in 1971, the Matigulu captured the Nyoni
estuary, which is deflected along the coast in the same
way. Thus the estuarine channel, which carries the
waters of the two rivers, stretches for about 14 km
parallel with the seashore. More recently (1978), a
flood in the Makigulu River caused a new exit channel
to break through the coastal dunes closer to the axis
of the river (see Fig. 14.9).

The ebb current is by no means constant in volume
or velocity. It obviously increases at spring tide and
during the discharge of flood waters which incidentally
transport the major part of the fluvial sediment. As
the volume of water within the estuary basin increases,

a hydraulic gradient develops and augments the
strength of the ebb current which tends to cut a more
direct route from the main axis of the estuary. It may
merely erode the tip of the sandspit on the updrift
side of the longshore current or it may break through
the sandspit and form a new exit channel. When the
ebb current direction is changed in this way there is
no scouring action to keep the old estuary mouth
open and it is rapidly closed either by fluvial sediment
or longshore transport so that a blind channel remains.
On South African coasts such blind inlets are to be
seen at St Lucia Bay, at the Mlalazi estuary, the
Beachwood mangrove swamp adjacent to the Mgeni
estuary and particularly at Keurbooms estuary in
Plettenberg Bay. Here the mouth migrates along the
shore and is at present near the centre of the old exit
channel which is closed at its western end (see Fig.
2.13C). The original inlet of the Great Berg estuary
was also deflected by north-westerly gales so that it
ran south-westerly, parallel to the coastline before
opening into St Helena Bay. Since there is an important
fishing harbour on the bank of the estuary a new
mouth was cut opposite the harbour and this has been
stabilised by training walls; the old inlet has closed
and the channel to it is silting up. In brief, the inlets
of many South African estuaries migrate along sandy
shores, while the smaller ones close in the dry season.
But not all of them. A study of those that remain
open shows that the inlet has migrated until the mouth
opens in the lee of a rocky promontory.

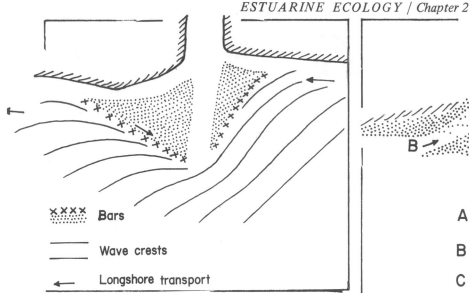

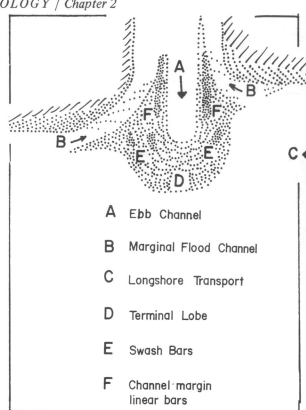

Figure 2.14
Effect of inlet form on wave propagation (after Hayes & Goldsmith, 1970).

A   Ebb Channel

B   Marginal Flood Channel

C   Longshore Transport

D   Terminal Lobe

E   Swash Bars

F   Channel·margin
    linear bars

Figure 2.15
Diagram of ebb tidal delta (after Hayes, 1977).

The effects of wave refraction and tidal range on the ebb tide delta

The presence of the ebb tide delta with its shoals and channels adjacent to an inlet, causes refraction of approaching waves so that the end of the crestline nearest the shoal curves to reach it parallel to its outer margin and the direction of sediment transport is affected (Fig. 2.14). Since the direction of longshore transport is also determined by the predominant wave approach, there tends to be a larger accumulation of sediment on the downdrift side of the ebb tide delta and the shoals are asymmetrical with respect to the inlet (Hayes & Goldsmith, 1970). Moreover, the coastline on the downdrift side of the inlet usually protrudes seaward with the beach face normal to the oncoming waves.

Hayes (1977) focused attention on the effects of tidal range on the morphology of sand accumulation in both ebb tide and flood tide deltas. Following Davies (1964) with a slight modification to metric units, Hayes classifies estuaries as microtidal if the tidal range is less than 2 m, mesotidal if the tidal range is 2-4 m and macrotidal if the tidal range exceeds 4 m. He states: 'River deltas and barrier islands are most common in microtidal areas, tidal deltas and tidal inlets are most common in mesotidal areas and linear sand ridges (sand ridges deposited on a shallow continental shelf by tidal currents), tidal flats and salt marshes are most common in macrotidal areas'. He shows that barrier islands enclosing sheltered bays are common in the Gulf of Mexico. The barrier islands are interrupted by tidal inlets with associated tidal deltas, while in the sheltered bays there are projecting river deltas and other features. Similar barrier islands

and protected sounds extend along much of the Atlantic seaboard of the United States and elsewhere but in these areas wave energy is seldom as great as it is on the coasts of South Africa. Although there is geological evidence that barrier islands existed in the past and may still be seen on the coast of Mocambique, they do not exist on South African coasts at present. The distributive forces on open coasts are too strong to allow fluvial sediments to be deposited as prograding deltas and the shoals that are formed off the mouths of estuaries are planed off below low tide levels.

Postma (1967) has shown that in estuaries and other inlet channels, the tidal currents are seldom in phase with the times of high and low tide. The maximum ebb current occurs late in the tidal cycle near low water, and the maximum flood current occurs before high water so that as the water level starts rising, a strong ebb current is still flowing out of the estuary. This erodes the main estuarine channel and as the water level rises further, the flood current flows in over the shallows on one or both sides of the ebb channel and part of the sediment derived from both the ebb and the flood channels is deposited in the quieter areas between the channels as sand ridges or linear bars. The sediment transported along the ebb channel is deposited in the sea as what Hayes (1977) terms the terminal lobe. On either side of this, he describes broad sheets of sand termed swash platforms and on these, wave action forms isolated swash bars. 'Marginal flood channels usually occur between a swash platform and the adjacent updrift and downdrift beaches' (Hayes, 1977). These features are illustrated in Fig. 2.15. The degree to which the ebb tide delta is offset from the inlet in the downdrift direction depends on the strength of the longshore current.

Hayes reports that tidal inlets and associated tidal deltas are most common in mesotidal areas where the barrier islands are short. He illustrates this by reference to the decreasing length of the barrier islands towards the east on the shorelines of The Netherlands and Western Germany where the tidal range increases. On the coast of Denmark the barrier islands lengthen again as the tidal range decreases to the microtidal category. As noted earlier, there are barrier islands and protected sounds on the coasts of southern Mocambique. At the mouths of Delagoa Bay and Inhambane Bay, where the spring tide range is about 3 m, there is a series of short stubby barrier islands and intertidal shoals comparable to those off the German coast. Within both bays, there is a complex of shoals forming the ebb tide deltas of the estuaries. The flood tide delta of Morrumbene estuary will be discussed in chapter 4.

REFERENCES

BRUNN, P. 1966. Tidal inlets and littoral drift. *Stability of coastal inlets.* **24.** Skipnes Offsetrykkeri, Trondheim.

BRUNN, P. & F. GERRITSEN 1960. *Stability of coastal inlets.* North Holland Publ. Co, Amsterdam.

CLOETE, C.E. & W.D. OLIFF (eds) 1976. South African marine pollution survey report 1974-1975. *S.Afr. Natl Sci. Programmes* Rep 8, CSIR, Pretoria.

DAVIES, J.L. 1964. A morphogenetic approach to world shorelines. *Zeitschr. Geomorph.* **8:** 127-142.

DOODSON, A.T. & H.D. WARBURG 1941. *Admiralty manual of tides.* HM Stationery Office, London.

FROMME, G.A.W. 1977. *Empirical relationship between sand grain size and settling velocity.* Natl. Res. Inst. Oceanol., Stellenbosch, S.Afr. (unpublished internal report).

GALVIN, C.J. 1970. A gross littoral drift rate formula. Abstr. 140, *Proc. 12th Conf. Coastal Engin., Washington, September 1970.* ASCE, Washington.

HAYES, M.O. 1977. Morphology of sand accumulation in estuaries: an introduction to the symposium. *In:* L.E. Cronin (ed), *Estuarine research, vol. II: geology and engineering.* Academic Press, New York. 587pp.

HAYES, M.O. & V. GOLDSMITH 1970. Offset coastal inlets. *Proc. 12th Conf. Coastal Engin., Washington, September 1970.* Vol II. ASCE, Washington.

HJULSTRÖM, F. 1935. The morphological activity of rivers as illustrated by River Fyris. *Bull. geol. Inst. Upps.* **25**(3).

IPPEN, A.T. 1966. *Estuary and coastline hydrodynamics.* McGraw-Hill, New York.

KING, C.A.M. 1962. *Oceanography for geographers* (2nd ed). Edward Arnold, London. 337pp.

LANE, E.W. 1947. Report of the sub-committee on sediment terminology. *Trans. Am. Geophys. Union* **28**(6).

MORGANS, J.F.C. 1956. Notes on the analysis of shallow water soft substrate. *J. anim. Ecol.* **25:** 367-377.

POSTMA, H. 1967. Sediment transport and sedimentation in the estuarine environment. *In:* G. Lauff (ed), *Estuaries.* Am. Ass. Adv. Sci., Washington.

SCHULZE, B.R. 1965. *The climate of South Africa. Part 8: General Survey.* WB **28**, Govt. Printer, Pretoria.

SHIPLEY, A.M. 1964. Some aspects of wave refraction in False Bay. *S.Afr. J. Sci.* **60:** 115-120.

US ARMY CORPS OF ENGINEERS 1973. *Shore protection manual.* Dept. of Army, Corps of Engineers, Washington.

Chapter 3
Estuarine currents, salinities and temperatures

J.H. Day

Department of Zoology, University of Cape Town

The current pattern and the distribution of salinity in an estuary are related. Both are basically due to the ratio between river flow and the volume of tidal water flowing in and out of the estuary mouth. Thus an estuary may be river dominated or marine dominated. This basic pattern is modified by wind drift, wave action, Coriolis force and the shape of the estuary basin so that all these factors must be examined in more detail.

RAINFALL AND RIVER FLOW

Although river flow is primarily determined by the area of the drainage basin and the annual precipitation, it is modified by many factors. The seasonal distribution of the rain or snowfall, the temperature, the rate of evapotranspiration, the permeability of the soil, the percentage of vegetation cover and the slope of the watershed are all important. To some extent they are interrelated.

In polar latitudes the precipitation is low; for example, the drainage basin of the Yukon in Alaska receives an average precipitation of 320 mm yr^{-1}. During the winter the river is frozen and the bottom water is clear and slow-flowing. When the thaw comes in spring, the ice floes erode the banks and the river carries muddy fresh water to the sea. During the summer the river water gradually clears and saline water moves upstream as a bottom current. In this case and in the fjords of mountainous areas, currents and salinities in the estuaries are primarily related to temperature changes.

In temperate regions there are many areas where the precipitation is between 650 and 750 mm per year. In the lowlands of most parts of Europe, the precipitation is fairly evenly distributed through the year, evaporation is low and the soil is permeable and well-covered with vegetation. The water percolates slowly into the soil and the river flow is continuous throughout the year. The salinity regime in the estuaries is fairly constant and the water, where not polluted, is clear though peat-stained. Conditions around Buenas Aires, Melbourne and Christchurch are similar. In South Africa, the Tsitsikama coast between George and Humansdorp has a similar even rainfall, but the coastal plateau is dissected by deep gorges. However, the forests limit runoff and erosion and the streams flow continuously although they are heavily stained with peat. The salinity in the estuaries varies little through the year.

Cape Town and Perth have a similar annual rainfall in the 650-890 mm range but the rains fall in winter and the summers are dry. The largest rivers such as the Berg in the Western Cape and the Swan in West Australia have catchments in well-watered mountain areas. The rivers carry muddy floods in winter so that the salinity in the estuaries is very low at that season. In summer the rivers flow slowly over the coastal plains, much of the silt is deposited and saline water penetrates well inland along the river bed.

In very arid areas such as Peru, Western Australia north of Perth, and in Namibia and Namaqualand in southern Africa, the rainfall is minimal and may fail completely in many years, while the rate of evaporation is always high. Normally there is only a slow seepage through the sands in the river beds and the water is often brackish. The occasional floods fill the river channels but they are of brief duration and such rivers have no estuaries. Even the mighty Orange River which drains a third of southern Africa, carries silt-laden fresh water to the sea in summer but is reduced to a trickle or may be even blocked by a sandbar during the dry season. The rate of flow is so changeable and the salinity changes are so abrupt that no estuary has developed. The exit to the sea is classified as a river mouth.

The subtropical coasts of the Transkei, Natal and southern Mocambique have a high rainfall of 1 000 to 1 250 mm yr^{-1}, most of which falls in summer. In the hilly country of the Transkei and southern Natal, the gradients of the rivers are so steep that the watersheds of the rivers are seriously eroded and chocolate brown floods occur every summer. The estuaries of the longer rivers are thus short and heavily silted and the salinity changes markedly with the seasons. Some of the shorter coastal rivers have cut deep ravines through the hills in past geological times and they now carry little silt; their estuaries are up to 35 m deep and the water is well stratified. Other Transkeian estuaries lie in ancient flood plains and are well-developed.

The coastal plain broadens in northern Natal and Mocambique and the land is low-lying with a range of sand dunes along the seashore. The rivers meander over the plains with many marshes and some even form permanent lakes which have obviously been cut off from the sea during Recent geological times for they are still slightly saline. Others broaden to form shallow estuarine lagoons before they cut a narrow exit channel to the sea. Salinities vary with the seasons and during droughts the smaller estuaries close while the very large St Lucia lakes become hypersaline.

TIDES IN ESTUARIES

The current patterns of estuaries are complex due to the stratification or various degrees of mixing of salt and fresh water. It is thus simpler to deal first with tides in homogeneous channels such as marine inlets and completely mixed estuaries and return later to consider the effects of density differences.

As a tidal wave enters a shallow channel of uniform width, its energy is decreased by friction. The length of the tidal wave decreases and so too do its height, velocity and the relative duration of the ebb and flow. The length of the tidal wave is a function of the depth of the channel. The mean depth of the tidal reach

(which includes the tidal river beyond the limit of saline intrusion) is difficult to determine but if it is 3 m the length of the tidal wave is about 242 km and if the mean depth is 1 m the length of the tidal wave is about 140 km (King, 1962). If the length of the tidal reach exceeds this, more than one tidal wave passes up the channel as in the Amazon river. Dyer (1973) states that if the tidal reach exceeds one quarter the length of the tidal wave, the maximum tidal velocity upstream occurs at the time of high tide, slack water occurs at mid-tide and the maximum velocity seaward occurs at low tide. If the tidal reach is less than one quarter the length of the tidal wave (as in all South African estuaries), the velocity decreases before high tide so that slack water occurs at high tide and the maximum velocity is near mid-tide. If the energy of the tidal wave is completely dissipated along the tidal reach so that there is no reflection of energy, the tidal range and the velocity of the current decreases towards the head of the estuary. As a result, the times of high and low water are delayed more and more as the distance from the sea increases. Moreover, as the tidal velocity depends on the depth of the channel and the depth below the crest of the tidal wave is greater than the depth below the trough, the high tide progresses more rapidly upstream than the low tide. Thus the duration of the flood tide is less than the duration of the ebb even when the river is not flowing. It is thus important to determine the times of high and low water along an estuary before arranging any sampling programme. As the discharge of the river increases, so does the duration of the ebb and the water may still be flowing out of the mouth when the tide is rising in the sea. In a heavy flood the water flows into the sea at all states of the tide.

Thus far it has been assumed that the channel is of uniform width. Most estuaries in southern Africa are constricted by sandspits at the mouth and then widen into a lagoon before the upper reaches narrow again towards the head of the estuary. In broad sections, both the velocity and range of the tide is reduced while in narrow sections the velocity and range is increased. When the northern half of Richards Bay was being modified to form a deep-water harbour, the southern half or lagoon was separated off by a berm to form a nature reserve and a sea inlet was cut through the sand dunes as shown in Figure 3.1. The tidal regime as determined by the National Institute for Oceanology (NRIO) shows the reduction in tidal range and the lag in the times of high and low water at that time, although very different tidal conditions developed when the sanctuary changed (see chapter 14).

A different tidal regime develops in Knysna estuary. Here the tidal range decreases in the broad lagoon and then increases in the narrow channel through the upper reaches as shown in Table 3.1 and Figure 3.2, based on data from NRIO reported by Anderson (1976).

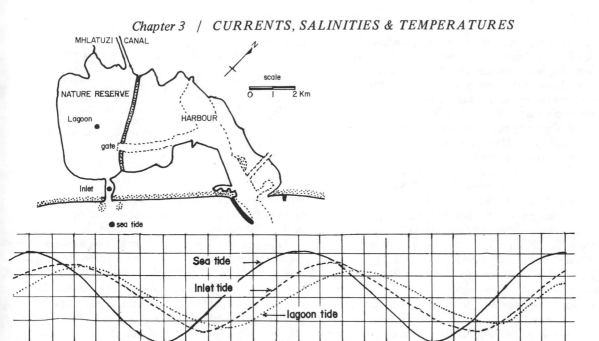

Figure 3.1
Tides in Richards Bay nature reserve when first opened (based
on NRIO observations).

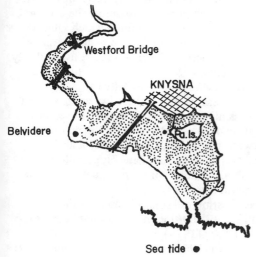

Table 3.1
Tidal observations in Knysna estuary during spring tide,
October 14, 1976

Position	Distance from sea km	Channel width km	Tidal range m	Lag at HWS min	Lag at LWS min
Sea mouth	–	–	1,83	0	0
Paarden Is.	4,2	1,3	1,78	35	75
Belvedere	9,8	0,8	1,89	50	90
Westford Br.	15,5	0,1	1,92	60	100

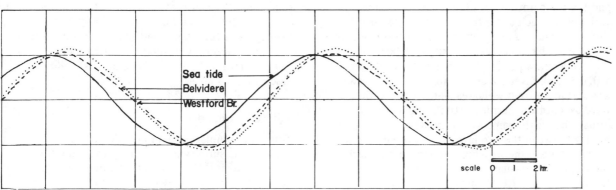

Figure 3.2
Tides in Knysna estuary (based on NRIO observations).

The tidal wave which enters the deep channel between Knysna Heads where the channel is about 0,14 km wide has a surface velocity of 1,2 m.sec^{-1} but slows down in the lagoon. Opposite Paarden Island where the channel is 1,3 km wide, the surface velocity decreases to about 0,5 m.sec^{-1} and the range is reduced. Opposite Belvedere the channel width decreases and the tidal range increases. In the upper reaches at Westford Bridge the width of the channel is reduced to 0,1 km and the tidal range is 1,92 m. It may also be seen that the tidal lag at LWS is much greater than the lag at HWS.

If an estuary has a wide mouth and the channel width decreases progressively upstream so that it is effectively funnel-shaped, the tidal range increases dramatically. There are no estuaries of this type in southern Africa but Perkins (1974) records that at the mouth of the Bristol Channel which is 64 km wide, the average tidal range is 6,06 m. About 128 km upstream near Worcester where the Severn estuary is only 0,8 km wide, the range is 10 m. Due to the decreasing depth along the Severn, the velocity of the incoming tidal wave decreases but towards high tide the depth below the crest increases and the front of the wave moves more rapidly and becomes steeper until it moves upstream as a wall of water and a tidal bore is formed. Bores are rare and are most likely to occur at spring tides in those sections of an estuary where the slope of the bed and the constriction of the channel are suitable and then disappear where the estuary broadens and the bed is more level (King, 1962). They are known not only in the Severn but also in sections of the Trent in England, the Bay of Fundy in Canada, the Hoogly in India and elsewhere. The most impressive is in the Tsientang Kiang in China where a wall of water over 4 m high advances upstream at almost 8 m.sec^{-1}.

THE EFFECT OF SALINITY DISTRIBUTION ON ESTUARINE CURRENTS

The density of sea water depends on both the salinity and the temperature but in estuaries where the salinity variation is much larger than in the sea, temperature differences have a relatively minor effect on density. River water has an average density of 1,001 g.ℓ^{-1} at 20° while sea water of 35 ‰ salinity has a density of 1,026 5 g.ℓ^{-1}. River water thus tends to flow seaward as a surface current while sea water tends to flow landward as a bottom current until its level over the estuary bed is the same as that in the sea. If there is no mixing between the two layers, the *salt wedge* extends landward along the shelving bed until its tip reaches the head of the estuary. As the tide rises and falls the tip of the salt wedge advances and retreats so the length of the estuary changes. The volume of estuarine water between high and low tide levels is known as the

Table 3.2
Salinity terminology and the reaches of an ideal estuary

Salinity range (‰)	Grade	Reaches in an ideal estuary
0,1 – 0,5	limnetic	river
0,5 – 5,0	oligohaline	head of estuary
5 – 18	mesohaline	upper reaches
18 – 25	polyhaline	middle reaches
25 – 30	polyhaline	lower reaches
30 – 40	euhaline	mouth
Over 40	hypersaline	

tidal prism and as it increases from neap to spring tide or with the discharge of the river so does the velocity of the tidal current.

Although in theory there may be no mixing between the surface layer and the salt wedge below, this must be extremely rare. Usually there is some degree of eddy diffusion and turbulent mixing at the interface. This may result in a mixed layer at intermediate depth so that there is a vertical stratification of salinity. If mixing proceeds further there may be merely a gradual increase of salinity with depth, while if turbulence is extreme the salinity will be homogeneous from surface to bottom.

Different estuaries and even different sections of the same estuary differ greatly in salinity. Many different terms have been proposed to define the different grades of salinity (or chlorinity upon which salinity was originally based) but the Venice system has now been universally adopted. Carriker (1967) used salinity grades and, following Day (1951) related the grades to the various reaches of an ideal estuary.

The various reaches of a real estuary are based not only on the salinity but also on the current velocity and the nature of the substrate so that the correspondence between the reaches of an estuary and the salinity range is only approximate. Hypersaline conditions are only liable to occur in the upper reaches of estuaries in arid countries.

The classification of estuaries based on the current system and the distribution of salinity has been developed by Pritchard (1955), Cameron and Pritchard (1963), Pritchard (1967) and others. The main factors which affect the current system and the distribution of salinity are the ratio of river discharge to tidal flow, the degree of turbulence and the width and depth of the estuary basin. A mathematical treatment will be found in Dyer (1973). The account that follows is a descriptive summary.

Salt wedge estuaries

This is the simplest system in which there is no mixing between the surface layer of river water and the sea water below it. It is a theoretical condition which may occur where a calm, slow-flowing river meets the sea where there is little or no tidal action. The surface of the salt wedge is horizontal and at midtide it corresponds to mean sea level but as the tide rises the tip of the wedge advances and as the tide falls it retreats so that there are slow currents within the wedge although there is no net motion over a complete tidal cycle. The velocity of the surface layer is determined by the gradient between the head of the river and the sea surface and varies with the discharge of the river. When the system is in equilibrium, the volume of sea water entering during the flood tide plus the volume of river water that has accumulated in the estuary basin during the whole tidal cycle must be discharged to the sea during the same period. This discharge is not necessarily limited to the fall in tidal level, for the surface water will continue to flow into the sea until the tidal level in the sea rises to correspond with that in the estuary mouth. As the width of an estuary usually increases towards the mouth the maximum velocity of the ebb current decreases downstream but there is no decrease in velocity with depth from the surface to the discontinuity between the surface and the bottom layer. If the width of the estuary exceeds 0,5 km, Corioli's force becomes significant and the interface between the two layers will be tilted laterally. In the northern hemisphere the depth of the surface layer will be greatest near the right bank looking downstream and the thickness of the bottom layer will be greatest near the left bank. The reverse will be true in the southern hemisphere.

Such a complete separation of surface fresh water and bottom sea water does not occur in the estuaries of southern Africa. It may occur in some fjords in winter when there is little fresh water outflow. In the fjords that have been investigated however, there are one or more layers of mixed water at intermediate depths.

Highly stratified estuaries

As the velocity of the surface layer of river water increases, the friction at the interface with the lower layer will increase too and internal waves will develop. If these waves break, salt water will be trapped in the upper layer. The process is called *entrainment* and the salinity and volume of the upper layer will increase progressively towards the sea. The salinity of the salt wedge is not reduced for the mixing is limited to the upper layer but there is now a slow net landward movement in the salt wedge and its surface slopes upward towards the sea to resist the friction with the upper layer. Corioli's force acts as before. This type of circulation is called a *two layered flow with entrainment* and occurs in deep estuaries with a high ratio of river flow to tidal flow (see Figure 3.3A).

Although some estuaries in southern Africa, such as the Mtamvuna, are well stratified it is unlikely that this is due to entrainment alone. According to Pritchard (1967), this type of circulation occurs in the mouth of the Mississippi and according to Dyer (1973) it also

occurs in the Velar estuary in Bengal. It seems likely that it occurs in some fjords during summer when the melt water from glaciers causes an increased flow of fresh water. Most fjords have a sill at the mouth and the exchange between the fjord and the sea is limited to layers above the sill. If the sill is shallow the deeper water in the fjord becomes stagnant and may become anoxic but many sills are over 50 m deep and during the summer fresh water outflow causes a compensating inflow of sea water at a deeper level which refreshes the system.

Partially mixed estuaries

As turbulence at the interface between the surface layer and the bottom layer increases, mixing develops in both layers. The turbulence may be increased by different factors; it increases when the ratio between tidal flow and river flow approaches 1:1 and when there are irregularities in the channel bed. It is accentuated in shallow estuaries where the volume of the tidal prism is large in comparison with the total volume of the estuary basin. In broad estuaries the fetch of the wind may be sufficient to generate waves of sufficient height to cause turbulence at the interface (see Figure 3.3B).

According to the degree of turbulent mixing, the vertical salinity gradient may show a halocline with a sudden increase of salinity between the two layers or the salinity may increase evenly with depth to the bottom of a shallow estuary or until the isohaline part of the bottom current is reached in a deep estuary.

The addition of saline bottom water to the upper layer not only increases its salinity but also its volume until it may be many times the volume of river water that entered the estuary (Pritchard, 1967). Moreover, its velocity is increased as it conveys the river water to the sea. Similarly the downward mixing from the upper layer decreases the salinity of the lower layer and increases its volume and during the flood tide its landward velocity is increased to replace the losses to the upper layer. Pritchard (1967) estimated that in the estuary of the James River the net seaward flow in the upper layer was 20 times the discharge of the river and the net landward flow in the lower layer was 19 times the river flow. These net flows are of course masked by the much larger volumes transported up and down the estuary by the flood and ebb tides.

The velocity of the upper layer is greatest at the surface and decreases with depth until the interface with the lower layer is reached. Here there is a layer of no net motion over a tidal cycle and below this the velocity of the lower layer increases until it is affected by friction with the estuary bed.

The salinity of the upper layer increases downestuary and the salinity of the lower layer decreases up-estuary to the tip of the salt wedge so that over

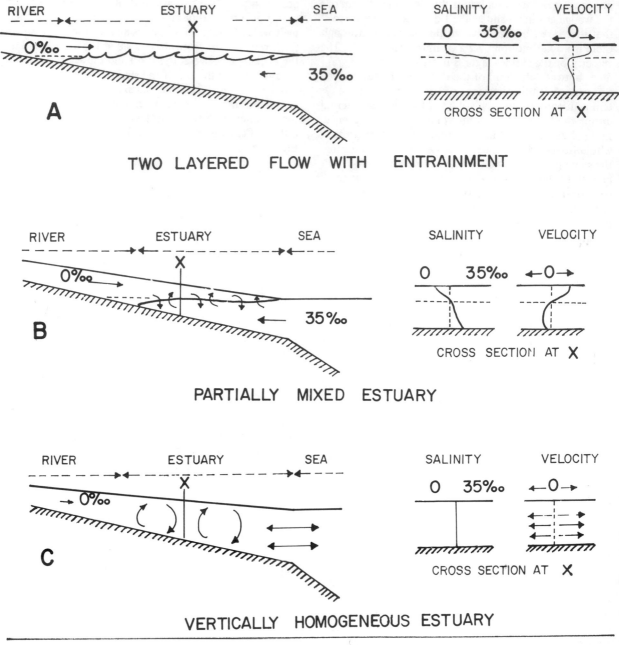

RIVER   ESTUARY   SEA   SALINITY   VELOCITY

X

0‰ →

← 35‰

0   35‰   ← 0 →

CROSS SECTION AT X

A

TWO LAYERED FLOW WITH ENTRAINMENT

RIVER   ESTUARY   SEA   SALINITY   VELOCITY

X

0‰ →

← 35‰

0   35‰   ← 0 →

CROSS SECTION AT X

B

PARTIALLY MIXED ESTUARY

RIVER   ESTUARY   SEA   SALINITY   VELOCITY

X

→ 0‰

0   35‰   ← 0 →

CROSS SECTION AT X

C

VERTICALLY HOMOGENEOUS ESTUARY

Figure 3.3
Salinity and current distribution in three types of estuaries.

most of the estuary there is a continual increase of salinity towards the sea. At the head of the estuary where the river water enters at the surface, there is vertical stratification and the isohalines are horizontal but thereafter the isohalines slope up towards the sea. As the degree of mixing increases they become steeper until they may be nearly vertical. A section across the main channel of a narrow estuary would show a series of horizontal isohalines but in a broad estuary Coriolis force makes them slope down towards the right bank in the northern hemisphere and towards the left bank in the southern hemisphere as has been shown by Pritchard (1952). Pritchard (1967) classes such estuaries as 'moderately stratified' and quotes Chesapeake Bay, the Savannah River estuary and several others as examples. Dyer (1973) refers to them as partially mixed estuaries and mentions the James estuary, the Mersey, the Thames and Southampton Water as some of the many estuaries of this type.

Vertically homogeneous estuaries

While complete separation of river water and sea water marks one theoretical limit, complete mixing to form a vertically homogeneous estuary marks the opposite extreme (see Figure 3.3C). This system is the result of turbulent mixing from surface to bottom and this is most likely to occur in shallow estuaries with a large tidal range so that the ratio of tidal inflow to river discharge is of the order of 10:1 or greater. The limit of saline intrusion which marks the head of the estuary advances upstream with the flood tide and retreats with the ebb but over many tidal cycles the net flow downstream would tend to drive the salt towards the mouth. According to Dyer (1973), the balance in narrow estuaries is maintained by mixing along the sides of the estuary. The high salinity water which advances upstream with the flood tide is trapped in embayments and backwaters and bleeds back into the main channel as the tide falls. It may be added that in arid countries the evaporation from intertidal flats increases the salinity of this seepage water. In broad estuaries where Corioli's force is effective, the high salinity water flows in along the left bank of northern hemisphere estuaries and the low salinity water of the ebb tide flows out along the right bank so there is a salinity difference across the estuary. There is also a tendency for separate ebb and flow channels to develop between the shoals. In such cases the circulation is horizontal rather than vertical but complete lateral separation of salinities merely represents a theoretical limit. Normally there is a gradual increase in salinity from one bank to the other as well as along the length of the estuary so that the isohalines are oblique to the axis of the estuary. Pritchard (1955) shows that this type of circulation occurs in the wide lower reaches of the Delaware and Raritan estuaries in New Jersey.

Periodically closed estuaries

These are the bodies of saline water referred to by Day (1951) as blind estuaries and termed estuarine lagoons by Jennings & Bird (1967). They may be open to the sea at frequent intervals or they may be closed for a year or more. There are many such estuaries in southern Africa, Western Australia and other arid areas but as yet their hydrodynamics have not been studied in detail.

As explained in chapter 2, bars develop very rapidly at the mouths of estuaries opening on the sandy shores of high energy coasts. While the river is flowing strongly the outflow of the tidal prism keeps the mouth scoured and open but during periods of low flow the bars become shallower, the sandspits lengthen and eventually the mouth closes. Tidal effects then cease within the estuary and the only circulation is that due to the inflow of the river and the stress of the wind. In broad shallow lagoons, the fetch of the wind may be sufficient to ensure complete mixing from surface to bottom but more usually salinity and temperature gradients become marked and in deep estuaries stratification develops and the bottom layers become stagnant and may be depleted of oxygen. The salinity depends on the ratio between evaporation and river discharge plus direct precipitation. If the gains of fresh water exceed the losses, the salinity decreases and the level of the estuary rises until it exceeds the height of the bar. The initial trickle of imprisoned water rapidly develops into a torrent and the bar bursts open. When the level of the estuary has fallen, sea water enters and a normal salinity regime is reestablished. Occasionally the loss of saline water continues at intervals until the estuary becomes hyposaline. If the losses by evaporation exceed the gains of fresh water, the level of the estuary falls and the salinity increases until the estuary becomes hypersaline.

Hypersaline estuaries

These are estuaries in which the salinity exceeds 40 ‰. This may occur even while the mouth is still open. When the level of the estuary falls below mean sea level there is a net inflow of sea water over a tidal cycle, and as the 35 ‰ sea water flows along the channel it evaporates and the salinity increases. The salinity at the mouth remains fairly constant with the inflow and outflow of the tides so that hypersaline conditions develop further up the estuary and become most marked in the upper reaches. Usually the mouth closes during the process and salinities may then reach 90 ‰ or more. Since the density of sea water is less than that of hypersaline water, sea water lies on the surface and the denser hypersaline water lies in the deeper channels. When the mouth is open it can only escape to the sea if the channel slopes seaward and the bar at the mouth is sufficiently deep. When the mouth is closed and tidal currents are absent the only circulation is that due to the wind but in broad shallow lagoons this may be sufficient to ensure complete mixing from surface to bottom. The classic example of a hypersaline estuary is the Laguna Madre in Texas described by Hedgpeth (1947, 1967). South African examples will be described later.

SALINITY PATTERNS IN THE ESTUARIES OF SOUTHERN AFRICA

The basic concepts of estuarine circulation introduced by Ketchum (1951), Pritchard (1952, 1967), Stommel (1953), Bowden (1963) and Dyer (1973) were largely based on a studies of estuaries in low-lying countries where the precipitation was high and the estuaries long and permanently open to the sea. Conditions in southern Africa are more variable. Details are given in chapter 14. Apart from the Tsitsikama coast where the rainfall is fairly uniform throughout the year, many drainage basins have a markedly seasonal rainfall with winter rains in the Western Cape and summer rains in the Transkei, Natal and Mocambique. Thus the river discharge varies greatly and at one season the estuary is river dominated and at another it is marine dominated. Flash floods are common but the river may be diminished to a trickle in the dry season and the estuarine salinity increases in consequence. Many of the smaller estuaries close in the dry season and the ratio of evaporation to river flow then determines whether the salinity decreases or increases to abnormally high values. The estuaries of southern Africa thus provide a rewarding field of study but in fact little attention has been paid to circulation patterns and salinity structure. Many of the records were made by biologists primarily interested in the distribution of the estuarine fauna or chemists monitoring pollution. In many cases these records are fragmentary or made without due regard to the lag in the times of high and low tide along the length of the estuary. Examples of estuaries which have been studied in some detail are quoted below. As there are many differences between permanently open and periodically closed estuaries the two classes will be considered separately.

Permanently open estuaries

As shown in chapter 14, the Msikaba is one of the very few deep and well stratified estuaries in southern Africa. It opens some 30 km north of Port St Johns. The coast here is precipitous and the estuary flows for 3,8 km through a steep canyon. While making a plankton survey, Woolridge (1976) recorded the salinity on various occasions between 1970 and 1973. The salinities during July and November are shown in Table 3.3.

During the dry winter, littoral drift builds up the bar until the mouth is only 0,5 m deep. Inside the estuary the surface salinity is high and there is a marked halocline at 2-3 m. During summer the increased outflow cuts a deeper channel through the bar so that it was 2,5 m deep in the summers of 1970 and 1971. The salinity records indicate that the high salinity surface water in July was being replaced by low salinity water in November and that sea water of 35 ‰ was flowing in along the bottom to fill the deeper parts of the estuary to within 4 or 5 m of the surface.

While studying the infratidal zonation in the same estuary in 1973, Blaber, *et al* (1974) found that the maximum depth in the Narrows about 2 km from the mouth was 35 m. At this position they made a detailed study of the salinity, temperature, light penetration and the percentage saturation of oxygen at different depths and presented their data in their figure 3. They show that the percentage saturation of

Table 3.3
Salinities in Msikaba estuary in July and November 1971

Distance from sea (km)	0,5		1,7		2,9		3,8	
Depth (m)	5		25		8		1,5-5,0	
Salinity (‰)	July	Nov	July	Nov	July	Nov	July	Nov
0 m	20	4	20	2	22	1	6	1
1 m	29	27	20	2	23	5	25	1,5
2 m	34,5	31	25	21,7	32	23	32,5	10
3 m	35	33	34,5	29	34	34		25
4 m	35	33,5	34,5	34	34	35		
5 m	35	35	34,5	35	34	35		
8 m			34,5	35	34	35		
16-23 m			35	35				

oxygen in the deeper layers changed during the year. In January it decreased from 100 % at the surface to less than 5 % below 12 m; in June it decreased gradually from 60 % at 12 m to 48 % at 33 m. By October the deep layers had again become depleted of oxygen so that the percentage saturation was less than 5 % below 12 m.

This additional information confirms that the surface layers are stratified and that the deeper layers have a homogeneous salinity although it is less in October than in January and June. Salinity values suggest that the stratification is limited to the first 5 m while oxygen values suggest that changes extend down to 12 m. Due to bad weather the oxygen observations in April were limited to the top 14 m and no salinities were recorded. Nonetheless the increase in oxygen concentration between 10 and 14 m in April from the low values recorded in January and the much higher values in June suggest that when the bar deepens, well-oxygenated sea water flows in along the bottom. By June it has mixed with the old water below 6 m, increasing both the oxygen content and salinity. By October the oxygen content below 12 m has again been depleted and the deep water remains stagnant until the next flood deepens the mouth.

Vertical stratification and oxygen depletion of the bottom layer is known to occur in other deep estuaries in southern Africa. It occurs in the temporarily blind Mtamvuna estuary described in chapter 14 also in Lake Sifungwe in the Kosi Bay system and in Swart Vlei, both of which will be described in chapter 15 on coastal lakes.

Many estuaries in southern Africa are completely mixed near the mouth, partially mixed in the middle reaches and stratified at the head but the length of each section is very variable. Many factors are responsible, including the ratio of river discharge to tidal flow and the depth of the estuary basin.

The Mlalazi estuary on the Zululand coast is typical of many shallow estuaries in southern Africa. Salinities were first recorded by Hill (1966) and a further set of records was made by the NIWR and edited by Oliff (1976) as part of the first (unpublished) report of the S. African National Marine Pollution Monitoring Pro-

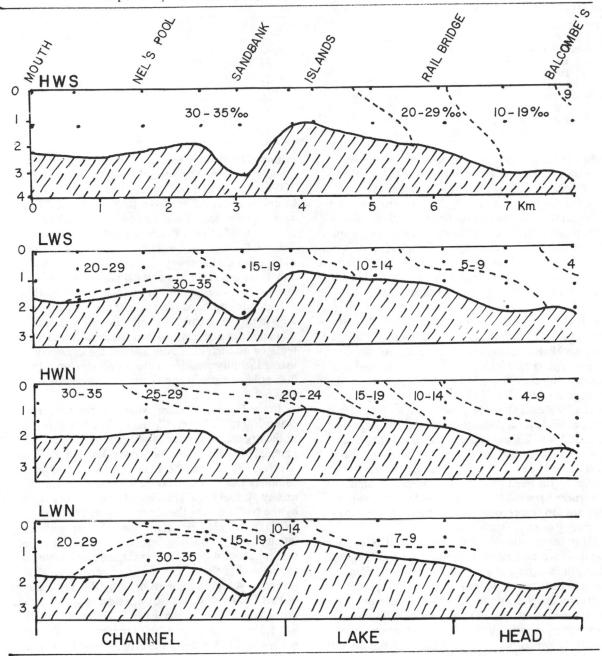

Figure 3.4
Isohalines in Mlalazi estuary at different tides (modified from Hill, 1966).

gramme. The two sets of records agree in broad outline but differ in detail. Hill's figure 3, which shows salinities at high and low tide of neaps and springs is reproduced above.

The estuary is 7 to 9,3 km long depending on river discharge, 100 to 250 m broad and 1 to 3,5 m deep at HWS. The actual mouth is constricted to 30 m between sandspits. This is important for, as Hill remarks, the volume of water that enters during the flood tide of

springs cannot all escape before the next tide starts rising in the sea. For this reason the lowest water level in the estuary occurs at low neap tide. It is only after a flood has temporarily widened and deepened the mouth that the water level is lowest at low spring tide. Salinities are affected in the same way. At low neap tides salinities of less than 10 ‰ extend half way down the estuary and the isohalines slope up very gently towards the sea. At high spring tide salinities of more than 30 ‰ extend more than 5 km from the mouth and further up, the isohalines slope fairly steeply. The isohalines in Hill's figure are not drawn at close intervals but the text suggests that the water

Table 3.4
Salinities in Mngazana estuary

		0,3	1,9	2,8	3,7	4,6 (causeway)	5,1	6,5
Distance from sea (km)		0,3	1,9	2,8	3,7	4,6 (causeway)	5,1	6,5
Depth at LW (m)		2,3	2,5	2,5	1,8	1,5	3,0	0,1
Salinities in July	Surf.	35	35	35	32	31	15	3
(‰) LW	Bot.	35	35	35	32	30	30	–
HW	Surf.	35	35	35	34	33	15	4
	Bot.	35	36	35	33	31	30	–
Salinities in Dec.	Surf.	36	22	24	20	20	5	3
(‰) LW	Bot.	36	32	30	30	29	26	–
HW	Surf.	35	35	35	32	32	8	–
	Bot.	35	35	35	35	34	24	–

in the lower reaches is homogeneous and the NIWR records of 27 May 1971 confirm this and show that completely mixed water of 34,5 ‰ extends to the surface 3 km from the mouth. At this point there is a 1 m depression in the estuary bed and high salinity water remains here during the ebb. Even during a spate when water of 2-3 ‰ flows over the surface, the water in the depression is stratified with a salinity of 30 ‰ at the bottom. The NIWR records show that during heavy floods the high salinity water is washed out of all depressions and is not replaced until the next high spring tide. It appears that after floods the salt wedge takes several days to reach the head of the estuary.

In the Mlalazi, complete mixing in the lower reaches is most extensive at high spring tides and is facilitated by turbulence in the narrow mouth. In Morrumbene estuary, Day (1974) reported complete mixing at all states of the tide in the broad shallow lagoon which extends for a distance of 10 km. The large tidal range of 2,4 to 2,6 m at springs and the strong tidal currents between the many shoals are probably responsible. In the quiet channels through the mangrove swamps to the head of the estuary there are salinity differences between the surface and the bottom but there are too few records to say whether there is partial mixing or stratification.

Many other estuaries along the coasts of Natal, the Transkei and the Eastern Province as far south as Port Elizabeth have been studied by the South African National Institute of Water Research. A long series of reports on 'Marine disposal of effluents' and the first two annual reports for 1975 and 1976 on 'Marine pollution surveys' edited by Oliff (1976) contain much valuable information on water quality but salinity observations were merely incidental to the study of pollution. Nonetheless they show that all the estuaries of large rivers are surprisingly short and shallow. During the summer rains these large rivers are often in flood and heavily silted. The estuary basins are filled with mud and the saline water is swept out to sea.

The smaller estuaries whose drainage basins are limited to the well-watered coastal belt of Natal and the Transkei are more stable and better developed. Those which open on unprotected sandy beaches are often closed by sandbars in the dry winters but if the mouth opens under the lee of a rocky headland which diverts littoral drift, it shallows but remains open. Mngazana estuary in the Transkei at 31°45'S is an

example. Its ecology has been studied by Branch & Grindley (1979). The estuary is 6,5 km long, 100-200 m wide and 1,5-2,5 m deep but there is a shallow causeway about 4,8 km from the sea. The tidal range at springs decreases from 1,5 m at the mouth to 1,0 m near the head of the estuary. As will be seen from Table 3.4, the salinity regime is remarkably constant as far up as the causeway. During the dry season in July there is no sign of a vertical salinity gradient. Such a gradient is present at low tide in summer but at high tide undiluted sea water extends to the surface 2,8 km up the estuary. A comparison of high and low tide salinities suggests that half the estuary water is replaced on each tidal cycle. Above the causeway a marked salinity gradient is always present since the high salinity water at the bottom is trapped by the causeway.

In the Eastern Province, where the rainfall is normally 500 to 700 mm.yr^{-1}, the smaller estuaries close in the dry season but the larger ones such as the Keiskamma, the Great Fish, the Kowie, Kariega, Bushmans, Sundays and the Swartkops all remain open. The most complete sets of records were made in the Swartkops; one by McLachlan & Grindley (1974) and one made by the NIWR during the spring tides of February 1975 (Oliff, 1976). His records show a very uniform salinity of about 33,7 ‰ from the surface to the bottom for about 4 km from the sea and from there on to the head of the estuary there are vertical salinity gradients. A report commissioned by the city engineer of Port Elizabeth shows that the tidal range at springs drops from 1,62 m in the sea to 1,3 m in the mouth and 1,2 m at a distance of 9,3 km upstream. During the severe flood of 1971 the estuary rose 1,5 m but the tidal range was hardly affected. Possibly the flood waters spread so widely over the marshes that the increased depth in the channel had negligible effects on the tidal range.

The mouth of Knysna estuary is stabilised between rocky headlands and the depth on the bar is 4 m. Within the headlands the estuary broadens into a lagoon but the main channel deepens in places to 14 or 16 m and is mainly 5 m deep as far as the rail bridge. The estuary is 19 km long and a general account of seasonal changes in salinity along the whole estuary is given by Day, et al (1952) who found no vertical salinity gradients in the lower half of the estuary. Grindley (1978) quotes more detailed observations made by the Sea

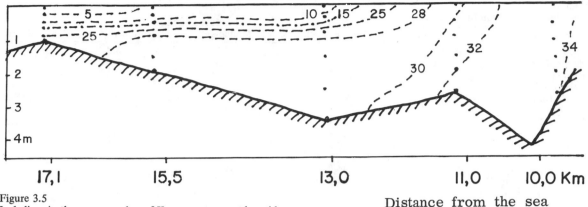

Distance from the sea

Figure 3.5
Isohalines in the upper reaches of Knysna estuary at low tide
(after Korringa, 1956).

Fisheries Branch in August 1976 and April 1977. These show high salinities decreasing from about 35,3 near the mouth to between 32 and 33 ‰ half way up the estuary. At some stations the salinity increases slightly with depth but if there are salinity gradients they are ephemeral. In the upper reaches the vertical salinity gradients become very marked. The most complete sets of records in this section are those of Korringa (1956). His low tide records taken on 24 January 1953 are shown as isohalines in Figure 3.5.

It will be seen that the isohalines are almost horizontal in the narrow channel extending from 17,1 to 13 km from the sea but become steeper 11 km from the sea. This indicates that the degree of mixing increases down estuary until the water becomes completely mixed in the lower reaches. Korringa's second set of records taken over high tide on 2 March 1953 has not been reproduced here but shows steeper isohalines extending to 19 km up the estuary indicating that at high tide mixing extends almost the whole length of the saline intrusion.

A consideration of the changing width and depth distribution along the estuary shows that the lower 10 km must contain well over 80 % of the total volume of the estuary basin and the fact that salinities in excess of 32 ‰ reach the surface throughout this section indicates that the estuary is marine dominated. Since the currents during spring tides may reach velocities of 0,5 to 1,1 m.sec^{-1} and the channel is beset by rocky promontories at the mouth and many shoals elsewhere, the currents are turbulent and homogeneous salinity distribution is to be expected. Further, these conditions in the lower half of the estuary are maintained throughout the year. Even in the upper reaches where the water is only partially mixed, Korringa's low and high tide records show that river discharge has only a brief effect on the mean salinity. Although the rainfall at 830 mm.yr^{-1} is fairly high, the area of the drainage basin is only 640 km^2 and the total volume of river water which reaches the estuary must be of the order of 5 % of the inflow from the sea.

Moreover much of the rain seeps into the forest soil and reaches the estuary slowly and continuously so that floods and marked changes of salinity are uncommon.

The estuaries of small rivers which drain the lowlands of the southern Cape close in dry summers but those of large rivers such as the Breede, the Great Berg and the Olifants, which have catchments in the mountains, remain open. The Great Berg meanders for about 300 km across the flat coastal plain to St Helena Bay. Over the last 50 km the fall is only 1 m and during dry summers estuarine water originally extended over this distance while fresh water reached the sea in winter. In 1952, Voëlvlei dam was built and 1,4 m^3.sec^{-1} was pumped into the river to limit saline water to the last 20 km and thus provide fresh water for farmers higher up. The Fisheries Development Corporation (FISCOR) report F22-1 of 1973 mentions the cutting of a new mouth to provide better access to the fishing harbour at Laaiplek in 1966 and an increased discharge of 2,1 m^3.sec^{-1} from Voëlvlei dam in 1967. Even so, the FISCOR surface salinity records show that in the dry summer of 1967-68 saline water extended 38,4 km from the sea. A study of the vertical salinity structure would be interesting.

The Breede (or Breë) River flows from the mountains around Worcester to San Sebastian Bay. Unpublished observations made by parties of biologists from the University of Cape Town over several years are summarised in Table 3.5.

The tidal range is not greatly reduced even at Malgas some 40 km upstream. The lag in the time of high tide is considerable and as usual, the lag at low tide is much longer. The salinity records are very incomplete but the main trends are clear. During the winter the water is effectively fresh from surface to bottom to within 10 km from the sea but further down the estuary the salinities are very variable. After a spate the surface layer at low tide may be almost fresh to the sea, but during the flood tide high salinity water enters along the bottom and when the river discharge

37

Table 3.5
Hydrographic conditions in Breede estuary

Position		Mouth	Port Beaufort	Moddergat	Karool's Kraal	Malgas
Distance from sea (km)		0	2,0	2,7	6,8	40
Midtide depth (m)		4,3	6,0	3,8	7,0	1,5
Spring tide range (m)		1,8	1,6	1,2-1,6	1,2	1,2
Lag at HWS (min)			0		37	120
Lag at LWS (min)			80		100	215
Salinity (‰)						
4.7.51 HWS	top		8,7		0,2	0,1
	bot.		26,8		0,1	0,1
(winter) LWS	top		0,2		0,1	0,1
	bot.					
4.7.74 HWS	top		34,0	34,0		
	bot.		34,0	34,0		
LWS	top			6,0	1,0	0
	bot.				4,0	0
8.2.52 HWS	top		34,4		28,1	1,8
(summer)	bot.		34,4		28,6	2,0
LWS	top		25,9		15,7	2,0
	bot.		27,3		21,9	2,0
8.1.74 HWS	top		34,0	35,0	31,0	0
& 16.1.80	bot.		34,2	35,0	33,0	0,5
(summer) LWS	top		31,2			
	bot.		30,1			

returns to normal, 34 ‰ salinity water may extend up to the surface 2,7 km from the mouth. During the summer when the river discharge is low, the estuary extends over the whole 40 km to Malgas. According to residents high salinity 'sea water' may reach this point but only low salinities were actually recorded.

Periodically closed estuaries

Umgababa estuary described in chapter 14 is typical of the many shallow estuaries which close periodically. The mouth is either closed or flows as a trickle over the beach during dry months. It bursts open during heavy rain and the level of the estuary then drops about a metre. Salinity and temperature records made by the NIWR and presented by Oliff (1976) are shown in Table 3.6.

It is interesting to see that salinity gradients are maintained along the estuary when the mouth is closed and that they are better marked in the rainy summer than the dry winter. Vertical gradients are also better marked in summer except at station 4 in the broad lagoon where wind action causes better mixing. The bottom at station 5 just inside the bar is below low tide in the sea and the high salinity indicates seepage through the sand. The temperature records will be discussed later.

The Hermanus Lagoon or Klein Rivier estuary described in chapter 14, illustrates the effect of early or late closure of the mouth. It is usually closed in summer and open in winter but in spring and late autumn it opens and closes depending on the rains. Only surface salinities were recorded by Scott *et al* (1952) but these show that when the mouth is open, tidal exchange maintains a normal salinity gradient from 0,4 ‰ at the head to 18 or 33 ‰ at the mouth depending on the

Table 3.6
Salinity and temperature records in Umgababa estuary

Station		5	4	3	2	1
Distance from sea (km)		0,25	0,75	1,55	2,25	2,95
Depth (m)		2,6	1,2	1,6	1,4	1,5
Salinity (‰)						
1.2.76	Surf.	16,2	16,5	5,0	2,5	0,8
	Bot.	32,7	16,5	12,0	7,5	0,8
17.6.76	Surf.	11,4	9,8	9,6	10,2	1,8
	Bot.	31,6	11,4	11,0	11,0	12,6
Temperature (°C)						
1.2.76	Surf.	26,2	27,4	29,4	29,2	27,2
	Bot.	25,2	27,2	29,4	28,8	27,2
17.6.76	Surf.	17,6	17,4	18,4	18,4	17,0
	Bot.	21,4	18,7	18,5	19,4	19,9

state of the tide. If the mouth closes early in spring while the river is still flowing, the salinity remains low through the summer with 0,5 to 2,3 ‰ at the head increasing to about 28 ‰ at the bar. If the mouth closes at the end of spring the water level in the estuary has already fallen and evaporation boosts the salinity near the bar to 38,5 ‰ in February. By April, salinities up to 40 ‰ were recorded 1,5 km further up, showing that a reversed salinity gradient had developed in the lower reaches. The rains in May reduce the surface salinities, and when the winter rains raise the level still further, the mouth bursts open or the sandbar is cut by farmers whose lands may be flooded. The level of the lagoon falls about a metre within a day and when the tide flows in, a normal salinity regime is re-established. It would be interesting to record vertical salinities to learn what changes occur in the deeper levels.

The deep Mtamvuna estuary in the summer rainfall area of southern Natal often closes in winter. Records made by the NIWR will be found in Oliff (1976). Like the Msikaba and other estuaries nearby, the Mtamvuna flows through a deep canyon and its circulation is

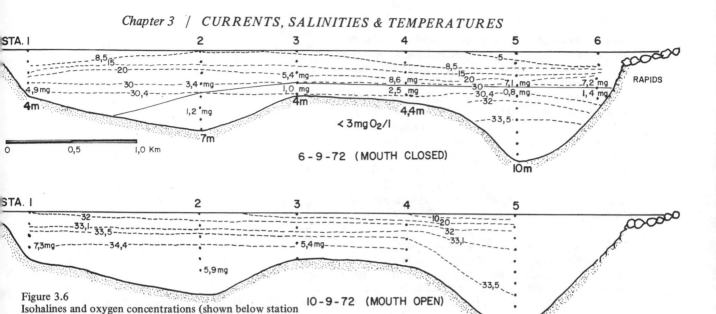

Figure 3.6
Isohalines and oxygen concentrations (shown below station numbers) in Mtamvuna estuary (data from Oliff, 1976).

hardly affected by the wind. It is 5,25 km long, 30-80 m wide and 4-10 m deep but it shallows and broadens near the mouth where a sandbar often blocks the exit to the sea. When the NIWR party first visited the estuary on 6 September 1972 the mouth had been closed for seven to ten days. Salinity, temperature and dissolved oxygen concentrations were measured and then a channel was dug through the sandbar. The outflowing water deepened and widened the channel overnight and the level of the estuary dropped 1 m. Seawater entered with the flood tide and a complete set of records was made three days later. Isohalines and oxygen concentrations based on the two sets of records are shown in Figure 3.6.

While the mouth is closed, river water flows over the surface gradually mixing with the saline water below so that the 15 ‰ isohaline lies at a depth of 1-2 m. There is a halocline between 2 and 3 m and below this salinities gradually increase from 30 to 33,5 ‰ in the deep 10 m depression. Dissolved oxygen concentrations gradually decrease from the surface to the bottom of the mixed layer but below the halocline they are less than 3 mg.ℓ^{-1} along the whole length of the estuary suggesting that the bottom layer remains unrefreshed for long periods.

The records made three days after the mouth had been opened show that sea water had flowed in along the bed of the estuary mixing with the resident water and increasing both its salinity and oxygen concentration. Meanwhile the surface water had escaped to the sea. As a result, the 32 ‰ isohaline had risen to 1 m and below this the salinity had increased to 34,4 ‰ as far up as station 4 but the high salinity water had not flowed over the rise to reach the 10 m depression.

Hyposaline conditions occur briefly in many estuaries but in a few small closed estuaries in Natal they persist for many months. The Manzimtoti and the Fafa are examples that have been investigated by the NIWR. The unpublished records are given by Oliff (1976) and Begg (1978). In the Manzimtoti, salinities are usually low, the minimum recorded being 0,15 ‰ at the head and 3,09 ‰ at the mouth in November and the maximum 1,13 to 15,6 ‰ over the length of the estuary in July. Vertical salinity gradients extend all along the estuary and as tides are absent these must be due to the inflow of the river and the turbulence caused by the wind. It is surprising that complete mixing does not result in this 1,0 m deep estuary.

The Fafa lagoon nearby is very similar to the Manzimtoti but a weir has been built across the lowest part of the sandbar to maintain a minimum depth of 1,3 m in the lagoon for recreational purposes. The long series of records made by the NIWR (Oliff, 1976) shows the effect of the weir on salinity. Tidal exchange has been reduced and sea water can only enter when waves slop over the weir at high tide. For most of the year the salinity in the lagoon is less than 5 ‰, and during the heavy rains in summer the river rises, breaks through the sandbar and runs fresh to the sea. The bar reforms rapidly and the salinity rises slowly but if the bar breaks again during late autumn rains, the river discharge does not persist long enough to reduce the salinity to its normal low value and a salinity of 20,75 ‰ was recorded in May 1972. As in the Manzimtoti, a vertical salinity gradient extends all along the estuary.

Milnerton Lagoon in the suburbs of Cape Town illustrates the extremes of salinity that can occur in a small estuary. As described by Millard & Scott (1954), the Diep River flows into Reit Vlei, a marshy area near Milnerton Race Course. Below the Blaauwberg bridge the marshes are almost fresh during the winter rains but become hypersaline in summer and may even dry up to a salt pan in autumn. From here on the estuary flows into a lagoon. At the start of winter rains the mouth of the lagoon is closed but when the water level rises to a height of 1,5 m above mean sea level, the bar bursts. The lagoon level drops about 1 m and a tidal range of 0,1 m is established. The mouth remains open through the winter and spring but as the river flow decreases the bar builds up and closes in summer and often remains closed through autumn. During this period evaporation exceeds precipitation plus river flow and the level of the lagoon drops to 0,3 m below mean sea level. Seasonal changes of salinity derived from the monthly records of chlorinity given by Millard & Scott for the years 1948 to 1951 are shown in Table 3.7.

During winter and spring a normal salinity gradient exists; salinities at the mouth remain fairly high due to tidal exchange but fall to hyposaline values along the rest of the estuary. In summer sea water slops over the top of the bar at high tide or seeps through the sand and evaporates as it flows towards the head of the estuary so that a reversed salinity gradient develops. In the autumn salinities up to 348,6 ‰ occur in the marshes above Blaauwberg Bridge and the area may dry out as a salt pan. In May 1951, early rains flushed out the area and the salinity dropped to 1,6 ‰. While the mouth is open, tidal currents maintain a vertical salinity gradient but when the bar closes the only circulation is that due to the wind. The salinity becomes uniform from the surface to the bottom but this may be due to evaporation of the less saline surface layer.

The St Lucia system in Zululand shows that hypersaline conditions develop slowly in a large estuary during years of drought. The system was described in detail by Day, *et al* (1954) and the fresh water budget has been determined by Hutchison (1976).

During good rains the level of the lake rises above sea level and a slow gravity flow maintains a normal salinity gradient through the lake and down the channel. Tidal action is limited to the last few kilometres and the only circulation in the lake is that due to the wind but this is sufficient to ensure mixing from surface to bottom. During the dry winter months the lake level falls below mean sea level and sea water flows along the channel and into the lake, increasing its salinity. The sea water evaporates as it flows slowly for 60 km through the channel and the lake so that the salinity increases progressively. During droughts, a reversed salinity gradient is established. The salinity in False Bay at the northern end of the lake rose to 52,6 ‰ in July 1948. There was a flood in April 1949 and the lake level rose about a metre and washed out most of the salt water so that the salinity in False Bay decreased to 25 ‰. In 1950 the rains were again below normal and the salinity increased to 45 ‰ and the mouth of the estuary closed. Since 1951 the mouth has been kept open by dredging. During the good rains of 1956, 1961 and 1963 the lake level was well above

Table 3.7
Seasonal changes in salinity in Milnerton Lagoon from June 1948 to May 1951 (based on data from Millard & Scott, 1954) (bar shown as o — open, c — closed)

Season	Rainfall mm	Bar	Salinity ‰			
			Mouth	Old bridge	KG fort	Saline marsh
Win. 48	277	o	34,3	8,1	1,8	1,8
Spr. 48	127	o	7,8	1,9	1,4	1,4
Sum. 48/49	41	c	44,9	48,4	51,8	148,1
Aut. 49	138	c	54,2	59,6	61,4	27,6
Win. 49	221	o	23,2	3,4	1,3	1,8
Spr. 49	145	o	25,3	8,7	2,0	1,9
Sum. 49/50	15	o-c	46,1	48,1	53,6	189,1
Aut. 50	140	c	56,0	59,6	59,6	348,6
Win. 50	323	o	17,8	12,5	3,3	2,5
Spr. 50	267	o	22,8	4,2	1,9	1,7
Sum. 50/51	53	c	39,1	39,8	42,2	127,4
Aut. 51	237	c-o	34,4	31,3	31,3	dry-1,6

sea level and salinities were below 20 ‰; Millard & Broekhuysen (1971) record salinities of 5,5 to 10,9 ‰ in False Bay in 1964 and 1965. Part of the fresh water in the rivers entering the lake is now used for irrigation and Champion (1976) records that the salinity of False Bay increased from 23 ‰ in 1966 to 102 ‰ in 1970 and then decreased to 20 ‰ in 1973. Hutchinson (1976) compares virgin conditions up to the mid-1960's with conditions between the late 1960's and early 1970's and concluded that the mean annual runoff has decreased from $364 \times 10^6 m^3$ to $295 \times 10^6 m^3$ due to irrigation and afforestation. He also developed a mathematical model to simulate lake levels and salinities which shows good agreement with recorded lake levels and salinities.

FLUSHING TIME AND RESIDENCE TIME

To quote Dyer (1973): 'The flushing time is the time required to replace the existing fresh water in the estuary at a rate equal to the river discharge'. The converse is the residence time. A knowledge of the net flow through an estuary, or a section if it, is obviously important in pollution studies while the residence time will determine the period available for the growth of estuarine plankton populations. Mathematical methods for calculating the flushing time have been developed by a number of physical oceanographers particularly Ketchum, Pritchard and Bowden and references to their papers as well as a general review will be found in Dyer (1973). The brief account below merely summarises the main principles.

The fraction of fresh water method. Fresh water is used as a tracer and the difference in salinity between the estuary water and that of the sea is an indication of the proportion of fresh water in any section. The data required are: R the rate of river flow; d the volume of the estuary section; S_s the salinity of the sea water entering the estuary; S_n the mean salinity of the estuary section. Assuming a steady state, the fraction of fresh water in the section is

$$f = \frac{S_s - S_n}{S_s}$$

and the total volume of fresh water in the section is $Q = fd$. The flushing time through the section is $T = Q/R$. The flushing time through the whole estuary is the sum of the values of T for all the sections. The flushing time through the lower reaches is greater than that through the upper reaches where the current velocity is lower; also the flushing time obviously decreases during periods of low flow. Flushing time may be measured in absolute units or in the number of tidal cycles.

The tidal prism method. In estuaries in which river water and sea water are completely mixed the flushing time may be obtained from a knowledge of the volume of the estuary at high tide and low tide. The difference between the two is the intertidal volume or the tidal prism. Part of this is due to the river discharge over a tidal cycle and part is due to the inflow of sea water during the flood tide. On the ebb tide the tidal prism flows out to sea and the fresh water content is due to the river discharge. Thus if V is the low tide volume and P is the intertidal volume then the flushing time in tidal cycles is

$$T = \frac{V + P}{P}.$$

Usually there is incomplete mixing of river and sea water in the estuary and this method then gives a shorter flushing time than other methods.

The modified tidal prism method. Ketchum (1951) improved the method of dividing the estuary into segments corresponding to the average distance moved by a drop of water during the flood tide. As explained by Dyer (1973) the uppermost segment O at the head of the estuary has a low tide volume V_o and an intertidal volume P_o equal to the discharge of the river R during a tidal cycle. Thus the high tide volume is $V_o + P_o = V_o + R$. The next segment is defined so that its low tide volume V_1 is equal to the high tide volume of segment O, ie $V_1 = V_o + R$. During the flood tide the intertidal volume P_1 is added, so that the high tide volume is $P_1 + V_o + R$. Further segments are defined in the same way so that the low tide volume of segment n or $V_n = V_o + R + \sum_1^{n-1} P$. P_n fills it to the high tide level.

If mixing is complete by the time of high tide, the volume of mixed water which moves into the next segment is the intertidal volume and the fraction is

$$\frac{P_n}{P_n + V_n} = r_n$$

which is termed the exchange ratio while the fraction that remains is $1 - r_n$. $1/r_n$ is the flushing time in tidal cycles. It will be noted that each segment received

volume R of fresh water per tidal cycle to add to the fractions remaining from previous tidal cycles. Dyer calculates that the total volume of river water which accumulates in segment n is $Q_n = R/r_n$ and the amount that flows out is:

$$R(r_n + r_n(1 - r_n) + r_n(1 - r_n)^2 \ldots r_n(1 - r_n)^{m-1} = R$$

From this the flushing time for the segment $1/r_n$ may be calculated. The flushing time for the whole estuary is the sum of the flushing times of all the segments.

If the water in a segment is not completely mixed the exchange ratio must be multiplied by $\hbar/\bar{H}$ where $\bar{H}$ is the average depth of the mixed layer and $\hbar$ is the average depth of the segment. If the salinity of the sea water that flows into the mouth of the estuary is known the salinity in each segment may be calculated and compared with the observed salinity as a check on the accuracy of the flushing time. The method has been used to determine the flushing time in several British and North American estuaries and the best agreement between observed and calculated salinities has been found in large estuaries which are well mixed and in which many segments may be defined.

Wood (1979) proposes a modification of Ketchum's segmental model. He points out that the complete displacement of the high tide content of segment i to form the low tide content of segment i + 1 during the ebb tide is known to be unrealistic. Mixing between the displaced and displacing water must occur. During the ebb phase 'each segment contains some of its original high-tide water, and some of the displaced water penetrates into the next but one downstream segment (0 into 2, 1 into 3, etc)'. Similar mixing across the segmental boundaries occurs in the flood phase. On this basis, Wood develops an average dispersion coefficient. He goes on to show that when Ketchum's model is modified in this way, it gives a closer approach to Ketchum's observations of the salinity distribution in Raritan estuary than does the original model.

The simplest method of measuring the flushing or residence time is to record the movement of a dye such as fluorescin as it moves through the various sections of an estuary. Due to vertical differences of salinity and incomplete mixing, the method gives only approximate values. The movement of fluorescin through a calibrated model of Knysna estuary containing fresh water was measured by the National Institute for Oceanology of the CSIR. Two tests were made, one to simulate the movement of sea water through the mouth and the other to simulate the downflow of fresh water from the head of the estuary. The results were reported as the residence time in different sections during spring and neap tides by Anderson (1976) and are reproduced in Table 3.8.

The rapid replacement of water between the Heads and the Rail Bridge agrees with the rapid currents and the high salinities in this section. Between Belvedere and Westford Bridge the residence time is longer than expected and may reflect the difference between the

Table 3.8
Residence time at positions in Knysna estuary

Position	Distance from sea km	Residence time in tidal cycles	
		Spring tide	Neap tide
The Heads	0,64	<1	<1
Thesens Jetty	4,6	1	1
Rail Bridge	5,4	1	27
Belvedere	9,8	15	37
Westford Bridge	15,5	24	67
Charlesford Rapids	18,9	36	104

uniform density fresh water used in the model and the stratified salinities in the real estuary.

ESTUARINE TEMPERATURES

Since an estuary is an area of mixing between river water and the sea, estuarine temperatures are initially determined by the ratio of tidal inflow to river discharge, and then modified by solar heating and evaporative cooling.

Rivers are usually cooler than the sea in winter and warmer in summer but if they arise in high mountains and flow rapidly to the sea, they may remain cooler throughout the year. Fjords and estuaries in high latitudes show surprisingly large temperature changes. Saelen (1967) records that in the Hardangerfjord the surface temperature falls to 3,8 ° in January and rises to 14,9 ° in August although the temperature of the salt water below 150-200 m remains below 7 ° throughout the year. The surface of the St Lawrence freezes in winter and during the spring blocks of ice float down the estuary eroding the banks and piling up in the shallows. During summer the surface temperature is over 18 °.

In the southern and western Cape, winter temperatures do not fall below 9 ° even in the Great Berg which has a high mountain catchment for the river flows for more than 150 km over the coastal plain. Summer temperatures rise to 27 °. The Breede has a slightly smaller range of 11 to 24 °. There are not many records of seasonal changes along the whole length of an estuary but two may be quoted. In Knysna estuary, Day *et al* (1952) record mean winter and summer temperatures of 12,6 to 27,7 ° at the head of the estuary and 15,3 to 20,8 ° at the mouth where tidal exchange maintains the estuarine temperature close to that of the sea which normally has a seasonal range of 13,5 to 19,9 °. In the subtropics the seasonal range decreases and in Morrumbene estuary at 23°43'S on the coast of Mocambique, Day (1974) reported seasonal changes of 18,8 to 28 ° at the head and 21 to 25 ° at the mouth. Local sea temperatures vary from 19 to 25 °.

Where coastal upwelling occurs, sea temperatures vary rapidly with onshore and offshore winds. On the west coast of southern Africa summer temperatures may fall from 15 to 10 ° within a day. In the warmer waters along the Tsitsikama coast the changes are less frequent but more dramatic and summer temperatures may drop from a norm of 19,9 to 11 ° numbing the fish along the shore. Occasionally the cold water enters Knysna estuary. Day et al (1952) record that after a prolonged spell of southeasterly wind, the high tide temperature in the mouth dropped from 21,8 to 11,5 ° in a day, driving marine fish up the estuary. Over the next few days the cold water penetrated more than 3 km mixing with the estuarine water and causing a fall from 22 to 17,4 °. Perkins (1974) records similar changes in an Oregon estuary.

In deep estuaries, seasonal changes are mainly limited to the surface layers. The Msikaba estuary is unusually deep and Blaber et al (1974) record a maximum depth of 35 m. Surface temperatures range from 14,1 ° in winter to 28 ° in summer but throughout the year the temperature below 16 m remains between 19,6 and 21,5 °. The low salinity surface water floats over the more saline deep water even when the surface temperatures are lower. When they are higher they accentuate the density gradient so that the thermocline and halocline tend to coincide.

In shallow estuaries with wide intertidal flats, solar radiation and evaporative cooling cause marked temperature changes. Hedgpeth (1967) records that in the shallow Laguna Madre in Texas, where river flow is of minor importance, the temperature varies from 11-16 ° in winter to 30 ° in summer. Gunter (1967) states that summer temperatures in other Texas estuaries may reach 40 °. In a severe winter on the other hand, mush ice forms along the shores of bays causing heavy mortalities to the fauna. This must be the widest temperature range recorded. Temperatures may also change very rapidly in shallow estuaries. In the lower York River in Virginia where the depth is 1,5 m, McHugh (1967) states that bottom temperatures change 4,4 ° in a day. In Langebaan Lagoon which has no river inflow, Day (1959) recorded that cooling on intertidal flats during a winter's night caused a drop from 14 ° in the afternoon to 9 ° in the early morning and solar radiation in summer caused a rise from 17,5 to 25 ° as the flood tide spread over the same sand flats. In the subtropical Morrumbene estuary, Day (1974) recorded 32 ° in the flood tide advancing over the sand flats while it was only 27 ° in the main channel. In isolated pools on sunlit mudflats the temperature reached 35 ° but it was only 26 ° in pools shaded by mangroves. This was the same as the air temperature and suggests that in the humid atmosphere under mangroves and probably also in dense salt marshes, evaporative cooling has little effect. When the tide falls and the water drains off the tail end of a shallow bank it forms a plume of warm water in the channel. That's where you will find the wise anglers on a summer evening.

There are few systematic temperature records in

small estuaries that close in the dry season. The most complete are those in the Umgababa estuary reported by Oliff (1976). Umgababa is typical of many blind estuaries on the south coast of Natal and the salinity and temperature records were reproduced earlier in Table 3.4. The main difference between this and an open estuary is the lack of a clear temperature gradient along the length of the estuary either in winter or summer. Winter temperatures at the head of the estuary are rather higher than the 14 ° or 15 ° in earby open estuaries probably due to the slower river flow. Similarly the surface temperature at the mouth is 17,6 ° whereas 20 ° would be expected with tidal exchange. Nonetheless, there are vertical temperature differences in both summer and winter and it will be seen that in winter the warmer water is at the bottom suggesting that the salinity gradients established when the mouth was open still determine the vertical density gradients. Bottom temperatures at station 5 near the sandbar are 25,2 ° and 21,4 ° which are close to sea temperatures in summer and winter and like the bottom salinities discussed earlier, indicate seepage through the sand.

The temperature records from closed estuaries in the Transkei and the Cape are incomplete. In a blind lagoon near the mouth of the Bashee estuary, January temperatures were 30 ° at the head and 26 ° at the sandbar. In the West Kleinemond estuary in the Eastern Cape, Hill (unpublished) reports a summer range of 27 to 22 ° and a winter range of 12 to 17 °. In Hermanus Lagoon, Scott *et al* (1952) conclude that the annual range is 12 to 28 °. The records from Milnerton Lagoon are liable to be affected by sea temperatures from autumn to spring but in the upper reaches at King George Fort the records of Millard & Scott (1954) indicate a range of 12 to 27,1 °.

From all these records it is concluded that the surface temperature gradient along the estuary which was established when the mouth was open do not persist when the mouth closes but the vertical gradients do, for the density differences depend on salinity. The seasonal temperature changes are similar to, but somewhat smaller than, those in an open estuary.

REFERENCES

ANDERSON, F.P. 1976. *Knysna Lagoon model investigation Part I: Main report; Part II: Appendix.* Coastal engineering and hydraulics division, National Research Institute for Oceanology, Stellenbosch (typescript, unpublished).

BEGG, G.W. 1978. *The estuaries of Natal.* Rep. 41, Natal Town and Regional Planning Commission, Pietermaritzburg.

BLABER, S.J.M., HILL, B.J. & A.T. FORBES 1974. Infratidal zonation in a deep South African estuary. *Mar. Biol.* 28: 333-337.

BOWDEN, K.F. 1963. The mixing processes in a tidal estuary. *Int. J. Air. Water Poll.* 7: 343-356.

BRANCH, G.M. & J.R. GRINDLEY 1979. The ecology of South African estuaries, Part XI: Mngazana, a mangrove estuary in Transkei. *S.Afr. J. Zool.* 14(3): 149-170.

CAMERON, W.M. & D.W. PRITCHARD 1963. Estuaries. *In:* M.N. Hill (ed), *The sea.* Vol. 2, Wiley, New York.

CARRIKER, M.R. 1967. Ecology of estuarine benthic invertebrates: a perspective. *In:* G. Lauff (ed), *Estuaries.* 442-485. Am. Ass. Adv. Sci., Washington.

CHAMPION, H.F.B. 1976. Recent prawn research at St Lucia with notes on the bait fishery. *In:* A.E.F. Heydorn (ed), *St Lucia Scientific Advisory Council workshop – Charters Creek, February 1976.* Natal Parks Bd, Pietermaritzburg.

DAY, J.H. 1951. The ecology of South African estuaries. Part I: General considerations. *Trans. roy. Soc. S.Afr.* 33: 53-91.

DAY, J.H. 1959. The biology of Langebaan Lagoon: a study of the effect of shelter from wave action. *Trans. roy. Soc. S.Afr.* 35: 475-547.

DAY, J.H. 1974. The ecology of Morrumbene estuary, Mocambique. *Trans. roy. Soc. S.Afr.* 41: 43-97

DAY, J.H., MILLARD, N.A.H. & G.J. BROEKHUYSEN 1954. The ecology of South African estuaries. Part 4: The St Lucia system. *Trans. roy. Soc. S.Afr.* 34(1): 129-156.

DAY, J.H., MILLARD, N.A.H. & A.D. HARRISON 1952. The ecology of South African estuaries. Part 3: Knysna, a clear open estuary. *Trans. roy. Soc. S.Afr.* 33: 367-413.

DYER, K.R. 1973. *Estuaries: A physical introduction.* Wiley, London, New York. 140pp.

FISCOR 1973. Berg River salinity. *Report F22-1, Fish. Devel. Corp. S.Afr. Cape Town* (34pp typescript).

GRINDLEY, J.R. 1976. *Report on ecology of Knysna estuary and proposed Braamekraal marina.* School Environ. Studies, Univ. Cape Town (123pp typescript).

GRINDLEY, J.R. 1978. *Environmental effects of the discharge of sewage effluents into Knysna estuary.* School Environ. Studies, Univ. Cape Town (62pp typescript).

GUNTER, G. 1967. Some relationships of estuaries to the fisheries of the Gulf of Mexico. *In:* G. Lauff (ed), *Estuaries.* Am. Ass. Adv. Sci., Washington.

HEDGPETH, J.W. 1947. The Laguna Madre of Texas. *Trans. 12th North American Conf.* p 364.

HEDGPETH, J.W. 1967. Ecological aspects of the Laguna Madre, a hypersaline estuary. *In:* G. Lauff (ed), *Estuaries.* 408-419. Am. Ass. Adv. Sci., Washington.

HILL, B.J. 1966. A contribution to the ecology of Umlalazi estuary. *Zool. Afr.* 2: 1-24.

HUTCHINSON, I.P.G. 1976. The hydrology of the St Lucia system. *In:* A.E. Heydorn (ed), *St Lucia Scientific Advisory Council workshop – Charters Creek, February 1976.* Natal Parks Bd, Pietermaritzburg.

JENNINGS, J.N. & E.C. BIRD 1967. Regional geomorphological characteristics of some Australian estuaries. *In:* G. Lauff (ed), *Estuaries.* Am. Ass. Adv. Sci., Washington.

KETCHUM, B.H. 1951. The exchanges of fresh and salt water in tidal estuaries. *J. mar. Res.* 10: 18-38.

KING, C.A.M. 1962. *Oceanography for geographers* (2nd ed). Edward Arnold, London. 337pp.

KORRINGA, P. 1956. Oesterteelt in Suid-Afrika. Hidrographiese, biologiese en oestrologiese waarnemings in die Knysnastrandmeer met antekenings oor toestande in ander Suid-Afrikaanse waters. *Invest. Rep.* 20, *Dept. Industries (Sect. Fish.) S.Afr.* 1-94.

McHUGH, J.L. 1967. Estuarine nekton. *In:* G. Lauff (ed), *Estuaries.* Am. Ass. Adv. Sci., Washington.

McLACHLAN, A. & J.R. GRINDLEY 1974. Distribution of macrobenthic fauna of soft substrate in the Swartkops estuary, with observations on the effects of floods. *Zool. Afr.* 9(2): 211-233.

MILLARD, N.A.H. & G.J. BROEKHUYSEN 1970. The ecology of South African estuaries. Part 10: St Lucia; a second report. *Zool. Afr.* 5(2): 277-307.

MILLARD, N.A.H. & K.M.F. SCOTT 1954. The ecology of South African estuaries. Part 6: Milnerton estuary and the Diep River, Cape. *Trans. roy. Soc. S.Afr.* **34**: 279-324.

OLIFF, W.D. (ed) 1976. (unpublished) *National marine pollution monitoring programme. First annual report* (509pp typescript); *Second annual report* (172pp typescript). NIWR Durban (includes many progress reports by the South African National Institute for Water Research).

PERKINS, E.J. 1974. *The biology of estuaries and coastal waters.* Academic Press, London. 678pp.

PRITCHARD, D.W. 1952. Estuarine hydrography. *Adv. Geophy.* **1**: 243-280.

PRITCHARD, D.W. 1955. Estuarine circulation patterns. *Proc. Am.* Soc. *Civil Eng.* **81**(77).

PRITCHARD, D.W. 1967. Observations of circulation in coastal plain estuaries. *In:* G. Lauff (ed), *Estuaries.* Am. Ass. Adv. Sci., Washington.

SAELEN, O.H. 1967. Some features of the hydrography of Norwegian fjords. *In:* G. Lauff (ed), *Estuaries.* Am. Ass. Adv. Sci., Washington.

SCOTT, K.M.F, HARRISON, A.D. & W. MACNAE 1952. The ecology of South African estuaries. Part 2: The Klein River estuary, Hermanus. *Trans. roy. Soc. S.Afr.* **33**: 283-331.

STOMMEL, H. 1953. The role of density currents in estuaries. *Proc. Minn. Int. Hydrol. Conf.*

WOOD, T.A. 1979. A modification of existing simple segmented tidal prism models of mixing in estuaries. *Estuarine coastal mar. Sci.* 8(4): 339-348.

WOOLDRIDGE, T. 1976. The zooplankton of the Msikaba estuary. *Zool. Afr.* **11**: 23-44.

Chapter 4

Estuarine sediments, turbidity and the penetration of light

J.H. Day

Department of Zoology, University of Cape Town

INTRODUCTION

An introduction to marine sedimentology was given in chapter 2. The grading of sediments was described and the effect of currents and turbulence on the erosion, transportation and deposition of sediments. The formation of the outer bar off the mouth of an inlet and the cross-sectional stability of the mouth itself were discussed. In the terminology adopted by Hayes (1977) the outer bar corresponds to the ebb tidal delta. The formation of the inner bar or flood tidal delta remains to be considered.

This chapter is written for ecologists interested in the environmental conditions in estuaries. Sedimentologists and civil engineers who require a more detailed account are referred to Dyer (1979). His first chapter reviews the basic principles of estuarine hydrography and sedimentation, while later chapters deal with methods of measurement and analysis of the results.

Many of the basic principles of estuarine sedimentation were discovered by Dutch geologists and engineers concerned with reclamation works and the conditions on the enormous tidal flats or wadden around the Zuider Zee. A most valuable review of this work was given by Postma (1967). He builds on the hypothesis of van Veen (1950) that the creeks that drain the mudflats during the ebb take a meandering course while the flood channels take a more direct course as the tide rises so that the two often run side by side or intersect leaving accumulations of sediment which form ridges between the two channels. Postma's own studies show that due to scour lag and other factors, fine particles (<0,1 mm) accumulate near the high tide mark.

Investigations in the Danish wadden over 30 years have been reviewed by Schou (1967). Some of the wadden are as much as 15 km wide with a number of sandy islands offshore and a complex series of flood and ebb channels across the mudflats. The raised areas between the channels are overgrown by salt marsh vegetation which increases their stability and facilitates further sedimentation. The construction of a 9 km berm wall to the island of Rømø on the outer edge of the wadden provided an opportunity to determine the factors which influence the deposition or erosion of sediments and the rates at which these processes occur.

In the past there has been much discussion in the geological literature as to whether estuarine sediments are of marine or terrestrial origin. In the light of present knowledge the arguments and counter arguments seem of little importance for there is now no doubt that both types of sediment are present in most estuaries, On high energy coasts where littoral drift is heavy, a flood tidal delta of marine sand may largely occlude the inlet and if the tides have a high amplitude the marine sand is carried rapidly to the head of the estuary. Conversely on low energy coasts where littoral drift is minimal, little marine sand enters the estuary. The estuary basin may remain deep or, if the drainage basin of the river is eroded and the estuary is river-dominated, the estuary basin is largely filled with fluvial sediments. These represent extreme cases and normally there is a preponderance of fine silts and clays of fluvial origin at the head of the estuary grading to medium or corase sand of marine origin at the mouth. Many workers have stressed that the distribution of sediments and the location of shoals and channels resulting from sorting and resorting during many years of normal river flow may be washed away by a single severe flood and a completely new system established.

The physico-chemical process of flocculation where fresh and salt waters mix has been discussed by Postma (1967), Dyer (1972), Burton (1976) and others. The process will be outlined later. Suspended floccules form the turbidity maximum in the upper reaches of estuaries and Inglis & Allen (1957) give an account of the resulting soft mud in the upper reaches of the Thames. Silting has been a problem for many centuries since the old London Bridge was built in 1215 in the days of King John. It was not until 1832 when the present bridge with longer spans and less obstructive supports was built that the tidal range above the bridge increased by 25 % and scoured a deeper channel. Nonetheless dredging was still essential. The spoil was dumped in the Black Deep at the mouth of the Thames. When it was learnt that much of the spoil was carried back to its original site by the flood tide flowing along the bottom, the spoil was carried far out to sea.

Modern reviews of estuarine sedimentology will be found in Postma (1967), Dyer (1972) and many engineering texts. Perkins (1974) considers the biological implications of sedimentation, erosion and turbidity. Jennings & Bird (1967) review the geomorphology of many Australian estuaries while Kulm & Byrne (1967) give a detailed account of the sedimentology of Yaquina Bay, an estuary on the Oregon coast. In southern Africa there have been only a few studies of estuarine sediments by geologists. Orme (1974) has discussed the sediments of Natal estuaries and van Heerden (1976) has analysed the distribution of sediments in Lake St Lucia. Flemming (1977) has made a thorough study of the sedimentology of Langebaan Lagoon. It is particularly interesting that his findings confirm that the lagoon never formed part of an estuary but that it was, and still is, a sheltered arm of the sea. For the rest, the observations of estuarine sediments have been made by biologists and chemists whose prime interest was the distribution of the fauna and flora or the degree of pollution. Obviously much remains to be done by skilled sedimentologists.

Where relevant, the observations on estuarine sediments in southern Africa will be interpreted in the light of overseas researches.

MARINE SEDIMENTS

Coastal plain estuaries are drowned valleys originally carved into the bed rock during Pleistocene regressions of sea level and subsequently filled to varying degrees by Pleistocene and Recent sediments. Such sediments were, and still are being derived from the sea and from the drainage basins of the rivers. It is convenient to deal with the marine sediments first.

The volume of sediment that accumulates near an inlet depends on the supply of sediment, the geology of the coast, the slope of the sea bed and the strength of the distributive forces. Little or no sediment accumulates on steep rocky coasts and the outer bar remains deep. For example, the bar at the entrance to Knysna estuary, which opens between precipitous cliffs is 4 m deep and hardly any marine sediment enters the estuarine channel (Chunnett, 1965).

On gently shelving sandy coasts large quantities of marine sediment are available within the surf zone. The formation of sandspits and the shoals which form the outer bar or ebb tide delta have been described in chapter 2. The flood tide flowing in through the estuary mouth forms a corresponding flood tide delta or inner bar.

Within the estuary channel wave energy is greatly reduced and the main forces at work are the tidal currents and seasonal floods of the river. The form and location of the shoals and channels are determined by these forces and by the shape of the estuary basin. Hayes (1977) stresses the importance of tidal range and in his figure 4 he illustrates the features that commonly occur in a broad estuarine sound enclosed by a bay-mouth bar. Such sounds are common on the Atlantic and Gulf coasts of the United States. Here the flood tide entering an inlet through the bar rapidly loses velocity so that the marine sediment it transports is deposited close to the inlet and the flood tide delta formed is quite separate from the river delta formed on the opposite side of the sound. Similar flood tide and river deltas were formed in Richard's Bay in Natal before it was modified to form a deep water harbour and may sometimes be seen in broad lagoons but in more elongated estuarine basins the sediments from the two sources are mixed.

An idealised flood tide delta is illustrated in Hayes' figure 13 which has been reproduced here as figure 4.1. Its characteristic features include the main flood ramp which is covered with flood-oriented sand waves and ends in an ebb shield. The main flood channel diverges on either side of the ramp to form smaller flood channels which shallow in the direction of flow and may form spillover lobes. The outer edges of the whole shoal are sharply defined by long ebb spits which separate the flood and ebb channels.

There are many variations of such an ideal flood tide delta. It is seldom symmetrical and often one of the ebb channels silts up and eventually closes so that the ebb shield unites with the sand flat on that side of the estuary. This has occurred in the estuary of the Breede river in the Cape. There is no ebb channel

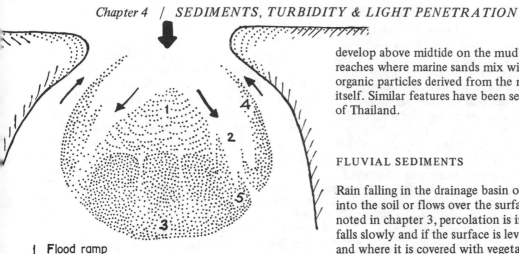

1 Flood ramp
2 Flood channel   4 Ebb spit
3 Ebb shield   5 Spillover lobe

Figure 4.1
Ideal form of flood tide delta (from Hayes, 1977).

between the right bank of the estuary and the flood ramp so that as the tide falls, the two are seen to be united as a single large sandflat covered with flood oriented sand ripples.

If no lagoon is formed inside the sandspits at the mouth and the width of the estuary decreases slowly upstream, the velocity of the flood current is maintained and marine sediments are transported well into the estuary basin. The flood tide delta becomes elongated and fragmented by spillover lobes while the finer marine sediments are carried further upstream to mix with the fluvial silt. Hayes has stressed that as the tidal range increases from microtidal estuaries with a range of 0-2 m, through mesotidal estuaries with a range of 2-4 m to macrotidal estuaries with a range above 4 m, tidal currents become more and more dominant. The banks of the estuary form extensive mudflats and the sandy deposits are mainly concentrated in the middle of the channel as elongated sandy shoals. In South Africa where the tidal range is less than 2 m all the estuaries would be classed as microtidal and in many cases constriction of the mouth by sandspits and rock promontories decreases the range even further. Thus the mudflats on the banks are not extensive except in the Swartkops in the Eastern Province where the coastal plain is very flat and flood waters extend the width of the estuary. The muddy banks are further stabilised by salt marsh vegetation. In Mocambique where the tidal range increases to 3 m or more, the broad muddy banks are stabilised by mangroves. The Morrumbene estuary described by Day (1974), is an example. Here the flood tidal ramp and the ebb shield are united to the right bank to form a sandflat over 2 km wide and a complex of linear sandy shoals develops further upstream. Extensive mangrove swamps

develop above midtide on the mudbanks in the upper reaches where marine sands mix with the silt and organic particles derived from the river and the estuary itself. Similar features have been seen in the estuaries of Thailand.

FLUVIAL SEDIMENTS

Rain falling in the drainage basin of a river either sinks into the soil or flows over the surface as run-off. As noted in chapter 3, percolation is increased if the rain falls slowly and if the surface is level and permeable, and where it is covered with vegetation. The seepage water that eventually reaches the river in this way carries little sediment apart from colloidal suspensates particularly humus. Such rivers are common in the lowlands of north-western Europe, on the Atlantic coast of the United States, in the Western Cape, northern Zululand and southern Mocambique. Run-off is increased by the opposite conditions. If the rainfall is markedly seasonal with droughts followed by heavy rains as occurs in the uplands of the Eastern Cape, the Transkei and Southern Natal, the grass may die and the run-off over the bare soil cuts erosion furrows down the hillsides and carries a heavy sediment load to the river. Overgrazing and tilling on steep slopes has the same effect. Much depends on the geology of the drainage basin. The weathered slates and shales in the lower reaches of the Olifants River in Namaqualand, the Breede River in the southern Cape and in the Sundays and Fish River valleys of the Eastern Cape form silts and clays which are carried by flood waters. Granite weathers slowly to form both sand and clay while sandstone and quartzite form only clean sand, as is evident along the Tsitsikama coast. During normal flow the larger particles are deposited along the river banks and only the lighter and finer particles including silt, clay, plant debris and humus are carried down to the estuaries. During floods the river overflows its banks and much of the sediment is deposited on the flood plain, particularly if it is overgrown by vegetation. Alexander (1979) regards the denudation of the vegetation along the river banks as a more important cause of siltation in estuaries than erosion in the river catchment. Swamps are very efficient filters and even silt may be deposited so that relatively clear water reaches the estuaries. This was once the case in the Mfolozi which flowed into St Lucia Bay. As recorded by Day *et al* (1954), the swamps in the lower Mfolozi were canalized in 1918 and the alluvial soil was planted to sugar cane. By 1947 St Lucia Bay and the lower channel of St Lucia estuary was filled with mud. As further sediment was deposited in the Mfolozi, the level of the river rose and levees were formed along its banks until they were higher than the rest of the flood plain. In recent floods the levees were breached and the cane lands were inundated. Much the same cycle of events has occurred in Richards Bay. Before it was divided to form a deep water harbour and a sanctuary for wild life, Millard & Harrison (1954) recorded that the Mhlatuzi flowed through swamps before emptying into Mhlatuzi Lake which, although shallow and muddy, had a stable depth of 1-2 m. When the harbour was constructed in 1974, the Mhlatuzi River was diverted so that it flowed into the sanctuary and, despite the advice of a panel of ecologists, a canal was cut through the swamps. Since then floods have filled most of the sanctuary with mud and it is only a few centimetres deep at low tide (Emanuel, 1977).

As floods subside and a river returns to its normal level it erodes the banks of the flood plain, particularly on the outer sides of bends where the channel is deepest and the sediment originally deposited in floods moves slowly onwards to the estuary. Marginal vegetation slows this process and protects the banks, but farmers cast envious eyes on the rich alluvium. In Natal the banks are cleared and cane is planted to the very edge of the river. Heydorn (1977) gives a graphic account of the erosion which has resulted along the Mhlanga River, north of Durban.

When the river approaches sea level its velocity is influenced by the tides even beyond the limits of saline intrusion. Coarse sand and pebbles from higher up the river are only transported into the tidal reaches during floods and medium to fine sand is deposited during normal flow. Silt, clay and colloidal humus remain in suspension until fresh and salt water mix at the head of the estuary. In some estuaries such as the Knysna, Swartkops, Keiskamma, Mngazana and Mtamvuna, the head of the estuary is fixed by rapids but in others it varies greatly with the discharge of the river or the state of the tide. At the head of Knysna estuary Day *et al* (1952) reported the bottom sediments as gravel and coarse sand changing to soft silt in the upper reaches. The sediments at the head and upper reaches of Swartkops estuary described by Macnae (1957) are similar. The upper reaches of the Berg and Breede estuaries are soft mud for many kilometres and Dyer (1972) reports that the mud reaches of the Thames extend for 10 to 40 km below London Bridge. In fact the sediments in the upper reaches of most estuaries in the world are soft mud. In contrast to this, Flemming (1977) has reported that the sediments in Langebaan Lagoon are deficient in both silt and clay particles and that the distribution of very fine sand with particle diameters between 0,063 and 0,125 mm is very limited. It has long been known that the salinity of the lagoon shows little variation from sea water values since there is no fresh water inflow. The nature of the sediment thus confirms that the lagoon is a sheltered inlet of the sea and not an estuary.

FLOCCULATION

Silt, clay and colloidal particulates such as humus are carried in suspension in river water. The physicochemical changes that occur when fresh and salt water mix have been reviewed by Postma (1967), Dyer (1972) and Burton (1976). Briefly, silt and particularly clay particles bear negative surfaces charges due to the adsorption of anions, particularly OH⁻ and to cation substitution in the crystal lattice and to broken bonds at the edges of the particle. These negative charges are balanced by a double layer of hydrated cations. The thickness of this double layer depends mainly on the ionic concentration of the water in which the particles are suspended. River water usually has a low electrolyte content and the charges on the particles repel one another. Estuarine water has a high electrolyte content so that the repulsive charges diminish and when the particles collide they unite to form a large spongy network or floccule. The flocculation of silt particles, most types of 'humus' and the clay minerals illite and kaolinite mainly occurs in salinities of 1-4 ‰, but montmorillonite flocculates slowly as the salinity increases to full strength sea water. Sewage which contains large carbohydrate and protein molecules as well as polyvalent metallic ions, promotes flocculation while bacteria and other organic particles act as binding agents. As more and more particles are added to the floccule it grows until its diameter may exceed 0,5 mm. Over 90 % of the spongy floccule is water and it sinks in still water at about 0,4 cm.sec⁻¹. Settling velocities vary with the nature of constituent minerals; they also increase as the floccules grow and decrease at higher salinities and lower temperatures.

Flocculation starts at the head of the estuary and while the floccules grow they drift down into the upper reaches where the water becomes very turbid so that this stretch is known as the turbidity maximum.

In the St Lucia system, the turbidity maximum occurs at the mouths of the Hluhluwe and Nyalazi Rivers entering False Bay and at the mouth of the Mkuze in North Lake. In both cases the salinity is usually very high, but in the Breede estuary where the increase in salinity is more gradual, maximum turbidities extend from Malgas to Karrols Kraal, a distance of 30 km. Scott *et al* (1952) report that when the mouth of Hermanus Lagoon is closed, the increase in turbidity is clearly correlated with increases in salinity up to 10 ‰ in the channel through which the Klein River enters the lagoon. However the sediment settles rapidly and the main part of the lagoon is clear though peat stained. The flocculation of the various types of 'humus' demands further investigation.

The particles sink as turbulence decreases so that the concentration of suspended particles increases with depth and deposition occurs at the slack of the tide. Much of the material is resuspended by ebb and flood currents and thus moves up and down the estuary. Postma (1967) reports that if floccules are carried upstream by the flood tide into the fresh water of the tidal river they disaggregate into their constituent particles only to reaggregate when the tide turns and the particles reach saline water. The floccules that are carried downstream by ebb currents are deposited at slack tide but as they are the lightest particles in the area, they are the first to be resuspended and carried back towards the upper reaches by the rising tide. The concentration of suspended particles thus increases with the velocity of the tidal currents. Dyer (1972) reports that due to lag effects, the peak concentration in the upper Chesapeake occurs one or two hours after the tidal currents have passed their greatest velocity.

If the sinking floccules are very concentrated with particle densities of 10 g.ℓ^{-1} or more, they sink as layers which drift along the bottom as fluid mud. Dyer (1972), quoting Inglis & Allen (1957), states that in the Thames particle concentrations reach 100 000 ppm. Fluid mud was also reported in Richards Bay by Millard & Harrison (1954) before it was dredged to form a deep water harbour and it also occurs in Milnerton lagoon (Millard & Scott 1954). It has even been reported over muddy bottoms in the open sea when heavy swells pass over the area. Zoutendyk, in a personal communication, stated that when charting the soling grounds in Mossel Bay as a scuba diver, he had normal underwater vision when standing upright, but could not see his hand before his face when he bent down.

When the floccules finally settle and further sediment is deposited on top, the pressure of the overburden slowly expels the interfloccule water and the floccule structure finally breaks down (Dyer, 1972). As the pressure increases the mud is consolidated and as the shear strength increases it becomes more difficult to erode. Silt that settles on intertidal banks loses water by evaporation and towards the high tide mark the mud is well-consolidated, but below mid-tide it remains soft and thixotropic so that tidal currents erode a salting cliff at this level.

SUBTIDAL SEDIMENTS

From the very variable sediments at the head of an estuary and the mud in the upper reaches, particle sizes increases towards the mouth. Further generalizations are of little value for the grade of sediment differs from one estuary to another and there are changes in any estuary where the channel broadens or narrows and the tidal currents swing from one bank towards the other. It is hoped that the distribution of subtidal sediments in the several estuaries described below will provide a guide to the type of changes that occur under different conditions.

Knysna estuary, described by Day *et al* (1952) is a well-developed estuary but by no means typical of most estuaries in southern Africa. The river discharge, which is usually less than 3 m^3.sec^{-1} is clear though peat-stained and only a little fine sand and silt is transported. The mouth is deep and opens on a rocky shore so there are no sand spits. Grindley (1976), after summarising earlier reports and a recent study by the National Research Institute for Oceanology, quotes a statement by Chunnett (1965) that practically no new sediment has been derived from the sea or the river during the last 100 or 150 years. Thus the observed changes in the location of the channels and the size of the shoals have been due to the reworking of the original estuarine sediments. The road cuttings along the banks of the estuary show strata of clay, sand, gravel and cobbles deposited during Pleistocene and Recent periods and similar sediments must extend to a considerable depth below present sea level. Krige (1927) reported that when the railway bridge was built across the lagoon bed, rock was not reached in borings to 80 ft (24 m). Before the bridge was built sailing ships loaded timber at Belvedere, 3 km further upstream. These channels are now too shallow for commercial navigation and there is little doubt that the many shoals and mudbanks along the solid embankment leading to the bridge spans across the main channel have grown in the last 50 years. Further mudbanks have been deposited along the causeway to Paarden Island and personal observations over the last 10 years show that the new road bridge at Ashford has deepened and stabilized the channel under the bridge spans but caused the deposition of more sediment along the solid embankment. Thus all the obvious redistribution of sediments appear to be due to engineering works. Similar shoals and mudbanks have been deposited along the embankments leading to the bridge spans across the estuaries of the Great Brak, the Keurbooms, the Swartkops, the Bushmans and several other estuaries (Day, unpublished). It is interesting that when the rubble was dredged from beneath one of the spans of the bridge across the Bushmans estuary and the channel was thus deepened in 1976, further sediment accumulated in the undredged channel below the other span (Weaver, 1979).

The distribution of particle sizes along the bed of Knysna estuary agrees with the general pattern found in European and American estuaries as reviewed by Postma (1967). While there is a gradual increase in particle size towards the sea, the swifter channels are floored with coarser sediments and the broader stretches between Belvedere and Leisure Isle are covered with sandy mud in depressions and cleaner medium sand elsewhere. Between Leisure Isle and the sea where the currents reach their maximum velocity of 1,2 m.sec^{-1}, the bed of the channel is coarse sand, shells and gravel.

The major shoals and channels shown in Figure 4.2 illustrate some of the features reported by Hayes (1977). Between the mouth and the railway bridge

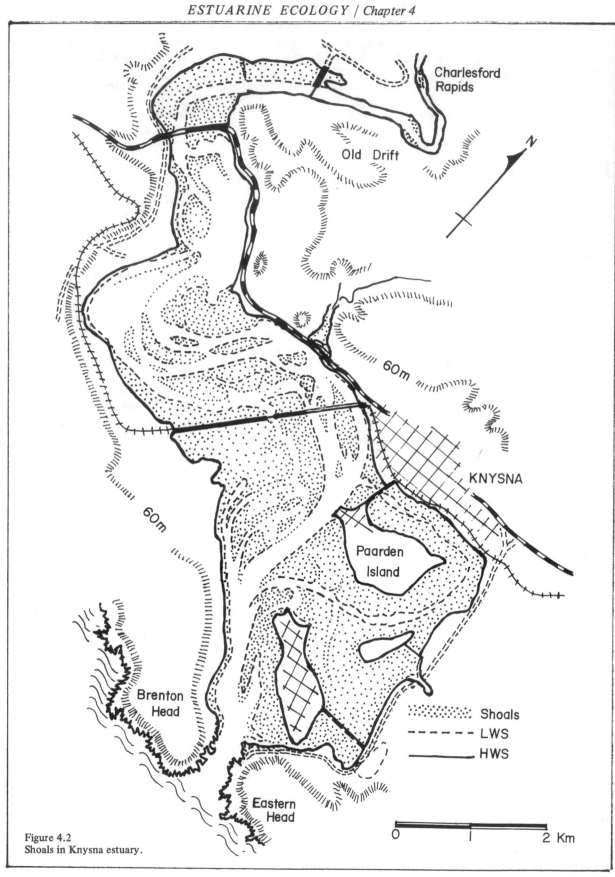

Figure 4.2
Shoals in Knysna estuary.

there is only one major channel which carries both flood and ebb tide currents. It hugs the right bank past Leisure Isle and then swings across to the left bank to pass under the spans of the railway bridge. One may speculate that Leisure Isle which is very flat and sandy was originally formed as part of the flood tide delta. The sand flats on its seaward face show flood oriented sand ripples which characterise a flood ramp. There is a minor flood channel which shallows in the direction of flow and along its western margin there is an elevated sandspit which slopes very steeply into the main channel. A spillover lobe has developed immediately north of Leisure Isle. Elongated shoals have formed under the span of the railway bridge and further up towards Belvedere there is a complex of tapered shoals with intervening flood and ebb channels. The sandy margins of the shoals slope fairly steeply whereas the mudbanks slope more gently illustrating the dependence of slope on particle size. The grading of sediments on intertidal banks will be discussed later.

In contrast to the slow rate of sedimentation in Knysna lagoon and other estuaries on the Tsitsikama coast, the estuaries of the large rivers draining the uplands of the Transkei and southern Natal receive enormous quantities of fluvial sediment during summer floods. The Kei, the Bashee, the Mzimvubu, the Mzimkulu and the Mkomazi are extreme examples. The sea is discoloured for many kilometres along the coast and from the air, successive outflows may be traced at deeper and deeper levels. Much of the sediment is deposited in the estuary basins as soft mud. Port Shepstone at the mouth of the Mzimkulu was the second port in Natal from 1883 to 1901, but the silting became so bad that coasters had to wait months outside the bar and the port was finally limited to fishing boats. Port St Johns at the mouth of the Mzimvubu was used by coasters until 1940, but when a party of biologists surveyed the estuary in 1950, most of the port was filled with mud, the maximum depth was less than 3 m and the shoals in mid-channel were only knee deep at low tide. During the low flow period in winter, part of the silt is carried out to sea by the ebb tide leaving the sand so that the bottom becomes firmer and the water less turbid. Oliff (1976) reports that during August 1974 the mean particle size on the bottom of the Mzimkulu varied from 0,61 to 0,33 mm, which is surprisingly coarse. Under such conditions the fauna and flora regenerates in spring only to be smothered during summer.

The rate of sedimentation in small estuaries along the same coasts is much lower. Their drainage basins are mainly in the bush-covered coastal strip where there is little erosion. Some of these estuaries flow through narrow ravines and the water is surprisingly deep. Almost all of them have narrow mouths between sandspits and may close from time to time unless they are sheltered from littoral drift by rocky promontories. In the Mtamvuna, described by Oliff (1976 unpub-

lished) and by Cloete & Oliff (1976), marine sand has not been transported beyond the temporarily blind mouth and the rest of the estuary basin is 4 to 10 m deep. Apart from the bar which is composed of 63 to 67,9 % coarse sand, the estuary bed is covered by fine sediments which include 75,4 to 93,8 % of silt and clay-sized particles brought down by the river. It is suggested that these sediments are deposited very slowly for the water is normally clear and even after the fairly heavy rains in January 1951 Secchi readings were 8 to 10 cm as compared to 2-3 cm in the nearby Mzimvubu. It is surprising that marine sand has not spread from the bar along the bed of the estuary, but possibly the narrow inlet through the bar is never deep enough to allow the flood current to reach erosion velocity.

The Msikaba estuary described by Blaber *et al* (1974) and by Wooldridge (1976), is a very similar estuary with precipitous banks and a maximum depth of 35 m, the greatest depth recorded in any estuary in southern Africa. As in the Mtamvuna, there is a narrow inlet through the sand bar but in this case it appears to be permanently open and deepens to 2,5 m during the summer so that marine sand is transported into the lower reaches of the estuary. Nonetheless, most of the estuary basin is floored with fluvial silt transported by the low salinity surface water in summer. Light penetration is reduced in this season but it increases to a maximum of 16 m in autumn as the floccules are deposited on projecting rocky ledges along the vertical banks or on the estuary bed.

The Mngazana described by Macnae (1963), Wooldridge (1977) and Branch & Grindley (1979) is the best developed estuary in the Transkei. Like the Mtamvuna, it is small and there is little erosion in the 275 km² coastal watershed, so that even during the summer rains the river is fairly clear. The estuary lies in a wide valley floored with Pleistocene and Recent sediments and the mouth is permanently open in the lee of a rocky headland. There is a large sandspit on the northern shore and marine sand has been transported upstream to form small sandy beaches and shoals in the lower reaches and to mix with estuarine sediments elsewhere. The channel is fairly uniform in width, the banks gradually becoming steeper upstream, while the bed material of the estuary consists mainly of fine sand and silt (Md $\phi \leqslant 2,0$) changing to softer silt and finally coarse sand, pebbles, and silt at the head. There are extensive mud flats overgrown with mangroves in the backwaters and tributaries. Dr Branch's records show that in such sediments the median particle diameter is about 0,1 mm and there is 7,4-11,0 % organic matter.

Many of the small estuaries which are closed by sandbars for months at a time broaden behind the sandbars to form lagoons. Hermanus lagoon described by Scott *et al* (1952) is a well-known example in the Western Cape and the Umgababa described by Oliff

(1976) is a typical example in Natal. There are many others. In no case has the sediment distribution been recorded but personal observations show that the bottom is soft silt where the river enters and then changes through varying mixtures of silt and medium sand to clean sand at the bar. The bottom contours are very constant and practically no change has been observed over the last 25 years. When the water level rises during the rainy season a channel up to a metre deep is eventually cut through the bar and a good deal of the soft silt is carried out to the sea but estuarine sand is not eroded. Neither does material from the sandbar get carried into the lagoon at high tide. In the dry season the bar closes and the film of riverine silt is slowly replaced. On the margins of the lagoon the silt is continually stirred up by wind-generated waves until it settles in deeper water.

The Gouritz, the Breede, the Berg and the Olifants are the four largest rivers in the southern and western Cape and of these, the estuary of the Breede is the best developed. The Breede flows for more than 100 km over the coastal plain and during the high flow period in winter it carries silt and clay derived from weathered slates and shales. The estuary is up to 40 km long when saline intrusion reaches Malgas in summer, but only 10 km in winter when fresh water extends from surface to bottom at Karools Kraal. Thus the head and upper reaches vary in position and the sediments along all this 30 km stretch are a muddy mixture of silt and clay with fine sand increasing downstream. At Port Beaufort, 3 km from the mouth, the sediment changes from muddy sand to clean, medium to coarse sand derived from the sea. There is evidence that since 1952 when the depths along the main channel and the position of the shoals were charted, the marine sand has been extending further and further upstream. By 1974 the large sandy shoal near the mouth had expanded and was dissected by minor channels; most of the original mud flats opposite Port Beaufort were covered with a surface layer of clean sand with a prominent ridge along the edge of the channel. This marine sand is obviously transported by the flood tide as a bed load and the channel bed is marked with sand waves and smaller ripples whose orientation show their movement upstream. The midchannel sandbank opposite Moddergat, 4 km from the mouth has grown until in 1978 it was awash at extreme low tide. It is probable that this upstream movement of marine sand is due to the decreased discharge of the river following the construction of dams and irrigation works in the drainage basin.

Swartkops estuary near Port Elizabeth is typical of several estuaries in the rather dry Eastern Province of the Cape. Macnae (1957) gives a general account of its ecology and the unusually broad intertidal salt marshes. McLachlan & Grindley (1974) give a very complete account of the physical characteristics of the water and the nature of the sediments at low spring tide.

Oliff (1976) adds a few more records of the sediments on the bed of the lower channel. As in several other estuaries, saline intrusion is limited by a causeway, in this case at Perserverance, about 16 km from the mouth. In consequence, the bottom sediment there is coarse to medium sand with a small percentage of subsieve grades. In the muddy upper reaches the main sediment is fine sand with 13,4 to 40 % of subsieve particles. From Swartkopsvillage to the National Road Bridge near the mouth, the bottom sediments are fine sand with less silt. In contrast to this, the backwaters, as in Tippers Creek, become increasingly muddy with over 67 % of silt and organic detritus at the upper end. The mouth below the bridge is clean fine to medium sand.

The Mlalazi, as described by Hill (1966) and Oliff (1976), is one of the few comparable estuaries in Zululand although the intertidal flats are not extensive and the head of the estuary is not fixed by a causeway. As shown by the salinity records and the analyses of sediments along the estuary, the tip of the salt wedge extends from 7 to 9,3 km from the sea according to the river flow. Between these two points the bottom sediments are either eroded to coarse sand with little silt or may be covered with mud containing more than 40 % silt. The muddy upper reaches with 53,8 % silt extend on to within 6 km from the sea and thereafter the middle and lower reaches are floored with medium sand and a decreasing percentage of subsieve grades. The mouth as usual is clean medium to coarse sand.

The immensely thick layers of silt and clay which form the bed of the St Lucia lakes were paralleled only by the sediments in Richards Bay before it was modified to form a deep-water harbour and a sanctuary for wild life. Oliff (1976) gives an analysis of the Richards Bay sediments and shows that over most of the bay between 75 and 98 % were subsieve grades with diameters averaging 35 microns. The water was seldom more than 1 m deep and the constant winds stirred up the soupy sediments so that Secchi disc readings were usually between 10 and 20 cm. The St Lucia lakes are very similar but much larger. Van Heerden (1976) has analysed a series of sediment cores. Some 28 m of clay have been deposited in the southern part of False Bay and there are alternating layers of clay, silt and sand in the northern part. In Hells Gates where False Bay opens into North Lake, in North Lake itself, and in South Lake, the series of sediments is similar but the thickness decreases from 13 to 7 m. Van Heerden concludes that the fine sediment particles flocculated where the four rivers enter the saline lakes. Deposition occurred in the vicinity during periods of low flow but the floccules were carried further over the lakes during floods. At such times the salinity is much lower and the depth of the lake is greater. Turbulence caused by wind generated waves rapidly stirs up the floccules when the lake level is low and surface drift has carried part of the sediment to the southern end of the lake.

INTERTIDAL BANKS

In addition to the increasing size of sediment particles along the estuary bed there are often differences between the subtidal sediments in the channels and the intertidal sediments on the shores. The contours of the banks also change. Both have an effect on the water content of the intertidal sediments and thus on the vertical zonation of the fauna and flora. There is an interplay of many factors. Storms and floods, even if infrequent, are of major importance. During normal flow there are changes in the velocity of tidal currents at different depths; the main channel may swing from one bank towards the other; wind-generated waves may affect windward shores and some banks may be well-protected by aquatic vegetation while others are bare and easily eroded.

Estuarine currents often reach peak velocity at midtide and the erosion of the shore is maximal at this level. Slack water occurs at high and low tide, although deposition of suspended material may be 1-2 hours later due to the settling lag (Dyer, 1972). The strongest currents are in the centre of the channel but decrease in the shallows and it is here that sediment accumulates towards high tide. Tidal flats are often extensive in the upper reaches where fine sediment is deposited and these are rapidly stabilised by the roots of salt marsh vegetation. The stems and leaves further reduce current velocities and as deposition continues, the elevation of the salt marsh increases. The lower edge of the salt marsh vegetation is near the midtide mark and below this the naked mud is eroded to form a salting cliff extending down to about low tide of neaps where a broader or narrower shelf of soft mud extends down to the level of low spring tide. The subtidal sediments are often coarser and firmer and slope more steeply to the channel bed. Where the estuary bends and the channel deepens towards the outer bend the high tide flats are eroded and the remains of the salting cliff is higher while on the opposite bank deposition exceeds erosion and the whole intertidal bank slopes gently and evenly into the channel. In the lower reaches where the estuary broadens the main channel may be well offshore and the more sandy banks are less easily eroded. Nonetheless, floods carve away the sand at higher levels and the flood tide deposits material at lower levels until the contour of the bank is often concave with broad flats of silty sand above low tide. At the mouth where the currents are stronger and the sand is coarser, the banks are steeper. The contours of the intertidal banks at a series of surveyed transects along Knysna estuary are illustrated by Day (1967) and those along Morrumbene estuary are shown in Day (1974).

Table 4.1
Sediment distribution on intertidal banks at Linga-Linga, Morrumbene estuary

Shore	Tidal level	Median particle diameter mm	Percentage silt
Exposed shore (no vegetation)	HWS	0,23	0,02
	LWN	0,19	0,02
	LWS	0,24	0,02
Sheltered shore (sea grasses from LWN to LWS)	HWS	0,20	0,02
	LWN	0,19	0,77
	LWS	0,21	3,30

The data in Table 4.1 were recorded on transects near the mouth of the Morrumbene estuary.

In broad stretches of these and other estuaries the fetch of the wind generates waves which erode fine sediments from upper tide levels on exposed windward shores so that the sediment changes from coarser particles at high tide mark through finer ones at low tide of neaps to coarser ones at low spring tide. This effect is reduced by sea grasses growing at and below low tide.

Soft mud accumulates in coves and backwaters of all estuaries and covers extensive areas of the wadden between the Netherlands and Denmark. Postma (1967) points out that mud is much less porous than well-sorted sand and as the tide falls V-shaped ebb channels are formed which unite and deepen towards the edges of the banks. The rising tide flows along these channels eroding the finer particles from the deeper levels where the currents are stronger and transporting them higher up the bank. Due to settling lag, the particles still move forward after the current has fallen below the critical suspension velocity. The ebb tide carries the particles down again but while the current in the drainage channel is increasing to reach erosion velocity, the water level over the whole bank is falling and with it, the strength of the ebb current so that the particles are transported a shorter distance down the bank than they were carried up by the flood. Over many tidal cycles the finest particles accumulate at the top of the shore and the particle size increases towards low tide.

PERMEABILITY AND THE DEPTH OF THE WATER TABLE

Intertidal banks are populated by a wide range of plants and animals and their density and distribution is related to the contour of the bank, the nature of the sediment, its water content and other factors. While the water over mud banks drains off the surface as the tide ebbs, much of it sinks below the surface of porous sand flats increasing the oxygen content of the interstitial water and catching organic particles and plankton on the surface as on a filter paper. Such sand banks are thus rich feeding areas. However, capillarity may draw water to the surface where it evaporates and the salinity increases or the water table may fall and the surface layers become very dry. Much depends on the size of the particles and the degree of sorting denoted by Qdϕ.

The ratio of interstitial water to solid particles is known as the *void ratio e* = volume of water to volume of solids. The *porosity* of the sediment is equal to e/1 + e and Bruce (1928) has shown that the porosity of an artificial sediment composed of uniform glass spheres is 25,96 % of the total volume regardless of the diameter of the spheres. Natural grains are rarely spherical and do not pack so well; thus well-sorted sand with particle diameters of 0,11 to 0,09 mm holds 44,7 % water and slightly coarser or finer sediments hold a little less. In poorly sorted sands the smaller particles tend to fill the spaces between the larger grains and the water content drops to about 20 %. Mud with a high silt or clay content has a high water content of over 80 % when first deposited but as it becomes compacted by evaporation or overburden pressure, the water content is greatly reduced.

The rate at which water seeps through a sediment or the *velocity of percolation*

$$V_s = \frac{1 + e}{e} pi,$$

where *p* is the coefficient of permeability and *i* is the hydraulic gradient. As the tide falls and the sandbanks and mudflats emerge, the hydraulic gradient increases and the water table in the banks sinks at a rate equal to the velocity of percolation. At the time of low tide, the *permanent water table* even in well-sorted sands well away from the water's edge is above that in the channel although it may be deep below the surface of the sand. On relatively impermeable mudbanks, the water table is permanently at the surface and shallow pools are common. The contour of the banks also affects the depth of the water table. Day (1974) has shown that in an evenly sloping shore of well-sorted sand with a negligible silt content, the permanent water table at HWS was below 30 cm and 22,5 cm at LWN. On a nearby concave bank which sloped steeply down to LWN the permanent water table was equally deep at HWS, but the water was seeping out on the surface at LWN. On banks of poorly sorted sand with a Qdϕ value of 0,8 or more, the water table at LWN was only 1-3 cm below the surface and on mudbanks containing more than 20 % silt and a Qdϕ value of more than 1,3 the water table at LWN never sank below the surface. This may be partly due to capillary action since the width of the capillary channels between the particles decreases with particle size and the degree of sorting. Below the permanent water table, water movements are reduced and as decomposing organic matter is usually present and the oxygen content of the interstitial water is low, a black layer of iron sulphide develops.

The penetrability of the sediment also depends on the porosity and water content of the sediment. When pressure is applied to a *dilatant* sediment there is an increased resistance to shear. Part of the interstitial water is driven out but the sand grains remain and the pore spaces are reduced so that the sand is compacted and more difficult to penetrate. In *thixotropic* sediments with a high water content such as quicksands and unconsolidated mud, there is a decreasing resistance to shear. The particles suspended in the interstitial water are displaced from the area under pressure and penetration is facilitated. The penetrability of a sediment is obviously important to burrowing animals.

TURBIDITY AND LIGHT PENETRATION

The penetration of light in the photosynthetic range of 380-720 mμ is important in the productivity of phytoplankton and the depth to which attached plants can grow. Most of the published work deals with measurements in the sea or lakes for salinity appears to have negligible effects on light transmission. Photometers fitted with a series of filters have been used to measure the attenuation of the various wave bands with depth. The extinction coefficient

$$\lambda = \frac{2,3}{d} \log_{10} I_o/I_d$$

where I_o is the intensity of the wave band at the surface and I_d is the intensity at depth *d*. A review of the energy flux of sunlight, its variations with the altitude of the sun and the degree of cloud cover, the percentage of light reflected from a calm or wave-rippled sea surface and the extinction coefficients of the various wave lengths will be found in Harvey (1955). A more recent review by Strickland (1965) includes comments on more sophisticated instruments, a discussion of the effect of the various wave lengths on plant pigments and an estimate of the light intensity at the compensation depth. This is 'probably somewhat less than 0,005 ly.min^{-1} for most natural populations near the bottom of the euphotic zone'.

The subsurface attenuation of light is due to absorption by water molecules and dissolved organic substances particularly 'gelbstof' and other humates, and to scattering or reflection by suspended particles including plankton and silt or clay. Red light is absorbed more rapidly than blue light but is less affected by scattering. Thus the maximum transmission in clear oceanic water is at 480 mμ, while in turbid estuarine and coastal water it is between 500 and 600 mμ. It is generally accepted that the compensation depth for phytoplankton corresponds to the depth where the surface intensity decreases to 1 %. In oceanic water this is about 100 m, in coastal water it is of the order of 15 m and in estuarine water it is seldom as much as 3 m. Quasim (1973) reports that in Cochin Backwater near Bombay, 1 % of the surface light reaches 2-6 m, but in the estuaries of southern Africa it is usually less.

In estuaries where the turbidity varies so much, photometers have seldom been used and a 30 cm Secchi disc which is less accurate and merely measures the penetration of visible light is a more convenient instrument. The relation between the Secchi depth D, the extinction coefficient of white light and the quantity of suspended particles is thus important. According to Perkins (1974) $\lambda = 1,7/D$ in clear sea water but in turbid estuarine water the approximate relation is $\lambda = 1,5/D$. Harvey (1955) states that about 16,7 % of the surface light reaches the Secchi depth. Thus the Secchi depth which is much less than the depth reached by 1 % of the surface illumination is assumed to be sufficient for photosynthesis. Allanson & van Wyk (1959) used both a Secchi disc and a photometer in Lake Sibayi and in Lake Nhlange in northern Zululand. In Lake Nhlange when the Secchi depth was 1,0 m, 1 % of the surface illumination reached 6 m and in Lake Sibayi when the Secchi depth was 2,8 m they determined that 1 % of the surface illumination reached 12 m. It should be noted that neither lake contains an appreciable quantity of suspended silt or plankton. Francis-Boeuf (1943) has related the Secchi depth to the weight of silt suspended in the surface water. The relationship is only approximate but is useful when the Secchi depth is less than 1,0 m.

The following data may be quoted:

Silt content of surface layers (g.ℓ^{-1})	0,27	0,26	0,22	0,13	0,11
Secchi depth (m)	0,35	0,35	0,45	0,60	1,0

Dyer (1972) has reviewed reports of the suspended sediment concentrations in several estuaries. The concentrations are greatest in the upper or mud reaches of estuaries and increase with depth and the discharge of the river. Maximum values are related to high tidal velocities and surface concentrations are greatest during the ebb although, due to lag effects, they occur after the current has reached its peak velocity. Obviously there will be similar variations in turbidity and light penetration.

Only a few of the many Secchi disc readings made in the estuaries of southern Africa are worth discussion. Some are isolated records while others do not show any unusual features although most of them indicate higher turbidities than have been reported from unpolluted European and American estuaries. The highest turbidities have been recorded in the estuaries of large rivers draining the Transkei and southern Natal. Unpublished records made during the January floods in the Bashee, the Mzimvubu, the Mzimkulu and the Mkomazi gave Secchi disc readings of 5-10 cm at all states of the tide even at the mouth. During low river flow in winter the water is much clearer and Secchi disc readings in the Mzimkulu reported by Oliff (1976) show Secchi depths of 1,3 m in the middle reaches and 1,4 m at the mouth. Smaller estuaries draining the coastal strip of the same region are clearer. The most complete set of readings are those made in the Mnga-

Table 4.2
Secchi disc readings (in m) in Mngazana estuary

Distance from sea (km)	Mouth	1,3	3,0	6,0	7,4	8,0
July 1975 HW reading	4,0	3,6	1,1	0,6	0,6	0,5
LW reading	0,9	1,0	1,0	0,5	0,5	0,5
Dec. 1975 HW reading	3,5	3,0	1,0	0,7	0,5	0,4
LW reading	0,8	0,8	0,9	0,5	0,4	0,1

zana. Dr George Branch has kindly permitted me to reproduce part of his data.

As will be seen, the turbidity decreases from high to low tide and from the mouth to the upper reaches of the estuary. Similar changes have been reported in the Morrumbene estuary by Day (1974) and in the Mlalazi estuary by Hill (1966). Hill's figure 5 illustrates the changes over a spring tide cycle in the broad middle reaches of the estuary. As the flood tide moves along the bottom the Secchi depth and transparency of the surface water increases slowly as the suspended particles sink. Due to lag effects, the maximum transparency occurs after high tide. During the ebb the turbidity increases rapidly and the minimum Secchi reading was recorded an hour before low tide. Over the whole tidal cycle the Secchi depth varied from 52 to 24 cm.

Unpublished Secchi disc records in the Breede estuary in the southern Cape show maximum turbidities during winter rains and minimum turbidities during low flow periods in summer as might be expected. The Secchi disc readings varied from 15-20 cm over the whole estuary after a flood in July 1951, and from 53 cm at the head of the estuary to 150 cm at the mouth during February 1952. Tidal variations were very obvious near the mouth. The clear 'blue' sea water entering with the flood tide flows under the more turbid estuarine water and at Port Beaufort, 2 km from the mouth, the Secchi disc readings changed from 130 cm at low tide to 150 cm at high tide. The change when the turbid surface water retreats with the ebb and advances with the flood is sharply defined. Many other parameters apart from turbidity change abruptly in such areas and Australian ecologists at Cronulla have referred to these areas as frontal systems.

When an estuary is closed by a sandbar, the clarity of the water increases for the only turbulence is that due to the slow inflow of the river and wave action in the shallows. In Hermanus lagoon Scott *et al* (1952) reported Secchi depths increasing from 10-50 cm where the river entered to 3,0 m in the centre of the lagoon. In the blind Umgababa estuary in Natal the Secchi depth was 60 cm in January 1950 and only 8 cm in the large open Mkomazi estuary nearby.

Changes in turbidity due to the wind are well-illustrated by a comparison of Lake St Lucia in Zululand and the Lagoa Poelela in southern Mocambique. Both are saline and atidal. Secchi depths recorded in St Lucia, which is 1-2 m deep, by Day *et al* (1954) show Secchi readings of 18 cm on windy days and up to 84 cm on calm days. In the Lagoa Poelela where the depth is 30-35 m, Hill *et al* (1975) record that 3 % of the surface illumination reached 22 m.

REFERENCES

ALEXANDER, W.J.R. 1979. Sedimentation of estuaries: causes, effects and remedies. *Abstract 4th S.Afr. natl. oceanog. Symp. Cape Town, July 1979.*

ALLANSON, B.R. & J.D. VAN WYK 1969. An introduction to the physics and chemistry of some lakes in northern Zululand. *Trans. roy. Soc. S.Afr.* **38**: 217-240.

ANON. 1973. Berg river salinity. Report **F22-1**. *Fish. Corp. S.Afr.* Cape Town (typescript).

BLABER, S.J.M., HILL, B.J. & A.T. FORBES 1974. Infratidal zonation in a deep South African estuary. *Mar. Biol.* **28**: 333-337.

BRANCH, G.M. & J.G. GRINDLEY 1979. The ecology of South African estuaries. Part XI: Mnagazana, a mangrove estuary. *S.Afr. J. Zool.* **1**(3): 149-179.

BRUCE, J.R. 1928. Physical factors on the sandy beach. Part 1. Tidal, climatic, and edaptic. *J. mar. biol. ASS. UK* **15**: 535-565.

BURTON, J.D. 1976. Basic properties and processes in estuarine chemistry. *In:* J.D. Burton & P.S. Liss (eds), *Estuarine chemistry.* Academic Press, London.

CHUNNETT, E.P. 1965. Siltation problems in Knysna Lagoon. *CSIR Report No. MEG.353:* 1-25.

CLOETE, C.E. & W.D. OLIFF 1976. South African marine pollution survey report 1974-1975. Rep 8. *S.Afr. natl. sci. Program,* CSIR, Pretoria.

DAVIES, J.L. 1964. A morphogenetic approach to world shorelines. *Zeitz. f. Geomorph.* **8**: 127-142.

DAY, J.H. 1967. The biology of Knysna estuary, South Africa. *In:* G. Lauff (ed), *Estuaries.* Am. Ass. Adv. Sci., Washington.

DAY, J.H. 1974. The ecology of Morrumbene estuary, Mocambique. *Trans. roy. Soc. S.Afr.* **41**: 43-97.

DAY, J.H., MILLARD, N.A.H. & G.J. BROEKHUYSEN 1954. The ecology of South African estuaries. Part 4: The St Lucia system. *Trans. roy. Soc. S.Afr.* **34**(1): 129-156.

DAY, J.H., MILLARD, N.A.H. & A.D. HARRISON 1952. The ecology of South African estuaries. Part 3: Knysna, a clear open estuary. *Trans. roy. Soc. S.Afr.* **33**: 367-413.

DYER, K.R. 1972. Sedimentation in estuaries. **In:** R.S.K. Barnes & J. Green (eds), *The estuarine environment.* Applied Science, London.

DYER, K.R. 1979. *Estuarine hydrology and sedimentation.* Cambridge Univ. Press. 230pp.

EMANUEL, A. 1977. Conservation at Richards Bay. *Omgewing, Environment, RSA* **4**(12): 5-7 Dept. Planning and Environment, Pretoria.

FLEMMING, B.W. 1977. Distribution of recent sediments in Saldanha Bay and Langebaan Lagoon. *Trans. roy. Soc. S.Afr.* **42**(3-4): 317-340.

FRANCIS-BOEUFF, C. 1943. Physico-chimie du milieu fluvio-marin. *CR Soc. Biogeogr.* **169-170**: 19-26.

GRINDLEY, J.R. 1976. *Report on the ecology of Knysna estuary and proposed Braamekraal Marina.* (Unpublished). School of Environmental Studies, Univ. Cape Town. 128pp.

HARVEY, H.W. 1955. *The chemistry and fertility of sea waters.* Cambridge Univ. Press.

HAYES, M.O. 1977. Morphology of sand accumulation in estuaries: an introduction to the symposium. *In:* L.E. Cronin (ed), *Estuarine research, 2. Geology and engineering.* Academic Press, New York.

HEYDORN, A.E.F. (ed) 1976. *St Lucia Scientific Advisory Council workshop – Charters Creek, February 1976.* Natal Parks Bd, Pietermaritzburg.

HEYDORN, A.E.F. 1977. Agriculture and earthworks – death knell of Natal's estuaries. *African Wild Life* **31**(6): 27-30.

HILL, B.J. 1966. A contribution to the ecology of Umlalazi estuary. *Zool. Afr.* **2**: 1-24.

HILL, B.J., BLABER, S.J.M. & R.E. BOLTT 1975. The limnology of Lagoa Poelela. *Trans. roy. Soc. S.Afr.* **41**: 263-271.

INGLIS, C.C. & F.H. ALLEN 1957. The regimen of the Thames estuary as affected by currents, salinities and river flow. *Proc. Inst. Civil Eng.* 7: 827-868.

KRIGE, A.V. 1927. An examination of the Tertiary and Quaternary changes of sea level in South Africa in favour of Recent world-wide sinking of ocean level. *Ann. Univ. Stellenbosch* **5**: 1-18.

KULM, L.D. & J.V. BYRNE 1967. Sediments in Yaquina Bay, Oregon. *In:* G. Lauff (ed), *Estuaries.* Am. Ass. Adv. Sci., Washington.

MACNAE, W. 1957. The ecology of the plants and animals in the intertidal regions of the Zwartkops estuary near Port Elizabeth, South Africa – Parts I and II. *J. Ecol.* **45**: 113-131; 361-387.

MACNAE, W. 1963. Mangrove swamps in South Africa. *J. Ecol.* **51**: 1-25.

McLACHLAN, A. & J.R. GRINDLEY 1974. Distribution of macrobenthic fauna of soft substrata in the Swartkops estuary, with observations on the effect of floods. *Zool. Afr.* **9**(2): 211-233.

MILLARD, N.A.H. & A.D. HARRISON 1954. The ecology of South African estuaries. Part 5: Richards Bay. *Trans. roy. Soc. S.Afr.* **34**(1): 157-179.

MILLARD, N.A.H. & K.M.F. SCOTT 1954. The ecology of South African estuaries. Part 6: Milnerton estuary and the Diep River, Cape. *Trans. roy. Soc. S.Afr.* **34**: 279-324.

OLIFF, W.D. (ed) 1976. *S.African National Marine Pollution Surveys.* Annual report 1, pp 1-509; Second annual report; section C, estuarine surveys pp 1-146. (Unpublished typescript) NIWR, Durban.

ORME, A.R. 1974. Estuarine sedimentation along the Natal coast, South Africa. *Tech. Rep.* **5**, *Office of Naval Research USA.* 1-53.

PERKINS, E.J. 1974. *The biology of estuaries and coastal waters.* Academic Press, London and New York.

POSTMA, H. 1967. Sediment transport and sedimentation in the estuarine environment. *In:* G. Lauff (ed), *Estuaries.* Am. Ass. Adv. Sci., Washington.

QUASIM, S.Z. 1973. Productivity of backwaters and estuaries. *In:* B. Zeitzschel (ed), *The biology of the Indian Ocean.* Chapman & Hall, London.

SCHOU, A. 1967. Estuarine research in the Danish moraine archipelago. *In:* G. Lauff (ed), *Estuaries.* Am. Ass. Adv. Sci., Washington.

SCOTT, K.M.F., HARRISON, A.D. & W. MACNAE 1952. The ecology of South African estuaries. Part 2: The Klein River estuary, Hermanus. *Trans. roy. Soc. S.Afr.* **33**: 283-331.

STRICKLAND, J.D.H. 1965. Production of organic matter in the primary stages of the marine food chain. *In:* J.P. Riley & G. Skirrow (eds), *Chemical Oceanography.* 1. Academic Press, London and New York.

VAN HEERDEN, I.L. 1976. The geology of Lake St Lucia and some aspects of its sedimentation. *In:* A.E.F. Heydorn (ed), *St Lucia Scientific Advisory Council workshop – Charters Creek, February 1976.* (Typescript). Natal Parks Bd, Pietermaritzburg.

WEAVER, A.v.B. 1979. The effects of flow restriction on selected grain size parameters of the sediments in the Bushman's River estuary. *Abstract 4th (S.Afr.) natl. oceannogr. Symp. Cape Town, July 1979.*

WOOLDRIDGE, T. 1976. The zooplankton of Msikaba estuary. *Zool. Afr.* **11**: 23-44.

WOOLDRIDGE, T. 1977. The zooplankton of Mngazana, a mangrove estuary in Transkei, Southern Africa. *Zool. Afr.* **12**: 307-322.

Chapter 5
The chemistry and fertility of estuaries
J.H. Day

Department of Zoology, University of Cape Town

INTRODUCTION

This brief account is presented as one facet of the environmental conditions in estuaries. It includes a summary of chemical records in a few polluted European estuaries and some chemical records from estuaries in South Africa, few of which are as yet polluted. Far more comprehensive accounts of the chemistry of European and North American estuaries are available as was evident in the recent workshop on the biogeochemistry of estuarine sediments (Goldberg, 1978). However, many of the reactions in the biogeochemical cycle are controversial and a discussion of the evidence would make this chapter too long. In attempting to condense and present only those features which are essential for an ecological appreciation, some dogmatic statements have been made. I am grateful to Dr M.J. Orren and Dr G.A. Eagle of the National Institute for Oceanology for eliminating the more obvious innaccuracies; those that remain are my own. I am also grateful to Mr W.D. Oliff of the National Institute for Water Research for making available his unpublished data on estuaries along the east coast of southern Africa.

SEA WATER

Since estuaries are formed by a mixing of sea and river water certain aspects of the chemistry of sea water

and river water provide a necessary introduction to the chemistry of estuaries. The proportions of the major ions in sea water are constant in all oceans whereas those in river water are variable and where the two mix, a number of changes occur. Thus estuarine water is not simply diluted sea water. The chemical species may change, some particulate substances become more soluble while some dissolved substances are precipitated and are absorbed or adsorbed on particles of silt, clay or organic matter. Other elements show conservative properties and their concentrations vary directly with dilution as measured by the salinity of the sample. Moreover, estuaries are very shallow in comparison to the sea and are much richer in living and dead organic matter. As a result, there is a rapid flux of materials between members of the biological community, the water and the sediments. Rapid decomposition of organic detritus on the estuary bed leads to changes in pH and Eh in the subsurface layers so that inorganic compounds are reduced with resulting changes in solubility.

The chemistry of sea water has been reviewed by Riley & Skirrow (1965, 1975) and Riley & Chester (1971). Accounts of estuarine chemistry will be found in Phillips (1972) and Burton & Liss (1976). Much of the material in this chapter has been drawn from these sources.

Sea water is a solution of many inorganic salts, atmospheric gases, traces of organic matter and in addition there is a small but variable concentration

Table 5.1
Major constituents of sea water of 35 $‰$ salinity (g.kg^{-1} sea water). After Riley & Skirrow (1975)

Na+	10,773	B	0,004
Mg^{2+}	1,294	Cl$^-$	19,344
Ca^{2+}	0,412	SO$_4{}^{2-}$	2,712
K+	0,399	Br$^-$	0,067
Sr^{2+}	0,008	F$^-$	0,0013
		HCO$_3$	0,142

of particulate material. The salinity, originally defined as the total mass of inorganic solids dissolved in 1 kg of sea water, is typically 35 $‰$ but ranges from 33 to 37 $‰$ in the open ocean. Sea water contains practically every known element in solution but most of them occur in such minute concentrations that the 11 major constituents comprise over 99,6 % of the dissolved solids. Estimates differ slightly and the concentrations of the major ions in Table 5.1 are quoted from Riley & Skirrow (1975).

The constancy of the major constituents led to the classical method of determining salinity by the volumetric titration of chlorinity but the more modern method is by determining the conductivity with a salinometer. In estuaries where the salinity may vary rapidly with the tide, approximate values are conveniently obtained in the field with a pocket refractometer. Where the salinity is below 5 $‰$, the proportion of the major constituents is affected by the variable composition of the river water and the ionic strength is usually expressed in conductivity units or in terms of chlorinity.

The carbon dioxide system, boric acid and pH

All the atmospheric gases are dissolved in sea water in concentrations related to their partial pressures in the atmosphere and their solubilities. The ratio of the partial pressures of $N_2:O_2:CO_2$ in the atmosphere is 2400:630:1 but in the sea, the ratio of the dissolved gases is 28:19:1 and their exact concentrations at saturation depend on the temperature and salinity. The dissolved CO_2, whether it be derived from the atmosphere, from respiration of aquatic organisms or bacterial decay is in equilibrium with carbonates and bicarbonates:

$$CO_2 + H_2O \rightleftharpoons H_2CO_3 \rightleftharpoons HCO_3^- + H^+ \rightleftharpoons 2H^+ + CO_3{}^{2-}.$$

In well-oxygenated sea water, HCO_3^- is the main species and undissociated H_2CO_3 and dissolved CO_2 together represent 1 % or less of the system. Also sea water contains cations of alkali metals in excess of the equivalent anions derived from strong acids and this excess base or total alkalinity, is in equilibrium with the carbon dioxide system. Expressed in milliequivalents per kilogram of sea water the total alkalinity ranges from about 2,2 to 2,5. The significance of this equilibrium is that when HCO_3^- is absorbed by plants during photosynthesis, it is immediately regenerated from H_2CO_3 so that it is never a limiting factor for plant growth. The same is true in estuaries.

H_3BO_3 is also weakly dissociated, and the presence of both the H_2CO_3 and the H_3BO_3 systems give sea water a limited buffering action. In the open sea, the pH varies from 7,8 to 8,3 according to the absorption of HCO_3^- during photosynthesis and production during respiration or bacterial decomposition so that pH gives an approximate measure of biological activity. The same processes also affect the concentration of dissolved oxygen and plant nutrients. In dense growths of algae, the pH may rise to 9,6 and the dissolved oxygen may become super-saturated during the day but at night the pH may fall to 6,8 and oxygen may fall to very low concentrations. At such low pH values the concentration of dissolved CO_2 increases and in organic deposits it is reported to affect the chemical species of some of the rarer elements. Riley & Chester (1971) discuss the equilibria of the carbon dioxide and boric acid systems in sea water and state that the effect of boric acid on the carbon dioxide system is only significant above pH 8. They reproduce tables of the first and second dissociation constants (pK_1 and pK_2) prepared by Lyman. I am indebted to Professor B. Allanson for drawing my attention to these and more recent studies. Allanson & Rudd (1978) provide revised tables of pK_1 and pK_2 for chlorinities of 1 to 10 $‰$ which are reuired for the determination of the total available carbon in oligohaline water.

Oxygen

The solubility of oxygen in sea water is a function of its partial pressure and the temperature and salinity of the water. A table of saturation values at different temperatures and salinities is given by Riley & Skirrow (1975) in terms of cm^3dm^{-3} ($ml.ℓ^{-1}$) and Kester (1975) gives a table in units of $\mu moles.kg^{-1}$. Selected values have been extracted and shown with the corresponding values in $ml.ℓ^{-1}$ in Table 5.2.

Photosynthesis supplements the oxygen absorbed from the atmosphere so that the whole euphotic zone is at times supersaturated. Below this the gradual decay of plankton and faecal material falling through the water depletes the oxygen concentration and an oxygen minimum layer develops in the tropics. In the southeastern Atlantic, its core is at 300-500 m. Part of this flows along the edge of the continental shelf of South West Africa and Visser (1969) reports that when it spills onto the shelf, oxygen concentrations fall to between 0,2 and 0,42 $ml.ℓ^{-1}$ (or 0,29-0,60 $mg.ℓ^{-1}$). When this water wells up inshore near Walvis Bay, plankton and fish are trapped and mass mortalities occur. Similar but less extreme conditions may occur as far south as St Helena Bay.

Deoxygenation may occur wherever bottom water is cut off from the surface circulation by haloclines or thermoclines. It is well-known in the Black Sea and the deeper parts of the Baltic. It occurs in deep lakes where surface cooling in winter is not sufficient to

Table 5.2
Saturation values for dissolved oxygen in μ mols.kg^{-1} seawater and in ml.ℓ^{-1} at selected temperatures and salinities (modified from Kester, 1975)

Temperature	S = 0 ‰	S = 16 ‰	S = 28 ‰	S = 35 ‰
0 °C	456,4 (11,3)	404,0 (10,0)	368,7 (9,1)	349,5 (8,6)
5	398,8 (9,8)	354,4 (8,7)	324,4 (8,0)	308,1 (7,6)
10	352,6 (8,7)	314,6 (7,8)	288,9 (7,1)	274,8 (6,8)
15	315,1 (7,8)	282,3 (7,0)	259,9 (6,8)	247,7 (6,1)
20	284,2 (7,0)	255,5 (6,3)	235,9 (5,8)	225,2 (5,6)
25	263,3 (6,5)	237,3 (5,9)	219,6 (5,4)	209,9 (6,2)
30	236,7 (5,8)	214,2 (5,3)	198,8 (4,9)	190,3 (4,7)

cause convection currents deep enough to refresh the hypolimnion. It also occurs in many fjords with a sill at the mouth which prevents exchange with oceanic water. Unpolluted estuaries are normally well-oxygenated from surface to bottom by turbulence due to tidal currents although in deep estuaries such as Chesapeake Bay, Taft & Taylor (1976) report that during summer the dissolved oxygen concentration may fall to nil below 6 m but this is re-oxygenated by convection currents in autumn. Low oxygen concentrations develop in deep depressions in an estuary bed where high salinity water tends to remain, while low salinity water flows over the surface. Three examples may be quoted from southern Africa. Boltt & Allanson (1975) and Oliff (1976) investigated the Kosi Bay system in Zululand. The estuary is formed by a series of deep lakes linked by shallow channels. Lake Nhlange is the largest and the deepest, reaching 24 m in one area. The salinity is low, varying from less than 1 ‰ during the summer rains, but normally it is 3-4 ‰. The bottom water is well-oxygenated in winter, but in some summers the concentration near the bottom may fall to 1-2 mg $O_2.\ell^{-1}$. Boltt & Allansom suggest that this may be due to the inflow of organic material from the surrounding swamps and its decomposition during the warm season. Lake Mpungwini (reported as Sifungwe), the next in the series, is 18 m deep in one area and usually has a well-marked salinity stratification at 8-10 m. Oliff reported that in 1971 there was anoxic water from 12 m to the bottom with traces of H_2S in some areas. When river discharge increased, much of the high salinity water was flushed out and in 1976 the anoxic water was limited to a small area in the deepest part of the lake.

The Msikaba estuary in southern Natal, studied by Blaber *et al* (1974) and Wooldridge (1976) has a maximum depth of 35 m in the middle reaches but shallows to less than 2 m at the mouth which is almost blocked by a sandbar. There is a well-marked halocline and during periods of normal river discharge the low salinity surface water flows out of the mouth, leaving the high salinity bottom water unrefreshed. Blaber *et al* report that the water below 10 m gradually loses oxygen and may be reduced to 5 % saturation below 12 m in summer but increases to 50 % saturation in winter. Wooldridge (1976) reports that heavy floods cut a channel up to 2,5 m deep through the bar.

Much of the surface water is flushed out and the sea water can enter as a bottom current to refresh the water in the deeper parts of the estuary to some degree before the bar builds up again.

The observations reported by Oliff (1976) on the Mtamvuna estuary at the southern border of Natal confirm and extend the observations in the Msikaba. The Mtamvuna has a maximum depth of 10 m and the mouth is periodically closed by a sandbar. When surveyed in September 1972, the mouth was closed and the estuary had a marked halocline at 2 m with a salinity of 30 ‰ and less than 20 % oxygen saturation below. Details are shown in figure 3.6.

When the NIWR investigators cut through the sand bar the level of the estuary fell about 1 m until sea water entered with the rising tide and flowed along the bed of the estuary. Within three days it had reached the rise at Station 4. The oxygen saturation of the bottom water increased to more than 90 %. It is interesting to note how rapidly the bottom water may be refreshed through a narrow channel and it would be interesting to know the rate at which an unpolluted estuary such as this loses oxygen.

Observations in other estuaries in southern Africa where tidal currents are restricted or absent, indicate that oxygen concentrations in the deeper layers depend on the ratio of area to depth and the circulation due to the wind. The St Lucia lakes are beyond the influence of tidal action but the area is over 3 000 km^2 and the depth is 1-2 m. Wind action is sufficient to prevent the formation of thermoclines and the water is well-oxygenated to the bottom. The Umgababa estuary in Natal, the Kleinemond estuary in the Eastern Cape and the Klein River estuary near Hermanus are all closed by sandbars for long periods but they are relatively broad and shallow and there is seldom evidence of substantial oxygen depletion even in the deepest areas.

Oxygen depletion by pollution

The decomposition of organic material in an estuary is a normal process and the introduction of domestic sewage merely adds to the quantity of matter that must be oxidised. The important factors are the rate of input of organic matter, its dilution and distribution by tidal currents, and the balance of oxygen absorption and replacement. Assuming the ratio of elements in the organic matter to be similar to that in a mixed plankton sample which is 106C:16N:1P (Flemming, 1940), the overall reaction during total decomposition to CO_2, NO_3^- and PO_4^{3-} according to Head (1976) is:

$$(CH_2O)_{106}(NH_3)_{16}H_3PO_4 + 138O_2 \rightarrow 106CO_2 + 122H_2O + 16HNO_3 + H_3PO_4$$

Head quotes studies of the oxidation of sewage in the Thames estuary and says: 'The carbonaceous oxygen demand of much of the sewage entering the estuary was found to be between 2,8 and 3,0 $gO_2.gC^{-1}$ which is slightly above the theoretical value of 2,67 for oxidation of carbohydrate. Most of the nitrogenous oxidation consisted of the oxidation of ammonium nitrogen and thus required the theoretical consumption of oxygen of 4,7 $gO_2.gN^{-1}$. On this basis one gram of dry sewage would require 2,89 g or 4,13 ℓ of oxygen. This would exhaust the oxygen contained in 575 ℓ of fully saturated water with a temperature of 15 ° and a salinity of 16 ‰. This gives some estimate of the dilution required for domestic sewage. Normally oxygen is absorbed from the atmosphere during the process of decomposition. The CO_2 formed is taken up by the alkali reserve so that there is only a minor decrease in pH while the PO_4^{3-} and the NO_3^- add to the fertility of the estuary. In brief, a small amount of sewage, if chlorinated to kill pathogens and comminuted to break up the faeces will cause no damage.

Decomposition of organic matter by aerobic bacteria is a slow process and some organic compounds are more rapidly oxidised than others which may take weeks even under favourable conditions (high temperature). The oxygen absorbed during various stages may be measured by different techniques and over different periods. The biological oxygen demand is usually measured over five days (5 day B.O.D.); it may be measured by the oxygen absorbed by acid permanganate during four hours (4 hour O.A), or the oxygen absorbed by permanganate in an alkaline medium in 30 minutes which is a measure of the easily oxidisable organic content. The oxygen absorbed from acid potassium dichromate is known as the chemical oxygen demand (C.O.D.). Details of methods will be found in manuals of water chemistry but the salinity of estuarine water introduces difficulties which do not occur in fresh water.

If the organic pollution is introduced too rapidly for oxygen replacement, the DO falls and when it reaches 10 % of saturation the oxidation of NH_3 ceases. When it drops further, many aquatic animals die, adding to the B.O.D. load and eventually the water becomes anoxic. Anaerobic bacteria then develop and absorb oxygen, first from NO_3^- reducing it through NO_2^- to NH_3 or N_2. Finally oxygen is absorbed from SO_4^{2-} reducing it to H_2S which combines with iron oxides to form FeS and when the available iron is exhausted, H_2S is liberated. The nitrogen and sulphur

cycles will be discussed in more detail later. When toxic H_2S spreads through the water column all animals die increasing the B.O.D. load further.

The anoxic water spreads along the estuary reducing the DO concentration far from sewage outfalls. The intensity of pollution in the Thames estuary has been studied over 80 years (Barrett, 1972). During periods of low flow in 1947, the Thames was anoxic for 20 miles and during the 1950's H_2S was evolved. Since then the sewage treatment plants have been enlarged and improved and the effect on the Thames has been monitored. It was found that the intensity of pollution along the Thames could best be measured by oxygen sag curves in which the mean DO concentration at mid-tide over a month is plotted against the mean river flow. A more sensitive measure of pollution 'is obtained by considering the seaward end of the sag-curve, for although the reduction in polluting load may not result in much change in the sag-curve minimum, it should in any event produce a shortening of the reach of low oxygen content' (Barrett, 1972). Since 1964 the anaerobic reach has disappeared and migratory fish pass up and down the estuary.

Both domestic and industrial wastes contain surfactants which reduce the DO concentration of the water. They are used to emulsify fats and oils in industry and are essential constituents of domestic detergents. Both anionic and non-ionic surfactants cause foaming at the water surface and they also reduce the exchange coefficient of oxygen by 20 % (Perkins, 1974). In addition, domestic detergents contain ployphosphates which, on decomposition, add to the fertility of the water. The first surfactants developed were resistant to bacterial decomposition and toxic to many organisms but the more modern ones are biodegradable and less toxic.

Oliff (1976) reports that apart from the small Salt (or Black) River which drains the industrial area of Cape Town and receives the effluent from one of the sewage plants, no large estuary in South Africa is known to be seriously deoxygenated as a result of pollution, although this does occur in several rivers. The Mgeni estuary in Durban receives the effluent from the Seekoei sewage works and is moderately polluted as judged by counts of faecal bacteria. However, the sewage absorbs oxygen in the maturation pond and the estuary itself is well-oxygenated. The Mhlatuzi river receives the effluent of a sugarcane mill and cane fibre reaches the bed of Richards Bay, but the water above is not deoxygenated. A 'marina' or cottages in a caravan park is being developed on the Fafa lagoon but a series of tests by the NIWR showed that the river water was well-oxygenated. There is minor pollution of the Mzimkulu at Port Shepstone but no deoxygenation. Pollution from Uitenhage tanneries and wool-washing plants flows into the Swartkops river but deoxygenated water seldom reaches the estuary.

Table 5.3
Comparison of some major constituents in average river water compared with values for sea water (modified from Burton & (Liss, 1976)

Constituent	River water			Sea water (S = 35 %)	
	Concen. (mg.ℓ^{-1})	% of total dissolved material		Concen. (mg.ℓ^{-1})	% of salinity
Na^+	6,3	5		10 770	31
K^+	2,3	2		399	1
Mg^{2+}	4,1	3,5		1 294	4
Ca^{2+}	15,0	12,5		412	1
Fe^{3+}	0,7	<1		<0,01	–
Cl^-	7,8	6,5		19 340	55
SO_4^{2-}	11,2	9		2 712	8
HCO_3^-	58,4	49		140[a]	0,4
SiO_2	13,1[b]	11		<0,1->10[c]	–

a. inorganic carbon expressed as HCO^-
b. Si expressed as SiO_2
c. variable with location and depth

Table 5.4
Total dissolved solids in zones of the Berg River* (data from Harrison & Elsworth, 1958)

Zone	Geological formation	Total dissolved solids (mg.ℓ^{-1})	
		Winter (rains)	Summer (dry season)
II	Table Mountain sandstone	10	40
IIIA	Cape granite	19	78
IV	Malmesbury slates	45	584
V	Sand, limestone, gravel	68	381

* Analysis of main ions omitted

Table 5.5
Main constituents and total dissolved solids in ten large unpolluted rivers in Natal (ppm)

	TDS	Na^+	K^+	Ca^{2+}	Mg^{2+}	SO_4^{2-}	Cl^-	CO_2^*
Dry season	458,4	62,3	2,1	27,7	21,9	23,0	105,6	3,1
Rainy season	212,6	31,9	1,6	12,8	11,1	10,7	42,1	5,3
% TDS (mean of both seasons)		14,3	0,7	6,0	5,0	5,0	21,4	1,6

* Inorganic carbon in solution expressed as CO_2

RIVER WATER

In contrast to sea water, river water has a very low and extremely variable concentration of dissolved solids. Moreover, the proportions of the various salts in solution also differ from those in sea water. Riley & Chester (1971) list the concentrations of all the dissolved elements in both environments and Burton & Liss (1976) present a shorter list of some of the main constituents in average river water and sea water based on the work of Livingstone (1963). A modified version of the table is shown here as Table 5.3.

Livingstone estimated the average concentration of dissolved solids in the rivers of the world as 120 mg.ℓ^{-1} as compared with a concentration of 35 000 mg.ℓ^{-1} in typical sea water. He emphasised that the total salt concentration in river water is extremely variable even within a single river, and quotes the concentration in the Moreau River in Canada as varying from 160 to 3 400 mg.ℓ^{-1} during a year; further, the concentration in an average river of one continent may differ markedly from the average in other continents. The variability depends on many factors. The geology of the drainage basin is obviously important. In rivers draining well-leached soils the concentration of salts is relatively low and stable but may show the effect of sea salt particles transported by the atmosphere and carried down by precipitation (Gibbs, 1970). Where weathering is in progress, the salt concentration is high and increases during periods of low river flow. As large rivers receive tributaries draining different geological formations the distinctions disappear and the total concentration of dissolved solids rises. These changes are shown by the different zones of the Great Berg River in the Cape Province investigated by Harrison & Elsworth (1958) (see Table 5.4).

A more representative analysis of total dissolved solids and the relative concentrations of the main constituents has been obtained by averaging the values in 10 large rivers in Natal. The original data were obtained by the NIWR and are contained in volume 13 of the Natal Town and Regional Planning Report (Brand *et al*, 1967, Archibald *et al*, 1969 and Kemp *et al*, 1976). The rivers selected were the Mtamvuna, Mzimkulu, Mkomazi, Matigulu, Mlalazi, Mhlatuzi, Mfolozi, Hluhluwe, Mkuze and Pongola ranging from the Transkei border to northern Zululand. The stations selected were those closest to the heads of the estuaries, while those in polluted rivers such as the Mgeni and Tugela or polluted stretches as in the Mhlatuzi were omitted. Average values are shown in Table 5.5.

In the dry season the total concentration of dissolved solids is double that in the wet season (summer) but even in the wet season the concentration is almost double the value of 120 mg.ℓ^{-1} given by Livingstone (1963) for the world average. A comparison of individual constituents shows many differences. In particular the concentrations of Na^+ and Cl^- are much higher possibly due to proximity to the sea while the concentration of Ca^{2+} is much lower. Although cretaceous deposits are present in Zululand, most of the coastal belt is covered by Recent sands.

In comparison to sea water, both the world average river water and the Natal rivers show marked differences in the relative proportions of the main constituents. As stated by Burton & Liss (1976), world average river water is dominated by calcium and bicarbonate while sea water is dominated by chloride and sodium and to a lesser extent by sulphate and magnesium. These differences from sea water are not so well-marked in Natal rivers.

Table 5.6
Estimated average concentrations of some dissolved minor constituents in global river water supply and in sea water (from Burton & Liss, 1976)

Constituent	Concentration ($\mu g.\ell^{-1}$)	
	River water	Sea water
Lithium	3	170
Fluoride	100	1 300
Barium	10	10
Uranium	0,3	3,3
Molybdenum	1	11
Chromium	1	0,2
Manganese	7	1
Copper	7	1
Zinc	20	2,5
Antimony	1	0,3
Lead	3	0,03

Table 5.7
Concentration of organic carbon in natural waters (based on Head, 1976)

Concentration of organic carbon ($mg.\ell^{-1}$)	River	Estuary	Coastal sea
Dissolved	10-20(50)	1-5(20)	1-5(20)
Particulate	5-10	0,5-5	0,1-1,0
Total	15-30(60)	1-10(25)	1-6(21)

Figures in parentheses are extreme values.

The minor constituents dissolved in river water and sea water

After stressing the variability of trace elements dissolved in river water, Burton & Liss quote the average values obtained by Turekian in comparison with the concentrations in coastal sea water. They are shown in Table 5.6.

The variations in concentration in individual rivers are very wide particularly in regard to antimony, cromium, cobalt and molybdenum. Burton & Liss point out that the concentration of uranium in Indian rivers may vary by three orders of magnitude. Human influences also have important effects. Uranium is leached from phosphatic fertilizers, lead is derived from anti-knock fuels and 30-40 % of the sulphate carried by rivers is due to the burning of fossil fuels and the transport of sulphur dioxide in the atmosphere.

The transport of particulate matter

Only part of the trace metals is carried in solution in river water, and five times more is transported in particulate form adsorbed on particles of silt, clay and organic matter. Iron, aluminium, manganese and silicon in particular are carried in this way. Burton & Liss (1976) quote the estimates of Gibbs (1973) of the relative percentages of iron, nickel, cobalt, chromium, copper and manganese transported either in solution or in particulate form.

The concentration of suspended solids and organic matter varies greatly with the climate and the rate of flow of the river. In arid areas with occasional heavy rains, the precipitation reaches the river mainly as run-off carrying large amounts of suspended matter whereas in areas of steady precipitation, the vegetation is well-developed and much of the water percolates into the soil. The height of the water table in the soil rises and the lower levels, whose residence time may be measured in years, eventually reaches the river. Such seepage water carries little suspended matter but more dissolved solids including soluble and colloidal organic matter. The contrast between arctic and tropical climates is equally striking. Gibbs (1978) shows that the Yokon river carries 1 000 mg sediment.ℓ^{-1} during the spring thaw, but only 1 mg.ℓ^{-1} in winter when the river is flowing under ice. In the Amazon, the Javari tributary carries 81 mg.ℓ^{-1} of suspended solids during the season of high discharge and 40 mg.ℓ^{-1} during the season of low flow. The similarity between rivers in Alaska where variations in sediment load are due to temperature changes, and the rivers in arid parts of South Africa where variations are related to seasonal changes in rainfall, is very striking. Harrison & Elsworth (1958) give a range in the lower reaches of the Berg River from 37 mg.ℓ^{-1} (as SiO_2) during the dry summer to a maximum of 966 mg.ℓ^{-1} (as SiO_2) during a winter flood. Extremes of turbidity occur in the Transkei and southern Natal where the water is so muddy during summer floods that a Secchi disc is invisible below 5-10 cm. During the dry winter months the Secchi depth is of the order of 1-2 m.

Organic matter

River water carries large amounts of organic matter. During floods, much is in the form of plant litter picked up from the flood plain, but in normal flow most of the material is organic detritus derived from riverine vegetation, marshes and polluted areas. The detritus includes both particulate and dissolved organic matter, much of it in colloidal form. The fraction retained by a 0,5 μm filter is termed particulate organic matter (POM) while that which passes through is termed dissolved organic matter (DOM). Both are usually estimated by their carbon content, referred to as POC and DOC.

Organic matter in the sea is discussed by Riley & Chester (1971), Williams (1975) and Head (1976). Head reports that both dissolved organic carbon and particulate organic carbon in suspension decrease from rivers through estuaries and coastal seas to oceanic waters. Part of his Table 1 is reproduced above as Table 5.7.

The dissolved organic carbon in estuaries may be absorbed directly by heterotrophic micro-organisms or it may be precipitated. In mixing with estuarine water which has a much higher ionic strength and a high and stable pH, the colloidal material flocculates, absorbs soluble material and is eventually deposited. Details will be discussed later.

One of the important forms of organic matter, both in fresh water and the sea is 'humus' or Gelbstoff. Its nature and formation has been discussed by Riley & Chester (1971), Phillips (1972) and Head (1976). It has long been recognised that humic material may be divided into four fractions according to its solubility in water, alcohol and alkali. Each of these fractions represents groups of compounds of similar molecular mass, with fulvic acid including the lighter substances and humic acids including the heavier substances (Head, 1976). According to Phillips (1972), 'humic materials are complicated but seem to consist of large, three-dimensional polyanions composed of aromatic nuclei with phenolic and carboxyllic functional groups linked together by various bridging groups some of which contain nitrogen'.

In marshy soils organic matter is broken down into simpler substances by anaerobic decomposition. Some of these later recombine to form humic materials. They are leached out of the soil in the company of ferrous ions and these when oxidised, form insoluble ferric compounds which are adsorbed onto the humus. The resulting colloid may be precipitated or remain in suspension depending on the pH of the water. Acid streams are often stained brown with humus; Harrison & Elsworth (1958) report that 'streams running down the seaward slopes of the Western Cape coastal ranges are strongly peat-stained while those running off the landward slopes are normally lightly stained, if at all. Thus the head waters of the Great Berg River though acidic are not normally deeply coloured except during heavy rains and flooding'. In contrast, the Palmiet River which rises in the same mountains is heavily peat-stained and 'the dark colour persists in the Palmiet during the dry season when it disappears in the upper Great Berg'.

The brown colouring of humus is not always associated with ferric iron. Part may be precipitated where river water mixes with salt water in an estuary while part remains in suspension and is discharged to the sea (Scott *et al* 1952, Burton & Liss 1976). Here it mixes with humic material derived from algae. According to Riley & Chester (1971) this is formed from phenolic compounds excreted by brown algae such as *Laminaria* and *Fucus*. These compounds combine in sea water with algal carbohydrates and proteins to form a polymer with a yellow colour and originally distinguished as 'Gelbstoff'. Head (1976) quotes statements of several workers that humic materials chelate trace metals and make them available to organisms and limit the toxic effect of copper, and probably other metals.

CHEMICAL CHANGES IN ESTUARIES

It has been emphasised that not only is the concentration of dissolved salts in typical sea water 35 000 mg.ℓ^{-1} as compared with 120 mg.ℓ^{-1} in typical river water, but the ratio of the major ions is also very different. Sea water is buffered on the alkaline side of neutrality and the Eh is about 400 mV. According to Burton & Liss (1976) 'sea water salts contribute some 98 % of the dissolved material at a salinity of 5 ‰ and so, in general terms, the compositional characteristics become dominant early in the mixing process'. These changes have marked effects on the solubility and the chemical species of the materials transported by the river; the changes in colloidal materials and trace metals are particularly important.

The suspended material which reaches the estuary includes many particles in the colloidal range of 1 nm to 1 μm. In fresh water they carry negative charges but these are neutralised in salt water. According to Pravdic in Burton (1976) the charge inversion generally occurs at a salinity of 2 ‰ and there is a net positive charge at 6 ‰. Between salinities of 2 and 5 ‰ the suspended particles are no longer mutually repelled and they flocculate. As the floccules drift, they gradually increase in size as new particles of silt, clay and organic matter are added and nutrients and trace metals in solution are adsorbed. The turbidity in the upper reaches of the estuary increases to a maximum and the floccules start sinking at the slack of the tide and mix with coarser sediments and decomposing organic matter on the muddy bottom. This is the beginning of the 'nutrient trap' in estuaries. The floccules may be resuspended as the flood tidal current increases and are carried further upstream in the saline bottom current and then rise up into fresh water due to turbulence. Here the floccules disaggregate into their constituent particles only to flocculate again when they are carried downstream by the ebb tide into more saline water. The net landward movement of the salt wedge along the bottom of the estuary maintains the finest particles which have the maximum absorptive area high up the estuary. This is another aspect of the nutrient trap. A third one is the exidised surface layer of the bottom sediment which largely inhibits the upward diffusion of soluble phosphate contained in the deoxidised lower layer of the sediment.

When sea water and river water mix in an estuary, some dissolved constituents derived from each source may behave as though simple dilution were occurring, so that there is a linear relationship between the concentration of the dissolved constituent and changes in salinity. Thus the concentration of some sea water constituents decreases with decreasing salinity while the concentration of some river water constituents decreases with increasing salinity. Such behaviour is termed conservative. If reactions occur in the estuary so that the ratio between the dissolved concentration of an ion or compound and its particulate phase is altered, the behaviour is said to be non-conservative and there is no longer a linear relationship with changes in salinity. The test of conservative behaviour is obviously most useful for those constituents whose relative concentration differs widely from river water to sea water.

Riley & Chester (1971) and Phillips (1972) have noted that the chemical species of trace elements are affected by high concentrations of other ions and quote as examples the species of cadmium, magnesium and copper in sea water. Burton (1976) deals in detail with the activity of elements in solution. He points out that whether or not an element behaves conservatively during estuarine mixing, its activity, which represents the effective concentration involved in chemical reactions, will vary with changes in salinity and composition. These effects are expressed as the activity coefficient for the species concerned. The relationships may be expressed as:

$$a_{xn} = f_{xn} C_{xn}$$

where a, f and c are respectively the activity, the activity coefficient and the concentration of the ion x^n. In solutions of high ionic strength such as sea water, the relation between activity and concentration can depend, not only on ionic strength but also on the specific composition of the solution, including single ions and complexes. The relation between the activity of an element and its total concentration is given by the total activity coefficient. Burton quotes calcium as an example:

$$a_{Ca^{2+}} = f_{Ca^{2+}} (total) \, C_{Ca^{2+}} (total)$$

The effects of ionic strength on the activity of the free ion are expressed by:

$$f_{Ca^{2+}} (total) = f_{Ca^{2+}} (free) \, y$$

where ý is the fraction of the total calcium content as unassociated ions. In sea water the free ion activity coefficient has been estimated as 0,28 but only 87 % of the calcium ions behave as free ions. The value of $f_{Ca^{2+}}$ (total) is thus 0,28 x 0,87 = 0,24. 'Clearly, in an estuarine system the value of $f_{Ca^{2+}}$ (total) will vary with ionic strength and with changing composition' (Burton, 1976). Burton also tabulates the total activity coefficients of the major ions in sea water as determined by several workers. The values are in good agreement and provide a basis for calculating the speciation of these constituents in the decreased salinities of estuaries. It would be interesting to know whether they provide an equally valid basis in hypersaline estuaries and what the limiting salinities are for major constituents.

The speciation of the minor elements is more difficult to model. Burton summarises the main problems. Among them he emphasises that:

1. The unstable exidation states are due to continuous renewal, biological cycling and the slow rates of oxidation.

2. Complexes with organic matter may be important but the bulk of the organic compounds are poorly defined and changes in their composition may occur during estuarine mixing.

3. There is uncertainty regarding the redox potential of oxygenated sea water. The pE has been taken as 12,5 but it has recently been suggested that it may be about 8,5.

4. The stability constants of all the associations which may occur are not available.

For these reasons the earlier work on the speciation of trace elements may be misleading. 'Since the toxicity of dissolved elements and their roles in processes such as bioaccumulation and inorganic sedimentary interactions may be critically affected by their chemical forms, the problems of speciation form an important focus for further research' (Burton, 1976).

The surface layers of estuarine sediments are usually oxygenated so that iron particles are in the Fe(III) form but the subsurface layers of the sediment are often anoxic and the iron is reduced to Fe(II). The metals absorbed on iron such as Mn and Co, may then be released into the interstitial water. Zinc, lead and cadmium are precipitated as sulphides in anoxic sediments but when the sediment is eroded and becomes oxidised these metals are released and may be absorbed by organisms. Manganese is normally precipitated as the hydrated oxide but is released in anoxic sediments and if the water column is also polluted and anoxic, manganese is carried as soluble organic complexes.

The effects of the estuarine mixing as tested by conservative or non-conservative behaviour by many workers has been discussed by Liss (1976). His voluminous list of references is not repeated here and only an outline of the main findings is summarised below.

Most of the major constituents of sea water including Na^+, K^+, Ca^{2+} and SO_4^{2-} are conservative. Tests on Mg^{2+} yield conflicting results. F^- is conservative but in sea water up to 50 % of fluoride forms the ion pair MgF^+. It is of interest that the maximum acceptable concentration of fluoride was 5,5 to 6,9 $\mu g.F^-$ per litre (Connell & Airey, 1979). Boron is non-conservative and in some estuaries 25-30 % may be removed by clay minerals. The soluble fraction of silicon was thought to exist in colloidal form but apparently it is mainly absorbed on alumina. It is conservative in some estuaries but in others between 20 and 30 % is removed by the coagulation of the suspended particles. In the Scheldt estuary, vigorous growth of diatoms in summer removes almost all the soluble silicon and much less is removed in winter (Wollast & Peters, 1978). Molybdenum is also partly absorbed by growing organisms but this is balanced by desorption from particulate material. Colloidal aluminium hydroxide which is one of the most abundant constituents of river water, flocculates in salinities of 2 ‰ and a very low concentration remains but further admixture with sea water which contains a higher concentration, causes the aluminium concentration to rise in the lower reaches of an estuary. Iron and manganese are non-conservative for, as noted earlier, the hydrated oxide particles flocculate in estuaries. Cobalt and nickel are conservative but the evidence in regard to zinc is conflicting.

Heavy metals in rivers and estuaries

Worldwide alarm has been raised by the increasing load of pollutants including domestic sewage, pesticides, petroleum products and heavy metals carried by rivers and estuaries to the sea. Minimata disease caused by methyl mercury taken up by algae and invertebrates, eaten by fish and thus passed to man in Japan has caused concern particularly in Sweden and Denmark where high concentrations of mercury have been found in flatfish. Cadmium levels in the Baltic are rising dramatically and this metal is concentrated by shellfish. Copper, which is an essential plant nutrient in trace concentrations, is toxic in higher concentrations and causes fish poisoning in the West Indies to an extent that many fish are not fit for human consumption.

Quite apart from the heavy pollution in rivers and estuaries draining large industrial cities in western Europe, the United States and Japan, where such pollution kills plankton, fish and invertebrates, the urgent question is how much pollution the sea can absorb. Is the input of heavy metals balanced by deposition on the sea bed or are the concentrations in the water rising? International committees have called for the monitoring of pollution in seas and estuaries to determine base line concentrations in unpolluted areas and to locate the main impact areas. The work is now in progress.

Interest is focused on toxic metals including Hg, Pb, Cd, As and Cr. A detailed review, with references to the many papers which contain conflicting results will be found in the chapter by de Groot, Salomons and Allersma in Burton & Liss (1976). Most of the heavy metals are transported in fine suspended sediments and many workers base their estimates on analyses of the concentrations in particulate material <63 μm or <16 μm or <2 μm in diameter. Förstner & Müller (1974) who estimate concentrations in <2 μm sediments, quote the concentrations in various rivers in Western Germany and report that less than 0,1 % of the heavy metals are associated with iron and manganese oxides. According to de Groot *et al*, the cadmium particles are mainly $CdCO_3$ and CdS, while those of lead are $PbCO_3$ and $Pb(OH_2)$ or complexes of both. Zinc is carried as $ZnCO_3$ and $Zn(OH)_2$, while arsenic is carried as arsenates, although in polluted rivers, both arsenic and mercury tend to combine with organic sulphur compounds and clay minerals. Copper is absorbed more by humus than by clays and the importance of humic acids as metal carriers is stressed by many workers. All of these heavy metals are carried by rivers draining industrial areas in alarming concentrations. Moreover, the concentrations are increasing and the future looks bleak.

In the slow-flowing upper reaches of estuaries, the suspended particles are deposited and in the reduced lower layers of the sediment where many metals form insoluble sulphides, others form soluble complexes with organic matter. De Groot *et al* quote reports that the organic complexes of Co, Ni, Cu and Zn may be five times more concentrated in the interstitial water than in the free water column. Some bacteria methylate Hg and As and the methylated mercury may be taken up by other organisms. Filter feeding animals are particularly important for not only do they scavenge many metal particles, but the glycoproteins on the filtering organs also act as chelating agents. Enzymes in the gut break down the complexes of useful metals such as Fe, Mn, Co, Cu, Zn and Mo and these are assimilated while other metals remain chelated and pass out with the faeces.

The concentration of trace elements by organisms against the concentration gradient is discussed by Riley & Chester (1971) who quote the relative concentration of different groups of elements in the periodic table given by Bowen (1966). In regard to metals they state: '(iv) the orders of affinity of cations for organisms is 4+ and 3+ elements $>$2+ transition metals $>$2+ Group IIA metals $>$1+ Group I metals'. Further '(v) In general the heavier elements in a particular Group of the periodic table are taken up more strongly than the lighter'.

Organisms vary in their ability to concentrate elements. Riley & Chester remark that the 'lower' organisms concentrate trace elements more than 'higher' ones. Some families concentrate a particular element more than others and there are even differences between individual species. Thus some ascidians concentrate vanadium 10^6 times while others concentrate the related metals niobium or titanium. Molluscs tend to concentrate cadmium and mercury particularly in the digestive gland, and they concentrate other trace metals as well. Common and widely distributed molluscs such as oysters and mussels have thus been used to compare the concentration of trace metals in different areas. Watling & Watling (1974, 1975) have used this method in South African estuaries and lagoons.

As noted earlier, part of the metal content of the sediments in the upper reaches of polluted estuaries is desorbed. However, much remains trapped in the mud and causes problems if the area is reclaimed for agriculture, as happened in Holland. The rest of the sediment is slowly transported by tidal currents towards the sea. In the process, the mud mixes with coarser marine sediments which contain a lower percentage of heavy metals. According to de Groot *et al* in Burton & Liss (1976), the evidence as to whether the decrease in the heavy metal content of the sediments along the length of an estuary is due to the chemical process of desorption, or to the physical process of mixing, is still conflicting.

Table 5.8
Mean concentrations of heavy metals in estuaries and shallow seas around South Africa

Concentrations in filtered water ($\mu g.\ell^{-1}$)	Hg	Cu	Cd	Pb	Zn	Co	Ni	Cr
Kosi Bay	0,10	0,24	0,08	0,45	1,69	0,48	1,28	–
Richards Bay 1974 (undeveloped)	0,17	1,70	0,15	1,87	–	–	–	–
Richards Bay 1976 (harbour)	0,80	0,91	0,08	2,25	0,17	2,46	4,56	–
Umgababa	0,82	1,81	0,07	2,52	1,9 2	9,65	–	–
Mzimkulu	0,06	1,45	0,02	2,93	6,53	–	–	–
Bashee December 1975	0,11	8,46	0,27	6,01	11,43	–	–	–
Swartkops December 1975	0,11	5,62	0,27	0,71	10,79	–	–	–
Concentrations in sediment ($\mu g.\ell^{-1}$)								
Kosi Bay	–	–	–	–	–	–	–	–
Richards Bay 1974 (undeveloped)	0,22	9,88	0,57	24,2	98,4	–	–	–
Richards Bay 1976 (harbour)	0,02	26,70	1,00	17,90	102,70	–	–	86,20
Umgababa	0,55	2,55	0,32	4,45	6,35	1,60	5,01	11,00
Mzimkulu	0,01	9,43	0,69	8,14	76,60	–	–	–
Bashee June 1975	0,01	9,56	0,49	14,23	17,00	–	–	9,55
Swartkops February 1975	0,05	6,53	1,08	23,40	19,70	–	–	42,90
Concentrations in filtered sea water ($\mu g.\ell^{-1}$)								
ECOR stations 1-9 off Durban	0,15	0,32	0,47	0,68	0,70	1,05		
Concentrations in $\mu g.g^{-1}$ in shallow marine sediments off South Natal								
ECOR stations 1-5	<0,01	1,64	1,23	8,81	9,70	8,02	–	7,63
M stations 1-3; O stations 1-3								
Concentrations in Saldanha Bay sediments (ppm)								
(Willis, Fortuin & Eagle, 1977)	–	<1	–	–	7	<3	<2	17

The concentration of trace metals in the estuaries and seas around South Africa is being determined by the South African Committee for Marine Pollution as part of the international monitoring programme. Preliminary results from the coast between Zululand and Algoa Bay are contained in internal reports from the NIWR edited by Oliff (1976) and a summary of the findings has been published by Cloete & Oliff (1976). They report concentrations of Hg up to 0,88 $\mu g.\ell^{-1}$ in sewage outfalls around Durban, but these outfalls flow directly into the sea and the outfalls into Durban Bay do not contain more than 0,55 $\mu g.\ell^{-1}$. Further work is in progress and the concentrations in Cape seas and estuaries, as recorded by the National Institute for Oceanology, are awaited. The concentrations of trace metals in the Wilderness Lakes and in Knysna estuary have been recorded by Watling (1977) and by Watling & Watling (1977) in unpublished reports. Although there is evidence of contamination by man, the values are not significantly high. The concentrations in Saldanha Bay and Langebaan Lagoon prior to the loading of iron ore from the new deep water harbour have been recorded by Fourie (1976), Willis *et al* (1977) and Watling & Watling (1974). There was no indication of metal pollution.

The concentrations of several metals in filtered water and in unsorted sediments reported by Oliff (1976) have been used to obtain the mean values of each shown in Table 5.8. As stated by Oliff, there have been difficulties in obtaining samples without contamination when filtering silt-laden samples. Also, there were losses while frozen water samples were transported many hundreds of miles to the Durban

laboratory. For these reasons some of the records are anomalous and these have been excluded when determining the means shown in Table 5.8. The records from any particular estuary are too few to show reliable trends even where the whole length of the estuary has been sampled and the means have been derived from all the records. When comparing one estuary with another, Kosi Bay appears to have the lowest concentrations in the water column possibly because it is fed by streams from marshes in a thinly populated area. The Umgababa has the highest concentration of mercury and cobalt which is interesting as the estuary is often closed by a sandbar. Copper is highest in the Bashee and cadmium is highest in the Swartkops, also lead, possibly due to the hot water effluent from a power station. Zinc is high in the Richards Bay sediments. When the Bay was dredged to form a deep water harbour and the effluents from aluminium and fertilizer factories flowed into the bay, increased concentrations of mercury, copper, lead, cadmium and nickel appeared. Oliff presents evidence that if the ratios of these heavy metals to iron are examined, there appear to be two groups. 'Mercury, copper, cadmium and lead form one [group] with a low dependence on iron. Zinc, chromium, cobalt and nickel form a second group with a high dependence on iron'. Dependence on manganese is not mentioned but otherwise there is general agreement with results obtained from Europe. In passing it may be noted that high concentrations of fluoride were recently found in the effluent from the fertilizer factory but tests did not indicate toxic effects to fishes. It should be emphasised that none of these South African

Table 5.9
Amounts of nitrogen and phosphorus in runoff from an average area (from Hobbie, 1976)

Sewage	N(kg.ha^{-1}.yr^{-1})	P(kg.ha^{-1}.yr^{-1})
Human wastes	6,6	0,8
Detergents	–	0,4
Street runoff	0,7	0,1
Industrial wastes	0,7	0,1
	8,0	1,4
Agricultural land and forest		
Arable land	2,3-5,8	0,1-0,5
Meadows and grasslands	4,3-13,3	0,1-0,5
Forests	1,0	0,1
	8,6-20,1	0,3-1,1
Total	16,6-28,1	1,7-2,5

estuaries is polluted to the extent of those draining industrial areas in Western Europe. Comparisons are difficult for the concentrations of metals in the water column are not quoted for European estuaries and as noted earlier, the concentrations given for European sediments are related to fine particles only, whereas the South African concentrations are related to the total sediments which may include all particle sizes. This may explain the higher concentration in Richards Bay where there is a high percentage of fine particles. De Groot *et al* quote concentrations in the estuarine sediments 16 μm in diameter of polluted German rivers that are two to three orders of magnitude higher than those in the sediments of Natal and Eastern Cape estuaries, which are in fact, unpolluted. Nonetheless, as shown in Table 5.8 both the waters and sediments of Natal estuaries have a higher concentration of heavy metals than the waters and sediments of coastal seas.

NUTRIENT CONCENTRATIONS IN ESTUARIES

The supply of trace metals required by organisms as co-enzymes has been discussed earlier. In addition, some algae are partly heterotrophic and can absorb simple organic substances in solution and it is well-known that peridinians flourish in late summer when exudates from the main summer plankton population are available. Hellebust (1967) estimates that about 10 % of the carbon absorbed by phytoplankton in photosynthesis is secreted in the form of carbohydrates. Many benthic unicellular plants and animals including flagellates and diatoms, absorb these and other simple organic compounds released by bacterial decomposition. In addition, many of these organisms, particularly blue-green algae and bacteria, excrete vitamins such as biotin, thiamine and vitamin B_{12} (cyanocobalamin), required by planktonic algae. Stewart (1972) quotes a table from Droop showing the requirements of the various divisions of algae for these vitamins. The growth promoting properties of humic acids have

been discussed earlier and it is widely reported that the transport of all these substances to neritic seas promotes phytoplankton growth.

It remains to consider the supply of nitrogen and phosphorous. Although there are reserves of both in the organically rich sediments and living organisms, the original source of nitrogen and phosphorous compounds is from rivers with smaller supplies from the sea as well as traces of nitrogen from rain. In all estuaries, there is a slow loss of nutrients to the sea in the form of organic detritus and migratory animals.

Perennial plants in all environments tend to abstract nutrients from dying leaves and retain them in the living stems and roots so that minimal amounts are lost with the dead material. Hobbie (1976) illustrates this by reference to a New Hampshire forest where 'more nitrogen and phosphorous fall in the rain than leave the forest in streams . . . In the same forest, 1 900 grams of phosphorous per hectare reach the forest floor as leaf fall but only 21 grams of phosphorous per hectare leave in the streams'. The rest is mineralised by soil bacteria and recycled through the roots.

Hobbie (1976) also quotes a table of N and P values from Vollenweider which shows how much the terrestrial input into rivers depends on the land use. See Table 5.9 above. He points out that the range of nutrients from farmlands, meadows and grasslands reflects the differing amounts of fertilizer reaching the streams. Obviously, uncultivated areas will provide much less and the contrast with the large amounts from industrial cities will be more marked.

In rivers, the organic matter from the land is decomposed and much of the nutrient content is absorbed by the aquatic plants or the sediments. The soluble forms of nutrients in the river water, as defined by the concentration in the filtrate passing through a 0,5 μm filter, have been reported in many forms. Nitrogen compounds include NH_4^+ which is oxidised to NO_2^- and then NO_3^- as well as small amounts of amino acids and colloidal organic nitrogen compounds. The individual ions may be recorded or the total nitrogen including the organic compounds may be recorded as kjeldahl nitrogen. The soluble inorganic phosphate derived from the decomposition of organic compounds is recorded as such but some colloidal organic phosphorous passes into the filtrate and this, plus the inorganic phosphate is recorded as total phosphorous. The units used include g.m^{-2}, g.m^{-3}, ppm, mg.ℓ^{-1}, μg.ℓ^{-1} or the element content may be shown as μg at P.ℓ^{-1}. International committees now suggest that soluble inorganic orthophosphate should be reported as soluble reactive phosphate, abbreviated as SRP. Useful conversion tables will be found in Riley & Skirrow (1965, 1975) and Riley & Chester (1971). A table of various constants and conversion factors is also given at the end of this book.

Burton & Liss (1976) give the concentration of the

various elements in solution in an average river of the world as compared with that in surface sea water. Included in their list is the total combined nitrogen (mainly nitrate) as 226 μg.ℓ^{-1} and phosphorous (?SRP) as 20 μg.ℓ^{-1}. It is interesting to compare these values with those recorded in the lower reaches of some unpolluted and polluted South African rivers. The means of the seasonal values in the Great Berg River in the winter rainfall area of the Western Cape, have been derived from the records in Harrison & Elsworth (1958) and are shown in terms of μg.ℓ^{-1} of nitrate nitrogen (excluding nitrite and ammonia) and μg.l^{-1} SRP in Table 5.10.

The nitrate nitrogen records are of the same order as those for world average rivers and the winter wet season maximum is well marked. As Harrison & Elsworth remark, the SRP (phosphate) values are unusually low except for the rise in the dry summer season. The soils of the Western Cape have a low phosphorous content.

Data for Natal rivers were abstracted from Volume 13 of the Natal Town and Regional Planning Report (Brand *et al*, 1967, Archibald *et al*, 1969, Kemp *et al*, 1976). The means of stations in the lower reaches of 10 large unpolluted rivers ranging from the Mtamvuna on the southern border of Natal to the Pongola on the northern border are shown in Table 5.11 below. Values are quoted for the rainy summer season and the dry winter season. For comparison, the means for four polluted rivers near Durban namely the Mlazi, Umbilo, Mbokintwini and Mhlatuzana are included in Table 5.11.

The nitrate nitrogen values, even in the polluted rivers are not abnormally high but the SRP values in both unpolluted and polluted rivers are considerably higher than the world average possibly due to the fertilizers used in the cane fields which cover much of the coastal belt. Why the values of nitrate fall in the dry season both in the Western Cape and Natal, while the values of phosphate rise, is unknown.

The nitrogen cycle in estuaries

The several chemical reactions which comprise the nitrogen cycle in estuaries are the same as those in the sea but the organisms concerned, and the situations in which they take place, often differ (Wood, 1965). In the sea, the release of ammonia by the decomposition of proteins and other complex organic nitrogen compounds contained in dead plankton and semidigested faecal material falling through the water, is a slow process. Also the oxidation of ammonia to nitrite and nitrate may take months in the cold deep layers of the sea. The reserves of inorganic nitrogen that accumulate are vast, but they only become available to plants in the lighted surface layers in areas of upwelling or when convection currents develop in autumn and winter. In estuaries, decomposition takes place on

Table 5.10
Seasonal changes of nitrate and soluble reactive phosphate in the Great Berg River (data from Harrison & Elsworth, 1958)

	Autumn	Winter	Spring	Summer
μg.NO$_3$N.ℓ^{-1}	170	250	90	90
μg.SRP.ℓ^{-1}	1	–	3	60

Table 5.11
Means of dissolved nitrate and soluble reactive phosphate (SRP) in 10 unpolluted large rivers and four polluted small rivers in Natal

	Rainy season	Dry season
Unpolluted rivers		
μg.NO$_3$N.ℓ^{-1}	220	140
μg.SRP.ℓ^{-1}	48,9	68,5
Polluted rivers		
μg.NO$_3$N.ℓ^{-1}	340	100
μg.SRP.ℓ^{-1}	277	603

shallow muddy bottoms and the inorganic nitrogen compounds released may be absorbed immediately even in anoxic soils by the roots of vascular plants. Alternatively, they may diffuse into the shallow water where they are available to phytoplankton and benthic algae. The whole estuarine ecosystem is not only richer in bacteria, plants and animals, but also more concentrated in space. The nitrogenous excretions of animals (even urea) are immediately available to the surrounding plants and the flux of nitrogen compounds is correspondingly more rapid, particularly in the warm months. Hobbie (1976) states that 'nitrate is almost completely removed from solution' as the water in the Pamlico estuary passes through to the sea; urea excreted by some aquatic animals 'is completely turned over every day during the summer but during the winter the turnover time is 200 days. Ammonia in the same estuary may cycle even faster since the algal photosynthesis during a single day in August requires 231 metric tons of nitrogen while only 5 metric tons of nitrate nitrogen and 100 metric tons of ammonia nitrogen were present. This implies that ammonia may have a turnover time of half a day'. Conversely, low concentrations of nitrogen in the water column may not mean that plants lack nitrogen but that it is absorbed as soon as it is liberated.

As noted, the main source of nitrogen in an estuary is the river. Apart from those polluted rivers where decomposition has already occurred, the fluvial supply is mainly in the form of organic detritus carried during the rainy season. Further, detritus is formed from the plants and animals in the estuary itself. Some estuaries gain nitrogen from the sea. This applies particularly to marine-dominated estuaries such as the Ythan in Scotland where the water is replaced each tidal cycle, and to deep fjords where nutrient-rich sea water from below the euphotic zone flows in over the sill to replace the outflow of light, oligohaline surface water. In general, however, estuaries supply more nitrogen to the sea than they gain and, as

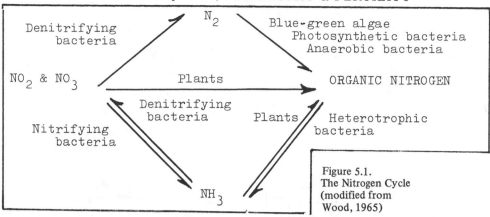

Figure 5.1.
The Nitrogen Cycle
(modified from
Wood, 1965)

in the case of rivers, it is mainly transported as organic detritus carried out at the end of the growing season or as migratory animals. Small amounts of both ammonia and nitrate are also received in rain. Estimates vary widely, but according to Vaccaro (1965), the mean for the USA is 0,06 μg at $N.\ell^{-1}$ (= 0,84 μg $NO_3^-N.\ell^{-1}$). If the rainfall is 1 000 mm.yr^{-1}, this is only 600 mg.m^{-2} yr^{-1} which is unimportant.

Molecular nitrogen is the most abundant form of nitrogen. It is dissolved in all natural waters, but it can only be metabolised by blue-green algae in the presence of sunlight and by anaerobic bacteria in richly organic muds. Lipschultz *et al,* 1979 report nitrogen fixation associated with four species of submerged angiosperms in Chesapeake Bay but it remains uncertain whether the fixation is due to the angiosperms or to associated organisms. Blue-green algae (or Cyanophyceae) such as *Oscillatoria* and *Lyngbya,* are common on estuarine mud particularly among salt marsh vegetation and some genera are epiphytic. Capone & Taylor (1977) report that *Calothrix* which grows on turtle grass *(Thalassia)* in Florida fixed 4-5 mg.m^{-2} day^{-1} when the water was clear. According to Vaccaro (1965), nitrogen fixation by blue-green algae is associated with photosynthesis since 'nitrogen acts as an alternative acceptor to carbon for hydrogen released photochemically from water'. During the process, part of the amino acid formed dif-

fuses into the surrounding water, adding to its fertility. Fixation of nitrogen by anaerobic bacteria in the sediment requires the energy obtained by the decomposition of 1 gram of carbohydrate for the production of 1 mg nitrate, which means that it can only occur in organically rich sediments. *Desulphovibrio* has been reported by Zuberer & Silver (1976) to fix nitrogen in organically rich mangrove sediments in the absence of ammonia. It is implied that the fixation occurred in close association, possibly in the rhizosphere around mangrove roots. Haines *et al* (1977) report that the rate of nitrogen fixation of all types in a Georgia *Spartina* marsh is 'probably about 5,8 g N.m^{-2}yr^{-1}'. Nitrate and nitrite may also be reduced by denitrifying bacteria to molecular nitrogen. Haines *et al* estimate the losses by denitrification in the Georgia *Spartina* marshes as approximately 79 g.N.m^{-2}yr^{-1}. This is greater than the gain by nitrogen fixation but it may be pointed out that nitrogen fixation occurs when the concentration of ammonia is negligible and denitrification occurs when nitrate is available; this would imply that it increases the fertility of the soil.

The nitrogen cycle in estuaries and the sea is summarised in Figure 5.1 omitting the transfer from plants to animals.

The nitrogen compounds in solution in the water column in South African estuaries have been recorded

Table 5.12

Concentrations of NO_3^-N and NH_3^-N (in $\mu g.\ell^{-1}$) and the mean of both forms of nitrogen in the water column of several South African estuaries

Salinity range	5-19 ‰			20-35 ‰		
Estuary	NH_3^-N	NO_3^-N	Mean	NH_3^-N	NO_3^-N	Mean
Kosi Bay (winter)	145	4,5	75	143	47	95
Richards Bay (winter)	—	—	—	30	194	112
Richards Bay (summer)	27,8	26,3	27,1	57,1	130,7	93,9
Mlalazi (winter)	7,4	101	54,2	1,0	246,5	124
Mgeni (winter)	760	1 400	1 080	759	1 455	1 107
Mzimkulu (winter)	67	151	109	—	—	—
Mtamvuna (winter)	240	60	150	300	60	180
Bashee (winter)	21	13	17	—	—	—
Swartkops (summer)	165	404	285	77,7	132	104
Great Berg* (spring)	—	164			172	
Olifants* (summer)		485			379	

* only NO_3 recorded and salinities in winter below 2,85 ‰ due to floods.

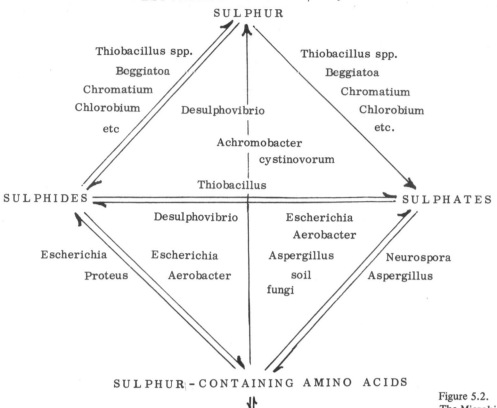

Figure 5.2.
The Microbial Sulphur Cycle
(from Wood, 1965)

in terms of nitrate, nitrite and ammonia nitrogen and as Kjeldahl nitrogen in different estuaries. Detailed records at various stations in a number of estuaries between Port Elizabeth and Kosi Bay are given by Oliff (1976) but the data are too voluminous to be presented here. The mean concentrations of nitrate and ammonia nitrogen have been extracted for two salinity ranges and these, with the average of both forms of inorganic nitrogen are presented in Table 5.12.

Two estuaries are obviously polluted. The Mgeni receives the effluent from one of the sewage works in Durban and it is evident that much of the ammonia has already been oxidised to nitrate and as the Mgeni is a short estuary, it carries much nitrogen to the sea. The Swartkops receives pollution from Uitenhage but more than half the nitrogen is absorbed in the long estuary and the discharge to the sea is normal. The other estuaries are unpolluted and the values show the range to be expected in the Transkei and Natal. In all of them, the nitrogen values increase in the higher salinities possibly due to release from organic detritus. Kosi Bay and the Mtamvuna show little of the ammonia oxidised to nitrate while Richards Bay and the Mlalazi show the reverse. The Olifants estuary shows high nitrate values possibly due to fertilization of vineyards on the banks of the river, or inflow of the nutrient-rich Benguela current.

Even the means of the original records show many variations from high to low salinities and from season to season. Before the reasons for these variations can be explained, more detailed records will be necessary. It is suggested that the work be confined to a few different types of estuaries and that the samples be obtained at known states of the tide in the different reaches of the estuaries and in each of the four seasons.

The sulphur cycle

The sulphur cycle is important biologically in that it is concerned with the conversion of insoluble phosphate in the sediment to soluble reactive phosphate which may be released into the water column. Also, in highly polluted estuaries, elemental sulphur may be formed or toxic H_2S released. The sulphur cycle is important geochemically in that as the redox potential falls, some metals form insoluble sulphides while others are released in soluble form. Aston & Chester in Burton & Liss (1976), give a table in which the concentration of many elements in oxidised and anoxic sediments is shown. They conclude 'that Ni, Co, Cr, V, Ba, Sr, Pb, Zn and Y have similar concentrations in both oxidising and anoxic sediments but Mn, Cu, Sc, Zn and Mo tend to be relatively enriched in anoxic sediments'. Mercury is mainly associated with suspended organic matter.

Table 5.13
Total inputs and outputs of nutrients in tons N or P to the Pamlico estuary during the year 1971-1972 (after Hobbie, 1976)

	PO_4^{3-}	NO_3^-	NH_4^+
Input	715	2 804	795
Output	459	1 425	744
Net gain to estuary	256	1 379	51

When this sediment is deposited and reduced in the lower layers, most of the inorganic mercury is converted to toxic methyl mercury and lost from the sediment to the water column.

A detailed account of the sulphur cycle will be found in Wood (1965) who summarises the many reactions facilitated by bacteria and fungi in a diagram reproduced in Figure 5.2. It will be seen that the sulphate may be taken up by micro-organisms as sulphur containing amino-acids or converted by anaerobic bacteria to sulphides which may in turn be reoxidised to elemental sulphur or sulphate, so that the whole sulphur cycle is reversible. This depends on the redox potential and the presence of suitable organic substrates for the bacteria concerned.

The oxidised surface layers of organic sediments contain a large community of micro-organisms with aerobic bacteria breaking down organic detritus, absorbing oxygen and producing simpler organic substrates such as the lactate, acetate and pyruvate required by other organisms. While the Eh exceeds 200 mV, the hydrated oxides of iron are in the brown ferric form and sulphate is not reduced. In the deeper layers of the sediment where the oxygen concentration falls to low levels and the Eh is about 100 mV, sulphate is reduced to sulphide by *Desulphovibrio* and species of *Thiobacillus* (Wood, 1965) and according to the equation:

$$(CHO)_{106}(NH_3)_{16}H_3PO_4 + 53SO_4^{2-} \rightarrow 106CO_2 + 106H_2O + 16NH_3 + H_3PO_4 + 53S^{2-}.$$

According to Berner (1967) the sulphide then forms H_2S although HS^- is the main species:

$$S^{2-} + H^+ \rightleftharpoons HS^- + H^+ \rightleftharpoons H_2S.$$

In this reducing environment, the hydrated ferric oxide is also reduced to ferrous oxide and combines with H_2S:

$$2HFeO_2 + 3H_2S \rightarrow 2FeS + S^0 + 4H_2^0.$$

The ferrous sulphide is black and crystallises slowly and the presence of this black layer in the sediment has long been recognised as indicating hypoxic or anoxic conditions. Berner investigated even deeper layers of the sediments in Chesapeake Bay where the pH falls from 7 to 4, providing conditions for the bacterially produced elemental sulphur to combine with FeS to form FeS_2 or pyrite which is grey green. Thus the colour sequence from the surface of the sediment downwards, is yellowish brown, then black and in deep cores, there are grey green bands of pyrite.

In grossly polluted estuaries where the dissolved oxygen has been absorbed and the nitrates reduced to ammonia, the surface sediment is black with sulphide (Barrett, 1972). When no further iron is available, toxic H_2S is liberated.

The phosphorus cycle

Although estuaries contain reserves of phosphorus trapped in the mud, they receive more from the river than is transported to the sea. Hobbie (1976) shows the nutrient budget for a year in the Pamlico estuary (see Table 5.13).

In the oxidised surface layer where bacterial action is most rapid, the SRP immediately combines with hydrated ferric oxide to form insoluble ferric phosphate or, as Liss (1976) more cautiously puts it, 'some type of Fe(III) phosphate complex'. Some phosphate also forms insoluble calcium phosphate, but according to Wood (1965) this phosphate can be fairly easily extracted by a variety of micro-organisms particularly in acid conditions. As the ferric phosphate is buried under fresh layers of sediment, the iron is reduced and the soluble phosphate is liberated into the interstitial water. As it diffuses upward to the oxidised layer, it is reprecipitated so that the sediment acts as a phosphate trap. Pomeroy (1970) estimated that the reserve in a Georgia salt marsh is sufficient for 500 years growth of *Spartina*.

The roots of vascular plants penetrate down into the reduced sediments and absorb the soluble phosphate; indeed sea grasses such as *Zostera*, *Halophila* and *Posidonia* will only grow in such sediments and one may speculate that the same is true of *Spartina* and many mangroves. McRoy *et al* (1972) have worked out the phosphorus budget for *Zostera*. It seems that the eelgrass absorbs phosphate both through its leaves and its roots. One transport system carries the phosphorus from the leaves to the roots and the other transports the phosphorus absorbed by the roots in the sediment through the stems and leaves to the water column as shown in figure 5.3.

Thus *Zostera* acts as a 'phosphate pump' supplying excess SRP to the organisms in the surrounding water. *Spartina* is also reported to be a phosphate pump and so are *Ruppia* and *Potamogeton*.

The release of SRP also occurs during gross pollution and Knox & Kilner (1973) report that when the overlying water of the Avon-Heathcote estuary in New Zealand lacks oxygen, phosphate is released from the sediment. In unpolluted estuaries there is a slow release of SRP due to the erosion of the sediments by strong currents or the burrowing activities of animals, but in contrast to nitrogen compounds whose concentration decreases down the length of the estuary, the concentration of phosphate in the water remains fairly constant, although the value differs from one estuary to another.

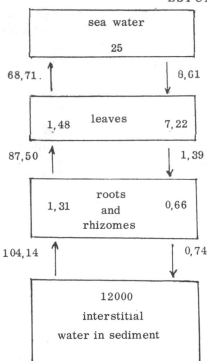

Figure 5.3. Calculated daily flux of phosphorus (in $\mu g.\ell^{-1}$) through 1 gram dry weight of *Zostera* (modified from McRoy *et al*, 1972)

Table 5.14
Soluble reactive phosphate in the water ($\mu g.SRP.\ell^{-1}$) (after Pomeroy *et al*, 1975)

Initial concentration	Final concentration
0	22,3
15,5	22,6
31,0	27,9
77,5	27,6
133,2	27,0
260,1	49,9

of 3-9 μg $SRP.gm^{-1}$ in estuarine sediments from Hong Kong. They report that the clay mineral kaolinite is the best phosphate absorbant and that maximum absorption occurs at about 2 ‰ salinity where the principal species of phosphate is $H_2PO_4^-$ and is less at higher salinities where the principal species is HPO_4^{2-}.

Phosphorus concentrations in South African estuaries

Most of the observations have been made as part of the marine pollution survey by officers of NIWR or NRIO. The intention was to obtain the range of concentrations in polluted and unpolluted estuaries and it is obvious that more detailed studies of certain estuaries will be required.

A brief summary of a few of the many records in the estuaries is given in Table 5.15 and concentrations in coastal seas are given for comparison.

Most of the records were made during brief synoptic surveys and even when they are condensed into two salinity ranges, the data are too few to draw reliable conclusions regarding changes along the estuaries or with the seasons. Such changes are most obvious with SRP records; the total P records, while indicating the maximum concentration of phosphorous in solution do not show the amount available to phytoplankton. The mouth of the Salt (= Black) River, the outfall of the Strandfontein sewage ponds, the upper reaches of Swartkops and the Mgeni are obviously polluted. Only the Swartkops shows that the phosphorus is absorbed along the estuary (as noted for nitrogen compounds). It is doubtful if much phosphorus is absorbed in the Mgeni for the estuary is very short. The remaining unpolluted estuaries show the range of SRP to be expected in the Cape, the Transkei and Natal, but more research will be required to explain why such a wide range occurs.

The concentration of PO_4-P in coastal seas shows the enriching effect of upwelling in the Benguela current at St Helena Bay and Saldanha Bay as compared with the lower values on the south and east coasts. Occasional upwelling off Knysna and neighbouring coasts, occurs during prolonged south-east winds in summer and the enriched water is known to enter Knysna estuary. The records off Durban made at the innermost station of the ECOR transect worked by the NIWR, indicate that phosphorus concentrations

Stefansson and Richards (1963) found concentrations of about 37 μg $SRP.\ell^{-1}$ in the Tamar estuary. Pomeroy *et al* (1965) felt that the absorption of phosphate by the sediment was due to microbial activity and poisoned one sample of sediment with formalin as a control. While the poisoned and healthy sediments were settled, there was no difference in the phosphate uptake from the overlying water. However when the samples were shaken and the sediments suspended, the healthy sediment took up far more phosphate showing that biological activity was important. To test the buffering action which must include both uptake and release from the sediment, Pomeroy *et al* added aliquots of sediment to estuarine water containing various concentrations of SRP and then mixed. After standing for an hour the concentrations in the overlying water were determined. His results are quoted in Table 5.14 below.

It will be seen that where the initial concentration ranged from 0-15,5, the sediment released SRP but above an initial concentration of 31 $\mu g.\ell^{-1}$ the sediment removed SRP from solution thus acting as a buffer, maintaining the concentration in the water between 22 and 28 $\mu g.\ell^{-1}$ until its absorptive capacity was exceeded.

The absorptive capacities of different sediments vary, the buffer concentration ranging from about 27-28 in the Tamar to 37 in the Columbia estuary. Stirling & Wormald (1977) found an equilibrium concentration

Table 5.15
Mean phosphorus concentrations in two salinity ranges of the surface waters of some South African estuaries and coastal seas (total P records marked with an asterisk, SRP records unmarked. All values in $\mu g.P.\ell^{-1}$).

Estuary and month	Salinity 4-20 ‰	Salinity 20-35 ‰	Reference
*Olifants (Feb)	96	84	Gledhill *et al* (unpublished)
*Great Berg (Oct)	27,8	58,1	”
Mouth of Salt River	–	464,6	Oliff (1976)
Strandfontein sewage outfall	–	743,4	”
*Knysna (Feb)	–	10,6	Korringa (1956)
Knysna (Apr)	–	25,6	Grindley (1978)
Knysna (Aug)	–	32,2	”
Swartkops (Feb + Dec)	593	157	Oliff (1976)
* Bashee (June)	2,5	–	”
* Bashee (Dec)	–	20	”
* Mtamvuna (Sept)	8,7	–	”
* Mzimkulu (Aug)	13,5	–	”
* Mgeni (May + Aug)	104	–	”
Mlalazi (June)	10,6	11,4	”
* Richards Bay (Nov + Feb)	8,7	14,7	”
Richards Bay (May + Aug)	–	30,8	”
St Lucia Lake (Oct + Mar)	30	15,0	Johnson (1976)
St Lucia Lake(Apr + Sept)	18	–	”
*Kosi Bay (Aug)	6	7,5	Oliff (1976)

Seasonal changes in coastal seas in $\mu g.P.\ell^{-1}$

	Spr.	Sum.	Aut.	Win.	
St Helena Bay	20,4	26,0	27,9	27,3	Clowes (1954)
Saldanha Bay	22,6	31,9	46,5	46,8	Henry *et al* (1977)
Off Knysna	–	9,3-21,7[a]	–	–	Korringa (1956)
Off Durban	10,0	14,0	12,5	13,0	Oliff (1976)

a. Temporary upwelling

along the Natal coast are about half those in the Benguela current. Both the west and east coast records show minimal concentrations in spring.

The nitrogen:phosphorus ratio

The depletion of either nitrogen compounds or SRP or both of these, has been reported to limit plant growth in many areas. In the open ocean the correlation between low concentrations of these nutrients and the decrease in phytoplankton production is convincing, but whether the decrease in the phytoplankton maximum in shallow seas is due to lack of nutrients, limited light penetration or to grazing by zooplankton is controversial. The crucial question is what concentration of available nitrogen and phosphorus is necessary for sustained healthy growth of phytoplankton. The analysis of mixed phyto- and zooplankton samples by Flemming (1940) gave a ratio of C:N:P of 106:16:1 and this 16:1 ratio has been quoted by many workers.

Harvey (1955) on the other hand, reported that his diatom cultures required a ratio of 9:1. Ketchum (1939), by means of enrichment experiments in cultures, concluded that a minimum of 47 μg $NO_3 - N.\ell^{-1}$ and 15 μg $SRP.\ell^{-1}$ was required by phytoplankton for healthy growth. This gives a ratio of N:P of 4,4:1. Strickland (1965), after a discussion of these results and his own experiments, concludes that Ketchum's estimates of minimal requirements were too high. In regard to nitrogen he says: 'we have only a rough indication of the limiting concentration of nitrate. It is probably safe to assume that it is less than 50 and probably less than 10 μg $NO_3 - N.\ell^{-1}$. In regard to Ketchum's minimum of 15 μg for phosphorus he says 'it is doubtful if this figure has any general applicability and the 'limiting' phosphorus concentration in some cultures has been found by the present author to be less than one third of this amount'. Elsewhere, Strickland reports that unicellular algae can accumulate reserves of nitrogen as protein and phosphorus as polyphosphate. Thus they can continue to divide several times when external concentrations have already fallen below the minimum necessary for growth.

As noted earlier, there are large reserves of phosphate in estuarine sediments which are available to the roots of vascular plants and Jefferies (1972) confirms the view of many workers that phosphorus is not a limiting factor. Algae, and in particular phytoplankton, must depend on the concentration of phosphorus in the water column which may be 'buffered' at various levels. Nitrogen has more often been reported to limit the growth of phytoplankton in estuaries. Bearing in mind the rapid turnover of nitrogen reported by Hobbie (1976), low concentrations of nitrogen compounds in the water column lose their significance. Nonetheless, Johnson (1976) has shown by enrichment experiments, that the production of phytoplankton in Lake St Lucia may be increased by the addition of nitrate.

According to Taft & Taylor (1976), the phytoplankton in deep estuaries such as Chesapeake Bay may be limited by either phosphorus or nitrogen deficiencies at different seasons. In May 1971, the whole water column was well-oxygenated and the phosphate was not released from the bottom sediments. As the temperature rose in August, bacterial activity increased so that the bottom water became anoxic and phosphate was released; in autumn, convection currents carried oxygen down to the bottom and there was no further phosphate release. In May, the phosphate concentration in the water was 5,11 μg $SRP - P.\ell^{-1}$ and in August it increased to between 17,04 and 25,56 μg $SRP.\ell^{-1}$. The ratio of N:P in the phytoplankton increased from 14,5:1 in winter to between 34 and 55:1 in spring and then fell to between 12 and 29:1 in summer. Taft & Taylor suggest that these seasonal changes in the N:P ratio mean that phosphate was limiting in spring and nitrogen was limiting in summer.

Although Strickland (1965) is too cautious to state limiting values of nitrogen and phosphorous, his evidence implies that limiting values of 20 μg $NO_3 - N.\ell^{-1}$ would not be unreasonable. If this is accepted as a working hypothesis when considering the concentrations reported in South African estuaries, only the Bashee would be limited by nitrogen and phosphorus. It must be noted, however, that most of the records have been made in winter and lower concentrations would be expected during the growing season in spring and summer.

REFERENCES

ALLANSON, B.R. & S. RUDD 1978. New tables for the first and second apparent dissociation constants, k_1 and k_2 of carbonic acid for estuarine and brackish waters. *Estuar. coastal mar. Sci.* 8(5): 495-498.

ARCHIBALD, C.G.M., COETZEE, O.J., KEMP, P.H., PRETORIUS, S.J. & R.R. SIBBALD 1969. The rivers of northern Natal and Zululand. *Water quality and abatement of pollution in Natal rivers. Part 4.* Natal Town and Regional Planning Commission, Pietermaritzburg.

BARRETT, M.J. 1972. The effects of pollution on the Thames estuary. *In:* R.S.K. Barnes & J. Green (eds), *The estuarine environment:* 119-122. Applied Science, London.

BERNER, R.A. 1967. Diagenesis of iron sulphide in Recent marine sediments. *In:* G. Lauff (ed), *Estuaries.* Am. Ass. Adv. Sci., Washington.

BLABER, S.J.M., HILL, B.J. & A.T. FORBES 1974. Infratidal zonation in a deep South African estuary. *Mar. Biol.* 28: 333-337.

BOLTT, R.E. & B.R. ALLANSON 1975. The benthos of some South African lakes. Part III: The benthic fauna of Lake Nhlange, Kwazulu, South Africa. *Trans. roy. Soc. S.Afr.* 41: 241-262.

BOWEN, H.J.M. 1966. *Trace elements in biochemistry.* Academic Press, London.

BRAND, P.A.J., KEMP, P.H., PRETORIUS, S.J. & H.J. SCHOONBEE 1967. Survey of the Three Rivers region. *Water quality and abatement of pollution in Natal rivers.* Part 2. Natal Town and Regional Planning Commission, Pietermaritzburg.

BURTON, J.D. 1976. Basic properties and processes in estuarine chemistry. *In:* J.D. Burton & P.S. Liss (eds), *Estuarine chemistry* 1-36. Academic Press, London.

BURTON, J.D. & P.S. LISS (eds) 1976. *Estuarine chemistry.* Academic Press, London. 229pp.

BUTLER, E.I. & S. TIBBITS 1972. Chemical survey of the Tamar estuary. I: Properties of the waters. *J. mar. biol. Ass. UK* 52: 681-699.

CAPONE, D.G. & B.F. TAYLOR 1977. Nitrogen fixation (acetylene reduction) in the phyllosphere of *Thalassia testudinarum. Mar. Biol.* 40: 19-28.

CLOETE, C.E. & W.D. OLIFF (eds) 1976. South African marine pollution survey report 1974-1975. *S.Afr. natl. Sci. Prog. Rep.* 8, CSIR, Pretoria.

CLOWES, A.J. 1954. The South African pilchard *(Sardinops ocellata).* The temperature, salinity and inorganic phosphate content of the surface layer near St Helena Bay, 1950-1952. *S.Afr. Div. Sea Fish., Invest. Rep.* 16.

CONNELL, A.D. & A.D. AIREY 1979. The chronic effects of fluoride on the estuarine amphipods *Grandidierella lutosa* and *G. lignorum. Abstracts 4th S.Afr. natl. oceanog. Symp.,* Cape Town, July 1979.

FLEMMING, R.H. 1940. The composition of plankton and units for reporting population and production. *Proc. 6th Pacif. Sci. Congr. Calif. 1939* 3: 535-540.

FOURIE, H.O. 1976. Metals in marine organisms from Saldanha Bay and Langebaan Lagoon prior to industrialisation. *S.Afr. J. Sci.* 72: 110-113.

FORSTNER, U. & G. MULLER 1974. *Schwermetalle in Flüssen und Seen.* Springer-Verlag, Berlin. 225pp.

GIBBS, R.J. 1970. Mechanisms controlling world water chemistry. *Science, NY* 170. 1088-1090.

GIBBS, R.J. 1973. Mechanism of trace metal transport in rivers. *Science, NY* 180: 71-73.

GLEDHILL, W.J., FRICKE, A.M., GREENWOOD, P.J. & M.J. ORREN (in preparation). Preliminary chemical and biological studies in two South African west coast estuaries.

GOLDBERG, E.D. (ed) 1978. *Biogeochemistry of estuarine sediments.* UNESCO, Paris.

HAINES, E., CHALMERS, A., HANSON, R. & B. SHERR 1977. Nitrogen pools and fluxes in a Georgia salt marsh. *In:* M. Wiley (ed), *Estuarine processes Vol II:* 241-254. Academic Press, New York.

HARRISON, A.D. & J.F. ELSWORTH 1958. Hydrobiological studies of the Great Berg River, Western Cape Province. Part I: General description, chemical studies and main features of the fauna and flora. *Trans. roy. Soc. S.Afr.* 35 (3): 125-226.

HARVEY, H.W. 1955. *The chemistry and fertility of sea water.* Cambridge Univ. Press, Cambridge, 224pp.

HEAD, P.C. 1976. Organic processes in estuaries. *In:* J.D. Burton & P.C. Liss (eds), *Estuarine chemistry:* 54-85. Academic Press, London.

HELLEBUST, J.A. 1967. Excretion of organic compounds by cultured and natural populations of marine phytoplankton. *In:* G. Lauff (ed), *Estuaries.* Am. Ass. Adv. Sci., Washington.

HENRY, J.L., MOSTERT, S.A. & N.D. CHRISTIE 1977. Phytoplankton primary production in Langebaan Lagoon and Saldanha Bay. *Trans. roy. Soc. S.Afr.* 42: 383-398.

HOBBIE, J.E. 1976. Nutrients in estuaries. *Oceanus* 19(5): 41-47.

JEFFERIES, R.L. 1972. Aspects of salt marsh ecology with particular reference to inorganic plant nutrition. *In:* R.S.K. Barnes & J. Green (eds), *The estuarine environment* 61-85. Applied Science, London.

JOHNSON, I.M. 1976. Studies of the phytoplankton of the St Lucia system. Pap. 9. *In:* A.E.F. Heydorn (ed), *St Lucia Scientific Advisory Council workshop — Charters Creek, February 1976.* Natal Parks Bd, Pietermaritzburg.

KEMP, P.H., CHUTTER, F.M. & D.J. COETZEE 1976. The rivers of southern Natal. *Water quality and abatement of pollution in Natal rivers.* Part 2. Natal Town and Regional Planning Commission, Pietermaritzburg.

KESTER, D.R. 1975. Dissolved gases other than CO_2. *In:* J.P. Riley & G. Skirrow (eds), *Chemical Oceanography,* 2nd ed, Vol 1: Academic Press, London.

KETCHUM, B.H. 1939. The absorption of phosphate and nitrate by illuminated cultures of *Nitzschia closterium. Am. J. Bot.* 26: 339-.

KNOX, G.A. & A.R. KILNER 1973 (unpublished). *The ecology of the Avon-Heathcote estuary.* Zoology Dept, Univ. Canterbury, Christchurch, NZ.

KORRINGA, P. 1956. Oesterteelt in Suid Afrika. Hidrographiese, biologiese en oesterologiese waarnemings in die Knysnastrandmeer, met aantekeninge oor toestande in ander Suid-Afrikaanse waters. *S.Afr. Dept. Industries (Sect. Fish.) Invest. Rep.* 20: 1-94.

LEACH, J.H. 1971. Hydrology of the Ythan estuary with reference to distribution of major nutrients and detritus. *J. mar. biol. Ass. UK* **51**: 137-157.

LIPSCHULTZ, F., CUNNINGHAM, J.J. & J.C. STEPHENSON 1979. Nitrogen fixation associated with four species of angiosperms in the central Chesapeake Bay. *Estuar. coastal. mar. Sci.* **9**: 813-818.

LISS, P.S. 1976. Conservative and non-conservative behaviour of dissolved constituents during estuarine mixing. *In:* J.D. Burton & P.S. Liss (eds), *Estuarine chemistry:* 93-130. Academic Press, London.

LIVINGSTONE, D.A. 1963. Chemical composition of rivers and lakes. *Prof. Pap. US Geol. Surv.* **440G**, 64pp.

McROY, C.P., BARSDATE, R.J. & M. NEBERT 1972. Phosphorus cycling in an eelgrass *(Zostera marina L)* ecosystem. *Limnol. Oceanog.* **17**: 58-67.

OLIFF, W.D. (ed) 1976 (unpublished). South African National Marine Pollution Program. First annual report 509pp. Typescript; second annual report 172pp. Typescript. (Includes many separate progress reports from 1969 onwards by the National Institute for Water Research, Durban; first under the heading *Natal Rivers Research Fellowships,* later as reports for the CSIR on *Marine disposal of effluents* and from 1974 onwards, as reports on *National marine pollution surveys (east coast section)* to the S. African national committee for environmental sciences.

PERKINS, E.J. 1974. *The biology of estuaries and coastal waters.* Academic Press, London. 678pp.

PHILLIPS, J. 1972. Chemical processes in estuaries. *In:* R.S.K. Barnes & J. Green (eds), *The estuarine environment:* 33-50. Applied Science, London.

POMEROY, L.R. 1970. The strategy of mineral cycling. *Ann. Rev. Ecol. Systems* **1**: 171-190.

POMEROY, L.R., SMITH, E.E. & C.M. GRANT 1965. The exchange of phosphate between estuarine water and sediments. *Limnol. Oceanog.* **10**: 167-172.

RILEY, J.P. & R. CHESTER 1971. *Introduction to marine chemistry.* Academic Press, New York and London. 465pp.

RILEY, J.P. & G. SKIRROW 1965. *Chemical oceanography Vol 1.* Academic Press, New York and London. 712pp.

RILEY, J.P. & G. SKIRROW 1975. *Chemical oceanography* (2nd ed) *Vol 2.* Academic Press, New York and London. 647pp.

SCOTT, K.M.F., HARRISON, A.D. & W. MACNAE 1952. The ecology of South African estuaries. Part 2: The Klein River estuary, Hermanus. *Trans. roy. Soc. S.Afr.* **33**: 383-331.

STEFFANSSON, U. & F.A. RICHARDS 1963. Processes contributing to nutrient distributions off the Columbia River and the Strait of Juan de Fuca. *Limnol. Oceanog.* **8**: 394-410.

STEWART, W.D.P. 1972. Estuarine and brackish waters — an introduction. *In:* R.S.K. Barnes & J. Green (eds), *The estuarine environment:* 1-9. Applied Science, London.

STIRLING, H.P. & A.P. WORMALD 1977. Phosphate/sediment interaction in Tolo and Long harbours, Hong Kong and its role in estuarine phosphate availability. *Estuarine coastal mar. Sci.* **5**: 631-642.

STRICKLAND, J.D.H. 1965. Production of organic matter in the primary stages of the marine food chain. *In:* J.P. Riley & G. Skirrow (eds), *Chemical oceanography Vol I:* 477-610.

TAFT, J.L. & W.R. TAYLOR 1976. Phosphorus dynamics in some coastal plain estuaries. *In:* M. Wiley (ed), *Estuarine processes.* Vol 1: 79-89. Academic Press, New York.

VACCARO, R.F. 1965. Inorganic nitrogen in sea water. *In:* J.P. Riley & G. Skirrow (eds), *Chemical oceanography* Vol 1: 365-408. Academic Press, London.

VISSER, G.A. 1969. Analysis of Atlantic waters off the west coast of southern Africa. *S.Afr. Div. Sea. Fish. Invest. Rep.* **75**.

WATLING, R.J. 1977. Trace metal distribution in the Wilderness lakes. *Special report FIS 147, National Physical Research Laboratory,* CSIR, Pretoria (unpublished).

WATLING, R.J. & H.R. WATLING 1974. Environmental studies in Saldanha Bay and Langebaan Lagoon. I: Trace metal concentrations in selected molluscs and algae. *FIS Special Rep.* **70**: 1-77 (unpublished).

WATLING, R.J. & H.R. WATLING 1975. Trace metal studies in Knysna estuary. *Environment RSA* **2**(10): 5-7.

WATLING, R.J. & H.R. WATLING 1977. Metal concentrations in surface sediments from Knysna estuary. *Special FIS Rep.* **122**, CSIR, Pretoria (unpublished).

WILLIAMS, P.J. de B. 1975. Biological and chemical aspects of dissolved organic material in sea water. *In:* J.P. Riley & G. Skirrow (eds), *Chemical oceanography* (2nd ed) Vol 2: 301-363. Academic Press, London, New York.

WILLIS, J.P., FORTUIN, H.G.G. & G.A. EAGLE 1977. A preliminary report on Recent sediments in Saldanha Bay and Langebaan Lagoon. *Trans. roy. Soc. S.Afr.* **42**: 497-509.

WOLLAST, R. & J.J. PETERS 1978. Biogeochemical properties of an estuarine system: the River Scheldt. *In:* E.D. Goldberg (ed), *Biogeochemistry of estuarine sediments.* pp 279-293. UNESCO, Paris.

WOOD, E.J.F. 1965. *Marine microbial ecology.* Chapman & Hall, London, 243pp.

WOOLDRIDGE, T. 1976. The zooplankton of the Msikaba estuary. *Zool. Afr.* **11**: 33-44.

ZUBERER, D.A. & W.S. SILVER 1976. Mangrove associated nitrogen fixation. *Proc. Internatl. Symp. Biol. Management Mangroves, Honolulu 1974.* Vol 2. Univ. Florida Publ.

Chapter 6

The estuarine flora
J.H. Day

Department of Zoology, University of Cape Town

INTRODUCTION

This chapter is divided into two sections. The first part, which is based on international researches and recent reviews, outlines the composition of an estuarine flora and the factors which affect distribution, biomass and primary productivity. The second part reviews what is known of the flora of estuaries in southern Africa. The phytoplankton is discussed in chapter 8. It may be said at once that the benthic microflora is practically unknown but the macrophytes have been described and their distribution in a number of estuaries has been recorded. Studies of biomass and primary production are in progress and there is a general discussion at the end of the chapter.

Although a number of marine algae and sea grasses extend into estuaries, the bulk of the flora has been derived from the land or fresh water. Almost all the divisions (or phyla) of the plant kingdom are represented with the probable exception of the Bryophyta. In spite of this, the macrophytes belong to relatively few genera, many of which are widely distributed. Chapman (1960) recognised nine estuarine floristic regions around the world; Chapman (1977) describes nine groups of saltmarsh formations and six groups of mangals. The microflora, judging by the reviews of Wood (1965) and Perkins (1974), is even more widely distributed.

Micro-organisms are easily suspended by tidal currents and later deposited in calmer areas so that the distinction between planktonic and benthic forms is often arbitrary. Wood (1965) and Hobbie (1976) have stressed that the whole estuarine ecosystem is closely knit with a rapid flux of materials between autotrophs and heterotrophs. The composition of the benthic and epiphytic communities, whether on rock, sand, mud or on the leaves of sea grasses and salt marsh vegetation, is determined by the nature, tidal level and permanence of the substrate. By way of example, rock surfaces between tide marks may be covered by a muddy felt formed by blue-green algae, diatoms, flagellates, sarcodina and ciliates as well as larger plants and animals.

Macroscopic algae are never abundant in coastal plain estuaries and are usually restricted to the cleaner rocks or man-made structures at the mouth. Due to the turbidity of the water, they seldom extend more than a metre below the low tide mark and indeed, light penetration rather than nutrient deficiencies appears to be the main factor limiting the growth of submerged plants and phytoplankton. Even in clear estuaries, the compensation depth for phytoplankton is seldom more than 5 m (exceptionally 20 m) and in turbid estuaries it may be much less than a metre. Moreover, the phytoplankton is continually drifting out to sea and unless the residence time of the estuarine water is prolonged, the growth of the lush marginal vegetation becomes the main source of primary production. Although there are exceptions to any generalisation regarding estuaries it is widely agreed by Odum and many others that estuarine productivity is much higher than that of the open sea and is comparable to good agricultural land.

Sandbanks at the mouth are so unstable that the microflora is scanty and macrophytes are absent. On the quieter sandflats away from the channel margin, the felt of micro-organisms in the sand provides sufficient stability for the seedlings of rooted plants to establish themselves, and once established, the roots bind the sediment and the shoots and leaves decrease current velocities and promote the deposition of more silt and organic detritus. As more species grow in such silty soils, a vertical zonation is established at different tidal levels depending on the period of submergence, the water potential of the soil and the resistance of different species to water loss or to the accumulation of salt in metabolising tissues. Jefferies (1972) reports that direct osmotic effects are of secondary importance so that the flora changes slowly along the estuary and at comparable tidal levels it is more closely related to the nature of the substrate than to the salinity.

As shown in chapter 5, estuaries are rich in nutrients mainly derived from the rivers. The soluble phosphate (SRP) trapped in the anoxic sediments is absorbed by the roots of vascular plants. The excess over requirement may be released to the water column so that *Zostera* and *Spartina* have been called phosphate pumps. Phosphate is seldom a limiting nutrient; inorganic nitrogen compounds however, are required in larger quantities and although some bacteria and blue-green algae fix molecular nitrogen and there is rapid recycling, nitrogen deficiencies do occur during seasons of rapid growth. This has been reported particularly in phytoplankton populations but Jefferies (1972) suggests that it also occurs in dense stands of salt marsh plants. Many of them are perennials which start growing earlier than the annuals and competition for nitrogen may limit the growth and distribution of the latter.

The species composition of an estuarine flora is related both to coastal sea temperatures and to the atmospheric temperature and humidity. Most of the macroscopic algae and sea grasses are derived from the sea and coastal currents determine their geographic distribution. In contrast to this, most of the emergent marginal plants including both salt marsh plants and mangroves, are derived from the land and their distribution is related to atmospheric temperatures and humidities. Details will be discussed later.

BACTERIA

This brief account, based largely on the work of Wood (1965), is no more than an introduction to the bacterial flora of estuaries. The work of Zobell and his colleagues (reviewed in Zobell (1946)) put marine microbiology on a firm basis; Wood (1965) extended our knowledge and his many observations in Australian estuaries and coastal seas show the close similarity between marine and estuarine bacteria. Indeed, there appear to be no bacteria restricted to estuaries and Scholes & Schewan (1964) doubt that there is even a distinct marine bacterial flora. One is left with the impression of a single worldwide population which is highly adaptable. Wood (1965) quotes *Desulphovibrio* as an example of environmental tolerance. It was isolated from estuarine sediments with a pH range of 6,8-8,4, a salinity of 19-30 ‰, and a temperature range of 14-23 °. In culture, it tolerated a pH range of 3,8 to 11,0, salinities between 1 and 24 ‰ and a temperature range of 10-65 °. In spite of this adaptability, some strains of *Desulphovibrio* proved to be sensitive to even small changes of temperature and salinity.

Estuaries receive bacteria from many sources and the survival of pathogens is of practical importance. Wood (1965) quotes the work of Ketchum (1952), who assessed: 'the relative effects of dilution, bactericidal action of sea water and predation in reducing the population of coliform bacteria in the tidal estuary of the Raritan River . . .' He concluded that 'bactericidal action is the most important factor followed by predation and dilution'. Ciliates are particularly important as predators of coliform bacteria. Wood notes that 'bacteriophages are not numerous in oceans and phages of terrestrial organisms must be of terrestrial origin'. He states that 'coliform organisms do not persist long in estuarine waters so that their presence indicates recent contamination'. More recent work shows that this is only true in well-illuminated water so that turbidity becomes important.

The normal bacterial population includes a few autotrophs; these do not contain true chlorophyll and use H_2S or thiosulphates as hydrogen donors. Few marine and estuarine parasitic forms are known and most of those described are species of *Pseudomonas* which cause fish diseases. The rarity of bacterial diseases among algae is possibly due to the production of antibiotics. Thus the great majority of the heterotrophs are free living, and among them, aerobic forms greatly outnumber the facultative or obligate anaerobes which are normally active in the subsurface layers of the sediment.

Forbisher (1968), who gives a valuable account of the various orders and suborders of bacteria and allied organisms, states that they grow best when attached to surfaces. He considers that when attached, the extra-cellular enzymes they secrete and the products of the reaction are less liable to diffuse away from the cell body. This is supported by later workers. Thus bacterial counts in the water column tend to increase with the number of particles in suspension and bacterial counts in the sediment are far greater than those in the water column. The counts increase in the finer sediments where the surface/volume ratio is greater. Wood (1965) quotes estimates made in Lake Macquarie, an unpolluted estuary in New South Wales. Viable aerobes and facultative anaerobes numbered between

5×10^1 and 1×10^4 in the water column compared with 3×10^5 and $6,5 \times 10^6$ in the sediment, representing an increase of two orders of magnitude quite apart from bacteria concerned with the sulphur cycle and other anaerobes. Very similar results were obtained in estuarine bays along the coast of Texas. The numbers in the surface layers of the sediment exceed those in the deeper layers but the rate of change depends on the percentage and nature of organic substrates and the porosity of the sediment. In Port Hacking in New South Wales, Wood noted that where new sediment rich in organic matter was deposited due to a current diversion, the bacterial count was $5 \times 10^7.ml^{-1}$ at the surface, $8,5 \times 10^7.ml^{-1}$ at 0,3 m and then decreased to $4 \times 10^7.ml^{-1}$ at 0,9 m.

Bacteria take part in a variety of chemical reactions many of which have been summarised in chapter 5. The decomposition of complex organic compounds contained in the dead bodies of plants and animals, and the eventual formation of simple compounds, takes place in a series of stages. Senescent leaves are attacked by fungi before they fall and autolysis helps to break down the tissues. Thereafter suites of organisms are concerned, since those that take part in the earlier stages are seldom specific in their substrate requirements but they provide the simpler organic compounds required by the bacteria that follow. Aerobic bacteria break down complex carbohydrates to simple acids such as lactic, pyruvic and butyric as well as some carbon dioxide. Cellulytic bacteria, including species of *Vibrio* and *Cytophaga,* break up long carbohydrate molecules into sections of one or a few hexose units. Proteins are hydrolysed to small peptides or individual amino acids and inorganic phosphate is rapidly split off from organo-phosphates. These products may be absorbed immediately by other members of the microbial community on the sediment surface, such as fungi, protozoa and unicellular algae many of which are facultative heterotrophs. Alternatively, as more sediment accumulates on top, the Eh falls and anaerobic bacteria become active. The simple carboxyllic acids provide the substrates needed by sulphate reducers such as *Desulphovibrio* and denitrifying bacteria such as *Pseudomonas* which reduce nitrates and nitrites to ammonia or molecular nitrogen. Wood doubts that the reduction of carbon dioxide to methane and hydrogen, which occurs in fresh water marshes, results in the evolution of these gases in estuaries for the products of the reduction are taken up in the sulphur cycle. Both the sulphur and nitrogen cycles have been outlined in chapter 5.

YEASTS AND FUNGI

Perkins (1974), who reviews the estuarine yeasts and fungi, quotes numerous examples of saprophytic, symbiotic and parasitic species. An account of the estuarine yeasts is also given by van Uden (1967) who suggests that 'yeast populations attain their highest densities in inland waters and have their lowest densities in the open sea'. Nonetheless, yeasts are far less abundant than bacteria, and although Wilson (1960) quotes several genera which are semiparasitic on algae, fishes and invertebrates in addition to the common saprophytic forms in organic detritus. He notes that their biology is poorly known.

The estuarine fungi include representatives of all classes; the Phycomycetes, Ascomycetes and the Fungi Imperfecti predominate, while Myxomycetes (or Mycetozoa) are uncommon and the Basidiomycetes are virtually absent. As in the case of the bacteria and yeasts, there appears to be little difference between terrestrial and marine forms. All estuarine species are euryhaline but judged by the production of fruiting bodies most of them prefer low salinities (Perkins, 1974).

The Phycomycetes are mainly saprophytic, some genera attacking wood in association with the shipworm *(Teredo)* and the gribble *(Limnoria)* which suggests that they provide proteins required by those wood-borers although the association is controversial (Wood, 1965). The Saprolegniaceae are largely parasitic causing diseases in fishes and attacking the eggs of crustacea as well as a variety of algae including diatoms. The Myxomycete *Labrynthula macrocystis* and the Pyrenomycete *Lulworthia halina* were suspected of causing the rapid decline of many *Zostera* beds in Europe and North America during the 1930's. Recent studies by Rasmussen, noting that both species may be found on healthy plants, suggest that the real cause of the decline was a temperature increase with the fungi attacking the moribund plants. Since *Zostera marina* occurs over a wide temperature range the suggestion obviously needs confirmation.

ALGAE

All taxonomic divisions of the algae are represented in estuaries, and although most estuarine algae particularly the macroscopic forms are immigrants from the sea, numerous unicellular algae are derived from fresh water. This is particularly true of the Chlorophyceae. Wood (1965) gives an interesting account of the distribution and nutrition of many unicellular marine and estuarine algae. The dinoflagellates may be holozoic or holophytic or may be facultative heterotrophs; some benthic diatoms are partly heterotrophic and so are many of the green flagellates. *Euglena* and allied genera may become so abundant on polluted mudbanks, that they form green patches particularly where the salinity is fairly low. In other algae, the organic requirements may be limited to substances they cannot synthesise themselves and are thus auxotrophs. As noted in chapter 5, Stewart (1972) quotes the vitamin requirements of the several divisions of algae. Many of the necessary vitamins are produced by bacteria and blue-green algae growing on muddy sediments (Wood 1965, Perkins 1974). Silicate and trace metal deficiencies, which have been reported to limit the growth of diatoms and other phytoplankters in open oceans have not been reported as limiting factors in estuaries. Gutknecht & Dainty (1969) give a detailed review of the ionic requirements of marine algae including a number of species which penetrate estuaries.

The most obvious, though not necessarily the most abundant planktonic algae are diatoms, while naked peridinians are more common than armoured forms. Little is known of the nanoflagellates except that the minute Cryptophyceae, Coccolithophorineae and Chrysomonadaceae all occur in estuaries. Although they escape through the meshes of plankton nets, chlorophyll extraction techniques show that the nanoflagellates often form the major part of the phytoplankton biomass (Wood, 1965). Nanoplankton in general is reported to be responsible for about 90 % of phytoplankton production in estuaries. The benthic microalgae include the same groups with an increase of blue-green algae (Myxophyceae) and euglenoids (Heterokontae). Although there are no clear cut divisions between the planktonic and benthic populations, the planktonic diatoms include more centric forms and the pennate genera often have spines or other flotation mechanisms, while the benthic pennate genera often have pectinoid sheaths for attachment to sand grains, rocks and sea grasses. Some have a raphe or slit so that they are mobile and can migrate through surface sediments (Wood, 1965).

Estuarine plankton will be discussed in more detail in chapter 8. According to Perkins (1974) who quotes the species of planktonic algae in the Tamar estuary from Mommaerts (1969), there are basically three populations. First there are neritic marine species which drift in and out of the estuary with the tides and tend to dominate the higher salinities near the estuary mouth. Second, there are a few freshwater species carried by river water; these survive for a while in low salinities. Third, there are autochthonous estuarine species in the middle and upper reaches. Although these are both eurythermal and euryhaline, they only become abundant if the estuarine water has a long residence time for in most estuaries the phytoplankton drifts out to sea too rapidly for a large population to develop. Photosynthesis is restricted by turbidity to the surface layers and seldom extends below 3 m. As noted, primary production by phytoplankton in shallow coastal plain estuaries is seldom as important as that due to benthic plants.

Many benthic diatoms grow on rocks or as epiphytes on the leaves of *Zostera, Thallasia, Posidonia, Ruppia* and other sea grasses. These epiphytic diatoms together with filamentous algae such as *Ectocarpus, Chaetomorpha* and *Polysiphonia* as well as blue-green algae and a number of small sedentary animals, form a furry covering on the older leaves known as periphyton or 'Aufwuchs'. This provides a rich source of food for invertebrates and a number of fishes. Mullets graze on the periphyton but a number of herbivorous fishes crop the leaves of the sea grasses. Blaber (1974) has shown that the sparid fish *Rhabdosargus holubi*, which packs its stomach with the leaves of *Zostera* or *Ruppia* is only capable of digesting the periphyton presumably because the fish lacks the strong cellulase necessary to digest the tough cellulose covering the *Ruppia* leaves. Bell *et al* (1978) have made similar observations on the Monacanthid fish which feed in *Posidonia* beds in Australia.

A similar diverse community of microscopic algae, animals and bacteria grows on sand banks and muddy sediments particularly in salt marshes where the organisms form mats which bind the sediment grains. The diatoms are famous for their vertical migrations, particularly the naviculoid genera and the same is true of dinoflagellates. They are partly heterotrophic and live within a few millimetres of the sediment surface and migrate up or down with the tides, the intensity of illumination and the water content of the sediment. The migration of diatoms is discussed by Perkins (1974). It seems that the basic response is to light. During the day, they move up so as to project from the sediment surface and are exposed by the ebbing tide if the light intensity is not too high. As the flood tide advances over the sand flat, 'tidal shock' transmitted through the sediment causes them to retreat into the soil before they are submerged. Thus golden brown or green patches may be seen on exposed intertidal flats due to diatoms, dinoflagellates, euglenoids or occasionally, the turbellarian *Convoluta* with its green symbiont *Platymonas convolutae*.

Macroscopic algae are not well-represented in estuaries. Apart from marine algae which drift into estuaries after storms and survive for a while, the permanent residents are limited to a small number of widespread genera which can withstand the turbidity, deposition of silt and the absence of wave action. Horizontal surfaces are rapidly covered with silt and the encrusting species are limited to the vertical surfaces of piles in swift channels. Coralines including encrusting 'lithothamnia' are virtually absent. Among the Chlorophyta the common genera are *Enteromorpha, Ulva, Cladophora, Chaetomorpha* and *Codium*. The gas-filled tubules of *Enteromorpha* allow it to float in quiet backwaters and mats of tubules attached as epiphytes may be so dense as to prevent light from reaching submerged vegetation. *Ulva* grows on stones, gravel and shells but is easily detached and proliferates while floating until it is cast ashore by wind. In organically enriched lagoons such as the shallow and polluted Avon-Heathcote estuary at Christchurch in New Zealand studied by Knox & Kilner (1973), the rotting mounds of *Ulva* cast ashore are a public nuisance. In

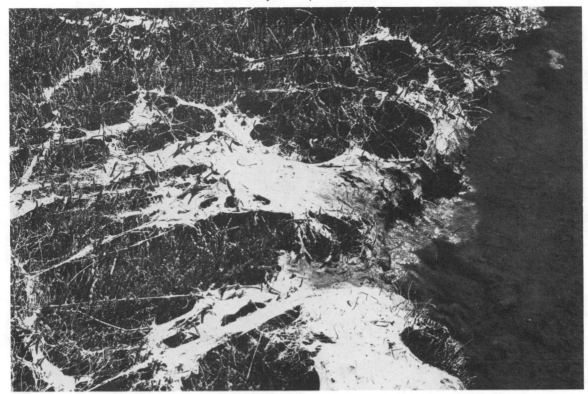

Matted *Chaetomorpha gracilis* drying at edge of a saltmarsh pool (Photo: J.H. Day)

many estuaries tufts of *Cladophora* and *Codium* fringe the sides of rocks and bridges and *Codium tenue* often becomes detached and continues to grow on shallow sandy bottoms.

Among the Phaeophyta, species of *Fucus, Pelvetia* and *Ascophyllum* are restricted to the Northern Atlantic where large plants are abundant on rocky slopes and smaller plants of *Fucus* and *Pelvetia* mix with vascular plants in salt marshes and contribute substantially to salt marsh production. In the tropics and the southern hemisphere, there are smaller amounts of *Sargassum* on hard substrates while blobs of *Colpomenia* and fans of *Dictyota* often occur on sea grasses such as *Posidonia* and *Thalassodendron. Hormosira banksii* is particularly common on rocks and oyster rafts in Australian estuaries.

The Rhodophyta are represented mainly by small species; *Polysiphonia, Ceramium* and *Laurencia* often growing as epiphytes on sea grasses while species of *Bostrychia* and *Caloglossa* grow on salt marsh plants and the boles and roots of mangroves. *Porphyra* grows at much the same level but is seldom common even at the mouths of estuaries. The largest and the most important alga commercially, is *Gracilaria* which has a small holdfast and is easily detached and then grows as long, thin branching strings which drift along the bottom until anchored in the sand of clear lagoons and sheltered marine bays. In South Africa it is the main

source of agar and is abundant in Langebaan Lagoon and Saldanha Bay (Simons, 1977). Dr Nigel Christie discusses its productivity in chapter 7. It is also found in Swartkops estuary (Macnae, 1957) and Hedgpeth (1967) records it in the Laguna Madre of Texas.

VASCULAR PLANTS

The vascular plants include the giant fern *Acrostichum* found in tropical and subtropical estuaries and a number of halophytic angiosperms. These form two groups; the sea grasses adapted to complete submergence at or below low tide, and the herbaceous salt marsh plants and mangroves growing at and above mid-tide and adapted to varying periods of emergence. The mangrove trees are of course perennials but it is somewhat surprising that the majority of the sea grasses and low-growing saltmarsh plants are perennials too with recorded life spans of 10-40 years. Only a few genera eg *Salicornia*, are annuals. It is convenient to deal with the low-growing saltmarsh plants separately.

Many genera of sea grasses have a worldwide distribution in sheltered bays and estuaries with *Thalassia, Thalassodendron, Cymodocea, Posidonia* and *Halodule (= Diplanthera)* and several other genera in warm and tropical waters and *Zostera, Halophila, Ruppia, Potamogeton* and *Zannichellia* extending into temperate

Zostera capensis and *Spartina maritima* (Photo: J.H. Day)

estuaries. Many of these genera are euryhaline but their extension into low salinities is very varied. *Posidonia, Thalassia, Cymodocea* and *Thalassodendron* prefer salinities above 20 ‰ while *Zostera* and *Halodule* tolerate 10 ‰; *Ruppia* and *Potamogeton* prefer low salinities and *Zannichellia* extends into fresh water and is seldom found in salinities above 10 ‰. Exceptions will be discussed later.

The main limit to growth is light and abrasion by sandladen water in more turbulent habitats. In heavily silted estuaries, *Zostera* is absent; in many estuaries it extends no more than a metre below low tide but in clear estuaries it may extend to 6 m and according to Ranwell (1972) it grows at the surprising depth of 30 m off the coast of California. *Halophila ovalis* is a widely distributed associate of *Zostera* and usually grows in the upper fringe, while *Posidonia, Thalassia* and some species of *Cymodocea* extend below the *Zostera* level. In Australia, *Posidonia australis* extends to 5 or 10 m while *Zostera capricorni* is usually restricted to 1-2 m, but may reappear with *Halophila* at the lower fringe of *Posidonia* at 9-12 m (Larkum, 1973).

All the sea grasses have well-developed rhizomes in the sediment from which arise longer or shorter stems bearing long, flexible leaves which, in the case of *Zostera,* may reach over a metre. However, its leaves vary greatly; where it grows at or above low tide level and is subject to wave action, the leaves are short and nar-

row but where it grows in deeper water or in more sheltered situations, the leaves are much longer. As a result the ecotypes have received different specific names.

Sea grasses grow best in muddy sediments which are often anoxic below the surface and in the case of *Zostera* at least, oxygen is conveyed to the roots through a series of intercellular spaces while nutrients are translocated through the xylem which opens at the apices of the leaves. As noted in chapter 5, McRoy *et al* (1972) have shown that phosphate is carried in both directions and the excess over requirement received from the roots, is released by the leaves. Possibly this is one reason why the older leaves of *Zostera, Thalassia* and *Posidonia* are often densely covered with epiphytic micro-organisms among which, blue-green algae that can fix molecular nitrogen may be present. Lipschultz *et al* (1979) states that nitrogen fixation is associated with four species of submerged angiosperms in Chesapeake Bay, and that the contribution of the rhizomes may be significant. The genera concerned are *Potamogeton, Myriophyllum, Ruppia* and *Elodea.*

At the end of the growing season most of the food reserves of the sea grasses are translocated to the rhizomes and the old leaves are attacked by fungi and bacteria. Eventually they disintegrate and add to drifting organic detritus. At the beginning of spring the

Sarcocornia perennis and *Juncus kraussii* (foreground) and *Zostera capensis* (Photo: J.D. Day)

rhizomes grow and the main method of reproduction is vegetative; new leaves appear and sometimes flowers but they are rare and seedlings are not readily established in new areas. If the rhizome system in existing sea grass beds is damaged by human activities such as hauling boats over the shallows, dredging, explosions or the movement of heavy vehicles, the bare areas take years to recover. This has been reported in *Posidonia* and *Zostera* beds in Australia by Wood (1965) and by Kirkman & Parker in the CSIRO Fisheries and Oceanography report for 1974-77. Experimental verification in regard to *Thalassia* is given by Zieman (in Wood & Johannes, 1975).

THE SALTMARSH VEGETATION

Low growing halphytic grasses, sedges and succulents cover the muddy banks of estuaries at and above mid-tide levels in temperate regions and mangroves occupy similar levels in more tropical estuaries and sheltered bays. Where the two occur together, the canopy of the mangrove trees shades out the low growing saltmarsh plants and they are restricted to the fringes and open glades.

Widely distributed genera include *Spartina, Puccinellia, Triglochin, Salicornia, Sarcocornia (Arthrocnemum), Limonium, Juncus, Suaeda, Sporobolus* and many

others at extreme high tidal levels. *Phragmites* extends from fresh water marshes into low salinities.

The nutrient requirements of saltmarsh plants have been studied by Piggott (1969) and Tyler (1971) among others, and it is generally agreed that there is no lack of phosphate in the sediment but ammonia or nitrate nitrogen is the limiting nutrient during periods of rapid growth. Competition develops between slow growing and rapidly growing species and Piggott points out that opportunistic annuals such as *Salicornia europea* and *Suaeda maritima* which grow in bare patches in the marsh may be stunted by competition from early growing perennials nearby. The same is true of *Salicornia meyeriana* in South Africa.

The halophytic angiosperms are euryhaline and the changes in the vegetation are more closely related to the nature of the soil than to changes in salinity along an estuary. But this does not apply to the changes in vegetation at different tidal levels. The transport of nutrients from the roots to the leaves demands a continuous flow of water through the plant and its eventual loss by transpiration. In order to absorb water from soil solutions with the same salinity as the sea, the osmotic pressure of the sap must exceed 20 bars. Ranwell (1972) reports that the average suction pressure for salt marsh plants is 32,6 bars, 65 % of which is due to chloride. The soil at high spring tide level is only inundated every 14 days and during neap tides

Triglochin sp (right front), *Displyma crassifolium* (long pale leaves), *Chenolea diffusa* (short broad leaves) and *Sarcocornia* sp (beaded stems) (Photo: J.H. Day)

Juncus kraussii (right) in mixed *Sarcocornia* and *Limonium* marsh (Photo: J.H. Day)

Limonium linifolium in flower (Photo: J.H. Day)

the soil water may become hypersaline. Jefferies (1972) who reviews the problems due to high salinity, has emphasised the need for more research on the matrix potential of soils. He says (p 66): 'Some accounts of the availability of water in saline soils for plant growth only stress the lowering of the free energy of water as a result of osmotic forces and neglect the role of the matrix potential in lowering the water potential of soils'. Thus the nature of the soil as well as its salinity must be considered and the clay content decreases the availability of water to the plant. As might be expected oil pollution has a similar effect (Baker in Cowell, 1971).

In order to absorb the necessary nutrients and water through the cortical cells of the roots without also absorbing an excess of salt, a high degree of ion selectivity and an efficient sodium pump to extrude the salt are necessary. The cortical cells of the roots of *Plantago* and *Triglochin* have a high K^+ to Na^+ ratio (Jefferies, 1972) but nonetheless some salt is absorbed which leads to ionic imbalance in metabolising tissues and a decreased growth rate. Sodium is excreted from special salt glands in the young leaves of *Spartina* and *Limonium* or is allowed to accumulate in non-metabolising cells until the leaves are shed and the sodium lost. Succulence in *Salicornia* and *Sarcocornia* also dilutes the high concentration of salt.

Although loss of water by transpiration is essential, it is restricted to the minimum and many halophytes in the upper marsh are xerophytic and the stomata may be small, reduced in number or closed for most of the day. Alternatively the leaf area may be reduced and during hot dry summers the leaves are shed early.

According to their ability to grow in waterlogged soils at midtide or to maintain adequate water and tolerate excessive salt at extreme high spring tide mark, the dominant species of saltworts occupy different tidal levels and as more silt is deposited and the marsh level rises, the dominants change. Anomalies may be due to many factors; the soil may change, drainage may be improved by the development of new channels and some species tolerate grazing by rabbits, cattle or even hippopotami better than others. Nonetheless saltmarshes tend to be stable over the years and Jefferies (1972) has estimated that *Limonium vulgare* at Stiffkey in Norfolk, is over 40 years old and Pierce (1979) estimates that *Spartina maritima* in Swartkops estuary, South Africa may reach 14 years. Since the stands of these perennials are dense it is not surprising that sexual reproduction is infrequent and asexual reproduction is common.

The distribution of saltmarsh plants at different tidal levels has been discussed by many authors including Jefferies (1972) and Perkins (1974). Although the zones and associations have been described as a Spartinetum, a Salicornetum and a Limonetum, etc, others

Dead mangroves at Beachwood Creek, Durban, 1974 (Photo: P. Berjak)

have noted many variations in vertical distribution. Zedlers (1977) found that in a Californian estuary dominance changed with tidal level but found no realistic associations or zones. The same genera do not always occupy the same position in different climates and at a species level, only the sequence within a floristic region is worth consideration; the dominant species at different levels in the estuaries of southern Africa will be described later.

MANGROVES

An excellent account of the mangroves in the Indo-west-Pacific region is given by Macnae (1968). Chapman (1977) discusses the mangroves of all oceans. Mangrove swamps or 'mangals' as Macnae prefers to call them, are associations of halophytic trees, shrubs, palms and occasional creepers growing on tropical and subtropical coasts. Outliers extend into temperate regions and there are small stands in 37 ° south latitude in Westernport Bay in Victoria and at Auckland in New Zealand. Mangroves are limited to calm waters and grow best on the muddy shores of sheltered bays and estuaries where the rainfall exceeds 1 500 mm per year and where temperature does not fall below 20 °; under such conditions they may form dense forests with trees up to 40 m high as occurs in New Guinea. The biogeography of mangroves is discussed by Chapman (1975) who says there are about 68 species with a distribution related to continental drift. In South East Asia there are some 30 genera but only a few are widespread through the Indo-Pacific; among them are *Avicennia, Bruguiera, Rhizophora, Ceriops, Lumnitzera, Sonneratia* and the estuarine fern *Acrostichum.* In dry climates, mangroves are sparsely developed and may be restricted to fresh water seepage areas or estuaries.

In the cooler climates of New South Wales and southern Africa there are only five or six species, West Africa has six (Chapman (1977), Moll & Werger (1978)) and the coasts of tropical America have about ten species with *Avicennia, Rhizophora* and *Laguncularia* as the dominant genera.

As usual, nitrogen is the main limiting nutrient during periods of rapid growth. Nitrogen-fixing blue-green algae and photosynthetic bacteria grow on the surface of the sediments. Zuberer & Silver (1975) found the sulphate reducing bacterium *Desulphovibrio,* living in close association with the roots of *Rhizophora mangle.* Kimball & Teas (1975) found that the rate of nitrogen fixation varied from 1,13 to 12,93 $\mu g.m^{-2}.hr^{-1}$ in Florida.

Mangroves have surprisingly shallow roots for such large trees and as the roots grow in anoxic soils, oxygen is absorbed by the lenticels on the pneumatophores of genera such as *Avicennia, Sonneratia* and *Bruguiera* or on the prop-roots of *Rhizophora.* The lenticles can only absorb oxygen from the air and when they are submerged the oxygen supply to the roots is soon exhausted. Periodic emergence of the lenticels is thus essential and normally occurs with the rise and fall of the tides. Macnae (1963), Breen & Hill (1969) and Berjak *et al* (1977) have shown that if the water level is stabilized, the mangroves whose pneumatophores or prop-roots are submerged soon die while those growing above water level survive. Oil pollution has the same effect; if the bulk of the lenticels are covered with oil the tree dies and if they are clean the tree survives.

Some mangroves produce normal seeds which develop after they fall on damp soil. According to Steinke (1975) the seeds of *Avicennia* float in the water for three or four days before they germinate so that they are widely distributed and implantation may occur at

Regenerating *Avicennia marina* saplings at same area of Beachwood Creek, 1979 (Photo: P. Berjak)

Bruguiera gymnorrhiza at Kosi Bay (Photo: J. Grindley)

any tidal level. The seeds of *Rhizophora, Bruguiera* and allied genera, germinate while still attached to the parent tree and a long, cigar-shaped hydrocotyle is formed as food reserves accumulate. The propagules hang down vertically and may be implanted in the mud where they fall, or they may drift away in the sea so that coastal distribution is determined by the direction of the currents.

Avicennia is usually the pioneer of the mangrove association for it has a wide temperature and salinity range. The seedlings even grow in sand above low neap tide level and as silt and organic matter accumulate around their roots and the foliage provides shade, other genera are established. At their preferred mid-tidal level *Rhizophora* and *Bruguiera* tolerate more shade, and grow more rapidly than *Avicennia* so that in dense stands the latter dies out at this level and persists only at lower and higher levels of the mangrove swamp. *Rhizophora* requires well-saturated soils but cannot tolerate the very saline soils near high spring tide. Its prop-roots support it on the margins of creeks and channels and where erosion occurs, the lower fringe of *Avicennia* falls and *Rhizophora* is left on the margin. On accreting shores Macnae (1968) and Saenger *et al* in Chapman (1977) give many examples of vertical zonation in dense mangrove forests. Although the difference in level between the lower fringe at about mid-tide and the upper fringe at the level of extreme high spring tide is seldom more than a metre or two, the slope of the shore is so gentle that the mangrove swamp may extend more than 5 km from the channels of the estuary and may cover many thousands of hectares. Vertical zonation is related to the period of exposure as measured by the number of tides that innundate a particular area each year (Walter, 1973).

Sonneratia alba often forms the lower fringe on the shores of sheltered bays and the lower reaches of estuaries for it is a stronger competitor than *Avicennia* but is restricted to salinities of 25-35 ‰. Next comes a band of *Avicennia marina,* then *Bruguiera* and thickets of *Ceriops tagal* with *Rhizophora mucronata* tracing the course of seepage channels and waterlogged areas and finally the upper fringe of *Avicennia* with *Lumnitzera* in the driest areas and *Nypa* palm in wetter climates. It is emphasised that this is a simplified and idealised sequence in the Indo-West-Pacific; in southeast Asia many more genera and species complicate the sequence, if indeed there is one. In more temperate climates there may be only one or two species, and in young or poorly developed mangrove swamps no sequence may be discernible.

The vertical range of each species is related, not only to its ability to tolerate shade as in the case of *Rhizophora* and *Bruguiera* but also to the salinity of the soil and its matrix potential as noted for saltmarsh plants. At the high spring tide mark, the salinity of the interstitial water in the soil may exceed 50 ‰ in

Well-grown *Avicennia marina* at Mlalazi (Photo: J.H. Day)

the dry season and fall to almost fresh water values in heavy rains when surface pools stand for days on end (Sasekumar, 1974). Many species of mangroves are xerophytic and Walter (1973) states that some develop suction pressures in their leaves of more than 40 atmospheres so that they can absorb nutrients from the roots. Also the leaves are often succulent and Joshi *et al* (1975) has shown that transpiration may be reduced by closing the stomata except for brief periods in the morning before maximum temperatures develop.

As in all halophytes, the main problem is to maintain the correct ratio of inorganic ions in metabolising tissues. Scholander (1968) states that there are three basic strategies for resisting high salinities although they are not mutually exclusive. They are:

Salt exclusion, eg *Rhizophora, Ceriops, Bruguiera*

Salt excretion, eg *Avicennia, Aegiceras, Acanthus*

Salt accumulation, eg *Sonneratia, Lumnitzera, Exoecaria*

Scholander reported a process of ultrafiltration by the roots of salt excluders. In salt excretors the concentration of salt in the xylem sap is similar to that in sea water and the excess sodium in young leaves is excreted by special salt glands; for example in *Avicennia* the excretion has a salinity of 41 ‰ and drips off the leaves on humid days. According to Berjak (1978) the salt glands of mature leaves disappear and the salt is eliminated through abaxial stomata. She describes the subcellular mechanisms concerned. In salt accumulators the excess salt is stored in the bark particularly in pneumatophores. In all mangroves, sodium remains in the dying leaves and is lost when they fall; indeed, there is a tendency for leaves to fall early during droughts.

THE ESTUARINE PLANTS OF SOUTHERN AFRICA

There are only a few studies of the microscopic algae of estuaries in southern Africa. Planktonic algae will be discussed in chapter 8 and benthic microalgae are poorly known.

Although the dominant macrophytes have often been recorded by biologists primarily concerned with the distribution of the estuarine fauna, very few workers have studied the estuarine flora in detail. Field work has been facilitated by the generic keys provided by Levyns (1966) and the more detailed account of the Chenopodiaceae by Tolken (1967). Similar accounts of the Potamogetonaceae, Cyperaceae and halophytic Graminae are urgently needed.

An account of the salt marsh vegetation in several Cape estuaries has been given by Scott *et al* (1952) for Hermanus lagoon, Millard & Scott (1954) for Milnerton lagoon, by Grindley (1976, 1978) for an area of Knysna lagoon and Macnae (1957) and Pierce (1979) for the Swartkops estuary. The halophytes of subtropical estuaries between the Transkei and Mocambique have been studied by Macnae (1963), Ward (1976) and Branch & Grindley (1979). Some notes on the vegetation of Morrumbene estuary are given by Day (1974). Inhaca Island, though not estuarine illustrates the effects of fresh water seepage, and accounts of the distribution of the saltmarsh plants, mangroves and sea grasses are given by Macnae & Kalk (1958, 1962) and Mogg (1963). Work on biomass and primary productivity is in progress; in chapter 7, Dr Nigel Christie deals with several of the species which grow in Langebaan Lagoon.

Although there is an overlap between the subtropical estuarine flora of southern Mocambique and the temperate flora of Cape estuaries, the latter flora does not include mangroves or tropical sea grasses and it is convenient to deal with the two separately. There is little change between the halophytes of the eastern, southern and western Cape Province as far as the Olifants estuary. The mouth of the Orange River on the Atlantic coast has virtually no estuary and the estuarine area at Sandwich harbour in Namibia, has not been studied. In the present state of our knowledge any attempt at a phytogeographical study of the halophytes of southern Africa would be worthless. Far more records are obviously needed and the identification of existing records needs to be checked in any case. Nonetheless, the known distribution of dominant or characteristic macrophytes (but by no means all the species) is given in the appendix as a basis for further work. It is based on the papers mentioned earlier, supplemented by the records in the estuarine surveys referred to in Day (1974) and my own and Professor Grindley's unpublished records from several estuaries. These are catalogued in the Zoology Department of the University of Cape Town.

THE MACROPHYTES OF SUBTROPICAL ESTUARIES

Many of the Indo-west Pacific mangroves recorded by Macnae (1968), and Chapman (1975) do not extend to the subtropical coasts of Africa (Moll & Werger, 1978). Nonetheless, the environmental features such as the shelter from wave action, rainfall, the number of tidal inundations, the matrix potential of the soil, the effect of shade and the seepage of fresh water from higher levels apply equally well in southern Africa. Mogg (1963) in particular, has stressed the importance of fresh water seepage and the downwards extent of the root system of species living near high tide.

Far more species have been recorded in the salt marshes of Inhaca Island than along the banks of the Morrumbene estuary studied by Day (1974). This is due to the greater intensity of collecting at Inhaca and the only mangrove which is recorded at Morrumbene but not at Inhaca is *Sonneratia alba*. Although the vertical distribution of mangroves and other halophytes is the same at the two localities, the sequence of zones or associations is better marked at Inhaca.

In southern Mocambique the rainfall is about 1 000 mm.yr^{-1} in the coastal belt, the tidal range is 3,0-3,3 m and shade temperatures seldom rise above 30 ° or fall below 19 °. Under these conditions the mangrove swamps are not as well developed as those in the East Indies, north eastern Madagascar or northern Mocambique where the rainfall and temperatures are higher.

The mangroves and low growing halophytes extend down from the extreme level of high tide of springs to about mid-tide although there are many local variations possibly due to fresh water seepage. The vertical sequence of individual species or associations at Inhaca

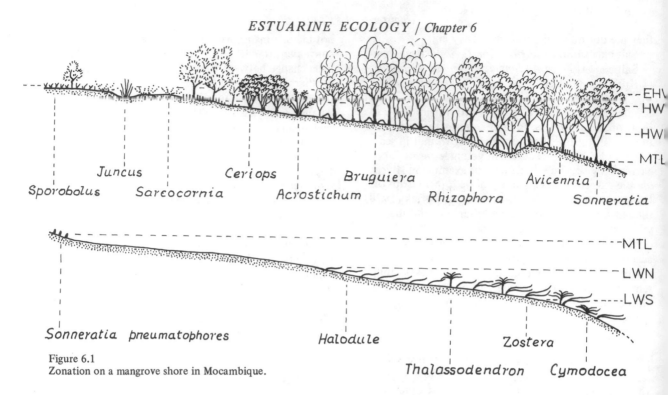

Figure 6.1
Zonation on a mangrove shore in Mocambique.

and in the lower reaches of Morrumbene estuary are briefly as follows:

1. From extreme high tide of springs (EHWS) to mean high water of springs (MHWS) there is a diverse and variable association of halophytes illustrated by Macnae & Kalk (1958) and Mogg (1963). It may be reduced to salt encrusted flats in dry clayey soils which are colonised by stunted bushes of *Avicennia marina,* tufts of *Digitaria littoralis* and strongly halophytic species of *Salicornia* or *Sarcocornia.* In better drained sandy areas *Sporobolus virginicus* may form a sward extending to much lower levels; growing in it are *Salicornia perrieri, Sesuvium portulacastrum* and better grown trees of *Avicennia.* The grass *Stenotaphrum secundatum* prefers moister areas and will grow under shade with several other halophytes where the coastal scrub encroaches on EHWS.

2. *Sporobolus* and *Salicornia perrieri* extend down to MHWS which is usually marked by *Chenolea diffusa* in drier areas or *Juncus kraussii* in wetter soils. *Sarcocornia decumbens* or *S. indica* replace the *Salicornia* and form creeping mats where the mangroves are scattered, but these and other saltworts are shaded out where the mangroves become dense. *Lumnitzera racemosa* marks the upper fringe of the mangroves on well-drained sandy slopes and *Avicennia marina* elsewhere.

3. Where the mudflats between HWS and midtide are very broad and covered with mature forests of mangroves the classical zones or bands described earlier may be discerned. As shown in figure 6.1, this includes an upper fringe of *Avicennia,* thickets of *Ceriops,* then *Bruguiera* with *Rhizophora* in seepage channels and a lower fringe of *Avicennia* and *Sonneratia.* More usually these genera are mixed and only the upper and lower fringes can be distinguished The pneumatophores of *Avicennia* in the lower fringe and sometimes the boles of these and other trees are covered with a mosslike association of algae including *Bostrychia* cf *Binderi, Caloglossa leprieuri* and other genera. In open glades, *Acrostichum aureum, Phragmites australis* and *Triglochin striata* often form dense stands in seepage areas, and *Sporobolus virginicus* forms a sward on the drier sandy slopes.

4. Below the mangrove belt there is usually a bare zone from about midtide to LWN. This is eroded and steep where the tidal currents sweep along the shore but broader and flatter where the main channel is offshore. Sea grasses may appear at LWN or their upper limit may hardly extend above LWS. As Day (1974) has shown, vertical distribution on estuarine shores is largely determined by the depth of the water table in the sediment. On the steeper, upper slopes it may sink below 30 cm and on the flatter, lower slopes it may ooze out on the surface so that even sandy shores may be permanently wet from LWN down. Only on evenly sloping shores is the vertical distribution of sea grasses and invertebrates closely related to tidal levels.

In southern Mocambique, the sea grass *Halodule uninervis* may extend from LWN to below LWS. *Thalassodendrum ciliatum* grows among the lower *Halodule* and extends to 5 or 10 m below LWS. *Cymodocea rotundata, C. serrulata, Zostera capensis* and *Halophila ovalis* grow near LWS, while *Syringodium isoetifolium* and *Thalassia hemprichii* both grow subtidally.

As usual, few macroscopic algae occur in estuaries. Although Day (1974) lists 12 genera from Morrumbene, the only really common species were *Enteromorpha intestinalis* and *Spyridia filamentosa* which grew as an epiphyte on *Zostera*.

The changes with decreasing salinity along Morrumbene estuary described by Day need not be detailed here. *Holodule* has a greater salinity range than the other sea grasses and was recorded in salinities of 10 ‰. Among the mangroves *Bruguiera* had the greatest salinity range and was found with *Acrostichum* among the fresh water mangroves *Barringtonia racemosa* and *Hibiscus tiliaceus* at the head of the estuary where the salinity varied with the tides from 10 ‰ to almost fresh water.

Changes in the macrophytes of Natal estuaries

As shown in the appendix, many tropical macrophytes disappear between southern Mocambique and the Transkei and South African species increase. Presumably this is due to decreased temperatures for the rainfall in the coastal belt at 1 000-1 250 mm.yr^{-1} is slightly higher than in southern Mocambique while water temperatures in winter may fall to 14 ° in southern Natal as compared to about 19 ° in Mocambique. However, there are several other factors which limit the macrophytes in particular estuaries.

As noted earlier, mangroves require a periodic change in water level to aerate the lenticels which supply oxygen to the roots. If tidal range is inhibited by the closure of the mouth, mangroves are absent. This happens in many small estuaries such as Mhlanga and Umgababa. It also occurred for four months in Kosi Bay and Breen & Hill (1969) record that the mangroves whose pneumatophores or prop-roots were submerged died while those at higher levels survived. Hill (1966) suggests that the same things happened in the Mlalazi prior to 1950. Bridges or causeways which restrict tidal flow have the same effect as seen in the Mgobezeleni estuary at Sordwana Bay (Bruton & Appleton 1975) and Beachwood near Durban (Berjak *et al* 1977). The harbour works at Durban destroyed most of the mangroves in the bay (Day & Morgans, 1956) and the same thing has happened recently in Richards Bay although new stands are appearing in the sanctuary area.

Heavy silting kills submerged vegetation such as *Zostera* and *Ruppia* as may be seen in the Mkomazi and the Mzimkulu, in Natal and the Mzimvubu, the Bashee and the Kei although smaller and clearer estuaries nearby have fringes of *Zostera* at the low tide mark. Small beds may occasionally be found in some blind estuaries such as the Umgababa.

High salinities have a devastating effect on the aquatic vegetation. The St Lucia lakes are hypersaline during long droughts and in the upper lakes the salinity may rise to over 100 ‰. Even when the salinity is above 50 ‰ all macrophytes apart from *Enteromorpha* disappear from below the water level and *Phragmites australis* is restricted to the banks where fresh water seepage occurs (Day *et al* 1954).

During good rains the high salinity water is flushed out and salinities between 7,2 and 8,3 ‰ were recorded throughout the lakes in the winter of 1964. The aquatic vegetation flourishes at such times, and Millard & Broekhuysen (1970) give a long list of the species present. As the salinity rises again in a dry year, the lakes provide natural laboratory in which the salinity tolerance of the various species may be observed. Ward (1976) incorporates earlier records with his own observations. When the salinity is below 10 ‰ *Bruguiera gymnorrhiza* saplings were found growing in 0,75 m of water confirming that this species has a lower salinity limit than other mangroves and that the saplings are not dependent on tidal range. *Zostera capensis, Ruppia spiralis* and *Potamogeton pectinatus* were abundant and *Najas marina* was present as well as the algae *Nitella* sp and *Enteromorpha intestinalis*. As the salinity rose, some species were restricted to areas of fresh water seepage or disappeared. *Najas* and *Nitella* were not recorded in salinities above 10 ‰; *Potamogeton* survived in 26,5 ‰ but died in 33,5 ‰; *Zostera* died when the salinity rose above 45 ‰ but *Ruppia spiralis* was found living in 50 ‰.

The vertical sequence of halophytes in Mgazana

Mgazana estuary in the Transkei has the best developed mangrove swamp south of Inhaca at 26 °S. It has been described by Macnae (1963) and a more detailed account of its ecology has been published by Branch & Grindley (1979). A brief description will be found in chapter 14.

Extensive salt marshes and mangroves border the main channel and the two tributaries. *Chenolea diffusa* marks the highest tidal levels in the drier and more sandy areas and a sward of *Sporobolus virginicus* extends down from above. *Juncus kraussii* grows at the same level as *Chenolea* in wetter areas and with it is *Stenotaphrum secundatum* with knee-high bushes of *Sarcocornia pillansiae* here and there. Below the *Chenolea* and *Juncus* at HWS, there are carpets of *S. pillansiae, S. natalensis* and *S.perennis* more or less in sequence downwards with *Triglochin striata* and *Cotula coronopifolia* in waterlogged areas. Fields of *Phragmites australis* mark areas of low salinity.

The mangroves grow on the more gentle slopes between HWS and LWN. There are only four species but where the mangroves are well-developed, they occupy the same habitats as described earlier. The upper fringe of *Avicennia* grows as scattered trees in

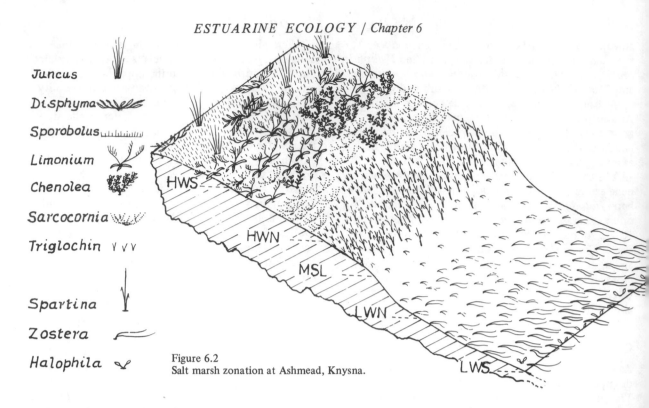

Juncus
Disphyma
Sporobolus
Limonium
Chenolea
Sarcocornia
Triglochin
Spartina
Zostera
Halophila

Figure 6.2
Salt marsh zonation at Ashmead, Knysna.

the carpet of *Sarcocornia* spp, then follows the main stand of *Bruguiera* which is dense and may grow up to 7 m high with *Rhizophora* along the seepage channels or on eroding banks and finally the lower fringe of *Avicennia*. Below the mangroves the mud may be bare for there is no sign of *Spartina*. *Zostera* grows well between LWN and LWS with patches of *Halophila* here and there; neither extend into deeper water or on to sandbanks.

The last few *Avicennia* and *Bruguiera* extend south of the Bashee estuary but they are poorly developed, and West (1944) has reported them from Kei mouth. This is the southern border of the Transkei and although saplings have been reported from Gonubie and Nahoon near East London by Steinke (1972) and Moll & Werger (1979), they do not persist.

The macrophytes of Cape estuaries

The average summer maximum temperature is of the order of 27 ° and the corresponding winter minimum is about 12 ° in the southern Cape or 10 ° in the Western Province. As shown in the appendix, several subtropical species are absent and two temperate species, *Limonium scabrum* (or *L. linifolium*) and *Spartina maritima* became abundant. There are, of course, many other temperate species but the records are too scattered to state definitely that they do not reach the Transkei. Again, the flora of west coast estuaries is so poorly known that it is not possible to say more than that there are no striking differences from the south coast.

The most complete account of zonation is that given by Macnae (1957) for the Swartkops estuary. While the dominants in other estuaries are not always the same, nor the associations so well-defined, the general sequence in many estuaries is similar to that in the Swartkops.

Between extreme high tide of springs and the mean level of high spring tide there is a diverse vegetation in which several species may be locally dominant. Macnae describes a sward of *Sporobolus virginicus* in drier areas or *Stenotaphrum secundatum* in moister soils with scattered bushes of *Sarcocornia pillansiae, S. decumbens* and *S. natalensis*. At a lower level but still above MWS, Macnae describes a limonetum with *Limonium linifolium* as dominant and several other species including *Chenolea diffusa* and *Juncus kraussii* as common to abundant. In other estuaries, *Limonium* may be no more than one of several common genera. In particular areas, *Sarcocornia pillansiae* may be dominant and *Chenolea* more common than *Limonium* at HWS. Grindley (1976, 1978) describes the Ashmead area in Knysna Lagoon with *Limonium* and *Chenolea* as co-dominants but hardly extending above HWS, and, at higher levels, *Sporobolus, Juncus, Sarcocornia pillansiae, Carpobrotus edulis* and *Disphyma crassifolium*. A general impression is shown in figure 6.2. Boucher & Jarman (1977), in an article on the terrestrial vegetation around Langebaan Lagoon, define two of the marsh communities as 'Juncus kraussii Dense Sedgelands' at higher levels and '*Chenolea-Salicornia* Dwarf Succulent Shrublands' below high tide. *Salicornia meyeriana* is one of the few annuals in the salt marsh.

Between HWS and HWN, *Sarcocornia perennis* carpets extensive areas in many estuaries, mixing with *Limonium* and *Chenolea* at higher levels and *Spartina maritima* below. In waterlogged depressions it is mixed with, or largely replaced by *Triglochin bulbosum* or *T. striata.*

The rice grass *Spartina maritima* is one of the most conspicuous features of Cape salt-marshes. It grows best on mudflats between HWN and midtide and covers many hectares in Knysna, Swartkops, Keiskamma and several other estuaries, although it is absent from blind estuaries. Large stands are transected by seepage channels which are stabilised by blue-green algae and fringed with *Triglochin bulbosum* or *Cotula coronopifolia*. Tufts of the alga *Bostrychia vaga* grow at the bases of the *Spartina* and *Limonium* plants.

Between the lower edge of the *Spartina* fields at midtide, and *Zostera* at low tide there is usually a bare muddy slope but in sheltered coves and backwaters, the slope is very gradual and the *Zostera* and *Spartina* meet.

The width of the band occupied by *Zostera capensis* varies with the current velocity, the nature of the substrate and the clarity of the water. It is absent from clean mobile sands at the estuary mouth and from very turbid estuaries but in clearer waters there is a fringe of small, thin-leafed plants above LWS on the mudbanks and in very clear estuaries such as Knysna and the Hermanus Lagoon, luxuriant beds with long leaves extend down to about 2 m.

Close-growing carpets of *Halophila ovalis* are common in the upper fringe of the *Zostera* belt and their matted rhizomes often support the burrows of the pistol shrimp *Alpheus crassimanus*. Isolated plants of *Codium tenue* grow here and closely cropped plants of *Ulva* sp grow on dead shells and pebbles. Small epiphytes including pennate diatoms and filamentous algae, form the periphyton on the older leaves which persist longest in the backwaters. Scott *et al* (1952) have recorded *Ectocarpus, Polysiphonia, Rhodocorton* and *Rhizoclonium* in Hermanus Lagoon but they have not been identified elsewhere. As noted earlier, the periphyton appears to be more important than the *Zostera* itself to herbivorous fishes. Day *et al* (1952) found *Zonaria tournefortii* as abundant in the deeper *Zostera* beds in Knysna but this seems to be an isolated occurrence. Similarly, Macnae (1957) reported *Gracilaria verrucosa* in a backwater of Swartkops estuary but it is mainly confined to Saldanha Bay and Langebaan Lagoon and has recently been decimated by the silt resulting from dredging(Simons, 1977).

As the salinity decreases to about 10 ‰ near the head of an estuary, many of the halophytes mentioned earlier decrease in abundance. *Phragmites australis* appears at the water's edge and there are clumps of *Scirpus venustulus* among the *Juncus, Cotula* and *Triglochin*. The submerged vegetation changes to an upper band of *Ruppia spiralis* above deeper *Zostera*

beds at salinities of about 20 ‰ and in lower salinities *Ruppia* and *Potamogeton pectinatus* become more abundant and *Zostera* disappears. *Scirpus globiceps* grows as tufts in sandy areas and *Zannichellia* replaces *Ruppia* in salinities below 5 ‰ in both Milnerton Lagoon and Swartkops estuary. The macroscopic alga *Chara* seems to be limited to the still waters of brackish lakes such as Swartvlei.

BIOMASS AND PRIMARY PRODUCTIVITY

Estuaries receive plant material from the land carried down the river and material drifting in from the sea with the tides. Leach (1971) suggests that these sources are more important than autochthonous production in the Ythan estuary. In deep clear estuaries, phytoplankton appears to be more important and in broad shallow estuaries there are important contributions from submerged sea grasses, benthic micro-algae growing on the mud flats and either salt marsh vegetation or mangroves. Few estimates of autotrophic biomass or total primary production have been made in such a complex ecosystem. Nonetheless the biomass or primary production of some of the most important plants has been assessed in terms of dry weight, ash-free dry weight, grams of carbon per square metre or kilocalories or kilojoules of energy. It is suggested that conversion factors be given since they differ from one species to another. Moreover, it is not always clear whether gross or net production has been measured and Teal (1962) has estimated that net production in salt marshes may be only 20 % of gross productivity.

Although estuarine plankton will be discussed in chapter 7, it is convenient here to compare phytoplankton production with that of benthic plants. Head (1976), after quoting estimates from several fjords and coastal plain estuaries, suggests a net production of 100-500 $g.C.m^{-2}.yr^{-1}$. Writing of the deep, slow-flowing waters of Chesapeake Bay, Hobbie (1976) states that 'annual primary production (of phytoplankton) may reach several hundreds of grams of carbon per square metre'. From the work of Oliff (1976) it would appear that the net production in the turbid estuaries of southern Africa is lower, possibly in the range of 100-200 $g.C.m^{-2}.yr^{-1}$. As will be shown, the net production of salt marsh vegetation is much higher than this.

Joint (1978) estimated the net production of benthic micro-algae in the Lynher estuary near Plymouth as 143 $g.C.m^{-2}.yr^{-1}$ as compared with 81,7 $g.C.m^{-2}.yr^{-1}$ for phytoplankton. The benthic algae which photosynthetise when exposed by the falling tide, start increasing a month earlier than the phytoplankton and are presumably not affected by the turbidity of the water. Joint also quotes estimates from several other estuaries varying from a minimum of 31 $g.C.m^{-2}.yr^{-1}$ in Ythan estuary to 200 g.C.m^{-2}.yr^{-1} in a Georgia saltmarsh. It would appear that 100 g.C.m^{-2}.yr^{-1} from benthic microalgae may be expected in many estuaries. Estimates from southern Africa would be most valuable.

Macrophytic algae are seldom abundant in estuaries although species of *Fucus* and allied genera are common in salt marshes in the northern hemisphere. Jefferies (1972) states that benthic algae contribute 25 % to the maximum standing crop of the salt marsh community in Norfolk but no separate estimates of algal production is available. Species of *Ulva* and *Enteromorpha* are abundant in very calm shallow estuaries, and in pools in salt marshes. For *Enteromorpha* Leach (1971) gives the mean biomass as 150 g dry mass m^{-2} in the Ythan estuary; Joint (1978) records a bloom of 95 g.m^{-2} in the Tamar; Knox & Kilner (1973) record a maximum biomas of 47 g dry mass m^{-2} of *Enteromorpha* and 176 g dry mass m^{-2} of *Ulva* in the Avon-Heathcote estuary in New Zealand.

The main production of vascular halophytes in salt marshes that has been measured is in the form of leaves and shoots which arise from rhizomes or rootstocks each year. In *Zostera, Posidonia* and *Spartina*, the older leaves die and fall away while new leaves are growing. Where it is not possible to determine growth rates directly, the peak standing crop above ground is used and a loss factor is added to determine annual net production. Sand-Jensen (1975) has determined the loss factor of *Zostera marina* in Denmark as 3,8 times the maximum standing crop which gives an annual net production of 838,8 g dry mass m^{-2}.yr^{-1}. If the same loss factor is applied to the standing of *Z. marina* in the North Carolina estuary recorded by Penhale (1977), the net production is 399 g dry mass m^{-2}.yr^{-1}. *Z. capricorni* in Australia and *Z. capensis* in South Africa appear to be similar to *Z. marina*. An analysis of the data of Clough & Attiwill (1975) indicates that the net production of *Z. capricorni* in Westernport Bay, Victoria is 633 g dry mass m^{-2}.yr^{-1}. Grindley (1978) records the standing crop of *Z. capensis* in Knysna estuary as 206 g dry mass m^{-2}, while Dr Christie's records in Langebaan Lagoon give a mean of 217 g dry mass m^{-2}. If Sand-Jensen's loss factor is applicable to *Z. capensis* the net annual production would be 782,8 and 824,6 respectively which is of the same order as the net production of other species of *Zostera*. It is very likely that in turbid estuaries, the annual production would be much less.

There are more records of the net production of *Spartina alterniflora* than any other salt marsh plant. Hatcher & Mann (1975) have compared the production in various salt marshes along the Atlantic and Gulf coasts of the United States using a loss factor of 25 % where none had been recorded. They record a net production of 289 g.C.m^{-2}.yr^{-1} in Nova Scotia and quote a minimum of 133 g.C.m^{-2}.yr^{-1} in New Jersey and a maximum of 1 153 g.C.m^{-2}.yr^{-1} in Georgia. In spite of local variations possibly due to low nutrient concen-

Table 6.1
Annual mean net productivity (g.dry mass.m^{-2}.yr^{-1}) (after Jefferies, 1972)

Location	Principal species	Net produc.	Reference
Georgia	*Spartina alterniflora*	3 700	Odum (1959) Pomeroy (1959)
N.Carolina	*S.alterniflora*	650	Williams & Murdoch (1969)
Bridgewater Bay, UK	*S.anglica*	960	Ranwell (1961)
Norfolk, UK	*S.anglica*	980	Jefferies (1972)
"	*Limonium vulgare*	1 050	"
"	*Salicornia* spp	867	"

Table 6.2
Primary productivity (in gC.m^{-2}.yr^{-1}) in shallow estuaries on the Atlantic coast of USA (after Hobbie, 1976)

	Georgia saltmarsh	Beaufort (NC) estuary
Salt marsh	700	256
Submerged plants	–	300
Attached micro-algae	–	75
Mud algae	420	–
Phytoplankton	–	66

Table 6.3
Mean biomass of shoots and roots of saltmarsh plants at Ashmead, Knysna estuary (from Grindley, 1978)

Species	Live biomass g.m^{-2}	Dry biomass g.m^{-2}	Ash-free biomass g.m^{-2}
Limonium scabrum	3 160	1 475	1 059
Chenolea diffusa	5 240	954	630
Sarcocornia decumbens	6 032	858	601
Triglochin bulbosa/ striata	12 460	1 935	1 229
Spartina maritima	15 816	3 464	1 614
Zostera capensis	1 757	206	67,5*

* *Zostera* patchy and the ash-free biomass of a more luxuriant quadrat was 238,4 g.m^{-2}.

trations, it is apparent that net production increases towards the subtropics.

Jefferies (1972) compares the production of *S. alterniflora* with *S. anglica, Limonium vulgare* and *Salicornia* spp growing in estuaries in southern England. His Table 3 is reproduced above as Table 6.1 but it may be noted that, since the units differ, the records are not directly comparable with those of Hatcher & Mann.

Hobbie (1976) compares estimates of primary production in Georgia and North Carolina. The enormous intertidal salt marshes along the Georgia coast are drained by a network of muddy creeks in which macroscopic vegetation is virtually absent. The salt marshes at Beaufort in North Carolina are more varied and the shallows at low tide and below have a rich growth of *Zostera* covered with periphyton or 'attached microalgae'. See Table 6.2.

Although the units used by Jefferies & Hobbie are different, their estimates are of the same order of magnitude. A round estimate of 500-1 000 g.C.m^{-2}.yr^{-1}

for salt marsh vegetation seems justified; this is comparable to the net annual production of good agricultural land.

Estimates of primary production in Langebaan Lagoon (which is not estuarine) and the coastal lakes of southern Africa will be discussed in later chapters. It is convenient here to consider the mean biomass of common salt marsh plants at one station in Knysna estuary reported by Grindley (1978). The data are presented in Table 6.3 and represent the biomass of the main halophytes (shoots and roots) along a transect from HWS to LWS. The samples were taken in November at the end of spring and must be close to peak biomass from which annual production could be assessed if the losses of dead material were known. Such data are urgently needed.

As noted earlier, the production of *Zostera capensis* is comparable to records of *Z. marina* if the same loss factor is used. The loss factor of *Sarcocornia* (= *Arthrocnemum*) is very small and the dry biomass recorded in Knysna is similar to the net production of *Salicornia* spp given by Jefferies (1972). The biomass of *Spartina maritima* is comparable to the maximum value for *S. alterniflora* reported from Georgia, and the biomass of *Limonium scabrum* is higher than the value for *L. vulgare* quoted by Jefferies. There are no comparable records for *Chenolea* and *Triglochin*. The biomass records of the common species in the Ashmead transect are thus as high or higher than the records discussed earlier. Whether the Ashmead transect represents the mean biomass in other salt marshes in Knysna is doubtful; experience suggests that it is better grown than most but a number of comparable transects would be required to confirm this.

The salt marshes of Knysna are more extensive and the plants appear more luxuriant than in most other estuaries in the Cape. Professor D. Baird has kindly given me an estimate of the biomass of salt marsh plants in Swartkops estuary. The total area of the salt marsh is 363 217 ha of which 32 % is covered by *Spartina maritima*. The mean standing crop (above ground) of the common plants in the marsh in g (dry mass) m^{-2} is: *Spartina maritima* (live) 546,8 and (dead) 94,8; *Triglochin bulbosum* 210,0; *Sarcocornia perennis* 1 290,2; *Chenolea diffusa* 889,0; *Limonium linifolium* 563,3. Most of these values are considerably lower than those from the Ashmead transect in Knysna quoted in Table 6.3 but it should be noted that the Knysna values include roots as well as shoots.

Professor Baird and P.E.D. Winter have also determined the annual net primary production of *Spartina* in Swartkops estuary. Their values as quoted by Grindley & Heydorn (1959) are: 'Net primary production in energy terms was 10 380 $kJ.m^{-2}.yr^{-1}$ but of the 602 g. $m^{-2}.yr^{-1}$ produced, it is estimated that only about 376 $g.m^{-2}.yr^{-1}$ accumulates as detritus. The remainder is apparently deposited elsewhere or carried out to sea'.

Many other estuaries such as the Breede, the Bushmans, Sundays and Kowie have relatively steep banks and the salt marsh vegetation is confined to a narrow strip. Little vegetation develops on the sand banks near the mouth and the estimate by Ranwell (1972) that three-fifths of the salt marsh production comes from the middle reaches of an estuary is supported by observations in South Africa.

Although the biomass of mangrove trees in terms of organic carbon is much greater than that of salt marsh vegetation, the annual production of leaf litter is not very different. Clough & Attiwill (1975) estimated that the biomass of stunted *Avicennia* in Westernport Bay, Australia was 8 600 g dry mass m^{-2} above ground and 14 600 g dry mass m^{-2} for roots. It is well-known that perennial plants including halophytes, abstract proteins from the leaves before they fall and it is significant that Clough & Attiwill found twice as much nitrogen and phosphorous in the roots as in the branches and leaves. Much of the organic carbon in the roots remains permanently in the sediment as peat and the wood in the branches is not available to consumers for years. Thus the litter fall is the usual estimate of annual production.

By analysing monthly records from litter baskets, Pool, Ludo & Snedaker (1975) estimate the litter fall in a Puerto Rican mangrove swamp as 838,6 g dry mass $m^{-2}.yr^{-1}$. They add that there is no obvious difference between the various species of mangroves or between the mangroves of Puerto Rico and Florida. They also compare the litter fall in mangrove swamps with that in terrestrial forest and scrub. The mangrove litter fall varied from 838,6 to 876 g dry mass $m^{-2}.yr^{-1}$ while that of terrestrial scrub and forest varied from 100 to 485 g dry mass $m^{-2}.yr^{-1}$. This suggests that the annual net litter productivity of mangroves is similar to that of salt marsh plants and twice that of terrestrial trees.

Appendix overleaf

APPENDIX: COMMON PLANTS IN THE ESTUARIES AND LAGOONS OF SOUTHERN AFRICA

	Morrumbene	Inhaca Island	Kosi Bay	St Lucia	Richards Bay	Mlalazi	Durban Bay	Estuaries near Umkomaas	Estuaries near Port Shepstone	Estuaries near Port St Johns	Mgazana	Estuaries near Bashee River	Keiskamma	Bushmans	Sundays	Swartkops	Knysna	Great Brak	Breede	Hermanus Lagoon	Milnerton Lagoon	Langebaan Lagoon	Berg River	Olifants River
	1	2	3	4	5	6	7	8	9	10	11	12	13	14	15	16	17	18	19	20	21	22	23	24
ALGAE																								
Bostrychia cf Binderi Harv.	●																							
Bostrychia scorpioides (GM) Moss											●		?				●							
Bostrychia vaga Hook & Harv.																	●							
Caloglossa leprieuri (Mont.) J.Ag.	●			●	●																			
Chatomorpha sp				●													●			●		●		●
Chara sp.																								●
Cladophora spp.				●			●										●			●	●			
Codium tenue Kutz.															●	●	●							
Ectocarpus sp		●																		●				
Enteromorpha bulbosa (Suhr.) Kutz.																	●		●	●	●	●	?	●
Enteromorpha intestinalis Linn.	●		●	●							●					●	●							
Gracilaria verrucosa (Huds.) Papenf.																●	●					●		
Lyngyba confervoides Gom.				●													●							
Polysiphonia incompta Harv.	●			●							●					●	●			●		●		
Rhizoclonium sp																				●			●	
Rhodocorton sp																				●				
Spyridia filamentosa (Wulf.) Harv.	●																							
Ulva spp							●			●	●					●	●			●		●		●
PTERIDOPHYTA																								
Acrostichum aureum Linn.	●	●		●	●	●																		
SPERMOPHYTA																								
Avicennia marina (Forsk.) Vierh.	●	●	●	●	●	●	●			●	●	●												
Barringtonia racemosa (Linn.) Lam.	●	●	●	●	●	●	●																	
Bruguiera gymnorrhiza (Linn.) Lam.	●	●	●	●	●	●	●	●		●	●	●												
Carpobrotus edulis N.E.Br.																	●		●					
Ceriops tagal (Perr.) Robinson	●	●	●																					
Chenolea diffusa Thunb.	●	●		●				●			●	●	●			●	●		●	●	●	●	●	●
Cotula coronopifolia Linn.											●	●	●			●	●		●	●		●	●	●
Cymodocea rotundata Ehrenb. & Hempr.	●	●																						
Cymodocea serrulata (R.Br.) Asch. & Magn.	●	●																						
Cyperus corymbosus Rottl.				●		●																		
Cyperus laevigatus (C.B.Cl.)				●		●																		
Cyperus natalensis Hochst.		●	●								●		●											
Cyperus textilis Thunb.										●						●	●			●		●		
Digitaria littoralis Stent.	●																							
Disphyma crassifolium Schwantes													●			●						●		
Filicia ficoides D.Cl.													●				●							
Galenia secunda (L.F.) Sond.													●			●					●			
Halodule uninervis (Forsk.) Asch.	●	●	●	●																				
Halophila ovalis (R.Br.) Hook	●	●					●				●				●									
Hibiscus tiliaceus Linn.	●	●		●	●	●	●				●													
Juncus kraussii Hochst	●	●	●	●		●	●	●	●				●	●	●	●			●	●	●	●	●	
Limonium linifolium Kuntze													●			●	●		●		●			
Limonium scabrum Kuntze																	●					●	●	
Lumnitzera racemosa Willd.		●	●																					
Naias marina Linn.			●	●																				

	Morrumbene	Inhaca Island	Kosi Bay	St Lucia	Richards Bay	Mlalazi	Durban Bay	Estuaries near Umkomaas	Estuaries near Port Shepstone	Estuaries near Port St Johns	Mgazana	Estuaries near Bashee River	Keiskamma	Bushmans	Sundays	Swartkops	Knysna	Great Brak	Breede	Hermanus Lagoon	Milnerton Lagoon	Langebaan Lagoon	Berg River	Olifants River
	1	2	3	4	5	6	7	8	9	10	11	12	13	14	15	16	17	18	19	20	21	22	23	24
Phragmites australis (Cav.) Trin.	●	●	●	●	●	●	●	●	●	●	●	●	●			●	●		●		●	●	●	●
Potamogeton pectinatus Linn.				●									●			●	●		●					
Puccinellia fasciculata Bickn.																	●	●						
Rhizophora mucronata Lam.	●	●	●	●			●	●		●	●													
Ruppia maritima Linn.	●		●	●													●					●	●	
Ruppia spiralis Linn.				●									●				●				●			
Salicornia meyeriana Moss.													●			●	●	●	●	●		●		
Salicornia perrieri Chev.	●	●		?									●			●	●		●					
Samolus porsus Thunb.									?		●		●			●	●		●					
Sarcocornia capensis (Moss) Scott												●	●			●				●				
Sarcocornia decumbens (Tolken) Scott		●									●		●			●	●							
Sarcocornia indica (Willd.) (Moq.) Scott	●	●		●																				
Sarcocornia natalensis (Bunge) Scott		●		●		●	●	●			●	●	●			●	●	●			●	●	●	
Sarcocornia perennis (Miller) Scott		●		●			●	●			●	●	●			●	●	●	●		●	●	●	
Sarcocornia pillansiae (Moss) Scott				●		●					●					●	●	●			●	●	●	
Scirpus globiceps C.B.Cl.																	●							
Scirpus litoralis Schrad				●		●	●	●		●							●							
Scirpus maritimus Linn.												●		●								●	●	
Scirpus nodosus Rottl.																						●	●	
Scirpus venustulus (Koenth.) Boeck																	●				●	●		●
Sesuvium portulacastrum Linn.		●																						
Sonneratia alba Smith	●																							
Spartina maritima (Curtis) Fernald													●	●		●	●		●		●			
Spergularia marginata (D.Cl.) Kitt								●												●				
Spergularia salina Presl.													●		●					●				
Sporobulus pungens						●	●										●			●		●		
Sporobulus virginicus Kunth.	●	●	●	●			●	●	●				●			●	●	●	●		●	●	●	
Stenotaphrum glabrum Trin.			●		●							●				●	●							
Stenotaphrum secundatum Kuntze		●		●				●					●			●			●					
Suaeda caespitosa W. Dod.													●									●		
Syringodium isoetifolium Dandy	●	●																						
Thalassodendron ciliatum (Forsk.)	●	●	●	●																				
Thalassia hemprichii (Ehr.) Asch.		●																						
Triglochin bulbosum Linn.			●	●	●	●	●	●					●			●	●				●	●	●	
Triglochin striata Ruiz & Pavon		●	●	●			●	●					●			●	●		●	●				
Zannichellia aschersoniana Graebn.															●						●		●	
Zostera capensis Setchell	●	●	●	●	●	●	●	●			●		●	●	●	●	●	●	●	●		●		●

REFERENCES

BALL, J.D., BURCHMORE, J.J. & D.A. POLLARD 1978. Feeding ecology of three sympatric species of leatherjacket (Pisces: Monacanthidae) from a *Posidonia* seagrass habitat in New South Wales. *Aust. J. Mar. Freshw. Res.* **29**: 631-643.

BERJAK, P. 1978. Subcellular adaptations in plants of extreme environments II. Towards understanding salt excretion in *Avicennia* spp. *Proc. Electron Micros. Soc. S.Afr.* **8**: 101-102.

BERJAK, P., CAMPBELL, G.K., HUCKET, B.I. & N.W. PAMMENTER 1977. *In the mangroves of southern Africa.* Natal Branch, Wildlife Soc. S.Afr., Durban. 72pp.

BLABER, S.J.M. 1974. Field studies of the diet of *Rhabdosargus holubi* (Steindachner) (Pisces: Sparidae) in a closed estuary. *J. Zool. Lond.* **173**: 404-417.

BOUCHER, C. & M.L. JARMAN 1977. The vegetation of the Langebaan area, South Africa. *Trans. roy. Soc. S.Afr.* **42**: 241-272.

BRANCH, G.M. & J. GRINDLEY (1979). The ecology of South African estuaries. Part XI: Mgazana, a mangrove estuary in Transkei. *S.Afr. J. Zool.* **14**(3): 149-170.

BREEN, C.M. & B.J. HILL 1969. A mass mortality of mangroves in the Kosi estuary. *Trans. roy. Soc. S.Afr.* **38**: 285-303.

BRUTON, M.N. & C.C. APPLETON 1975. Survey of Mgobezeleni lake-system in Zululand with a note on the effect of a bridge on the mangrove swamp. *Trans. roy. Soc. S.Afr.* **41**: 283-294.

CHAPMAN, V.J. 1960. *Saltmarshes and salt deserts of the world.* Interscience, London and New York.

CHAPMAN, V.J. 1975. Mangrove biogeography. *In:* G. Walsh, S. Snedaker & H.L. Teas (eds), *Proc. Internatl. Biol. Symp. Management Mangroves, Honolulu 1974.* Univ. Florida.

CHAPMAN, V.J. (ed) 1977. *Ecosystems of the world 1: Wet coastal ecosystems.* Elsevier, Amsterdam. 428pp.

CLOUGH, B.F. & P.M. ATTIWILL 1975. Nutrient cycling in a community of *Avicennia marina* in a temperate region of Australia. *Proc. Internatl. Symp. Biol. Management Mangroves, Honolulu 1974.* Vol 1: 137-146. Univ. Florida Publ.

COWELL, E.B. (ed) 1971. *The ecological effects of oil pollution on littoral communities.* Institute of Petroleum, London.

CSIRO *Fisheries and oceanography report for 1974-77,* Canberra.

DAY, J.H., MILLARD, N.A.H. & A.D. HARRISON 1952. The ecology of South African estuaries. Part 3: Knysna, a clear, open estuary. *Trans. roy. Soc. S.Afr.* **33**: 367-413.

DAY, J.H. & J.F.C. MORGANS 1956. The ecology of South African estuaries. Part 7: The biology of Durban Bay. *Ann. Natal Mus.* **13**: 259-312.

FROBISHER, M. 1968. *Fundamentals of Microbiology.* 8th ed. Saunders Co, Philadelphia, London. 629pp.

GRINDLEY, J.R. 1976. *Report on ecology of Knysna estuary and proposed Braamekraal marina.* School Environm. Studies, Univ. Cape Town. 133pp. Typescript.

GRINDLEY, J.R. 1978. *Environmental effects of the discharge of sewage effluent into Knysna estuary.* School Environm. Studies, Univ. Cape Town. 62pp. Typescript.

GRINDLEY, J.R. & A.E.F. HEYDORN 1979. Man's impact on the estuarine environment. *S.Afr. J. Sci.* **75**(12): 554-560.

GUTKNECHT, J. & J. DAINTY 1969. Ionic relationships of marine algae. *Oceanog. mar. Biol. Ann. Rev.* **6**: 163-200.

HATCHER, B.G. & K.H. MANN 1975. Above-ground production of marsh cord-grass *(Spartinon alterniflora)* near the northern end of its range. *J. Fish. Res. Board Can.* **32**: 83-87.

HEAD, P.C. 1976. Organic processes in estuaries. *In:* J.D. Burton & P.C. Liss (eds), *Estuarine chemistry:* 54-85. Academic Press, London.

HEDGPETH, J.W. 1967. Ecological aspects of the Laguna Madre, a hypersaline estuary. *In:* G. Lauff (ed), *Estuaries.* Am. Ass. Adv. Sci., Washington.

HILL, B.J. 1966. A contribution to the ecology of Umlalazi estuary. *Zool. Afr.* **2**: 1-24.

HOBBIE, J.E. 1976. Nutrients in estuaries. *Oceanus* **19**(5): 41-47.

JEFFERIES, R.L. 1972. Aspects of saltmarsh ecology with particular reference to inorganic plant nutrition. *In:* R.S.K. Barnes & J. Green (eds), *The estuarine environment.* Applied Science, London.

JOINT, I.R. 1978. Microbial production of an estuarine mudflat. *Estuarine coastal mar. Sci.* 7: 185-196.

JOSHI, G.V., JAMALE, B.B. & L. BOSHALE 1975. Ion regulation in mangroves. *Proc. Internatl. Symp. Biol. Management Mangroves, Honolulu, 1974.* Vol 2: 595-607. Univ. Florida Publ.

KETCHUM, B.H. 1962. Processes contributing to the decrease of coliform bacteria in a tidal estuary. *Ecology* **33**: 247-258.

KIMBALL, M.C. & H. TEAS 1975. Nitrogen fixation in mangrove areas of southern Florida. *In:* G. Walsh, S. Snedaker & H. Teas (eds), *Proc. Internatl. Symp. Biol. Management Mangroves, Honolulu, 1974.* Vol 2: 654-660. Univ. Florida Publ.

KNOX, G.A. & A.R. KILNER 1973. *The ecology of the Avon-Heathcote estuary.* Dept. Zoology, Univ. Canterbury, Christchurch. 358pp (typescript, unpublished).

LARKUM, A.W.D. 1973. The marine plants of Jervis Bay. *Operculum* 3(182): 43-46.

LEVYNS, M.R. 1966. *A guide to the flora of the Cape Peninsula* (2nd ed). Juta & Co, Cape Town. 310pp.

LIPSCHULTZ, F., CUNNINGHAM, J.J. & J.C. STEPHENSON 1979. Nitrogen fixation associated with four species of submerged angiosperms in the central Chesapeake Bay. *Estuar. coastal mar. Sci.* 9: 813-818.

MACNAE, W. 1957. The ecology of plants and animals in the intertidal regions of the Swartkops estuary near Port Elizabeth, South Africa. Parts I and II. *J. Ecol.* **45**: 113-131; 361-387.

MACNAE, W. 1963. Mangrove swamps in South Africa. *J. Ecol.* **51**: 1-25.

MACNAE, W. 1968. A general account of the fauna and flora of mangrove swamps and forests in the Indo-West-Pacific region. *Adv. Mar. Biol.* **6**: 74-270.

MACNAE, W. & M. KALK 1958. *A natural history of Inhaca Island, Mocambique.* Witwatersrand Univ. Press, Johannesburg. 163pp.

MACNAE, W. & M. KALK 1962. The ecology of mangrove swamps at Inhaca Island, Mocambique. *J. Ecol.* **50**: 19-34.

McROY, C.P., BARSDATE, R.J. & M. NEBERT 1972. Phosphorus cycling in an eelgrass *(Zostera marina L)* ecosystem. *Limnol. Oceanog.* **17**: 58-67.

MILLARD, N.A.H. & G.J. BROEKHUYSEN 1970. The ecology of South African estuaries. Part 10: St Lucia: a second report. *Zool. Afr.* **5**: 277-307.

MILLARD, N.A.H. & K.M.F. SCOTT 1954. The ecology of South African estuaries. Part 6: Milnerton estuary and the Diep River, Cape. *Trans. roy. Soc. S.Afr.* **34**: 279-324.

MILNE, H. & G.M. DUNNET 1972. Standing crop, productivity and trophic relations of the fauna of Ythan estuary. *In:* R.S.K. Barnes & J. Green (eds), *The estuarine environment.* Academic Science, London.

MOGG, A.O.D. 1963. A preliminary investigation of the significance of salinity in the zonation of species in salt marsh and mangrove swamp associations. *S.Afr. J. Sci.* **59**(3): 81-86.

MOLL, E.J. & M.J.A. WERGER 1978. Mangrove communities. *In:* M.J.A. Werger & A.C. van Bruggen (eds), *Biogeography and ecology of southern Africa.* Junk, The Hague.

MOMMAERTS, J.P. 1969. On the distribution of major nutrients and phytoplankton in the Tamar estuary. *J. mar. biol. Ass. UK* **49**(3): 749-766.

PENHALE, P.A. 1977. Macrophyte-epiphyte biomass and productivity in an eelgrass *(Zostera marine L)* community. *J. exp. mar. Biol. Ecol.* **26**: 367-387.

PERKINS, E.J. 1974. *The biology of estuaries and coastal waters.* Academic Press, London. 678pp.

PIERCE, S.M. 1979. *The contribution of Spartina maritima (Curtis) Fernald to the primary production of the Swartkops estuary.* MSc thesis, Rhodes University, Grahamstown.

PIGOTT, C.D. 1969. Influence of mineral nitrition on the zonation of flowering plants in coastal salt marshes. *In:* I.H. Rorison (ed), *Ecological aspects of mineral nutrition of plants.* Blackwell, Oxford.

POCOCK, 1955. Seaweeds of Swartkops estuary. *S.Afr. J. Sci.* **52**: 73-75.

POOL, D.J., LUGO, A.E. & S.C. SNEDAKER 1975. Litter production in mangrove forests of southern Florida and Puerto Rico. *Proc. Internatl. Symp. Biol. Management Mangroves, Honolulu, 1974.* Vol 1: 313-237. Univ. Florida Publ.

RANWELL, D.S. 1972. *Ecology of salt marshes and sand dunes.* Chapman & Hall, London. 258pp.

SAND-JENSEN, K. 1975. Biomass, net production and growth dynamics in an eelgrass *(Zostera marina L)* population in Vellerup Vig. Denmark. *Ophelia* **14**: 185-.

SASEKUMAR, A. 1974. Distribution of macrofauna on a Malayan mangrove shore. *J. anim. Ecol.* **43**: 51-69.

SCHOLANDER, P.F. 1968. How mangroves desalinate sea water. *Physiol. Plant.* **21**: 251-261.

SCHOLES, R.B. & J.M. SCHEWAN 1965. The present status of some aspects of marine microbiology. *Adv. mar. Biol.* **2**: 133-170.

SCOTT, K.M.F., HARRISON, A.D. & W. MACNAE 1952. The ecology of South African estuaries. Part 2: The Klein River estuary, Hermanus. *Trans. roy. Soc. S.Afr.* **33**: 283-331.

SIMONS, R.H. 1977. The algal flora of Saldanha Bay. *Trans. roy. Soc. S.Afr.* **42**: 461-482.

STEINKE, T.D. 1972. Further observations on the distribution of mangroves in the Eastern Cape Province. *J. S.Afr. Bot.* **3**: 165-178.

STEINKE, T.D. 1975. Some factors affecting the dispersal and establishment of propagules of *Avicennia marina* (Forsk) Vierh. *Proc. Internatl. Symp. Biol. Management Mangroves, Honolulu, 1974.* Vol 2: 402-414. Univ. Florida Publ.

STEWART, W.D.P. 1972. Estuarine and brackish waters — an introduction. *In:* R.S.K. Barnes & J. Green (eds), *The estuarine environment.* Applied Science, London.

TEAL, J.M. 1962. Energy flow in the salt-marsh ecosystem of Georgia. *Ecology* **43**: 614-624.

TOLKEN, H.R. 1967. The species of *Arthrocnemum* and *Salicornia* (Chenopodiaceae) in southern Africa. *Bothalia* **9**(2): 255-307.

TYLER, G. 1971. On the effect of phosphorus and nitrogen supplied to Baltic shore meadow vegetation. *Bot. Notiser.* **120**: 433-447.

VAN UDEN, N. 1967. Occurrence and origin of yeasts in estuaries. *In:* G. Walsh (ed), *Estuaries:* 306-310. Am. Ass. Adv. Sci., Washington.

WALTER, H. 1973. *Vegetation of the earth in relation to climate and ecophysiological conditions.* Springer-Verlag. Berlin. 237pp.

WARD, C.J. 1976. Aspects of the ecology and distribution of submerged macrophytes and shoreline vegetation of Lake St Lucia. *In:* A.E.F. Heydorn (ed), *St Lucia Scientific Advisory Council.workshop — Charters Creek, February 1976* (typescript). Natal Parks Bd, Pietermaritzburg.

WEST, O. 1944. Distribution of mangroves in the eastern Cape Province. *S.Afr. J. Sci.* **41**: 238-239.

WILSON, I.M. 1960. Marine fungi: a review of the present position. *Proc. Linn. Soc. Lond. (Bot.)* **171**: 53-74.

WOOD, E.J.F. 1965. *Marine microbial ecology.* Chapman & Hall, London. 243pp.

WOOD, E.J.F. & R.E. JOHANNES (eds) 1975. *Tropical marine pollution.* Elsevier, Amsterdam. 192pp.

ZEDLERS, J.B. 1977. Salt marsh community structure in Tijuana estuary, California. *Estuarine coastal mar. Sci.* **5**(1): 39-53.

ZOBELL, C.E. 1946. *Marine microbiology.* Chronica Botanica Press, Waltham, Mass, 240pp.

ZUBERER, D.A. & W.S. SILVER 1975. Mangrove associated nitrogen flixation. *Proc. Internatl. Symp. Biol. Management Mangroves, Honolulu, 1974.* Vol 2. Univ. Florida Publ.

Chapter 7
Primary production in Langebaan Lagoon

N.D. Christie

Department of Zoology, University of Cape Town

Present address: Ecological Consultants (Pty) Ltd., P.O. Box 258, Rondebosch

SUMMARY

The primary production of the main plant species was investigated in Langebaan lagoon during the period 1975-76. The species monitored were *Gracilaria verrucosa, Sarcocornia perennis, Spartina maritima, Zostera capensis, Typha capensis, Phragmites australis* and phytoplankton. The main nutrients in the lagoon water ie nitrates and phosphates showed clear seasonal patterns with nitrate declining down the length of the lagoon in summer while the pattern for phosphate was the opposite. Distribution and production of *Gracilaria* and phytoplankton was clearly correlated with nutrient concentrations and *Gracilaria* was shown to be the most important species in the ecology of the total lagoon. The saltmarsh genera *Spartina* and *Sarcocornia* reach maximum abundance in the southern half of the lagoon where nutrients in the marsh water were highest and where water currents were greatly reduced. The lagoon is a very important ecosystem and efforts should be made to conserve it in its natural unspoilt form.

INTRODUCTION

Langebaan lagoon is a sheltered arm of Saldanha Bay on the west coast of S. Africa, approximately 100 km north of Cape Town. The lagoon (Figure 7.1) is 14 km in length and about 3,5 km at its greatest width. It is completely marine with no inflowing river and draws all its sea water from Saldanha Bay which has a volume 20 times greater than the lagoon. In winter however, there is small seepage of fresh water due to rain.

In 1977, a symposium on research in the natural sciences of Saldanha Bay and Langebaan Lagoon was held at Saldanha. The symposium has been published in Volume 42 of the Transactions of the Royal Society of South Africa. The volume includes studies by Shannon & Stander on water replacement, current velocities and water chemistry; by Flemming on geology; by Willis *et al* on geochemistry; by Henry *et al* on phytoplankton production; by Grindley on zooplankton; by Boucher & Jarman on vegetation; by Simons on marine algae; by Christie & Moldan on benthic macrofauna; by Puttick on intertidal animals and by Summers on waders. Earlier papers on Langebaan Lagoon include Isaac (1956) on *Gracilaria* distribution and Day (1959) on animal distribution.

The aim of the present study is to describe the production of the phytoplankton, of *Gracilaria* and of several species of marsh plants in relation to the environmental factors to which they are exposed.

METHODS

The productivity of the main plant species in the lagoon was investigated over the year 1975-76 by two series of sampling trips 6-8 weeks apart. One series

Table 7.1
Sampling dates at lagoon and marsh stations and standing
crop of living *Typha* and *Phragmites* (grams dry mass per m²)

Date	Position of sample	Standing crop of *Typha*	Standing crop of *Phragmites*
1974 July		19,1	
1975 June	Marsh		1,72
July			13,5
Aug.	Marsh	53,0	129,4
Sept.	Marsh	76,8	338,4
Oct.	Lagoon	161,0	396,0
Nov.	Marsh	260,8	607,2
Dec.	Lagoon *(Gracilaria)*	435,8	964,0
1976 Jan.	Lagoon and Marsh	570,8	699,4
Feb.		747,6	818,8
Mar.	Lagoon and Marsh	716,2	
May	Lagoon and Marsh		
Aug.	Lagoon		
Sept.	Lagoon		

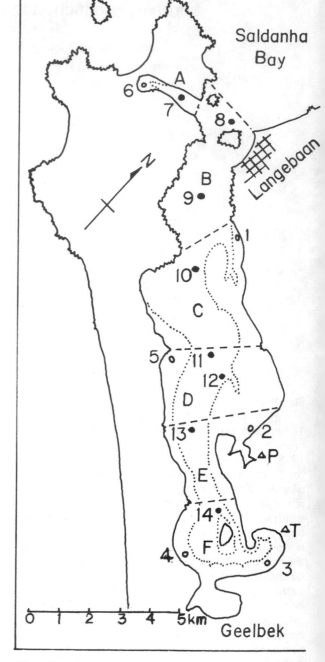

Figure 7.1
Map of Langebaan Lagoon showing lagoon sections A-F, shore
stations 1-6, lagoon stations 7-14, *Phragmites* station P and
Typha station T.

sampled the intertidal marshes and the other the sublittoral, hereafter referred to as the lagoon areas. Collection dates are given in Table 7.1.

The intertidal salt marshes include species such as
*Sarcocornia perennis, Scirpus triqueter, Chenolea dif-
fusa, Limonium scabrum, Triglochin bulbosum, Spartina maritima* and *Zostera capensis.*

Six shore stations, numbered 1-6, were visited on
each sampling trip and their positions are shown in
Figure 7.1. The plant species studied were *Spartina,
Sarcocornia, Zostera* and *Scirpus.* At some stations
there were two strains of *Spartina* ie tall and short and
of *Zostera* ie large and small. The presence of these
different strains was a direct result of close proximity
to stronger water movements and longer submergence
time, the larger or taller strains being found in channels
and on the edges of banks.

The standing crop was determined for each species
by removing the above ground shoots in two quadrats
of ⅛ m². The material from *Spartina* and *Scirpus* was
sorted into living and dead since these were clearly
distinguishable. Individual shoots of *Sarcocornia* do
not live more than a year but it was not possible to
separate dead from living material. *Sarcocornia* grows
by producing new bulbous segments along its shoots
while old ones narrow and are either absorbed into the
new material or die and pass into the lagoon system as
detritus. The dead material of *Zostera* was not easy to
collect since it is quickly removed by the tides. Hence
only living *Zostera* was collected.

In the laboratory the plants were dried to constant
weight at 50 °C and then ashed in a muffle furnace at
590 °C. The ash-free weights (equivalent to the organic
content) were then calculated. Energy values as kJoules
per gram dry mass (1 Calorie = 4,186 kJoules) were
obtained for samples in March and September using a
bomb calorimeter.

Production as a net value, ignoring loss by grazing
or decay, between sampling trips was calculated as an
energy unit in kJoules, ie increase in standing crop x

organic content x energy value. Calculations of production which make allowances for other factors are
explained with the results.

The marsh plants *Typha capensis* and *Phragmites
australis* were sampled and analysed by Dr R.W. Summers of the Percy Fitzpatrick Institute of African
Ornithology. The collection dates and the standing

crop are given in Table 7.1. The stations where the standing crop was monitored are shown in Figure 7.1 as T *(Typha)* and P *(Phragmites)*. The dry mass of the living material was obtained after removal from a 1 m^2 quadrat. The organic content and energy values in kJoules were calculated for the January sample only.

In the field, growth measurements of individual plants of *Spartina, Sarcocornia* and *Scirpus* were obtained by placing a numbered tag around the base of the shoots. During subsequent trips the shoots were remeasured. This method was tried for *Zostera* by placing the tag around the rhizome but this proved impractical due to continual loss of tags. Also the plants were disturbed by the planting of the tags. The methods of Zieman (1968) and Sand-Jensen (1975), ie of some form of clipping or marking also proved impractical because the *Zostera* found in Langebaan lagoon is extremely small. The mean leaf-widths for the small strain were 2 mm while values for the large strain were 3-4 mm. Only values for standing crop were obtained.

Aerial photography conducted during 1975 was used to obtain detailed maps of the lagoon and surrounding marsh vegetation. The surface areas covered by individual species were calculated. These figures were used in conjunction with net productivity values at each station to give net production values for the whole lagoon.

In addition, the air and water temperatures were recorded and surface water samples taken. Some of the water samples were immediately cooled and frozen in a portable freezer for later analysis of nitrate (Strickland & Parsons, 1960), nitrite and silicate (Strickland & Parsons, 1968), soluble reactive phosphate or SRP (Murphy & Riley, 1962) and total phosphorus (Menzel & Corwin, 1965). Samples were also taken for chlorophyll analysis (Strickland & Parsons, 1968) and salinity analysis.

During the March sampling trip, sediment from the habitat of each species was obtained for analysis of texture and organic carbon content (Morgans, 1956).

Table 7.1 gives the collection dates for the lagoon samples which were undertaken using a small boat. Station positions 7-14 are also shown in Figure 7.1.

Phytoplankton productivity was measured at each station using the light and dark bottle technique (Gaarder & Gran, 1927) and the Winkler analysis for the determination of oxygen. Measurements were made at depths of 1, 2 and where possible, 3 m by suspending the bottles from interconnecting transparent perspex rods. The water for the bottles was collected from the appropriate depth by a diver using a reversing bottle. Two bottles at each depth were fixed immediately to determine the original oxygen content of the water. The remaining bottles, ie two light and two dark, were left in situ for two to three hours. Stations were worked in pairs and the eight stations were completed in two days, weather permitting, otherwise an extra

day was required. In calculating productivity over a daily 24 hour basis, allowances were made for daylight intensity variations. Daylight readings were obtained from the Met Office at D.F. Malan Airport.

Water samples were analysed in the same way as the marsh water samples. The samples were all taken within one hour of each other starting from station 14 and working towards the lagoon entrance.

The dominant sublittoral macrophyte in the lagoon is *Gracilaria verrucosa*. This is a red alga which is commercially important for the extraction of agar. The standing crop of *Gracilaria* was obtained at each station during trips in December 1975 and August 1976, by using 2 x ¼ m^2 quadrats. The samples were treated in the same way as those of marsh plants to give the ash-free mass and energy content. Gross production was calculated by using the light and dark bottle technique with small pieces of *Gracilaria* inside the light bottles. The amount of oxygen evolved by the plants was converted into carbon using the conversion ratio of 1 ml O_2 = 0,536 mg.C. This was again converted into dry grams of plant material using a ratio of 26 % as given by Lapointe *et al* (1976).

The lagoon was divided into several cross-sectional areas (Figure 7.1), and from the aerial photographs, the amount of *Gracilaria* in hectares in each area was determined. Using the above data, production rates were calculated.

ENVIRONMENTAL CONDITIONS

The lagoon is quite shallow with an average depth of 1-2 m and a maximum depth of 6 m (Flemming, 1977). Since it was never an estuary, merely an enclosed arm of the sea with negligible fresh water inflow (normal rainfall 253,1-269,9 mm per year), all the sediments and nutrients are derived from the sea. No silt or clay was noted by Flemming and this was confirmed in the present study. The finer grades of sand are to be found on the western side of the lagoon which has the weakest water currents.

During tidal exchange, half of the lagoon water passes into Saldanha Bay (Shannon & Stander, 1977). This means that the richer nitrate and phytoplankton water of the bay moves in and out of the lagoon and values at Langebaan village vary with the state of the tide.

Figure 7.2 shows the monthly concentration of total phosphorus in each section of the lagoon. Some months have been omitted for clarity, but they do not alter the general trend. The concentration of SRP was generally 80-90 % of the total phosphorus value. In all months, the concentrations increased towards the southern end of the lagoon at sections E and F where they reached a maximum in October and January possibly due to the release of phosphate from *Zostera* and the decay of detritus. At the lagoon

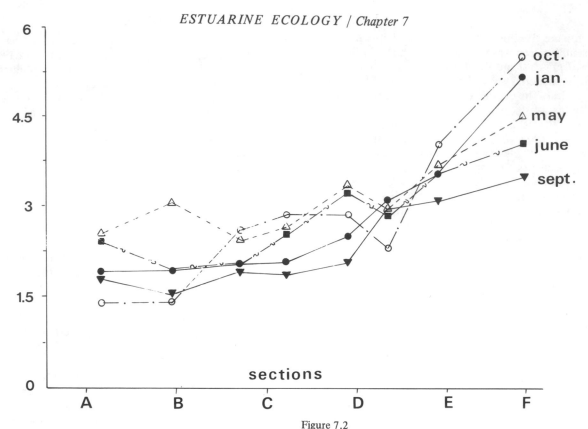

Figure 7.2
Concentration of total phosphorus (μg at. P.ℓ^{-1}) in lagoon water.

entrance (section B), values were low in September and October, possibly due to uptake by *Gracilaria* and phytoplankton and were lowest in May at the approach of winter.

An account of phosphorus recycling by *Zostera* is given by Barsdate & Nebert (1972) and by *Spartina* by Woodwell & Whitney (1977).

The distribution of nitrate nitrogen in the lagoon shown in Figure 7.3 revealed an opposite pattern to phosphate and decreased towards the southern end of the lagoon. At the entrance, the concentration in January was 10,15 μg at.ℓ^{-1} on a flood tide and in March it was 4,0 μg at.ℓ^{-1} on an ebb tide. This corresponds with the inflow of upwelled nutrients from the cold Benguela current in summer.

However, the nitrate in January (mid-summer) is utilised at a greater rate than in other months and its January value in the southern half of the lagoon is the lowest for the year. Summer is the peak production time for phytoplankton and *Gracilaria* which remove nitrate rapidly from the lagoon. This will be discussed in greater detail later. In contrast, during winter when production is low the concentration of nitrate hardly alters down the length of the lagoon.

The concentration of silicate in the lagoon in general is less than that found in the marsh pools. Concentrations of silicate in lagoon water have not been tabulated here but in summer there is an increase from station 8 at the lagoon mouth (January 18,28 μg at ℓ^{-1}, March 18,99 μg at.ℓ^{-1}) to station 14 at the southern end

Table 7.2
Nutrient concentrations (μg at.ℓ^{-1}) in salt marsh pools in June and January

Lagoon section	A	C	D	E	F	F
Shore station	6	1	5	2	3	4
June 1975						
Total P	3,65	4,68	4,30	4,95	6,08	5,43
SRP-P	2,85	4,65	3,81	4,67	5,72	4,99
NO$_3$-N	0,11	0,12	0,44	0,47	4,05	0,70
Silicate-Si	24,18	30,77	23,08	35,82	58,46	58,24
January 1976						
Total P	3,10	2,47	2,23	3,15	9,52	5,58
SRP-P	2,52	2,05	1,10	2,28	8,48	4,82
NO$_3$-N	1,29	0,40	0,0	0,28	0,18	0,70
Silicate-Si	31,82	23,42	24,31	30,72	70,94	64,09

Table 7.3
Water temperatures in the lagoon and salt marsh pools (records from lagoon stations 9 and 12 omitted)

Lagoon section	A	B	C	D	E	F
Lagoon station	7	8	10	11	13	14
Lagoon water						
Oct. 1975	16,5	15,6	18,0	19,2	20,5	22,5
Jan. 1976	21,0	14,9	18,0	21,0	22,2	24,0
Mar.	19,8	17,5	18,1	21,2	21,0	21,0
May	14,7	15,4	13,5	18,3	18,3	18,5
Aug.	14,8	14,0	13,5	13,5	14,3	14,8
Sept.	15,5	15,1	15,9	16,0	16,0	16,8
Marsh water						
Shore station	6	—	1	5	2	3,4
Jan.	24,0	—	26,3	27,2	30,5	26,1
Aug.	21,5	—	16,0	19,0	15,3	20,7

where it reaches 51,61 μg at.ℓ^{-1} during January. Winter (June) levels are virtually constant down the length of the lagoon with a mean of 35 μg at.ℓ^{-1}. Summer values are less than those in winter probably due to uptake and utilisation by phytoplankton.

The nutrients present in the marsh water for June and January are given in Table 7.2. The level of nutrients was generally high with the exception of nitrate. Station 3 at the south-east corner of the lagoon always had the highest concentration of SRP and total phosphorus. These values were highest during the summer months of November to March (SRP 9,14-8,48 μg at.P. ℓ^{-1}. This part of the lagoon has the greatest concentration of *Zostera* and *Spartina* which release phosphate to the surrounding water. During summer, when growth and decay is fastest, phosphate levels reach their highest.

Nitrate concentrations in salt marsh pools show the opposite pattern. During June (winter) when plant growth was at its slowest, the maximum value of 4,05 μg at $NO_3-N.\ell^{-1}$ was recorded at station 3. With the approach of summer, values fell progressively due to utilisation of the nitrate and then increased again during autumn. Monthly values (not recorded in Table 7.2) were: August 2,37; September 2,02; November 0,35; January 0,18; March 1,00 and May 1,67. Table 7.2 shows the January value. Values at other marsh stations were slightly higher during January with a maximum of 1,20 μg at.ℓ^{-1} at station 6 and a minimum of 0.0 μg at. ℓ^{-1} at station D.

A summary of water temperatures in five sections of the open lagoon and in salt marsh pools in three sections of the lagoon is given in Table 7.3.

The water in the shallow lagoon is much warmer than the water of Saldanha Bay in summer. Temperature records at the lagoon entrance (section B) taken at various states of the tide show lower temperatures during the flood tide and higher temperatures during the ebb. The January record of 14,9 ° shows cool bay water entering the lagoon and sections C, D and F show how the water temperature increases to 24,0 ° at the southern end of the lagoon. In March the temperature range is reduced with 17,5 ° at the entrance and 21,0 ° at the southern end. In winter (August), there is little difference between the bay water and that of the lagoon although the lagoon water is slightly cooler. Thus the bay water in section B is 14,0 ° and the lagoon water in sections C and D are 13,5 °. The records at station 7 in lagoon section A show the temperatures in Riet Bay, a blind arm of the lagoon near the entrance. The bay is shallow and solar radiation raises the temperature above that in the entrance to the lagoon so that the range from summer to winter is 21,0 to 14,7 °.

The temperatures in salt marsh pools during the day are naturally higher than those in the open lagoon. In section D, the January temperature in the salt marsh was 30,5 ° compared to 21,0 ° in the lagoon. Even in winter (August) the marsh temperature was 15,3 ° as compared to 13,5 ° in the lagoon. The higher marsh temperatures would increase the rate of growth of the plants and the rate of bacterial decay of the detritus.

Salinities in the lagoon are given in Figure 7.4. Salinities were virtually constant in Section B at the lagoon mouth ranging from 34,5 to 35,0 ‰. In section F at the southern end of the lagoon, October to January recorded a steady increase to a maximum of 38,04 ‰ due to evaporation. Winter (August) recorded a fall to 33,92 ‰ in the same section due to inflow of fresh water from rain. Records of salinity in the salt marshes have not been included in this paper but all marsh stations except station 1 in Riet Bay showed a marked increase in salinity from winter to summer with the highest values in excess of 43 ‰ in January. This is directly correlated with water temperature and solar radiation.

CHLOROPHYLL AND PHYTOPLANKTON

The concentrations of chlorophyll a in the lagoon surface water are given in Table 7.4. Mere traces of chlorophyll b were found during some months and these are not recorded.

Section B at the entrance, generally had the highest

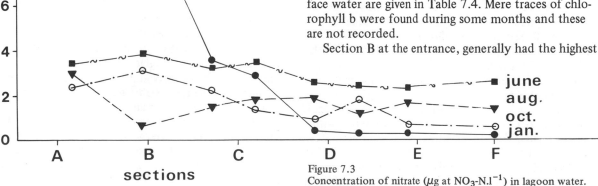

Figure 7.3
Concentration of nitrate (μg at $NO_3-N.l^{-1}$) in lagoon water.

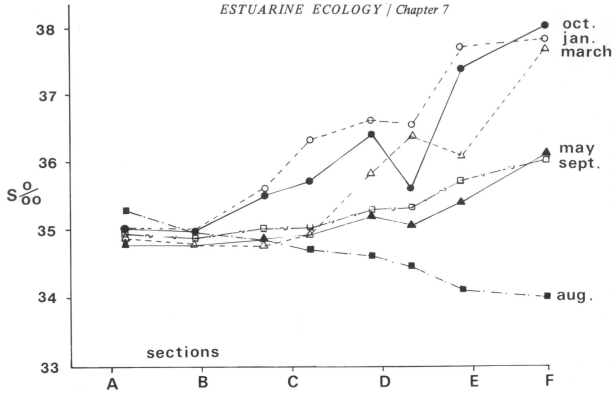

Figure 7.4
Salinity of lagoon water (‰).

Table 7.4
Chlorophyll a values ($\mu g.\ell^{-1}$) in the surface water of the lagoon

Lagoon section	A	B	C	D	E	F
Lagoon stations	7	8,9	10,11	12	13	14
Oct.	0,68	5,07	1,23	2,44	1,26	2,05
Jan.	0,78	2,00	2,98	1,28	2,41	2,26
Mar.	1,71	4,63	2,73	2,39	4,16	2,10
May	0,81	9,48	4,27	5,94	2,49	1,96
Aug.	0,52	2,42	1,50	1,26	1,29	1,31
Sept.	0,94	2,33	1,34	2,02	1,40	1,72

Table 7.5
Gross production of phytoplankton in $mgC.m^{-3}.day^{-1}$ as the mean of samples from 1, 2 and 3 m in each lagoon section throughout the year

Lagoon section	A	B	C	D	E	F
Lagoon station	7	8, 9	10, 11	12	13	14
Oct.	32	195	154	131	108	77
Jan.	43	314	300	171	63	46
Mar.	37	255	193	189	133	141
May	31	161	121	101	76	52
Aug.	34	103	97	56	67	41
Sept.	57	130	100	93	64	56

values with values declining towards the southern end of the lagoon. Section A in Riet Bay always had the lowest values. Although the concentration of nutrients is similar in sections A and B as shown by Figures 7.2 and 7.3, there is little circulation in Riet Bay while the rich phytoplankton in Saldanha Bay flows in and out of section B. Second, the sediment in Riet Bay is mainly angular quartz grains (Flemming, 1977) and the reflection of the sun from these grains in the shallow water with the maximum depth of 3 m may cause an inhibition effect.

Winter values of chlorophyll a in the lagoon (August and September) as would be expected, were low with a much smaller gradient in winter than in summer down the length of the lagoon.

Chlorophyll values were also recorded in marsh water but have not been reproduced here. No chloro-phyll b was found. Chlorophyll a values were highest in winter and autumn, the highest value being 7,30 μg. ℓ^{-1} in August in the salt marsh at station 3 in lagoon section F, where there is seepage of fresh water. It is suspected that the chlorophyll values are contaminated by decay products of marsh plants after the peak standing crop in summer.

The gross production of phytoplankton in lagoon water as $mgC.m^{-3}.d^{-1}$ is given in Table 7.5 where a general pattern is observed. Gross production values are means of the results obtained at each sampling depth. On sunny days, results from the 2 or 3 m depth range were higher than at 1 m while cloudy days showed the reverse pattern. Station 7 in section A (Riet Bay) consistently had the lowest values which correlates with the low concentration of chlorophyll a (Table 7.4).

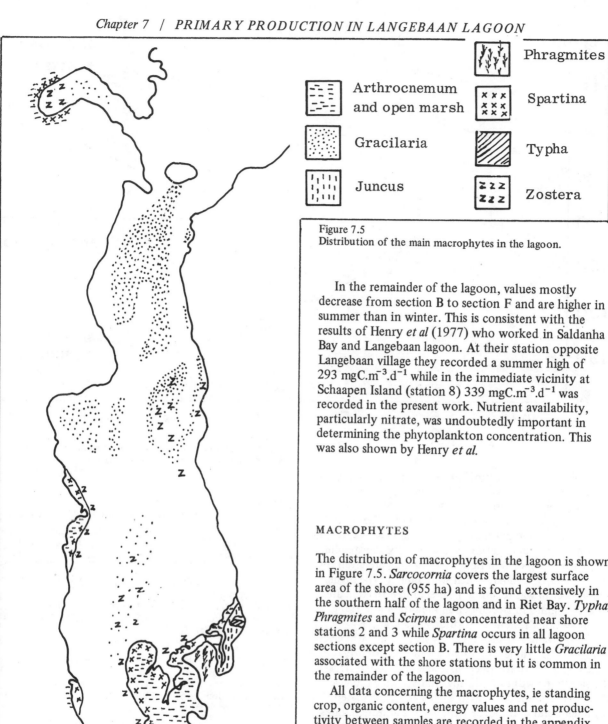

		Phragmites
	Arthrocnemum and open marsh	Spartina
	Gracilaria	Typha
	Juncus	Zostera

Figure 7.5
Distribution of the main macrophytes in the lagoon.

In the remainder of the lagoon, values mostly decrease from section B to section F and are higher in summer than in winter. This is consistent with the results of Henry *et al* (1977) who worked in Saldanha Bay and Langebaan lagoon. At their station opposite Langebaan village they recorded a summer high of 293 mgC.m^{-3}.d^{-1} while in the immediate vicinity at Schaapen Island (station 8) 339 mgC.m^{-3}.d^{-1} was recorded in the present work. Nutrient availability, particularly nitrate, was undoubtedly important in determining the phytoplankton concentration. This was also shown by Henry *et al.*

MACROPHYTES

The distribution of macrophytes in the lagoon is shown in Figure 7.5. *Sarcocornia* covers the largest surface area of the shore (955 ha) and is found extensively in the southern half of the lagoon and in Riet Bay. *Typha*, *Phragmites* and *Scirpus* are concentrated near shore stations 2 and 3 while *Spartina* occurs in all lagoon sections except section B. There is very little *Gracilaria* associated with the shore stations but it is common in the remainder of the lagoon.

All data concerning the macrophytes, ie standing crop, organic content, energy values and net productivity between samples are recorded in the appendix. This is not given here but is filed in the Zoology Department of the University of Cape Town. The important details are extracted and discussed in this report.

Gracilaria verrucosa

Gracilaria is mainly found in the open lagoon away from the marshes and occurs extensively in section B where 162 hectares were recorded. It is the dominant sublittoral macrophyte although large areas are also exposed at low spring tides.

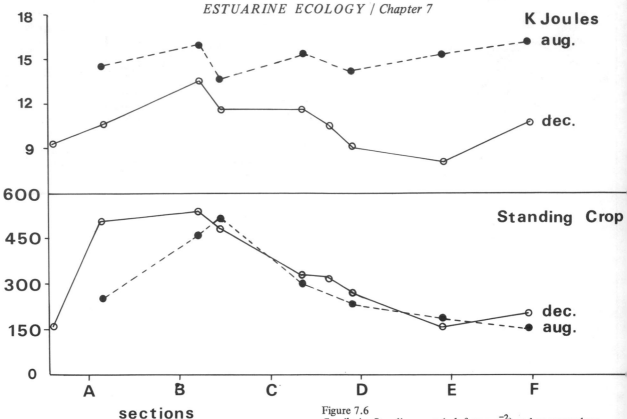

Figure 7.6
Gracilaria: Standing crop (ash free g.m^{-2}) and energy values (kJoules per gram (dry mass)$^{-1}$ in each lagoon section.

Table 7.6 gives the number of hectares of *Gracilaria* in each cross-sectional area. The amount present declines from the north to the south of the lagoon where only 0,1 ha was found in section F. The total area in the lagoon covered by *Gracilaria* was 342,625 ha.

A corresponding pattern was shown by the standing crop (ash free g.m^{-2}) in figure 7.6. Results are given for August and December, which were similar. The highest standing crop was recorded near station 8 (lagoon area B) at 543 g.m^{-2} in December and fell to below 200 g. m^{-2} at stations 12 and 13 (lagoon areas E and F).

Energy values as kJoules g(dry mass)$^{-1}$ are also given in Figure 7.6. Values for August were consistently higher than those for December and probably reflect the slower growth rates during winter. Table 7.6 gives the gross production rates at each station calculated from light and dark bottle experiments and these rates again reflect a similar pattern to standing crop and energy values. Production rates are gross in comparison to the net values given for other species where grazing and plant decay was considered to lead to a loss of material. This was not applicable in the case of *Gracilaria* where total production per unit of time was calculated.

Gross production per month (Table 7.6) as a percentage of standing crop was highest in the northern half of the lagoon at stations 7-10 for both August and December. The maximum production of 98,6 % per

Table 7.6
Gross production of *Gracilaria* in each section of the lagoon in August and December

Lagoon section	A	B	C	D	E	F
Lagoon station	7	8,9	10	11	13	14
August						
Gross production (% standing crop)	60,2	79,5	80,8	63,2	52,4	30,1
% organic matter	75,6	80,1	74,3	72,6	68,3	56,6
Energy production (10^3 kJ.m^{-2} month^{-1})	2,78	7,82	5,03	2,72	2,21	1,28
December						
Gross production (% standing crop)	66,7	88,4	62,4	40,4	30,8	18,5
% organic matter	56,1	51,9	55,4	56,8	60,4	61,4
Energy production (10^3 kJ.m^{-2} month^{-1})	6,19	11,4	3,32	1,71	0,63	0,65
Area of *Gracillaria* (ha)	18,1	161,5	130,6	29,5	2,9	0,1
Energy production in August per lagoon section (10^9 kJ.month^{-1})	0,5	12,64	6,57	0,80	0,06	0,01
Energy production in December per lagoon section (10^9 kJ.month^{-1})	1,12	18,43	4,34	0,50	0,02	0,01
Total production for lagoon (10^9 kJ.month^{-1})	August – 20,57			December – 24,41		
	Total for year = 269,88 (10^9 kJ)					

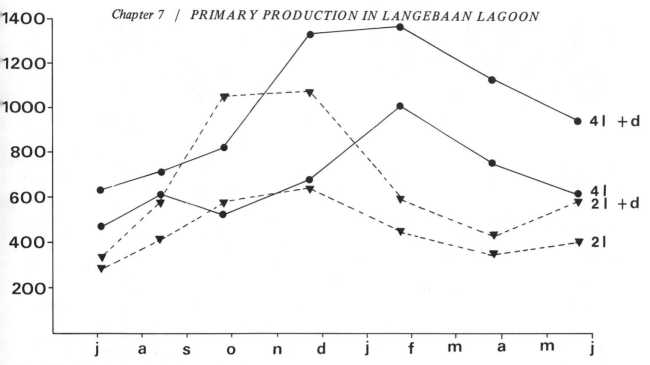

Figure 7.7
Spartina: Monthly standing crop of tall shoots (ash free g.m^{-2}) at stations 2 and 4 for live material (l) and live plus dead material (l + d).

Table 7.7
Spartina: standing crop of live and dead shoots in ash free g. m^{-2} also annual net production in 10^3 kJoules of the maximum standing crop of live shoots and annual net production of live, live + dead and live + dead + loss per lagoon section in 10^9 kJoules

Lagoon section	A	D	E	F
Shore station	6	5	2	4
Maximum standing crop g ash free g.m^{-2}				
Tall live shoots	711	661	649	1 016
Tall dead shoots	404	250	476	660
Short live shoots	335	272	339	366
Short dead shoots	47	199	222	251
Mean standing crop of live shoots g ash free g.m^{-2}				
Tall shoots	440	398	436	683
Short shoots	224	203	241	250
Annual net production per m^2 of maximum standing crop (10^3 kJoules)				
Tall live shoots	15,16	13,84	14,61	20,08
Short live shoots	6,02	5,59	7,88	7,52
Annual net production per lagoon section (10^9 kJoules)				
Shore area of	2,75	31,87	173,96	48,75
Spartina (ha)				
Tall live shoots	0,21	2,21	12,70	4,89
Tall live + dead	0,33	3,04	22,0	8,06
Tall live + dead + loss	0,55	5,75	33,02	12,71
Short live shoots	0,08	0,89	6,85	1,83
Short live + dead	0,09	1,54	11,34	3,08
Short live + dead + loss	0,14	1,56	11,65	3,11

Totals: live = 29,66; live + dead = 49,48;
live + dead + loss = 68,44

month occurred at station 8 in December.

The energy production for August and December in each lagoon cross-sectional area, and the approximate mean for the year are also given. Most of the production occurred in section B with 12,64 and 18,43 x 10^9 k-Joules for August and December respectively. Production at stations 11, 13 and 14 in the southern lagoon and at station 7 in Riet Bay were negligible. The annual total was estimated at 269,88 x 10^9 kJ and was calculated as six times the winter value in August, plus six times the summer value in December. The availability of nutrients, particularly nitrate, appears to be one of the most important growth regulating factors. The area covered decreased from the lagoon mouth to the southern end (Table 7.6). Section E has, on average, a higher water temperature which is more favourable for growth, as confirmed by unpublished culture experiments by the author and Mr S.G. Pheiffer. Water depths in the different sections where *Gracilaria* grows are comparable but nitrate values show great variation (Figure 7.3).

That nitrate was a limiting factor was supported by the colour of the *Gracilaria* which in sections E and F, was yellowish. Near Schaapen Island with maximum available nitrate, the *Gracilaria* was almost black. This colour steadily changed down the lagoon through brown and brown-green to yellow.

The nitrate availability also explains the observed production rates. In summer, when the highest nitrate concentration enters the lagoon, *Gracilaria* production is maximal in section B. By the time the water reaches the south, phytoplankton plus the attached macrophytes have depleted the nitrate so that *Gracilaria* production is much lower. The reverse occurs in winter. At the southern end of the lagoon *Gracilaria* production is higher in winter than it is in summer, although it is still much lower than at the northern entrance. This is probably due to the presence of nitrate released from decaying plant material in the marshes.

Spartina maritima

The monthly values of the standing crop of the tall strains of live, and live plus dead *Spartina* at stations 2 and 4 (sections E and F) are shown in Figure 7.7 in terms of g (ash free) m^{-2}. Peaks in the two components coincide fairly closely. The tall strain of *Spartina* recorded a maximum live standing crop of 1 016 g (ash free).m^{-2} at station 4 during January. The peak for all stations was in summer between November and January. Similarly the maximum of 366 g (ash free) m^{-2} for the short strain was recorded at station 4 in January.

The mean standing crop is also given in Table 7.7 for both strains. Again the highest values were recorded at station 4 indicating that growth conditions were most favourable at the station. However, new shoots appear throughout the year. A comparison of 50 tall shoots from each station collected in June, supports the observation that station 4 has the highest biomass. The mean length of the shoots from station 4 was 90 cm and the ash free mass was 1,55 g. The shoots from stations 2, 5 and 6 were shorter (50-70 cm) and at equivalent lengths, they were slightly lighter (0,5-1,0 ash free g). Shoots from station 2 were the lightest but labelled shoots showed that they grew slightly faster.

The values for standing crop of *Spartina* may be compared with values given by other workers. Kirby & Gosselink (1976) reported a peak live standing crop of 1 018 g (dry mass).m^{-2} in a Louisiana marsh. Nixon & Oviatt (1973) on Rhode Island gave a mean standing crop of 433-1 383 g (dry mass).m^{-2} while Udell *et al* (1969) gave 664-111 g (dry mass).m^{-2} for all *Spartina* on a Long Island marsh. A mean standing crop of 387 g (dry mass).m^{-2} in Mississippi marshes was given by Gabriel & de la Cruz (1974). To compare these results, the standing crop at Langebaan lagoon must be converted into grams dry mass. The mean percentage of organic matter is 77,0 and 76,0 for tall and short strain, respectively. Using these data with the mean ash free g.m^{-2} values of live plus dead material at each station, the lagoon mean annual standing crop is 635 g (dry mass).m^{-2} for tall *Spartina*. The corresponding value for short *Spartina* is 302 g (dry mass).m^{-2}.

Net production of live shoots was calculated in kJ. m^{-2}. Important components of these calculations were the organic contents and energy values which fluctuated throughout the year (Appendix). The highest organic content of approximately 80 % was found in winter with values falling to nearly 70 % during summer.

Net production was calculated by Milner & Hughes (1968) as the difference between maximum and minimum standing crops. However, since the vast majority of the shoots do not live more than a year, the minimum standing crop value will also turn over and a truer reflection of production would be given by using the maximum standing crop only. These values are given in Table 7.7 for annual production per m^2 and per shore region.

Kirby & Gosselink (1976) compare various methods of measuring production. In addition to the Milner & Hughes (1968) method, they employed the methods of Smalley (1959) and Wiegert & Evans (1964). Smalley's method makes allowance for dead plant material to which is added the production of live plants. The method of Wiegert & Evans accounts for all losses except that by grazing. Dead plants and pieces of dead plants are lost between sampling trips. Assuming no transport of undecayed leaves and shoots from the marsh, an estimation of this loss can be obtained by measuring the loss of similar material kept in gauze bags in situ on the marsh. The loss rate, multiplied by the mean dead standing crop gives the total loss which is added to the live and dead standing crops.

Kirby & Gosselink gave peak live standing crops of 1 018 and 788 g (dry mass).m^{-2} for tall and short *Spartina alterniflora* respectively. Using the Wiegert & Evans method, their annual production was 2 645 and 1 323 g (dry mass).m^{-2} for the tall and short strain respectively. These figures gave ratios of production to peak standing crop of 2,6 for the tall strain and 1,7 for the short strain.

The Langebaan lagoon samples of *Spartina maritima* were separated into living and dead material. The percentage of dead to live material ranged from 14,0 % for the short strain at station 6, to 73,3 % for the large strain at station 2. This dead material makes a considerable contribution to the detritus and therefore to the productivity of the lagoon. The kJoules produced by the live plus dead material for each shore region is indicated in Table 7.7.

In the present study, the loss rate of dead material between sampling trips was not determined. Use is therefore made of the ratios calculated by Kirby & Gosselink of actual net production to peak live standing crop as previously given. This calculation increases the mean productivity estimate by 38,32 %. It is interesting that the increases for the tall strain are all in excess of 50 % while the increases for the short strain are much less. The large additional values for the tall strain are due to the fact that they experience greater water movement than the short strain, and hence a greater loss of material between sampling trips.

As a mean, dead material accounted for 29 % of the contribution of *Spartina* to the organic material in the lagoon. Similarly the loss factor was 27 %. The final estimate for the total net production of *Spartina* in the whole lagoon was 68,44 x 10^9 kJ.

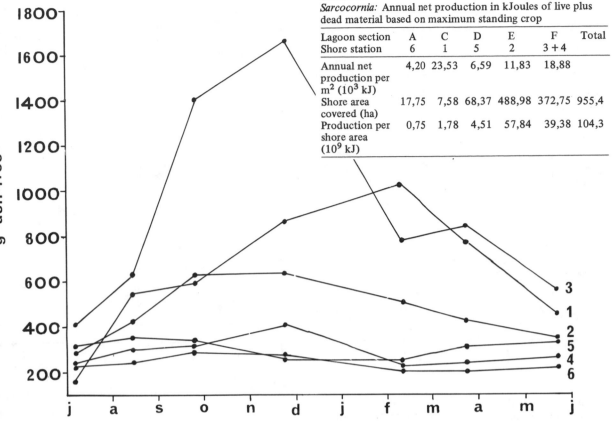

Table 7.8
Sarcocornia: Annual net production in kJoules of live plus dead material based on maximum standing crop

Lagoon section	A	C	D	E	F	Total
Shore station	6	1	5	2	3 + 4	
Annual net production per m^2 (10^3 kJ)	4,20	23,53	6,59	11,83	18,88	
Shore area covered (ha)	17,75	7,58	68,37	488,98	372,75	955,4
Production per shore area (10^9 kJ)	0,75	1,78	4,51	57,84	39,38	104,3

Figure 7.8
Sarcocornia: Monthly standing crop (ash free g.m^{-2}) at each shore station.

Sarcocornia perennis

The monthly standing crop in ash free g.m^{-2} is shown in Figure 7.8. No distinction was made between live or dead material since this was not practical for this species. The highest standing crop of 1 651 g.m^{-2} was found at station 3 during November. At station 1 at the northern end of the lagoon there was a peak of 1 014 g.m^{-2} but the area of *Sarcocornia* there was only 7,58 ha (Table 7.8). Large seasonal differences were observed at stations 1-3, whereas the standing crop at the remaining stations was more constant.

The net production was measured in the same way as *Spartina* and annual values for each lagoon section are given in Table 7.8. The energy values were similar to *Spartina* with a mean of 13,98 x 10^3 kJ.m^{-2}. Minimum production was at stations 4 and 6, which had values of 4,20 and 6,37 respectively.

Labelling of individual shoots indicated that constant mass was maintained although the shoot had increased in length. Shoots are composed of bulbous segments which narrow towards the base while new ones are produced at the apex. Production was therefore much higher than that calculated purely from har-

vesting. The value of 104,3 x 10^9 kJ can be doubled since this was the mean ratio of production to standing crop calculated from labelling. The production rate of *Sarcocornia* (208,6 x 10^9 kJ) for the whole lagoon is therefore very similar to that of *Spartina* since the total production for *Spartina* was 33 % of the value for *Sarcocornia* while occupying only 26,9 % of the surface area of the latter.

Scirpus triqueter

This species was only found in the south-east of the lagoon at station 3. Shoots reached a maximum height of 1,5 m in January and, like *Spartina,* new shoots appeared throughout the year.

The shoots when mature start to die and turn brown from the apex towards the base in a linear relationship between live and dead material. Often dead material had broken off before the entire plant died and this material must be included in productivity calculations.

Figure 7.9 shows the monthly standing crop of live and dead material. Live material reached a peak of 1 348 g (ash free).m^{-2} in November from a low of 212 g in June. Dead material also peaked in November at 908 g (ash free).m^{-2}. The large loss observed in August, was probably due to strong winter storms.

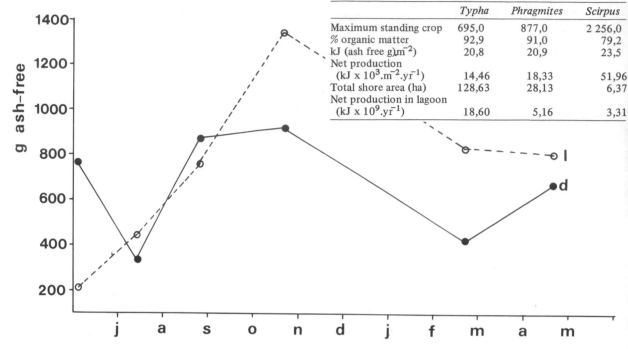

Table 7.9
Standing crop (ash free g.m^{-2}), energy data and net production for live *Typha* and *Phragmites* and live plus dead *Scirpus*

	Typha	*Phragmites*	*Scirpus*
Maximum standing crop	695,0	877,0	2 256,0
% organic matter	92,9	91,0	79,2
kJ (ash free g).m^{-2})	20,8	20,9	23,5
Net production (kJ x 10^3.m^{-2}.yr^{-1})	14,46	18,33	51,96
Total shore area (ha)	128,63	28,13	6,37
Net production in lagoon (kJ x 10^9.yr^{-1})	18,60	5,16	3,31

Figure 7.9
Scirpus: Monthly standing crop (ash free g.m^{-2}) for live (l) and dead material (d) at station 3.

Table 7.9 gives the productivity calculations. The mean energy content of 23,5 kJ per ash free gram of living material was the highest recorded for any marsh plant. Annual net production for *Scirpus* in an area of 6,37 hectares was 1,98, 1,33 and 3,31 kJ x 10^9 for live, dead and total material, respectively. The true figure is probably somewhat higher, as no allowance was made for loss of dead material between sampling trips.

Typha capensis and *Phragmites australis* (data from Summers)

These species are the largest in the marshes and are found only in marshes near stations 2 and 3 where there is fresh water seepage. They have mean lengths during December of 311 cm *(Phragmites)* and 166 cm *(Typha).* In comparison, *Scirpus* only reaches 150 cm and the tallest *Spartina* is just over 55 cm.

Typha and *Phragmites* both have a clearly defined growth pattern. The first shoots appear in June and reach peak standing crop in February and December, respectively. The standing crop for each sampling trip is given in Table 7.1. As stated previously, the organic matter ratio and energy values were only obtained for January and these have been used in calculating net production per m^2 and annual production per shore region shown in Table 7.9. Production was 14,46 and 18,33 kJ x 10^3.m^{-2}.yr^{-1} for *Typha* and *Phragmites* respectively, and therefore much lower

than for *Scirpus*. This was due to the fact that although the plants were much taller than *Scirpus,* their densities per m^2 were lower. Mean plant densities were *Typha* 19, *Phragmites* 22 and *Scirpus* 900.

Mason & Bryant (1975) made allowances for plants which disappeared between sampling trips, ie as the total standing crop increased during the growing season the number of plants decreased. No such allowances were made for the Langebaan samples since the number of plants either increased or remained stable.

Mason & Bryant report a net productivity for *Phragmites communis* in the Norfolk Broads of 1 080 g (dry mass).m^{-2} in 1972 and 551 g (dry mass).m^{-2} in 1973. These values were 14,6 and 5,1 % greater than the peak standing crop. Similar values for *Typha angustifolia* were 1 445 and 1 515 g (dry mass).m^{-2}. Except for the 1973 *Phragmites* values, these values are higher than those found in Langebaan lagoon.

Zostera capensis

The general distribution of *Zostera* and other macrophytes is indicated in Figure 7.5, while Table 7.10 gives the area in hectares covered by *Zostera*, both on intertidal shores, and submerged sandbanks. As will be seen, the maximum shore cover (71,3 ha) is in section E and the whole shore area of *Zostera* is

112

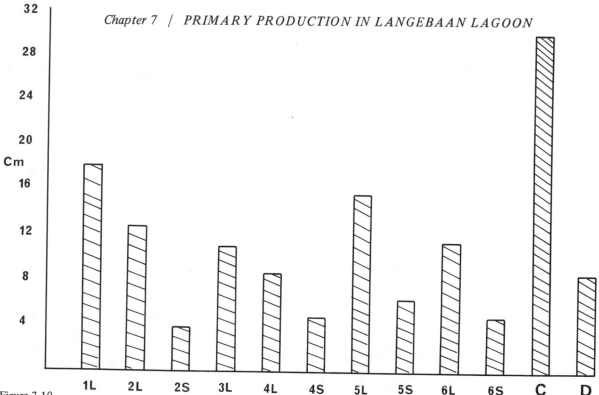

Figure 7.10
Zostera: Mean length (cm) of the main leaves of small (s) and
large (l) plants at shore stations 1-6 and sandbanks in lagoon
sections C and D.

Table 7.10
Annual net production of *Zostera* in 10^9 kJoules on the shores
and sandbanks in each lagoon section based on maximum standing crop of leaves.

Key: (1) Lagoon section and shore station number
     (2) Leaf size
     (3) Area of *Zostera* (ha)
     (4) Maximum standing crop ash free $g.m^{-2}$
     (5) Annual net production (10^3 kJoules.m^{-2})
     (6) Annual net production per *Zostera* area
        (10^9 kJoules)

(1)	(2)	(3)	(4)	(5)	(6)
Intertidal shores					
A sta. 6	large	1,05	340	7,11	0,07
	small	4,2	108	1,05	0,04
C sta. 1	large	4,625	190	4,00	0,19
D sta. 5	large	0,2	170	2,07	<0,01
	small	0,8	32	0,35	<0,01
E sta. 2	large	14,275	119	1,82	0,26
	small	57,10	94	1,41	0,81
F sta. 3	small	22,625	151	2,31	0,52
F sta. 4	large	4,325	299	4,07	0,18
	small	17,30	83	1,18	0,20
	Total	129,35		Total	2,28
Infratidal sandbanks					
A		–	–	–	–
B		4,875	36,4	0,65	0,03
C		49,375	36,4	0,65	0,32
D		17,375	131,84	2,18	0,38
E		19,125	131,84	2,18	0,42
F		54,875	151	1,81	1,02
	Total	146,15		Total	2,17

129,35 ha. The largest area on sandbanks is in section
F (54,9 ha) and the area on all sandbanks is 146 ha,
so that the total area in the lagoon covered by *Zostera*
is 275 ha. It tends to be concentrated at the southern
end of the lagoon where the ground is very flat and
there is a good deal of seepage water.

The leaves arising from the rhizomes in the sediment, are liable to be abraded by waves on exposed
shores so that different areas on the shore have large
or small leaves and even the leaves on submerged banks
vary, possibly due to the strength of the tidal currents
and the fertility of the sediment. The longest leaves on
the shores were recorded in sheltered bays (eg section
A) while those on submerged banks were longest in
section C (see Figure 7.10). The mean width of large
leaves was 3-4 mm while the small leaves averaged 2 mm.

The maximum standing crop for large and small
leaves in each section of the lagoon is given in Table
7.10. The maximum standing crop of the large-leaf
strain on the shore, was at station 6 where 340 g (ash
free).m^{-2} was recorded in January. The values on the
sandbanks were respectively 151 in section F, 131,84
in sections D and E and 36,4 in sections B and C. As
may be seen, the standing crop increased in more sheltered areas of the lagoon such as Riet Bay (section A)
and near Geelbek (section F) at the southern end of
the lagoon.

The leaves of *Zostera* are belts of living tissue and
as they are continually being abraded and lost at the
tip an estimate of production based on the weight of
the living leaves alone would be a severe under-estimate.

113

As mentioned earlier, an attempt to mark plants by labelling the rhizomes so that the growth rates of their leaves could be determined was unsuccessful. The annual net production was thus based on the maximum standing crop data shown in Table 7.10 and an allowance was made for loss by abrasion.

Sand-Jensen (1975) in Denmark, was able to mark a line on individual leaves chemically and thus to determine the growth rate accurately. He found that the maximum leaf production of 7,9 g (dry mass).m^{-2} day^{-1} occurred in mid-June and coincided with the period of maximum radiation. Net production per annum was 3,8 times the maximum leaf standing crop of 226 g (dry mass).m^{-2}. This ratio of 3,8 was applied to *Zostera capensis* in Langebaan although the growth rates could well be higher than for *Zostera marina* in Denmark. The annual net production based on the weight of living leaves per m^2 is shown in Table 7.10 in units of 10^3 kJ and for each lagoon section, it is shown in units of 10^9 kJ. The annual net production for all intertidal shores is 2,28 kJ x 10^9 and that for submerged sandbanks is 2,17 x 10^9 kJ. Sand-Jensen's loss factor of 3,8 would give values of 8,86 and 8,25 and on this basis the annual net production for the whole lagoon is 17,11 kJ x 10^9. Even allowing for such a factor (which is higher than the one for *Spartina*), it would appear that *Zostera* leaves contribute little production to the whole lagoon.

The mean standing crop in the whole lagoon is 135,3 g (ash free).m^{-2}. In comparison to this, Penhale (1977) found that in the salt marshes of North Carolina the average standing crop was 105 g (dry mass).m^{-2}. Although the units are different, the *Zostera* in Langebaan has a higher standing crop.

CONCLUSIONS

Although the reduction of light intensity limits primary production in many estuaries, this does not hold in Langebaan Lagoon where the water is clear and the depth is seldom more than 3 m. Indeed, the reduction of production in the surface layer of Riet Bay suggests that the high light intensity may inhibit production. The main limiting factor appears to be nutrient concentration, particularly nitrate, which decreases down the length of the lagoon in summer. The production of phytoplankton decreases in the same way. *Gracilaria* provides a better example for not only does the biomass and production decrease markedly towards the southern end of the lagoon, but the colour of the plants also changes from almost black at the entrance to yellow at the southern end.

While the production of vascular plants in the salt marshes may be affected by the nutrients in the water, they absorb most nutrients through the roots. It has been shown that their production is highest at the southern end of the lagoon. Here tidal currents are minimal and there are extensive marshes where decay and recycling is concentrated.

Sarcocornia has the highest annual production of 31 kJ x 10^3.m^{-2}. Comparable figures for *Spartina* and *Gracilaria* are 20 and 17 respectively. However, when the area occupied by each species is taken into account, it is evident that *Gracilaria* is the most important plant in the whole lagoon with an annual gross production of 270 x 10^9 kJ per year. *Sarcocornia* produces 208 x 10^9 kJ and *Spartina* 68. The remaining macrophytes that have been assessed, namely *Zostera*, *Scirpus*, *Typha* and *Phragmites* are less important in the lagoon ecosystem.

Acknowledgements

My thanks go to the following for their valuable assistance. Mrs K. Trotter for the chemical analyses; Mrs M. Jarman for the aerial photography; Commander Meyer of the SA Navy for research facilities at Langebaan; Dr G.M. Branch and Professor J.H. Day for their help and advice, and the Department of Planning and the Environment for funding the project.

REFERENCES

BARSDATE, R.J. & M. NEBERT 1972. Phosphorus cycling in a eelgrass *(Zostera marina* L*)* ecosystem. *Limnol. Oceanog.* 17: 58-67.
BOUCHER, C. & M.L. JARMAN 1977. The vegetation of the Langebaan area, South Africa. *Trans. roy. Soc. S.Afr.* 42: 241-272.
CHRISTIE, N.D. & A. MOLDAN 1977. Distribution of benthic macrofauna in Langebaan lagoon. *Trans. roy. Soc. S.Afr.* 42: 273-284.
DAY, J.H. 1959. The biology of Langebaan lagoon: a study of the effect of shelter from wave action. *Trans. roy. Soc. S.Afr.* 35: 475-547.
FLEMMING, B.W. 1977. Distribution of recent sediments in Saldanha Bay and Langebaan lagoon. *Trans. roy. Soc. S.Afr.* 42: 317-340.
GABRIEL, R.C. & A.A. DE LA CRUZ 1974. Species composition, standing crop, and net primary production of a salt marsh community in Mississippi. *Chesapeake Sci.* 15: 72-77.
GRINDLEY, J.R. 1977. Zooplankton of Saldanha Bay and Langebaan lagoon. *Trans. roy. Soc. S.Afr.* 42: 341-370.
HENRY, J.L., MOSTERT, S.A. & N.D. CHRISTIE 1977. Phytoplankton primary production in Langebaan Lagoon and Saldanha Bay. *Trans. roy. Soc. S.Afr.* 42: 383-398.
ISAAC, W.E. 1956. The ecology of *Gracilaria confervoides* (L) Grev. in South Africa with special reference to its ecology in the Saldanha-Langebaan lagoon. *Proc. 2nd Inst. Seaweed Symp.:* 173-185.
KIRBY, C.J. & J.G. GOSSELINK 1976. Primary production in a Louisiana gulf coast: *Spartina alterniflora. Ecology* 57: 1052-1059.
LAPOINTE, B.E., WILLIAMS, L.D. & J.C. GOODMAN 1976. The mass outdoor culture of macroscopic marine algae. *Aquaculture* 8: 9-21.
MASON, C.F. & R.J. BRYANT 1975. Production, nutrient content and decomposition of *Phragmites communis* Trin. and *Typha angustifolia* L. *J. Ecol.* 63: 71-95.

MENZEL, D.W. & N. CORWIN 1965. The measurement of total phosphorus in seawater based on the liberation of organically bound fractions by persulphate oxidation. *Limnol. Oceanog.* **10**: 280-286.

MILNER, C. & R.E. HUGHES 1968. Methods for the measurement of the primary production of grassland. *1-B-P Handbook No 6 Blackwell Sci. Publ., Oxford* 70p.

MORGANS, J.F.C. 1956. Notes on the analysis of shallow water soft substrata. *J. Anim. Ecol.* **25**: 367-387.

MURPHY, J. & J.P. RILEY 1962. Determination of phosphate in natural waters. *Anal. Chim. Acta* **27**: 31-36.

NIXON, S.W. & C.A. OVIATT 1973. Analysis of local variation in the standing crop of *Spartina alterniflora. Bot. Mar.* **16**: 103-109.

PENHALE, P.A. 1977. Macrophyte-epiphyte biomass and productivity in an eelgrass *(Zostera marina L)* community. *J. exp. mar. Biol. Ecol.* **26**: 211-224.

PUTTICK, G.M. 1977. Spatial and temporal variations in inter-tidal animal distribution at Langebaan lagoon, South Africa. *Trans. roy. Soc. S.Afr.* **42**: 403-440.

SAND-JENSEN, K. 1975. Biomass, net production and growth dynamics in an eelgrass *(Zostera marina L)* population in Vellerup Vig. Denmark. *Ophelia* **14**: 185-201.

SHANNON, L.V. & G.H. STANDER 1977. Physical and chemical characteristics of water in Saldanha Bay and Langebaan lagoon. *Trans. roy. Soc. S.Afr.* **42**: 441-459.

SIMONS, R.H. 1977. The algal flora of Saldanha Bay. *Trans. roy. Soc. S.Afr.* **42**: 461-482.

SMALLEY, A.E. 1959. The growth cycle of *Spartina* and its relation to the insect populations in the marsh. *Proc. Salt Marsh Conf. Sapelo Island Georgia.*

STRICKLAND, J.D.H. & T.R. PARSONS 1960. A manual of seawater analysis. *Bull. Fish. Res. Bd. Can.* **125**. 185pp.

STRICKLAND, J.D.H. & T.R. PARSONS 1968. A practical handbook of seawater analysis. *Bull. Fish. Res. Bd. Can.* **167**. 309p.

SUMMERS, R.W. 1977. Distribution, abundance and energy relationships of waders (Aves: Charadrii) at Langebaan lagoon. *Trans. roy. Soc. S.Afr.* **42**: 483-496.

UDELL, H.R., ZARUDSKY, J., DOHENY, T.E. & P.R. BURK-HOLDER 1969. Productivity and nutrient values of plants growing in the salt marshes of the town of Hempstead, Long Island. *Bull. Torrey bot. Club* **96**: 42-51.

WIEGERT, R.G. & F.C. EVANS 1964. Primary production and the disappearance of dead vegetation on an old field in south eastern Michigan. *Ecology* **45**: 49-63.

WILLIS, J.P., FORTUIN, H.H.G. & G.A. EAGLE 1977. A preliminary report on the geochemistry of recent sediments in Saldanha Bay and Langebaan Lagoon. *Trans. roy. Soc. S.Afr.* **42**: 497-509.

WOODWELL, G.M., WHITNEY, D.E., HALL, C.A.S. & R.A. HOUGHTON 1977. Flax pond ecosystem study. Exchanges of phosphorus between a salt marsh and the coastal waters of Long Island Sound. *Mar. Biol.* **41**: 1-6.

ZIEMAN, J.C. 1968. A study of the growth and decomposition of the seagrass, *Thalassia testudinum.* M.Sc. thesis, Univ. Miami. 50p.

Chapter 8

Estuarine plankton
J.R. Grindley

School of Environmental Studies, University of Cape Town

INTRODUCTION

Plankton comprises those organisms, both animal and plant, that live suspended in water and whose powers of locomotion are so weak that they drift with water currents. Plankton in estuaries include both plants referred to as *phytoplankton* and small animals or *zooplankton*. Those organisms with an entirely planktonic life history are referred to as *holoplankton* while vast numbers of temporary plankton or *meroplankton* also appear in estuaries. This latter category includes the larvae of benthic invertebrates, fish and other nektonic organisms. There are also many *tychopelagic* species in estuaries which live in the plankton or on the bottom with equal facility. Those species that live in the surface layer may be referred to as *neuston*. Some plankton in estuaries develops and passes its whole life history in the estuary and is referred to as *autochthonous*. Other forms enter from the sea or from rivers and are described as *allochthonous*. Among these are many plankton organisms that normally occur in the coastal waters; they are referred to as *neritic* plankton.

Neritic marine plankton penetrates estuaries to various degrees. Some *stenohaline* forms are unable to withstand major changes in salinity. *Euryhaline* species however flourish in estuarine waters of reduced salinity or in hypersaline lagoons. The lower reaches of estuaries are commonly occupied by neritic plankton but the number of species becomes reduced further up an estuary. In the upper reaches, the plankton community may include characteristic *estuarine species* not normally found in the open sea or the fresh waters of rivers. Few fresh water species will tolerate even very low salinities so that *fresh water* plankton plays very little part in estuaries.

The first part of this chapter covers general aspects with an emphasis on new findings. An account of phytoplankton of estuaries is followed by a general account of zooplankton. In the latter part of the chapter, an account is given of the plankton of certain South African estuaries.

ESTUARINE PHYTOPLANKTON

Due to the fluctuating salinity and temperatures the phytoplankton tends to be both euryhaline and eurythermal. Like the marine phytoplankton of temperate seas, the phytoplankton of the lower reaches of estuaries is dominated by diatoms, and dinoflagellates are less abundant although they may be important at certain seasons. Small nanoflagellates are usually abundant in the upper reaches. The significance of phytoplankton in estuaries is somewhat controversial; while it may play a major role and be the basis of heterotrophic life in the open sea, this appears to be unusual in estuaries. In most estuaries, the primary production of phytoplankton is insignificant in comparison to that of attached plants and the organic detritus derived from them.

Permanent or temporary autochthonous populations of marine diatoms play an important role particularly in the lower reaches of estuaries. *Skeletonema costatum* is an abundant and widespread species and so are *Nitzschia closterium* and *Thalassiosira decipiens*. In South African estuaries, species such as *Coscinodiscus granii, Rhizosolenia setigera, Chaetoceros lorenziana, Biddulphia mobiliensis* and *Actinoptychus splendens* fall into this category. Some dinoflagellates such as *Prorocentrum micans* and *Peridinium* species are temporarily autochthonous in South African estuaries and elsewhere.

Such marine species are seldom important but forms such as *Asterionella japonica, Coscinodiscus concinnus* and certain species of *Chaetoceros, Biddulphia* and *Rhizosolenia* may be significant at times.

Autochthonous brackish water species are particularly important in estuaries. Among diatoms, *Chaetoceros danicum* is important in European estuaries, while *Chaetoceros subtilus, Nitzschia longissima* and *Melosira dubia* occur in South Africa. Many brackish water species are small enough to be considered nanoplankton, also many flagellates such as *Cryptomonas* spp, *Heteromastix longifilis, Emiliana huxleyi, Pyramimonas orientalis*. Extremely small blue-green algae such as *Synechococcus leopoldiensis* may be important and some dinoflagellates such as *Glenodinium* spp.

Benthic or littoral diatoms do appear in the plankton, particularly when strong turbulence created by tidal currents puts them into suspension. Common species are *Achnanthes brevipes, Navicula distans, Nitzschia angularis* and *Triceratium* spp.

Red water caused by blooms of dinoflagellates has been recorded in estuaries including the St Lucia lakes and the Swartkops estuary.

Several papers have been published on diatoms from South African estuaries including Archibald (1966), Cholnoky (1960, 1962, 1963, 1968), Giffen (1963, 1966), Johnson (1976), Masson & Marais (1975). Studies of nanoplankton have been carried out by Henrici & Pienaar (1975, 1976). Studies of primary productivity have been carried out by Robarts (1976), Henry *et al* (1977) as well as scientists of the National Institute for Water Research (NIWR) including Connell, Hemens & Oliff who have recorded their results in reports edited by Oliff (1976, 1978).

Environmental controls

Whereas in clear tropical ocean waters the euphotic zone may extend below 100 m, in estuaries it is usually limited to 10 m or less. The turbidity, often limits the euphotic zone to a few centimetres. In such conditions, it is probable that the phytoplankton spends only short periods in the euphotic zone in turbulent conditions and must survive for periods below the compensation depth.

Various nutrients, including organic compounds

Table 8.1
Distribution of diatom species in an estuary with a stable salinity gradient (based on Kawamura, 1966)

Salinity (‰)	Dominant forms
2-5	*Anabaenopsis* sp, *Microcystis* sp, *Synedra ulna, Melosira varians.*
9-10	*Anabaena flos-aquae, Melosira varians, Chaetoceros* sp, *Biddulphia* spp, *Coscinodiscus* sp.
16	Euglenoids
20	*Melosira varians, Chaetoceros debilis, Ditylum brightwelli,* Peridinians.
24-31	*Skeletonema costatum, Rhizosolenia longiseta, Biddulphia aurita, Ditylum brightwelli,* Dinophyceans.

are required for phytoplankton growth. A number of other environmental factors also act as constraints. Zooplankton can rapidly reduce their numbers while tidal exchange can physically remove plankton from an estuary (Grindley, 1977b). Salinity and temperature and environmental parameters also determine the species composition, their abundance and productivity.

Riley (1967) has suggested that standing crop and productivity is determined by the availability of nutrients, the penetration of sunlight and the grazing pressure of filter feeders and that salinity does not appear to be an important factor. Where the estuarine water is reasonably transparent, a rich phytoplankton may develop. Its nature depends largely on its position in the estuarine system. In the Navesink estuary in New Jersey, Kawamura (1966) distinguished zones dominated by different species at particular salinities. A zone dominated by euglenoids occurred at salinities below 20 ‰. This was followed by a zone in which *Rhizosolenia* was dominant between 20 and 22 ‰. *Cerataulina bergonii* was dominant between 22 and 25 ‰ while the outer estuary was dominated by an assemblage of dinoflagellates. These included species of *Peridinium* and *Glenodinium*. Finally the diatom *Skeletonema costatum* was dominant in the open water beyond the mouth of the estuary.

On the basis of work in other areas, Kawamura (1966) suggested the generalized scheme for regions with a fairly stable salinity gradient shown in Table 8.1. However such generalizations are not reliable, and many investigators have found that particular estuaries have distinctive phytoplankton communities. Wood (1964) found that a number of Australian estuaries had their own peculiar communities of phytoplankton and that dinoflagellate species dominated several estuaries.

Riley (1967) discussed the role of depth and circulation pattern in determining the nature and productivity of phytoplankton. In all rapidly flowing estuaries, the phytoplankton population has little time to grow before it is swept out into the sea so that the residence time of the water in an estuary is very important. In deep estuaries including fjords, the water is usually stratified and inflowing sea water at the bottom carries in nutrients. River water on the surface may be nutrient-poor, particularly with regard to phosphorus and nitrogen compounds which are often the limiting factors for phytoplankton growth. Inflowing sea water is usually not very fertile when coming from the levels of marine phytoplankton growth, where nutrient supplies are reduced but it may be nutrient-rich in late autumn or winter when phytoplankton is at a minimum. Salt water which flows slowly in near the bottom, may have a high residence time which increases its significance. In deep estuaries, the proportion of benthic plants decreases so that competition between phytoplankton and benthic plants for nutrients is decreased. In shallow estuaries, sea and river water are well-mixed, and phosphates, silicates and iron are in good supply and organic nutrients including chelators and vitamins such as thiamine are largely derived from land drainage. Cobalamin which is essential for many neritic and estuarine species of phytoplankton including *Skeletonema costatum,* is derived from biogenic sources in the estuary itself. The limiting factor is usually nitrogen compounds.

Nitrogen and phosphorus are particularly important. Many estuaries receive sewage effluents in which the nitrogen to phosphorus ratio would lead one to expect that nitrogen might be a limiting nutrient. However, Taft & Taylor (1976) suggest that in Chesapeake Bay high N to P ratios in the water *and* in the phytoplankton show that phosphorus regulates phytoplankton productivity during the spring. The cycle of posphorus within this estuary is particularly interesting in that the maximum orthophosphate (SRP) concentrations within the water column occur in the summer. At this time, inorganic nitrogen also appears to be a major regulating factor. Taft & Taylor (1976) also show that phosphomonoesters may be a source of phosphorus for estuarine phytoplankton. Many phytoplankton species produce alkaline phosphatase enzymes when intracellular phosphate levels decline below the threshold value. These enzymes hydrolyze extracellular organic phosphorus monoesters to release orthophosphate ions. There is a rapid recycling of animal excretions in shallow estuaries and decomposition products on the bottom become available to the phytoplankton by turbulent mixing. The nitrogen and phosphorus flux through estuarine plankton depends to some extent on exchange processes between the sediment and the overlying water. Vertical transport of soluble forms of nitrogen and phosphorus are particularly important. In their study of nutrient availability and utilization in Namaquasett Bay, Furnas *et al* (1976) stress the dynamic nature of nutrient cycling and the problems of establishing nutrient budgets. The replenishment of nitrate in the water column may not be met by the excretion rates of zooplankton and benthos. However, it is now recognized that urea is a significant source of nitrogen for phytoplankton. Webb & Haas (1976) have shown that in the York

River estuary at times, urea supplied most of the nitrogen required by the phytoplankton. Nanophytoplankton being responsible for about 80 % of the urea uptake. Riley (1946) proposed a mathematical model of phytoplankton production in aquatic ecosystems considering solar energy, nutrients and zooplankton grazing. In his model, he assumed that each environmental parameter affected primary production independently. This assumption does not agree with Liebig's law of the minimum.

Takahashi *et al* (1973) attempted to overcome these difficulties in a study of the Frazer River estuary in Canada. Phytoplankton photosynthesis was estimated by a simulation model. The bits of information considered included photosynthetic responses of phytoplankton and actual measurements of each environmental factor. Only one environmental factor was considered to limit photosynthesis at any given time and place. The results showed which factors actually limited phytoplankton photosynthesis at different times. Temperature limited primary productivity only near the surface during the winter and nitrate was limiting during two months in the summer. During the rest of the year, light was the main limiting factor near the surface and throughout the year at depth. The consideration of nutrient uptake was based on Michaelis-Menten kinetics which provide a universal physiological basis for the model. However, the model can not predict changes in the structure of the phytoplankton community. A further refinement based on temperature optima might make this possible. This might be a significant improvement since changes in community structure affect the size-selective grazing of herbivorous zooplankton. The dominant phytoplankton species involved at different times were *Skeletonema costatum, Thalassionema nitzschioides, Thalassiosira nordenskioldi, Corethron hystrix, Chaetoceros convolutus* and a species of *Rhizosolenia.*

Other factors

In shallow estuaries the grazing pressure of benthic invertebrates is added to that of the zooplankton. The competition for nutrients between phytoplankton and attached macrophytes is not well-understood. However, estuaries with large *Zostera* beds and broad salt marshes are generally poor in phytoplankton. Floods are obviously more serious for phytoplankton than for rooted vegetation and in shallow estuaries, flushing occurs periodically. In all estuaries, much depends on the penetration of light and thus on turbidity. Indeed, there is broad agreement that the seasonal changes in light intensity are more closely correlated with turbidity than with the intensity of insolation. Photosynthesis is inhibited by very high as well as low light intensities and the effect of turbidity may thus depend on the intensity of the incident light and the depth of the estuary. Many phytoplankton species are adapted to the wave lengths least affected by turbidity.

The existence of a clearly defined surface micro-layer of phytoplankton in estuaries has been recognized. Greater concentrations of organisms, of at least several orders of magnitude, may occur. It is now evident (Manzi *et al*, 1977) that the differences are not only in abundance but also in the nature of communities at the surface. The surface microlayer generally exhibits lower diversity, evenness and richness than subsurface communities.

The effects of enclosing an estuarine area by dykes has been described by Bakker (1964). The changes in the resulting lake-like area known as the Veerse Meer in Holland, involved the elimination of many polyha-line species and the gradual appearance of many brackish water species among which flagellates and dinoflagellates were the most important groups. Stand-ing crops increased greatly but this was probably affected by eutrophication as well as closure. As stand-ing crops increased, diversity declined and the originally stable marine environment became an unstable brackish water lake with periodic blooms. Some allochthonous marine species now enter through a sluice from the Oosterschelde and persist for various periods depending on conditions. *Skeletonema costatum* tolerated salini-ties as low as 6 ‰ and several species flourished when salinities rose to 10 ‰ in summer (Bakker, 1972).

One of the problems in Dutch estuaries that have been closed in the Delta Plan, is the development of blooms of blue-green algae when the water becomes stagnant. Possibilities of preventing this have been dis-cussed by Peelen (1969). In the Brielse Maas which was dammed in 1950, cyanophycean blooms deve-loped although they fail to develop in sea water.

Blooms

Seasonal changes in the standing-crop of phytoplankton have been studied in several temperate American estua-ries. In general, phytoplankton production is low from mid-autumn to mid-winter due to the turbulence and the resulting turbidity. The concentration of nutrients increases and the flowering of phytoplankton may start as early as mid-winter which is much earlier than the spring flowering in the open sea. Riley (1967) reports that in Long Island Sound, where the salinity range is 24-29 ‰ and the temperature varies from 2-3 °C in February, to 23 °C in August, the flowering starts in January or February. The observed increase in the standing stock shown by chlorophyll measurements was in good agreement with theoretical values. In these, the increases in chlorophyll are positively correlated with increases in light intensity less the losses due to respiration which increase with temperature. As Riley remarks, other factors should be taken into conside-ration but the agreement between observed and com-puted results shows that the main factors have been identified. Recent studies show that humic compounds originating from land drainage have a stimulating effect on the growth of phytoplankton in estuaries (Prakash, 1971).

In the brackish waters of Chilka Lake in India, two blooms occur annually (Patnaik, 1973). A peak of phytoplankton abundance occurs in summer with *Chaetoceros* spp and *Asterionella japonica* as the domi-nant forms. A second brief peak occurred in winter as the salinity and nutrient concentrations increased.

Declines in abundance

Declines in abundance after plankton blooms may be caused by one or more factors, including grazing by zooplankton, depletion of essential nutrients and self-shading through cell abundance. This latter effect can cause increased turbidity of the water reducing light intensity for the whole phytoplankton community and hence inhibiting growth (Riley, 1967).

The decline from the spring maximum has been attributed by some (Marshall *et al*, 1934, Cushing, 1975) to grazing by zooplankton and by others includ-ing Riley (1967), to exhaustion of nutrients. In Long Island Sound, it appears to be due to depletion of nutrients for the increase in zooplankton does not occur until after the phytoplankton maximum has passed. During the summer, phytoplankton produc-tion is very variable but generally high, possibly due to rapid regeneration of nutrients and rapid turnover. The eventual decline appears to be related to a reduc-tion in nutrient availability.

A factor which may tend to inhibit phytoplankton growth is the production of tannins by certain plants including mangroves (Smayda, 1970). Perhaps this is one of the factors which account for the smaller num-bers of diatoms in mangrove estuaries, but the high turbidity in most mangrove estuaries may be more important.

In Chesapeake Bay it has been found that adult copepods feed on a variety of suspended particles including phytoplankton cells and particulate detritus. Feeding habits result in a gradation of grazing pres-sures (Richman *et al*, 1977). The largest particles are selected first after which the copepods switch succes-sively to the biomass peaks of the smaller size categories. The results suggest considerable flexibility in copepod feeding behaviour which cannot be explained solely by the mechanism of a fixed sieve. Comparisons of the feeding behaviour of *Eurytemora affinis*, *Acartia tonsa* and *A. clausi* show that the *Acartia* species have greater capabilities for taking large particles; this may be asso-ciated with modifications of their mouth parts for raptorial feeding. More work is needed on the mouth-part morphology and feeding behaviour of estuarine copepods to clarify selective grazing mechanisms. Opportunistic feeding behaviour occurs when particles are taken out of proportion to their numbers, but from size classes forming biomass peaks. This maximizes energy retrieval per unit of energy expenditure. At the same time, such selective feeding helps to maintain the diversity of the phytoplankton species.

Adaptations of estuarine species

Riley (1967) has noted that the phytoplankton of estuaries tends to be quantitatively abundant but limited as to the number of important species. The small number of species that can succeed in the variable estuarine environment, leads to the dominance of those few. Over the whole year, many species appear in estuaries but most of them are in small numbers and are found at the mouth of the estuary. Very few species are always common. Riley (1967) reports only 13 that form 5 % or more of the population of Long Island Sound in one month or another.

Skeletonema costatum is one of the best adapted and most successful diatom in estuaries in many parts of the world. It has been found to be a dominant species in Block Island Sound (Riley, 1952), Great Pond (Hulbert, 1956), Long Island Sound (Conover, 1956), Chesapeake Bay (Patten *et al*, 1963), James River estuary (Marshall, 1967), Cape Fear river estuary (Carpenter, 1971), Tamar estuary (Mommaerts, 1969), Valparaiso Bay (Avaria, 1965), Cananeia estuary (Kutner, 1976) and many others. *S. costatum* appears to be a eurythermal species probably as a result of the existence of a series of geographical races (Braarud, 1962). Experiments carried out in Sao Paulo, showed that it grew best at 24 °C (Kutner, 1976). Clones of *S. costatum* from the Guianas coast were unable to grow below 25 °C while clones collected near Cape Cod grow very well at 3-4 °C (Hulbert & Guillard 1968). Although temperature appears to have a strong influence on the annual cycle of abundance, it is probable that land drainage provides the nutritive substances causing blooms. According to Curl & McLeod (1961) phosphates and nitrates are probably the factors that determine the standing stock of *S. costatum* while Smayda (1973) showed that nitrogen and sometimes silica were elements limiting the growth of this species. Riley (1967) has suggested that apart from the depletion of essential nutrients the end of a bloom may be caused by grazing by zooplankton or by self-shading. *C. costatum* is sometimes said to be a meroplanktonic species (Braarud, 1962), for it may live on the bottom during part of the year. It forms spores that sink to the bottom and germinate later under more favourable conditions (Wood, 1971).

Nanoplankton

Nanoplankton may play a very important role particularly in the upper reaches of estuaries. The relative importance of the larger forms of phytoplankton and of nanoplankton varies greatly. The nature of the plankton appears to depend upon the environmental conditions while competition for nutrients appears to determine the succession. Dinoflagellates and nanoflagellates favour high temperatures and may form blooms in the early autumn in the upper reaches of calm, slow-flowing estuaries.

In the Cananeia estuary in Brazil, Kutner (1976) found that microflagellates were most abundant at the inner stations while diatoms were less numerous there. Micro-flagellates did not show a seasonal variation whereas diatoms were at a minimum in winter and a maximum in summer. Since temperatures are high throughout the year river flow may be more important seasonally. The nanoplankton was composed of micro-flagellates and small diatoms and made up an average of 87,2 % of the phytoplankton. The most abundant diatom was *Skeletonema costatum*.

Rhode *et al* (1958) pointed out the importance of nanoplankton in primary production, noting that small algal cells are generally more effective producers of organic matter by virtue of their volume to surface ratio. They have great powers of nutrient utilization and the capacity for rapid multiplication.

Hulbert (1970) has shown that in some very shallow estuaries including Moriches Bay near New York and Salt Pond at Woods Hole phytoplankton cells may exceed 10^9 litre^{-1}. The dominant corganisms mainly nanoplanktonic sink more slowly than neritic marine species and monopolization of the nutrient supply appears to aid the dominance of these forms. Nutrients are kept down to almost undetectable levels continually so diatoms introduced from the sea are reduced and also tend to sink out of the shallow water.

The phytoplankton which passed through 35 μm mesh (nanoplankton) were responsible for 89,6 % of the phytoplankton productivity in Chesapeake Bay (McCarthy *et al*, 1974). On one summer cruise the nanoplankton was found to be responsible for 100 % of the primary productivity. At that time the <10 μm fraction was responsible for 94 % of the productivity.

Biomass of phytoplankton

Estimates of phytoplankton biomass are difficult since the seston or suspended organic particles, includes detritus and benthic microalgae eroded from the banks, as well as true phytoplankton. Odum & de la Cruz (1967) when measuring the seston in a creek draining a Georgia salt marsh, estimated that only 10 % was plankton, of which two-thirds to three-quarters was phytoplankton, ie about 7 % of the total mass of seston. Thus measures of pigment concentration or primary productivity are usually more reliable than measures of biomass. Calculations of biomass are sometmes made by cell volume calculations on the basis of cell counts and the geometry of the cells of each species.

Primary productivity

Primary productivity may be calculated from the rate of oxygen production or nutrient uptake using dark and light bottles, or from the rate of uptake of ^{14}C. Primary productivity has also been estimated on the basis of chlorophyll concentrations in relation to light penetration. However, Vollenweider (1971) has criticised this approach, pointing out that chlorophyll content is only one of a series of factors affecting productivity. With the nutrient uptake method, 'luxury consumption' where more nutrients are absorbed than are required for immediate needs, can cause inaccuracies. The ^{14}C method is widely used but the values that are obtained are neither gross nor nett production but are believed to be near the nett value.

Phytoplankton productivity in estuaries, varies greatly. In high latitudes, light intensity may be critical while in tropical areas other factors such as seasonal nutrient or salinity fluctuations may be more important. Furnas *et al* (1976) found an annual carbon production in the temperate Narragansett Bay of 308 gC. m^{-2} of which 42 % occurred during July and August.

In the Cochin Backwater, a tropical estuary in India, phytoplankton production far exceeded the rate of consumption by zooplankton herbivores (Qasim, 1970). Gross production ranged from 400 to 1 500 mg C.m^{-2}.d^{-1}. Qasim suggested that excess production found its way through 'alternate pathways' in the food web of the estuary. He suggested that shrimps, mullet and benthic organisms, perhaps via detritus, become the major phytoplankton consumers in a tropical estuary. During the monsoon, when salinity is in the 10-20 ‰ range, peak gross production was 280 gC.m^{-2}.y^{-1}. Nett production was 124 gC.m^{-2}.y^{-1}. Consumption by zooplankton was only 30 gC.m^{-2}.y^{-1} so it appears that most of the production adds to the detritus in the estuary (Qasim, 1973).

In the Parramatta estuary and Sydney Harbour in Australia, (Revelante & Gilmartin 1978) primary production, chlorophyll a and cell counts increase dramatically towards the upper reaches (Figure 8.1). In the inner estuary, values as high as 175 mgC.m^{-3}.h^{-1} and 19 mg chlorophyll a m^{-3} have been recorded. The production and crop data form three distinct sets (Table 8.2): As a result of 'cultural eutrophication' (pollution), the entire estuary had nutrients in excess of either micro- or nanoplankton demand. The marked gradients in production rates and crops therefore reflect biological and/or physical cropping pressures. Nanoflagellates increasingly dominate the phytoplankton communities towards the inner estuary.

Cell densities in excess of 110×10^6 cell per litre were recorded in the upper estuary. Nanoplankton: microplankton primary production ratios increased from near 1 in the outer estuary to ratios in excess of 4,2 in the inner estuary. Shannon-Weaver diversity indices decreased systematically in the estuary from 4,0 near the mouth to 2,4 near the head and were

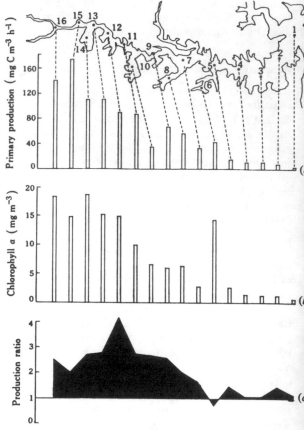

Figure 8.1
(a) Primary production rates, (b) chlorophyll a standing crop, and (c) nanoplankton/microplankton production ratios in the Parramatta Estuary and Sydney Harbour (from Revelante & Gilmartin, 1978).

Table 8.2
Sydney Harbour phytoplankton productivity (from Revelante & Gilmartin, 1978)
Key: (1) Region of estuary
  (2) Approximate salinity range (‰)
  (3) Mean rate of ^{14}C primary production (mg.C.m^{-3}.h^{-1})
  (4) Mean standing crop of chlorophyll a (mg chlorophyll a.m^{-3})

(1)	(2)	(3)	(4)
Outer estuary	34-35	11	1,8
Middle estuary	33-34	54	8
Inner estuary	29-33	127	17

inversely related to community biomass.

Measurements of the primary productivity of the hyposaline Fafa lagoon on the Natal coast were made by the NIWR using both the oxygen method and carbon-14 method. In all cases, the oxygen method gave higher results than the corresponding values for the carbon-14 method although the trend was similar. As noted, the carbon-14 method gives results which approach nett productivity. Values obtained with the ^{14}C method ranged from 0,32 to 7,50 mg.m^{-3}.h^{-1} in the lagoon. With the oxygen method, values ranged from 1,65 to 29,93 mg.m^{-3}.h^{-1}.

The pigments present within cells are more reliable as an indicator of the standing crop of phytoplankton organisms. Pigment analysis can be a very useful measure of biomass. The quantity of pigment present varies with the abundance of phytoplankton cells and Sakamoto (1966) has suggested a division of lakes on the basis of chlorophyll concentration as follows:

Oligotrophic 0,3-2,5 mg chlorophyll m^{-3}
Mesotrophic 1-15 mg chlorophyll m^{-3}
Eutrophic 5-140 mg chlorophyll m^{-3}

In the Fafa lagoon, the NIWR have recorded a maximum of 4,07 mg chlorophyll m^{-3}, suggesting a mesotrophic condition on this basis (Oliff, 1976).

In many South African estuaries, primary productivity is limited by the extreme turbidity caused by suspended silt. The lagoon area of Richards Bay was shallow and muddy so that the penetration of sunlight was very limited. Measurements by the NIWR revealed that phytoplankton production was usually less than 100 mg.C.m^{-3}.day^{-1}. After the division of Richards Bay into a northern harbour area and a southern 'sanctuary' area with a separate mouth, clear sea water entered the sanctuary with the tides. Productivity increased to 40 mg.C.m^{-3}.hr^{-1} or about 320 mg.C.m^{-3} day^{-1} in February 1976 presumably as a result of the improved light penetration.

It is not possible to draw conclusions on the productivity of estuarine waters by studying plankton samples collected with an ordinary plankton net, for the nanoplankton, including the smallest diatoms and unarmoured flagellates escape through its meshes. Even centrifuging is not altogether satisfactory for it causes many small flagellates to break up. Benthic microflora including diatoms, fungi and bacteria may also play an important part. For example, benthic diatoms like *Gyrosigma, Navicula, Pleurosigma* and *Triceratium* are found on the bottom mud of estuaries and may be much more common than is indicated by plankton samples and play a major role in primary production.

In the Kungsbacka Fjord, which is a moderately polluted estuary in Sweden, the rate of primary production was about 100 gC.m^{-2}.y^{-1} (Olsson & Olundh, 1974). The rate was highest in June when 25 gC.m^{-2} was recorded. The relationship between primary production and environmental factors was investigated using Spearmans rank correlation coefficient and a significant positive correlation was found between temperature and primary production.

ZOOPLANKTON OF ESTUARIES

Composition

A very wide variety of holoplanktonic and meroplanktonic organisms appear in the zooplankton of estuaries.

As in the open sea, the zooplankton tends to be more diverse than the phytoplankton.

The holoplanktonic fauna is dominated by small species of copepods although most of the taxonomic groups present in neritic seas may be found in estuaries particularly if marine dominated and salinities are high Chaetognaths, ctenophores, and ciliates including Tintinnidae may be common and large rhizostomid yelly-fish are seasonally abundant but planktonic foraminifera, hydroid medusae, euphausids, salps and larvaceans are usually scarce. Mysids and amphipods may be very abundant but as these spend much of their time on the bottom it is sometimes doubted that they should be included among the true plankton. Meroplanktonic forms including invertebrate larvae are seasonally abundant, particularly the larvae of barnacles, polychaete worms and decapod crustaceans.

Several genera of Copepoda are characteristic of estuaries in different parts of the world. Species of *Eurytemora, Acartia, Pseudodiaptomus* and *Tortanus* are particularly well-represented. The mouth area is usually occupied by euryhaline marine species of *Paracalanus, Centropages, Oithona, Pseudocalanus, Temora* and many harpacticoids including *Euterpina* and *Harpacticus*. Australian estuaries include a number of other groups such as *Boeckella, Gladioferens* and *Sulcanus*.

Many peracarid crustaceans such as Mysidacea, Amphipoda, Isopoda, Tanaidacea and also the Cumacea play an important role although they are really tychopelagic. Species of *Mesopodopsis* and *Gastrosaccus* are important among the mysids in South African estuaries while the most abundant cumacean is *Iphinoe truncata*. Species of Amphipoda and Isopoda are listed in Tables 8.5 and 8.6.

Pelagic eggs, larvae and juveniles of several species of fish appear in many samples. The larvae of invertebrates make up a large part of the plankton in the spring and summer. They include zoea larvae of crabs, decapod 'mysis' larvae, juvenile panaeids and polychaete and mollusan larvae.

In almost all estuaries the greatest species diversity occurs among neritic forms near the mouth. Various marine species penetrate to different degrees into the lower reaches, while the upper reaches are often dominated by estuarine species. The zooplankton of estuarie may thus be divided into four components on the basis of their salinity tolerance. (1) A stenohaline marine component penetrating only into the mouth (eg *Corycaeus* spp), (2) a euryhaline marine component penetrating further up the estuary (eg *Paracalanus* spp), (3) a true estuarine component comprising species confined to estuaries (eg *Pseudodiaptomus hessei*), (4) a fresh water component comprising species normally found in fresh water (eg *Diaptomus* spp).

Table 8.3
Distribution of Copepoda

ESTUARINE ECOLOGY / *Chapter 8*

Copepod species	Walvis Lagoon	Sandvis Lagoon	Luderitz Lagoon	Orange	Olifants	Verlorenvlei	Berg	Langebaan Lagoon	Kleinriviervlei*	Breede	Knysna	Swartkops	Bushmans	Bashee	Mngazana	Msikaba	Mzimkulu	uMgababa	Durban Bay	Mlalazi	Richards Bay	St Lucia	Kosi
Acartia (Paracartia) africana		●	●				●	●	●	●	●	●	●	●									
Acartia (Paracartia) longipatella		●	●	●			●	●				●	●	●									
Acartia (Arcartiella) natalensis										●	●	●	●	●			●		●	●	●	●	●
Calanoides carinatus	●	●	●	●			●	●				●	●	●					●				
Centropages brachiatus	●	●	●				●	●	●														
Centropages chierchiae												●		●	●				●	?	●		
Centropages furcatus																			●	●	?	●	●
Clausidium sp.		●	●					●				●											
Clausocalanus furcatus											●												●
Corycaeus spp								●						●	●	●	●	●	●	●	●	●	●
Ctenocalanus vanus							●																
Euterpina acutifrons			●				●	●			●	●		●	●				●	●	●	●	●
Halicyclops spp	●	●		●		●	●	●	●		●	●	●	●	●		●	●	●	●	●	●	●
Harpacticus ? gracilis							●	●	●		●		●						●	●	●		
Hemicyclops sp		●				●	●	●											●				
Nannocalanus minor							●	●															
Oithona brevicornis/nana	●	●	●				●												●		●	●	●
Oithona plumifera							●														●		
Oithona similis		●					●	●	●		●										●		●
Paracalanus aculeatus							●												●		●	●	●
Paracalanus crassirostris	●	●	●		●		●	●			●	●	●	●		●			●		●		●
Paracalanus parvus	●	●	●				●	●	●		●	●		●			●		●				●
Porcellidium sp	●				●			●						●									
Pseudodiaptomus stuhlmanni																			●	●	●	●	●
Pseudodiaptomus hessei		●	●	●	●		●	●				●	●						●	●	●	●	●
Pseudodiaptomus nudus		●	●				●			●	●		●	●					●		●		
Rhincalanus nasutus			●				●	●															
Saphirella stages	●	●	●		●		●	●			●											●	
Temora turbinata														●	●		●		●	●			●
Tortanus capensis					●		●	●				●											
Tegastes sp	●	●					●	●				●									●	●	

? Indicates uncertain identification   * Also called Hermanus Lagoon.

Copepoda

Very many species of Copepoda have been recorded from South African estuaries but most of them are occasional records of marine species of little significance to estuarine ecosystems. Table 8.3 is a selected list of the important species. Their recorded distribution in a series of 23 selected estuaries is indicated. It can be seen that a large number of species including *Euterpina acutifrons*, *Halicyclops* spp, *Harpacticus gracilis*, *Oithona similis*, *Oithona nana/brevicornis*, *Paracalanus aculeatus*, *P. crassirostris*, *P. parvus*, *Porcellidium* spp, *Pseudodiaptomus hessei*, *P. nudus*, *Saphirella* stages and *Tegastes* spp occur in estuaries right around the coast. This sort of distribution is very different from that of many intertidal benthic organisms which have distinctively limited geographical distributions. Some species, however, do appear to be limited to estuaries on certain parts of the coast. *Ctenocalanus vanus* and *Nannocalanus minor* were only recorded on the west coast and *Tortanus capensis* only in the south. *Acartia natalensis*, *Centropages Chierchiae*, *Centropages furcatus*, *Clausocalanus furcatus*, *Pseudodiptomus stuhlmanni* and *Temora turbinata* were recorded only on the east coast. *Corycaeus* spp were mainly on the east coast. A number of species including *Paracartia africana*, *Paracartia longipatella*, *Calanoides carinatus*, *Centropages brachiatus*, *Clausidium* sp, *Hemicyclops* spp, *Oithona plumifera* and *Rhincalanus nasutus* appear mainly on the west and south coasts. There are no exclusively west coast estuarine copepod species, which is also true of macrobenthic invertebrates and fishes (see discussion in chapter 14).

Species of *Pseudodiaptomus* were dominant in many South African estuaries. Descriptions of the species in southern African waters have been published (Grindley, 1963) and their zoogeography discussed (Grindley, 1969). *Pseudodiaptomus hessei* occurs in estuaries right around the coast of South Africa while *Pseudodiaptomus stuhlmanni* (= *P. charteri*) occurs predominantly in the subtropical estuaries of Zululand and Mozambique.

Table 8.4
Distribution of Mysidacea

Mysid species	Walvis Lagoon	Sandvis Lagoon	Luderitz Lagoon	Orange	Olifants	Verlorenvlei	Berg	Langebaan Lagoon	Kleinriviervlei*	Breede	Knysna	Swartkops	Bushmans	Bashee	Mngazana	Msikaba	Mzimkulu	uMgababa	Durban Bay	Mlalazi	Richards Bay	St Lucia	Kosi
Gastrosaccus brevifissura		•			•			•	•	•	•	•	•	•	•	•					•	•	
Gastrosaccus gordonae								•													•	•	
Gastrosaccus psammodytes		•	•								•				?					•			?
Heteromysis spp															•								
Leptomysis tattersalli								•	•														
Mesopodopsis africana								•		•					•					•	•		
Mesopodopsis slabberi	•	•	•	•				•	•		•		•	•	•	•					•	•	
Mysidopsis similis			•					•	•												•		
Rhopalopthalamus terranatalis							•	•	•		•	•	•	•	•					•	•		
Siriella dayi								•							?								
Tenagomysis natalensis									?												•	•	

? Indicates uncertain identifications   * Also called Hermanus Lagoon

In the Swartkops and Sundays River estuaries, *Acartia natalensis* and *Acartia longipatella* showed a clear seasonal pattern of succession (Wooldridge, 1979). *A. natalensis* appeared in greatest numbers in summer while *A. longipatella* was most abundant in winter. The cycle of dominance appeared to be regulated by temperature, salinity and competition between the two species. *A. natalensis* is more tolerant of reduced salinity, and the replacement of *A. longipatella* by *A. natalensis* starts in the upper reaches and spreads seawards.

On the west coast the contrast between the cold sea and the warm estuaries and lagoons, limits the penetration of west coast neritic species. The true estuarine species are widely distributed in west coast estuaries since temperatures in the estuaries are comparable to those elsewhere (see also chapter 14). On the east coast, the differences between the neritic environment and that of the estuaries is more favourable to the penetration of neritic species. Owing to the presence of the cool northward flowing counter current, species that might be considered typical of the west coast such as *Calanoides carinatus* have been recorded as far east as Richards Bay. On the north coast of Natal, where conditions are sub-tropical, *Pseudodiaptomus stuhlmanni* replaces *P. hessei* except in the isolated coastal lakes Sibaya and Nhlange. It seems possible that the presence of *P. hessei* in these relict estuarine systems is related to Pleistocene incursions (see also chapter 15). However, the present-day contrast between these cool deep lakes and the warm shallow estuaries inhabited by *P. stuhlmanni* appears to be a more likely explanation.

Saphirella stages are common in most South African estuaries. The name *Saphirella* continues to be used although Nicholls (1944) and Gurney (1945) indicated that they probably represent stages of *Hemicyclops* species.

Mysidacea, Amphipoda and *Isopoda*

In Table 8.4, the distribution of species of Mysidacea in 23 South African estuaries is recorded. It may be seen that both *Gastrosaccus brevifissura* and *Mesopodopsis slabberi* are very widely distributed. *Mesopodopsis africana* and *Rhopalopthalamus terranatalis* occur mainly in estuaries on the south and east coasts. Most of the other species occur irregularly.

In Table 8.5, the distribution of Amphipoda in plankton samples from South African estuaries is recorded. Griffiths (1976) provides much more extensive distribution records of amphipods in South African estuaries, but most of the species do not appear in the plankton. *Austrochiltonia subtenuis, Melita zeylanica* and *Paramoera capensis* are widely distributed, while *Corophium triaenonyx* and species of *Grandidierella* were limited to east coast estuaries.

In Table 8.6 the distribution of Isopoda appearing in plankton samples from the same 23 estuaries is recorded. *Eurydice longicornis* was widely distributed but a number of species appear to occur mainly in east coast estuaries. *Paridotea ungulata* appeared on the west and south coasts only.

Mangrove estuaries

Many of the mangrove estuaries on the coast of Mozambique, Natal and Transkei include some distinctive and characteristic benthic communities. This raises the question as to whether there are also distinct plankton communities associated with mangroves. Neritic marine plankton, sometimes with more than a hundred species of zooplankton and many species of phytoplankton, enter with the tides. This high diversity usually corresponds to a low biomass while further up such estuaries, the characteristic estuarine plankton with a higher

Table 8.5
Distribution of Amphipoda in plankton samples

Amphipod species	Walvis Lagoon	Sandvis Lagoon	Luderitz Lagoon	Orange	Olifants	Verlorenvlei	Berg	Langebaan Lagoon	Kleinriviervlei*	Breede	Knysna	Swartkops	Bushmans	Bashee	Mngazana	Msikaba	Mzimkulu	uMgababa	Durban Bay	Mlalazi	Richards Bay	St Lucia	Kosi
Aora typica								●							●								
Austrochiltonia subtenuis				●		●	●	●						●		●			●				
Caprella equilibra								●							●								
Corophium acherusicum								●				?											
Corophium triaenonyx												●	●		●	●		●	●	●			●
Gitanopsis ? pusilla								●							●								
Grandidierella bonieroides												?		?	●	?		?			●	●	
Grandidierella lignorum												●		●	●	●			●				
Grandidierella lutosa															●			●					
Jassa falcata								●							●								
Lysianassa ceratina								●		●					●								
Melita zeylanica							●	●		●	●			●	●				●	●	●		
Paramoera capensis								●	●	●	●	●	●						?	●	●	●	
Orchestia ancheidos								●											●	?			
Urothoe elegans								●															?

? Indicates uncertain identifications
* Also called Hermanus Lagoon

Table 8.6
Distribution of Isopoda in plankton samples

Isopod species	Walvis Lagoon	Sandvis Lagoon	Luderitz Lagoon	Orange	Olifants	Verlorenvlei	Berg	Langebaan Lagoon	Kleinriviervlei*	Breede	Knysna	Swartkops	Bushmans	Bashee	Mngazana	Msikaba	Mzimkulu	uMgababa	Durban Bay	Mlalazi	Richards Bay	St Lucia	Kosi
Anilocra capensis (parasitic)																				●			
Caprella cicur																	●						
Capellina longicollis																							
Cirolana fluviatilis										?		?			●	●			●				?
Cirolana hirtipes								●															
Corallana africana																●			●				
Cryptoniscid larvae								●				●	●	●		●	●			●	●		
Cyathura carinata														●	●					●	●	●	
Eurydice longicornis	●							●	●		●		●		●	●							
Exosphaeroma hylecoetes								●	●	●	?		●										
Exosphaeroma truncatitelson								●															
Gnathia sp								●									●	●	●				
Leptanthura laevigata																				●			
Paridotea ungulata		●		●			●	●			●												
Pontogeloides latipes								●															●
Sphaeroma annandalei												?							?				
Sphaeroma terebrans																					●		
Sphaeroma spp																						●	
Synidotea setifer												?						?	●				
Synidotea variegata																					●	●	?

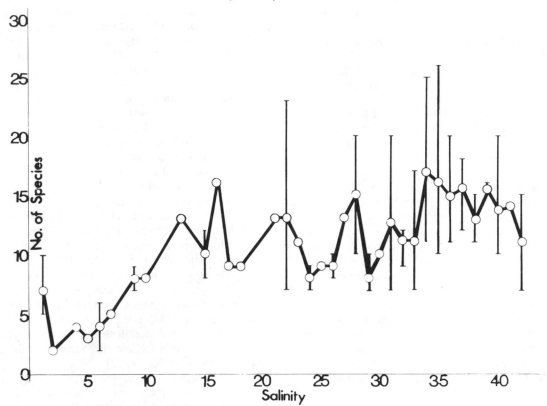

Figure 8.2
Zooplankton diversity (species richness) of samples taken in
the Swartkops estuary in 1967-68. Means and ranges are indicated for each salinity (Grindley, 1976a).

biomass but a low diversity occurs. These distribution
patterns appear to be determined by the extent of tidal
exchange and are apparently unrelated to the position
of mangrove areas. The composition of the zooplankton
is similar to that elsewhere and there do not appear to
be any species restricted to mangrove estuaries, at least
in Southern Africa.

Mysidacea do seem to be more important than elsewhere however, and may make up more than 75 % of
the zooplankton. *Gastrosaccus brevifissura* and *Mesopodopsis africana* are common, and the latter reached
over 10 000 m^{-3} in the Mngazana estuary (Wooldridge,
1977). Some species of Copepoda also became very
abundant at times. *Pseudodiaptomus hessei* reached
42 700 m^{-3} in Richards Bay (Grindley & Wooldridge,
1974) while *Acartia natalensis* reached 10 600 m^{-3} and
Oithona similis reached 123 000 m^{-3} in the Mngazana
estuary (Wooldridge, 1977).

Diversity

In almost all estuaries, the greatest diversity occurs
near the mouth where a wide range of neritic species
appear. In the lower reaches, various marine species
penetrate to different degrees. If species diversity is
plotted against salinity for a particular estuary, it is
found that peak diversity occurs at near 35 ‰ and
that diversity decreases with greater variations of
salinity. Below 2 ‰ an increase in diversity may occur
as a result of the penetration of the fresh water component (Figure 8.2).

Although many planktonic species occur in estuaries
right around the coast of southern Africa, there is an
increase in maximum diversity in the sub-tropical
estuaries in the north-east. There is, however, no close
correlation between water temperature and maximum
diversity. Estuaries are richest in species when they are
permanently open to the sea and marine dominated.

The changing communities of zooplankton along
the length of Langebaan Lagoon are indicated in
Figure 8.3. The frequency of occurrence of a series of
important zooplankton organisms from the open waters
of Saldanha Bay (area A) through the mouth of the
laboon (B) to the head of the lagoon (E) is portrayed.

Vertical migration

The vertical migration of plankton in the sea is well-
known. In many estuaries, there is hardly any zooplankton in the water during the hours of daylight. The
settled volume of plankton obtained in surface hauls is
commonly less than one-tenth of that found at night.
Owing to differences in the vertical migration behaviour
of different species, their relative numbers may also

change markedly (Figure 8.4). Thus it is important to have an understanding of the behaviour of the different species to make a reasonable assessment of their relative numbers.

Studies have been made of the diurnal fluctuations of zooplankton at the surface in six different estuaries and lagoons in South Africa (Grindley, 1972). Study sites included Langebaan, Milnerton, Kleinriviervlei (Hermanus Lagoon), Swartvlei, Swartkops and Lake Sibaya. Hourly plankton samples and recordings of environmental data were made over a 24 hour period at each locality. Vertical migration appeared to occur in almost all of the species of zooplankton observed and under a variety of conditions. Differences in migration pattern were observed between species. *Pseudodiaptomus hessei* appears at the surface in large numbers only during the night, whereas *Paracartia africana* also appears in fairly large numbers during the day. Some taxa including Cumacea, Ostracoda and 'Mysis' larvae of Decapoda only appear at the surface for part of the night. No differences in behaviour between the adult males and females or copepodite stages could be detected in *Pseudodiaptomus hessei* but in *Paracartia africana* higher propotions of adult males and females migrate than do copepodite stages. Studies using a plankton pump, revealed that the numbers of *Pseudodiaptomus* increase throughout all depths at night, suggesting that they rise off or from within the bottom sediments. Sampling of superficial bottom mud with a sled-like instrument indicated that some *Pseudodiaptomus* actually burrowed into the layers of mud and detritus on the bottom during the day, and this was confirmed in the laboratory (Grindley, 1972).

The patterns of vertical migration of most species observed in the course of this work differed from the usual type. The plankton rose *after* dark, reached peak concentrations at the surface near the middle of the night, and descended *before* dawn. This pattern of behaviour was best explained by an endogenous rhythm. This has since been confirmed by laboratory experiments in the case of *Pseudodiaptomus hessei* (Grindley, 1972). A diurnal fluctuation in activity was obtained in·laboratory experiments in constant darkness and in constant light, although true migrations were not manifest. It thus appears that although their migrations are phased by the diel cycle and their magnitude is affected by moonlight and other factors, there is an endogenous rhythm of activity.

An apparatus described by Grindley (1964) was designed to carry out experiments on vertical migration in the field in such a manner as to study the effects of salinity layering. The apparatus was in the form of a one metre-long tube built of perspex and nylon. It could be separated into three compartments by two sliding trap doors which enabled the separation of components of the plankton swimming upwards, downwards, or remaining at the same level under

various conditions. It was found that *Pseudodiaptomus* migration was largely inhibited by water of lower salinity on the surface. Nevertheless, *Pseudodiaptomus* is at times found living naturally in unstratified water of lower salinity (Grindley, 1964).

Vertical migration must be of profound significance to have evolved in almost all planktonic animals. The survivial value, to estuarine planktonic animals, of inhibition by low surface salinity is probably in helping them to maintain their position in an estuary. In estuaries there is usually a net surface outflow of low salinity water, and a compensating influx of saline water along the bottom which maintains the salinity gradient. By vertical migrations, the planktonic animals drift alternately upstream and downstream tending to maintain their position in the estuary. When floods occur, however, the planktonic animals would be in danger of being swept out to sea if they rose to the surface. The observed inhibition of migration produced by a strong salinity discontinuity would prevent this.

This behavioural mechanism would reduce the coefficient of reproduction required to maintain a zooplankton population in an estuary for a given exchange ratio of water escaping seaward at each tidal cycle (Grindley, 1964, 1972).

Various stages of development also behave differently in vertical migrations. Juveniles often occur at higher levels than adults, particularly ovigerous females. Among benthic larvae, the late larval stages which are about to metamorphose and settle on the bottom prefer lower levels in the water as in the oyster *Crassostrea virginica* (see chapter 9).

These general remarks about vertical migration, provide an introduction to the problem of how zooplankton remains in an estuary whose waters, over a complete tidal cycle, are flowing continually into the sea. Of course during a river flood, much of the plankton *is* swept out but it reappears surprisingly quickly. Conversely if the flow is sluggish and the residence time of the estuarine water is long, planktonic populations build up to high densities. In well-stratified estuaries, marine plankton invades the estuary in the saline bottom current and spreads along the estuary with the flood tide. During the ebb, the ability to remain on, or actually in the bottom sediments, may allow it to spread rapidly along the length of the salt wedge and by upward migration to sample the water above until it reaches its salinity optimum.

Tidal dispersion

The patterns of distribution of zooplankton found, frequently bear little relation to the distribution of the physical and chemical parameters conventionally investigated, such as salinity, temperature, dissolved oxygen, and the concentrations of various nutrients. Tidal exchange appears to be the single most important factor controlling distribution of plankton. It would seem

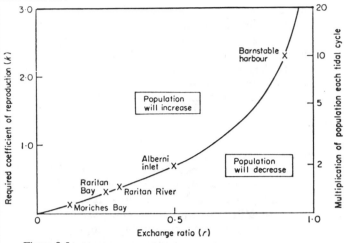

Figure 8.5
The coefficient of reproduction required to maintain steady state of endemic populations in estuaries as a function of the exchange ratio (from Perkins,1974 after Ketchum).

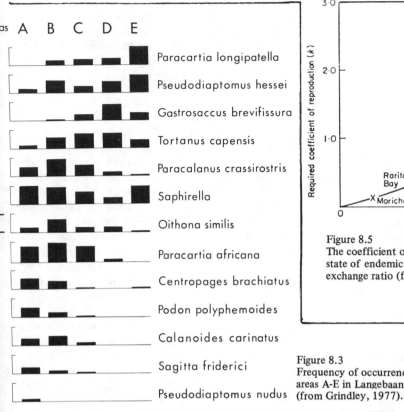

as	A	B	C	D	E	
						Paracartia longipatella
						Pseudodiaptomus hessei
						Gastrosaccus brevifissura
						Tortanus capensis
						Paracalanus crassirostris
						Saphirella
						Oithona similis
						Paracartia africana
						Centropages brachiatus
						Podon polyphemoides
						Calanoides carinatus
						Sagitta friderici
						Pseudodiaptomus nudus

Figure 8.3
Frequency of occurrence of some planktonic organisms in the areas A-E in Langebaan Lagoon (areas A-E defined in the text) (from Grindley, 1977).

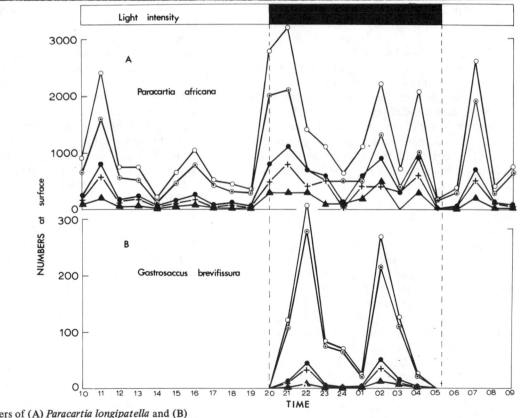

Figure 8.4
Changes in the numbers of (A) *Paracartia longipatella* and (B) *Gastrosaccus brevifissura* at the surface over a 24 h period at Schrywershoek in Langebaan Lagoon. Adult males – black triangles, adult females – black crosses, total adults – black circles, juvenile stages – centred circles, total – open circles (from Grindley, 1977).

129

that estuarine zooplankton can only survive at a point in the estuary where the rate of tidal replacement is not too great. Neritic plankton penetrating with the tides appears to be limited in a related but opposite manner largely to the mouth and channel area. Thus the ultimate survival of the estuarine plankton depends on individual reproduction rates and the exchange rate of the water as indicated in Figure 8.5.

The boundary between neritic and estuarine communities and the areas of concentration of each, appear to persist irrespective of the salinity regime prevailing. The boundary of course shifts with the state of the tide pushing up-estuary with the flood tide until high water and falling back towards the mouth with the ebb. It would seem that estuarine zooplankton can only survive where the residence time of the water is long enough for the generation time of a particular species. Margalef (1967) has shown experimentally that if the residence time of the water is too short, or a particular species is not adapted to maintain station in an estuary, that species cannot multiply effectively.

Salinity tolerance

Salinity tolerance is *one* of the most important factors limiting the distribution of estuarine plankton. To understand the extent to which distribution is limited by salinity tolerance, survival and acclimation experiments have been carried out with various species of estuarine plankton. Species of *Pseudodiaptomus* are well-known for their ability to inhabit a wide range of salinities (Marsh, 1933). *P. hessei* has been found in water from less than 1 to 74 ‰, while *P. stuhlmanni* found from less than 1 to 75 ‰. Survival experiments indicate peak survival for both species near 35 ‰ but with a wide spread of tolerance reaching above 70 ‰ in both *P. hessei* and *P. stuhlmanni*. The fact that they are seldom found in sea water near the mouths of estuaries but dominate the hypo- and hyper-saline parts of estuaries, indicates that factors other than salinity preference, influence their distribution. Probably they cannot compete with the wide range of species occurring in sea water but in higher and lower salinities where there is little competition, they flourish.

In Australian estuaries *Gladioferens imparipes* lives in waters with a wide range of salinity. Studies in the Swan River estuary of Western Australia (Hodgkin & Rippingale, 1971) revealed that their distribution was limited by other factors and that they only occurred in water of low salinity. They were seldom present where either of two less euryhaline species, *Sulcanus conflictus* or *Acartia clausi* were abundant. These two species prey on the larvae of *Gladioferens*. When these predators are present in large numbers they may eliminate the nauplii of *Gladioferens*. *Gladioferens* is then restricted to parts of the estuaries where environmental conditions are unfavourable to *Sulcanus* or *Acartia*.

Some interesting results were obtained in experiments with specimens of *P. hessei* from Lake Nhlange (salinity 3 ‰) and Lake Sibayi (chlorinity 0,1 ‰). Both groups showed peak survival between 5 and 15 ‰ but they would not survive in salinities above 25 ‰. In becoming adapted to the exceptionally low salinities of their relict estuarine environment, their range of tolerance has been reduced. However, it was found by acclimation experiments that they could be readapted to sea water, indicating that their low-salinity adaptation was physiological rather than a fixed genetic change. In acclimation experiments *P. stuhlmanni* were able to extend from fresh water to over 70 ‰ (Grindley, unpublished).

As noted in chapter 14 and discussed in chapter 3, there is a reduced flow of fresh water into the northern lakes of the St Lucia system during droughts, while evaporation steadily increases their salinity. During March 1969 the salinity reached a maximum of 89 ‰ in northern False Bay while in March 1970 the salinity reached over 100 ‰ in the eastern part of North Lake. These extreme conditions resulted in the death of almost all living organisms including the plankton, over wide areas. In March 1970, a series of plankton samples were taken along a transect from Charters Creek in South Lake to Lister Point in False Bay covering a salinity range from 40 to 76 ‰. The maximum salinity in which various species could survive under field conditions was determined in this way.

Table 8.7 shows the ranges of salinity in which various species of Copepoda have been found in South Africa. It is to be emphasized that these are not salinity tolerances, but merely their recorded range in natural conditions. It must be recognised that the absence of particular species above or below a particular salinity does not indicate that it cannot tolerate a greater range. Liebig's Law of the Minimum holds although the ultimate limit may be biological and related to competition, predation or feeding.

In 1969, the hypersaline conditions in the St Lucia system resulted in various secondary ecological disturbances affecting the plankton (Grindley & Heydorn, 1970). Red water, due to *Noctiluca scintillans* appeared in July 1969 when the salinity was reduced by rains. It is probable that this example of red water in an estuarine system was initiated by the organic nutrients released in the decay of the organisms killed by the hypersaline water. In the absence of their normal predators, chironomid pupae became an important component of the plankton and the exceptionally large swarms of midges that followed, caused further ecological disturbances on land.

Detritus

A general discussion of organic detritus will be found in chapter 16. In recent years many workers have found that suspended detritus may play an important role in

Table 8.7
Recorded salinity range of Copepoda

Copepod species	0	10	20	30	40	50	60	70	80
Acartia (Paracartia) africana									
Acartia (Paracartia) longipatella									
Acartia (Arcartiella) natalensis									
Calanoides carinatus									
Centropages brachiatus									
Centropages chierchiae									
Centropages furcatus									
Clausidium sp									
Clausocalanus furcatus									
Corycaeus spp									
Ctenocalanus vanus									
Euterpina acutifrons									
Halicyclops spp									
Harpacticus ? gracilis									
Hemicyclops sp									
Nannocalanus minor									
Oithona brevicornis/nana									
Oithona plumifera									
Oithona similis									
Paracalanus aculeatus									
Paracalanus crassirostris									
Paracalanus parvus									
Porcellidium sp									
Pseudodiaptomus stuhlmanni									
Pseudodiaptomus hessei									
Pseudodiaptomus nudus									
Rhincalanus nasutus									
Saphirella stages									
Temora turbinata									
Tortanus capensis									
Tegastes sp									

the feeding of estuarine zooplankton. In most examples, the evidence is indirect or circumstantial.

Heinle & Flemer (1975) in their study of *Eurytemora affinis* in the Patuxent estuary, found no pronounced spring phytoplankton maximum. Gross algal production was many times (5-75) smaller than the carbon requirements of the *Eurytemora* population. However, the water was always highly turbid with detrital carbon (POC) present in large amounts. They suggested that organic detritus may constitute an important link between primary and secondary production in the plankton as well as in other sections of estuarine ecosystems. Heinle *et al* (1976) suggested that the mysid *Neomysis americana* may be an important estuarine detritivore but as yet there was not direct evidence. They did, however, indicate that *Eurytemora affinis* can use the micro-organisms on detrital particles as food.

A phytoplankton maximum occurs in the Cochin Backwater during mid-summer but Wellershaus (1976) has shown that there is no corresponding zooplankton maximum. There is a clear peak of Copepoda in the autumn and minor peaks of other zooplankton in the spring. It would appear that the zooplankton feeds on sources other than endogenous phytoplankton. Abundant organic matter is supplied by the rivers particularly during the monsoon season and this is probably their main food. It is suggested that the zoo-

plankton may play a useful role in cleaning the water of particulate sewage in this way.

In the mesohaline zone of the Westerschelde estuary in the Netherlands, the dominant copepod *Eurytemora affinis* is abundant throughout most of the year. In the nearby mesohaline lake Veere, *Acartia tonsa* is the dominant, but its abundance is limited to summer. Bakker *et al* (1977) suggest that the continuous supply of suspended detritus in the Westerschelde, explains the prolonged high biomass of *Eurytemora* while the seasonal availability of phytoplankton limits the summer abundance of *Acartia*.

The importance of organic detritus in estuarine ecosystems is widely recognized but there have been doubts about its importance to suspension feeders and few direct demonstrations of its nutritional importance to estuarine copepods or other plankton. The potential value of organic detritus and its attached micro-organisms is great (Newell, 1965, Odum & de la Cruz, 1967). The detritus food web contributes stability to estuarine ecosystems by making the energy fixed seasonally by primary producers available to consumers throughout the year as suggested by Riley (1963).

Heinle *et al* (1977) investigated the significance of detritus as food for estuarine copepods. A variety of detrital foods derived from *Spartina* and other marsh plants were fed to the copepods *Eurytemora affinis* and *Scottocalanus canadensis*. Simulation of natural

Table 8.8
Potential food resources for harpacticoid copepods (production is given as $g.C.m^{-2}.day^{-1}$. Biomass and detritus are given as $g.C.m^{-2}$)
    Autotrophic biomass was calculated from chlorophyll by using a ratio of carbon to chlorophyll of 30:1. Heterotrophic biomass was calculated from ATP by using a ratio of carbon to ATP of 280:1 and subtracting autotrophic biomass. Detritus was calculated from total organic carbon by subtracting autotrophic and heterotrophic biomasses. All calculations are rounded to two significant figures (from Sibert *et al*, 1977).

Depth in sediment (cm)	Primary production $(g.C.m^{-2}.day^{-1})$	Autotrophic biomass $(g.C.m^{-2})$	Heterotrophic activity per hour	Heterotrophic biomass $(g.C.m^{-1})$	Detritus $(g.C.m^{-2})$
0 to 1	0,079	3,2	0,12	10	160
1 to 2		1,8	0,13	10	240
2 to 5		3,9	0,063	10	580
Total	0,079	9,0	0,088*	31	980

* Depth-weighted average

detritus in laboratory feeding experiments is difficult because the nature of the microbial population appears to be critical. Autoclaved or sterilized detritus resulted in poor survival and no egg production. If a rich and abundant microbiota was present they did well. Ciliated protozoans appear to be particularly important in the transfer of detrital energy to copepods. When fed on a protozoan infusion, egg production was as high, if not higher, than for copepods fed on algal cultures. Protozoa were particularly numerous when cultured in an infusion of naturally dried *Spartina*.

The food of many harpacticoid copepods is the bacteria associated with organic detritus. Sibert *et al* (1977) investigated the feeding of harpacticoid copepods in the Nanaimo estuary. Mixed assemblages including *Tisbe furcata, Harpacticus uniremus, Harpacticus spinulosus* and *Dactylopodia crassipes* were collected from the estuary. They were presented with a food mixture labelled with $^{14}CO_2$ (for autotrophs) and ^{3}H-glucose (for heterotrophs). The copepods were found to ingest heterotrophically processed carbon sources about nine times as rapidly as carbon from autotrophic sources. The potential food resource of harpacticoids was also studied by several methods. Autotrophic biomasses were estimated from chlorophyll; heterotrophic biomasses from adenosine 5'-triphosphate (ATP), and detritus from total organic carbon measurements. Autotrophic primary productivity was measured by $^{14}CO_2$ uptake and heterotrophic activity by ^{14}C-glucose incorporation in sediment slurries (Table 8.8).

It is clear that the biomass and activity of the heterotrophic-fed populations are much greater than those of the autrotrophic population and that there is a large pool of detrital organic carbon.

The possible detrital pathways which may be involved in the utilization of organic detritus and associated micro-organisms by copepods are shown in Figure 8.6.

In theory, direct consumption of detritus (seven in Figure 8.6) should lead to maximum food-chain efficiency but Heinle *et al* (1977) suggest that this pathway is unimportant. They found that detritus is acceptable by *Scottocalanus canadensis* if combined with algal cells (8). With *Eurytemoera affinis*, Heinle *et al* have

shown that a 2-stage route (1 and 4) via bacteria appears possible; protozoa, which are a good food source may be used as food which implies either a 2-stage (3 and 5) or 3-stage (1, 2, 3) route. It is probable that many of the routes shown in Figure 8.6 are important and that combinations of pathways provide an important stabilizing effect on the trophic relations in the estuary. The pulsed seasonal production of detrital carbon from tidal marshes results in similarly pulsed production of zooplankton in some estuaries apparently without a time lag (Heinle *et al*, 1976). Feeding studies revealed that detritus from the marshes is rapidly incorporated into higher trophic levels. The resultant timing of cpoepod production in this way, indicates that year-to-year variations in detrital production may be important in the survival of larval fish in particular years, since harpacticoid copepods are the principal food of chum salmon larvae.

Predation

Zooplankton predators may appear in large numbers when their prey is abundant. Kremer (1976) studied the late summer increase of the ctenophore *Mnemiopsis leidyi* in Narragansett Bay. Their biomass increases by several orders of magnitude, reaching a peak of 60 g wet mass m^{-3} followed by a rapid decline in autumn.

Day (1967) suggested that plankton is not a major source of food in South African estuaries. He noted that few invertebrates and fishes in estuaries were known to be plankton feeders. However, the vertical migrations of zooplankton take them down to, or even into the bottom so that they may provide food for bottom feeders as well. *Pseudodiaptomus stuhlmanni* appears in the gut contents of penaeid prawns in the St Lucia lakes (Grindley, unpublished).

Whitfield (1969) studied the food resources for fish in Mhlanga estuary and found that only 1,8 % of the common species depend on zooplankton which has an annual standing crop of 1,7 $kJ.m^{-3}$ (0,08 $g.m^{-3}$). However, Heeg & Blaber (1979) concentrating on filter-feeding fish, found that *Gilchristella aestuarius, Hilsa kelee* and *Thryssa vitrirostris* in the St Lucia lakes fed

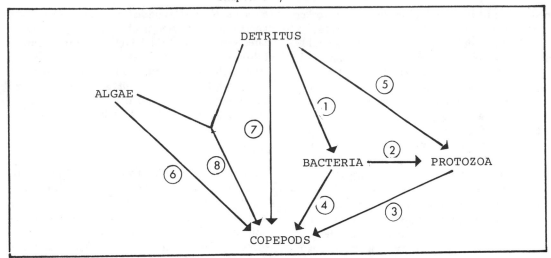

Figure 8.6
Possible pathways involved in utilization of organic detritus and associated micro-organisms (after Heinle *et al*, 1977).

largely on zooplankton. *Pseudodiaptomus stuhlmanni* formed more than 70 % of the calorific value of the zooplankton during 70 % of the study period. The mysid *Mesopodopsis africana* was second in importance in terms of its calorific contribution. Heeg & Blaber pointed out that the zooplankton, and *P. stuhlmanni* in particular, supports not only the filter feeding teleosts, but indirectly the predatory fish *Argyrosomus hololepidotus* and *Elops machnata* which feed on them. *P. stuhlmanni* thus plays an important role in the food web of the St Lucia ecosystem.

Seasonal changes

Phytoplankton in estuaries usually becomes most abundant in spring, while the zooplankton usually reaches its peak in late spring and summer. These basic patterns are frequently disrupted by flooding or by the opening and closing of estuary mouths and such changes may override the basic seasonal pattern. Although there are changes in abundance, the species composition of the plankton varies only slightly with the seasons. Considering the great changes that have taken place in the St Lucia system during the past 30 years'(Grindley, 1976), it is remarkable that there has been so little change in the plankton. It is evident that estuarine plankton is remarkably tolerant of environmental changes. Indeed, there is little evidence of long-term changes in the plankton of any South African estuary.

In many parts of the world, survival through the cold winter months is a major problem for estuarine plankton. Davis (1976) investigated the overwintering strategies of copepods in some northern Norwegian fjords. *Calanus finmarchicus* and *Pseudocalanus minutus* survived through the winter as juveniles, maturing and

producing eggs in early spring. *Acartia longiremis* occurred only as adult females in the winter and these old females produced eggs in spring. *Oithona helgolandica* was seen in all stages throughout the winter but there was no evidence of egg production. *Microsetella norvegica* remained through the winter only as adults with no egg production until spring.

Models

Plankton systems in the open sea are particularly well-suited to modelling. Phytoplankton production and the transfer of energy to herbivorous zooplankton is the sort of system where the rates of change are clearly related to measurable factors. However, estuarine conditions make relationships more difficult to clarify. By keeping most of the variables constant and studying the changes in only one or two parameters this complexity can be reduced in experimental situations. Growth and respiration rates can be established in this way but unfortunately behavioural effects can not be dealt with as readily. The activities of animals in their natural environment are not as easy to model rationally.

Gordon Riley (1946) expressed the rate of change of abundance of marine phytoplankton per unit of time (ie reproductive rate) in terms of photosynthesis and respiration rates and their environmental determinants. Riley's model is:

$$\frac{dP}{dt} = P (Ph - Re - G),$$

where Ph is the instantaneous rate of production of organic material by photosynthesis, Re is the instantaneous rate of loss of organic material by respiration, and G is the grazing rate. Far more sophisticated models have been developed in recent years.

In principle there is little difference between the procedures involved in establishing a model and those used in designing an experiment. A hypothesis is established and must be tested. Riley and his co-workers found that the seasonal variations in the amount of both phytoplankton and zooplankton in the sea could be simulated by an expression involving environmental limiting factors and physiological coefficients determined in laboratory experiments. Riley's models were reasonably successful and fair agreement was obtained between calculated and observed changes.

In estuaries stratification, turbulent mixing, flushing and limited residence time all modify the basic biological variables. Many early numerical models of estuarine hydrological systems involved vertically integrated equations which were useful for studying tidal dynamics but of little value in simulating estuarine circulation. Models using two-dimensional grids arranged vertically, or full three-dimensional models can overcome these limitations. Numerical models can only be as good as the data used to set them up and to verify them. Nevertheless, mechanistic simulation models involving both biological and hydrological variables can be valuable tools in the study of estuarine ecosystems.

As noted earlier, the biomass of the ctenophore *Mnemiopsis leidyi* in Narragansett Bay, shows marked seasonal variations, reaching a peak of 15-60 g wet mass m^{-3} in the late summer. Kremer (1976) established a computer simulation model of the dynamics of individual growth and population biomass. Zooplankton concentration (food) and temperature were the two forcing functions of the model. The model's prediction of the population increase and peak biomass agreed well with observations. Reduced ctenophore growth and depressed fecundity explained the stabilization of the population in late summer when its zooplankton prey was depleted. However the model was inadequate to explain the autumn decline. Hydrological flushing, predation and metabolic demands must be postulated here. Nevertheless the model clarified the pulsed nature of the population dynamics of the ctenophore. Not all plankton models relate to production cycles. The effect of tidal exchange on the distribution of estuarine plankton is particularly interesting. Margalef's work and recent studies (Grindley, 1978a), indicates that this is the most important factor involved. In most estuaries there is a point above which the residence time of the water is sufficient for estuarine plankton to survive while nearer the sea the rapid tidal exchange results in this area being dominated by neritic plankton. For each species there is a point in the estuary where there is a balance between the exchange ratio of water and the coefficient of reproduction required to maintain the population. Population increase can occur upstream while nearer the mouth tidal dispersion will produce a decrease.

The effects of tidal exchange have been investigated by means of a physical model. With the co-operation of the South African National Research Institute for Oceanology (NRIO) and the Fisheries Development Corporation a physical model of the Knysna estuary

was made by NRIO and has been used for simulation experiments. As noted in chapter 3, the model estimated the extent of tidal exchange and residence time at different points in Knysna estuary using fluoresceine dye as an indicator. Dye was later used at the top of the estuary to represent estuarine plankton; here the dilution indicated the rate of plankton dispersion by tidal exchange (Grindley, 1977b).

Biomass

The biomass of zooplankton is often significantly higher in the upper reaches of estuaries where the diversity is low and estuarine species predominate. Zooplankton species may reach several thousand per cubic metre. Thus *Pseudodiaptomus stuhlmanni* reached 42 700 m^{-3} in the unmodified Richards Bay in November 1971 (Grindley & Wooldridge, 1974).

Zooplankton biomass may be estimated from settled volumes of zooplankton. Although the settled volume is generally regarded as an unreliable measure of plankton biomass, the relative uniformity of size of organisms in estuarine plankton gives a fairly consistent packing density. A significant correlation was found between settled volume and biomass (dry mass) with a correlation co-efficient significant at the one per cent level. When tested, this relationship was found to be reliable for samples from Richards Bay and with samples from elsewhere except where such samples were dominated by large shrimps or gelatinous organisms. Using this method, determinations of zooplankton biomass have been found to range up to several hundred milligrams per cubic metre of water.

Zooplankton biomass may vary very greatly from time to time and from place to place even within one estuary. Thus single figures or means are of limited value. The range of values recorded for estuaries around the South African coast are as follows:

East coast estuaries:	1,2 – 1 200 mg.m^{-3}
South coast estuaries:	1,0 – 112,7 mg.m^{-3}
West coast estuaries:	1,4 – 1 014,5 mg.m^{-3}

However, the majority of biomass figures fall between 10 and 100 mg.m^{-3} (Grindley, unpublished).

Secondary productivity

Quantitative studies are now providing some understanding of zooplankton secondary productivity in estuaries. The rate of recovery from major floods indicates high levels of secondary productivity at such times. On this basis, a preliminary assessment of nett secondary zooplankton production in Richards Bay gave a maximum value of 12 mg (dry mass) m^{-3}.day^{-1} or approximately 4,4 g.m^{-3} y^{-1}. This gives a P/B ratio of 0,04 per 24 hours or a P/B ratio of 13 per year (Grindley & Wooldridge, 1974).

Studies of growth rate in relation to salinity and water temperature with *Pseudodiaptomus hessei* and studies of the rate and frequency of egg production have been completed but are not yet published.

In recent years it has become clear that the productivity of estuaries is not dependent on the same factors as operate in the open sea or fresh water. Plankton samples in most estuaries, indicate large amounts of suspended detritus, while the phytoplankton is relatively sparse. Day (1951) emphasized that attached plants and organic detritus derived from them formed the food of the bulk of the estuarine fauna. Although photosynthesis by plants may provide the primary source of food, an assessment of the amount of food available to animals is not a simple matter. A discussion will be found in chapter 16, but it may be noted here that many animals do not feed directly on the plants or on detritus but rather on the bacteria and other micro-organisms growing on the detritus. The estuarine zooplankton appears to play a role in the utilization of this secondary bacterial production either directly or by feeding on the protozoa and other micro-organisms which live on or with the bacteria. A knowledge of the primary productivity of any aquatic ecosystem is a prerequisite for a real understanding of the secondary productivity of estuaries. Studies of the zooplankton are only beginning to provide some understanding of the secondary productivity of estuaries. Studies of the inter-relationships of phytoplankton, bacteria, protozoa and zooplankton are required. Little is yet known of the relative importance of bacteria, protozoa, nanoplankton, phytoplankton, zooplankton and detritus in the bionomics of estuaries.

THE PLANKTON OF SOUTH AFRICAN ESTUARIES

Twenty years ago practically nothing was known of the plankton of South African estuaries and the dominant plankton species had not even been described. In more recent years, research at several universities and institutes has progressed rapidly and today the plankton is reasonably well-known. This research has also provided additional information on the present condition of the estuaries which is of practical importance for planning and conservation.

The plankton in 95 estuaries around the coast of southern Africa from Angola to Mozambique has been surveyed (Grindley, 1970a). A series of stations was established in each estuary from the mouth often to the limit of saline influence. All sampling was done at night. Various nets were used as well as a Clarke-Bumpus sampler, and basic hydrological data were recorded. Certain estuaries were periodically sampled to investigate seasonal fluctuations and their annual variations. Experimental studies have been made of certain aspects of the biology and behaviour of selected plankton species. Vertical migration behaviour, salinity

tolerance, development, feeding and responses to light have also been investigated.

In the following pages, accounts of the plankton of eight South African estuaries and lagoons are given. These examples have been selected to illustrate features of particular interest as well as general distribution patterns common to many southern African estuaries.

Olifants estuary

The Olifants estuary is a large permanently open estuary on the west coast with a catchment in the winter rainfall area. A general account is given in chapter 14.

The flood tide brings dense phytoplankton into the estuary from the rich upwelling waters of the Benguela current. Neritic diatoms are abundant in estuarine plankton samples but no study has yet been made of the phytoplankton as a whole.

Near the mouth, typical marine forms are dominant. The larvae of polychaetes, cirripedes, decapods, molluscs and fish appear as well as neritic copepods such as *Calanoides carinatus* and *Centropages brachiatus* typical of the Benguela current. Euryhaline marine forms such as *Oithona brevicornis* and *Paracalanus crassirostris* are common in the lower reaches and penetrate further up the estuary.

The upper reaches also support large numbers of estuarine copepods such as *Pseudodiaptomus hessei* and *Paracartia longipatella*. Other Crustacea occurring in significant numbers, are the mysids *Gastrosaccus brevifissura* and *Mesopodopsis slabberi* as well as various Amphipoda, Isopoda and Cumacea. Details have not yet been published although some points have been included in general papers (Grindley, 1978c, 1979).

Langebaan Lagoon

Langebaan lagoon has been described in many publications and a map and summarized account will be found in chapter 14. It is a 15 km long, shallow marine inlet, part of Saldanha Bay on the west coast of South Africa. It is not an estuary but its fauna includes many estuarine species as do many well-sheltered embayments.

Monthly measurements of phytoplankton primary production were made at six stations in Saldanha Bay and Langebaan lagoon using the light and dark bottle (oxygen) method (Henry *et al,* 1977). The gross primary production showed a marked seasonal variation from 261,6 mg.C.m^{-3}day^{-1} in winter through 675,4 mg. C.m^{-3}day^{-1} in spring to a maximum of 885,3 mg.C.m^{-3}. day^{-1} in summer. Production rates ranged from 611,0 mg.C.m^{-3}day^{-1} at the entrance of the lagoon to 162,6 mg.C.m^{-3}day^{-1} at a station in the middle of the lagoon where nutrient levels were apparently limiting. In Saldanha Bay, which receives nutrient-rich upwelled water from the Benguela current, mean 'total chloro-

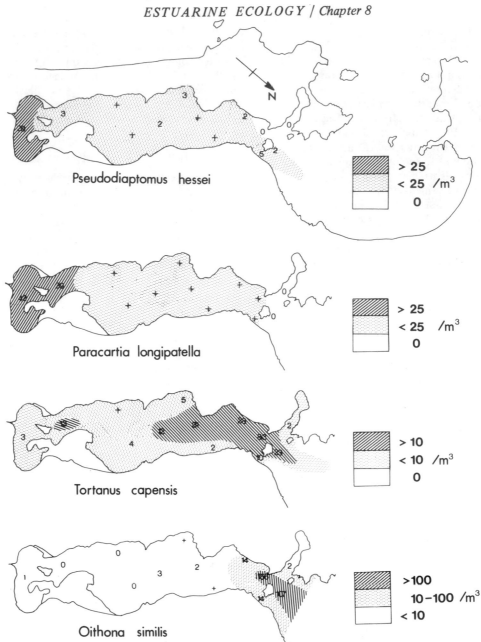

Figure 8.7
Distribution of four species of Copepoda in Langebaan Lagoon
and Saldanha Bay (from Grindley, 1977).

phyll' values ranged from 4,9 to 32,9 mg chlorophyll
m⁻³. A correlation between the values for total chlorophyll and gross primary production established for the
lagoon, was used in conjunction with these chlorophyll
values and water transparency measurements from Saldanha Bay to make a rather doubtful estimate of the
gross primary production for the bay. The estimates
ranged from 364,9 mg.C.m⁻²day⁻¹ in winter, after
which a steady increase took place until the late summer values reached 2 040,7 mg.C.m⁻²day⁻¹. The highest
value was in autumn, the estimated value for May
being 3 301,2 mg.C.m⁻²day⁻¹ (Henry *et al*, 1977). These

values are based on the correlation established for
stations in Saldanha Bay and Langebaan lagoon and
the latter is dominated by nanoplankton flagellates
while Saldanha Bay is dominated by diatoms. Possible
differences between the productivity of different forms
per milligram of chlorophyll have not been taken into
consideration in these estimates. Moreover, Vollenweider (1971) points out that chlorophyll content is
an unreliable measure because of the number of other
factors concerned.

The zooplankton of Langebaan lagoon and Saldanha
Bay was described by Grindley (1977a) on the basis of
127 plankton samples taken between 1946 and 1976.
Different zooplankton communities occur in the diffe-

rent parts of the Saldanha-Langebaan system which ranges from typical marine conditions in Saldanha Bay to sheltered conditions at the head of Langebaan lagoon. High-tide salt marsh pools may become hypersaline but the lagoon water retains a sea water range of salinity.

In Saldanha Bay, the plankton is dominated by species which occur in the neritic plankton of the west coast. The most abundant Copepoda are *Paracartia africana, Paracalanus crassirostris, Paracalanus parvus, Oithona similis,* various harpacticoids, *Centropages brachiatus* and *Calanoides carinatus.* A wide range of other planktonic invertebrates including the cladoceran *Podon polyphemoides,* the mysid *Mysidopsis similis* and the chaetognath *Sagitta friderici* may also be abundant at times. As is to be expected in a sheltered bay, large numbers of meroplanktonic larvae of benthic invertebrates occur. Among the most abundant, are cirripede, polychaete and decapod larvae.

In most parts of Langebaan lagoon, harpacticoid copepoda, and the calanoids *Paracartia africana, Paracalanus crassirostris, Saphirella* stages and *Oithona similis* are common and *Pseudodiaptomus hessei* and *Tortanus capensis* occur frequently. Many of the neritic species which occur in Saldanha Bay, penetrate into Langebaan lagoon to a limited extent. The mysid *Gastrosaccus brevifissura* which is common in much of the lagoon was absent in Saldanha Bay. Vast numbers of larvae of benthic invertebrates appear in the lagoon.

At the head of the lagoon, the dominant species are *Pseudodiaptomus hessei, Paracartia longipatella, Saphirella* stages, several species of harpacticoid copepods and the mysid *Gastrosaccus brevifissura.* These same species are characteristic of estuaries on the west and south coasts of South Africa. Cirripede nauplii and hydroid medusae are absent in this area. Organic detritus is also a prominent feature of plankton samples from this and other areas of Langebaan Lagoon.

The plankton of the central part of Langebaan lagoon is distinct from that of Saldanha Bay as well as from that of the area at the head of the lagoon in regard to species composition in terms of frequency and abundance, diversity and biomass. The zooplankton at the head of the lagoon is almost indistinguishable from that of estuaries. This is remarkable as it has been stressed that Langebaan lagoon is not an estuary. It has no inflowing rivers and salinity values are in the normal sea-water range or, in pools, slightly into the hyper-saline range (32,7-43,0 ‰). In these variable conditions at the head of the lagoon, estuarine species such as *Pseudodiaptomus hessei* and *Paracartia longipatella* appear to be better able to survive than most species common in the lagoon, while neritic marine species barely penetrate this area (Figure 8.7). Studies of plankton movements made it clear that tidal exchange transports the plankton extensively. The markedly reduced percentage occurrence of 'estuarine' species beyond this area indicates that they do not compete successfully except in areas where environmental

fluctuations limit the success of other species. The estuarine species need to maintain their position in areas where they can compete and musta avoid tidal dispersion to unfavourable areas. As noted earlier, the residence time of the water must be long enough to accommodate the generation time of the species that occur.

As elsewhere, vertical migration behaviour appears to play an important role. The results of two 24 hour series indicate that nearly all of the species of zooplankton studied *do* undertake vertical migrations. The zooplankton tends to move with the tidal currents but remains in water of similar salinity. It is not clear what controls the plankton community in the rest of Langebaan lagoon. They too must be limited by tidal dispersion both near the head of the lagoon where they tend to be swept into more variable salinities and at the mouth where they tend to be swept into the fully marine conditions of Saldanha Bay. However, it is not clear what features of the lagoon environment are significant. Copepods such as *Tortanus capensis,* mysids such as *Gastrosaccus brevifissura,* vast numbers of harpacticoid copepods and large numbers of meroplanktonic larvae occur in the middle reaches. Shallow conditions and the proximity of the bottom are probably important. Certainly the benthic crustaceans which are only temporarily planktonic such as gammarid amphipods and cumaceans may be affected more by the nature of the benthic environment. The paucity of planktonic diatoms in the lagoon and the relative abundance of nanoplanktonic flagellates and organic detritus are probably important but insufficient is known of the trophic relationships of the zooplankton to assess the significance of such factors. However, the 24 hour series of samples shows that there must be some environmental factors which favour the success of the plankton community within the lagoon and limit its success in Saldanha Bay. Various neritic marine species which are known to breed in Saldanha Bay penetrate only to a limited extend into Langebaan lagoon. Here the tidal dispersal limit must operate in reverse. The tidal transport of phytoplankton, nano-plankton and organic detritus must be significant for it has been estimated that 50 % of the lagoon water may be exchanged at spring tides (Shannon & Stander, 1977). Some data suggest that tidal import of marine phytoplankton may be significant in the nutrition of zooplankton at the entrance of the lagoon. Tidal export of organic detritus (or tidal import of marine detritus) may be even more important.

Wilderness lagoons

The environment and the ecology of Swartvlei in the series of lakes and lagoons associated with estuaries between Wilderness and Knysna are described by Professor Brian Allanson in chapter 15. The phytoplankton of the upper reaches of Swartvlei was composed mainly of nanoplankton (less than 60 μm). The dominant diatom was *Coscinodiscus lineatus* but other diatoms, dinoflagellates and other flagellates occurred in smaller numbers. The upper reaches of Swartvlei could be described as oligotrophic as total cell counts never exceeded $3,5 \times 10^5$ cells ℓ^{-1} (Robarts, 1976). Macrophytes including *Potamogeton pectinatus* are the major primary producers and are responsible for 86 % of the primary production (Howard-Williams, 1978).

Manoplankters have been shown to be more efficient primary producers than larger phytoplankton organisms in many warm waters. Small cell size limits the rate of sinking and the extent of self-shading which may limit the productivity of larger forms. Robarts (1976) has pointed out that flagellates including dinoflagellates have an adaptive advantage over less mobile diatoms or benthic plants in being able to migrate to optimum light and nutrient levels. In systems such as Swartvlei with poor light conditions, mobile flagellates thus have a selective advantage. Low light intensities may therefore not only limit photosynthesis but also play a role in selecting the type of primary producers which occur (see Figure 8.8).

Studies of zooplankton have been described in an unpublished paper (Grindley & Wooldridge, 1973), which recorded the occurrence of 45 species of plankton organisms. *Pseudodiaptomus hessei* was abundant through most of the Wilderness Lakes system, while *Acartia natalensis* was abundant in some areas. *Grandidierella lignorum* was the commonest amphipod in the zooplankton. In the channel linking Swartvlei to the sea, ostracods were very abundant also the cumacean *Iphinoe truncata*; *Upogebia africana* larvae were recorded from the Touw river estuary.

The vertical distribution of zooplankton in the 12 m deep central area of Swartvlei revealed some interesting features. *Pseudodiaptomus hessei* was absent from the surface waters in daylight and appeared mainly between 4 and 9 m below. However, they were absent in the deoxygenated water below 9 m in this meromictic system. *Pseudodiaptomus hessei* normally migrates to the bottom during the day but the deoxygenated bottom water of Swartvlei usually enforces a midwater daylight distribution. These copepods were only able to reach the bottom after equinoctial spring tides induced an influx of sea water which resulted in aerobic conditions at 12 m.

Acartia natalensis occurred from the surface to the bottom with maximum numbers between 5 and 7 m. Their presence in surface waters in daylight is not unusual as several species of *Acartia* appear to tolerate high light intensities. Small numbers occurred down in the deoxygenated zone. The ability of this species to penetrate completely deoxygenated waters has also been observed in Mpungweni (*olim Sifungwe* lake) in the Kosi system. However, they appeared in largest numbers in the bottom water when conditions were aerobic.

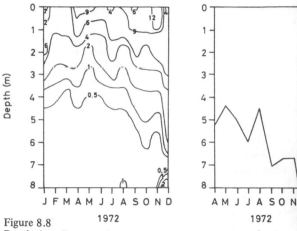

Figure 8.8
Depth-time diagram of primary productivity (mg.C.m^{-3}.h^{-1}) in the pelagic zone of the upper reaches of Swartvlei during 1972: right, the 1 % level of white light penetration (from Robarts 1976).

A species of *Halicyclops* occurred only in the lower layers of Swartvlei between 5 and 12 m. They were commonest one metre from the bottom in the deoxygenated zone and were absent on the bottom when it was aerobic.

Studies of the horizontal distribution of zooplankton sampled in a grid of 30 stations in the surface waters of Swartvlei, revealed interesting comparisons with the deoxygenated zone. This zone was present only in the deeper areas. It was found that *Acartia natalensis* was the dominant species over the deoxygenated zone in the central areas of the vlei while *Pseudodiaptomus hessei* was the dominant species in the marginal areas of Swartvlei where there was no deoxygenated zone (Figure 8.9). The lowest plankton biomass was recorded over the deep areas. All stages of *Pseudodiaptomus hessei* in reduced numbers were recorded over the deep water. Males, females (including both ovigerous and non-ovigerous forms) and juveniles were affected. The amphipod *Grandidierella lignorum* was also concentrated in the plankton over the shallow marginal areas and almost absent from central areas.

Groenvlei, unlike the remainder of the Wilderness lakes has no connection to the sea and the zooplankton is correspondingly different although the water is brackish with a salinity of 2-3,5 ‰. *Mesocyclops leuckarti* is the dominant copepod and *Diaptomus capensis* is present, while *Acartia natalensis* and *Pseudodiaptomus hessei* are absent. Cladocera including *Moina* and *Daphnia* and Ostracoda and Amphipoda are present. Comparison with Lake Sibaya, where the salinity is less than 1 ‰ is interesting, for *Pseudodiaptomus hessei* is abundant there, and other estuarine relicts also occur. These differences would appear to be related to the estuarine origin of Lake Sibaya in the Terminal Pleistocene, as discussed in chapter 15.

A detailed study of the ecology of the zooplankton of the Wilderness Lakes has been completed recently

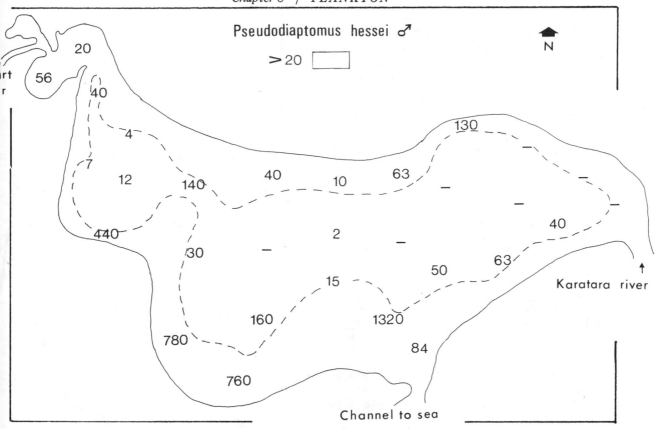

Figure 8.9
Distribution of *Pseudodiaptomus hessei* males in the surface waters of Swartvlei.

in the form of an unpublished thesis (Coetzee, 1978).

Swartkops estuary

The Swartkops estuary near Port Elizabeth has been studied more intensively than most other South African estuaries. A general account of the estuary with references to detailed studies is given in chapter 14.

The phytoplankton studies include those of Chol-noky (1960) and Giffen (1963). Studies of the nano-plankton were carried out by Henrici & Pienaar (1975, 1976) using the transmission electron micrscope. Their studies revealed a diverse nanoplankton flora; the best represented algal classes being the Haptophyceae, Chrysophyceae, Prasinophyceae and Cryptophyceae. The species identified from the Swartkops estuary are summarized in Table 8.9. Scale-bearing nanoplankters were well-represented and the Prasinophyceae and some Haptophyceae were shown to withstand con-siderable ranges of salinity (1-28 ‰). Although total chlorophyll determinations indicated that there were higher concentrations of micro-algae at the head of the estuary than near the mouth, there was, however, a greater diversity of scale bearing nanoplankton species at the mouth and this region was particularly rich in species of *Chrysochromulina*, which appear to be

Table 8.9
Nanoplankton species occurring in the Swartkops estuary (after Henrici & Pienaar, 1976).

	Summer	Winter
Prasinophyceae		
Heteromastix longifilis	●	●
Platymonas sp	●	
Pyramimonas ? amylifera	●	●
Pyramimonas orientalis		●
Pyramimonas disomata		
Pyramimonas sp	●	
Chrysophyceae		
Apidinella spinifera	●	
Paraphysomonas butcheri	●	
Sphaleromantis marina	●	
Haptophyceae		
Chrysochromulina ephippium	●	
Chrysochromulina ericina		●
Chrysochromulina regularis		●
Chrysochromulina strobilus		●
Prymnesium parvum		●
Prymnesium ? saltans	●	
Imantonia rotunda	●	
Hymenomonas carterae	?	●
Emiliana huxleyi	?	●

139

Table 8.10
Larval fish families and species sampled in the Swartkops estuary (from Melville-Smith & Baird, 1979)

Family	Species	Common name
Clupeidae	*Gilchristella aestuarius*	estuarine round-herring redeye,
	Etrumeus teres	round-herring
Engraulidae	*Stolephorus commersonii*	tropical anchovy
Hemiramphidae	*Hemiramphus far ?*	spotted halfbeak
Soleidae	*Heteromycteris capensis*	Cape sole
	Solea bleekeri	blackhand sole
Syngnathidae	*Syngnathus acus*	longnose pipefish
Sciaenidae	*Argyrosomus hololepidotus**	kob
Monodactylidae	*Monodactylus falciformis*	Cape moony
Pomadasyidae	*Pomadasys* sp	grunters
Sparidae	*Rhabdosargus* sp	stumpnose
Atherinidae	*Hepsetia breviceps*	Cape silverside
Gobiidae	More than one species	gobies
	Psammogobius knysnaensis	Knysna sandgoby
Blenniidae	More than one species	blennie
	Omobranchus woodi	kappie blenny
Clinidae	*Clinus superciliosus ?*	super klipfish
Platycephalidae	*Platycephalus indicus*	bartail flathead
Elopidae	*Elops machnata*	tenpounder

* very doubtful — Editor

unable to withstand fluctuations in salinity. Temperature was also thought to be important in the occurrence of nanoplankton species as judged by seasonal changes. Certain species such as *Pyramimonas* sp could be classed as summer forms while winter species included *Pyramimonas orientalis;* others again, such as *Heteromastix longifilus* were present throughout the year. Although productivity studies have not yet been undertaken it was clear that the nanoplankters were significant primary producers in Swartkops estuary. This applies particularly to the Prasinophyceae.

The zooplankton of the Swartkops estuary was studied for eight years (Grindley, 1976a). During the first year and a half, monthly sampling was carried out at a series of 13 stations through the estuary and in later years only at two stations which were found to be representative of the lower and upper reaches of the estuary. Zooplankton abundance (as shown in Figure 8.10) varied by nearly two orders of magnitude during the eight year period. Also a seasonal cycle was evident with an early summer peak of abundance and occasional smaller peaks in autumn.

The copepod *Pseudodiaptomus hessei,* although absent near the mouth, occurred throughout the rest of the estuary being particularly abundant in the upper reaches with peak numbers between October and December. They breed throughout the year and there was a peak of egg production in September. Species of *Acartia* included *Acartia clausi, Acartia filosa, Paracartia longipatella* and *Acartia natalensis. Acartia* species were present almost throughout except after severe floods when they appeared to be unable to re-establish themselves in the low salinities prevail-

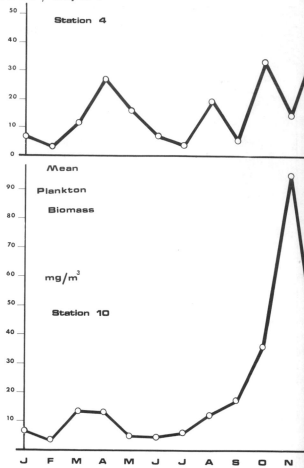

Figure 8.10
Mean monthly zooplankton biomass for two stations in the Swartkops estuary for the period 1967 to 1975. Station 4 is in the lower reaches and Station 10 in the upper reaches (from Grindley, 1976a).

ing during these periods. Peak numbers were obtained in the late spring and summer. Several species of *Oithona* occurred, including *O. similis, O. nana, O. brevicornis* and *O. plumifera* and all were most abundant in the lower estuary.

The mysid *Gastrosaccus brevifissura* was present throughout the year and virtually throughout the estuary, but it was most common in the upper reaches during late spring and summer. There was a peak of egg production in spring and juveniles became most abundant in January while later peaks in numbers of juveniles suggested three broods per year. *Gastrosaccus psammodytes,* which normally lives in sandy beaches appeared in small numbers near the mouth. Another mysid, *Mesopodopsis slabberi* occurred throughout the estuary and throughout the year. A distinct seasonal cycle was apparent with peak numbers between October and December. Floods and consequent low salinity appeared to cause downstream shifts of the population.

Three species of amphipods namely *Corophium*

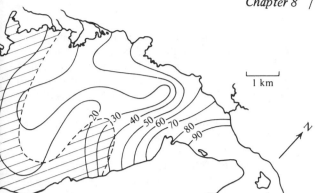

Figure 8.11
Distribution of juveniles and ovigerous females of the copepod *Pseudodiaptomus stuhlmanni* in Richards Bay in January 1970. The percentage of juveniles in the population is indicated by contours. The area where ovigerous females make up more than 30 % of the population is hatched (from Grindley & Wooldridge, 1974).

triaenonyx, Melita zeylanica and *Paramoera capensis* were common in the plankton with maximum numbers in mid-summer. The shrimp *Palaemon pacificus* (elsewhere considered as part of the benthos) occurred throughout the estuary at all times of the year but they were most abundant in spring. The single peak of numbers suggests that there might be only one prolonged breeding season per year.

Few fish breed in estuaries, but eggs of some species were obtained. Eggs ranged from 0,6 to 1,0 mm with the commonest forms 0,65 and 0,90 mm in diameter. Studies of larval fish have been made by Melville Smith & Baird (1979) (see Table 8.10).

A clear relationship between species diversity of plankton and salinity was evident. Species richness was greatest near 35 ‰. The number of species decreased towards the maximum salinity of 42 ‰. From 35 to 13 ‰ there was a gradual decrease beyond which there was a rapid decrease to 2 ‰. Below this again, there was an increase due to the appearance of fresh water species.

The patterns of distribution of phytoplankton and zooplankton in the Swartkops estuary show some measure of permanence in relation to the geography of the estuary despite fluctuating conditions. Except after severe floods, the usual distribution pattern of neritic plankton dominating the lower reaches and estuarine plankton dominating the upper reaches persisted under various environmental conditions. As in most of the other estuaries investigated, tidal exchange appears to be the most important factor controlling the distribution of plankton.

Msikaba estuary

The Msikaba estuary in Transkei is possibly the deepest

estuary on this coast with a maximum recorded depth of 35 m. This deep estuary in a narrow gorge is described in chapter 14.

The phytoplankton has not yet been studied but an account of the zooplankton is given by Wooldridge (1976). It was found that the well-developed stratification had a major influence on the distribution of the zooplankton. Marine species penetrated far up the estuary in the high salinity bottom water. Copepod species including *Centropages furcatus, Corycaeus africanus, Euterpina acutifrons, Canthocalanus pauper* and *Macrosetella gracilis* were recorded. The oligohaline species appeared in the surface waters in the upper reaches, but were not a major component of the zooplankton. The euryhaline component including estuarine species of copepods such as *Pseudodiaptomus hessei, Acartia natalensis* and *Paracartia longipatella* were relatively well-established near the mouth. There was evidence that as usually occurs, they maintained their position within the estuary by vertical migration between inflowing bottom and outflowing surface waters.

As described in chapter 3, the narrow mouth and shallow sand-bar limits turbulence due to tidal currents with in this deep narrow estuary and tidal mixing of river and marine water is thus minimized, and a stable, stratified system is established in the middle reaches. Tidal mixing does occur near the mouth and it is in this area that the true estuarine plankton is best developed. In most estuaries, this community reaches maximum abundance in the upper reaches. In this quiet deep estuary the marked stratification changes the normal pattern of distribution.

Mysidacea were represented by five species of which *Mesopodopsis africana* was the most abundant. Isopoda included *Cirolana fluviatilis* and *Eurydice longicornis,* while Amphipoda included *Corophium triaenonyx* and *Melita zeylanica.*

Zooplankton was most abundant in the late winter (August) and spring when a maximum biomass of 35 mg.m^{-3} was measured. Values fluctuated considerably, but seasonal means were similar with 16 mg.m^{-3} in winter and 15 mg.m^{-3} in summer.

Richards Bay

Richards Bay is a large shallow estuary on the east coast in which a commercial harbour has recently been constructed. The northern half has been dredged and is now separated by a berm from the southern sanctuary area which was a separate mouth. The biota and environmental conditions before modification of the Bay and the changes that have occurred in the sanctuary since then are summarized in chapter 14.

A study of the plankton was made as part of the overall environmental survey before the construction of the harbour (Grindley & Wooldridge, 1974). The phytoplankton was studied by the NIWR. They found that turbidity in Richards Bay limited the penetration of sunlight to such an extent, that phytoplankton production was usually less than 100 mg.C.m^{-3}day^{-1}.

The zooplankton was rich and two different plankton communities were apparent (Figure 8.11). In the channel area from the mouth for about 4 km, neritic plankton was predominant. This community included the larvae of many invertebrates: Copepoda such as *Centropages* spp, *Canthocalanus pauper, Pseudodiaptomus nudus, Eucalanus monachus, Temora turbinata, Lucicutia flavicornis, Oithona fallax, Oithona rigida, Corycaeus* spp, *Oncaea* spp, *Euterpina acutifrons, Macrosetella gracilis, Microsetella norvegica* and *Monstrilla* sp. The mysid *Gastrosaccus gordonae;* Ostracoda including *Cypridina vanhoeffeni* and the Ctenophore *Pleurobrachia pileus.*

In the shallow lagoon area which covered most of the upper part of the estuary, a true estuarine plankton occurred, dominated by the copepods *Pseudodiaptomus stuhlmanni, Acartia natalensis* and *Oithona brevicornis.* Mysids including *Mesopodopsis africana,* the amphipod *Grandidierella bonnieri,* the shrimps *Acetes erythraeus, A. natalensis* and the larvae of other invertebrates.

The variable and occasionally low salinity of the lagoon area severely limited the number of species able to exist there. As a result, the species diversity was low. While over 150 species of zooplankton were recorded from the channel area, less than 80 were recorded from the lagoon and most of the latter did not occur regularly. The few species of true estuarine plankton mentioned above, tended to dominate the plankton of the lagoon at all times.

While species diversity was significantly higher in the channel area, the numbers and biomass of zooplankton was significantly higher in the lagoon area. The most abundant species in the lagoon was the copepod *Pseudodiaptomus stuhlmanni* which occurred in numbers up to 42 700 m^{-3}, while *Acartia natalensis* occurred in numbers up to 186 000 m^{-3}.

The biomass of zooplankton in Richards Bay varied from 4 to 344 mg.m^{-3}. An estimate of secondary productivity of zooplankton on the basis of rate of recovery following a flood, indicated a maximum nett value of 12 mg.m^{-3}.d^{-1} (ie approximately 4,4 g.m^{-3}.d^{-1}). This gives a P/B ratio of 0,4 per 24 hours or a P/B ratio of 13 per year.

It appeared that the distribution of zooplankton in Richards Bay was determined primarily by the rate of tidal exchange in different areas. The distribution of the major communities remained essentially similar at all seasons despite a wide range of prevailing environmental conditions.

St Lucia

The complex of estuarine lakes making up the St Lucia system constitute the largest estuary in South Africa.

Details of this system are given in chapter 14, where the remarkable salinity fluctuations including periods of hypersalinity during droughts are described.

The phytoplankton of St Lucia is predominantly autochthonous and diatoms contribute most to phytoplankton volume. Small blue-green algae, flagellates and dinoflagellates are also present. Allochthonous marine and allochthonous fresh water elements also play a limited role at times.

The diatoms of St Lucia were described by Cholnoky (1968) and his ecological notes provided background information on the species recorded. He noted that there was evidence from diatoms in the sediments that hypersaline and hyposaline phases had occurred in earlier times just as in recent years. Grindley & Heydorn (1970) included some observations of phytoplankton in their description of the red water and associated phenomena during the high salinity of 1969-70.

Johnson (1976) studied the phytoplankton from 1973 to 1975. She found organisms varying in size from the large centric diatom *Coscinodiscus granii* (2×10^6 μm^3) to the minute blue-green alga *Synechococcus* sp ($4 - 10$ μm^3). Allochthonous fresh water species of *Cosmarium* and *Scenedesmus* were occasionally found at river mouths, and allochthonous marine species such as *Asterionella japonica* and species of *Thalassiothrix* and *Ceratium* were sometimes present near the mouth. Such allochthonous species were of minor significance in the system as a whole. Temporary autochthonous populations introduced from the sea, but capable of limited proliferation contributed significantly to the system. Neritic marine diatoms such as *Rhizosolenia setigera* and *Rhizosolenia stolterforthii* and some other species in lower numbers were introduced by a net inflow of sea water at times of low lake levels. Both species of *Rhizosolenia* are recognized as being eurythermal and euryhaline and in St Lucia they occurred at salinities and temperatures between 5-24 ‰ and 25,0-28,5 °C.

The permanently resident autochthonous population included *Coscinodiscus granii* which is usually considered a neritic marine species. In St Lucia, it ranged from 16-50 ‰ salinity and 18-30 °C. Autochthonous brackish water species included *Pleurosigma delicatulum, Nitzschia longissima, Chaetoceros subtilis* and two species of the blue-green alga *Synechococcus.* Dinoflagellates and other flagellates occurred in small numbers.

The largest plankton volumes occurred in the northern areas which have the most extreme salinity ranges and also the major river inputs. During 1975, phytoplankton volumes were lower than in the previous years which might have been related to the lower salinities (8-24 ‰) or to nutrient limitation. Algal bioassay indicated that P was the primary limiting nutrient in the first half of the year while both N and P were limiting in the latter half. Nutrient limitation may not have been important in the previous years for it had been suggested that the mass mortalities which occurred during the preceding hypersaline period must have caused massive nutrient release (Grindley & Heydorn, 1970). *Pleurosigma delicatulum* was the dominant species in the phytoplankton in the late summer each year with maximum abundance at 28 °C. This species was largely responsible for the peak volumes during this period.

Studies of the zooplankton of the St Lucia system have been carried out by Grindley (1976b) as part of a programme of research on the plankton of southern African estuaries. The period covered by the samples included in this study was nearly 30 years as samples were collected from 1948 to 1974 at irregular intervals. The samples available thus cover a number of major and significant environmental changes including opening and closing of the mouth and great salinity fluctuations.

Several different zooplankton communities appear in the St Lucia system. The allochthonous marine component includes stenohaline marine forms such as species of *Corycaeus* in the mouth area. Euryhaline marine species such as *Paracalanus* species may penetrate somewhat further. Most of the system, however, is dominated by a typical autochthonous estuarine zooplankton. An allochthonous fresh water community of species including *Diaptomus* and *Cyclops (sensu lato)* occurs in the mouths of the rivers and may penetrate the lakes during periods of low salinity. The true estuarine plankton which occupies mot of the system, is dominated by the copepods *Pseudodiaptomus stuhlmanni* and *Acartia natalensis*. Other copepods include *Oithona* and *Halicyclops* species and a number of harpacticoids. The most abundant mysids are *Mesopodopsis africana* and *Gastrosaccus brevifissura*. In addition, there are species of Isopoda, Amphipoda, Ostracoda, Cumacea and Tanaidacea in smaller numbers, and the larvae of many groups of invertebrates. Small numbers of fish eggs and fish larvae appeared in some samples.

During the periods of excessively high salinity between 1969 and 1971, most of the plankton disappeared in some northern parts of the St Lucia system and only the few species referred to in the section on salinity tolerance, survived. Apart from this, approximately the same community of estuarine plankton species occurred throughout the St Lucia system except near the mouth. Considering the extreme range of salinity conditions (0-115 ‰) during the period of observations, and the occasional closing of the mouth, it is remarkable that the plankton composition remained so stable. Various seasonal and longer term changes in total biomass and relative abundance were observed. For the 1967 series, the mean ash-free dry biomass was 19 mg.m^{-3} and for the 1969 series in the Charter's Creek area, 37 mg.m^{-3}. This would appear to be about an order of magnitude less than the biomass of phytoplankton calculated on the basis of the cell volume figures presented by Johnson (1976).

Except where salinities exceeded 75 ‰, the species composition of the zooplankton did not change very significantly and it is suggested that the ecological fluctuations which have been observed in higher plants and animals, in the St Lucia system must be looked for as direct salinity effects on those organisms rather than indirect effects in the lower trophic levels.

Kosi Estuary

The Kosi estuary just south of the border between Mocambique and South Africa, includes a series of interlinked basins of which the largest is Lake Nhlange. Details of this estuary system are given in chapter 14. The zooplankton of the Kosi estuary was studied by Grindley in 1967 (unpublished).

Allochthonous marine plankton dominates the mouth area and lower estuary (30-36 ‰). Twenty-two species of Copepoda were recorded of which *Corycaeus africanus, Oithona rigida, Paracalanus parvus, Euterpina acutifrons* and juvenile calanoids were the most abundant. Meroplanktonic larvae included the usual neritic groups, and there were small numbers of other tychopelagic groups.

Further up the estuary at salinities below 30 ‰, only 13 species of Copepoda appeared of which the commonest were species of *Paracalanus* and Harpacticoida. At 15 ‰ and below, *Acartia natalensis* and *Pseudodiaptomus hessei* became the dominant copepods and a number of other estuarine organisms were common. The cumacean *Iphinoe truncata* was remarkably common in the reed-fringed channels linking the separate basins. Various amphipods including *Corophium triaenonyx, Melita zeylanica, Grandidierella* sp, *Urothoe serrulidactylus* and *Austrochiltonia subtenuis* appeared as well as Tanaidacea and the mysid *Mesopodopsis africana*.

The 20 m deep basin called Mpungweni (Sifungwe in earlier reports) is meromictic and the deeper layers are at times completely anoxic and suffused with hydrogen sulphide. *Acartia natalensis* was particularly abundant on the surface here as was the mysid *Mesopodopsis africana*. The latter were found to be most abundant at a depth of about 8 m just above the anoxic zone. Both *Acartia natalensis* and *Mesopodopsis africana* were found to be active in the anoxic zone which raises interesting questions regarding their respiratory metabolism.

In the 30 m deep Lake Nhlange, the plankton is dominated by *Pseudodiaptomus hessei* while two species of cyclopoid copepods appeared only in small numbers; *Mesopodopsis africana* was present as were the amphipods *Corophium triaenonyx, Grandidierella* sp and *Austrochiltonia subtenuis* while meroplanktonic forms appeared in small numbers.

REFERENCES

ARCHIBALD, R.E.M. 1966. Some new and rare diatoms from South Africa. 2. Diatoms from Lake Sibaya and Lake Nhlange in Tongaland (Natal). *Nova Hedwigia* 12 (suppl.): 477-495.

AVARIA, P.S. 1965. Diatomeas y silicoflagelados della Bahia de Valpariso. *Rev. Biol. Mar.* 12: 61-119.

BAKKER, C. 1964. Planktonuntersuchungen in einem hollandischen Meersesarm vor und nach der Abdeichung. *Helgol. wiss. Meeresunters.* 10: 456-472.

BAKKER, C. 1972. Milieu en plankton van het Veerse Meer, een tien jaar oud Brakwatermeer in Zuidwest Nederland. *Med. Hydrobiol. Veren.* 6: 15-38.

BAKKER, C., PHAFF, W.J., VAN EWIJK-ROSIER, M. & N. DE PAUW 1977. Copepod biomass in an estuarine and a stagnant brackish environment of the SW Netherlands. *Hydrobiologia* 52: 3-13.

BAKKER, C., PHAFF, W.J., VAN EWIJK-ROSIER, M. & N. DE PAUW 1977. Copepod biomass in an estuarine and a stagnant brackish environment of the SW Netherlands. *Hydrobiologia* 52(1): 3-13.

BRAARUD, T. 1962. Species distribution in marine phytoplankton. *J. Oceanogr. Bol. Japan.* 20th Annivers. Vol. pp 628-649.

BRADFORD, J.M. 1976. A new species of *Bradyidius* (Copepods, Calanoida) from the Mngazana Estuary, Pondoland, RSA, and a review of the closely related genera *Pseudotharybis. Ann. S.Afr. Mus.* 72: 1-10.

CARPENTER, E.J. 1971. Annual phytoplankton cycle of the Cape Fear River Estuary, North Carolina. *Chesapeake Sci.* 12: 95-104.

CHOLNOKY, B.J. 1960. Beiträge zur kenntnis der Ökologie der Diatomeen in dem Swartkops-Bache nahe Port Elizabeth (Südost-Kaapland). *Hydrobiologia* 16: 229-287.

CHOLNOKY, B.J. 1962. Beiträge zur Kenntnis der Südafrikanischen Diatomeenflora. 3. Diatomeen aus der Kaap-Provinz. *Revista de Biologia* 3: 1-80.

CHOLNOKY, B.J. 1963. Beiträge zur Kenntnis des marinen Litorals von Südafrika. *Botanica Marina* 5: 38-83.

CHOLNOKY, B.J. 1968. Die Diatomeenassoziationen der Santa-Lucia-Lagune im Natal (Südafrika). *Botanica Marina* 11 (suppl.): 1-121.

COETZEE, D.J. 1978. *A contribution to the ecology of the zooplankton of the Wilderness lakes.* pp 1-167. PhD thesis, University of Stellenbosch.

CONNELL, A.D. 1974. Mysidacea of the Mtentu River estuary. *Zool. Afr.* 9: 147-159.

CONNELL, A.D. & J.R. GRINDLEY 1974. Two new species of *Acartia* (Copepoda, Calanoida) from South African estuaries. *Ann. S.Afr. Mus.* 85: 89-97.

CONOVER, R.J. 1956. Oceanography of Long Island Sound, 1952-1954. VI. Biology of *Acartia clausi* and *A. tonsa. Bull. Bingham Oceanog. Coll.* 15: 156-233.

CONOVER, S.A. 1956. Oceanography of Long Island Sound, 1952-1954. IV. Phytoplankton. *Bull. Bingham Oceanog. Coll.* 15: 62-112.

CURL, H., jr & G.C. McLEOD 1961. The physiological ecology of a marine diatom *Skeletonema costatum* (Grev.) Cleve. *J. Mar. Res.* 19: 70-88.

CUSHING, D.H. 1975. *Marine ecology and fisheries.* Cambridge University Press, Cambridge. 278pp.

DAVIS, C.C. 1976. Overwintering strategies of common planktonic copepods in some north Norway fjords and sounds. *Astarte* 9: 37-42.

DAY, J.H. 1951. The ecology of South African estuaries. Part 1. A review of estuarine conditions in general. *Trans. roy. Soc. S.Afr.* 33: 53-91.

DAY, J.H. 1967. The biology of the Knysna estuary, South Africa. pp 397-407. *In:* G. Lauff (ed), *Estuaries.* Am. Ass. Adv. Sci., Washington.

DAY, J.A. 1976. Southern African Cumacea. Part 2. Family Bodotriidae, subfamily Bodotriinae. *Ann. S.Afr. Mus.* 75: 159-290.

FURNAS, M.J., HITCHCOCK, G.L. & T.J. SMAYDA 1976. Nutrient-phytoplankton relationships in Narragansett Bay during the 1974 summer bloom. pp 118-133. *In: Estuarine processes* Vol 1. Academic Press, New York.

GIFFEN, M.H. 1963. Contributions to the diatom flora of the estuaries of the eastern Cape Province. *Hydrobiologia* 21: 201-265.

GIFFEN, M.H. 1966. Contributions to the diatom flora of South Africa. 3. Diatoms of the marine littoral regions at Kidd's Beach near East London, Cape Province, South Africa. *Nova Hedwigia* 13: 245-292.

GREEN, J. 1968. *The biology of estuarine animals.* Sidgwick & Jackson, London. 401pp.

GRIFFITHS, C. 1976. *Guide to the marine amphipods of southern Africa.* SA Museum, Cape Town. 106pp.

GRINDLEY, J.R. 1963. The Pseudodiaptomidae (Copepods, Calanoids) of southern African waters, including a new species *Pseudodiaptomus clarteri. Ann. S.Afr. Mus.* 46: 373-391.

GRINDLEY, J.R. 1964. Effect of low-salinity water on the vertical migration of estuarine plankton. *Nature* 203: 781-782.

GRINDLEY, J.R. 1969. The quaternary evolution of the Pseudodiaptomidae. *S.Afr. Archaeol. Bull.* 24: 149-150.

GRINDLEY, J.R. 1970a. The plankton of South African estuaries, pp H1, 1-16. *In: Oceanography in South Africa.* CSIR, Pretoria.

GRINDLEY, J.R. 1970b. The role of fresh water in the conservation of South African estuaries. *In: Water for the future.* RSA Water Year 1970 Convention.

GRINDLEY, J.R. 1972. The vertical migration behaviour of estuarine plankton. *Zool. Afr.* 7: 13-20.

GRINDLEY, J.R. 1976a. The plankton of the Swartkops estuary. *In: South African National Conference on Marine and Fresh Water Research, Port Elizabeth, 1976.* CSIR, Port Elizabeth.

GRINDLEY, J.R. 1976b. Zooplankton of St Lucia. *In:* A.E.F. Heydorn (ed), *St Lucia Scientific Advisory Council Workshop meeting – Charters Creek 15-17 February 1976.* Paper 12. Natal Parks Bd, Pietermaritzburg.

GRINDLEY, J.R. 1977a. The zooplankton of Langebaan Lagoon and Saldanha Bay. *Trans. roy. Soc. S.Afr.* 42: 341-369.

GRINDLEY, J.R. 1977b. Residence time tests. *In: Knysna Lagoon model investigation* 1: 28-38. NRIO, Stellenbosch.

GRINDLEY, J.R. 1978a. A model study of the role of tidal exchange in the control of estuarine plankton distribution. *S.Afr. Ass. Adv. Sci. (Abstracts),* Stellenbosch, July 1978.

GRINDLEY, J.R. 1978b. A new species of *Tortanus* (Crustacea, Copepoda) from South Africa. *Ann. S.Afr. Mus.* 74: 219-228.

GRINDLEY, J.R. 1978c. A community structure model for the estuaries and lagoons of the desert west coast of southern Africa. *In: Internatl. Conf. Ecol., Jerusalem, Israel.* Intercol, Jerusalem.

GRINDLEY, J.R. 1978d. The zooplankton of mangrove estuaries on the east coast of southern Africa. *In: Internatl. Conf. Ecol., Jerusalem, Israel.* Intercol, Jerusalem.

GRINDLEY, J.R. 1979. The plankton of west coast estuaries. *In: Fourth (S.Afr.) Natl. Oceanog. Symp., Cape Town, July 1979.* SANCOR, Cape Town.

GRINDLEY, J.R. & G.A. EAGLE 1978. *Environmental effects of the discharge of sewage effluent into Knysna estuary.* School of Environmental Studies, Univ. Cape Town. 62pp.

GRINDLEY, J.R. & A.E.F. HEYDORN 1970. Red water and associated phenomena in St Lucia. *S.Afr. J. Sci.* 66: 210-213.

GRINDLEY, J.R. & T. WOOLDRIDGE 1973. *The plankton of the Wilderness lagoons.* Port Elizabeth Mus., Port Elizabeth. 21pp.

GRINDLEY, J.R. & T. WOOLDRIDGE 1974. The plankton of Richards Bay. *Hydrobiol. Bull.* (Amsterdam) 8: 201-212.

GURNEY, R. 1945. Some notes on the copepod genus *Saphirella. Ann. Mag. nat. Hist.* (Ser. 11) **11**: 825-829.

HEEG, J. & S.J.M BLABER 1979. The biology of filter feeding teleosts in Lake St Lucia, Zululand. *In: Fourth (S.Afr.) Natl. Oceanogr. Symp., Cape Town, July 1979.* SANCOR, Cape Town.

HEINLE, D.R. & D.A. FLEMER 1975. Carbon requirements of a population of the estuarine copepods. *Eurytemora affinis. Mar. Biol.* **31**: 235-247.

HEINLE, D.R., HARRIS, R.P., USTACH, J.F. & D.A. FLEMER 1977. Detritus as food for estuarine copepods. *Mar. Biol.* **40**: 341-353.

HEINLE, D.R., FLEMER, D.A. & F. USTACH 1976. Contribution of tidal marshlands to mid-Atlantic estuarine food-chains. *In: Estuarine Processes* Vol **2**. pp 309-320. Academic Press, New York.

HEINRICI, R. & R.N. PIENAAR 1975. A preliminary study of the nanoplankton from the Swartkops estuary, Port Elizabeth. *Proc. Electron Micros. Soc. S.A.* Vol **5**: 51-52.

HENRICI, R. & R.N. PIENAAR 1976. Observations on scale bearing nanoplankton from the Swartkops estuary, Port Elizabeth. *In: S.A. Natl. Conf. Mar. Fresh Water Res. Port Elizabeth, 1976.* pp 1-22. CSIR, Port Elizabeth.

HENRY, J.L., MOSTERT, S.A. & N.D. CHRISTIE 1977. Phytoplankton production in Langebaan Lagoon and Saldanha Bay. *Trans. roy. Soc. S.Afr.* **42**: 383-388.

HILL, B.J. 1966. A contribution to the ecology of the Umlalazi estuary. *Zool. Afr.* **2**:1-24.

HODGKIN, E.P. & R.J. RIPPINGALE 1971. Interspecies conflict in estuarine copepods. *Limnol. and Oceanog.* **16**: 573-576.

HOWARD-WILLIAMS, C. 1977. The distributions of nutrients in Swartvlei, a southern Cape coastal lake. *Water S.Afr.* **3**: 213-217.

HOWARD-WILLIAMS, C. 1977. The distribution of nutrients in Swartvlei, a southern Cape coastal lake. *Water S.Afr.* **3(4)**: 213-217.

HULBERT, E.M. 1970. Competition for nutrients by marine phytoplankton in oceanic, coastal and estuarine regions. *Ecology* **51**: 475-484.

HULBERT, E.M. & R.R. GUILLARD 1968. The relationship of the distribution of the diatom *Skeletonema tropicum* to temperature. *Ecology* **49**: 337-339.

HULBERT, G.M. 1956. Phytoplankton of Great Pond, Massachusetts. *Biol. Bull.* **110(2)**: 157-168.

JOHNSON, I.M. 1976. Studies on the phytoplankton of the St Lucia systems. *In:* A.E.F. Heydorn (ed), *St Lucia Scientific Advisory Council Workshop – Charters Creek, February 1976.* Natal Parks Bd, Pietermaritzburg.

KAWAMURA, T. 1966. Distribution of phytoplankton populations in Sandy Hook Bay and adjacent areas in relation to hydrographic conditions in June 1962. *Tech. Pap. Bur. Fish. Wildl. Wash.* **1**: 1-37.

KORRINGA, P. 1956. Oyster culture in South Africa. *Invest. Rep. Div. Sea Fish. S.Afr.* **20**: 1-86.

KREMER, P. 1976. Population dynamics and ecological energentics of a pulsed zooplankton predator, the ctenophore *Mnemiopsis leidyi. In: Estuarine Processes* Vol **1**: 197-215. Academic Press, New York.

KUTNER, M.B. 1976. Seasonal variation and phytoplankton distribution in Cannanéia region, Brazil. PhD thesis. *Publ. 361, Inst. Oceanog.* University of São Paulo.

MANZI, J.J., STOFAN, P.E. & J.Z. DUPUY 1977. Spatial heterogeneity of phytoplankton populations in estuarine surface microlayers. *Mar. Biol.* **41**: 29-38.

MARSH, C.D. 1933. Synopsis of the calanoid crustaceans, exclusive of the Diaptomidae, found in fresh and brackish waters, chiefly of North America. *Proc. US Natl. Mus.* **82**: 1-58.

MARSHALL, H.G. 1967. Plankton in James River Estuary, Virginia. 1. Phytoplankton in Willoughby Bay and Hempton Roads. *Chesapeake Sci.* **8**: 90-101.

MARSHALL, S.M., NICHOLLS, A.G. & A.P. ORR 1934. On the biology of *Calanus flamarchicus.* V. Seasonal distribution, size, weight and chemical composition in Loch Striven in 1933, and their relation to the phytoplankton. *J. mar. biol. Ass. UK* **19**: 793-827.

MASSON, H. & J.F.K. MARAIS 1975. Stomach content analyses of mullet from the Swartkops estuary. *Zool. Afr.* **10**: 193-207.

MᶜCARTHY, J.J., TAYLOR, W.R. & M.E. LOFTUS 1974. Significance of nanoplankton in the Chesapeake Bay estuary and problems associated with the measurement of nanoplankton productivity. *Mar. Biol.* **24**: 7-16.

MELVILLE-SMITH, R. & D. BAÍRD 1979. Abundance, distribution and species composition of fish larvae in the Swartkops estuary, Port Elizabeth. *In: Fourth (S.Afr.) Natl. Oceanog. Symp., Cape Town, July 1979.* SANCOR, Cape Town.

MOMMAERTS, J.P. 1969. On the distribution of major nutrients and phytoplankton in the Tamar estuary. *J. mar. biol. Ass. UK* **49**: 749-765.

NATIONAL INSTITUTE FOR WATER RESEARCH 1976. *Reports on South African Estuaries.* NIWR, Durban.

NEWELL, R. 1965. The role of detritus in the nutrition of two marine deposit feeders, the prosobranch *Hydrobia ulvoe* and the bivalve *Macoma balthica. Proc. zool. Soc. Lond.* **144**: 25-45.

NICHOLLS, A.G. 1944. Littoral Copepoda from South Australia (ii): Calanoida, Cyclopoida, Notodelphyoida, Monstrilloida and Caligoida. *Rec. S.Aust. Mus.* **8**: 1-62.

NOBLE, R.G. & J. HEMENS 1978. *Inland water ecosystems in South Africa – a review of research needs.* S.Afr. Natl. Sci. Programme Rep. **34**: 1978.

ODUM, E.P. & A.A. DE LA CRUZ, 1967. Particulate organic detritus in a Georgia salt marsh – estuarine ecosystems. pp 383-388. *In:* G. Lauff (ed), *Estuaries.* Am. Ass. Adv. Sci., Washington.

OLIFF, W.D. (ed) 1976. *National marine pollution monitoring program.* First ann. Rep. (509pp typescript). Second ann. Rep. (172pp typescript). NIWR, Durban.

OLSSON, I. & E. OLUNDH 1974. On plankton production in Kungsbacks Fjord, an estuary in the Swedish west coast. *Mar. Biol.* **24**: 17-28.

PATNAIK, S. 1973. Observations on the seasonal fluctuations of plankton in the Chilka Lake. *Indian J. Fish* **20(1)**: 43-55.

PATTEN, B.C., MULFORD, R.A. & J.E.WARINNER 1963. An annual phytoplankton cycle in the lower Chesapeake Bay. *Science* **4**: 1-20.

PEELEN, R. 1969. Possibilities to prevent blue-algal growth in the delta region of the Netherlands. *Verh. Internat. Verein. Limnol.* **17**: 763-766.

PRAKASH, A. Terrigenous organic matter and coastal phytoplankton fertility. *In:* J.D. Costlow, jr. (ed), *Fertility of the sea.* Vol 2: 351-368. Godion & Breach, New York.

QASIM, S.Z. 1970. Some problems related to the food chain in a tropical estuary. *In:* J.H. Steele (ed), *Marine Food Chains* pp 45-51. Oliver & Boyd, Edinburgh.

QASIM, S.Z. 1973. Productivity of back waters and estuaries. *In:* B. Zeitschel (ed), *The biology of the Indian Ocean, ecological studies 3.* pp 143-154. Springer-Verlag, Berlin.

REVELANTE, N. & M. GILMARTIN 1978. Characteristics of the microplankton and nanoplankton communities of an Australian coastal plain estuary. *Aust. J. mar. Freshw. Res.* **29**: 9-18.

RICHMAN, S., HEINLE, D.R. & R. HUFF 1977. Grazing by adult estuarine calanoid copepods of the Chesapeake Bay. *Mar. Biol.* **42**: 69-84.

RILEY, G.A. 1946. Factors controlling phytoplankton populations on George's Bank. *J. mar. Res.* **6**: 54-73.

RILEY, G.A. 1967. The plankton of estuaries *In:* G. Lauff (ed), *Estuaries.* pp 316-326. Am. Ass. Adv. Sci., Washington.

RILEY, G.A. 1952. Phytoplankton of Block Island Sound, 1949. *Bull. Bingham Oceanog. Coll.* **13**: 40-64.

RILEY, G.A. 1963. Organic aggregates in sea water and the dynamics of their formation and utilization. *Limnol. Oceanogr.* **8**: 373-381.

RHODE, W., VOLLENWEIDER, R.A. & A. NAUWERCK 1958. The primary production and standing crop of phytoplankton. *In:* A.A. buzzati-Travesso (Ed), *Perspectives in Marine Biology.* pp 299-322. Univ. Calif. Press, Berkeley.

ROBERTS, R.D. 1976. Primary productivity of the upper reaches of a South African estuary (Swartvlei). *J. exp. mar. Biol. Ecol.* **24**: 93-102.

SAKOMOTO, M. 1966. The chlorophyll amount in the eutrophic zone in some Japanese lakes and its significance in the photosynthetic production of phytoplankton communities. *Bot. Magazine, Tokyo.* **79**(932-933): 77-78.

SHANNON, L.V. & G.H. STANDER 1977. Physical and chemical characteristics of water in Saldanha Bay and Langebaan Lagoon. *Trans. roy. Soc. S.Afr.* **42**: 441-459.

SIBERT, J., BROWN, T.J., HEALEY, M.C., KASK, B.A. & R.J. NAIMAN 1977. Detritus-based food webs: Exploitation by juvenile chum salmon. *Science* **196**: 649-650.

SMAYDA, T.J. 1970. Growth potential bioassay of water masses using diatom cultures: Phosphorescent Bay (Puerto Rico) and Caribbean waters. *Helgoland. wiss. Meeresuntersuch.* **20**: 172-194.

SMAYDA, T.J. 1973. The growth of *Skeletonema costatum* during a winter spring bloom in Narragansett Bay. Rhode Island. *Norw. J. Bot.* **20**: 219-247.

TAFT, J.L. & W.R. TAYLOR 1976. Phosphorus dynamics in some coastal plain estuaries. *In: Estuarine Processes* Vol 1: 79-89. Academic Press, New York.

TAKAHASHI, M., FUJII, K. & T.R. PARSONS 1973. Simulation study of phytoplankton photosynthesis and growth in the Frazer River Estuary. *Mar. Biol.* **19**: 102-116.

TATTERSALL, O.S. 1952. Report on a small collection of Mysidacea from estuarine waters in South Africa. *Trans. roy. Soc. S.Afr.* **33**: 153-188.

VOLLENWEIDER, R.A. 1971. *A manual on methods for measuring primary production in aquatic environments.* (IBP Handbook 12). Blackwells Sci. Publ., Oxford.

WEBB, K.L. & L.W. HAAS 1976. The significance of urea for phytoplankton nutrition in the New York River, Virginia. *In: Estuarine Processes* Vol 1: 90-102. Academic Press, New York.

WELLERSHAUS, S. 1976. Some aspects of the plankton ecology in the Cochia backwater (a South Indian estuary). *German Scholars on India* **2**: 341-360. Nachiketa Publications, Bombay.

WHITFIELD, A.K. 1979. A quantitative study of the trophic relationships within the fish community of the Mhlanga estuary. *In: Fourth (S.Afr.) Natl. Oceanog. Symp., Cape Town, July 1979.* SANCOR (CSIR), Pretoria.

WOOD, E.J. 1971. Phytoplankton distribution in the Caribbean region. *In: Symposium on Investigations and Resources of the Caribbean Sea and Adjacent Regions, Curacáo, 18-26 November 1968.* pp 399-410. UNESCO, Paris.

WOOD, E.J.F. 1964. Studies in microbial ecology of the Australasian region. V. Microbiology of some Australian estuaries. *Nova Hedwigia* **8**: 461-527.

WOOLDRIDGE, T. 1976. The zooplankton of Msikaba estuary. *Zool. Afr.* **11**: 23-44.

WOOLDRIDGE, T. 1977a. The zooplankton of Mgazana. A Mangrove estuary in Transkei, southern Africa. *Zool. Afr.* **12**: 307-322.

WOOLDRIDGE, T. 1977b. A new species of *Laicyclops* (Copepoda, Cyclopoida) from estuaries in Transkei, southern Africa. *Ann. S.Afr. Mus.* **73**: 361-372.

WOOLDRIDGE, T. 1979. Aspects of the ecology of two estuarine species of *Acartia* (Crustacea, Copepoda). *In: Fourth (S.Afr.) Oceanog. Symp., Cape Town, July 1979.* SANCOR, CSIR, Pretoria.

Chapter 9

The estuarine fauna
J.H. Day

Department of Zoology, University of Cape Town

INTRODUCTION

Those features which apply to all elements of an estuarine fauna will be dealt with briefly for fuller discussions will be found in many reviews including Spooner & Moore (1940), Day (1951), Remane & Schlieper (1958), Baas Becking & Wood (1955), Lauff (1967), Green (1968), Barnes & Green (1972) and Perkins (1974).

Estuaries are richer in nutrients than either fresh waters or the sea so that the flora is highly productive and animal life is prolific. Surprisingly few fresh water animals invade estuaries and most of the species are of marine origin but even these include only a small fraction, probably less than a tenth of those found in the sea nearby. Thus an estuary is rich in individuals but poor in species. Sanders (1968) has correlated low species diversity with a stressed environment and one of the obvious stress factors is the variable salinity. In normal estuaries the salinity is reduced and varies with the tide; in hypersaline estuaries the salinity may increase to over 100 ‰ but tidal changes, if present at all, are small and the salinity increases slowly as river flow decreases and evaporation proceeds. Both the absolute value of the salinity and its variability must be considered for Bassindale (1943) has shown that the rate of change is as important as the magnitude of the change. Thus the fauna of the Elbe estuary studied by Caspers (1959) where the salinity never exceeds 18 ‰ and varies with the tide is much poorer in

species than in the Baltic sea nearby recorded by Remane & Schlieper (1958). The tidal changes in the Baltic beyond the Skagerack are negligible and the salinity changes gradually from fresh water in the Gulf of Finland to sea water values at the Skagerack, a distance of more than 500 km. Although Pritchard (1967) does not include the Baltic in his definition of an estuary, the distribution of the Baltic fauna in relation to salinity is often quoted in discussions of the faunistic components of an estuary. It is satisfying to give exact salinity limits to different elements of the fauna but these are not really applicable in estuaries.

The ratio of major ions in estuarine water is the same as that in the sea and to this extent marine animals are preadapted to life in estuaries, but in variable salinities osmotic problems arise. Plant cells being protected by tough cellulose cell walls and provided with a vacuole which can vary in volume are capable of resisting osmotic swelling or shrinkage within a wide range of salinity. They are not subject to the direct effect of osmosis but their tolerance to high salinities is limited by the accumulation of ions in their cells and the reduced availability of water (Jefferies, 1972). Animal cells which lack these protective devices may absorb ions in dilute media and exclude or excrete them in hypertonic environments and within limits they can maintain the internal osmotic pressure so as to prevent excessive inflow or outflow of water. In addition there are many other behavioural, physiological or anatomical adaptations. Beadle (1972) has made the interesting point that triploblastic animals protect the vital mesodermal tissues by the circulation of body fluids which act as a buffer between the internal cell concentrations and those in the external medium. Many animals avoid extreme external changes, others have impervious integuments, or passively tolerate some internal changes in concentration and then osmoregulate to eject excess osmotic inflow of water. Discussion of these mechanisms is beyond the scope of this book and readers are referred to reviews by Potts & Parry (1964), Dehnel (1967), Beadle (1957, 1972), Newell (1970, 1976) and Kinne (1964). Salinity and temperature adaptations of estuarine animals in southern Africa are reviewed by Dr Burke Hill in chapter 11.

THE COMPONENTS OF AN ESTUARINE FAUNA

Although the composition and distribution of an estuarine fauna is due to the combined effect of many factors, salinity tolerance is a convenient means of separating most of the components. The term *stenohaline* indicates a limited salinity range while *euryhaline* indicates a wider range; thus there are stenohaline and eruyhaline marine species and euryhaline and stenohaline fresh water species. The exact salinity range of any single species is affected by the stage of development, the period of acclimation, the rate of change of salinity and the temperature so that the terms stenohaline and euryhaline are not given absolute values. Within these limits the several components of an estuarine fauna are set out below. They are based on those adopted by Day (1951) and amended by Remane (1958), Carriker (1967), Green (1968) and Perkins (1974).

1. *The stenohaline marine component.* This includes species which do not tolerate salinities which differ greatly from sea water. Day (1951) suggested a range of 25-35 ‰ but later work in hypersaline estuaries suggests that the range should be increased to 25-40 ‰. This agrees with the limits proposed by Carriker. However, Remane, Green and Perkins put the lower limit at 30 ‰ and imply, but do not state, that the upper limit is 35 ‰. Most of the animals which live in the sea are stenohaline and relatively few penetrate estuaries.

2. *The euryhaline marine component.* This includes species with a much wider salinity range. Day (1951) suggested 5-35 ‰ but later experience in St Lucia Lake which occasionally reaches salinities of over 100 ‰ shows that marine species such as the bivalve *Eumarcia paupercula* and the crab *Hymenosoma orbiculare* occur in salinities over 50 ‰ and are known in other estuaries in salinities of 15 ‰. Boltt (1975) suggested that the upper lethal limit is about 60 ‰. Hedgpeth (1967)

records the marine barnacles *Balanus eburneus* and *B. amphitrite* as living in Laguna Madre in salinities up to 80 ‰. Forbes & Hill (1969) showed that *Hymenosoma orbiculare* not only lives but breeds in the fresh water of Lake Sibayi in Zululand. Such wide salinity ranges may be exceptional but they suggest that an ability to osmoregulate may be used in both hyposaline and hypersaline waters. Remane (1958) subdivided the euryhaline marine component in the Baltic since he found many species in salinities of 15 ‰, fewer in 8 ‰, fewer still in 3 ‰ and one or two below 3 ‰. As Green (1968) has remarked: 'It is doubtful if these groups can be applied to the more variable salinities of estuaries'. Probably a salinity range of 5-50 ‰ would apply to most euryhaline marine animals if the change extended over more than a week.

3. *The true estuarine component.* These are species of marine origin but restricted to estuaries and have not been recorded from the sea or fresh water. They are obviously euryhaline and may be found all over an estuary from oligohaline to hypersaline conditions between salinities of 2 and 60 ‰. There are only a few such species and as they are relatively important in low salinities. Green (1968) has referred to them as the brackish water component.

4. *The euryhaline fresh water component.* This includes a few species derived from rivers which tolerate salinities higher than 0,5 ‰ which is the accepted limit between fresh and estuarine water. The component includes a small number of aquatic oligochaetes, insects of the groups, Ephemeroptera, Hemiptera, Diptera and Coleoptera and a few Gastropods. Remane has subdivided Baltic representatives into those that penetrate to salinities of 3, 8 and above 8 ‰. In the variable salinity of estuaries a general range of 0,5-8 ‰ with exceptional species extending to 18 ‰ appears sufficient. Boltt (1975) reports large numbers of chironomids (mainly a species of *Polypedilum*) from Lake St Lucia when the salinity was between 70 and 80 ‰ and the rest of the fauna was greatly reduced. It would seem that some insects at least can not only tolerate hypersaline conditions but can multiply rapidly when competition and predation are reduced.

5. *The terrestrial component.* Green (1968) and Perkins (1974) have distinguished a small terrestrial component which seems to be indifferent to salinity. In most estuaries it is limited to a few land anthropods such as kelpflies, staphilinid and carabid beetles, small centipedes, mites and wolf spiders which feed in the drift line along the estuary. In mangrove swamps Macnae (1968) mentions ants, mosquitoes, fire flies and even several vertebrates which may be added to this component.

6. *The migratory component.* This includes those spe-

cies which actively migrate out of estuaries during part of their life cycle. Most penaeid prawns, the larger portunid crabs and the majority of pelagic fishes enter estuaries as juveniles for food and shelter and return to the sea to spawn. Many then remain there but others enter estuaries again seasonally on feeding excursions. A few fishes breed in estuaries and return to fresh water while anadromous species such as salmon and catadromous species such as eels of the genus *Anguilla* use estuaries as highways on their migrations between fresh waters and the sea. The embryonic and larval stages of many species are less tolerant of low and variable salinities than the adults and many active swimmers move into higher salinities to spawn even if they do not reach the sea. The American blue crab *Callinectes sapidus* which ranges almost to fresh water, mates in low salinities and gravid females migrate to salinities above 20 ‰ until their embryos hatch (Darnell, 1959 in Green, 1968).

Fish-eating and wading birds are an anomalous group. Although Milne & Dunnet (1972) have shown that they are very important in the economy of estuaries they are not really a component of the estuarine fauna. They are indifferent to salinity and fly from one estuary to another or from estuaries to inland waters or the sea according to the local abundance and availability of food. A full discussion of the role of birds in estuaries will be found in chapter 13.

The importance of calm water

Stephenson (1947) showed that the fauna and flora on the rocky coasts of South Africa changes progressively with increasing shelter from wave action. Lewis (1964) has shown the same on British shores. Day & Morgans (1956) and Day (1959) investigated the fauna in the sheltered waters of Durban Bay and Langebaan Lagoon where the salinity is the same as the sea and found that 70-80 % of the species in these calm waters do not occur on wave-washed coasts, but the majority of them do occur in estuaries. The fauna and flora on well-sheltered rocks is very poor and includes relatively few of the algae and benthic invertebrates present on surf-beaten rocks. The reverse is true on sandy shores. The rich fauna on sheltered sand banks is completely different from the sparse fauna on open sandy shores. It must be concluded that the estuarine fauna and flora is composed of calm water species. It is not that the euryhaline species prefer reduced salinities; they live and flourish in estuaries in spite of variable salinities.

Where wave action is reduced, stable banks of sandy mud containing organic detritus are deposited. This occurs not only on the shores of lagoons and estuaries but also at increasing depths in bays and the open sea. It is reasonable to suppose that the marine components of the estuarine fauna have been derived from the fauna of soft sediments on the sea bed below 20 m on low energy coasts and below 30 or 40 m on high energy coasts. Day *et al* (1971) surveyed the distribution of 619 species of benthic invertebrates on a transect from the open sandy shore of North Carolina to a depth of 200 m. Not one of the few species found on the sandy shore extended below 3 m. Earier workers had reported that 95 of the species that occurred in depths of 10-100 m on the transect were present in the sheltered estuarine sounds nearby, either between tide marks or 1-2 m below. There could be no doubt that the bulk of the estuarine fauna on soft sediments had been derived from the bed of the sea and the size of the specimens suggested that they grew more rapidly in the rich estuarine mud. In my later and still unpublished study of Merimbula estuary in New South Wales, a few of the species found on rocky shores outside the estuary extended to similar habitats in the estuary but none of the species on the muds and sands of the estuary was present on the barren sandy beaches outside. As further evidence it may be noted that Wallace (1975) found that many of the fish and penaeid prawns which enter Zululand estuaries as juveniles occur as adults at depths of 30-40 m in the sea.

THE DISTRIBUTION OF THE BENTHIC MACROFAUNA

For practical work it is convenient to subdivide the benthon into size groups. Animals that are retained in a sieve with meshes of 1,0 mm are referred to as the macrobenthos; those that pass through this mesh but are retained by a mesh of 0,1 mm are termed the meiofauna and those that pass through this mesh are termed the microfauna. It should be noted that these size groups are not universally adopted; Perkins (1974) quotes slightly different mesh sizes and Green (1968) uses the term microbenthos to include all animals that pass through a 1 mm screen. Sieves are never completely selective in any case.

In general the microfauna includes the Protozoa; the meiofauna consists largely of nematodes, ostracods and harpacticoid copepods while the macrofauna includes the larger members of other invertebrate phyla.

The estuarine macrofauna includes only a few species of the many insects, oligochaete worms and molluscs which occur in fresh water. Many marine invertebrates are also poorly represented. There are no corals even in tropical estuaries and there are few species of sponges, hydroids, anemones, ectoprocts, echinoderms or ascidians. Polychaete worms are always abundant. Amphipods are abundant in Cape estuaries but seem to decrease towards the tropics while the reverse is true of bivalves and gastropods although this may be merely a reflection of the larger number of species in tropical estuaries. Caridean shrimps are common in all

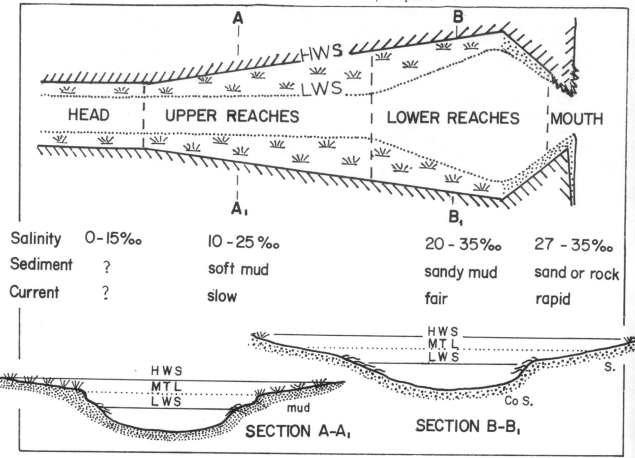

Figure 9.1
Plan and sections of an ideal estuary (modified from Day, 1951).

estuaries but penaeid prawns and crabs become very abundant in the tropics.

An estuary is a complex environment and along its length there are changes in wave energy, salinity, temperature, current velocity, substrate and aquatic vegetation. The macrobenthos is subject to the combined effect of all of these as well as other secondary factors and it is stressed that it is extreme and not average conditions which limit distribution and abundance. If any single factor exceeds the range of tolerance of a species that is the limiting factor, be it an abiotic physical factor such as salinity or a biotic factor such as suitable food or competition in its broadest sense.

If environmental factors changed evenly along the whole estuary, the fauna would change progressively. It would not be possible to divide an estuary into regions each with a greater homogeneity of species than in the overlap between them. Possibly this does occur in some estuaries but in most of them there are longer stretches where environmental conditions and thus the benthic fauna changes slowly and intervening stretches where the rate of change is more rapid. Day (1951) divided an ideal estuary into four reaches, the mouth, the middle reaches, the upper reaches and the head. Later workers have used the term 'lower reaches' instead of 'middle reaches' and for several reasons

'lower reaches' is more appropriate and is used here.

Marked changes in salinity, current velocity, the nature of the substrate and the changes in the aquatic vegetation indicate the approximate boundaries of the four reaches. The method is subjective but practical and improves with experience. The length of any reach varies from one estuary to another. The mouth region (Figure 9.1) is usually short and well-defined, the lower reaches with gently shelving banks of muddy sand are longer while the upper reaches with extensive weedbeds on the mudbanks and a steep slope or eroding salting cliff above low tide are usually the longest stretch of an estuary. The head of the estuary is the most variable stretch of all. If the gradient of the inflowing river is steep, the head of the estuary is short, the substrate is coarse sand and the salinity varies greatly with the tide. If the river is slow flowing and muddy, the head of the estuary grades imperceptibly into the upper reaches and the seasonal changes in salinity indicate that what was part of the upper reaches in the dry season becomes the head of the estuary or even the non-saline tidal river in the wet season.

The salinity changes and the distribution of the components of the benthic macrofauna in four estuaries

Table 9.1

The distribution of the benthic macrofauna along four estuaries in southern Africa

Reach	Salinity %oo	Faunistic components				Total records of species
		Stenohaline marine	Euryhaline marine	True estuarine	Fresh water	
Mlalazi						
Mouth	35-25	13 %	87 %	0	0	38
Lower reaches	35-14	7 %	88 %	3 %	2%	59
Upper reaches	29-6	0	91 %	4 %	4%	54
Head of estuary	24-0,5	0	65 %	22 %	13%	23
Total species		9(11 %)	67(80 %)	5(6 %)	3(4 %)	84
Mngazana						
Mouth	35-30	55 %	41 %	4 %	0	56
Lower reaches	35-19	36 %	57 %	6 %	1%	117
Upper reaches	30-5	0	79 %	21 %	0	24
Head of estuary	20-1	0	45 %	55 %	0	11
Total species		73(35 %)	114(55 %)	20(9,5 %)	1(0,5 %)	208
Knysna						
Mouth	36-34	54 %	45 %	2 %	0	179
Lower reaches	35-29	36 %	54 %	10 %	0	199
Upper reaches	26-19	9 %	72 %	19 %	0	64
Head of estuary	14-1	0	42 %	36 %	14 %	33
Total species		139(45 %)	137(44 %)	27(9 %)	7(2 %)	310
Breede						
Mouth	35-27	37 %	61 %	2 %	0	57
Lower reaches	34-6	8 %	79 %	11 %	0	73
Upper reaches	29-1	0	83 %	17 %	17 %	24
Head of estuary	2-0,1	0	0	17 %	83 %	6
Total species		24(24 %)	63(62 %)	9(9 %)	5(5 %)	101

along the coast of southern Africa are tabulated above. Morrumbene estuary although extensively sampled (Day, 1974) was omitted for the distinction between the stenohaline and euryhaline components of many tropical species is unknown. Several other estuaries were omitted either because they have been incompletely sampled or because they have unusual features such as a periodically closed mouth or become hypersaline at intervals. The estuaries of the Mlalazi, Mngazana, Knysna and Breede rivers were selected for they provide a wide range of conditions but they are all normal, well-developed estuaries, and all four are illustrated and briefly described in chapter 14. The Mlalazi is a subtropical estuary in Zululand with a narrow mouth opening on a sandy shore. There are narrow belts of mangroves along the banks, rather small *Zostera* beds and few rocks. The data on environmental changes and the distribution of the fauna given by Hill (1966) has been supplemented with additional records made by the NIWR (Oliff, 1976). Mngazana is a subtropical estuary in the Transkei with a mouth opening between a sandy beach and a rocky headland (Branch & Grindley, 1979). There are extensive mangrove swamps and *Zostera* beds in the lower reaches, and eroding mud banks and boulders, further up. Knysna estuary on the warm-temperate Cape coast has a wide mouth between rocky headlands where wave action is appreciable and there are swift tidal currents; the salinity remains high well up the estuary and there are extensive salt marshes and *Zostera* beds. Environmental data have been extracted from Day *et al* (1952)

and the faunistic records have been revised.

The Breede estuary is 300 km further west along the southern Cape coast. The river flow is strong, and a long stretch of the estuary is oligohaline. There are outcrops of slates along the banks but most of the substrate is soft mud; sand banks are only present within 4 km of the mouth, which is constricted by a sandspit so that wave action in the estuary is negligible. All data have been extracted from unpublished records made by myself and other members of the Zoology Department of the University of Cape Town.

The euryhaline fresh water component and the true estuarine component have been defined earlier. The euryhaline and stenohaline components should rightly be distinguished by an experimental determination of the salinity ranges of the various species. Failing this they have been distinguished by their distribution in many estuaries where the salinity variations were known. Those marine species which do not penetrate beyond the mouths of more than 20 estuaries that have been surveyed are regarded as stenohaline and those which extend into reduced or hypersaline reaches are regarded as euryhaline.

The benthic macrofauna of all four estuaries has been well-sampled; further collecting and a better identification of sponges, nemerteans, oligochaetes and insect larvae would certainly increase the number of records but it would be unlikely to change the percentage of the four components except possibly the fresh water component. This has few species compared to the numbers listed by Green (1968) for British estuaries.

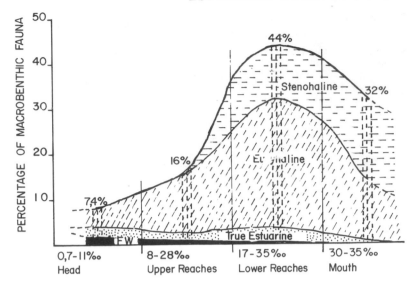

Figure 9.2
Distribution of faunistic components
and percentages of total
macrobenthos in the four reaches
of an estuary.

On the other hand the total number of macrobenthic species is much larger. It may be surprising to see that Mlalazi with 84 species has the poorest macrobenthic fauna for subtropical estuaries are potentially rich as shown by the 378 species recorded by Day (1974) in Morrumbene estuary. Mngazana with 209 species is the richest estuary in the Transkei. Knysna with 310 macrobenthic species is unusually rich for a temperate estuary but the rocks at the mouth, and the extensive sandbanks, salt marshes and *Zostera* beds provide a great variety of habitats.

While there are obvious differences in the percentages of the four components along the four estuaries, they all show the same trends. The stenohaline marine component is highest at the mouth and decreases to nil at the head of the estuary. It is particularly poor in Mlalazi and as the salinity is high at the mouth the poor fauna must be due to the shifting sands. In Knysna estuary the stenohaline marine component forms 45 % of the total macrobenthos. This is due to the inward extension of many seashore species on the wave-washed rocks at the mouth and the high and stable salinity of the lower reaches. The euryhaline marine component forms a major part of the macrobenthos of all four estuaries; only in the Breede does it stop at the upper reaches for the salinity at the head of this estuary is always below 3 ‰. The true estuarine component forms 6-9,5 % of the total macrobenthos. It is absent at the mouth of the Mlalazi but it extends along the whole length of the other three estuaries and becomes increasingly important in the upper reaches where the euryhaline marine component is reduced. The euryhaline fresh water component is even smaller (0,5-5 % of the macrofauna) but would be larger if the insect larvae could be identified and included in the analysis.

When discussing the distribution of components along an estuary many workers including Hedgpeth (1967), Beadle (1972) and Barnes (1974) have repro-

duced the graph of Remane & Schlieper (1958) or modifications of it. As noted earlier, it should not be applied to estuaries where the salinity chages with the tide. It may be added that Remane's diagram does not separate the stenohaline and euryhaline marine components or the changes in species richness along an estuary. Figure 9.2 attempts to do this. It has been derived from the mean values of the four sets of data in Table 9.1.

The percentage of the whole macrobenthic fauna recorded in each of the four reaches of the estuary is shown by histograms. These have been joined by smoothed curves to illustrate the gradual changes from one reach of the eastuary to the next. The greater diversity in the lower reaches is obviously due to the increased number of euryhaline species. The stenohaline marine species decrease from the mouth but a few are still present in the upper reaches where the salinity maximum is 28 ‰. The true estuarine and euryhaline fresh water species extend over most of the estuary.

MAINTENANCE OF STATION AND SELECTION OF HABITAT

Due to the seaward drift of estuarine water, all planktonic organisms including the eggs and larvae of benthic animals tend to be carried down the estuary and out to sea. Further, the adult population may be decimated by floods. How the population maintains its station in the estuary and the particular habitat where it normally occurs has been discussed by many workers. Reviews will be found in Carriker (1967), Newell (1970) and Lockwood (1976).

Most estuaries are stratified to some degree with a seaward current of lower salinity water near the surface, a landward current of higher salinity near the bottom and a level of no net motion over a complete tidal cycle at the halocline. All tidal currents increase from neaps to springs and the discharge to the sea increases during the rainy season. Adaptations to maintain station in an estuary are related to all these factors and several different strategies have been adopted.

Some animals have no planktonic stage while in others it is abbreviated. Amphipods and isopods retain their eggs in brood pouches until they reach the adult form and the adults themselves are active swimmers. The larvae of *Arenicola* spp and *Nereis diversicolor* are not planktonic. Several families of polychaetes (eg Eunicidae, Capitellidae and Maldanidae) undergo early development in gelatinous bags attached to the substrate and so do gastropods such as *Natica, Notarchus* and *Siphonaria.* As a result the planktonic stage is abbreviated. Forbes (1973) has shown that the burrowing prawn *Callianassa kraussi* also has an abbreviated larval stage and is distributed at the juvenile benthic stage. Carriker (1967) has suggested that estuarine species have a shorter planktonic stage than marine species to reduce the drift to the sea.

The timing of egg release is also important. Carriker noted that the clam *Mercenaria mercenaria* tends to spawn at low neap tide in summer when river discharge is minimal and the rising tide carries the eggs upestuary. How far this applies to other estuarine animals is unknown but egg release is related to lunar cycles in many cases and is often triggered by a rise in temperature such as occurs when the rising tide floods over sun-warmed banks. Day (1974) recorded a temporary increase from 27 to 32 ° as the tidal water slowly inundated on the sand banks of Morrumbene estuary. It is presumed that this is widely applicable in summer.

Zooplankton, including planktonic larvae, rise towards the surface at night and sink during the day and the vertical movement may be limited by salinity stratification. Carriker (1967) summarises many reports on the larvae of *Crassostrea virginica* as an example. The early larval stages are uniformly distributed through the water column but the older, eyed larvae are more concentrated on or near the bottom during the ebb. They are stimulated by the increased salinity and turbulence of the flood tide to greater activity and tend to rise off the bottom and be carried upestuary. Carriker also notes that the immature larvae of several gastropods and bivalves tend to congregate near the halocline where net drift is minimal but when mature, the larvae of *Mya* like those of *Crassostrea*, sink below the halocline.

The work of Bousfield (1955) on barnacle larvae in Miramichi estuary illustrates the way in which three species maintain station in different reaches of the estuary. *Balanus balanoides* and *B. crenatus* extend from the sea into the mouth of the estuary. The larvae of *B. balanoides* always remain near the surface and drift in and out of the estuary with the tide. *B. crenatus* larvae sink deeper as they develop and are more liable

to return to the estuary. *B. improvisus* is mainly estuarine and its larvae also sink as they develop and may be carried upestuary beyond their spawning sites. However, they are restricted by low salinities and do not extend as far in rainy summers as they do in dry ones.

As yet the only study of larval recruitment in southern Africa is that of McLachlan (1974). He showed that after the bivalve population of Swartkops estuary had been decimated by prolonged floods of 1971, larvae appeared in the lower reaches. Some species settled there while others were distributed higher up the estuary so that the main settlement of each species took place in the area where the adult population normally occurs. The strategy whereby the larvae regained their normal station in the estuary is unknown.

The settlement of larvae on hard surfaces

The final selection of the adult habitat by larvae that are ready to settle has been studied by many workers. Such larvae continually test the substrate with chemosensory and tactile organs; if a suitable habitat is found they metamorphose rapidly but if not, they can delay metamorphosis for long periods. The larvae of sedentary invertebrates which live on the solid surfaces of stones, dead shells or even algae have been studied for many years by Professors E.W. Knight-Jones and D.J. Crisp and their associates. References to their many papers will be found in a review by Newell (1970). Light intensity, current velocity and the texture of the surface are all important to the searching larvae but the main attractants are chemotactile stimuli. Experiments show that the attractive substance must be present on the solid surface, and not merely dissolved in the surrounding water. Extracts of the encrusting coralline *Lithothamnion* promote the settlement of *Spirorbis rupestris* which normally occurs on this substrate and extracts of *Fucus serratus* attract the larvae of *Spirorbis borealis* which is common on this alga. Settlement on rock may be illustrated by the experiments on *Balanus balanoides* which has been more thoroughly tested than other species of barnacle. Barnacles tend to grow in dense aggregations and at an early stage in the investigations it was found that *B. balanoides* larvae were attracted by extracts of barnacles and that a higher percentage of larvae settled where *B. balanoides* extracts were present than where extracts of other barnacles were used. Even dead shells were attractive to the larvae and it was eventually shown that the attractive substance is the protein arthropodin which is to some extent species specific. Once settled, the cypris larva explores the surface with its chemotactile antennules preferring rough surfaces to smooth ones, orientating itself in relation to the current and finally metamorphosing sufficiently far from neighbouring *B. balanoides* to allow for growth. Proximity to other species is disregarded so that the latter may be overgrown and in this way the dominance of *B.*

balanoides at a particular tidal level is established.

The settlement of larvae on soft sediments

Habitat selection on soft sediments is apparently guided by different stimuli. MacGinitie (1935) and many experienced field workers have stressed that species distribution in estuaries is related to 'the nature of the sediment'. To be more specific than this has proved very difficult. The many attempts to relate species distribution or abundance with the median particle size of the sediment have led to conflicting results. Better correlations have been obtained with the percentage of subsieve particles or the percentage of organic detritus but unfortunately even these parameters do not define the requirements of individual species.

Early experiments on the larvae of a burrowing polychaete were made by Day & Wilson (1934) and Wilson's many later experiments are summarised by Wilson (1955). He found that although the grade of the sediment is important, the nature of the micro-organisms is of greater significance. Work by Scheltema (1961) using veligers of the deposit feeder *Nassarius obsoletus* and by Gray (1966) using the archiannelid *Protodrilus symbioticus* indicated that the nature of the bacteria in the sediment was the main attractive factor. Moreover the attraction persisted even when the bacteria were killed so long as the organic film they had secreted on the sediment particles remained unaltered. Thus the prime selection of a suitable sediment is not made by the settling larvae but by particular strains of bacteria.

In a recent review, Stevenson & Erkenbrecker (1976) first make the point that the density of heterogenous microbial populations in estuaries is several orders of magnitude higher than those in open oceans. They go on to say that judging by the potential rate of uptake of organic substrates, four factors are of particular importance in determining the demise or survival of particular strains. These are fluctuations of salinity, competition for particular organic substrates, the nature of the surface to which the micro-organism is attached and above all its ability to remain dormant during severe conditions. Since smaller particles have a larger surface to volume ratio, the density of bacteria increases in finer sediments. Further research is obviously needed but is evident that the strains of bacteria and the settlement of the benthic larvae which they attract are ultimately dependent on salinity distribution, the grade of sediment, its organic content and the tidal level. These parameters in turn determine the permeability of the sediment, the depth of the permanent water table between tide marks, and the depth within the sediment where the oxygen concentration falls to such low levels that anaerobic bacteria develop and black ferrous sulphide is formed.

Although the nature of the sediment changes continually along the length of an estuary in relation to current velocities it is convenient to distinguish three basic types: clean well-sorted sand, sandy mud and soft mud with a high organic content. Clean sands and soft muds have a poor fauna while sandy mud in its many grades has the richest fauna in the estuary (Day 1959, 1967); Branch & Grindley (1979).

Clean, well-sorted sands are porous and as the tide rises well-oxygenated water percolates down to the permanent water table and seeps out at lower levels. Towards the high tide mark the water table sinks to 30 cm or more at the time of low tide and as the surface dries out temperatures and salinities fluctuate widely with atmospheric conditions. Under such harsh conditions high tide sands are very barren. Below mid-tide where the surface remains moist there are a few members of the epifauna particularly amphipods which forage over the surface when covered by the tide and hide below the surface when it falls. At and below low tide, seepage of interstitial water must continue so long as there is a hydraulic gradient between the level of the water table in the sand and the free water surface in the channel. However, there is little silt in the sand and deep permanent burrows collapse. Shallow-burrowing suspension feeders such as *Cardium* and *Mactra* or heart urchins and cake urchins *(Echinocardium* and *Echinodiscus)* are sometimes abundant. *Mactra lilacea* has been obtained in 2-3 m at the entrance of Langebaan Lagoon in concentrations with a biomass of 114 gC.m^{-2} (Christie & Moldan, 1977) and *Cardium edule* at the mouth of Burry Inlet in Britain has a biomass of 120 g (dry mass) m^{-2} (Walne, 1972).

As the percentage of silt and clay particles in the sand increases from less than 1 % to between 2 and 20 %, deep burrows do not collapse and the infauna includes many species of polychaetes, bivalves and prawns. The organic content increases in the finer sediments and the epifauna of amphipods, isopods, small crabs and gastropods increases as well, particularly when *Zostera* or other sea grasses provide shelter. Most species (possibly 85-95 % of the fauna) are deposit feeders although micro-algae are inevitably included in the diet.

As the percentage of silt and clay increases above 30 % the fauna decreases possibly due to anoxic conditions in the sediment. In soft sloppy mud where the percentage of subsieve particles may rise to 80 % only a few specialised members of the epifauna can exist (McLachlan & Grindley, 1974).

Both Green (1968) and Eltringham (1971) have described the biology of species that live in soft sediments in British estuaries. A similar treatment is not possible in southern Africa. There are literally hundreds of species and the fauna changes between the subtropical estuaries of Mocambique to the Transkei and the temperate estuaries of the Cape. Many are apparently rare and others that are common in some localities have only been recorded a few times and their ecology is unknown. Attention will therefore be focussed on a few of the better-known species that live on hard or soft substrates and notes on the distribution and biology of the other characteristic species will be found in the Appendix. Day (1969) gives an illustrated key to the common plants and animals which live in estuaries and on open sea shores.

THE BENTHIC MACROFAUNA OF ESTUARIES IN SOUTHERN AFRICA

All hard surfaces in estuaries including rocks, concrete structures, wooden posts and mangrove roots are covered with silt apart from vertical surfaces in rapid currents. As a result, macroscopic algae are reduced to a few specialised genera such as *Ulva*, *Enteromorpha*, *Codium* and *Bostrychia*. The herbivores which feed on encrusting algae such as 'lithothamnia' and *Ralfsia* disappear and as the rock crevices near low tide are filled with silty sand the cryptic fauna of sponges, hydroids, ectoprocts and ascidians is limited to a few sites below low tide. Thus the rich fauna found on rocky seashores is reduced to a few species. Most of those above mid-tide are browsing gastropods which feed on lichens and the plankton and detritus left by the receding tide, or grapsid crabs and oniscid isopods which harbour in crevices and feed at lower levels when the tide is down. At mid-tide and below, the main forms are hard shelled barnacles and oysters. Marine mussels hardly extend inside the estuary mouth but one species of estuarine mussel *(Musculus virgiliae)* appears among the barnacles in low salinities, and amphipods and isopods are common among them.

The vertical zonation on rocky seashores has been described by Stephenson (1947, 1972) and the species that extend into estuaries maintain the same levels. Brief accounts will be found in Day *et al* (1952) and Day (1974). It may be noted in passing that species such as *Nodolittorina natalensis* and *Littorina kraussi* which occur in the supratidal spray zone on exposed rocky shores are absent from the calm waters of estuaries and the upward extension of *L. scabra* and *L. africana* var. *knysnaensis* is strictly limited to the level of high spring tide. The fauna on hard substrates below low tide is poorly known but Blaber *et al* (1974) have observed thick growths of sponges *(Grantessa ramosa)*, barnacles *(Balanus amphitrite)* and mussels *(Musculus virgiliae)* on the steep rocky walls of the Msikaba estuary and among them was *Ficopomatus (= Mercierella) enigmatica* which is normally found above low tide. They emphasise that all these rock-dwelling forms are limited to vertical rock faces for silt accumulates to such an extent on flat ledges that the burrowing prawn *Upogebia africana* is common.

The zonation on hard substrates of both subtropical and temperate estuaries is shown in Figure 9.3 and the biology of the better-known species is described below. The serpulid polychaete *Ficopomatus enigmatica* is known from warm temperate estuaries all over the world. Its recorded salinity range is 1,5 to 50 ‰, its spawning range is 5-30 ‰, its temperature range is 12-28 °C and its depth range is from low water of neaps to at least 32 m (Millard & Broekhuysen, 1970, Perkins, 1974, Blaber *et al*, 1974). It may be attached not only to rocks, wooden posts and mangrove roots but even to compact lumps of mud. It reaches maturity in six weeks at 20 ° and the settling larvae are obviously attracted by old tubes for it forms coralliform masses of twisted tubes over 10 cm thick. Indeed, Tebble (1953) reported that where estuaries in England are warmed by effluents from power stations the masses may grow so thick that they prevent the closure of sluice gates. Despite all this it has not been reported from the sea and its occurrence in the estuaries of southern Africa is sporadic. While abundant in Kosi Bay, it is absent from Morrumbene and it may be either rare or common in other subtropical estuaries. In the Breede estuary it is usually abundant but only dead tubes were found where *Pomatoleios kraussii* was found alive. In Milnerton Lagoon where *Pomatoleios* is absent *Ficopomatus* is always abundant. Possibly competition with related genera such as *Pomatoleios* and *Hydroides* may exclude it from the sea but its irregular occurrence in estuaries requires further investigation.

Of the several species of barnacles which are abundant on rocky coasts only *Chthamalus dentatus, Tetraclita serrata* and *Balanus algicola* enter estuaries and even these are only common at the mouth. *Balanus amphitrite* replaces them in calmer waters and more variable salinities; indeed it is the dominant species in harbours and estuaries in warm and tropical seas. The Australian *Elminius modestus* which is spreading rapidly in Europe has now been reported from Cape Town docks but is not known from local estuaries. Like most barnacles *B. amphitrite* is gregarious and aggregations are found on bridge supports and mangrove roots along the channel margins. According to Perkins (1974) it requires a temperature range of 17-18 ° for spawning and its salinity range is 12-55 ‰ but Millard & Broekhuysen (1970) did not find live specimens in salinities above 36 ‰ in St Lucia which periodically becomes hypersaline. In normal estuaries it extends from the mouth to the upper reaches and grows best at mid-tide although like several other species, it extends to slightly higher levels on shaded surfaces. In Cape and Transkei estuaries *B. amphitrite* is mixed with the endemic estuarine species *B. elizabethae* and as *B. amphitrite* decreases in reduced salinities, *B. elizabethae* becomes dominant and grows in sheets at the head of Knysna estuary where the salinity is normally 3,9-7,2 ‰. It has not been found at the head of Breede estuary where the salinity seldom exceeds 2 ‰ so that its salinity range appears to be similar to that of *B. pallidus stutsburi* in Lagos harbour studied by Sandison (1966).

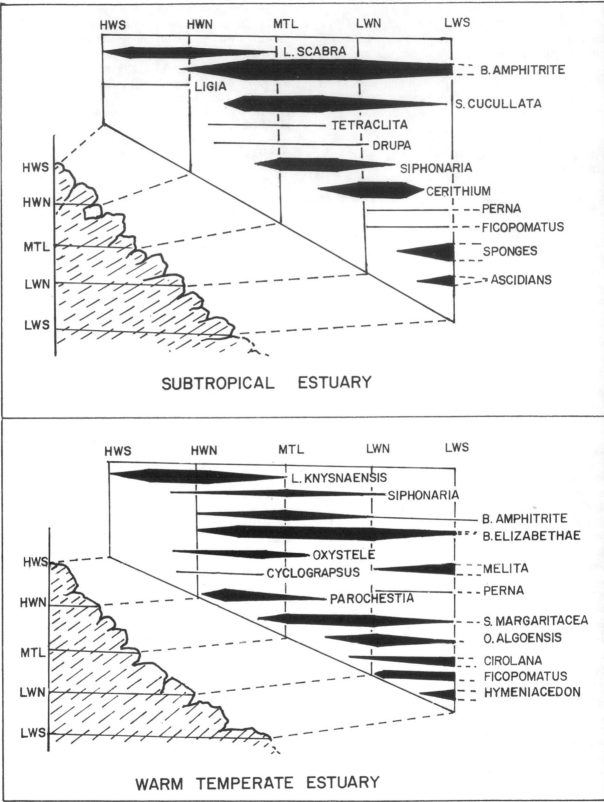

Figure 9.3
Zonation on rocks in the lower reaches of subtropical and
warm temperate estuaries in southern Africa.

Cyclograpsus punctatus is the common shore crab in the Cape and extends into estuaries, sheltering in rock crevices near high tide or burrowing in eroding mudbanks and salt marsh vegetation. Its reproduction and development have been described by Broekhuysen (1941), its osmoregulation has been investigated by Boltt & Heeg (1975) and its ecology has been described by Alexander & Ewer (1969). Although it is a scavenger and opportunistic carnivore it is partly herbivorous feeding above the water level and thus avoiding low salinities. Its main competitor in estuaries is the marsh crab *Sesarma catenata* which is restricted to sandy mud.

Oysters extend from the seashore into the lower reaches of estuaries on the warm eastern and southern coasts. The Indo-Pacific species *Crassostrea cucullata* (recently referred to the genus *Saccostrea*) is the main species in subtropical estuaries and is cultivated commercially in Australia but has not been tried in South Africa. Normally it forms a conspicuous white band on mid-tide rocks and the lower fringe of mangrove roots and has been recorded in salinities of 18-36 ‰ but dies when the water becomes hypersaline. The endemic *Saccostrea margaritacea* lives near low tide in the subtropics but extends to higher levels in the Cape even in situations where it is liable to be covered by sand. The shell elongates from the point of attachment to maintain the aperture above the sediment so that the whole shell may be curved like a parrot's beak and is often referred to as a 'pappagaaibek'. Wild oysters have been collected commercially for many years and the Fisheries Development Corporation have attempted to cultivate them in Knysna estuary. Korringa (1956) who described the ecology of *S. margaritacea* also gave an account of the early difficulties in cultivating oysters. The settlement of the first season's spat appeared promising but the shells did not grow and eventually proved to belong to the small 'weed' oyster *Ostrea algoensis* which is now known to be common or abundant in many Cape estuaries. Wild oysters were then brought into the estuary for fattening but were eaten by the 'mussel cracker' *Sparodon durbanensis*. They were then surrounded by fences on the sand banks but they were either buried by burrowing prawns or gathered by octopi to form nests. Eventually the European *Crassostrea angulata* was introduced and was reported to grow more rapidly than *S. margaritacea* but unfortunately the parasitic *Polydora ciliata* was introduced with it. Genade (1973) reported that the Japanese *Crassostrea gigas* was proving successful.

Apart from Korringa's account, nothing has been published on the factors that affect the reproduction of South African oysters although much may be contained in unpublished reports. Meanwhile local biologists have relied on the extensive published researches on the American *C. virginica* and the European *C. angulata* and *Ostrea edulis*. Taken together these make an interesting story. *C. virginica* has a prolonged breeding season in Florida and Texas when the temperature on the oyster bars is 25 ° or higher but on the cold coast of Maine breeding is restricted to a short period in summer when the temperature exceeds 20 °(Green, 1968). Similarly *O. edulis* breeds for extended periods in the warm Mediterranean in temperatures over 15 °, for shorter periods in England when the temperature is over 18 ° and for a brief period in the Norwegian fjords when the temperature in the oyster trays suspended below the surface film of fresh water reaches 25 °(Green, 1968). This restriction of the breeding season of warm water species in colder climates is particularly well-illustrated by oysters and barnacles and probably applies to other groups. According to a recent review by Bayne (1976) the actual emission of fertile eggs must be preceded by temperature increases over a longer period during which the slower processes of yolk accumulation and maturation of oocytes take place. Thus the sequence of temperature and salinity changes is as important as the temperature at which egg emission occurs.

Few of the gastropods on rocky sea shores extend into the mouths of estuaries and fewer still occur on silt-covered rocks in the lower reaches. In Cape estuaries the main species are generalised browsers living on the upper part of the shore, such as *Littorina africana* var. *knysnaensis*, *Oxystele variegata*, *Siphonaria oculus*, *S. capensis*, *S. aspera* and *Patella oculus*. *Thais dubia* is the main predator and feeds largely on barnacles. All these species are rare in reduced salinities except *Littorina africana* var. *knysnaensis* which extends into salinities of 4-7 ‰. Possibly it is only subjected to the upper part of this range as it lives near the high tide mark. As Broekhuysen (1940) has shown, it is very resistant to desiccation and tolerates a water loss of 22 %. In subtropical estuaries, *L. a. knysnaensis* is replaced by the Indo-Pacific *L. scabra* which tolerates temperatures of 30-15 ° and extends into salinities below 10 ‰. Unlike *L. africana* var. *knysnaensis* it is restricted to sheltered bays and estuaries and is common in mangrove swamps.

The fauna of soft sediments

Many of the reviews which deal with the fauna of soft sediments deal with both marine and estuarine environments and it is not always possible to distinguish the characteristic features of estuarine populations. Dahl (1952) makes an interesting comparison between the crustaceans which live on temperate and tropical shores; Eltringham (1971) deals with life on sand and mud and includes one chapter on estuaries; Newell (1970, 1976) deals with adaptations to life between tide marks; Perkins (1974) gives an integrated account of life in estuaries and shallow seas and focusses attention on the coast of Britain. Only Green (1968) and Lockwood (1976) limit their attention to the estuarine environment and the adaptations of the fauna. In southern Africa discussions of the zonation of the

Arenicola loveni (Photo: J.H. Day)

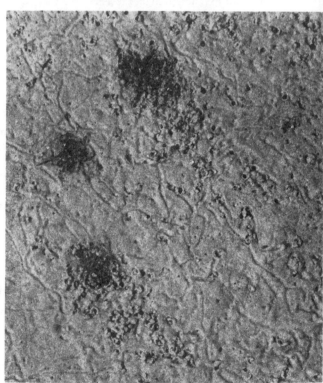

Assiminea globulus and feeding tracks (Photo: J.H. Day)

Loripes clausus (Photo: J.R. Grindley)

fauna and flora of soft sediments will be found in Day (1951, 1967, 1974) and Macnae (1957).

The distinction between the epifauna and the infauna is somewhat arbitrary for many members of the epifauna retire to deep burrows when not feeding and the shallow burrowers of the infauna are often exposed on the surface. The criterion adopted here is that the epifauna forages over the surface while members of the infauna feed within their burrows.

The epifauna may be divided into two groups. One group forages over the damp sediment above the receding water level and among them are talitrid amphipods and crabs of the families Ocypodidae and Grapsidae. They are capable of rapid locomotion, they can respire in air for limited periods and tolerate high temperatures and desiccation for considerable periods. With the notable exception of *Dotilla* and *Mictyris* which makes temporary burrows wherever they are overtaken by the rising tide, many members of this group retire to burrows in the upper part of the shore so that the position of the burrow is not necessarily an indication of the feeding area. The other group of the epifauna feeds below the water surface and is not capable of aerial respiration and rapidly succumbs to desiccation. This group includes several amphipods (particularly the family Haustoriidae), cirolanid and sphaeromid isopods, palaemonid shrimps, penaeid prawns, hymenosomatid and portunid crabs and almost all the gastropods. They forage widely over the intertidal banks while the tide is high and when it falls, slow-moving forms bury themselves in the damp sediment while the faster swimmers retreat to the shelter of *Zostera* beds. Below the level of *Zostera* and other sea grasses such as *Halodule* and *Thalassodendron* in more tropical estuaries, the bed of the estuary is surprisingly barren both of epifauna and infauna. This requires further investigation but the low percentage of organic detritus in the shifting sands of the swifter channels and the low oxygen content of the mud in the quiet upper reaches are probably important.

The separation of the two groups of epifauna is very marked on bare sandy shores. On mud banks where the upper levels are overgrown with salt marsh vegetation, the epifauna extends down with it. Often there is a step or salting cliff at mid-tide (Figure 9.4) but on evenly sloping banks the salt marsh vegetation meets the *Zostera* and the upper and lower groups of the epifauna overlap (Figure 9.5).

Members of the infauna can only feed and pump oxygenated water through their burrows when submerged are thus more abundant on the lower part of the shore. This is particularly true of suspension feeders such as *Mactra* and *Solen* but deposit feeders extend further up wherever interstitial water from high tide banks seeps over the surface. Day (1974) has shown that the upper limit of the fauna and sea grasses is more closely related to the depth of the water table than it is to tidal levels. This is illustrated in Figures 9.6 and 9.7.

Nemertea

Burrowing anemones, turbellarians and nemerteans are rare in estuaries but there is one very large nemertean, possibly the largest in the world. It is *Polybrachiorhynchus dayi* which reaches a length of half a metre and 2,5 cm in breadth. It is found burrowing in sandy mud although it can also swim like an eel. The striking feature is the large branching proboscis which can spread out to cover the palm of one's hand. It is thought to prey on *Upogebia* and the polychaete worms among which it occurs.

Polychaeta

Polychaete worms are abundant in estuarine muds. *Ceratonereis erythraeensis* is probably the most common and appears to be the ecological equivalent of *Nereis diversicolor* in Europe. It feeds on microscopic algae and surface deposits and extends from sheltered bays to the heads of estuaries and tolerates salinities below 10 ‰. *Diopatra cuprea* which is common on muddy bottoms at 30 m or more in the sea, extends into the shallows in subtropical bays and even above lowtide in Morrumbene estuary in salinities as low as 20 ‰. It makes sandy tubes beset with shells and often extends well above the surface. The gut contains no sand or detritus and as the head is provided with muscular tentacles and strong jaws it is obviously a carnivore. It has been seen to stretch out of its tube and take amphipods from the sediment surface and is suspected of feeding on moribund zooplankton as well. Most of the other polychaetes are deposit feeders for there are few suspension feeders in estuarine sediments. Subsurface deposit feeders include species of *Orbinia*, *Scoloplos*, *Capitella* and *Notomastus* and the well-known *Arenicola*. *A. loveni* which is one of the largest species of the genus reaches a length of 60 cm and a diameter of 2 cm. The mean mass is 41,3 g (Barham, 1979). The U-shaped burrows are common in sandy areas at low tide of neaps but for some unknown reason they do not occur below low spring tide. The burrows extend down below the permanent water table into black thixotropic deposits where the oxygen concentration must be low. Unpublished work by Professor A.C. Brown indicates that the respiratory adaptations of *A. loveni* resemble those of *A. marina*. *A. loveni* does not make worm casts but squirts out black watery sand from the tail shaft. It reaches sexual maturity at 10-15 g, breeds in mid-summer (Barham, 1979) and eggs have been found on warm damp sand at the edge of the rising tide which may allow them to be carried upestuary. The larvae are covered with sand grains so that they are not planktonic but drift over the bottom. Mr Gaigher of the Cape Department of Nature Conser-

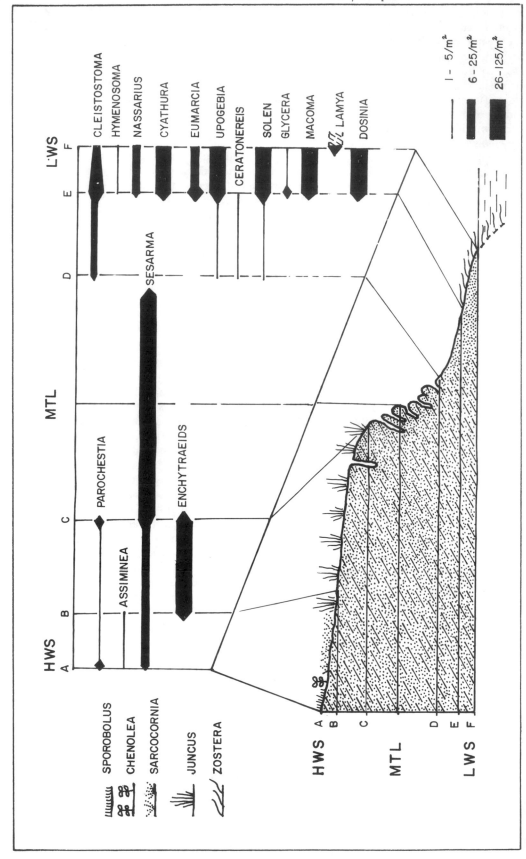

Figure 9.4
Transect at Ashford (Knysna) to show effect on zonation of a salting cliff (from Day, 1967).

vation informs me that although the adults are found above low tide, very young worms have been found on subtidal banks. How the juveniles return to the adult habitat is unknown.

Amphipoda

Melita zeylanica is the most abundant estuarine amphipod and indeed it is confined to estuaries and has been recorded in salinities up to 49 ‰ in St Lucia (Millard & Broekhuysen, 1970) and below 1 ‰ in the blind estuary at Hermanus where it may form 84 % of the catch in weed beds during autumn (Scott *et al,* 1952). *Corophium triaenonyx* extends into even lower salinities and has been reported by Boltt (1969) as part of the relict estuarine fauna of Lake Sibayi which is now fresh. Like the European *C. volutator* whose biology is summarised by Green (1968), it makes shallow burrows in mud often among stones, and filters organic particles from the excavated sediment and from the water currents which it draws through its burrows.

Talitrid amphipods of the genus *Orchestia* are part of the epifauna near high tide. *O. ancheidos* and *O. rectipalma* shelter under the cast-up weed on the sandy banks, forage down to mid-tide when the tide falls and eventually hop back to their refuge as the tide rises. Although the hopping movements of talitrids appear to be random, Newell (1970) reviews the evidence that the European species *Talitrus saltator* orientates itself by the sun or the moon and has an internal clock to correlate its migrations with the changing angle of the light source. It is an example of the many members of the epifauna that maintain station on the shore in the same way.

Decapoda: Natantia

Palaemonid shrimps shelter in weed beds below low tide. *Palaemon pacificus* is the most abundant and widespread species of the genus and feeds mainly on epiphytic algae. In subtropical estuaries it is replaced in low salinities by *Macrobrachium equidens* and *Caridina nilotica* both of which extend into fresh water although ovigerous females occur in estuaries. *P. pacificus* breeds both in estuaries and the sea and studies of the breeding of European palaemonids indicate that the fecundity tends to increase in higher salinities (Green, 1968).

Several species of penaeid prawns are sufficiently abundant along the coasts of Mocambique and Zululand to be fished commercially, although the standing stock is not very large. The adults of many penaeids breed in the sea and the juveniles enter estuaries for food and shelter during the spring and return to the sea in autumn as their gonads develop. Champion (1976) reviewed the findings of Joubert & Davies

(1966) and his own observations in Richards Bay and St Lucia in 1970 and 1976. Records of catch per unit effort show that *Penaeus indicus* which is the most important species commercially, and *Metapenaeus monoceros* which is rather small are the two most common penaeids in local estuaries preferring soft muddy bottoms and tolerating salinities over 60 ‰. *P. monodon* and *P. japonicus* are present in reduced numbers in hypersaline conditions, the former preferring weedbeds and the latter sandy bottoms. *P. semisulcatus* appears to be restricted to salinities below 40 ‰ and is mainly found in weed beds with *P. monodon.* All five species have been recorded in salinities below 10 ‰ and probably tolerate 5 ‰. They are reported to be omnivorous although individual food preferences have not been studied. In the mangrove swamps of Mocambique they forage over the mudbanks when the tide is high and drift back along the creeks to the main channel when the tide falls. African fishermen build reed-walled traps at the mouths of the creeks and on every suitable intertidal sandbank. The average catch per trap per night is seldom more than 1 kg, but the meshes of the traps built in the creeks are so fine that even juveniles of 3-4 cm are caught which is most uneconomic. Trawling in Inhambane and Delagoa Bay is more profitable and the South African Fisheries Development Corporation has undertaken pilot experiments in Natal to determine whether the prawns could be cultured economically. The choice of Amatigulu as a test area was unfortunate. Better results in a more suitable estuary might be successful.

Decapoda: Thalassinidea

As penaeid prawns decrease in abundance between Natal and the Cape, burrowing prawns become more abundant. Indeed the two thalassinidean prawns *Callianassa kraussi* and *Upogebia africana* are the dominant prawns in all South African estuaries and must contribute very largely to the total biomass of the macrobenthos.

Callianassa kraussi is a subsurface deposit feeder and makes branching burrows in sandbanks in sheltered bays, lagoons and estuaries including those that are blind for part of the year. Brown (1953) estimates densities of 300 m^{-2} in the blind Kleinemond estuary and Christie & Moldan (1977) record densities up to 83 m^{-2} in Langebaan Lagoon. Forbes (1973) has shown that there is only a short non-planktonic larval stage but the biology of *C. kraussi* is otherwise similar to that of *C. californiensis* described by MacGinitie (1935). It creates currents through its burrow with its paddle-shaped pleopods, loosens the sand from the walls of the burrow and winnows the suspended food particles through a mesh of long setae on its pereiopods. At intervals the excavated sand is blown out of the

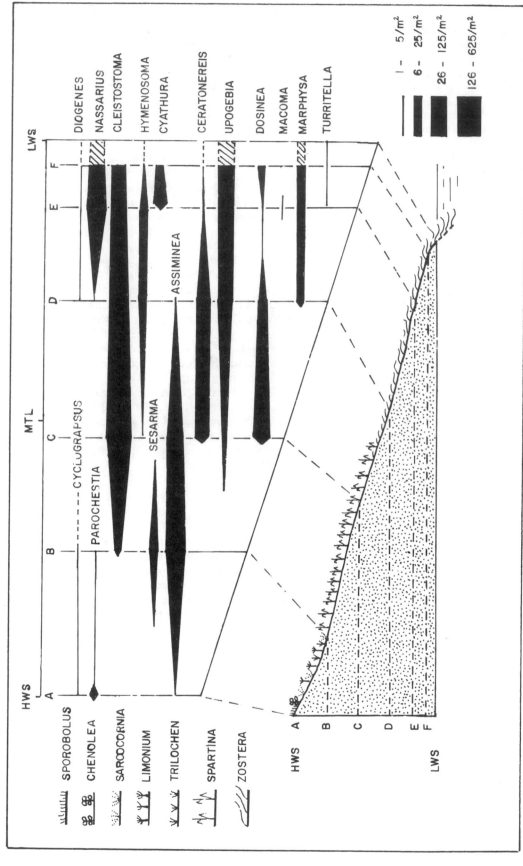

Figure 9.5
Transect at the rail bridge (Knysna) to show zonation on an even slope (from Day, 1967).

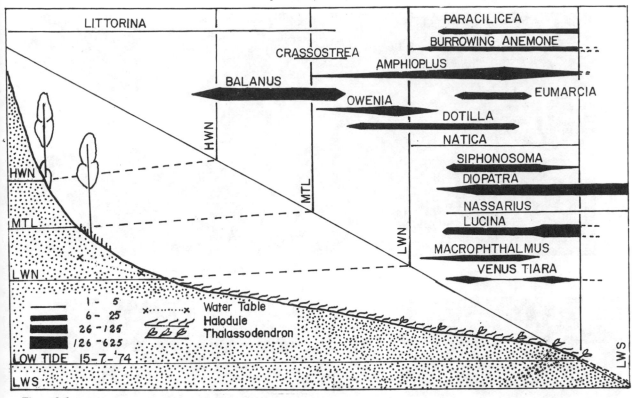

Figure 9.6
Transect across muddy sand at Linga-Linga (Morrumbene) with
water table at surface from LWN to LWS (from Day, 1974).

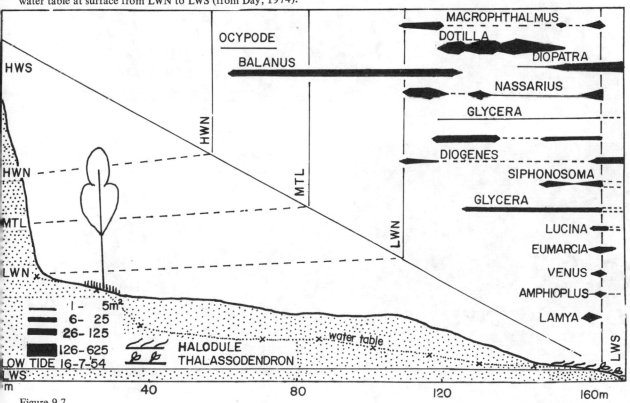

Figure 9.7
Transect across muddy sand at Rio Sambé (Morrumbene) with
water table at surface near LWN and below surface almost to
LWS (from Day, 1974).

Upogebia africana and burrows (Photo: J.H. Day)

Holes in dense *Callianassa* bed (Photo: J.G. Grindley)

Penaeus monodon (Photo: George Begg)

Callianassa kraussi (Photo: J.H. Day)

164

burrow so that it forms conical mounds on the surface. The subsurface sand eventually becomes so riddled with burrows that one may sink in up to the knees. There are subsidiary effects as well. The water currents through the burrows increase the oxygen content of the interstitial water in the sand and Dye (1976) has noted that the meiofauna whose depth range is normally limited by falling oxygen concentrations extends to much deeper levels in *Callianassa* beds. On the other hand the activities of the prawn must destroy the deep permanent burrows of other members of the infauna and McLachlan & Grindley (1974) suggest that competition with *Callianassa* limits the distribution of burrowing bivalves. Certainly neither *Arenicola loveni* nor *Solen capensis* occur where *Callianassa* is common but the epifauna and shallow-burrowing polychaetes do not appear to be affected.

Upogebia africana makes U-shaped burrows in sandy mud and feeds on detritus drifting over the surface with the tide and probably for this reason it does not occur in tideless estuaries which have been closed by sandbanks for long periods. The biology of *U. africana* is generally similar to that of *U. pugetensis* described by MacGinitie (1930) and further details are given by Hill (1967). Dr Burke Hill also reviews the effects of salinity, temperature and oxygen concentrations in chapter 11. Siegfried (1962) gives biometric data of prawns in the Uilenkraal estuary near Danger Point and estimates that breeding starts at 11 mm carapace length, while Hill states that in the warmer Kowie estuary breeding starts at 10 mm and that females produce two batches of eggs, one in early spring and one at the end of summer. These hatch in 24-30 days and Hill's data suggest that prawns start breeding when a year old, and die after 3-4 years.

Upogebia is the most important member of the infauna in low tide mudbanks and is used extensively for bait. The average wet weight is 1,7 g and as densities of 50-250 m^{-2} are commonly found in mud with 3-20 % of silt and clay, the total wet mass must be of the order of 255 g.m^{-2}. McLachlan & Grindley (1974) give a maximum wet mass of 450 g.m^{-2} in Swartkops estuary and Grindley (1976) quotes an ash-free biomass of 7,76 g.m^{-2} for the Ashmead area of Knysna estuary. In sediments with more than 20 % subsieve particles *Upogebia* is less common but it persists to the upper reaches of estuaries in salinities of 10 ‰ and in the Breede estuary it tolerates salinities of 2 ‰ for at least a week. While maximum densities occur between low tide of neaps and springs it becomes uncommon below the *Zostera* belt and its normal upper limit is about mid-tide.

Upogebia is preyed on by long-billed waders such as curlews and by many species of fish particularly the white steenbras *Lithognathus lithognathus* and the spotted grunter *Pomadasys commersonni*. These are often called 'blowers' for they hang vertically in the water and blow vigorously down one hole so that the prawn is ejected from the other. Bait collectors stamp on an inverted jam tin or use a prawn pump with the same effect. Cunning anglers cast drift bait where they see the tails of the blowers waving in the shallows.

Decapoda: Brachyura

Crabs of the families Ocypodidae, Grapsidae, Portunidae and Hymenosomatidae dominate the epifauna of subtropical estuaries. Day (1974) recorded 67 species in Morrumbene estuary, 18 species in southern Cape estuaries and four species in Atlantic coast estuaries. In British estuaries Green (1968) reports that the only common species is *Carcinus maenas*. It is to be expected that active animals with well-enclosed gill chambers such as the Ocypodidae and Grapsidae should be common on intertidal banks in the tropics and subtropics but it is difficult to understand why the Portunidae, Hymenosomatidae and other crabs which live below tide marks are so common in subtropical estuaries and so uncommon in cold temperate waters.

Ocypode ceratophthalmus which is common on surf beaches in the subtropics barely extends into the mouths of estuaries and is replaced on sheltered sandbanks with swarms of *Dotilla fenestrata*. Dahl (1952) suggests that *Ocypode* is the tropical equivalent of the temperate talitrid amphipods but in estuaries *Dotilla* or the Australian soldier crab *Mictyris* is a closer parallel. Armies of *Dotilla* forage up and down the damp sandbanks and disappear into temporary burrows when overtaken by the incoming tide. Like several other genera of the Ocypodidae (eg *Uca*, *Macrophthalmus* and *Cleistostoma*) they are deposit feeders, scraping up the surface of the sand with their flattened chelae, brushing out the edible particles with both the spoon-tipped and plumose setae on their maxillipeds and finally rejecting the remaining sand as pseudofaecal pellets. Macnae (1968) describes the maxillipeds of several species of *Uca* and *Macrophthalmus* and Icely & Jones (1978) give further details of the feeding of *Uca* spp.

Since there are thousands of *Dotilla* easily visible on the open sandbanks it might be expected that they would be subject to heavy predation. They are relatively slow moving and not very aggressive but the gulls take no notice of them and they have not been found in fish stomachs either. One can only presume that they are very distasteful. Since most species of *Uca* and *Macrophthalmus* and grapsid crabs of the genus *Sesarma* are associated with mangroves these will be discussed with the rest of the mangrove fauna.

Hymenosoma orbiculare feeds on detritus and microscopic algae and is common all around southern Africa both in the sea and in estuaries wherever the bottom is muddy sand. It occurs in False Bay at depths of 0-102 m where the temperature falls to 10 ° during upwelling and extends into subtropical estuaries where temperatures rise to 30 °. Millard & Broekhuysen (1970) record it in salinities up to 45,4 ‰ in St Lucia and Broekhuysen (1955) found it breeding in salinities of 5 ‰ in Cape estuaries. Apparently it may be acclimated to even lower salinities for Allanson *et al* (1966) record it as breeding in fresh water in Lake Sibayi where it forms part of the relict estuarine fauna. It is normally found half buried in the sand. In the *Zostera* beds in Knysna Day (1967) recorded densities of 6-25 m^{-2} and it is often found in fish stomachs.

Scylla serrata is the largest of Indo-Pacific portunid crabs (120-160 mm carapace width) and occurs in estuaries as far south as Cape Agulhas. Hill (1974, 1975, 1976, 1979) has made a detailed study of its ecology which is outlined below. Although it may be found in deep burrows at low tide it is subtidal and ranges over 100 m or more of muddy sand, preying on small mussels, gastropods and other crabs. It is obviously euryhaline and Hill (1979) reports a salinity range of 2 to 60 ‰; it leaves Lake St Lucia when the salinity rises beyond this. Nonetheless the females migrate to the sea to spawn like many other portunids. Hill reports that they leave Cape estuaries in the autumn and if the estuary mouth is closed by a sandbar, females cross the open sands at night to reach the sea. They have not been reported by trawlers working along the Cape coasts and presumably they migrate over deep muddy bottoms to reach the tropics. One was taken in 400 m off the Natal coast.

Mollusca: Pelecypoda

Almost all the bivalves on soft sediments are burrowers and only one small mytilid, *Lamya capensis* is common on the surface. It attaches itself to dead shells, weeds and algal carpets at, and below low tide and at times the byssal threads attach one shell to the next so that a carpet of mussels is formed. It extends into salinities below 10 ‰ in the upper reaches of all estuaries and is important in the diet of *Scylla* and many demersal fishes, particularly *Rhabdosargus* spp.

The burrowing bivalves include few suspension feeders, the most common being species of *Solen* which make deep vertical burrows and feed at the entrance. The united siphons which project above the surface are provided with a series of breaking planes so that they break when attacked by fish browsing along the bottom and leave the body intact. *Solen capensis* occurs in the cleaner sands and higher salinities near the mouths of Cape estuaries while the smaller *S. corneus* burrows in soft mud with 10-40 % subsieve particles and extends into salinities below 20 ‰ in the upper reaches of estuaries from Cape to Mocambique. The deposit feeders *Eumarcia paupercula, Macoma litoralis* and *Dosinia hepatica* have a similar habitat and distribution and McLachlan & Grindley (1974) record them in densities up to 40 m^{-2}

in the upper reaches of Swartkops estuary which they refer to as the middle reaches. They claim that competition with *Upogebia* restricts their distribution further down the estuary but their measurements of the organic nitrogen content in the sediment do not indicate that organic detritus is a limiting factor. It is also unlikely that the U-shaped burrows of *Upogebia* would interfere with the burrows of bivalves as do the ramifying burrows of *Callianassa*. As noted earlier, McLachlan (1974) studied the larval recruitment of these bivalves and he further estimated the growth rates of *S. corneus, M. litoralis* and *D. hepatica* and concluded that they live five to six years. Similar studies of other important elements of the estuarine fauna are needed before energy budgets in estuarine ecosystems can yield meaningful results.

Mollusca: Gastropoda

Relatively few gastropods are adapted to life in soft sediments. Apart from the group that lives in mangrove swamps, most of the species are small but they are sometimes present in enormous numbers. Green (1968) records *Hydrobia ulvae* as having a normal density of 5 000-9 000 m^{-2} in the Clyde estuary and densities of more than 15 000 m^{-2} in Denmark. The common species in the Cape is *Assiminea globulus.* Day (1967) recorded densities of 125-625 m^{-2} in Knysna estuary and Puttick (1977) found 30 000-85 000 m^{-2} (mixed with *Hydrobia* sp) in Langebaan Lagoon with a dry biomass of 12-26 $g.m^{-2}$ or 65 % of the total biomass in the surface layers of the shore. In subtropical lagoons and estuaries the main species is *Assiminea bifasciata* but it is only one of the several small gastropods on muddy sediments of mangrove swamps and the densities are much lower than those of *A. globulus* in the Cape. It also has a wider range. It extends from above the water level to a depth of 3 m in Lake St Lucia and has a salinity range of 8,3-37,6 ‰ (Millard & Broekhuysen, 1970). Blaber (1976, 1977) reports that it is important in the diet of several species of mullet and Branch & Grindley (1979) report that it is preyed on by *Scylla serrata*.

Assiminea and *Hydrobia* are but two examples of the many species of detritus feeders which comprise more than 85 % of the fauna of many estuaries. Newell (1965, 1970) gives a detailed account of the nutrition of *Hydrobia ulvae*. In brief, it ingests organic detritus and the accompanying micro-organisms, assimilates the organic nitrogen compounds and voids most of the carbohydrate in its faecal pellets. These are colonised by bacteria some of which fix atmospheric nitrogen to form amino acids and proteins. The enriched detritus is ingested by *Hydrobia* or other deposit feeders so that the nitrogen compounds are transformed from bacterial protoplasm into animal flesh and the carbohydrate content of the original detritus gradually decreases. As Newell remarks, it is not surprising that there is a good correlation between the biomass of deposit feeders and the organic nitrogen and carbon content of the deposits and that a variety of animals are capable of surviving for long periods on a diet of bacteria.

The species of *Nassarius* are mainly scavengers and two species, *N. arcularia* and *N. coronatus* are particularly common on muddy sand in Morrumbene estuary and both reach Durban Bay. They extend below low tide and if stranded on the shore lie half buried in the damp sand. *N. kraussiana* is abundant on more muddy banks at the *Zostera* level and extends all around southern Africa. Day (1967) has recorded densities of 125-625 m^{-2} in Knysna estuary and it is suspected that they are detritus feeders but this needs confirmation.

The large tectibranch *Notarchus leachii* is one of the few herbivores, feeding mainly on *Ulva* and microscopic algae. It is sometimes so common as to block the filters of intake ducts. The main predaceous molluscs on the sandbanks are species of *Natica* and *Polynices* which plough through the moist sand feeding on bivalves. *N. gualteriana* and *P. tumidus* are tropical species while *N. tecta* is endemic.

The fauna of mangrove swamps

The mangrove fauna is considered separately as it includes a characteristic group of species, some of which live on mud while others live on the hard surfaces provided by the roots and tree trunks. Moreover, it has been claimed that mangrove swamps have a specialised fauna although more recent work has shown that all the species that are found among mangroves either extend onto the mudbanks below the mangrove belt or occur elsewhere. Some are found on rocks and others live in dense saltmarsh vegetation. Even *Cerithidea decollata* which clusters on the mangrove trunks has been recorded by Macnae (1957) in the salt marshes of Swartkops estuary. As emphasised by Macnae (1968), there are no species completely restricted to mangroves. Nonetheless the same or closely related species dominate mangrove swamps over wide geographical regions and in this sense there is a characteristic mangrove fauna.

On the east coast of Africa mangrove trees extend from the tropics to the southern border of the Transkei but they are not present in all bays and estuaries. They grow on muddy shores between extreme high tide and mid-tide but they are absent from tideless estuaries that are closed by sandbars for long periods (Breen & Hill, 1969). The best developed swamps occur on gently sloping banks and may extend two kilometres or more from the main channel. It is in such dense and extensive swamps that the mangrove fauna is fully developed.

One of the best known mangrove swamps is on the sheltered side of Inhaca Island in Delagoa Bay. Over the years more than 1 000 species of marine invertebrates have been collected from coral reefs, open shores and mangroves around the island but surprisingly few of them were actually found in the mangrove belt. Macnae & Kalk (1958, 1962) describe the habits of 22 species. Macnae (1963) gives a comparative account of the mangrove swamps between Mtata River mouth at 32 °S in the Transkei and Kosi Bay at 27 °S on the northern border of Zululand and lists 41 species. Day (1974) records 31 species from mangroves in Morrumbene estuary as part of a total of 300 species between tide marks. Branch & Grindley (1979) report 33 mangrove species and a total benthic fauna of 208 species in Mngazana estuary at 31°42'S. Obviously the mangrove fauna represents only a small part of the total estuarine fauna but this may be expected since the mangrove belt is situated on the upper part of the shore which is always poor in species. Indeed, the mangrove belt harbours many more species and has a higher biomass than is found at the same level on open sunlit shores although detritus derived from the mangroves is largely responsible for the rich fauna found at lower levels (Day 1974, Branch & Grindley 1979).

The lower fringe of the mangrove fauna appears among the pneumatophores of *Avicennia marina* at about mid-tide. These pneumatophores are densely encrusted with *Balanus amphitrite* and *Saccostrea cucullata* while the moist surface of the muddy sand swarms with *Uca lactea annulipes. Uca urvillei* and *Macrophthalmus grandidieri* are restricted to shallow pools and seepage channels. Digging among the matted roots is difficult but the infauna appears to be very scanty at this level of the shore. No burrowing bivalves were found and only a few polychaetes, mainly *Dendronereis arborifera* and *Marphysa macintoshi,* but these extend up among the dense mangroves.

Among the mangrove trees the fauna starts to change. The barnacles and oysters extend onto the tree trunks and even the lower branches along the edges of the channels, but for some reason they do not extend into thickets. *Littorina scabra* extends throughout the mangrove swamp living between high tide of neaps and springs. It may even be found on the leaves. Macnae (1963) states that they feed on the leaves but this has not been observed and it is thought that they browse on the lichens and the film of detritus left by the tide as they do on rocks. On shaded trunks, the first aggregations of *Cerithidea decollata* appear and become more numerous in the dense thickets. These gastropods seldom feed in daylight but Brown (1971) states that they descend to the mud surface at night to feed on the microflora and organic deposits at intervals related to the cycle of spring and neap tides. More recent and as yet unpublished studies by Cockroft indicate that the juveniles remain on the mud surface and it is only the adults that climb which is reminiscent of many land snails. The hermit crab *Clibanarius padavensis* and the large predatory crab *Metopograpsus thukuhar*

Mounds of *Thalassia anomala* in Thailand (Photo: J.H. Day)

Uca urvillei (Photo: J.R. Grindley)

Sesarma meinerti (Photo: J.R. Grindley)

Uca lactea annulipes and pseudo-faecal pellets (Photo: J.H. Day)

Cerithidea decollata on a mangrove trunk (Photo: J.H. Day)

169

both climb amongst the tangled strut roots and the peculiar *Periophthalmus* which hunts insects climbs too, although it always remains within reach of standing water. Macnae (1968) who describes the biology of several Indo-Pacific species states that both *P. sobrinus* and *P. kalolo* occur in southern Africa and from his description of their habits *P. sorbrinus* is the more common.

As one hacks a transect through dense mangrove stands, the environment changes and the fauna with it. The canopy closes overhead and the light becomes dim. The air is still and oppressively humid. The black anoxic mud becomes waterlogged and water squirts from the crab holes at every step. As Table 2 in Day (1974) indicates, the percentage of silt and clay in Morrumbene mangals rises to over 20 % and the organic content to between 3,6 and 7,1 %. Branch & Grindley (1979) record up to 12 % in Mngazana mangroves; there is little smell of H$_2$S unless one digs down through the layer of matted roots.

The infauna is very restricted; *Marphysa macintoshi* and a few nereid polychaetes are occasionally found but there are no burrowing bivalves. Rather surprisingly groups of *Upogebia africana* and *Alpheu crassimanus* occur in water-logged areas for they normally occur near low tide levels. The dense shade and high humidity must reduce desiccation and organic detritus is obviously abundant, but the feeding time must be short so high up the shore and the specimens are smaller than those below the mangrove belt.

The surface of the mud is dominated by crabs, particularly species of *Uca* and *Sesarma*. The habits of the several species have been described by Macnae (1963) and Edney (1961, 1962) has determined the tolerance of *Uca* spp to water loss and variations of temperature and salinity. *Uca lactea* extends from the fringe of pneumatophores into shaded areas but not into dense thickets where *Uca chlorophthalmus* is common. *Macrophthalmus depressus* replaces *M. grandidieri* in muddy seepage channels and *Uca urvillei* occurs there as well. Species of *Sesarma* scavenge all through the mangroves. *S. guttata* is abundant in dense mangroves in Mocambique, feeding on mangrove seedlings, organic detritus and, when opportunity serves, on other members of the epifauna. It does not make permanent burrows, but merely scuttles away under the prop roots or into the nearest hole when disturbed. *S. catenata* has similar habits but is more common in mangrove swamps in the Transkei and extends into salt marsh vegetation in the Cape. Its ecology in Kowie estuary has been discussed by Alexander & Ewer (1969). *S. eulimene* extends into the upper fringe of mangroves, but is mainly found among reeds and sedges in low salinities and has been seen climbing grass stalks to feed on the seeds. The largest species, *S. meinerti,* burrows at extreme high tide and scavenges through the mangroves at night and drags mangrove leaves back to its burrow (Branch & Grindley, 1979).

Apart from *Cerithidea decollata* and *Littorina scabra,* the main gastropods are *Cassidula labrella* and species of *Melampus*. The large *Terebralia paulstris* is common in some mangrove swamps and reaches a biomass of more than 50 gm.m^{-2} in Sordwana but is absent from many other estuaries.

It might be expected that amphipods and isopods would be abundant among the rotting leaves on the mud, but they are not really common. Talitrids such as *Orchestia ancheidos* occur in the drift line at the high tide mark but there are few other species. The only common isopod is *Sphaeroma terebrans* which bores into rotting roots on the edges of eroding banks.

The diversity of the mangrove fauna decreases towards the head of the estuary with decreasing salinity (Day 1974, Table 5). Changes in the number of specimens per square metre, however, are not so well marked and appear to vary with the depth of the water table and the intensity of shade although this needs further investigation. Vertical zonation is outlined by Macnae & Kalk (1958), Millard & Broekhuysen (1970) and Day (1974). The zonation at Tinga-Tinga (Morrumbene estuary) is reproduced below. There are obvious changes at the lower and upper fringes of the mangrove belt and the fauna on the tree trunks occupies different levels. The epifauna on the mud banks which consist largely of crabs may be assigned to different levels according to whether the position of the burrow or the level over which the species feeds is regarded as more important.

THE ZOOGEOGRAPHICAL AFFINITIES OF THE ESTUARINE FAUNA OF SOUTHERN AFRICA

The distribution of benthic invertebrates around the coast of southern Africa has been discussed by many workers. Stephenson (1947, 1972) gives references to the earlier papers and discusses the distribution of the dominant species of algae and invertebrates on the rocky shores of South Africa. Ekman (1953) and Briggs (1974) deal with the fauna of the whole continental shelf of southern Africa. Others have dealt with specific groups from the shore to the edge of the continental shelf; Day (1967) dealt with the Polychaeta, Hulley (1972) with the skates and rays, Griffiths (1974) with the Amphipoda, Kensley (1974) with the Decapod Crustacea and Millard (1978) with the Hydrozoa. The most recent general review of the distribution of marine and estuarine plants and animals is that of Brown & Jarman (1978).

There is some disagreement regarding the boundaries of the faunistic provinces and the terminology to be used. This is largely due to the depth range considered since the inshore surface temperatures are higher and more variable than those on the outer shelf at depths of 50-150 m. Inshore temperatures between Inhambane at 24 °S and Port St Johns at 31 °S normally vary season

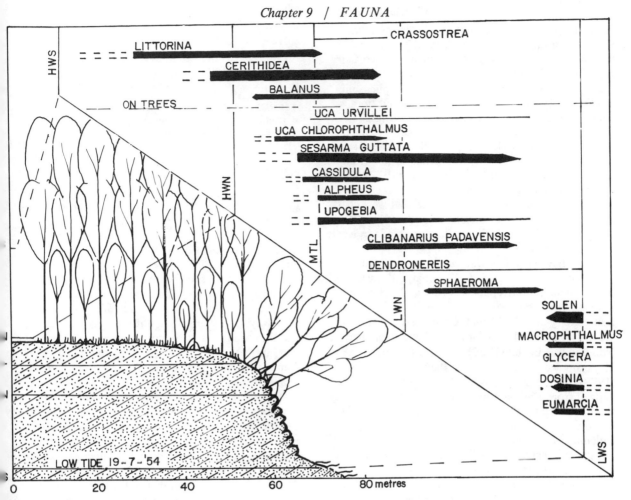

Figure 9.8
Transect through lower part of a mangrove swamp at Tinga-Tinga (Morrumbene) (from Day 1974).

ally between 25 and 18 ° and seldom fall below 16 ° at 75 m. Further to the south-west where upwelling may occur, temperatures are lower and more variable. Between East London at 33 °S and Cape Agulhas at 35 °S inshore temperatures vary from 25-11 ° with a mean of about 20-14 °. On the Atlantic coast from Cape Point at 34°30'S to Rocky Point at 19 °S where upwelling occurs frequently, inshore surface temperatures range from 18-9 ° with a mean of 14-12 °. There is thus a marked change between the south and the west coasts with an overlap between Cape Agulhas and Cape Point. But this does not apply on the outer shelf. Hulley (1972) gives a temperature range of 9-13 ° at 50-150 m from East London around the Cape to Walvis Bay. As a result, the fauna of the outer shelf shows little change between these points and may be regarded as a single Cape-Namaqua fauna (Day 1967, Griffiths 1974, Millard 1978).

Since the sea water that flows into estuaries with the flood tide is inshore surface water, the distribution of the marine components of the estuarine fauna should be more closely related to the distribution of the shore fauna than that of the outer shelf. All workers are agreed that tropical Indo-Pacific species decrease as they spread southwards down the coasts of South Africa and that endemic species increase. Whether these should be divided into subtropical endemics ranging from southern Mocambique to the Transkei and temperate endemics on Cape coasts is controversial and more careful identification and further records are needed to decide the issue. Stephenson (1947, 1972) regards the whole shore fauna of Natal including tropical and endemic species as essentially subtropical and points out that although reef corals are present they do not form coral reefs as they do in the tropics. Following on the work of Stephenson (1947) in Natal and more recent studies of Kalk (1959) and Day (1974) in southern Mocambique, it has become evident that apart from a tropical fauna around Inhaca Island (26 °S), the whole of this coast from Inhambane at 24 °S to the Port St Johns at 31 °S is subtropical. While tropical species are common, there are many subtropical endemics and an increasing number of temperate endemics towards the south. There is an overlap between the subtropical and warm temperate fauna on the Transkei coast and from there to Cape Agulhas the South Coast fauna is warm temperate. There is another overlap with the West Coast fauna between Cape Agulhas and Cape Point and from there on the West Coast fauna extends north to Namibia. Stephenson (1947, 1972) states that the shore fauna and flora is cold temperate while Ekman (1953) and Briggs (1974) regard the whole West Coast fauna as warm temperate since it lacks the species characteristic of the subantarctic. This is true, but the shore fauna differs from that on the South coast. As Stephenson (1972) and Brown & Jarman (1978) state, the matter is simply a question of terminology. Possibly the cold west coast should be referred to as the Namaqua province. Penrith & Kensley (1970) have shown that it extends to Rocky Point (18°59'S).

There have been few studies of zoogeographic distribution in estuaries. Day (1974) analysed the affinities of the fauna in Morrumbene estuary at 24°S. Similar analyses of the fauna of Mlalazi estuary at 29°S, Mngazana estuary at 31°42'S and Knysna estuary at 34°04'S are shown below. The records for Mlalazi were derived from the lists in Hill (1966) and Oliff (1976); those for Mngazana were extracted from Branch & Grindley (1979), while the records for Knysna estuary were derived from a revision of the lists in Day *et al* (1952) supplemented by the list of fishes in Grindley (1976). It should be noted that the subtropical and temperate endemics of southern Africa are combined in Table 9.2 but the tropical and 'other foreign' categories are the same as those defined in Day (1974).

It is evident that the percentage of tropical species decreases in more southern estuaries, while the percentage of endemic species increases. The change is very marked between Mngazana in the Transkei and Knysna in the Cape. This agrees with the findings of Stephenson (1947, 1972). Stephenson did not distinguish between endemic and tropical species in the East Coast biota and more detailed analyses of distribution must await more extensive records and taxonomic revisions of the molluscs in particular. It would be particularly interesting to know whether the distinction between subtropical and temperate endemics in southern Africa is valid. In the subtropical Morrumbene estuary, Day (1974) showed that there is a higher percentage of subtropical

Table 9.2
Zoogeographical distribution in four estuaries in southern Africa

	Total species recorded	Number analysed	Tropical species	Endemic species	Other foreign species
MACROBENTHIC INVERTEBRATES					
Morrumbene	378	340	65 %	21 %	14 %
Mlalazi	81	80	58 %	28 %	14 %
Mngazana	208	145	44 %	41 %	15 %
Knysna	310	268	19 %	50 %	31 %
FISHES					
Morrumbene	114	104	87 %	13 %	0 %
Mlalazi	57	54	78 %	15 %	7 %
Mngazana	63	52	67 %	21 %	12 %
Knysna	70	67	31 %	48 %	21 %

species in the upper reaches than at the mouth. A preliminary analysis of the Mlalazi fauna shows 13 % of subtropical species in the upper reaches and only 9 % in the mouth and lower reaches. In Knysna estuary temperate endemics decrease from 56 to 49 % in the same way. The changes are not large but they do suggest that the fauna of the upper reaches of estuaries is more influenced by terrestrial temperatures than is the fauna at the mouth which is affected by coastal sea temperatures.

The number of species recorded in any estuary obviously depends on the size and nature of the estuary and the intensity of collecting. Regional groups of estuaries may be expected to show a decrease in diversity between the tropics and cold temperate coasts. This has been discussed by Day (1974) and his Table 8 shows that while the number of benthic invertebrates in the estuaries of the southern Cape totalled 357, the fauna of all estuaries on the Atlantic coast of South Africa totalled 59. This is a much greater decrease in diversity than has been recorded in the sea. Day relates this very striking reduction in Atlantic coast estuaries to the drastic change in temperature at the mouths of these estuaries. This is particularly marked during periods of upwelling in summer. Estuarine temperatures are then 19,5 ° or higher but when the upwelled water flows into the estuary with the rising tide, temperatures may drop to 10,5 °. It is significant that the larvae of invertebrates and the juveniles of fishes are normally recruited from the sea in summer.

Appendix 9.1 follows on page 173

Appendix 9.1

CHARACTERISTIC MACROBENTHIC SPECIES IN ESTUARIES OF SOUTHERN AFRICA

Note: The following abbreviations are used:

Zoogeographic range

T — tropical, recorded north of 20 °S in the Indian Ocean
E — endemic to southern Africa south of 20 °S
C — cosmopolitan, recorded in northern and southern temperate zones
O — other foreign species recorded from scattered localities

Abbreviations for estuary names

Mor — Morrumbene, Mocambique
Kos — Kosi Bay, Zululand
RhB — Richards Bay, Zululand
Ml — Mlalazi, N. Natal
Umk — estuaries near Mkomazi
Shp — estuaries near Port Shepstone
St.J — estuaries near Port St Johns
Mng — Mngazana, Transkei
Bas — Bashee River, Transkei

Ham — Keiskamma River, E. Cape
Bmr — Bushmans River, E. Cape
Sun — Sundays River, near Port Elizabeth
Swa — Swartkops River, Port Elizabeth
Kny — Knysna, S. Cape
Bre — Breede River, S. Cape
Her — Hermanus Lagoon, S. Cape
Mil — Milnerton Lagoon, Cape Town
Brg — Berg River, W. Cape
Olf — Olifants River, W. Cape

Estuarine component (defined in text)

St.m — stenohaline marine
Est — restricted to estuaries
Eur — euryhaline marine
FW — euryhaline fresh water

Remarks

Brief notes are given for those species not discussed in the text.

Species	Zoogeography; estuarine range	Estuarine component	Remarks
PORIFERA			
Hymeniacedon perlevis (Mont.)	O; Bmr - Kny	St.m	On rocks or shells in mud at LWS and below.
CNIDARIA			
Hydractinia kaffraria Millard	T; St.L - Bre	Eur	Commensal on shells of *Nassarius kraussiana*.
Pseudactinia flagellifera (Hertw.)	E; Swa - Kny	St.m	On rocks and *Zostera* at LWS and below.
NEMERTEA			
Polybrachiorhynchus dayi Gibson	E; Mor - Bre	Eur	See text.
POLYCHAETA: ERRANTIA			
Ceratonereis erythraeensis Fauvel	T; Mor - Olf	Eur	See text.
Ceratonereis keiskama Day	E; Kos - Brg	Est	Common in mud in low salinities; omnivorous.
Dendronereis arborifera Peters	T; Mor - Bas	Eur	See text.
Diopatra cuprea (Bosc)	T; Mor	Eur	See text.
Glycera tridactyla Schm.	C; Mor - Bre	Eur	A subsurface predator in muddy sand.
Lumbrineris tetraura (Sch.)	C; Mng - Mil	Eur	A subsurface predator in muddy sand.
Marphysa macintoshi Crossl.	T; Mor - Mng	Eur	See text.
Marphysa sanguinea (Mont.)	C; Bmr - Bre	Eur	Omnivorous; density 6-125 m^{-2} in *Zostera* beds.
Nephtys capensis Day	E; Mng - Mil	Eur	A subsurface predator in muddy sand, LWS and below.
Neptys tulearensis Fauvel	T; Mor - Kny	Eur	A subsurface predator in muddy sand at LWS.
POLYCHAETA: SEDENTARIA			
Arenicola loveni Kbg.	E; Bmr - Her	Eur	See text.
Capitella capitata (Fabr.)	C; Mor - Olf	Eur	A detritus feeder; common in polluted mud.
Cirriformia tentaculata (Mont.)	C; Mor - Kny	Eur	A detritus feeder in mud.
Desdemona ornata Banse	E; Ml - Mil	Est	A small filter feeder in mud at LWS and below.
Ficopomatus (= Mercierella) enigmatica (Fauvel)	C; St.L - Mil	Est	See text.
Notomastus abberans Day	E; Mor - Mng	St.m	A subsurface deposit feeder.
Orbinia angrapequensis (Aug.)	E; Mng - Olf	St.m	A subsurface deposit feeder in sand.
Owenia fusiformis D.Ch.	C; Mor	St.m	A suspension and deposit feeder building tubes in sand at LWN and below.
Pomatoleios kraussii (Baird)	T; Mor - Bre	St.m	A filter feeder building calcareous tubes on mid-tide rocks.
Prionospio sexoculata Aug.	O; St.L - Her	Eur	A detritus feeder in mud at LWS and below.
Thelepus plagiostoma Schm.	C; Mor - Kny	St.m	A detritus feeder making sandy tubes under stones.
CIRRIPEDIA			
Balanus amphitrite Darwin	C; Mor - Her	Eur	See text.
Balanus elizabethae Brnrd.	E; Shp - Kny	Est	See text.
Balanus algicola Pilsbry	E; Mng - Olf	St.m	Common below mid-tide on west coast.
Chthamalus dentatus Krauss	T; Mor - Her	St.m	On rocks above mid-tide.

AMPHIPODA

Austrochiltonia subtenuis (Sayce)	E; Kos - Olf	Est	In weedbeds; salinity range 3-50 ‰.
Corophium triaenonyx Stebb.	T; Mor - Her	Est	See text.
Grandidierella bonnieroides (Steph.)	T; Mor - Mng	Eur	In mud, salinity range 10-50 ‰.
Grandidierella lignorum (Brnrd.)	E; St.L - Bre	Est	In mud at low tide and below; extends to fresh water.
Melita zeylanica Stebb.	T; Mor - Olf	Eur	See text.
Orchestia ancheidos (Brnrd.)	T; Mor - Brg	Eur	See text.
Orchestia rectipalma (Brnrd.)	E; St.L - Olf	Est	See text.
Urothoe pulchella Costa	O; St.J - Kny	Eur	Epifauna on sandbanks at LWN.

ISOPODA

Cirolana fluviatilis Stebb.	T; Kos - Bre	Est	In sand and under stones; salinity range 5-52 ‰; omnivorous.
Cyathura estuaria Brnrd.	E; St.L - Her	Est	Makes shallow burrows in mud; salinity range 2-50 ‰.
Exosphaeroma hylecoetes (Brnrd.)	E; Umk - Mil	Est	On muddy sand among *Zostera;* mainly herbivorous.
Ligia dilatata Brandt	E; Ham - Olf	Eur	On rocks near high tide; scavenger.
Sphaeroma terebrans Bate	T; Mor - Kny	Eur	See text.
Synidotea variegata Cllge.	T; Mor - St.L	Eur	In *Zostera* beds; herbivorous; salinity range 8-42 ‰.

MYSIDACEA

See Chapter 8 on plankton.

MACRURA

Alpheus crassimanus Heller	T; Mor - Bre	Eur	Makes communal burrows in *Zostera* beds.
Caridina nilotica Roux	T; Mor - St.J	FW	See text.
Macrobrachium equidens (Dana)	T; Mor - Bas	FW	See text.
Metapenaeus monoceros (Fabr.)	T; Mor - Ham	Eur	See text.
Palaemon pacificus (Stimps.)	T; Kos - Olf	Eur	See text.
Penaeus indicus M.Edw.	T; Mor - Kny	Eur	See text.
Penaeus japonicus Bate	T; Mor - Bre	Eur	See text.
Penaeus monodon Fabr.	T; Mor - Bas	Eur	See text.
Penaeus semisulcatus de Haan	T; Mor - Mng	Eur	See text.

ANOMURA

Callianassa kraussi Stebb.	E; Kos - Olf	Eur	See text.
Clibanarius longitarsus (de Haan)	T; Mor - Mng	Eur	Scavenging on mud banks near LWN.
Clibanarius padavensis (de Man)	T; Mor - St.J	Eur	See text.
Diogenes brevirostris (Stimps.)	E; Ham - Her	Eur	A scavenger and filter feeder moving over the shore with the tide.
Upogebia africana (Ortm.)	E; Mor - Olf	Eur	See text.

BRACHYURA

Cleistostoma algoense Brnrd.	E; Mng - Kny	Eur	A deposit feeder in sloppy mud at LWS.
Cleistostoma edwardsii McLeay	E; St.L - Bre	Eur	A deposit feeder between HWN and LWS on sandy mud.
Cyclograpsus punctatus M.Edw.	O; Kos - Olf	Eur	See text.
Dotilla fenestrata Hilg.	T; Mor - Bre	Eur	See text.
Hymenosoma orbiculare Desm.	E; Mor - Olf	Eur	See text.
Macrophthalmus boscii Aud.	T; Mor - Mng	Eur	A deposit feeder in sloppy mud; LWS and below.
Macrophthalmus depressus Rüppell	T; Mor - RhB	Eur	A deposit feeder in mangrove creeks.
Macrophthalmus grandidieri M.Edw.	T; Mor - Mng	Eur	A deposit feeder in muddy pools, LWN-LWS.
Metopograpsus thukuhar (Owen)	T; Mor - Bas	Eur	See text.
Rhyncoplax bovis Brnrd.	T; Kos - Swa	Est	In weed beds in salinities down to 4 ‰.
Scylla serrata Forsk.	T; Mor - Bre	Eur	See text.
Sesarma catenata Ortm.	E; Mor - Bre	Eur	See text.
Sesarma eulimene de Man	T; Mor - Bas	Eur	See text.
Sesarma guttata M.Edw.	T; Mor - Mng	Eur	See text.
Sesarma meinerti de Man	T; Mor - Bas	Eur	See text.
Thalamita crenata (Latr.)	T; Mor - Ml	Eur	Both *T.crenata* and *T.integra* Dana are common predators below low tide; ovigerous females in burrows.
Tylodiplax blephariskios (Stebb.)	E; Mor - Mng	St.m	In sloppy mud at LWS and below.
Uca lactea f. annulipes (M.Edw.)	T; Mor - Ham	Eur	See text.
Uca chlorophthalmus (M.Edw.)	T; Mor - Ham	Eur	See text.
Uca urvillei (M.Edw.)	T; Mor - Mng	Eur	See text.
Varuna litterata (Fabr.)	T; Mor - Her	Eur	Clings to drift wood in the sea and extends to fresh water.

MOLLUSCA: BILVALVIA

Dosinia hepatica (Lam.)	E; Mor - Bre	Eur	See text.
Eumarcia paupercula (Holten)	T; Mor - Kny	Eur	See text.
Lamya capensis (Kr.)	E; Mor - Her	Eur	See text.
Loripes clausus Phil.	E; Mor - Bre	Eur	Dead shells are abundant but live specimens are not very common; a deposit feeder in muddy sand.
Macoma litoralis (Kr.)	E; Mor - Kny	Eur	See text.
Macoma retrorsa Sow.	E; Mor - St.L	Eur	Abundant in mudbanks at LWS in Morrumbene.
Mactra lilacea Lam.	E; Mor - Kny		In Langebaan lagoon, it reaches a size of 10 cm and a density of 550 m^{-2}. Net production reaches the high value of 15 983 kJ.m^{-2}.yr^{-1} (Lucas, 1979).
Musculus virgiliae Brnrd.	E; Kos - Bre	Est	See text.
Ostrea algoensis Sow.	E; Mng - Bre	Eur	Common in rocks at LWS even in muddy crevices.
Perna perna (Linn.)	T; Mor - Her	St.m	Extends into estuary mouths on rocks at LWN and below.
Psammotellina capensis (Sow.)	E; Mng - Bre	Eur	Burrows in sandbanks in the lower reaches.
Solen capensis Fischer	E; Mng - Olf	St.m	See text.
Solen corneus Lam.	T; Mor - Kny	Eur	See text.
Saccostrea (= Crassostrea) (cucullata (Born))	T; Mor - Mng	Eur	See text.
Saccostrea margaritacea (Lam.)	E; Mng - Her	Eur	See text.

MOLLUSCA: GASTROPODA

Assiminea bifasciata (Nevill)	T; Kos - Bre	Est	See text.
Assiminea ?globulus (Conolly)	E; Shp - Olf	Est	See text.
Cassidula labrella (Desh.)	T; Mor - Ml	Eur	See text.
Cerithidea decollata (Linn.)	T; Mor - Swa	Eur	See text.
Haminea alfredensis (Kr.)	E; Swa - Her	Eur	A common tectibranch in *Zostera* beds; mainly herbivorous.
Littorina africana var. *knysnaensis* Phil.	E; Shp - Olf	Eur	See text.
Littorina scabra Linn.	T; Mor - Sun	Eur	See text.
Nassarius arcularius (Linn.)	T; Mor - Dbn	St.m	See text.
Nassarius coronatus (Brug.)	T; Mor - Dbn	St.m	See text.
Natica gualteriana Rechluz	T; Mor - Ham	Eur	See text.
Natica tecta Anton	E; Mng - Her	Eur	See text.
Nerita albicilla Linn.	T; Mor - Bre	Eur	A generalised browser on mid-tide rocks; extends further south in estuaries than on seashores.
Notarchus leachii Blainv.	E; Mor - Her	Eur	See text.
Oxystele variegata (Anton)	E; Mng - Olf	Eur	A generalised browser on rocks from HWN-LWN.
Patella oculus Born	E; Bmr - Bre	Eur	See text.
Polynices tumidus (Swains.)	T; Mor - Mng	St.m	See text.
Siphonaria aspera Kr.	E; Umk-Kny	Eur	See text.
Siphonaria capensis Q. & G.	E; Kos - Her	Eur	See text.
Siphonaria oculus Kr.	E; Umk-Bre	Eur	See text.
Terebralia palustris (Brug.)	T; Kos - Mng	Eur	See text.
Thais dubia (Kr.)	E; Bmr - Olf	Eur	A predator feeding mainly on barnacles.
Turritella capensis Kr.	E; Kny	Eur	Abundant on muddy sand at LWS and below in Knysna and Langebaan Lagoon; a filter feeder.

ECHINODERMATA

Amphipholis squamata (D.Ch.)	C; Mor - Kny	St.m	Fairly common in shell gravel and under stones.
Patiriella exigua (Lam.)	O; Mng - Bre	Eur	Fairly common on hard substrates from LWN down.
Echinocardium cordatum Pennant	O; Mng - Her	St.m	Burrows in sand from LWS to bed of estuary.
Echinodiscus bisperforatus Leske	T; Mor - Kny	St.m	Burrows in sand at LWS in the mouths of estuaries and at 3-5 m in Plettenberg Bay and Mossel Bay.

TUNICATA

Pyura stolonifera (Heller)	E; Bmr - Her	St.m	On rocks and hard substrates at LWS and below; abundant from LWS – 10 m on rocky seashores.

PISCES

Periophthalmus sobrinus Eggert	T; Mor - Mng	Eur	On mud and climbing on mangrove roots.

REFERENCES

ALEXANDER, J.J. & D.W. EWER 1969. A comparative study of some aspects of the biology and ecology of *Sesarma catenata* and *Cyclograpsus punctatus* with additional notes on *Sesarma meinerti*. *Zool. Afr.* 4: 1-35.

ALLANSON, B.R., HILL, B.J., BOLTT, R.E. & V. SCHULTZ 1966. An estuarine fauna in a fresh water lake in South Africa. *Nature* 209(5022): 532-533.

BAAS BECKING, L.G. & E.J. WOOD 1955. Biological processes in the estuarine environment. *Proc. K. ned. Akad. Wet.* 58: 160-181.

BARHAM, W.T. 1979. Spawning of *Arenicola loveni* Kinberg in the Heuningsnes River estuary, Bredasdorp. *S.Afr. J. Sci.* 75: 262-264.

BARNES, R.S.K. 1974. *Estuarine biology*. Edward Arnold, London: 76pp.

BARNES, R.S.K. & J. GREEN (eds) 1972. *The estuarine environment*. Applied Science, London. 133pp.

BASSINDALE, R. 1943. A comparison of the varying salinity conditions of the Tees and Severn estuaries. *J. Anim. Ecol.* 12: 1-10.

BAYNE, B.L. 1976. Aspects of reproduction in bivalve molluscs. *In:* M. Wiley (ed), *Estuarine Processes* Vol 1: 432-448.

BEADLE, L.C. 1957. Osmotic and ionic regulation in aquatic animals. *Ann. Rev. Physiol.* 19: 329-358.

BEADLE, L.C. 1972. Physiological problems for animal life in estuaries. *In:* R.S.K. Barnes & J. Green (eds), *The estuarine environment*. Applied Science, London.

BLABER, S.J.M. 1976. The food and feeding ecology of mullet in the St Lucia Lake system. *In:* A.E.F. Heydorn (ed), *St Lucia Scientific Advisory Council Workshop – Charters Creek, February 1976*. Natal Parks Bd, Pietermaritzburg. (Also): *Biol. J. Lin.. Soc. Lond.* 8: 267-277.

BLABER, S.J.M. 1977. The feeding ecology and relative abundance of Mugilidae in Natal and Pondoland estuaries. *Biol. J. Linn. Soc. Lond.* 8: 259-276.

BLABER, S.J.M., HILL, B.J. & A.T. FORBES 1974. Infratidal zonation in a deep South African estuary. *Mar. Biol.* 28: 333-337.

BOLTT, R.E. 1969. The benthos of some southern African lakes. Part II: The epifauna and infauna of the benthos of Lake Sibaya. *Trans. roy. Soc. S.Afr.* 38(3): 249-270.

BOLTT, R.E. 1975. The benthos of some southern African lakes. The Recovery of the benthic fauna of St Lucia Lake following a period of excessively high salinity. *Trans. roy. Soc. S.Afr.* 41: 295-323.

BOLTT, G. & J.HEEG 1975. The osmoregulation of three grapsoid crab species in relation to their penetration of an estuarine system. *Zool. Afr.* 10: 167-182.

BOUSFIELD, E.L. 1955. Ecological control of the occurrence of barnacles in the Miramichi estuary. *Bull. Natl. Mus. Can.* 137.

BRANCH, G.M. 1975. Mechanisms reducing intraspecific competition in *Patella* spp: migration, differentiation and terrestrial behaviour. *J. Anim. Ecol.* 44: 575-600.

BRANCH, G.M. & J. GRINDLEY 1979. The ecology of South African estuaries. Part XI: Mngazana, a mangrove estuary in Transkei. *S.Afr. J. Zool.* 14(3): 149-170.

BREEN, C.M. & B.J. HILL 1969. A mass mortality of mangroves in the Kosi estuary. *Trans. roy. Soc. S.Afr.* 28: 285-303.

BRIGGS, J.C. 1974. *Marine zoogeography*. McGraw Hill, New York.

BROEKHUYSEN, G.J. 1941. The life history of *Cyclograpsus punctatus* M.Edw.: Breeding and growth. *Trans. roy. Soc. S.Afr.* 28: 331-.

BROEKHUYSEN, G.J. 1955. The breeding and growth of *Hymenosoma orbiculare*. *Ann. S.Afr. Mus.* 41: 313-343.

BROWN, A.C. & N. JARMAN 1978. Coastal and marine habitats: *In:* M.J.A. Werger & A.C. van Bruggen (eds), *Biogeography and ecology of southern Africa*. W. Junk, The Hague.

BROWN, D.S. 1971. The ecology of Gastropoda in a South African mangrove swamp. *Proc. malac. Soc. Lond.* 39 (363): 263-279.

CARRIKER, M.R. 1967. Ecology of estuarine benthic invertebrates; a perspective. *In:* G. Lauff (ed), *Estuaries*. Am. Ass. Adv. Sci., Washington. 757pp.

CASPERS, H. 1959. Die Einteilung der Brakwasser Regionen in einem Aestuar. *Arch. Oceanog. Limnol. (Suppl.)* 11: 153-169.

CHAMPION, H.F.B. 1976. Recent prawn research at St Lucia, with notes on the bait fishers. *In:* A.E.F. Heydorn (ed), *St Lucia Scientific Advisory Council Workshop – Charters Creek, February 1976*. Natal Parks Bd, Pietermaritzburg.

CLOETE, C.E. & W.D. OLIFF (eds) 1976. *South African marine pollution survey report 1974-1975*. S.Afr. Natl. Sci. Program Rep. 8: CSIR, Pretoria.

DAHL, E. 1952. Some aspects of the ecology and zonation of the fauna on sandy beaches. *Oikos* 4: 1-27.

DAY, J.H. 1951. The ecology of South African estuaries. Part I: General considerations. *Trans. roy. Soc. S.Afr.* 33: 53-91.

DAY, J.H. 1959. The biology of Langebaan Lagoon: a study of the effect of shelter from wave action. *Trans. roy. Soc. S. Afr.* 35: 475-547.

DAY, J.H. 1967a. The biology of Knysna estuary, South Africa. *In:* G. Lauff (ed), *Estuaries*. Am. Ass. Adv. Sci., Washington. 757pp.

DAY, J.H. 1967b. *A monograph on the Polychaeta of southern Africa*. Parts I and II: Br. Bus. (nat. Hist.) London. 878pp.

DAY, J.H. 1969 (2nd ed 1974). *A guide to marine life on South African shores*. Balkema, Cape Town. 330pp.

DAY, J.H. 1974. The ecology of Morrumbene estuary, Mocambique. *Trans. roy. Soc. S.Afr.* 41(1): 43-97.

DAY, J.H., FIELD, J.G. & M.P. MONTGOMERY 1971. The use of numerical methods to determine the distribution of the benthic fauna across the continental shelf of North Carolina. *J. Anim. Ecol.* 40: 93-125.

DAY, J.H., MILLARD, N.A.H. & A.D. HARRISON 1952. The ecology of South African estuaries. Part 3: Knysna, a clear open estuary. *Trans. roy. Soc. S.Afr.* 33: 367-413.

DAY, J.H. & J.F.C. MORGANS 1956. The ecology of South African estuaries. Part 7: The biology of Durban Bay. *Ann. Natal Mus.* 13: 259-312.

DAY, J.H. & D.P. WILSON 1934. On the relation of the substratum to the metamorphosis of *Scolecolepis fuliginosa* (Claparède). *J. mar. biol. Ass. UK* 19: 655-661.

DEHNEL, P.A. 1967. Osmotic and ionic regulations in estuarine crabs. *In:* G. Lauff (ed), *Estuaries*. Am. Ass. Adv. Sci., Washington. 757pp.

DYE, A.H. 1976. Studies on the ecology of the estuarine meiofauna in South Africa. *Proc. 1st Interdisciplinary Conf. Mar. freshw. Res. S.Afr., Port Elizabeth, July 1976*.

EDNEY, E.G. 1961. The water and heat relations of fiddler crabs (*Uca* spp). *Trans. roy. Soc. S.Afr.* 36(2): 71-91.

EDNEY, E.G. 1962. Some aspects of the temperature relations of the fiddler crab (*Uca* spp). *Biometerology*, 79-96.

EKMAN, S. 1953. *Zoogeography of the sea*. Sidgwick & Jackson, London. 417pp.

ELTRINGHAM, S.K. 1971. *Life in mud and sand*. English Universities Press, London. 218pp.

FORBES, A.T. 1973. An unusual abbreviated larval life in the estuarine burrowing prawn *Callianassa kraussi*. *Mar. Biol.* 22: 361-365.

FORBES, A.T. & B.J. HILL 1969. The physiological ability of the marine crab *Hymenosoma orbiculare* to live in a subtropical freshwater lake. *Trans. roy. Soc. S.Afr.* 38: 271-283.

GENADE, A.B. 1973. A general account of certain aspects of oyster culture in Knysna estuary. *S.Afr. Natl. Oceanog. Symp. Cape Town 1973* (Abstracts pp 26-28).

GRAY, J.S. 1966. The attractive factor of intertidal sands to *Protodrilus symbioticus. J. mar. biol. Ass. UK* 45: 627-645.

GREEN, J. 1968. *The biology of estuarine animals.* Sidgwick & Jackson, London. 401pp.

GRIFFITHS, C.L. 1974. *The gammaridean and caprellid Amphipods of southern Africa.* PhD thesis, Univ. of Cape Town.

GRINDLEY, J.E. 1976. *Report on ecology of Knysna estuary and proposed Braamekraal Marina.* School of Environmental Studies, Univ. Cape Town (123pp typescript).

HEDGPETH, J.W. 1967. Ecological aspects of the Laguna Madre, a hypersaline estuary. *In:* G. Lauff (ed), *Estuaries.* Am. Ass. Adv. Sci., Washington.

HILL, B.J. 1966. A contribution to the ecology of Umlalazi estuary. *Zool. Afr.* 2: 1-24.

HILL, B.J. 1967. *Contributions to the ecology of the anomuran mud prawn Upogebia africana (Ortmann).* PhD thesis, Rhodes Univ., Grahamstown.

HILL, B.J. 1974. Salinity and temperature tolerance of zoeae of the portunid crab *Scylla serrata. Mar. Biol.* 25: 21-24.

HILL, B.J. 1975. Abundance, breeding and growth of the crab *Scylla serrata* in two South African estuaries. *Mar. Biol.* 32: 119-126.

HILL, B.J. 1976. Natural food, foregut clearance rate and activity of the crab *Scylla serrata. Mar. Biol.* 34: 109-116.

HILL, B.J. 1979. Biology of the crab *Scylla serrata* (Forskal) in the St Lucia system. *Trans. roy. Soc. S.Afr.* 44: 55-62.

HULLEY, P.A. 1972. The origin, interrelationships and distribution of southern African Rajidae (Chondrichthyes Batoidei). *Ann. S.Afr. Mus.* 60: 1-103.

ICELY, J.D. & D.A. JONES 1978. Factors affecting the distribution of the genus *Uca* (Crustacea: Ocypodidae) on an East African shore. *Mar. Biol.* 6: 315-326.

JEFFERIES, R.L. 1972. Aspects of salt marsh ecology with particular reference to inorganic plant nutrition. *In:* R.S.K. Barnes & J. Green (eds), *The estuarine environment.* Applied Science, London. 133pp.

JOUBERT, L.S. & D.H. DAVIES 1966. The penaeid prawns of the St Lucia Lake system. *Invest. Rep. Oceanog. Res. Inst. Durban* 13: 1-40.

KALK, M. 1959. The zoogeographical composition of the intertidal fauna at Inhaca Island, Mocambique. *S.Afr. J. Sci.* 55: 178-180.

KENSLEY, B.F. 1974. *A review of the southern African decapod Crustacean fauna.* PhD thesis, Univ. Cape Town.

KINNE, O. 1964. The effects of temperature and salinity on marine and brackish water animals. II Salinity and temperature – salinity combinations. *Ann. Rev. Oceanog. Mar. Biol.* 2: 281-339.

KORRINGA, P. 1956. Oyster culture in South Africa. Hydrological, biological and ostreological observations in the Knysna lagoon, with notes on conditions in other South African waters. *Invest. Rep. Div. Sea Fish. S.Afr.* 20: 1-85.

LAUFF, G. (ed) 1967. *Estuaries.* Am. Ass. Adv. Sci., Washington.

LEWIS, J.R. 1964. *The ecology of rocky shores.* English Univ. Press, London. 323pp.

LOCKWOOD, A.P.M. 1976. Physiological adaptation to life in estuaries. *In:* R.C. Newell (ed), *Adaptation to environment: essays on the physiology of marine animals.* Butterworths, London.

LUCAS, M.I. 1979. Growth and metabolism of *Mactra lilacea. Abstract. 4th (S.Afr.) Natl. Oceanog. Symp., Cape Town, July 1979.* CSIR, Pretoria.

MacGINITIE, G.E. 1930. Natural history of the mud shrimp *Upogebia pugetensis* (Dana). *Ann. Mag. Nat. Hist.* 10: 36-42.

MacGINITIE, G.E. 1935. Ecological aspects of a California marine estuary. *Am. Midland Naturalist* 16: 629-765.

MACNAE, W. 1957. The ecology of the plants and animals in the intertidal regions of the Swartkops near Port Elizabeth, South Africa. Parts I and II. *J. Ecol.* 45: 113-131; 361-387.

MACNAE, W. 1963. Mangrove swamps in South Africa. *J. Ecol.* 51: 1-25.

MACNAE, W. 1968. A general account of the mangrove swamps in the Indo-West Pacific region. *Advan. Mar. Biol.* 6: 73-270.

MACNAE, W. & M. KALK 1958. *A natural history of Inhaca Island, Mocambique.* Witwatersrand Univ. Press, Johannesburg.

MACNAE, W. & M. KALK 1962. The ecology of mangrove swamps at Inhaca Island, Mocambique. *J. Ecol.* 50: 19-34.

McLACHLAN, A. 1974. Notes on the biology of some estuarine bivalves. *Zool. Afr.* 9: 15-34.

McLACHLAN, A. & J.R. GRINDLEY 1974. Distribution of the macrobenthic fauna of soft substrata in the Swartkops estuary with observations on the effects of floods. *Zool. Afr.* 9: 211-233.

MILLARD, N.A.H. 1978. The geographical distribution of South African hydroids. *Ann. S.Afr. Mus.* 74: 159-200.

MILLARD, N.A.H. & G.J. BROEKHUYSEN 1970. The ecology of South African estuaries. Part 10: St Lucia, a second report. *Zool. Afr.* 5(2): 277-307.

MILNE, H. & G.M. DUNNET 1972. Standing crop, productivity and trophic relations of the fauna of the Ythan estuary. *In:* R.S.K. Barnes & J. Green (eds), *The estuarine environment.* Applied Science, London.

NEWELL, R.C. 1965. The role of detritus in the nutrition of two marine deposit feeders, the prosobranch *Hydrobia ulvae* and the bivalve *Macoma balthica. Proc. zool. Soc. Lond.* 144: 35-45.

NEWELL, R.C. 1970. *Biology of intertidal animals.* Logos Press, London. 555pp.

NEWELL, R.C. 1976 (ed) *Adaptation to environment: essays on the physiology of some marine animals.* Butterworths, London.

OLIFF, W.D. (ed) 1976. National marine pollution surveys. East coast section. Second annual report. *S.Afr. Natl. Inst. Water Res.* Durban (typescript). Note: This compilation includes many internal reports by officers of the NIWR. Summaries and references are given by Cloete & Oliff 1976.

PENRITH, M.-L. & B.F. KENSLEY 1970. The constitution of the intertidal fauna of rocky shores of South West Africa. Part 2: Rocky Point. *Cimbebasia (A)* 1: 243-268.

PERKINS, E.J. 1974. *The biology of estuaries and coastal waters.* Academic Press, London. 677pp.

POTTS, W.T.W. & G. PARRY 1964. *Osmotic and ionic regulation in animals.* Pergamon Press, New York.

PRITCHARD, D.W. 1967. What is an estuary: physical viewpoint. *In:* G. Lauff (ed), *Estuaries.* Am. Ass. Adv. Sci., Washington.

PUTTICK, G.M. 1977. Spatial and temporal variations in intertidal animal distribution at Langebaan Lagoon, South Africa. *Trans. roy. Soc. S.Afr.* 42: 403-440.

REMANE, A. & C. SCHLIEPER 1958. Die biology des Brack wassers. *Die Binnengewasser.* 22: 348.

SANDERS, H.L. 1968. Marine benthic diversity: a comparative study. *Am. Nat.* 103: 243-262.

SANDISON, E.E. 1966. The effect of salinity fluctuations on the life cycle of *Balanus pallidus stutsburi* Darwin in Lagos harbour, Nigeria. *J. Anim. Ecol.* 35: 363-378.

SCHELTEMA, R.S. 1961. Metamorphosis of the veliger larvae of *Nassarius obsoletus* (Gastropoda) in response to bottom sediment. *Biol. Bull. Wood's Hole* 120: 92-109.

SCOTT, K.M.F., HARRISON, A.D. & W. MACNAE 1952. The ecology of South African estuaries. Part 2: The Klein River estuary, Hermanus. *Trans. roy. Soc. S.Afr.* 33: 283-331.

SIEGFRIED, W.R. 1962. A preliminary report on the biology of the mudprawn *Upogebia africana*. *Invest. Rep. Cape Dept. Nature Conserv.* **1**: 1-24.

SPOONER, G.M. & H.B. MOORE 1940. The ecology of the Tamar estuary. VI: An account of the intertidal muds. *J. mar. biol. Ass. UK* **24**: 283-330.

STEPHENSON, T.A. 1947. The constitution of the intertidal fauna and flora of South Africa. III. *Ann. Natal Mus.* **11**: 207-324.

STEPHENSON, T.A. & A. STEPHENSON 1972. *Life between tide marks on rocky shores.* Freeman, San Francisco.

STEVENSON, L.H. and C.W. ERKENBRECKER 1976.

TEBBLE, N. 1953. A source of danger to harbour structures. *J. Inst. munic. Engrs.* **1953**: 1-7.

WALLACE, J.H. 1975. The estuarine fishes of the east coast of South Africa. I: Species composition and length distribution in the estuarine and marine environment. II; Seasonal abundance and migrations. *Invest. Rep. Oceanog. Res. Inst., Durban* **40**: 1-72.

WALNE, P.R. 1972. The importance of estuaries to commercial fisheries. *In:* R.S.K. Barnes & J. Green (eds), *The estuarine environment.* Applied Science, London. 133pp.

WELLS, G.P. 1945. The mode of life of *Arenicola marina* L. *J. mar. biol. Ass. UK* **26**: 170-207.

WILSON, D.P. 1955. The role of micro-organisms in the settlement of *Ophelia bicornis* Savigny. *J. mar. biol. Ass. UK* **34**: 513-543.

Chapter 10
Estuarine meiofauna
A.H. Dye and J.P. Furstenburg*

Department of Zoology, University of Port Elizabeth

INTRODUCTION

In 1942 Mare coined the term 'meiobenthos' to describe those organisms intermediate in size between the macrobenthos on one hand and the microbenthos on the other. This group, now often referred to as meiofauna, is approximately defined as those organisms passing a 1,0 or 0,5 mm sieve but being retained on a 0,045 mm sieve. Although the adult forms of the benthic groups occur in distinct habitats, some overlap of macro- and meiofauna occurs due to the presence of the larval stages of the former. McIntyre (1969) consequently divided the meiofauna into temporary forms, consisting of the larvae of macrofauna which will eventually outgrow the meiofauna size range, and the permanent meiofauna consisting of the Nematoda, Harpacticoida, Mystacocarida, Kinorhyncha, Ostracoda, Gastrotricha, Halacarida, Archiannelida, Tardigrada, Turbellaria and Rotifera. In addition to these, there are some specialized forms of Hydrozoa, Nemertina, Bryozoa, Gastropoda, the solenogasters, Holothuroidea, Tunicata, Priapulida and Sipunculoida (Coull 1973).

Since the early 1940's, a great number of papers dealing with meiofauna have appeared but few of these have been concerned with estuaries. This research was given a boost in the 1950's with the development of ecosystems energetics theory which brought about a realization of the potential importance of this ubiquitous group in terms of energy flow. Most estuarine

* Present address: University of Transkei, Umtata

meiofauna studies have been carried out in the Northern Hemisphere and these have followed three relatively independent courses (Coull, 1973):

a) Studies on distinct taxonomic groups (Capstick 1959, Wieser & Kanwisher 1961, Teal & Wieser 1966, Tietjen 1966, Barnett 1968, Coull & Vernberg 1970, Meyers & Hopper 1971, Wieser, Ott, Schiemer & Gnaiger 1974, Fava & Volkman 1975).

b) Studies of the relationships between environmental factors such as grain size, salinity and temperature and the meiofauna (Capstick 1959, Gordon 1960, Kinne 1964, Duff & Teal 1965, Johnson 1965, Jansson 1967b, Boaden 1968, Fenchel & Straarup 1971, Biggs & Flemmer 1972, Marsden 1973).

c) Quantitative analyses of the meiofauna and environmental correlations (Rees 1940, Mare 1942, Jansson 1968, Tietjen 1966, 1969, Damodoran 1974, Kurian 1974, Fava & Volkman 1975). Some of the more important reviews which have appeared in addition to the above are McIntyre, 1964, 1969, Swedmark, 1964, Gerlach, 1971, Coull, 1973.

Few of the above have attempted to analyse the observed distribution and fluctuations of the meiofauna in terms of environmental parameters in the long term. There is thus little data on topics such as organism-sediment relationships, composition, biomass and production. This is particularly true in southern Africa where there is a lack of knowledge of all aspects of estuarine meiofaunal ecology. This field of study is represented only by a recently completed

Table 10.1
Intertidal meiofaunal population densities in numbers per 10 cm^2.

Substrate	Area	Population	Reference
Sand	Baltic	500	Jansson (1968)
,,	,,	685	Smidt (1951)
,,	Blyth estuary	800*	Capstick (1959)
,,	Baltic	1 250*	Fenchel (1969)
,,	,,	968	Muus (in Coull 1973)
,,	Rhode Island	2 686	Tietjen (1969)
,,	Swartkops	960	Dye & Furstenberg (1977)
,,	Kromme	1 828	Dye (1977)
,,	Knysna	1 100	,,
,,	Berg River	187	,,
,,	Mgazana	2 000	Dye (1979)
Mud	Bristol Channel	5 955	Rees (1940)
,,	India	2 117	McIntyre (1968)
,,	Blyth estuary	501*	Capstick (1959)
,,	Swartkops	138	Dye & Furstenberg (1977)
,,	Kromme	60	Dye (1977)
,,	Knysna	794	,,
,,	Berg River	87	,,
,,	Mgazana	330	Dye (1979)
Salt marsh	New Jersey	749	Brickman (in Coull 1973)
,,	Georgia	12 400*	Teal & Wieser (1966)
,,	Massachussettes	1 785*	Wieser & Kanwisher (1961)
,,	Swartkops	45 000	Furstenberg (unpublished)
,,	Knysna	5 232	Dye (1977)

* nematodes only

study of the sand and mud flats of the Swartkops estuary (Dye 1977). Work is also in progress on the salt marsh meiofauna of this estuary (Furstenberg, unpublished). Periodic surveys of some of the West coast estuaries are being made as part of a pollution monitoring programme of the CSIR (Fricke, personal communication).

POPULATION DENSITY, COMPOSITION AND DISTRIBUTION

Published data indicate that meiofaunal population densities in non-estuarine beaches vary from 50 to 10 212 per 10 cm^2 (McIntyre, 1969). Densities between 60 and 2 250 per 10 cm^2 have been recorded from sandy beaches in Algoa Bay (McLachlan, 1977). Although great variation occurs, the average appears to be in the region of 1 000 individuals per 10 cm^2. A comparison of estuarine and non-estuarine areas, reveals that the former are considerably richer in meiofauna than the latter and this may be ascribed to a greater availability of food and a greater degree of shelter in estuaries. Analysis of available data leads to the conclusion that estuarine habitats may be divided into three on the basis of substrate and plant cover, ie sand flats, mud flats and salt marshes, and that these areas may be ranked in order of importance with the sand flats having the lowest meiofauna populations and the salt marshes the highest.

The average meiofauna density in sand flats appears to be 1 000 per 10 cm^2 (see Table 10.1) but some areas achieve this number in nematodes alone (Fenchel 1969). Studies in South Africa have shown meiofaunal populations to vary from 187 per 10 cm^2 (Berg River) to 1 828 per 10 cm^2 (Kromme River) (Dye, 1977). The average is still, however, close to 1 000 per 10 cm^2 and it appears that little change in density occurs between exposed sandy beaches and estuarine sand flats.

An increase in meiofaunal populations is usually found in mud flats (Table 10.1) and here the average is in the region of 3 000 per 10 cm^2 although densities as high as 6 000 per 10 cm^2 have been recorded (Rees, 1940). In all of the estuaries studied in this country, however, a marked decrease in meiofauna occurs in the mud flats. Table 10.1 shows that the average density in these areas is only 270 per 10 cm^2.

The salt marshes are characterized by high meiofaunal numbers. Teal & Wieser (1966), for instance, recorded 12 400 per 10 cm^2 in a Georgia salt marsh while Wieser & Kanwisher (1961) found 1 785 per 10 cm^2 in a Massachussettes salt marsh. Variation does, of course, occur and Brickman (in Coull 1973) recorded a total population of only 749 per 10 cm^2. The average of 5 000 per 10 cm^2 must still, however, be considered conservative. Data from a survey of the Knysna Lagoon (Dye, 1977) revealed an average meiofaunal density of 5 232 individuals per 10 cm^2 although numbers varied between 19 682 and 493 per 10 cm^2. Possibly the highest recorded meiofaunal density was found in the salt marshes of the Swartkops estuary where numbers between 30 000 and 65 000 per 10 cm^2 are common (Furstenberg, unpublished).

Although the meiofauna is a heterogenous group comprising a large number of taxa it appears that the finer sediments of estuaries do not support the diversity of meiobenthic species characteristic of coarser sediments. Nematodes are usually the cominant group accounting for more than 80 % of the total meiofauna. The taxonomic composition of the nematodes is related to feeding and is a characteristic of certain sediments. Wieser (1953) divided the nematodes into four groups based on buccal morphology. These groups are: a) selective deposit feeders, b) non-selective deposit feeders, c) epigrowth feeders and d) predators. The first two dominate in littoral fine sediments and in sub-littoral sands and muds. The epigrowth feeders occur in fine littoral sands and become dominant in muddy areas. Predators become dominant in coarse sands as well as in fine sands poor in deposits. (Perkins, 1974). Thus, in general, the sand flats of estuaries would be expected to have a predominance of deposit feeders and a significant percentage of epi-growth feeders while the muddy areas would be dominated by epi-growth feeders. Appendix 1 gives a list of the marine nematodes so far identified in Swartkops estuary.

The sub-dominant meiofaunal group is usually the harpacticoid copepods. These are more strongly influenced by factors such as dessication and oxygen availability. They thus occur near the surface of the sediment and towards the lower tidal levels. The population density decreases considerably in muddy areas due to their intolerance of low oxygen conditions.

The remainder of the meiofauna consists of a miscellany of taxa, particularly Oligochaeta and Ostracoda, occurring in low numbers. Perkins (1974) found ostracods and harpacticoids both to account for only 0,4 % of the total meiofauna in sandy areas of the Eden estuary with nematodes making up the remainder. At Whitstable, oligochaetes accounted for 0,1 %, ostracods for 10 %, harpacticoids for 7 % and nematodes for the remainder (Perkins, 1958). In the sand flats of Swartkops the meiofauna consists of nematodes (84 %), harpacticoids (11,5 %) and the remainder was made up of oligochaetes, polychaetes, flatworms and gastrotrichs. The composition of the mud flat community was similar with the addition of small numbers of amphipods and ostracods (Dye & Furstenberg, 1977). Some variation occurred in the other estuaries studied. For instance no harpacticoids were recorded in the Berg River, the bulk of the meiofauna being nematodes and the rest a mixture similar to that of Swartkops. Over 98 % of the total meiofauna of Knysna estuary consisted of nematodes with very low numbers of the other groups. Community composition in the lower reaches of the Kromme estuary was similar to that of Swartkops but harpacticoid copepods were absent in the mud flats.

Little data are available on the intertidal distribution of meiofauna in estuaries. Teal & Wieser (1966) found the highest densities at mid-tidal levels in a salt marsh (10 410 per 10 cm^2) and the lowest at high tide marks (980 per 10 cm^2). In Swartkops salt marshes, the intertidal distribution of meiofauna was found to depend on plant cover since the community consisted of 99 % plant parasitic nematodes. The highest numbers were found in areas dominated by *Triglochin* spp and *Chenolea diffusa* and the lowest in *Spartina maritima* areas. The same trend was found at Knysna.

In the muddy areas of the Swartkops estuary the highest density occurred at high tide (152 per 10 cm^2) and the lowest, 30 cm below low tide (63 per 10 cm^2). The same trend was found in the muddy areas of the Kromme estuary but in the Berg River the highest population density occurred at low water (Dye, 1977). Gray & Rieger (1971) reported the highest meiofaunal density at high tide in sand flats of Robin Hood's Bay, England (179 per 10 cm^2) and the lowest to occur at low tide (89 per 10 cm^2). They ascribed this to substrate properties with the very coarse substrate at low tide limiting the meiofauna. The opposite trend occurred in the Swartkops sand flats for the highest number occurred at low tide (776 per 10 cm^2) and the lowest at high tide (582 per 10 cm^2). This is due to desiccation which occurs at the higher tidal levels. A modification of this trend was found in the Kromme estuary where the highest population density occurred at mid-tide. Here the population is limited by desiccation at high water and by scouring at low water. This is a characteristic of the lower reaches of the Kromme estuary. A similar situation is found, though to a lesser extent, in the Berg River.

Interstitial meiofauna occupies an essentially three dimensional habitat and the vertical distribution of the various species depends on individual tolerance to environmental conditions. Most studies of intertidal estuarine areas have revealed that the meiofauna is most abundant in the top few centimetres (Rees 1940, Perkins, 1958, Teal & Wieser, 1966, Barnett, 1968). Penetration in sands is much greater and a more even distribution of the meiofauna can be expected. In Swartkops sand flats, the average depth of penetration is 34 cm ± 9 cm. Of the total population, 38 % occurred in the top 6,5 cm, 66 % occurred down to 13,0 cm and a total of 87 % was found down to 20,0 cm (Dye, 1977). In the mud flats the average maximum depth was 26 cm ± 7 cm with 56 % of the total occurring in the top 6,5 cm, 78 % down to 13,0 cm and 94 % down to 20,0 cm. This distribution was greatly affected by the presence of the prawn *Callianassa kraussi* the burrows of which allowed the penetration of oxygenated water into the sediment and caused the meiofauna to extend to a greater depth than would have been expected on the basis of sediment properties alone (Dye & Furstenberg, 1977). The nematodes have a greater vertical depth penetration than the other taxa and below 10 cm the meiofauna consists mainly of this group. Wieser & Kanwisher (1961) found nematodes penetrating to a depth of 12 to 14 cm in salt marshes with a loosely packed substrate. In more compact areas, the penetration was only 3 to 6 cm. This contrasts with the data from Swartkops salt marshes where, although the bulk of the meiofauna occurred down to 30 cm, significant numbers of nematodes were found down to 80 cm (Furstenberg, unpublished).

In certain circumstances the meiofauna may migrate to different depths to avoid unfavourable conditions (Perkins, 1958; Renaud-Debyser, 1963, Bush, 1966, Boaden, 1968, McLachlan, Erasmus & Furstenberg, 1977). Bush (1966) found a downward movement of the meiofauna after rain and Boaden (1968) found vertical migrations related to wave action. A seasonal rearrangement of the vertical distribution pattern was found by Renaud-Debyser (1963). McLachlan *et al* (1977) reported a vertical migration of 12 cm related to desiccation and temperature, such that the meiofauna moved downwards on the outgoing tide and upwards on the incoming tide. It is clear that there is no single stimulus for vertical migration. In well-drained beaches where some desiccation occurs this will be the most important stimulus. In poorly drained areas other factors such as oxygen depletion may cause vertical movements. No data are available on the vertical migrations of the meiofauna in estuaries apart from a study in the Swartkops estuary which revealed only a limited movement of ± 5 cm in response to desiccation in the sand flats (Dye, 1977).

A number of factors have been postulated to account for the observed distributions of meiofauna. However, as with vertical migration, no single factor is responsible but a number of generalizations may be made. Among the more important factors are sediment properties such as grain size, porosity and permeability. On these depend a host of other factors such as salinity and temperature variations, bacterial populations and oxygen availability.

Particle size exerts its effect through mechanical restrictions on the movement of animals. In coarse substrates, the pore spaces are large and the meiofauna is predominantly interstitial, that is, they slide between the sand grains. Fine substrates, although they have a higher overall porosity, have smaller pore spaces and the meiofauna is of the burrowing type. Wieser (1959) postulated a size barrier of 200 μm separating the sliders from the burrowers. McLachlan, Winter & Botha (1977) estimated the barrier to be 160 μm. Interstitial harpacticoids have been found in sands of median particle size 175 μm in the Swartkops estuary but at a median particle size of 130 μm, the harpacticoids were all of the burrowing type. This suggests that the lower limit for interstitial life, excluding nematodes, is closer to 160 μm than to 200 μm. The nematodes are capable of sliding through very small pores and their lower limit has been postulated to be 125 μm (McIntyre & Murison, 1973).

Fine substrates trap detritus and are characterized by high bacterial populations. This, in turn, limits oxygen and only those species capable of anaerobic survival, or tolerant to low oxygen conditions, can live in such substrates.

SEASONAL FLUCTUATIONS

A few long term studies of meiofauna in estuaries have revealed considerable seasonal changes in composition and abundance. For example, Jansson (1968) found a single peak of meiofauna in summer (18,3 °C) and a minimum in winter (4,5 °C) in the Baltic. The fluctuations were most pronounced among the Turbellaria, Gastrotricha, Oligochaeta, Harpacticoida and Halacarida but the nematodes maintained a relatively high population throughout the year. Muus (in Coull, 1973) also reported a seasonal fluctuation of meiofauna correlated with temperature. A similar finding was made by Perkins (1958). Barnett (1968) reported a spring peak for harpacticoids in an intertidal mud flat. It thus appears that in cool-temperate regions, temperature is the most important factor governing seasonal fluctuations. Data from a tropical estuary in India revealed a drastic decrease in meiofauna between July and September due to the monsoons (Damodoran, 1974). A seasonal study in the Swartkops estuary revealed an interesting mechanism of control, involving combinations of oxygen and temperature. The meiofauna in both the sand and mud flats, exhibited two peaks, one in spring and one in autumn. In summer (December/January) the meiofauna reached a minimum due to low oxygen conditions; another minimum occurred in winter (June/July) due to low temperatures. Between these two seasons, however, when both temperature and oxygen were in the middle of their ranges, the meiofaunal population increased (Dye & Furstenberg, 1977). It appears therefore, that the Swartkops represents a transition zone between areas where temperature is the dominant limiting factor and areas where oxygen assumes this role. The latter would be expected in tropical areas not experiencing extremes such as monsoons. In the absence of data from such areas, one may speculate that one peak will occur in winter when oxygen concentrations are high. The pattern found in the Swartkops estuary is entirely due to nematodes since the remainder of the meiofauna exhibited a random fluctuation not correlated with any of the measured environmental parameters. It is possible that the species composition of the meiofauna is not constant and that different species become dominant in response to different environmental conditions at various times of the year. This was found to be an important factor at Whitstable, Kent (Perkins, 1974). As with vertical distribution, the seasonal fluctuations of the meiofauna in the Swartkops estuary were affected by the presence of *Callianassa kraussi*. In areas where large numbers of the prawn occurred, the seasonal fluctuations of the meiofauna became random. This is due to the removal of an important limiting factor, namely oxygen. The pumping activity of the prawns alleviates to some extent the summer minimum of oxygen.

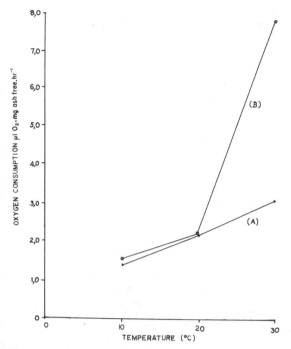

Figure 10.1
The effect of temperature on the oxygen consumption of nematodes from sandy (A) and muddy (B) areas of the Swart-kops estuary (after Dye, Erasmus & Furstenberg 1977).

MEIOFAUNA PHYSIOLOGY AND PRODUCTION

Many records are available on the oxygen consumption of estuarine nematodes. Wieser & Kanwisher (1961) measured respiration rates of a number of nematodes from a salt marsh and obtained results varying from 7,2 $\mu l.O_2$.mg (dry mass)$^{-1}$.hr^{-1} *(Axonolaimus spinosus)* to 1,38 $\mu l.O_2$.mg (dry mass)$^{-1}$.hr^{-1} *(Halichoanolaimus longicauda)* at 20 °C. There appears to be great individual variation as well as species variation. Data are also available on harpacticoid respiration. For instance, Coull & Vernberg (1970) measured the respiration of *Enhydrosoma propinquum* and *Longipedia helgolandica* and obtained values of 9,25 and 2,90 $\mu l.O_2$.mg (dry mass)$^{-1}$.hr^{-1} respectively at 20 °C. For *Asellopsis intermedia,* Lasker & Wells (1970) reported a respiratory rate of 3,8 $\mu l.O_2$mg (dry mass)$^{-1}$.hr^{-1} and 2,20 $\mu l.O_2$.mg (dry mass)$^{-1}$.hr^{-1} respectively. These values indicate that meiofaunal metabolism is about five times more active than that of macrofauna (Gerlach, 1971).

As far as responses to changes in temperature are concerned Wieser & Kanwisher (1961) concluded that nematodes do not regulate their metabolism. They found no evidence of compensation in animals collected in summer or winter. This conclusion appears to hold for animals from salt marshes and fine substrates. Figure 10.1 shows the effect of temperature

on the oxygen consumption of nematodes from sandy and muddy areas of the Swartkops estuary (after Dye *et al*, 1977). It can be seen that animals from muddy areas (B) do not regulate their metabolism and may be classed as temperature conformers. Individuals from the sandy areas (A), however, show considerable regulation up to 30 °C. Thus, the Q_{10} for mudflat nematodes is 2,48 and for sand flat nematodes it is 1,48. It appears that the sand flat types have adapted to the large diurnal fluctuations in temperature by becoming regulators, whereas the mudflat meiofauna, which does not experience such wide fluctuations, can best survive by being conformers.

Estimates of production of meiofauna are usually made by multiplying the standing crop biomass by a certain turnover. This in turn is obtained from studies of the life cycles of meiofaunal organisms (Gerlach, 1971). There is a great variation in life cycles with some organisms reproducing in as little as five days under laboratory conditions (Tietjen & Lee, 1972), While others require three months (Gerlach, 1971). Tietjen (1967) found that the life cycle of the nematode *Monhystera filicaudata* took between 24 and 35 days at 20 to 25 °C. Similarly, the life cyle of *M. disjuncta* and *Diplolaimella schneideri* takes 40 days (Chitwood & Murphy, 1964).

These studies were, however, carried out under optimal laboratory conditions and on the assumption that reproduction takes place immediately upon reaching sexual maturity. Since these conditions are rarely met in nature, the true life cycles can be expected to be somewhat longer than the above estimates indicate. In his summary of the subject, Gerlach (1971) estimated an annual turnover of 9. Prior to this, McIntyre (1969) had estimated a turnover of 10 per year. Recently an estimate of 6 per year was made by Arlt (1973). It appears, therefore, that an average life cycle of 44 days, or a turnover of 8 per year, is a reasonable estimate for meiofauna. Estimates of production based on oxygen consumption were made for nematodes from the Swartkops estuary. This method assumes an RQ of 0,85 (Hargrave, 1969) and is a simpler method than culturing nematodes for life cycle studies. Using this method in conjunction with *in situ* dark and light bottle measurements of total secondary production, it was determined that, of a mean total secondary production of 38,72 g.C.m^{-2}.yr^{-1} in the sand flats *meiofauna* accounted for 1,2 % or 0,46 g.C.m^{-2}.yr^{-1}. In the muddy areas the meiofauna accounted for 0,03 % of the production of 88 g.C.m^{-2}.yr^{-1} or 0,24 g.C.m^{-2}.yr^{-1}.

Estimates from the literature indicate great variation in the relative importance of the meiofauna. Some of these estimates are: 29 % (Wieser & Kanwisher, 1961), 15 % (Marshall, 1969) and 1,65 % (Smith *et al*, 1972). Thus, in terms of secondary production, the meiofauna of the sand and mudflat areas of the Swartkops estuary is relatively unimportant. The quantitative importance of the meiofauna in the salt marshes may be considerably greater since the annual production may be in excess of 40 g.C.m^{-2}.yr^{-1}.

Some indications of the qualitative importance of the meiofauna have come from recent studies of marine food webs. These studies have revealed that, in a laboratory system containing bacteria, meiofauna and the polychaete *Capitella capitata*, the total community respiration was three times higher when the system contained both meiofauna and *C. capitata* than with bacteria alone (Lee *et al*, 1974). This seems to indicate that the productivity of estuarine substrates is profoundly influenced by meiofauna even in areas where only modest populations exist.

Appendix 10.1
LIST OF MARINE NEMATODE GENERA SO FAR IDENTIFIED FROM THE SWARTKOPS ESTUARY

Note: To date, 52 genera have been identified of which the dominant one is *Theristus* and the subdominant ones are *Cobbia, Pomponema, Rhips, Leptogastrella* and *Axonolaimus.*

ORDER 1. ARAEOLAIMIDA
Fam. Axonolaimidae
Subfam. Axonolaiminae
Genus. *Axonolaimus*
Odontophora
n.g.
Subfam. Diplopeltinae
Genus. *Araeolaimus*

ORDER 2. MONHYSTERIDA
Fam. Linhomoeidae
n.g.
Fam. Monhysteridae
Subfam. Monhysterinae
Genus. *Cobbia*
Gonionchus
Leptogastrella
Mesotheristus
Paramonhystera
Theristus
Trichotheristus
Subfam. Xyalinae
Genus. *Xyala*
n.g.1
n.g.2
Fam. Siphonolaimidae
n.g.
Fam. Scaptrellidae
Genus. *Scaptrella*

ORDER 3. DESMODORIDA
Fam. Desmodoridae
n.g.
Subfam. Spiriniinae
Genus. *Onyx*
n.g.
Subfam. Metachromadorinae
n.g.
Subfam. Desmodorinae
Genus. *Desmodora*
Subgenus. *Croconema*
Fam. Ceramonematidae
Subfam. Dasynemellinae
Genus. *Dasynemoides*

ORDER 4. CHROMADORIDA
Fam. Comesomatidae
Subfam. Sabatieriinae
Genus. *Laimella*
Subfam. Dorylaimopsinae
Genus. *Dorylaimopsis*
Subfam. Comesomatinae
Genus. *Comesoma*
Fam. Chromadoridae
Subfam. Euchromadorinae
Genus. *Rhips*
Genus. *Actinonema*
Graphonema
Subfam. Hypodontolaiminae
Genus. *Hypodontolaimus*
Dichromadora
Nygmatonchus
Subfam. Chromadorinae
Genus. *Chromadorina*
Fam. Cyatholaimidae
Subfam. Pomponematinae
Genus. *Pomponema*
Subfam. Paracanthonchinae
Genus. *Paracanthonchus*
Paracyatholaimus
Subfam. Cyatholaiminae
Genus. *Metacyatholaimus*
Fam. Choniolaimidae
Genus. *Choniolaimus*
n.g.

ORDER 5. ENOPLIDA
Fam. Enoplidae
Subfam. Enoplolaiminae
Genus. *Enoplolaimus*
Mesocanthion
Subfam. Enoplinae
Genus. *Enoplus*
Subfam. Enoploidinae
Genus. *Enoploides*
Subfam. Trileptinae
Genus. *Fenestrolaimus*
Trileptium
Fam. Tripyloididae
Genus. *Bathylaimus*
Fam. Oncholaimidae
Subfam. Oncholaimellinae
Genus. *Viscosia*
Oncholaimellus
Oncholaimus
Mononcholaimus
n.g.
Fam. Enchelidiidae
Subfam. Eurystominae
n.g.

REFERENCES

ARLT, G. 1973. Zur produktionsbiologischen Bedeutung der Meiofauna in Küstengewassern. *Wiss. Zeitschr. Univ. Rostock* 22: 1141-1145.

BARNETT, P.R.O. 1968. Distribution and ecology of harpacticoid copepods of an intertidal mudflat. *Int. Rev. ges. Hydrobiol.* 53: 117-209.

BIGGS, R.G. & D.A. FLEMMER 1972. The flux of particulate carbon in an estuary. *Mar. Biol.* 12: 11-17.

BOADEN, P.J.S. 1968. Water movement — a dominant factor in intertidal ecology. *Sarsia* 34: 125-136.

BUSH, L.F. 1966. Distribution of sand fauna in beaches at Miami, Florida. *Bull. mar. Sci.* 16: 58-75.

CAPSTICK, C.K. 1959. The distribution of free-living nematodes in relation to salinity in the middle and upper reaches of the River Blyth estuary. *J. Anim. Ecol.* 28: 189-210.

CHITWOOD, B.G. & D.G. MURPHY 1964. Observations on two marine monhysterids, their classification, culture and behaviour. *Trans. Am. micros. Soc.* 83: 211-239.

COULL, B.C. 1973. Estuarine meiofauna. A review: Trophic relationships and microbial interactions. *In:* L.H. Stevenson & R.R. Colwell (eds), *Estuarine microbial ecology.* Belle Baruch Coastal Res. Inst., Univ. S.Carolina. pp 499-511.

COULL, B.C. & W.B. VERNBERG 1970. Harpacticoid copepod respiration: *Enhydrosoma propinquum* and *Longipedia helgolandica. Mar. Biol.* 5: 341-344.

DAMODORAN, R. 1974. Meiobenthos of the mud banks of the Kerala coast. *Proc. Indian Natl. Sci. Acad. 38B, (Biol. Sci.)* 3, 4: 288-297.

DUFF, S. & J.M. TEAL 1965. Temperature change and gas exchange in Nova Scotia and Georgia salt marsh muds. *Limnol. Oceanog.* 10: 67-73.

DYE, A.H. 1977. *An ecophysiological study of the benthic meiofauna of the Swartkops estuary.* MSc thesis, Univ. Port Elizabeth, South Africa.

DYE, A.H., ERASMUS, T. & J.P. FURSTENBERG 1977. An ecophysiological study of the benthic meiofauna of the Swartkops estuary. III. Partition of benthic oxygen consumption and the relative importance of the meiofauna. *Zool. Afr.* 13(2): 187-200.

DYE, A.H. & J.P. FURSTENBERG 1977. An ecophysiological study of the benthic meiofauna of the Swartkops estuary. II. The meiofauna: Composition, distribution, seasonal fluctuations and biomass. *Zool. Afr.* 13(1): 19-32.

DYE, A.H. 1979. Aspects of the ecology of meiofauna in Mngazana estuary, Transkei. *S.Afr. J. Zool.* 14: 67-73.

FAVA, G. & B. VOLKMAN 1975. *Tisbe* (Copepoda: Harpacticoida) species from the Lagoon of Venice. I. Seasonal fluctuations and ecology. *Mar. Biol.* 30: 151-165.

FENCHEL, T. 1969. The ecology of marine microbenthos IV. Structure and function of the benthic ecosystem, its physical factors and the microfauna communities with special reference to the ciliated Protozoa. *Ophelia* 6: 1-182.

FENCHEL, T. & T. STRAARUP 1971. Vertical distribution of photosynthetic pigments and the penetration of light in marine sediments. *Oikos* 22: 172-182.

GERLACH, S.A. 1971. On the importance of marine meiofauna for benthos communities. *Oceologia (Berl.)* 6: 176-190.

GORDON, M.S. 1960. Anaerobiosis in marine sandy beaches. *Science* 132: 616-617.

GRAY, J.S. & R.M. RIEGER 1971. A quantitative study of the meiofauna of an exposed beach at Robin Hood's Bay, Yorkshire. *J. Mar. biol. Ass. UK* 51: 1-19.

HARGRAVE, B.T. 1969. Epi-benthic algal production and community respiration in the sediments of Marion Lake. *J. Fish. Res. Bed. Can.* 26: 2003-2026.

JANSSON, B.O. 1967b. The significance of grain size and pore water content for interstitial fauna of sandy beaches. *Oikos* 18: 311-322.

JANSSON, B.P. 1968. Quantitative and experimental studies of the interstitial fauna of four Swedish sandy beaches. *Ophelia* 5: 1-71.

JOHNSON, R.G. 1965. Temperature variation in the infaunal environment of a sand flat. *Limnol. Oceanog.* 10: 114-120.

KINNE, O. 1964. The effect of temperature and salinity on marine and brackish water animals. II. Salinity and temperature salinity combinations. *Oceanog. mar. Biol. Ann. Rev.* 2: 281-339.

KURIAN, C.V. 1974. Ecology of benthos in a tropical estuary. *Proc. Indian Natl. Sci. Acad. 38B (Biol. Sci. 3, 4):* 156-163.

LASKER, R.J., WELLS, B.J. & A.D. McINTYRE 1970. Growth, reproduction, respiration and carbon utilization of the sand dwelling harpacticoid copepod *Asellopsis intermedia. J. mar. biol. Ass. UK* 50: 147-160.

LEE, J.J., TENORE, K., TIETJEN, J.H. & C. MASTROPAOLA 1974. An experimental approach towards understanding the role of meiofauna in a detritus-based marine food web. *Biol. Bull.* 147(2): 488-489.

MARE, MOLLY, F. 1942. A study of a marine benthic community with special reference to the micro-organisms. *J. mar. biol. Ass. UK* 55: 517-554.

MARSDEN, I.D. 1973. The influence of salinity and temperature on the survival and behaviour of the isopod *Sphaeroma rugicauda* from a salt marsh habitat. *Mar. Biol.* 21: 75-85.

MARSHALL, N. 1969. Trophic relationships in shoal benthic environments. *In:* J.H. Steele (ed), *Marine food chains.* Oliver & Boyd, Edinburgh.

McINTYRE, A.D. 1964. Meiobenthos of sublittoral muds. *J. mar. biol. Ass. UK* 44: 665-674.

McINTYRE, A.D. 1968. The meiofauna and macrofauna of some tropical beaches. *J. Zool., Lond.* 156: 377-392.

McINTYRE, A.D. 1969. Ecology of marine meiobenthos. *Biol. Rev.* 44: 245-290.

McINTYRE, A.D. & D.J. MURISON 1973. The meiofauna of a flatfish nursery ground. *J. mar. biol. Ass. UK* 53: 93-188.

McLACHLAN, A. 1977. Studies on the psammolittoral meiofauna of Algoa Bay. II. The distribution, composition and biomass of the meiofauna and macrofauna. *Zool. Afr.* 12: 33-60.

McLACHLAN, A., ERASMUS, T. & J.P. FURSTENBERG 1977. Migrations of sandy beach meiofauna. *Zool. Afr.* 12: 257-278.

McLACHLAN, A., WINTER, D. & L. BOTHA 1977. Vertical and horizontal distribution of sublittoral meiofauna in Algoa Bay, South Africa. *Mar. Biol.* 40: 355-364.

MEYERS, S.P. & B.E. HOPPER 1971. Nematological microbial interrelationships in estuarine biodegradetive processes. *In:* L.H. Stevenson & R.R. Colwell (eds), *Estuarine microbial ecology.* Belle Baruch Coastal Res. Inst. Univ. S. Carolina. pp 483-489.

PERKINS, E.J. 1958. The food relationships of the microbenthos with particular reference to that found at Whitstable, Kent. *Ann. Mag. Nat. Hist. (ser. 13)* 1: 64-77.

PERKINS, E.J. 1974. *The biology of estuaries and coastal waters.* Academic Press, London and New York.

REES, C.B. 1940. A preliminary study of the ecology of a mud flat. *J. mar. biol. Ass. UK* 21: 185-199.

RENAUD-DEBYSER, J. 1963. Recherches ecologiques sur la faune interstitielle des sable Bassin d'Arachon, île de Bimini, Bahamas. *Vie Milieu* 14: 463-550.

SMIDT, E.L.B. 1951. Animal production in the Danish Waddensea. *Meddr. Kommiss. Danmarks Fisk-og Havunders. (Ser. Fiskeri)* 11, 151.

SMITH, K.L., BURNS, K.A. & J.M. TEAL 1972. In situ respiration of the benthic communities in Castle Harbour, Bermuda. *Mar. Biol.* 12: 196-199.

SWEDMARK, B. 1964. The interstitial fauna of marine sand. *Biol. Rev.* 37: 1-42.

TEAL, J.M. & W. WIESER 1966. The distribution and ecology of nematodes in a Georgia salt marsh. *Limnol. Oceanog.* **11**: 217-222.

TIETJEN, J.H. 1966. *The ecology of estuarine meiofauna with special reference to the class Nematoda.* PhD thesis, Univ. Rhode Island, USA.

TIETJEN, J.H. 1967. Observations on the ecology of the marine nematode *Monhystera filicaudata* Allgen 1929. *Trans. Am. micros. Soc.* **86**(3): 304-306.

TIETJEN, J.H. 1969. The ecology of shallow water meiofauna in two New England estuaries. *Oecologia (Berl.)* **2**: 251-291.

TIETJEN, J.H. & J.J. LEE 1972. Life cycles of marine nematodes. Influence of temperature and salinity on the development of *Monhystera denticulata* Timm. *Oecologia (Berl.)* **10**: 167-176.

WIESER, W. 1959. The effect of grain size on the distribution of small invertebrates inhabiting the beaches of Puget Sound. *Limnol. Oceanog.* **5**: 121-137.

WIESER, W. & J. KANWISHER 1961. Ecological and physiological studies on marine nematodes from a small salt marsh near Woods Hole, Massachusettes. *Limnol. Oceanog.* **6**: 262-270.

WIESER, W., OTT, J., SCHIEMER, F. & E. GNAIGER 1974. An ecophysiological study of some meiofauna species inhabiting a sandy beach at Bermuda. *Mar. Biol.* **26**: 253-248.

Chapter 11

Adaptations to temperature and salinity stress in southern African estuaries

Burke J. Hill

Queensland Fisheries, P.O. Box 36, North Quay, Brisbane, Australia 4000

INTRODUCTION

Estuarine ecology in southern Africa is founded upon the survey work carried out by the University of Cape Town Zoology Department under the leadership of John Day. This work recorded the species of plants and animals occurring within estuaries, their distribution and relative abundance and led to certain important conclusions, the most significant one being that the fauna of southern African estuaries is largely a quiet water one similar to that of sheltered parts of the sea.

Following upon the surveys, estuarine research entered an experimental phase in which an attempt was made to assess the importance of various environmental factors to the fauna. The choice of factors investigated has been greatly influenced by work in the northern hemisphere and has concentrated on the effects of reduced salinity and elevated temperature. But to what extent are these parameters important in southern African estuaries? Has this emphasis been at the expense of other possibly more important aspects? Research in southern Africa has reached the phase where it is worth standing back and assessing the results so far obtained and then deciding in which direction we should now proceed. This chapter is therefore largely restricted to South African estuarine research. Readers are referred to reviews such as those of Beadle (1972), Kinne (1963, 1964) or Lockwood (1976) for a more general coverage of the adaptations of estuarine animals.

TOLERANCE TO DECREASED SALINITY

Most biologically oriented definitions of an estuary emphasize salinity since, as Day (1951) has pointed out, the main characteristic of an estuarine population is generally regarded as its ability to tolerate changes in salinity. Marine populations by contrast, cannot withstand low salinity and heavy mortalities have been recorded in areas where marine faunas have been exposed to a drop in salinity, for example in Jamaica by Goodbody (1961). This contrast is illustrated by the salinity tolerance of two species of the burrowing prawn *Upogebia* in South African waters (Figure 11.1). *U. africana,* which is found in estuaries, has a far greater osmoregulatory ability and is more tolerant of low salinity than *U. capensis,* a species which is restricted to the sea (Hill, 1971).

U. africana copes with low salinity by well-developed osmoregulatory powers but this is not an essential requirement for low salinity tolerance. The tubiculous polychaete *Ficopomatus enigmatica* can live in a range of salinities from less than 1 ‰ up to 55 ‰. Skaer (1974) found that over most of this range it is an osmoconformer, maintaining the blood slightly hyperosmotic to the medium, and only regulating significantly when in fresh or nearly fresh water. This polychaete illustrates that osmotic regulation is not a prerequisite in estuarine animals, provided that the cells can tolerate dilution of the body fluids.

Estuaries generally have a gradient of salinity along

their length with near seawater values at their mouths reducing to near freshwater at their heads. The distribution of many animals can be correlated with this salinity gradient although it is important to bear in mind that the relationship is not necessarily a causal one. Nevertheless, it is obvious that an animal with limited salinity tolerance would not be able to penetrate far up most estuaries. The sandshark *Rhinobatus annulatus* occurs in shallow waters in the sea on the east coast of southern Africa and, according to Smith (1953), 'enters estuaries in fair numbers in summer'. Bok (1973) exposed *R. annulatus* to low salinity and determined the salinity tolerance as well as plasma osmolality and composition. He found that the sandshark could not survive salinities below 26 ‰. In elasmobranchs, the blood is hyperosmotic to seawater, increased osmolality being obtained largely through the presence of urea which, in sandsharks in seawater, contributes about 40 % of the total osmotic pressure depending on the protein reserve available. Bok found that in lowered salinity, *R. annulatus* reduced the concentration of urea more than that of the major ions (Na$^+$ and Cl$^-$) indicating selective excretion of urea. Sandsharks could not cope with salinities below 26 ‰ in which the concentration not only of urea, but also of the major ions, dropped. Thus although able to enter the lower reaches of estuaries, *R. annulatus* will be effectively excluded from the upper reaches where low salinities are present.

Grapsoid crabs offer another example of the way in which salinity may control distribution within an estuary. *Cyclograpsus punctatus* is a common shore crab usually associated with rocks around high tide level (Broekhuysen, 1941) but it also enters estuaries. Alexander & Ewer (1969) found that *C. punctatus* and another grapsoid crab, *Sesarma catenata,* are both present in the lower reaches of the Kowie estuary, but that the latter species extends further up the estuary. A third crab, *Sesarma eulimene,* is dominant at the top of the Kowie estuary. Boltt & Heeg (1975) found that *C. punctatus* was not very tolerant of low salinity, the lowest salinity in which experimental animals showed no mortality over a 14 day period, being 20 ‰. *S. catenata* and *S. eulimene* were more tolerant, 100 % survival occurring in 7 ‰ and 2,7 ‰ respectively. The osmoregulatory abilities of the three species correspond to their salinity tolerance, with *S. eulimene* regulating most strongly and *C. punctatus* not being able to maintain hyperosmosticity. Boltt & Heeg pointed out that the salinity tolerance of the two species of *Sesarma* agrees well with their distribution, but that *C. punctatus* can apparently penetrate further up the Kowie estuary than its osmoregulatory ability would suggest. *C. punctatus* is primarily a shore crab which constructs its burrows near the top of the intertidal zone and spends long periods out of water (Alexander & Ewer, 1969). Boltt & Heeg suggested that this amphibious habit, coupled with a modest osmoregulatory ability enables

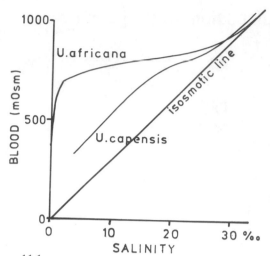

Figure 11.1
Blood osmotic pressure as a function of salinity in estuarine *Upogebia africana* and marine *U. capensis* (redrawn from Hill, 1971).

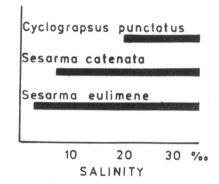

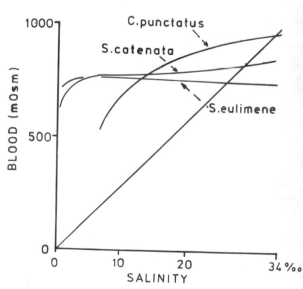

Figure 11.2
(a) Range of salinities tolerated for 14 days by three species of crabs. (b) Blood osmotic pressure curves of these crabs after four days in various salinities (redrawn from Boltt & Heeg,1975).

Table 11.1
Species of estuarine animals recorded in Lake Sibaya by Allanson *et al* (1974)

Polychaeta:	*Ceratonereis keiskama*	
Crustacea:	*Apseudes digitalis*	(tanaid)
	Corophium triaenonyx	(amphipod)
	Grandidierella lignorum	(amphipod)
	Orchestia ancheidos	(amphipod)
	Hymenosoma orbiculare	(crab)
Fish:	*Gilchristella aestuarius*	
	Hepsetia breviceps	
	Glossogobius giuris	

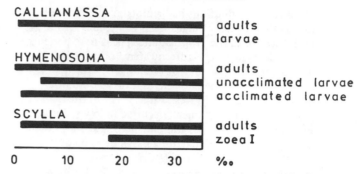

Figure 11.3
Salinity tolerance of adults and larvae of the prawn *Callianassa kraussi* and of the crabs *Hymenosoma orbiculare* and *Scylla serrata.*

C. punctatus to survive low salinity since they would be submerged for only brief periods at high tide when salinities are at their maximum. Thus, although salinity appears to be a factor in limiting the spread of *C. punctatus* up the estuary, behavioural considerations allow it to partially overcome a limited salinity tolerance. Day *et al* (1952) pointed out that in the Knysna estuary many marine animals which inhabit the upper levels of the shore penetrated further up the estuary than those which lived at the bottom of the shore.

Boltt & Heeg noted that most crabs which moulted in low salinities in their experiments died. Tolerance experiments should take into account the increased sensitivity to low salinity of crustacea at ecdysis. Hill (1971) found that *Upogebia africana* could survive in a salinity of 1,7 ‰ but individuals which moulted in this salinity died, probably due to uptake of dilute medium at ecdysis. The lowest salinity in which *U. africana* could successfully moult was 3,4 ‰. Although *U. africana* may postpone moulting if salinity is low, this delay cannot be continued indefinitely.

Most investigations of the effect of low salinity on South African estuarine animals have been restricted to the adult stage and it cannot be assumed that larval stages or juveniles will have an equivalent tolerance. Broekhuysen (1955) found adults of the crab *Hymenosoma orbiculare* in the Sandvlei estuary in winter when salinities dropped to between 1 and 5 ‰. Experimentally, he found however that the zoeae were killed by salinities below 5 ‰ (Figure 11.3) and he concluded that many of the larvae which hatched in winter would not survive. Hill (1974) similarly reported that the first zoeal stage of the crab *Scylla serrata* could not survive exposure to salinities below 17,5 ‰ and even at this salinity considerable mortality occurred. Adult *S. serrata* by contrast can tolerate salinities down to 2 ‰ (Hill, 1975).

Intolerance of larval or juvenile stages does not necessarily exclude adults of the species from occupying areas unfavourable for larvae. Forbes (1978) reports populations of the burrowing prawn *Callianassa kraussi* living in the upper reaches of two estuaries (Swartvlei and Keurbooms) at salinities below 5 ‰. Females in these areas never carried eggs although populations lower down in the estuaries had normal annual breeding cycles. Forbes found experimentally that 17 ‰ is the lowest salinity at which the eggs and larvae could undergo complete development but that postlarvae were far more tolerant. *C. kraussi* does not have a planktonic larval stage and dispersal is carried out by postlarval but subadult stages (Forbes, 1973). Since these stages are tolerant of low salinity they can spread into areas where breeding is not possible.

The ability of most South African estuarine animals to successfully cope with reduced salinity, is well-illustrated by the finding of several estuarine species in freshwater (135 ppm Cl⁻) in Lake Sibaya by Allanson *et al* (1966). The species include a polychaete *(Ceratonereis keiskama)*, several crustaceans, the tanaid *Apseudes digitalis*, the amphipods *Corophium triaenonyx*, *Grandidierella lignorum* and *Orchestia ancheidos*, the crab *Hymenosoma orbiculare* and several fish — *Gilchristella aestaurius*, *Hepsetia breviceps* and *Glossogobius giuris* (Allanson *et al*, 1974).

Boltt (1969) investigated the salinity tolerance and osmoregulatory ability of one of these species, the amphipod *G. lignorum.* He found that, provided he initially acclimated them to dilutions of seawater, animals from the Kowie estuary could be transferred to freshwater where they survived, moulted and grew. Boltt found that the osmotic concentration of the blood of *G. lignorum,* living naturally in low salinities, did not differ markedly from that of animals taken from seawater and exposed to equivalent dilutions. He concluded that no special physiological adaptation with respect to salinity tolerance and osmoregulation was necessary for estuarine *G. lignorum* to live in freshwater.

The finding of the crab *Hymenosoma orbiculare* in freshwater in Lake Sibaya was of interest because this species as stated above, was shown by Broekhuysen (1955) to have a larval stage intolerant of salinities below 5 ‰. Forbes & Hill (1969) undertook a comparative physiological study of *H. orbiculare* using crabs from Lake Sibaya and from estuaries. They found that *H. orbiculare* regulates hyperosmotically; in a salinity of 6,5 ‰ the blood has an osmolality equivalent to 31 ‰ seawater. Crabs from seawater put into lower dilutions had a blood osmolality corresponding to that of crabs from Lake Sibaya, and, as in the case of *C. lignorum,* adults appear to experience no problems in entering freshwater. Forbes & Hill investigated the breeding of estuarine *H. orbiculare,* and found, like Broekhuysen, that they could not obtain viable larvae in freshwater. If egg extrusion, incubation and hatching all occurred in low salinity, survival in low salinities was enhanced, but they could not obtain successful hatching in salinities below 1,7 ‰. Thus in the case of *H. orbiculare,* there appears to have been some change in the physiology of the zoeal stages which has enabled them to survive in freshwater in Lake Sibaya.

Tolerance to low salinity does not necessarily imply that the animal inhabits estuaries. The burrowing goby *Croilia mossambica* occurs in coastal lakes on the east coast of southern Africa at salinities from 8 ‰ down to freshwater. Blaber & Whitfield (1977) showed that *C. mossambica* can survive in seawater but despite this euryhalinity, does not occur in estuaries in the region. They suggested that its absence from estuaries is due to an intolerance of high temperature and water turbulence; it is therefore restricted to the quiet waters of coastal lakes.

It appears that many of the southern African estuarine animals so far investigated are euryhaline, and in many cases can survive in low salinities.

Most estuaries in South Africa are small, and this fact, coupled with occasional heavy falls of rain, results in their being prone to floods during which the whole estuary may experience salinities usually associated only with the head. Macnae (1957) pointed out that between half and one-third of the annual rainfall of the Swartkops estuary catchment frequently fell over a few days. He observed that the resulting floods appeared to have little effect on the animal life in the estuary. This suggests a remarkable ability on the part of estuarine animals to resist freshwater during floods. Although animals may exhibit a preference for a certain salinity range, unless there is some mechanism to cope with exposure to very low salinity, they cannot exist there indefinitely. The sea urchin *Parechinus angulosus* for example is not usually found in estuaries but it does occur in large numbers on rocky embankments in the mouth of the Kowie estuary. After a flood this population is completely eliminated and thousands of dead urchins can be found on the bottom of the estuary and on beaches adjoining the mouth. *P. angulosus* is clearly part of the stenohaline marine component and is incapable of tolerating reduced salinity.

Mobile animals which are not tolerant of low salinity can escape the effects of a flood by migrating into the sea. Some sessile animals can also avoid the consequences of a flood. According to Mclachlan & Erasmus (1974) the estuarine bivalves *Dosinia hepatica* do not regulate osmotically over the range 14 to 45 ‰. Below 14 ‰, they keep their shells tightly closed and, after

Table 11.2
Mean population density (numbers m^{-2}) of juvenile *Solen corneus* before, during and after a flood in the Kowie estuary. Sampling carried out 10 km up the estuary by means of a 0,225 m^2 van Veen grab.

Date	Bottom salinity ‰	Number of samples	Population density m^{-2}
April	30-32	35	98
May (flood)	0-1	22	89
September	20-22	30	35

23 days in distilled water, retain a salinity of 22 ‰ in their mantle cavities. Thus this species can successfully isolate itself from the surrounding water and can survive a flood. Mclachlan & Erasmus found that three other common bivalves, *Solen corneus, S. capensis* and *Macoma littoralis,* could not avoid low salinity in this way as their shells gape. These bivalves were found to be osmoconformers but were moderately tolerant of low salinity, the two species of *Solen* surviving down to about 14 ‰ and *M. littoralis* down to 7 ‰. This tolerance is however not sufficient to cope with floods and Mclachlan & Erasmus found that large scale mortalities occur at these times. If this is the case, how are populations of *Solen* maintained, and especially *S. corneus* which lives high up in estuaries? Mclachlan & Erasmus suggested that because salinity fluctuations are greatly reduced within the substrate, the bivalves would be insulated from low salinities to some extent. Mortality of *S. corneus* during a flood was confirmed in the Kowie estuary in May 1977 (Hill, unpublished data). Between 104 and 152 mm of rain fell in the catchment over two days and the Kowie river came down in flood. Four days later bottom salinities 8 km from the mouth were below 1 ‰ and 20 samples taken by means of a van Veen grab yielded large numbers of dead juvenile *S. corneus.* In September, a series of 30 grab samples in the same area indicated a population density of 35 juveniles m^{-2} whereas in April, just prior to the flood, the density had been 98 m^{-2} (35 samples). This suggests that about a third of the juveniles had survived. Adult *S. corneus* also survived in adjacent intertidal banks. Thus adults and juveniles can survive periods when the water overlying their burrows is well below their lethal limit.

The suggestion that benthic fauna may be protected from rapid salinity changes by the substrate, has been made by many authors although it is rarely checked. Thompson & Pritchard (1969) for example, suggested that *Callianassa californiensis* would probably be protected from salinity changes because of its burrowing habit. Forbes (1974) investigated this possibility for *Callianassa kraussi* which is abundant in many South African estuaries. Forbes showed experimentally that when *C. kraussi* was absent, the salinity of interstitial water of a sandy substrate was extremely stable. However, if *C. kraussi* were present there was rapid equili-

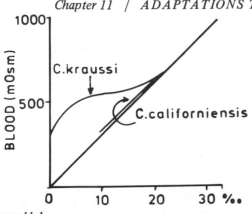

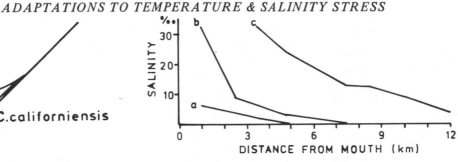

Figure 11.5
Salinities in the Kowie estuary on successive days following a
flood; (a) 1 day, (b) 3 days, (c) 7 days.

Figure 11.4
Blood osmotic pressure as a function of salinity in *Callianassa
kraussi* and *C. californiensis* (redrawn from Forbes, 1974).

bration between the interstitial and the overlying water.
Forbes concluded that the burrowing habit of *C. kraussi*
does not necessarily provide protection against sudden
changes in surface water salinity. *C. kraussi* is extremely
tolerant of low salinity, having been found down to
1 ‰. Forbes studied their osmoregulationa and found
that they were isosmotic from 35 ‰ down to 21 ‰.
Below this salinity, *C. kraussi* regulates hyperosmotically
maintaining the blood osmotic pressure at a level equivalent to between 17 and 21 % seawater. Forbes pointed
out that *C. kraussi* is more tolerant and osmoregulates
better than any other species of *Callianassa* so far investigated (Figure 11.4). He drew attention to the fact
that the same high tolerance applies to another South
African thalassinid, *Upogebia africana* which is also
more tolerant than species of *Upogebia* investigated in
other parts of the world. Forbes suggested that frequent
freshwater flooding of southern African estuaries would
act as a selective force favouring the improvement of
osmoregulatory ability.

Upogebia africana like *Callianassa* is a burrower and
the question arises as to whether its burrow offers any
protection from low salinity. Hill (1971) found that it
did not since the prawns would pump freshwater
through their burrows. Thus both *C. kraussi* and *U.
africana* rely upon their physiological capabilities to
survive floods, and are not significantly protected by
the burrowing habit.

Although the drop in salinity during a flood may
be large, it is usually of short duration and tidal action
soon reintroduces seawater into the estuary. Thus
within a few weeks or months the salinity regime will
be restored to preflood levels (Figure 11.5). One aspect
of floods has a much longer lasting effect, namely deposition of silt. Stephenson *et al* (1977) found that
changes in the fauna of a Queensland bay were detect-
able 14 months after a major flood and that this was
mainly due to deposition of silt. As shown by Mclachlan & Grindley (1974) and earlier workers, one of the
most important factors limiting macrobenthos distribu-

tion appears to be the type of substrate. Thus silt
deposited by floods may well have longer lasting
effects on the estuarine fauna than the relatively short-
lived salinity changes.

TOLERANCE TO HYPERSALINITY

Although the upper reaches of estuaries are usually
associated with decreased salinities, extended periods
of low or no rainfall can result in elevated salinities.
These rises in salinity are usually not large, seldom
exceeding 40 ‰. During the dry summers of the
South Western Cape, however, salinities may rise to
above 50 ‰ as found in the Heuningnes estuary by
Mehl (1973) and in Milnerton estuary by Millard &
Scott (1954). In contrast to these annual short periods
of hypersalinity, droughts cause extended periods of
highly saline conditions in the St Lucia system (Day
et al, 1954). According to Boltt (1975) salinities
throughout the St Lucia system were in excess of
40 ‰ in 1969 with a maximum of 89 ‰ in one area
(salinities over 100 ‰ have been recorded in the
northern lakes — editor).

Hypersaline conditions are very complex; apart
from changes in osmotic pressure, precipitation of certain salts occurs above 70 ‰ resulting in ionic changes.
Alterations also occur in alkalinity and pH and there
is a decrease in oxygen content and the specific heat
of water, resulting in more rapid temperature changes
(Copeland, 1967).Extremely little is known about the
effects of these changes on southern African estuarine
animals. Boltt (1975) stated that most benthic forms
tolerate salinities up to about 55 ‰. Blaber (1973,
1974) found that juvenile bream *(Rhabdosargus holubi)*
could tolerate salinities up to 70 ‰ and that even in
65 ‰, the osmotic pressure of the blood was not significantly different from the value in seawater. Wallace
(1975) stated that in St Lucia 10 species of fish had
their upper limits of distribution within the salinity
range 65 to 75 ‰ and recorded both *Mugil cephalus*
and *Rhabdosargus sarba* at salinities above 80 ‰. He
also reported *Sarotherodon mossambicus* and *Elops*

machnata as occurring above 110 ‰. The physiology of the tolerance of *S. mossambicus* to '200 % seawater' was described by Potts *et al* (1967). Although *M. cephalus* is estimated by Breuer (1957) to have an upper salinity tolerance of near 100 ‰, apparently no physiological studies of this tolerance have been undertaken.

Hill (1979) carried out experiments on the tolerance of the crab *Scylla serrata* to high salinity. *S. serrata* was found to have an upper limit of 60 ‰ but Hill pointed out that if the benthic animals on which it feeds are killed at lower salinities, *S. serrata* would probably move out of highly saline areas before their own lethal limit was reached. Hill (unpublished data) found that although *S. serrata* was a strong hyperosmoregulator in low salinities, when exposed to hypersaline conditions the crabs were nearly iso-osmotic, maintaining the blood slightly hyposmotic to the medium in the range 35 to 55 ‰.

Although extensive mortalities occur in the sessile or slow moving benthic fauna during periods of high salinity in St Lucia, the fish apparently are not killed to any great extent. Wallace (1975) showed that as salinities increased in the northern part of the system, some species moved southwards thereby avoiding extreme hypersaline conditions.

TEMPERATURE TOLERANCE

Kinne (1963) in an extensive review of the effects of temperature on marine organisms, pointed out that in most animals there are several different limiting temperatures. For example, breeding may occur over a narrower range than activities such as feeding and many of these activities may be suspended before lethal temperatures are reached. In most cases, the only temperature effect recorded is the temperature which causes death of some or all of the animals in a sample. In the case of high temperature tolerance, the experimental method affects the estimation. Read & Cummings (1967) showed that determining the upper lethal limit by continuously raising temperature, results in a considerable overestimation. In the case of the mussel *Mytilus edulis* they cited work in which an upper lethal limit of 40,8 °C was obtained if the temperature was raised at 1 °C every five minutes. Read & Cummings repeated these experiments and found an upper lethal limit of 30 °C if the temperature was raised at 1 °C every three and a half days, as compared to 26 °C if the animals were kept at a fixed temperature. Orr (1955) pointed out that it is of importance to realise that animals can survive short exposures to temperatures which would kill them on long exposure and that it is therefore necessary to compile time-temperature curves.

In South Africa, the most extensive investigation of upper temperature tolerance is that of Mclachlan &

Table 11.3
Upper lethal temperatures of the mussel *Mytilus edulis* estimated in three different ways (data from Read & Cummings, 1967)

Upper lethal temperature °C	Method
40,8	Temperature raised 1 ° every 5 minutes
30	Temperature raised 1 ° every 3,5 days
26	Animals kept at fixed temperature

Table 11.4
Upper lethal temperatures of bivalves exposed to slow (2 ° every 48 h) or rapid (1 ° every 10 minutes) heating regimes (data from Mclachlan & Erasmus, 1974)

Species	Upper lethal temperature °C	
	Slow heating	Rapid heating
Dosinia hepatica	39	45,5
Macoma litoralis	37	44
Solen corneus	–	44,5
S. capensis	–	41
Psammotellina capensis	–	42

Erasmus (1974) who studied five species of bivalves. Unfortunately they utilised the method of continuously raising temperature but they did try two heating regimes for some species. *Dosinia hepatica* and *Macoma littoralis* exposed to a temperature rise of 2 °C every 48 hours, died at 39 and 37 °C respectively. When heated rapidly (1 °C every ten minutes) the same species died at 45,5 and 44 °C respectively, illustrating the effect described by Read & Cummings (1967). Three other species — *Solen corneus*, *S. capensis* and *Psammotellina capensis* — were exposed to the rapid rate (1 °C 10 min^{-1}) only, and died at 44,5, 41 and 42 °C respectively. Although these results are not as satisfactory as time-temperature curves, they indicate that these five species of bivalve are well able to resist brief exposure to temperatures above 35 °C. Temperatures of this order have been reported by various authors as occurring in shallows in southern African estuaries, eg 35 °C in Morrumbene estuary (Day, 1974), 36 °C in the Swartkops estuary (Macnae, 1957), 38,5 °C in St Lucia (Millard & Broekhuysen, 1970) and 39 °C in Kosi estuary (Blaber, 1973).

McLachlan & Erasmus found that the shell of bivalves has little insulating effect, and the maximum difference between internal and external temperatures was 2,2 °C in the case of *Dosinia hepatica*. The burrowing habit of bivalves investigated by Mclachlan & Erasmus probably provides considerable protection against transient high temperatures. Rock dwelling species such as *Musculus virgiliae* and *Lamya capensis* as well as the common estuarine barnacle *Balanus amphitrite* by contrast, will be directly exposed to high temperatures in shallow waters in summer.

The burrow of the prawn *Upogebia africana* affords considerable protection against transient high temperatures. Hill & Allanson (1971) found that when water

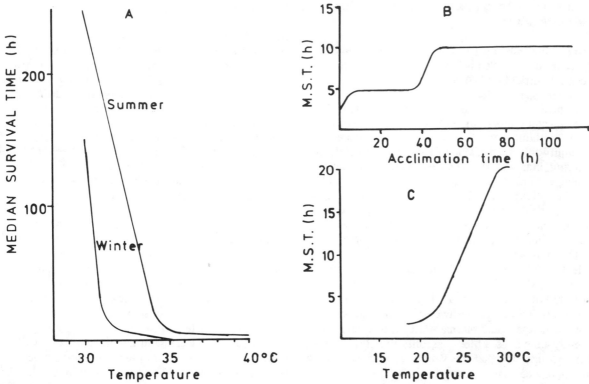

Figure 11.6
Median times of survival of *Upogebia africana*, (a) at various
temperatures in summer and winter, (b) collected in winter
and acclimated for differing periods at 24 °C and (c) after
58 h acclimation of winter prawns at a series of temperatures
(redrawn from Hill & Allanson, 1971).

above 32 °C was present above the burrow, the prawn
ceased pumping and closed down one of the entrances
to the burrow. The result was that the burrow temperature remained nearly 10 °C cooler than the overlying
water. Hill & Allanson compiled time-temperature
curves for *U. africana* and found that the upper lethal
limit for long term exposure was 29 °C in both winter
and summer. The prawns were able to resist temperatures above 29 °C for limited periods, the higher the
temperature the shorter the time. The resistance time
to temperatures above 30 °C was much greater in summer than in winter but it was possible to acclimate
winter prawns and to increase their resistance to a level
comparable to that of summer individuals. Seasonal
effects of this nature are frequently ignored by
research workers, but they may be quite large and
should be estimated. In addition, latitudinal effects
may occur; Forbes & Hill (1969) for example found
that specimens of the crab *Hymenosoma orbiculare*
collected from subtropical waters had a higher upper
lethal temperature than those from warm temperate
estuaries.

Blaber (1973) found that the teleost *Rhabdosargus
holubi* had an upper lethal limit of 30 °C and he suggested that this would probably restrict its northward

distribution. Whether or not high temperature does
restrict the spread of temperate south coast estuarine
animals into tropical east coast estuaries still remains
an unresolved question which has not been tested
experimentally. The absence of any reports of mass
mortality associated with high temperature, suggests
that if high temperature is a limiting factor, it usually
operates through mechanisms other than lethality. An
indication of the complexity involved, is given by the
effect of temperature on the prawn *Upogebia africana*.
Hill (1977) found that the length of the breeding season in *U. africana* was related to temperature and was
longer in warmer areas. An extended breeding season
was associated with smaller size in females possibly
due to diversion of energy into egg production. Female
U. africana are far smaller in Natal than in Cape estuaries as shown by the modal size of adults (8 mm versus
15 mm carapace length). Since the number of eggs
carried by *U. africana* females is related to size (Hill,
1977), smaller females carry fewer eggs. The crab
Hymenosoma orbiculare has also been found to breed
at a smaller size and to produce fewer eggs at the
northern end of its range in Natal and southern Mocambique than in Cape estuaries (Hill & Forbes, 1979).
Thus in tropical areas, a reduction in egg production
may occur, and might be a factor in limiting the northwards distribution of certain southern animals. Clearly
the effects of high temperature on estuarine animals
need a great deal more investigation which should not
be restricted to determining upper lethal temperatures.

INTERACTION BETWEEN LOW SALINITY AND
LOW TEMPERATURE

Between May and July 1976 a large scale mortality of
fish occurred in the St Lucia system. According to
Blaber & Whitfield (1976), at least 100 000 fish of
11 different species died. They reported that the fish
kills coincided with a period of very low salinity (1,0
to 3,5 ‰) and temperature (13 °C). Blaber & Whitfield pointed out that information on temperature
and salinity tolerance of one of the species, *Rhabdosargus holubi,* indicated that the combination of salinity
and temperature at St Lucia in winter of 1976 would
be lethal to this species. They suggested that the large
mortality was due either to a lethal salinity-tempera-
ture combination or to skin lesions which usually
follow osmoregulatory stress and rapidly become
infected by fungi.

Mehl (1973) found that the teleost *Lithognathus
lithognathus* showed similar lesions and died when
transferred to freshwater at a temperature of 9 to 14 °C.
Mehl also found that in freshwater, *L. lithognathus* had
higher haematocrit and plasma protein values and consequently an increased blood viscosity than fish in seawater. He suggested that this would retard vascular flow
resulting in a decrease in oxygen uptake and death
through asphyxiation.

The work of Allanson *et al* (1971) on *Sarotherodon*
(= *Tilapia*) *mossambicus* offers an alternative reason
for the death of marine teleosts in low salinity-low
temperature combinations. *S. mossambicus* occurs in
both freshwater and estuaries in the subtropical parts
of southern Africa but its southward distribution into
the Eastern Cape is restricted to estuaries. Allanson *et
al* found that osmoregulation of *S. mossambicus* breaks
down at 11 °C in salinities below 5 ‰. This breakdown
is associated with ultrastructural changes in the kidney
(Allanson & Cross, 1970). Sensitivity to low salinity
and low temperature is not limited to teleosts. Hill
(1979) reported that the 1+ year class of the crab
Scylla serrata in St Lucia was apparently killed at the
same time as the mortality of fish reported by Blaber
& Whitfield.

It·is probably of considerable significance, that the
South Western Cape is a winter rainfall area and thus
low temperatures and low salinity occur simultaneously
every year (Mehl, 1973, Millard & Scott, 1954, Scott *et
al,* 1952). These conditions could prevent many northern warm-water species from inhabiting southern
estuaries in winter. On the Natal coast, winter temperatures do not usually drop as low as those further
south and, in addition, the rainy season is in summer
when temperatures are high. Thus, except in St Lucia
where water exchange rates are extremely slow, Natal
estuaries do not experience the combination of low
temperature and low salinity. This may partly explain
Day's (1974) observation that there were more species
of crabs and fishes in tropical estuaries than in the
south coast ones.

CONCLUSIONS

Research into the effects of salinity on South African
estuarine animals has revealed that low salinity by itself is not an important factor. Many animals have a
tolerance well in excess of the levels encountered in the
areas in which they live. Indeed, many species have
been shown to be capable of living in freshwater. This
tolerance to low salinity may be related to the annual
low salinities which occur in the winter rainfall area of
the South Western Cape and also to the frequent floods
which occur along most of the east coast of southern
Africa. Animals do not live in estuaries because of low
salinity, they live there despite it.

Estuarine animals show a wide variety of ways in
which they cope with reduced salinity. Some such as
the sandshark, have a limited low salinity tolerance
and presumably leave the estuary in times of flood;
others like the polychaete *Ficopomatus enigmatica*
are exceptionally tolerant and are osmoconformers
from seawater to near freshwater. Most estuarine animals do osmoregulate but some, for example certain
bivalves, can isolate themselves from low salinity.

Although the decrease in salinity associated with a
flood may be drastic, it is only temporary; a much
longer lasting effect of floods is deposition of silt. The
quiet waters of estuaries are associated with fine sediments and this may be one of the most important factors in Day's hypothesis that the animals living in
estuaries largely represent a quiet water fauna. Silt
covers rocks limiting browsers, makes it difficult for
sessile rock-dwelling forms to settle, restricts sand-
dwellers and clogs the gills of many animals. Unfortunately no worker has yet attempted to investigate the
sheltered-water hypothesis but several studies indirectly
support it. Mclachlan & Grindley (1974) found that
substrate type was one of the most important factors
limiting distribution of the macrobenthos, salinity
playing only a small role. In Lake Nhlange — part of
the Kosi estuary system — Boltt & Allanson (1975)
found that while samples from sandy substrates were
rich in animals, those from muddy areas had only a
few animals. Lastly, Blaber (1976) has found large
differences in the species composition of mullet shoals
caught over different substrate types. Thus it may be
that a large part of the quiet water fauna does not
represent animals unable to live in rough water but
rather animals able to cope with, and probably to
utilise fine sediments. Clearly, substrate and its
influence on the fauna is a field requiring a great deal
of attention and those who venture into it should bear
Day's hypothesis in mind.

An important way in which low salinity may act as
a limiting factor is in combination with low temperature. This combination presents a severe physiological
stress to many animals and since it occurs annually in
winter in southern Cape estuaries and not infrequently
in Eastern Cape estuaries, it may act to prevent colo-

nisation of these estuaries by the warm water fauna. It is important to realise that it is the combination which is important, and low salinity or low temperature by themselves may not be a problem. More work in this field may partly explain the observation by Day (1974) that there are fewer species in temperate south coast as compared to subtropical east coast estuaries.

One of the few estuaries in which high salinity is a major problem is the St Lucia system. Salinities above 55 ‰ are lethal to a wide variety of organisms but the exact levels are known for remarkably few species. It would be of great value to have more information on the recovery of estuarine populations after a hyper-salinity event. Boltt (1975) gathered this information for some components of the benthic fauna but a great deal remains to be done. The physiology of tolerance to high salinity has received little attention. Consider-able experimental use could possibly be made of the hypersaline conditions which are deliberately created in evaporation ponds in the Eastern Cape for commer-cial salt production.

The southern African estuarine fauna appears to be well able to tolerate transient high temperatures and no natural mortalities have been reported. The real effect of high temperature is more subtle; there are strong indications that it may adversely affect fecun-dity in temperate species but the mechanisms are complex and obscure. Nevertheless, this is the only explanation so far offered whereby temperature may be responsible for the absence from tropical waters of many of the animals commonly found in temperate estuaries.

The recent interest in energy flow is beginning to yield new information on factors which may limit estuarine animals. The striking differences between the fauna of open and closed estuaries for example, may be related to greater food availability in estuaries which are permanently connected to the sea. Experi-ments by Forbes & Hill, for example, have shown that individuals of the sand prawn *Callianassa kraussi* trans-ferred into an open estuary have far more rapid growth rates than control animals left in a closed estuary.

Estuarine research in South Africa has tended to concentrate on animals in relation to one particular estuary or region — usually because of accessibility to the researcher. Future research should utilise a broader regional base if explanations are to be sought for the larger problems dealing with distribution of the fauna.

REFERENCES

ALLANSON, B.R. & R.H.M. CROSS 1970. Ultrastructural changes in renal proximal tubule cells of *Tilapia mossambica* following exposure to low temperature. *Septieme Congres International de Microscopie Electronique.* Grenoble. **197**: 611-612.

ALLANSON, B.R., BOK, A. & N.I. VAN WYK 1971. The influence of exposure to low temperature on *Tilapia mossambica* II: Change in serum osmolarity, sodium and chloride ion concentration. *J. Fish Biol.* **3**: 181-185.

ALLANSON, B.R., BRENNON, N.M. & R.C. HART 1974. The plants and animals of Lake Sibaya, Kwazulu, South Africa: A checklist. *Rev. Zool. Afr.* **88**(3): 507-532.

ALLANSON, B.R., HILL, B.J., BOLTT, R.E. & V. SCHULTZ 1966. An estuarine fauna in a freshwater lake in South Africa. *Nature* **299**: 532-533.

ALEXANDER, S.J. & D.W. EWER 1969. A comparative study of some aspects of the biology and ecology of *Sesarma catenata* and *Cyclograpsus punctatus. Zool. Afr.* **4**: 1-35.

BEADLE, L.C. 1972. Physiological problems for animal life in estuaries. *In:* R.S.K. Barnes & J. Green (eds), *The Estuarine Environment.* Applied Science, pp 51-60. London.

BLABER, S.J.M. 1973. Temperature and salinity tolerance of juvenile *Rhabdosargus holubi. J. Fish. Biol.* **5**: 593-598.

BLABER, S.J.M. 1974. Osmoregulation in juvenile *Rhabdosargus holubi. J. Fish. Biol.* **6**: 797-800.

BLABER, S.J.M. 1976. The food and feeding ecology of Mugi-lidae in the St Lucia lake system. *Biol. J. Linn. Soc.* **8**: 267-277.

BLABER, S.J.M. & A.K. WHITFIELD 1976. Large scale mor-tality of fish at St Lucia. *S.Afr. J. Sci.* **72**: 288.

BLABER, S.J.M. & A.K. WHITFIELD 1977. The biology of the burrowing goby *Croilia mossambica. Env. Biol. Fish.* **1**: 197-204.

BOK, A. 1973. *Aspects of ion and water metabolism in some elasmobranchs.* MSc Thesis, University of Port Elizabeth. 88pp.

BOLTT, R.E. 1969. *A contribution to the benthic biology of some southern African lakes.* PhD Thesis, Rhodes Univer-sity, Grahamstown.

BOLTT, R.E. 1975. The benthos of some southern African lakes Part V. The recovery of the benthic fauna of St Lucia following a period of excessively high salinity. *Trans. roy. Soc. S.Afr.* **41**: 295-323.

BOLTT, R.E. & B.R. ALLANSON 1974. The benthos of some southern African lakes Part III. The benthic fauna of Lake Nhlange, KwaZulu, South Africa. *Trans. roy. Soc. S.Afr.* **41**: 241-262.

BOLTT, G. & J. HEEG 1975. The osmoregulatory ability of three grapsoid crab species in relation to their penetration of an estuarine system. *Zool. Afr.* **10**: 167-182.

BREUER, J.P. 1957. Ecological survey of Baffin and Alazan Bays. *Publ. Inst. mar. Sci., Univ. Tex.* **4**: 134-155.

BROEKHUYSEN, G.J. 1941. The life-history of *Cyclograpsus punctatus:* breeding and growth. *Trans. roy. Soc. S.Afr.* **28**: 331-336.

BROEKHUYSEN, G.J. 1955. The breeding and growth of *Hymenosoma orbiculare. Ann. S.Afr. Mus.* **56**: 313-343.

COPELAND, B.J. 1967. Environmental characters of hyper-saline lagoons. *Contrib. Mar. Sci.* **12**: 207-218.

DAY, J.H. 1951. The ecology of South African estuaries Part I. A review of estuarine conditions in general. *Trans. roy. Soc. S.Afr.* **33**: 53-91.

DAY, J.H. 1974. The ecology of Morrumbene estuary, Mocam-bique. *Trans. roy. Soc. S.Afr.* **41**: 43-97.

DAY, J.H., MILLARD, N.A.H. & G.J. BROEKHUYSEN 1954. The ecology of South African estuaries Part IV. The St Lucia system. *Trans. roy. Soc. S.Afr.* **34**: 129-156.

DAY, J.H., MILLARD, N.A.H. & A.D. HARRISON 1952. The ecology of South African estuaries Part III. Knysna: a clear open estuary. *Trans. roy. Soc. S.Afr.* **33**: 367-413.

FORBES, A.T. 1973. An unusual abbreviated larval life in the estuarine burrowing prawn *Callianassa kraussi. Mar. Biol.* **22**: 361-365.

FORBES, A.T. 1978. Maintenance of non-breeding populations of the estuarine prawn *Callianassa kraussi* (Crustacea, Ano-mura, Thalassinidea). *Zool. Afr.* **13**: 33-40.

HILL, B.J. 1975. Abundance, breeding and growth of the crab *Scylla serrata* in two South African estuaries. *Mar. biol.* **32**: 119-126.

HILL, B.J. 1977. The effect of heated effluent on egg production in the estuarine prawn *Upogebia africana* (Ortmann). *J. exp. mar. Biol. Ecol.* **29**: 291-302.

HILL, B.J. 1979. Biology of the crab *Scylla serrata* in the St Lucia system. *Trans. roy. Soc. S.Afr.* **44**(1): 55-62.

HILL, B.J. & B.R. ALLANSON 1971. Temperature tolerance of the estuarine prawn *Upogebia africana. Mar. Biol.* **11**: 337-343.

HILL, B.J. & A.T. FORBES 1979. The biology of *Hymenosoma orbiculare* in Lake Sibaya. *S.Afr. J. Zool.* **14**: 75-79.

KINNE, O. 1963. The effects of temperature and salinity on marine and brackish water animals. I. Temperature. *Oceanog. mar. Biol. Ann. Rev.* **1**: 301-340.

KINNE, O. 1964. The effects of temperature and salinity on marine and brackish water animals. II Salinity and temperature-salinity relations. *Oceanog. mar. Biol. Ann. Rev.* **2**: 281-340.

LOCKWOOD, A.P.M. 1976. Physiological adaptation to life in estuaries. *In:* R.C. Newell (ed), *Adaptation to environment.* pp 315-392. Butterworths, London.

MACNAE, W. 1957. The ecology of the plants and animals in the intertidal regions of the Zwartkops estuary near Port Elizabeth, South Africa. Parts I & II. *J. Ecol.* **45**: 113-131; 361-387.

M⁽ᶜ⁾LACHLAN, A. 1974. Notes on the biology of some estuarine bivalves. *Zool. Afr.* **9**: 15-20.

M⁽ᶜ⁾LACHLAN, A. & T. ERASMUS 1974. Temperature tolerance and osmoregulation in some estuarine bivalves. *Zool. Afr.* **9**: 1-13.

M⁽ᶜ⁾LACHLAN, A. & J.R. GRINDLEY 1974. Distribution of soft substrata in Swartkops estuary with observations on the effects of floods. *Zool. Afr.* **9**: 211-233.

MEHL, J.A.P. 1973. Ecology, osmoregulation and reproductive biology of the white steenbras, *Lithognathus lithognathus. Zool. Afr.* **8**: 157-230.

MILLARD, N.A.H. & C.J. BROEKHUYSEN 1970. The ecology of South African estuaries Part X: St Lucia: a second report. *Zool. Afr.* **5**: 277-307.

MILLARD, N.A.H. & K.M.F. SCOTT 1954. The ecology of South African estuaries Part VI: Milnerton estuary and the Diep River, Cape. *Trans. roy. Soc. S.Afr.* **34**: 279-324.

ORR, P.R. 1955. Heat death. I Time-temperature relationships in marine animals. *Physiol. Zool.* **28**: 290-294.

POTTS, W.T.W., FOSTER, M.A., RUDY, P.P. & G. PARRY HOWELLS 1967. Sodium and water balance in the cichlid teleost *Tilapia mossambica. J. exp. Biol.* **47**: 461-470.

READ, K.R.H. & K.B. CUMMING 1967. Thermal tolerance of the bivalve molluscs *Modiolus modiolus. Mytilus edulis* and *Brachidontes demissus. Comp. biochem. Physiol.* **22**: 149-155.

SCOTT, K.M.F., HARRISON, A.D. & W. MACNAE 1952. The ecology of South African estuaries Part II. The Klein River estuary. *Trans. roy. Soc. S.Afr.* **33**: 283-331.

SMITH, J.J.B. 1953. *The sea fishes of southern Africa.* 565pp. Central News Agency, Cape Town.

SKAER, H. le B. 1974. The water balance of a serpulid polychaete, *Mercierella enigmatica.* I. Osmotic concentration and volume regulation. *J. exp. Biol.* **60**: 321-330.

STEPHENSON, W., COOK, S.D. & Y.I. RAPHAEL 1977. The effects of a major flood on the macrobenthos of Bramble Bay, Queensland. *Mem. Qld. Mus.* **18**: 95-119.

THOMPSON, L.C. & A.W. PRITCHARD 1969. Osmoregulatory capacities of *Callianassa* and *Upogebia. Biol. Bull.* **136**: 114-129.

WALLACE, J.H. 1975. The estuarine fishes of the East coast of South Africa. *Invest. Rep. oceanog. res. Inst., Durban* **40**: 1-72.

Chapter 12

Estuarine fishes

J.H. Day, S.J.M. Blaber and J.H. Wallace

Present addresses: *J.H. Day:* Department of Zoology, University of Cape Town. *S.J.M. Blaber:*
Department of Zoology, University of Natal, Pietermaritzburg. *J.H. Wallace:* Port Elizabeth Museum, Humewood

CLASSES OF FISHES IN ESTUARIES

Fishes are part of the nekton and in estuaries this
includes cuttle fish, swimming crabs, penaeid prawns,
crocodiles and occasional dolphins, turtles and dugongs.
Like the fishes, very few of these animals live permanently in estuaries but migrate either to the sea or to
the rivers.

Day (1951), McHugh (1967) and Perkins (1974)
have divided estuarine fishes according to their breeding and migratory habits into several groups or classes.
These are:

1. The largest group may be termed marine migrants.
As shown by Hartley (1940), Day (1951) and Gunter
(1961), the great majority of fishes found in estuaries
are the juveniles of species which breed in the sea and
enter estuaries to feed and shelter until their gonads
start to develop. They then return to the sea; some of
the adults are never seen in estuaries again, but a few
others visit estuaries seasonally to feed. Many families
are represented among the juvenile migrants. The Mugilidae, Clupeidae, Engraulidae, Atherinidae, Sciaenidae,
Sparidae and Pomadasyidae are the most important
and several other families might be added from the
long lists of fishes in West African estuaries given by
Pillay (1967a) and the even longer lists in Indian
estuaries given by Pillay (1967b). Many of the genera
listed by Pillay are present in the estuaries of southern
Africa and the Platycephalidae and Sillaginidae are
particularly common in Australian estuaries. Juvenile

marine migrants are common in tropical and temperate
estuaries all over the world but according to McHugh
(1967) they are not well-represented in high latitudes
and indeed, the number of species decreases in estuaries flowing into cold temperate seas (Day, 1974).

2. The anadromous fishes include the various species
of salmon *(Salmo* and *Oncorhynchus),* sturgeons
(Accipenser), the northern smelts *(Osmeridae),* the
shad *(Alosa),* the white perch *(Roccus americanus),*
the striped bass *(R. saxatilis)* and the lampreys (Petro-
myzontidae). In their migrations to and from their
breeding grounds in fresh water, these fishes dominate
the cold northern estuaries. McHugh (1967) states
that 20 million red salmon are taken annually from
one Alaskan river alone and that the salmon fishery
of Washington State is worth $100 million per year.
According to Korringa (1967), anadromous fish were
equally important in Europe at one time but now,
due to pollution and dam construction, they are scarce
or absent except in Norway and the western coasts of
Scotland and Ireland. Anadromous fish are virtually
absent in the tropics and as far as we are aware, there
are no endemic anadromous fish in the southern
hemisphere, although the lamprey *Geotria australis* is
known from Australia, New Zealand and South America.

3. The catadromous group includes the fresh water
eels *(Anguilla)* spp but not the many other eels found
in estuaries. After breeding in the depths of the Sar-
gasso Sea, the leptocephalus larvae of the North Atlan-

tic *Anguilla* migrate to eastern American coasts or those of western Europe and metamorphose to elvers. The elvers enter estuaries on their way to fresh water and are caught in great numbers. The fresh water eels of southern Africa are common as far south as Knysna. Crass (1964) states that the leptocephalus larvae take more than a year to drift in the South Equatorial Current past Madagascar to the African coast. *Anguilla mossambica* and *A. marmorata* are the main species in the Mocambique and Agulhas currents. After metamorphosis, large numbers of elvers migrate through the estuaries each summer. Strangely enough, fresh water eels have not been found in African rivers flowing into the South Atlantic; it has been suggested that there are no ocean deeps sufficiently close to the west coast of Africa in which they could breed.

4. There are only a few species which could properly be called estuarine residents in that they breed and complete their entire life cycle within an estuary. Mullets which are so common in estuaries, were at one time thought to breed there but it has been observed by many workers that if the estuary mouth is closed so that females with ripe roes cannot escape to the sea, the eggs are resorbed. The majority of residents are small species such as gobies, syngnathids, ambassids, atherinids, stolephorids and some clupeids. Accurate estimates of the proportion of residents in an estuary are difficult to obtain but Hedgpeth (1967) records nine species in a total of 70 species in the Laguna Madre; Wallace (1975a) estimates 10 among 86 species in St Lucia, and among the 58 common estuarine fishes in southern Africa listed in Table 12.1, 12 breed in estuaries. One of them, *Gilchristella aestuarius* can breed in fresh water as well. The resident species, like fresh water fishes, often show some evidence of reproductive specialisation. Male syngnathids carry the developing eggs in a brood pouch; gobies have sticky eggs and attach them to stones and guard them; sticklebacks and cichlids are famous for making nests and some of the latter are mouth breeders.

5. Finally, there is an anomalous group of fishes whose breeding habits and migrations do not fit any well-defined group. The American garfish *Lepidosteus spatula* and *Dorosoma cepidianum* breed in fresh water but live mainly in estuaries. The menhaden *Brevoortia*, which is one of the most important commercial fish on the Atlantic and Gulf coasts of the United States, differs from one species to another. *B. tyrannus* on the Atlantic coast actually breeds in the sea but migrates at an early stage into low salinities or even fresh water; according to Gunter (1967), it develops abnormally in full strength sea water. The Gulf species, *B. patronus* and *B. gunteri* both breed in the upper reaches of estuaries but the adults migrate to the sea. The African *Sarotherodon (olim Tilapia) mosambicus* is a fresh water fish but it is abundant in estuaries and can breed even in hypersaline water.

ENVIRONMENTAL ADAPTATIONS

Estuaries are characterised by a variable salinity, a temperature range greater than the sea and many have turbid water and a muddy bottom. In general, estuarine fishes are euryhaline, eurythermic and some of them have highly developed sensors similar to those developed by deep sea fishes. In the muddy estuary of the Ganges some fishes lack eyes.

We are concerned here with the ecological implications of adaptations by fishes to conditions of life in an estuary. Space does not permit a serious discussion of the physiological mechanisms whereby salinity and temperature adaptations are achieved. For such a discussion, the reader is referred to Parry (1966), the summary in Green (1968) and the more recent review in Kinne (1970, 1971). Experiments and observations on South African fishes and other estuarine animals are reviewed by Dr Burke Hill in chapter 11.

Salinity tolerance. There is palaeontological evidence that all fish evolved in fresh water and physiological studies confirm that the osmotic concentration of inorganic ions in the blood is less than that in sea water. This means that all marine fish are preadapted to decreased salinities. Although the skin is relatively impermeable to ions, the gills are of necessity permeable and many marine fish can tolerate reduced salinities if the change is gradual. For estuarine fish subject to tidal changes and sudden floods, the rate of change appears to be as important as the absolute concentration. Blaber (1973b) has shown that in the South African sparid *Rhabdosargus holubi,* the internal concentration of inorganic ions does not drop for 10 hours when the fish is subjected to abnormally low salinity. Probably the duration of the stress condition is important too although this requires further investigation.

Many estuarine species breed in the sea presumably because the eggs and embryonic stages are intolerant of salinity changes although the turbulence of the waves may also be important in keeping the eggs off the bottom. Juveniles, however, rapidly develop a greater tolerance to reduced salinity. Green (1968) quotes the case of the European herring. The eggs develop normally in salinities ranging from 6 to 52 ‰. The larvae have a wider range than this and tolerate salinities between 1,4 and 60 ‰ for over 20 hours while juveniles of 9 to 24 cm may be transferred direct from sea water to salinities of 6 to 40 ‰ without indications of stress. Green also states that flounders *(Pleuronectes flesus)* breed in the sea but their juveniles can live for years in fresh water. Experiments in southern Africa show that the adults of several species such as *Monodactylus falciformis, Gilchristella aestuarius, Hepsetia breviceps* and *Mugil cephalus* tolerate direct transfer from sea water to fresh. Similar results are quoted by McHugh (1967) in regard to North American species.

Tolerance to the hypersaline conditions which occur in some estuaries during droughts, has been recorded in relatively few species. Hedgpeth (1967) lists 70 species in the Laguna Madre of Texas of which 18 tolerate salinities as high as 60 ‰ and as low as 5 ‰. Wallace (1976) records 13 of the 88 species in Lake St Lucia, Zululand in salinities below 10 ‰ and above 60 ‰. Ten of them occur in salinities above 70 ‰, and among them is the fresh water fish *Sarotherodon mossambicus* which may be found in a salinity of 110 ‰ and can breed in 45 ‰. The recorded salinity tolerance of many individual species in the estuaries of southern Africa will be given later.

In spite of this wide salinity tolerance, it has been reported from many parts of the world that the greatest species richness is to be found at the mouth of an estuary. During floods the fish disappear for a few days and then return to the estuary. As might be expected, prolonged floods have a greater effect. Persistent low salinity conditions occur in the Blackwood estuary in south-western Australia (Hodgkin, 1978). All the rain falls in the three winter months and the whole estuary is flooded with fresh water and the fish population, apart from a few fresh water species, drifts out to sea. Due to the small tidal range, salinities of 5-10 ‰ persist until early summer and it is only then that juveniles from the sea start moving up the estuary. The lower reaches are well-populated in late summer but the fish disappear again when the estuary is flooded at the end of autumn. In brief, the fish population is very poor for most of the year.

Temperature tolerance. It is well-known that the upper reaches of an estuary are warmer than the sea in summer and cooler in winter. These seasonal changes have marked effects on the fish population particularly in high latitudes where temperature becomes more important than salinity. In cold climates where the whole surface of the estuary is frozen in winter, the fish population is reduced to a minimum. The anadromous species which dominate such estuaries, have returned to the sea and the overwintering juveniles and resident species remain inactive. They concentrate in the deeper parts of the estuary and show little growth. McHugh (1967) quotes a dramatic account of the seasonal changes in an Alaskan estuary. In April the ice starts to break up but the temperature is still low and there is nothing to be seen swimming in the clear water. Then very rapidly, as the light and temperature increase, plankton blooms develop and shoals of young herring and salmon fry appear with many small cottids and flounders congregating in the shallows. Seals come to attack the herring and salmon and by summer vast shoals of pink salmon are moving in and out of the creeks for weeks on end. Hundreds of puffins dive everywhere, blooms of large jellyfish appear and there are even occasional otters. And yet, by the late fall, everything has gone from the estuary and in the bay the water is cold, clear and apparently devoid of life. Soon a skin of ice appears and the winter has come.

In cold-temperate estuaries the seasonal changes are not so marked but they are equally important. As the turbidity due to the floods of winter or the spring thaw decrease, phytoplankton develops rapidly in large, slow flowing estuaries and submerged vegetation grows in the shallows. Juvenile fish enter from the sea as the water becomes warmer and the permanent residents grow rapidly. Immigration from the sea follows no fixed pattern but depends on the spawning season. This often occurs in the winter or spring so that juveniles may arrive at the estuary mouth from early spring to summer when food is plentiful. Predatory species follow the juveniles and by midsummer the estuary is well-stocked, with juveniles congregating in the warm upper reaches or sheltering in the weedbeds. As the temperatures fall in late autumn, the adults and most of the juveniles return to the sea but some juveniles remain in the estuaries for two or three years.

In warm and subtropical estuaries, the significance of temperature declines and salinity effects become more important. In such estuaries the rainy season may be in winter or summer or may be spread evenly through the year, and the immigration of juveniles is very variable although it usually occurs after a rise in temperature. Gunter (1967) states that juveniles enter the Mississippi with the increased outflow of estuarine water after the spring rains. In Natal, estuaries that have been closed by sandbars during the winter, burst open after the spring rains which start in September and juveniles enter before the mouths close again. Wallace (1975c) found that adults start spawning during late winter and early spring and suggests that this is an adaptation which produces juveniles inshore at the time of the year when the spring rains open river mouths and before the floods of high summer. Juveniles of most species are 2-4 cm long when they enter estuaries and early entries tend to be smaller than late entries which suggests that if the opening of the estuary mouth is delayed, the juveniles live and grow in the coastal shallows.

In the southern Cape Province, the main rains are in winter and the main influx of juveniles is in spring suggesting that the rising temperature is more important in the cooler climate. As noted, estuarine fish are eurythermic and tolerate a wider range of temperature than normally occurs in the estuaries they inhabit. There are no reports of fish being killed by seasonal changes of temperature in the temperate estuaries of the United States and Europe. In subtropical estuaries such as those of the Gulf of Mexico, however, where summer temperatures may exceed 30°, Gunter (1967) reports that in exceptionally cold winters mush ice may form in the shallows and fish die by the million as happened in 1962.

In tropical estuaries there are smaller seasonal

changes of temperature but there are marked changes in salinity. Pillay (1967b) reports that in India the estuaries vary from near fresh water during the monsoon rains to salinities of 30-33 ‰ in the dry season. There are no reports of mortality due to low salinities, rather there is a tendency for fresh water fish such as the Cichilidae to move down to the saline lagoons in the rainy season and back to the rivers in the dry season. The plankton develops at the end of the rains and marine migrants then become more abundant.

THE FEEDING HABITS OF ESTUARINE FISHES

A study of the diet of commercially important estuarine fish is necessary to evaluate the holding capacity of different estuaries. Detailed studies of the food requirements of the North American menhaden *(Brevoortia* spp), the channel bass *(Roccus saxatilis),* the European flounder *(Pleuronectes flesus)* and the Indo-Pacific Bombay duck *(Harpodon nehereus)* have already been made. A knowledge of the feeding habits of sport fish is possibly of greater economic importance since fish attract thousands of tourists to favoured angling resorts. The same applies to commercially successful aquaculture of mullets, bream, milk fish *(Chanos chanos)* and tilapia *(Sarotherodon mossambicus)* for the percentage survival often depends on correct food supplies for the early developmental stages. Beyond these commercial considerations is the theoretical importance of understanding the impact of migratory fishes on an estuary. The juveniles feed there and when grown transport not only organic matter in the form of fish flesh, but also pollutants such as mercury and persistent insecticides out of the estuary.

A particularly fascinating problem is the question of competition among fishes with what appear to be identical food requirements. Mullets are a case in point. They are classed as iliophagous or detritus feeders, although organic detritus has so many meanings for different workers that it is losing its value. Ten species of mullets occur in the St Lucia lakes in Zululand and Blaber (1976a) has analysed their feeding habits. Many of the mullets feed on the small gastropod *Assiminea bifasciata* but this is abundant and competition only becomes important when some resource is in short supply. However, the various species of mullet also take foraminifera, the large centric diatom *Aptinoptychus splendens,* small centric and pennate diatoms, filamentous algae, blue-green algae, plant fragments in varying quantities and sediment particles of different sizes. Mullets have villiform teeth on their pharyngeal pads and gill arches which presumably aid in sorting the required materials. They also have tough, gizzardlike stomachs in which the sediment particles grind up the hard shelled foods. Blaber has found that different species of mullet ingest different size ranges

of particles and suggests: 'that interspecific competition for food is reduced by substrate particle size selection, and perhaps by differences in feeding periodicity'. Marais & Baird (1979) have investigated the feeding habits and adaptations of mullet in the Swartkops estuary.

The feeding habits of fishes change with age. When the yolk-sac has been absorbed, most if not all juvenile fish pass through a stage of plankton feeding until their adult organs develop, when feeding becomes specialised to some degree. The permanent plankton feeders may be divided into two groups. Those with an efficient filtering mechanism of elongated gill rakers, include the Clupeidae, Engraulidae and Stolephoridae among estuarine species. All of them are shoal fish feeding near the surface at night or near the bottom in daylight. They are most abundant in broad, slow-flowing estuaries and lagoons where the long residence time of the water allows dense concentrations to develop. Blaber (1979) and Heeg & Blaber (1979) discuss the common filter feeding fish in Lake St Lucia. The dominant copepod *Pseudodiaptomus stuhlmanni* provides 70 % of the calorific value of the plankton although the mysid *Mesopodopsis africana,* the amphipod *Grandidierella lignorum* and veligers of *Assiminea bifasciata* are also important. The three abundant plankton feeding fish are *Gilchristella aestuarius,* juvenile *Thryssa vitrirostris* and *Hilsa kelee.* As is often the case, *Gilchristella* and *Thryssa* spawn in spring so that the juveniles are present when zooplankton is abundant. *Gilchristella* is a resident and spawns in the lake but *Thryssa* is unusual in that it enters the lake from the sea to spawn. It changes its diet as it grows and from a length of 10 cm preys mainly on juvenile *Gilchristella* and its own young but most of the adults return to the sea in winter. *Hilsa kelee,* like most estuarine fish, breeds in the sea and enters the estuary in summer feeding first on the large diatom *Coscinodiscus,* then changing to the abundant copepod *Pseudodiaptomus* in autumn and to veligers of *Assiminea* in winter. A variety of other plankters are taken depending on the local concentrations in the lake centre where *Hilsa* congregates.

The individual plankton feeders do not depend on a filter mechanism but snap up or suck in individual copepods, amphipods and mysids. Estuarine forms include Syngnathids, Ambassids and Monodactylids all of which are common in weed beds where amphipods are abundant and shelter from swift predators is available.

The presence of specialised herbivores in estuaries is doubtful although many fishes include benthic and ephipytic diatoms in their diet and will take filamentous algae particularly *Enteromorpha.* Others crop sea grasses such as *Zostera, Posidonia* and *Ruppia.* Typical adaptations are incisiform teeth and a long absorptive region in the gut for the food is not very nutritious. The rabbit fish *Siganus* spp is a typical example. Many mullets are partly herbivorous and juveniles of *Rhabdosargus* (particularly *R. globiceps* and *R. holubi*) crop the leaves of *Zostera* and *Ruppia.* According to Blaber (1974a), the pieces of leaf pass out of the rectum undigested. It would appear that the fish lacks the cellulase necessary to digest the thick cellulose covering the leaves. However, the leaves are usually covered with epiphytic diatoms whose surface is naked protoplasm and it is from these diatoms that nourishment is obtained and they contribute over 50 % of the food digested. Drift algae are also eaten and they are often covered with nutritious ectoprocts or other epizoans. It would be interesting to know whether any species of herbivorous fish (such as species of Scaridae or Monacanthidae) has developed cellulase itself or whether it depends on symbiotic micro-organisms as do ruminants. Possibly those that feed on filamentous algae such as *Enteromorpha* are able to masticate the relatively thin cell walls and release the protoplasm for digestion.

Several estuarine fishes are said to be detritus feeders suggesting that they live on decaying vegetable fragments and the community of bacteria and other small organisms that detritus supports. The case of mullets has been discussed earlier. Cichlids such as *Sarotherodon* also contain organic detritus in their stomachs as well as living plants and invertebrates while gobies feed mainly on small invertebrates.

The demersal fishes which feed largely on benthic invertebrates include many species of Blenniidae, Gobiidae, Sillaginidae, Sparidae, Platycephalidae, Sciaenidae, Lethrinidae, Lutjanidae as well as flat fishes and many other families in tropical estuaries. As noted earlier, the larval and early juvenile stages start off catching individual zooplankton organisms and then graduate to amphipods, mysids, isopods and shrimps. Those that develop molariform teeth such as the sparids, take hard-shelled prey including crabs, bivalves and gastropods. The less specialised fishes feed mainly on the epifauna although they also graze on the projecting siphons of bivalves, the fans of sabellid worms and they take gobies and other small fishes as they grow larger. Others develop special mechanisms for feeding on the infauna although there appear to be no hard and fast distinctions between one family of fishes and another. Like most carnivores in other environments, fishes are opportunistic feeders and change their diet according to the abundance and accessibility of the prey. When oyster culture was first started in Knysna Lagoon the oysters were set out in open trays. The first crop was eaten by the brusher *Sparodon durbanensis* although oysters are not its normal prey and brushers are seldom found in estuaries. The diet of common estuarine fish in southern Africa will be detailed later. Green (1968) summarises the work of Hartley (1940) in regard to the feeding of flat fish in the Tamar estuary at Plymouth. Three of them feed at the mouth of the estuary. The

brill *(Rhombus laevis)* feeds largely on gobies, the dab *(Limanda limanda)* nips off the crowns of the fanworm *Sabella penicillus* while the plaice *(Pleuronectes platessa),* which also feeds largely on polychaetes, has developed a pharyngeal pump to suck the complete worms out of their tubes. The flounder *(Pleuronectes flesus)* extends to lower salinities higher up the estuary and although its diet largely overlaps with that of the plaice, competition is reduced. The flounder also takes more mysids and fewer polychaetes. The diet of the flounder has been investigated in many other European estuaries. Green (1968) quotes reports that the diet of juvenile flounders includes mysids, shrimps, *Nereis diversicolor, Corophium volutator,* harpacticoid copepods, oligochaetes, chironomid larvae and small bivalves. It is obvious that the diet changes from estuary to estuary with the abundance of the prey.

Platycephalids which are particularly common in Australian estuaries, feed quite largely on crabs but also take gobies and other small fishes. *Terapon jarbua* which is widely distributed in Indo-Pacific estuaries, also feeds on crabs and amphipods but apparently its main diet is the scales of larger fish. Whitfield & Blaber (1978a) note that although the scale-eating habit has developed in a number of American fresh water characoids and African cichlids it has not been reported before in marine fishes. The Pomadasyidae, which are famous for the grunting sounds they make with their pharyngeal teeth, also have thick pouting lips and feed mainly on the infauna. Longhurst (1957) lists the diet of several fishes including *Pomadasys jubelini* from in the 'Sierre Leone River'. It feeds largely on burrowing polychaetes (particularly *Pectinaria sourei*) and cumacea. On occasion, it also takes large numbers of *Lingula.* As will be described later, *Pomadasys commersonni* in southern Africa feeds largely on burrowing prawns or penaeid prawns when the latter are abundant.

Predatory fishes which feed mainly on smaller fish in estuaries have been divided by Whitfield & Blaber (1978b) into two groups. 'Group one' predators in Lake St Lucia are high speed swimmers with lunate or forked tails and include the Zambesi shark *Carcharinus leucas,* the king fish *Caranx ignobilis,* the leervis *Lichia amia* and the barracuda *Sphyraena jello* although this is rare. Such fish are fast enough to feed on shoals of full-grown mullet and often attack in packs to break up the shoals of prey. Anglers have reported that sharks are occasionally stranded in the mouth of St Lucia estuary while attacking shoals of mullet entering the system. The mullet escape by leaping into the air and there is a report that a man was killed by a 5 kg *Mugil cephalus* which hit him in the kidneys.

The slower 'group two' piscivores often have truncate or rounded tails. The kob *(Argyrosomus hololepidotus)* and the ten-pounder *Elops machnata* are the most abundant species in this group and feed mainly on small plankton feeders such as *Thryssa vitrirostris* and *Gilchristella aestuarius*. The kob feeds on both prey according to their changes in density whereas the ten-pounder has a definite preference for *Gilchristella* when both prey species are abundant. During summer, when penaeid prawns are plentiful, both predators switch to this prey. Although the summary given here is a simplification of the several predators and their prey, it is evident that prey resources are shared so as to reduce competition and, as the abundance and availability of prey changes, so does the diet of the predators.

The amount of food eaten by a fish depends on its activity, its age, the water temperature and the calorific value of the food. Although fish often regurgitate their food when netted, the mass of the stomach contents is some indication of their rate of feeding. Whitfield & Blaber (1978b) quote a report by Todd (1915) that 90 % of the plaice caught in the North Sea in winter were empty while only 10 % were empty in summer. Longhurst (1957) found no seasonal change in the percentage of empty stomachs in a tropical estuary in Sierra Leone and Whitfield & Blaber themselves found no seasonal changes in the percentage of empty stomachs in St Lucia Lake. They conclude that while low winter temperatures in temperate waters may inhibit certain species from feeding, the relatively high winter temperatures in tropical and subtropical estuaries enable feeding to continue during all seasons. McHugh (1967) quotes observations on the amount of food eaten by fresh water trout. At a length of 2,5 cm and a temperature of 2,2 °, it eats 2,7 % of its body mass per day, while at 20 ° it eats 9,9 %. At a length of 25 cm and a temperature of 2,2 ° it eats 0,4 % of its body mass per day and at 20 ° it eats 1,7 %. The efficiency of conversion of food into fish flesh is estimated at 10 %. Blaber (1974a) estimated that a juvenile of the sparid *Rhabdosargus holubi* averaging 5 gm dry body mass eats 1,9 % of its body mass per day, but he himself regards his estimate as too low.

ESTIMATES OF BIOMASS

It would be of great interest to determine the total biomass of fish per hectare of an estuary and to follow the seasonal changes due to immigration, growth, mortality or emigration. Unfortunately, such an investigation is fraught with many difficulties, primarily because all the types of nets and other catching devices are selective to some extent so that random samples are almost impossible to obtain. McHugh (1967) discusses the matter in detail and quotes many statistics of commercial and sport fisheries in Chesapeake Bay. An analysis of his figures reveals that 173 kg.ha^{-1} are caught by commercial fishermen. Among the families included in the catch are Engraulidae, Clupeidae, Sciaenidae and Stromateidae. The sport fishery aimed mainly at striped bass, yields a further 7 kg.ha^{-1} so that

the total catch in 1962 was 180 kg.ha^{-1} which McHugh estimates as about 50 % of the annual increment. Korringa (1967) reports that before the Zuider Zee was reclaimed, the annual catch was 135 kg.ha^{-1}. This may be compared with 25 kg.ha^{-1} in the North Sea and 28 kg.ha^{-1} in the Japan Sea both of which are overfished. The summary by Milne & Dunnet (1972) of the extensive survey of the Ythan estuary near Aberdeen, includes an account of the seasonal changes in the population of gobies and flounders. Sea trout which pass in and out of the estuary are not included in the analysis. The biomass of gobies and flounders is shown graphically in terms of grams per square metre. If the monthly values are extracted and converted into kilograms per hectare, the seasonal changes are as follows:

Biomass (kg.ha^{-1})	Winter	Spring	Summer	Autumn
Gobies	3,5	1,6	2,4	6,9
Flounders	120	122	180	55
Both	123,5	123,6	182,4	61,9

The main contribution to the biomass is due to the flounders. The juveniles arrive in spring, feed rapidly in the summer and return to the sea in autumn. Although Milne & Dunnet list 23 species of fish in the estuary, the combined biomass of flounders and gobies alone is comparable with the estimates of McHugh in the Chesapeake and Korringa in the Zuider Zee in terms of kilograms per hectare. Similar estimates of the total fish population in an estuary in southern Africa are urgently needed.

These estimates of biomass in natural estuaries may be compared with production in estuarine fish farms. There are many of them in India and Indonesia, often in specially constructed fish ponds flooded with brackish water and stocked with small fishes and penaeid prawns. Pillay (1967b) refers to them as tambaks and 'bheris'. The annual production varies from 150 kg.ha^{-1} in poor ponds to 1 000 kg.ha^{-1} in well-managed ponds. The paddy fields of Bengal produce 100-200 kg of fish per hectare without decreasing the yield of rice.

THE ESTUARINE FISHES OF SOUTHERN AFRICA

The taxonomy of marine and estuarine fishes around southern Africa will be found in Smith (1949 and 5th edition 1965) which incorporates the earlier works of Barnard (1925, 1927) and many others. He also includes many biological notes on estuarine species but it is not always possible to distinguish observations made in estuaries from those made in the sea. The first purely biological study was that of Biden (1930) and although it is written for anglers it includes a wealth of information on game fish. The many later studies upon which this account is based, are given in the list of references.

The estuaries and lagoons covered here extend from the Morrumbene estuary on the coast of southern

Table 12.1 Zoogeographic distribution of 59 common estuarine fishes in southern Africa

Component	S.Mocambique	Natal	Transkei	S.Cape	Atlantic Cape	Total
Tropical	37 (80 %)	37 (70 %)	27 (59 %)	20 (49 %)	1 (6 %)	37
Subtropical	7 (15 %)	7 (13 %)	7 (15 %)	6 (15 %)	4 (25 %)	7
Cosmopolitan	2 (4 %)	3 (6 %)	3 (7 %)	3 (7 %)	3 (19 %)	3
S.African endemics	0	6 (11 %)	8 (17 %)	12 (29 %)	8 (50 %)	12
Totals	46	53	46	41	16	59

Mocambique at 23°40′S to the estuary of the Olifants River in Namaqualand at 31°31′S. An inspection of the mouth of the Orange River by Brown (1959) revealed that the river flow is so strong and the environment so changeable that no estuarine fauna is established there. The same appears to be true of the mouth of the Tugela River in Natal. Brief accounts of the environmental conditions in some 40 individual estuaries will be found in chapter 14.

Smith (1965) lists 1 500 species of fishes from the seas and estuaries around southern Africa. Day (1974) lists 114 species from Morrumbene estuary, Blaber (1978c) lists 124 species from the Kosi system, Wallace (1975b) lists 232 species from Natal estuaries generally and there are about 150 species in Cape estuaries. Many of these species are the same and some are mere strays that have been recorded in the mouths of estuaries. Probably about 300 species are estuarine for part of their life cycle, indicating that about 20 % of the fishes of southern Africa extend into estuaries or live there as residents. Since many estuaries have been incompletely sampled, it would be pointless to list them all. Wallace (1975a) selected 64 species as being common in Natal estuaries; the better-known ones and a few more that are particularly common in southern Mocambique or Cape estuaries are discussed here. In all, 59 species are discussed. Their zoogeographic distribution is analysed in Table 12.1 above.

Since this is a selected sample and not a random one, only the most obvious differences are discussed, but it may be noted in passing that Wallace's 64 species show the same trends. The majority of species in southern Mocambique, Natal and the Transkei extend north to the equator and are thus termed tropical. The percentage of these tropical species decreases sharply south of the Transkei. When the number of occurrences of tropical species is considered as well, this trend becomes more marked; indeed, many of the Cape records such as those of *Acanthopagrus berda* and *Arothron immaculatus* in Swartkops estuary and *Pelates quadrilineatus* in the Bushmans estuary are sporadic. Such species are found in some years but not in others, probably due to the transport of larvae from further north by the wandering Agulhas current which in some years flows close inshore. The subtropical component includes species which are common in southern Mocambique and Natal but do not reach the tropics. This component might be included either with the SouthAfrican endemics or with a more broadly defined tropical component, but until more is known of their

distribution it is advisable to keep these species separate as a group whose distribution may be centred on Mocambique, Madagascar and Natal. It may be noted that a number of estuarine invertebrates have the same range.

The cosmopolitan component includes *Mugil cephalus*, *Pomatomus saltatrix* and *Syngnathus acus* which extend all around southern Africa. *Argyrosomus hololepidotus*, included here in the endemic component, is known also from Australia; possibly it represents a southern hemisphere component. The South African endemics are defined as those species which are confined to South Africa and do not reach Mocambique. They are most common in the southern Cape and they form 50 % of the fishes in the Atlantic coast estuaries which have a very poor fauna. There are no species restricted to the Atlantic coast estuaries of southern Africa. In Day (1974) it was suggested that the sharp differences between the temperature of the cold Benguela current in summer and the warm temperature of the estuaries acted as a barrier to the migration of all but eurythermic species.

Considering the distribution of all components, there is no doubt that the estuaries of southern Africa may be grouped into three faunistic provinces. These are, a subtropical province extending from southern Mocambique to the Transkei as far as the Bashee or the Kei River, a warm temperate province extending from there to False Bay (Sandvlei estuary), and a cold temperate province on the Atlantic coast. These three provinces correspond with those of Stephenson (1947) for the intertidal biota. The interesting point is that the summer temperatures in the upper reaches of all these estuaries are 24-28 ° and thus differ very little all around southern Africa. Winter temperatures in the same upper reaches range from about 18 ° in the subtropics to about 11 ° in the southern Cape. Fish migrate into estuaries as the temperatures rise in spring and summer. Apparently it is the temperature of the coastal sea which determines the composition of the estuarine fauna.

Differences between estuaries. The gaps in the distribution of the 59 common species shown in Appendix 12.1 as well as records of abundance (not quoted here), show that the nature and density of the fish fauna changes from one estuary to the next. Part of this is undoubtedly due to inequalities of collecting for the number of samples and the nature of the nets used by different workers was not the same. But there are also important environmental differences between the estuaries. In those estuarine systems which include large saline lakes such as Kosi Bay and St Lucia, the long residence time of the estuarine water allows rich plankton swarms to develop and plankton feeders including *Gilchristella aestuarius* and *Thryssa vitrirostris* are abundant. Similar conditions develop in the sheltered waters of Delagoa Bay and Durban Bay and the same plankton feeders occur there. These forage fish as well as shoals of mullet and smaller numbers of other prey attract several types of piscivorous fish whose feeding habits have been described earlier. The breadth and depth of the lakes eliminates, or at least reduces, certain types of predation. Whitfield & Blaber (1978b, 1978c, 1978d), note that in weedy shallows the forage fish are safe from the large, fast swimming piscivorous fish but they are preyed on by other fish and by wading birds such as herons, egrets and by reed cormorants. In the open lake away from the sheltering weed beds, surface shoals are attacked not only by the swift piscivorous fish but also by diving birds, pelicans and white-breasted cormorants.

Many large estuaries have a poor fish fauna. Examples are the Mkomazi, Mzimkulu, Mzimvubu, the Bashee, the Kei and the Keiskamma. All of these carry heavy silt loads during the rainy season. While suspended, the silt reduces light penetration and when deposited it smothers the submerged vegetation and much of the benthic fauna. The productivity falls and the fish population is limited in consequence. The silt load decreases after the summer rains and the whole estuarine flora and fauna improves so that the fish population is richer in winter than in summer which is unusual.

Many of the smaller estuaries have relatively clear water and a rich fauna of small fishes. The Mngazana described by Branch & Grindley (1979) is an example. Gobies, small soles, mullet, *Ambassis, Monodactylus, Gilchristella, Hepsetia breviceps* and juvenile *Rhabdosargus* spp are common and the angling for larger fish such as kob, grunter, ten-pounder and leervis is good. Some of these estuaries are partially blocked by sandspits and large fish seldom enter unless the mouth is sufficiently deep for them to escape at low tide. Many more small estuaries are completely closed during the dry season, particularly those along the south coast of Natal, the Transkei and the Eastern Cape. In such periods fish are prevented from entering or leaving these estuaries.

Blaber (1973a, 1974b) has taken advantage of the fact to determine the size of the captive population of *Rhabdosargus holubi*, its rate of growth and the changes in the rate of mortality. In 1971 the West Kleinemond estuary in the Eastern Cape was closed for the six month period February to July. Weekly netting along the length of the estuary supplemented by marking and recapture data showed that the population of juveniles decreased from 55 360 in February to about 11 485 in July. This gives a mortality varying from 49 % in summer to 9 % in winter but averaging about 30 % per month over the whole six month period before the estuary mouth opened and the fish returned to the sea. In 1972 there was a new stock of 14 674 juvenile *R. holubi* when the estuary closed in April and about 12 000 when it reopened in November. In this year the total mortality over the nine month period was about 18 % and the monthly mortality, which was very low in mid-winter, averaged 2,5 %. No commercial fishing or angling takes place in the estuary and the mortality is almost certainly due to predation. The only piscivorous fish netted were a few leervis *(Lichia amia)*, and those examined were feeding mainly on young mullet and *Gilchristella*. There are also large crabs *(Scylla serrata)* but these are not suspected of catching healthy fish and feed mainly on the abundant small gastropods and bivalves. The mortality of *Rhabdosargus* is mainly due to fish-eating birds, particularly cormorants *(Phalacocorax africanus, P. carbo* and *P. capensis)*, the darter *(Anhinga rufa)* and the heron *(Ardea cinerea)*. The number of fish-eating birds was high in 1971 when the fish population was dense but much lower in 1972 when the fish population was sparse. The birds had moved to the neighbouring Kasouga estuary where the fish population was judged to be as dense as it was in W. Kleinmond during 1971. It would appear that when the fish density is high, mortality due to bird predation in W. Kleinemond averages 30 % per month, but when it is low the birds move to better stocked estuaries and the fish mortality falls.

Two other estimates of juvenile fish mortality in estuaries may be quoted. In Lake Macquarie estuary in New South Wales, Thomson (1959) estimated a long term mortality of *Rhabdosargus sarba* as 78 % but as the estuary was open, part of the decrease in population was thought to be due to emigration. In a Connecticut estuary, Pearcy (1962) recorded a monthly mortality of juvenile flounders as 31 %. The estimates of 80 and 78 % during the total estuarine stage and monthly mortalities of 30 % and 31 % in well-stocked estuaries to which many predators are attracted, and the much lower mortalities in poorly stocked estuaries are useful guides for further research.

The data given by Blaber (1973a, 1974b) may be analysed further to show the biomass changes in a juvenile population from the time it migrates into an estuary until it returns to the sea. The surface area of West Kleinemond is 25,5 ha but must increase somewhat after rains. When the mouth closed in February 1971, 55 360 *Rhabdosargus* with a modal length of 6 cm and a mean mass of 8 g were trapped in the estuary. This gives a biomass of 17,5 kg.ha^{-1} or 1,7 g.m^{-2}. After six months, predation had reduced the number to 11 485 but judging from Blaber's graph the mean length of the fish was then 12 cm and the mass 60 g so that the biomass of the population had increased to 27 g.ha^{-1} or 2,7 g.m^{-2}. Thus, in spite of the heavy predation, the biomass had increased. The mouth of the estuary opened in August and it is presumed that the bulk of the population migrated back to the sea. When the mouth closed again in April 1972 the enclosed population was 14 674 *Rhabdosargus* with a modal length of 6 cm and a mean mass of 8 g giving a biomass of 4,6 kg.ha^{-1} or 0,46 g.m^{-2}. When the mouth of the estuary opened again after nine months, this cohort had been reduced by predation to 12 000 but in the meantime the fish had grown to a mean length of 12 cm and a mass of 60 g which gives a biomass of 28,2 kg.ha^{-1} or 2,8 g.m^{-2}. Thus the lower density in 1972 had reduced bird predation to such an extent that the final biomass of the fish was slightly higher than in 1971.

These are approximate figures and there are other factors to be considered. There are of course many other species of fish in the estuary which would increase the estimate of total fish biomass. Further, the lower initial density of *R. holubi* in 1972 would decrease competition and might result in higher feeding rates and an eventual increase in the rate of growth.

The significance of estuaries to fish. As shown earlier, estuaries are far more productive than the open sea. In very large estuaries such as Chesapeake Bay, the commercial catch of fish is five times that in comparable areas on the continental shelf, and in addition, there are large catches of blue crab *(Callinectes)* and shellfish. Further, the outflow of organic detritus from an estuary supports the rich coastal fisheries at the mouth of the Chesapeake, the Mississippi, the Niger, the Ganges, the Nile and probably all the large estuaries. Day (1975) has described the effect of outpourings of mangrove detritus on the density of plankton near Nosy Bé in Madagascar. The point is that materials derived from estuaries eventually provide food for many fishes which never enter estuaries.

It was also shown earlier that about one-fifth of the 1 500 species of fish recorded from the seas of southern Africa occur in estuaries at some stage in their life cycle. They are derived from the coastal population but by no means all the coastal fishes enter estuaries. Those that live in specialised habitats such as the parrot fish, the chaetodons and the cardinals which live in coral reefs are restricted to the sea. So also are the majority of clinids and sparids which haunt rocky areas. The bulk of the marine fish which do migrate into estuaries are the juveniles of demersal feeders whose adults live in shallow areas along the coast. Little is known of the habits of these adults for they are difficult to sample adequately and much more work is needed.

Adults of species found in Mocambique estuaries are trawled on bottoms of sandy mud in or near Inhambane Bay (Day, 1974) or Delagoa Bay (Macnae & Kalk, 1958). In Natal, surf anglers regularly catch the adults of almost all the predaceous fish found in the estuaries and shoals of mullet are common. According to Wallace (1975a) the adults and juveniles of many estuarine fish have been trawled on sandy mud on the Tugela bank at depths of 30 to 40 m which is below the effect of wave action. Many tropical and subtropical species are also caught by anglers along the Transkei coast and Cape species such as *Lithognathus lithognathus* become more common there. In the southern Cape, angling on sandy shores is limited to sheltered areas like Plettenberg Bay where *Lithognathus, Argyrosomus* and *Lichia amia* are the main species whose juveniles enter estuaries. Somewhat surprisingly, *Rhabdosargus globiceps* whose juveniles feed in estuarine weedbeds change their feeding habits when they return to the sea and feed along the reefs (Talbot, 1955).

In South Africa, commercial fishing in estuaries is prohibited and the catch of estuarine dependent fish in the sea by commercial fishermen is not very important. Kob, Cape stumpnose, white steenbras and mullet are probably the most valuable although no statistics of landings are available. Sport fishing and the tourist attraction it provides are of much greater importance, not only in South Africa but in many other countries. Wallace & van der Elst (1975) quote de Sylva (1969) as stating that while the commercial fishery for estuarine dependent species in the United States was worth $75 million in 1965, the sport fishery for these species was worth over $331 million. Angling in South Africa is one of the major sports and is said to attract more enthusiasts than rugby. Thousands of anglers and their families visit coastal resorts every year. Van der Elst (1979) provides statistics that show how the number of anglers at St Lucia has increased over the years. Judged by the accommodation required at St Lucia, the number has increased from 50 000 in 1966 to 300 000 in 1976; judged by the value of bait sales about R18 000 was spent in 1966 and R63 000 was spent in 1975. During the Easter week-end, 400 anglers were recorded over a distance of 500 m near St Lucia mouth alone. St Lucia is an angler's paradise but it is only one estuary. The value of all estuaries for tourism has not been estimated but it must be very great.

South Africa and Namibia are relatively arid countries with few estuaries and many of them have been damaged by erosion and heavy siltation since the turn of the century. Others are closed by sandbars for long periods. St Lucia which is the largest estuarine system in southern Africa, has been damaged by both of these causes. More recently, the fresh water in the rivers entering Lake St Lucia has been so reduced by irrigation projects that the salinity in the upper reaches rose to over 100 ‰ in 1971 and the fish moved out of the lakes. In Durban Bay the feeding banks have been destroyed by harbour development and the water in the bay is sometimes so polluted that fish die by the million. Richards Bay has now been converted into a deep water harbour and the part cut off as a sanctuary has been so filled with silt that its value as a nursery

Acanthopagrus berda – river bream (Photo: George Begg)

for juvenile fish and prawns is insignificant. Langebaan Lagoon is already in danger of pollution from the ore loading plant in Saldanha Bay. Many of the smaller estuaries are now throttled by the solid embankments of bridges. It is urgent that future planning take into account the great economic value of the few estuaries that remain unspoilt.

BIOLOGICAL NOTES ON INDIVIDUAL SPECIES

Too little is known of the habits of the estuarine fishes to give a useful account of each and every one. Nor is there the space. The biological notes that follow are thumbnail sketches of the common species. The notes vary in length although all are condensed. Many are very brief indicating how little we know about even the common species. There is much to be learnt and many intriguing problems to be solved. From the evidence presented here it will be evident that the great majority of estuarine fishes tolerate salinities between 10 and 40 ‰ and some have an even wider salinity range. Thus a reduction of salinity is not an important deterrent to migration from the sea into an estuary. Gunter (1967) states that juveniles actively migrate into the mouth of the Mississippi against the outflowing current of estuarine water on the surface and are not passively carried in by the bottom current of sea water. This suggests that juveniles may be guided into the mouth of an estuary by the salinity gradient or at least by the gradient of some substance whose concentration changes from estuaries to the sea. Apparently the migration is not related to a change in diet for most juveniles continue to feed on plankton after they have entered

estuaries. Again, when the juveniles are approaching sexual maturity they migrate back to the sea. The intriguing problem is how they find their way. In linear estuaries where there is a gradual increase of salinity towards the mouth and an increase in tidal effects, these may act as guides. But the same migration occurs in saline lakes such as St Lucia Lake where there is no tidal action and the salinity gradient may either be seaward or towards the head of the estuary.

The material in the notes that follow has been culled from many sources both published and unpublished. To prevent constant repetition, only a few selected references are given and most of these have been quoted earlier. The detailed distribution in estuaries is left to the Appendix and it is emphasised that distribution in the sea is not necessarily the same. The location of all the estuaries mentioned is shown in Figure 14.3 on p 256 and the main characteristics of each are summarised in chapter 14.

Acanthopagrus berda (Försk.); River Bream. A tropical species recorded from Morrumbene estuary to the Swartkops but rare south of the Transkei. Breeds in shallow seas near the mouths of estuaries from May to August. Enters estuaries from July to December at a length of 1 to 5 cm but mainly over 3 cm; grows 10-12 mm per month and the bulk of the catch measures 4-16 cm long during February and March. Matures at 22 cm and spent adults up to 36 cm occur in estuaries. Recorded salinity range 0-72 ‰. From 2-6 cm the young stages feed on zooplankton, many amphipods, chironomid larvae, tanaids and small crabs. From 6-

Argyrosomus hololepidotus – kob (Photo: M.M. Smith)

12 cm many amphipods are still taken but the main food is bivalves, particularly *Lamya* also the siphons of *Solen*, a few gastropods *(Natica)*, gobies and weed. Larger fish take crabs, small fish and weed (? *Ruppia*) and adults feed on crabs, prawns, barnacles *(Balanus amphitrite)*, and a variety of bivalves and small fish and the stomachs are occasionally packed with cropped *Zostera* covered with epiphytes *(cf Rhabdosargus holubi)*. Van der Elst (1979) states that the catch of *A. berda* is declining.

Ambassis natalensis G. & T.; Slender Glassy. A tropical species common from Morrumbene estuary to the Bashee. Attains 9 cm and matures at 4,5 cm and spawns in estuaries from August to November. Feeds on zooplankton snapping up individual copepods, decapod larvae, fish larvae and amphipods; common in weed beds. Recorded salinity range 3,0-42,2 ‰. Both *A. natalensis* and *A. commersoni* are preyed on by *Caranx ignobilis*.

Amblyrhynchotes honckeni (Bloch); Evileyed Blaasop. A tropical and widespread species common on both rocky shores and in estuaries; recorded from Morrumbene estuary to Hermanus Lagoon, often appearing in cold water when other fish are scarce. Attains 15,5 cm but usually 6-10 cm. Matures at 8 cm and is nearly ripe in June and July. Feeds on small crustacea (mainly sphaeromid isopods) and crabs *(Hymenosoma)* and also takes bivalves such as *Lamya* and some *Zostera*. Flesh poisonous.

Argyrosomus hololepidotus (Lacep.); Kob. Known from the Atlantic, Mediterranean, Walvis Bay to Zululand and Australia. Recorded in estuaries from Kosi Bay to Heuningnes near Cape Agulhas. Attains 180 cm (75 kg) and breeds in the sea probably on the Agulhas Bank off Mossel Bay and matures at about 70 cm. Spawning not recorded but spent fish have been taken in Natal from June to October. Juveniles enter estuaries in summer at a length of 10 cm and larger sizes follow. In estuaries juveniles have undeveloped roes and adults are either spent or inactive. Juveniles feed mainly on shrimps, penaeid prawns, burrowing prawns, crabs and gobies; adults are bottom feeding piscivorous fish feeding on *Thryssa vitriristris*, *Gilchristella*, fingerling mullet and squid as they are not fast enough to catch adult mullet. Scuba divers have observed that when swift predators are attacking mullet shoals, the kob swims along the bottom snapping up damaged fish. In the sea there is evidence of Atlantic coast fish migrating south to the Cape in winter and on the east coast young kob move to Natal in the autumn and return to the southern Cape in summer. Salinity range 8-60 ‰; dead fish were found in St Lucia in higher salinities.

Arothron immaculatus (Bloch); Blackedged Blaasop. A tropical species common from Morrumbene estuary to the Transkei but only seen in small numbers from there on to the Sundays estuary. Found in estuaries from a length of 2 cm and attains 16 cm; mature at 12 cm and females with ripening ovaries found in estuaries in October but breeding area uncertain. Feeds on copepods at 2 cm but larger sizes feed mainly on bivalves particularly *Lamya* but also take a few *Assiminea*, *Dosinia* and crabs; cropped *Zostera* sometimes seen. Flesh poisonous. Recorded salinity range 8,1-37,6 ‰.

Caranx ignobilis (Försk.); Giant Kingfish. A tropical Indo-Pacific species and adults are taken along the coasts of Mocambique and Natal. Juveniles recorded in estuaries from Morrumbene to the Mzimvubu; both adults and juveniles present throughout the Kosi Bay system but those recorded in St Lucia between 4 and 45 cm long were all immature and were feeding mainly on *Thryssa*, gobies and penaeid prawns. Breeds in the sea and specimens up to a length of 116 cm are common just beyond the surf zone feeding on mullet. Recorded salinity range 3,0-37,7 ‰.

Caranx sexfasciatus Q. & G.; Bigeye Kingfish. A tropical Indo-Pacific species, recorded just beyond the surf zone of Mocambique and Natal, feeding on mullet. Juveniles in estuaries from Morrumbene to Kei mouth entering at a length of 4-8 cm and reaching 46 cm. They mature at about 35 cm and breed in the sea. From 4-14 cm they feed on zooplankton and fish larvae; larger ones take penaeid prawns and fish including gobies. Salinity range uncertain.

Carcharinus leucas (M. & H.); Zambesi Shark or Bull Shark. Distribution circumtropical, also common

Caranx ignobilis – kingfish (Photo: J. Ballard)

along the coasts of Mocambique and Natal, a few specimens even reaching Swartkops estuary. They often patrol the mouths of estuaries when floods make the water muddy and are then extremely dangerous to bathers. According to Bass *et al* (1973) and Bass (1976) this shark matures at 225 cm and reaches 300 cm. Mating occurs in coastal waters north of St Lucia and gravid females deposit about 12 young, each measuring 60-70 cm in the mouth of St Lucia and other estuaries. Juveniles feed in the estuary, occasionally visit the sea or ascend into fresh water. Bass (1976) reports that they grow very slowly and may take 20 years or more to reach maturity. Adults found only in the sea. Juveniles in St Lucia occur in salinities up to 47 ‰ and one was taken in 53 ‰. Sharks of 90-120 cm found mainly in the sea but smaller juveniles and larger adolescents move all over the estuaries. Bass *et al* (1973) reports sharks 1 120 km up the Zambesi. They feed on a wide variety of fish (mainly mullets), also penaeid prawns, when plentiful, large crabs *(Scylla serrata)* and occasionally mammals.

Clinus superciliosus (Linn.); Klipfish. An endemic species which occurs on rocky seashores from Swakopmund to the Kei mouth and is recorded in estuaries from the Bushman's River to Olifants River, ranging from 4,5 to 12 cm long; mainly found among stones and in weedbeds. It copulates and breeds in estuaries as well as in the sea. Matures at a length of 6,5 cm and embryos are present in the uteri from March to May. Feeds mainly on amphipods (particularly Talitridae and *Melita zeylanica*), isopods including *Paridotea* and *Exosphaeroma,* chironomid larvae, *Assiminea* and the crab *Hymenosoma;* large adults take gobies and hermit crabs as well.

Caffrogobius (olim Gobius) caffer (Gunther); Barehead Goby. Very like *C. nudiceps.* An endemic species common in rock pools on the seashore from Natal to the Atlantic coast and recorded in estuaries from St Lucia to the Olifants River, particularly in weedbeds, sometimes sharing the burrows of *Alpheus crassimanus.* Length up to 15 cm but ranging from 1,6

to 11,6 cm in estuaries. Males mature at 5,5 cm and females at 6,0 cm. Breeds mainly in winter both in the sea and in estuaries, attaching the eggs to shells and stones and guarding them. Juveniles up to 3 cm take zooplankton and later feed mainly on small crustacea including ostracods, tanaids, amphipods and isopods; adults feed on small crabs *(Cleistostoma* and *Hymenosoma), Upogebia* and amphipods. They are preyed on by many demersal fishes and wading birds. Replaced by other gobies in low salinities.

Oligolepis acutipennis (Val.), Sharptail Goby. Tail long and pointed. An Indo-Pacific species recorded only in estuaries from Morrumbene to Mngazana. Common in muddy shallows feeding on worms and small crustacea. Breeds in estuaries and guards its eggs. Recorded salinity range 9,1-35,2 ‰.

Diplodus sargus Linn.; Blacktail. A tropical species recorded all around southern Africa on rocky shores. Juveniles recorded in estuaries from Kosi Bay to Hermanus Lagoon. Attains a length of 30 cm but usually 6-20 cm in estuaries. Becomes mature at about 15 cm; ripening eggs were recorded in September but in February all gonads were inactive. Breeds in the sea and enters estuaries at 2-3 cm. Haunts rock ledges and weedbeds and often shoals with juveniles of *Rhabdosargus.* From 2-5 cm it feeds on zooplankton but then changes to polychaetes, amphipods, isopods, barnacles, small limpets, bivalve siphons and other members of the epifauna; some specimens have their stomachs packed with *Zostera* or filamentous algae. Biology of the adult not known. Salinity range uncertain, possibly 20-37 ‰.

Eleotris fusca (Bloch); Dusky Sleeper. A tropical Indo-Pacific species reaching the Transkei. Attains 15 cm but usually less than 10 cm long. Common in estuaries from Morrumbene to the Bashee River. An inactive fish found in weedbeds and under loose stones; probably omnivorous and breeds in estuaries.

Carcharhinus leucas – Zambesi shark (Photo: N.K. Nayogel)

Gilchristella aestuarius – whitebait (Photo: M.M. Smith)

Elops machnata (Försk.); Tenpounder. An Indian tropical species recorded in estuaries from Kosi Bay to Breede River. Often taken in the upper reaches of estuaries and tolerates salinities from 3,0-110 ‰. Spawns in the sea but area and season unknown. Attains 105 cm but size at maturity unknown. Juveniles enter St Lucia at 5-10 cm; larger fish abundant from April to October but roes not mature. A piscivorous feeder preying mainly on *Thryssa vitrirostris* and *Gilchristella aestuarius,* but also takes *Hyporamphus knysnaensis, Crenidens crenidens, Leiognathus equulus,* a few small mullet, gobies and penaeid prawns. No obvious seasonal migration observed.

Gerres acinaces Blkr.; Smallscale Pursemouth. A tropical Indo-Pacific species common in estuaries from Morrumbene to Mnganzana with a few extending on to Knysna. Juveniles of 4-12 cm are abundant but adults of 16-25 cm are rare. Breeds in the sea and developing roes found in specimens in Durban Bay in October and January. Juveniles of 4,5-6 cm feed on planktonic copepods and polychaete worms. As the protuberant jaws develop it feeds mainly on polychaetes and amphipods dug from the sand banks and specimens longer than 15 cm feed on small bivalves, polychaetes and gastropods *(Assiminea)* and an occasional *Upogebia.* Recorded salinity range 3-35 ‰. *Gerres rappi* and

three other species are also common in Natal estuaries.

Gilchristella aestuarius (G. & T.); Estuarine Round Herring. Recorded from lagoons, estuaries and fresh water from India and abundant from Delagoa Bay around the Cape to the Olifants estuary. Attains 7,5 cm and matures at 4,5 cm within one year. Swims in shoals and spawns in estuaries from October to December both in fresh water and in salinities up to 47,5 ‰. Larvae abundant in Swartkops estuary in summer (Melville-Smith & Baird, 1979). Recorded in salinities up to 52,6 ‰. Feeds on copepods, ostracods, mysids, amphipods and chironomid larvae but is a non-selective filter feeder. It is the main prey of *Elops machnata* and juvenile *Argyrosomus hololepidotus* but is also taken by many other piscovorous fish including *Hilsa kelee* and by birds.

Glossogobius giuris (Hamilton-Buchanan); Tank Goby. Snout elongated. A tropical Indo-Pacific species reaching Swartkops estuary. Occurs in estuaries and lagoons, and extends into fresh water. Abundant in southern Mocambique and Zululand particularly in muddy shallows and weedbeds. Reported to reach 50 cm in tropical waters but seldom more than 10 cm in Zululand. Matures at 5,5 cm and breeds in estuaries, a few ripe females being reported in January. Feeds on amphipods particularly *Melita zeylanica,* tanaids, chironomid larvae and smaller gobies. Recorded salinity range 0-42,2 ‰.

Lichia amia – leervis .

Hepsetia (olim Atherina) breviceps (Cuv.); Cape Silversides. A South African endemic present from St Lucia to Port Nolloth; not common in Zululand but becomes abundant in lagoons, sheltered bays and estuaries in the Cape, often in low salinities. Recorded salinity range 7,6-42,2 ‰. Attains 15 cm but usually 4-7 cm. Matures at 4,5 cm and breeds in estuaries. A zooplankton feeder swimming in shoals. Preyed on by many piscivorous fish and birds.

Hilsa kelee (Cuv.); Kelee Shad. A tropical Indo-Pacific species common in Delagoa Bay, St Lucia, Richards Bay and Durban Bay and just reaches the Transkei. Blaber (1979) has studied its biology in St Lucia. Attains 35 cm and matures at 15 cm; swims in shoals in open waters. Spawns in shallow seas from September to February and enters estuaries in late summer. Range in estuaries 3-17 cm. A non-selective plankton feeder, the stomach containing diatoms, veligers and copepods. Recorded salinity range 3,0-35,0 ‰. Preyed on mainly by *Otolithes ruber* and large carangids.

Lichia amia (Linn.); Leervis. Known from north-west Africa, rare from Walvis Bay to Table Bay but common from False Bay to Zululand. Both juveniles and adults recorded in estuaries from St Lucia to Sand-vlei (Muizenberg). Juveniles enter estuaries in summer when 4-8 cm long and Blaber (1974b) has shown that they grow from 9 to 20 cm between January and July. They are swift predators from the time they enter estuaries and feed on shrimps, penaeid prawns, juvenile mullet, *Hepsetia*, *Ambassis* and many other small fishes. Juveniles lurk under floating *Enteromorpha* mats and dash out at passing shoals but adults hunt in open waters. They tolerate low salinities of 10 ‰ and are common in the upper reaches of estuaries but the majority leave for the sea in autumn. According to Biden (1954) they probably spawn in coastal seas between Port Elizabeth and Cape Agulhas in December and January but fish with very large gonads reported in Plettenberg Bay in winter. Length at maturity about 55 cm. Adult fish up to a maximum of 25 kg and a length of 170 cm enter estuaries in the Cape after spawning but others migrate up the coast to Natal in May arriving in June with the sardine shoals and returning to the Cape in October and November.

Hyporhamphus knysnaensis (Smith); Knysna Half-beak. A South African endemic species extending from Natal to the southern Cape, and recorded in estuaries from Kosi Bay to Hermanus Lagoon. Attains 25 cm and matures at 9 cm; breeds in estuaries from October to January and juveniles recorded from November to March. Coetzee (1979) reports the volume of food organisms in the stomachs of fish caught in Rondevlei as 55,8 % amphipods and isopods, 20,9 % plants (mainly *Ruppia*), 12 % *Musculus virgiliae* and a few percent of other foods. Often seen leaping on the surface when chased by *Lichia amia* in Knysna Lagoon or by *Tylosurus leiurus* in Lake St Lucia. Recorded salinity range 5,0-42,2 ‰.

Johnius belengerii (Cuv.); Minikob. A tropical Indo-Pacific species common in Delagoa Bay, St Lucia and Richards Bay and reaches the Keiskamma estuary. Attains 27 cm and is trawled on the Tugela Bank in 30-40 m. Males mature at about 12 cm and the largest male recorded was 16,2 cm long; females mature at 15 cm and all large fish in estuaries are females up to 24,2 cm long, suggesting sex reversal. Breeding occurs in St Lucia and Richards Bay from September to February. It is demersal, living on shallow muddy bottoms and feeds mainly on crabs *(Hymenosoma)*, gobies and *Gilchristella* but also takes amphipods, mysids, bivalves and polychaetes. Recorded salinity range 3,0-55 ‰.

Leiognathus equulus (Försk.); Slimy. A tropical Indo-Pacific species extending to East London. Recorded in estuaries from Morrumbene to the Bashee River. Attains 30 cm, matures at 18 cm and breeds in the sea from September to April. Juveniles enter estuaries as small as 1,6 cm but mainly at 4-6 cm and

Lithognathus lithognathus (Photo: J. Wallace)

commonly reach 14 cm. The main food is zooplankton but they also take many benthic animals including amphipods, polychaetes and bivalves. Recorded salinity range 3,0-35 ‰. Preyed on by *Muraenesox bagio.*

Lethrinus nebulosus (Försk.); Blue Emperor. A tropical Indo-Pacific species common along the coasts of Mocambique and Zululand but rare further south and only strays reach East London. Maximum length 70 cm but usually 30-50 cm. Size at maturity unknown. Breeds in the sea and juveniles of 6-10 cm are abundant in Morrumbene estuary, Delagoa Bay and the mouth of Kosi Bay, but only a few reach Durban Bay suggesting the breeding area is further north. Juveniles feed on polychaetes, bivalves and small crustacea. Adults caught commercially off Inhambane Bay and Delagoa Bay.

Lithognathus lithognathus (Cuv.); White Steenbras. An endemic South African species extending from Natal to Namibia although the exact range on that coast is confused by the presence of *L. aureti.* Shoals are reported by Biden (1930, 1954) to congregate in warm sandy shallows of the southern Cape during summer and then migrate northwards up both the Atlantic and Indian ocean coasts in autumn and return in early summer. As shown by Mehl (1974) this species is a hermaphrodite reaching maturity in about the seventh year at about 50 cm long and may attain a maximum length of 180 cm and a weight of 22 kg. In estuaries the gonads are either immature or inactive. White Steenbras breed in shallow seas probably in July and August and spent males occur in September. Juveniles enter Cape estuaries with the spring rains at a length of 4 cm or less. Blaber (1974b) reports that they grow about 6 cm in the first year mainly during September to December and 8 cm in the second year and at the

end of two years they are about 20 cm long. Mehl (1974) tables length against age suggesting that fish (80-84 cm are 15 years old but the number of specim measured is small. Juveniles of 4-6 cm feed mainly on copepods; larger juveniles contain algae such as *Ulva* and *Enteromorpha,* many polychaetes including capitellids, spionids, nereids and *Ficopomatus,* amphipods, chironomid larvae and small bivalves. Large fish have a well-developed pharynx and a protrusible snout and, standing on their heads, blow down vertically into the sand to uncover or blow out burrowing prawns particularly *Upogebia* and *Callianassa* as well as the crab *Hymenosoma,* bivalves such as *Loripes* and *Tellina* and worms as large as *Arenicola.* Along sandy seashores they feed mainly on *Donax serra* and the crab *Ovalipes.* The White Steenbras is one of the prime gamefish in Cape estuaries. Mehl investigated the physiology and reports that they can live for a week in fresh water; Millard & Scott (1954) reported a salinity range of 1,8-57,8 ‰.

Lutjanus argentimaculatus (Försk.); River Snapper. A tropical Indo-Pacific species common in Mocambique and extending in decreasing numbers to the Transkei. Breeds in the sea and attains 90 cm but seldom more than 40 cm in estuaries. Juveniles reported from Morrumbene estuary to the Mngazana, mainly in rocky areas. An invertebrate predator commonly taking crabs and prawns.

Lutjanus fulviflamma (Försk.); Dory Snapper. A tropical Indo-Pacific species abundant in estuaries from Morrumbene to the Transkei with occasional specimens reaching Keiskamma estuary. Attains 30 cm and breeds in the sea probably among inshore reefs. Size at maturity unknown. Juveniles in estuaries are commonly between 5 and 16 cm and feed mainly on shrimps, small fish, crabs such as *Hymenosoma* and also on amphipods, and polychaete worms. It has not been

Monodactylus argenteus – moony (Photo: J. Wallace)

recorded from low salinities.

Monodactylus argenteus (Linn.); Natal Moony. A tropical Indo-Pacific species extending along rocky shores to Pondoland but rare in the Cape. Common in estuaries from Morrumbene to the Mtamvuna on the Natal south coast and a few reach the Breede River in the Cape. Breeds in the sea and matures at a length of 13 cm. Juveniles in estuaries feed mainly on zooplankton but are not filter feeders; they snap up individual copepods, amphipods, some filamentous algae and *Zostera.* According to Smith (1965) it enters fresh water.

Monodactylus falciformis Lacep.; Cape Moony. An Indo-Pacific species common along Cape coasts. Recorded in estuaries from Kosi Bay to Heuningnes near Cape Agulhas, extending into low salinities and even tolerating direct transfer to fresh water without ill effects. Breeding habits unknown; immature specimens ranging from 2,3-11,7 cm common in estuaries. From 2-7 cm it feeds on individual copepods and chrionomid larvae; larger ones feed mainly on amphipods and chironomids with occasional mysids, shrimps and a little *Zostera.*

Mugil cephalus Linn.; Flathead Mullet. A cosmopolitan species abundant in all temperate and tropical waters. Recorded in the estuaries of southern Africa from Kosi Bay to the Orange River. Attains 67 cm and breeds in the sea in inshore waters from May to September. Males mature at 45 cm and females at 49 cm. Some juveniles enter estuaries at 1-2 cm but most at 2,5-4 cm during June to October and spread along the estuary keeping mainly to the shallows and growing at 1,7-1,9 mm per month. They are the most abundant

mullet in St Lucia and remain there three or four years. Gonads start ripening in February and the fish congregate at the mouth in a prespawning condition and spawn in the breakers outside, possibly requiring turbulence to keep the eggs suspended. If trapped in the estuary by a sandbar at the mouth, the eggs are resorbed. Few spent fish re-enter the estuary. Blaber (1976) reports that within St Lucia their main food is the small gastropod *Assiminea bifasciata,* large foraminifera *(Rotalia* and *Quinqueloculina),* the centric diatom *Actinoptychus,* smaller diatoms and detritus and that they ingest sand grains with a mean size of 225 μ to grind the food in the gizzard. They feed mainly in the evening and continue at a slower rate at night. The main predators of the adults are the shark *Carcharinus leucas, Lichia amia, Caranx* spp and *Sphyraena* spp. Juveniles are preyed on by *Argyrosomus hololepidotus* and *Tylosurus* spp. Most of the mullet leave estuaries in September. Recorded salinity range 5,2-70 ‰; dead specimens were found in salinities of 75,6-80,0 ‰ and none were found alive in higher salinities.

Mugil (Liza) dumerili Steindachner; Groovy Mullet. A subtropical species extending from southern Mocambique to Mossel Bay and recorded in estuaries from Morrumbene to the Breede River, becoming more abundant in the Eastern Cape. Attains 36 cm, becomes mature at 20-28 cm and breeds in the sea near the mouths of estuaries from June to November. A few enter estuaries at a length of 2-5 cm during August and September but the major recruitment is at 5-8 cm from January to June. The species is abundant in St Lucia from July to November and according to Whitfield (1976) it is the third most abundant mullet forming 18,9 % of the mullet population. According to Blaber (1976) they feed mainly in the morning on the gastropod *Assiminea bifasciata,* foraminifera, the large diatom *Actinoptychus splendens* as well as smaller diatoms and detritus including sand grains with a mean diameter of 300 μ. Marais & Baird (1979) discuss the diet in the Swartkops estuary. Recorded salinity range is 3,0-52,6 ‰.

Mugil (Liza) macrolepis (A. Smith); Large-scale Mullet. A tropical Indo-Pacific species abundant along the coasts of Mocambique and Natal as far as Durban but in smaller numbers from there on to Port Alfred. Recorded in estuaries from Morrumbene to the Bushmans River and possibly extends to the Breede River. Attains a length of 35 cm and becomes mature at 20-28 cm. It breeds in the sea near the mouths of estuaries from May to November. Juveniles enter estuaries between 1 and 4 cm long; a few appear in May but most of them arrive between July and December and eventually become the second most abundant mullet in St Lucia forming 22,5 % of the mullet population. They feed mainly on small diatoms, and filamentous algae and take smaller amounts of foraminifera and the gastropod *Assiminea bifasciata.* Adults are preyed on by fast piscivorous fish including *Carcharinus, Lichia, Caranx* spp. and *Caranx.* Juveniles are preyed on by *Tylosurus* spp and to a lesser extent by *Argyrosomus* and *Elops.* The recorded salinity range is 0-72 ‰.

Mugil (Liza) richardsoni A. Smith; Southern Mullet. Very close to *Liza ramada* of Europe. An endemic species abundant on the Atlantic and southern coasts of the Cape but rare in Natal. Recorded in estuaries from the Mgazi near St Johns to the Olifants River. Attains 40 cm and is mature at about 25 cm. It breeds in the sea in summer and enters estuaries in February to July at a length of 2,5 cm or less. Marais & Baird (1979) report on the diet and feeding activities in the Swartkops estuary; the gut contains organic detritus and diatoms on sand grains. Recorded salinity range 5,6-59,4 ‰.

Mugil (Valamugil) cunnesius (Val.); Longarm Mullet. An Indo-Pacific species common in Zululand and reaching Durban in small numbers. Recorded in estuaries and sheltered bays from Delagoa Bay to Umtentwini River in southern Natal. Attains 29 cm and matures at about 22 cm, breeding in the sea near the mouths of estuaries. Both ripe and spent individuals are found in the mouth of St Lucia from November to June. Juveniles enter estuaries at a length of 2-4 cm, feeding on zooplankton. Adults feed mainly on diatoms, foraminifera and organic detritus and ingest sand grains with a mean diameter of 150 μ indicating that they feed on more muddy bottoms than other mullet. Recorded salinity range 5,9-65 ‰.

Mugil (Valamugil) buchanani Blkr.; Bluetail Mullet.

An Indo-Pacific species abundant along the shores of Mocambique and common as far as Durban, with a few extending on to Knysna. Only juveniles are recorded in estuaries from Morrumbene to Xora estuary in the Transkei. Attains 100 cm and matures at about 30 cm, spawning in the sea probably in spring to early summer. Juveniles enter estuaries mainly from February to July at a length of 2-5 cm. The main food is diatoms, *Ruppia* and foraminifera and the mean size of sand grains ingested is 200 μ. Recorded salinity range 20-55 ‰.

Mugil (Valamugil) seheli Försk.; Blue-spot Mullet. An abundant Indo-Pacific species in Mocambique but fewer in Natal. Recorded in estuaries and lagoons from Morrumbene to Durban Bay. Attains at least 45 cm. Breeding unknown. The main food is diatoms, foraminifera blue-green and filamentous algae and the mean size of sand grains ingested is 200 μ.

Myxus capensis (Smith); Freshwater Mullet. An endemic South African species extending from Kosi Bay to Knysna. Breeds in the sea, becomes fairly common in the upper reaches of many estuaries and extends into fresh water. Attains 36 cm (Crass, 1964), but size at maturity and breeding season unknown. Juveniles enter estuaries at a length of 1-2 cm while still feeding on zooplankton. Specimens over 3 cm feed mainly on organic detritus, pennate diatoms, foraminifera, filamentous algae and ingest fine to medium sand. Flying insects are taken from the water surface. *M. capensis* is the only mullet often caught by anglers; winged termites are the preferred bait although diced fish flesh and dough are commonly used. Recorded salinity range 0-34,5 ‰.

Pelates quadrilineatus (Bloch); Trumpeter. An Indo-Pacific species reaching the coast of Pondoland. Common in estuaries from Morrumbene to the Mlalazi and a few extending on to Bushmans River. Attains 20 cm and breeds in the sea but the breeding season and size at maturity unknown. Juveniles common in weedbeds in Morrumbene; omnivorous, cropping *Zostera* and taking small crustacea particularly amphipods and isopods.

Periophthalmus 'sobrinus' Eggert; Mudskipper. Classification doubtful; Smith (1965) refers the South African specimens to *P. cantonensis* Osbeck but Macnae (1968) who has studied many species in the mangroves of the Indo-Pacific, states that there are two species, *P. sobrinus* and *P. kalolo* in Mocambique which differ in the structure of their burrows. One or possibly both species extend from Morrumbene to Mngazana in the Transkei. Mudskippers spend much time out of water often climbing on mangrove roots, holding on by the pelvic sucker and grasping with their pectorals. They feed on insects, small crabs and amphipods. They breed in estuaries, the male digging a hole in the mud to which the female is attracted by a courtship display; the eggs are attached to the walls of the burrow and guarded. Mudskippers tolerate a wide range of salinity

Pomadasys commersonni – spotted grunter (Photo: George Begg)

and extend along the estuary wherever there are mangroves. Dr van Dyk has kept them for weeks in fresh water tanks (personal communication).

Platycephalus indicus (Linn.); Bartailed Flathead. An Indo-Pacific species, common along the Natal coast but rare further south although it reaches Mossel Bay. Recorded in estuaries from Morrumbene to Swartkops estuary. Attains 100 cm; matures at 42 cm and breeds in the sea from July to November. Occurs in estuaries from a length of 8-12 cm upwards and mainly found on muddy bottoms. A general predator feeding on shrimps, penaeid prawns, crabs (particularly *Hymenosoma*), and small fishes such as *Johnius belengeri, Solea bleckeri, Thryssa vitrirostris* and gobies. Recorded salinity range 8,3-55,0 ‰.

Pomadasys hasta Bloch; Javelin Grunter. A tropical Indo-Pacific species reaching the Transkei. Common in estuaries from Morrumbene to Mngazana. Attains 45 cm and breeds in the sea in June but size at maturity unknown. Juveniles 3-4 cm long enter estuaries in March and specimens up to 14 cm are common until November; smaller specimens up to 10 cm long feed mainly on mysids and other small crustacea but the larger sizes take many gobies. Salinity range 14-42 ‰.

Pomadasys commersonni (Lacep.); Spotted Grunter. This is an Indian Ocean species extending from the tropics to temperate waters. It is common in Natal and the Transkei and is the premier game fish in many estuaries; during spring the 'grunter run' in St Lucia attracts thousands of anglers. *P. commersonni* also forms the bulk of anglers' catches in Swartkops estuary. It extends to Southern Cape estuaries and odd specimens have been taken in Hermanus Lagoon and False Bay. It attains 92 cm and a weight of 10 kg but is commonly 2-4 kg. Males become mature at 26-28 cm

and females at 30-35 cm and breed in shallow seas from August to December. Juveniles 2-4 cm long enter estuaries from September to December and grow at the rate or 1,2 to 1,47 cm per month during the summer but more slowly in winter and at the end of the year they are about 20 cm long but still immature. At this stage they return to the sea. Young adults in a spent condition return to the estuaries but only during the summer months in Cape estuaries. At a length of 2-4 cm the juveniles are still feeding on zooplankton particularly on copepods but between 5 and 10 cm they take benthic animals such as polychaete worms, bivalves, crabs and shrimps. Larger fish take the same foods but in addition they blow holes in the sand to uncover larger prey including *Upogebia* and *Callianassa* and even the deep-burrowing *Solen*. Surprisingly few penaeid prawns are found in their stomachs although they take them voraciously when offered as bait. Recorded salinity range 0-74 ‰.

Pomadasys olivaceum Day; Piggy. An Indo-Pacific species extending to False Bay. More common in Cape than in Natal estuaries and extends as far as Hermanus Lagoon. Attains 30 cm, matures at about 15 cm and breeds in the sea probably in autumn. Juveniles found in rock pools and estuaries range from 6-15 cm long and are present in all months of the year. They feed mainly on amphipods with fewer polychaetes, chironomid larvae, amphipods, small crabs and *Upogebia*. *P. olivaceum* is an important forage fish in the Cape preyed on by elf, leervis and kob.

Pomatomus saltatrix (Linn.); Elf. This species which is cosmopolitan in temperate and tropical seas, has been recorded in estuaries from St Lucia to the Olifants

Pomatomus saltatrix – elf (Photo: C.L. Biden)

Rhabdosargus globiceps – Cape stumpnose (Photo: C.L. Biden)

River. Biden (1954) reports that it attains 105 cm and a weight of 18 kg on the Atlantic coast but only 9 kg off Natal coasts. It is a swift predaceous fish travelling in shoals along the Natal coast from August to November and along Cape coasts from September to March. It breeds in the sea and Biden suggests that it spawns in the Cape in December while van der Elst (1974) reports that it spawns in Natal from September to November although a ripe-running female 33 cm long was taken at Durban in January. Juveniles 2-3 cm long appear in the mouths of Natal estuaries and larger fish of about 15-20 cm are taken in the surf during May, June and July while adults are taken in August. The season in the Cape is from August to March. Shoals of adult fish move in and out of estuary mouths with the tide feeding on small mullet, *Ambassis, Gilchristella* and *P. olivaceum.* Juveniles of 15-20 cm extend all

over the estuaries presumably remaining there until the main shoals leave at the end of summer. Recorded salinity range 16,6-34,8 ‰. Van der Elst (1974) states that the Natal population is being seriously depleted by anglers and beach seines.

Psammogobius knysnaensis Smith; Knysna Sandgoby. An endemic South African species confined to lagoons and estuaries. Recorded from St Lucia to the Olifants River; rare in Natal but abundant in the Cape particularly in sandy shallows. Attains 6,5 cm but usually 3-5 cm. Size at maturity unknown. Breeds in estuaries and uses crab and prawn holes as temporary refuges. Feeds on worms, ostracods, amphipods, isopods, copepods and other small crustacea as well as chironomid larvae. Often found in the stomachs of demersal fish and wading birds.

Rhabdosargus globiceps (Cuv.); White Stumpnose. An endemic South African species which is rare in

215

Natal but becomes abundant in the southern Cape and extends along the Atlantic coast to Namibia (Sandwich Harbour) and is important to both commercial fishermen and anglers. Recorded in estuaries from the Bushmans to the Olifants rivers. Attains 50 cm but seldom more than 30 cm. Females mature in their third year at 22-24 cm and males at 27,5 cm. They breed in shallow seas probably during November and December and females greatly outnumber males. According to Talbot (1955) the juveniles enter estuaries at a length of 2 cm when they are still feeding on zooplankton and there appears to be a lag before they start feeding on amphipods, isopods, ostracods and algae such as *Enteromorpha.* At a length of 4 cm when the crushing molars start to develop they take many molluscs including *Assiminea, Lamya, Haminea* and the siphons of *Solen,* also crabs, *Callianassa* and shrimps. They often feed in weedbeds and pack their stomachs with cropped leaves of *Zostera* or *Ruppia* (see notes on *R. holubi*). Talbot stressed that filamentous algae are an important item in the diet. Fish of 6 cm grow at the rate of 5-10 mm per month or 6-12 cm per year. The juveniles remain in estuaries for two years and finally return to the sea when their gonads start to develop at the beginning of the third year and after this they do not return to estuaries. In the sea they haunt the reefs and feed mainly on mussels such as *Perna* and *Chloromytilus,* reef-worms *(Gunnarea)* and Talbot also reports amphipods, crushed barnacles and crabs in the stomachs. Biden (1954) reports that in the southern Cape they are abundant from May to October and come into sheltered waters such as Kalk Bay harbour at night to feed. In the summer months the shoals move up both the east and the west coasts and return in March. Salinity range unknown.

Rhabdosargus holubi (Steindachner); Cape Stumpnose. An endemic South African species abundant from Zululand to Knysna but less common in the Cape Peninsula. Juveniles recorded in estuaries from Kosi Bay to Milnerton Lagoon. Most of the work on the estuarine phase of the life cycle is due to Blaber (1973a, 1973b, 1974a, 1974b, 1974c) and further notes on breeding and the marine phase of the life cycle will be found in Wallace (1975a, 1975b) and Wallace & van der Elst (1975). *R. holubi* attains a length of 35 cm but is seldom more than 20 cm. It is mature at 15 cm and breeds in Natal seas from May to August. Juveniles enter estuaries at a length of 1-3 cm from July to September. Blaber's studies in regard to salinity tolerance, feeding, growth, juvenile population density, mortality and biomass have been referred to earlier and merely the highlights are given here. Blaber studied the captive population in the West Kleinmond estuary while the mouth was closed. The juveniles soon changed from feeding on zooplankton to benthic invertebrates, filamentous algae and the leaves of *Ruppia.* The leaves were not digested, only the epiphytic diatoms growing on them. *R. holubi* spread through the length of the estuary and experiments showed that it tolerates salinities of 0-70 ‰. First year fish grew 5 mm per month in winter and 10,3 mm per month in summer, the yearly growth rate being 60 mm. Juveniles return to the sea at a length of 13-14 cm so that the duration of estuarine life is two years. Adults do not return to estuaries. In February 1971, 55 000 juveniles entered the estuary but due to predation, the population fell to 11 000 by August. However, growth had increased the biomass from 17,3 kg.ha^{-1} to 27 kg.ha^{-1} during the six months. In April 1972 another cohort of 14 674 juveniles was trapped in the estuary. This had an initial biomass of 4,6 kg.ha^{-1} but in spite of predation, growth during nine months increased the biomass to 28,2kg. ha^{-1}. Thus the lower density in 1972 had reduced predation to such an extent that the final biomass was higher than in 1971. Predation in West Kleinmond was mainly due to cormorants, herons and egrets which migrated to better stocked estuaries when the density of *R. holubi* was low in 1972.

Rhabdosargus sarba (Försk.); Natal Stumpnose. This is an Indo-Pacific species known from Australia, the Indian Ocean and the east coast of Africa from the tropics to Knysna. It is abundant in Morrumbene and Natal estuaries but decreases in Cape estuaries and the last record is from the Great Brak. It is an important angling fish which attains a length of 72 cm and a weight of 10 kg in the sea but is seldom more than 25 cm in estuaries. It is a hermaphrodite and matures at a length of 26 cm and spawns in the sea in inshore waters from May to December. Juveniles of 2-3 cm enter Natal estuaries in August and September or sometimes as late as January when they are 3-6 cm. The monthly growth rate is 12 mm so that recruits entering in August reach a length of 18 cm in a year. They then return to the sea to spawn but spent fish revisit estuary mouths to feed from August to November. Fish under 10 cm long in Durban Bay feed on planktonic copepods, amphipods, shrimps and small crabs; from 10-15 cm they feed on small crabs and gastropods such as *Assiminea* and as the molar teeth grow, adolescent fish feed on bivalves such as *Eumarcia, Loripes* and *Dosinia* as well as crabs. Adult fish also take harder shelled molluscs such as *Nassarius* spp and *Perna* so that *R. sarba* is essentially a feeder on epifauna (Day & Morgans, 1956). The recorded salinity range is from below 10 to 80 ‰.

Sarotherodon (olim Tilapia) mossambicus (Peters); Largemouth Tilapia. A fresh-water fish abundant in tropical and subtropical east African rivers and extends into estuaries. Whitfield & Blaber (1979) report that it is common in closed estuaries but is absent from open estuaries, since it avoids swift currents and rapidly varying salinity. It shelters in weedbeds where it feeds on the rich organic detritus and nests on sandy bottoms. Recorded from Morrumbene to the Bushmans estuary and very abundant in the beds of *Zostera* or *Ruppia* in Lake St Lucia. Attains 30 cm but usually less than 20 cm. It matures at 18-20 cm and spawns both in rivers and estuaries and is a mouth breeder. Its main food is organic detritus which always contains microscopic organisms and it takes some living green 'weed' as well. Juveniles are abundant in winter months and are preyed on by herons, reed cormorants and crocodiles. Recorded salinity range 0-116 ‰.

Scomberoides tala (C. & V.); Needlescaled Queenfish. A tropical Indo-Pacific species reaching East London. Juveniles ranging from 8-19 cm are fairly common in estuaries from Morrumbene to the Umtamvuna River. It attains 75 cm and breeds in the sea; size at maturity unknown but all fish below 32 cm are immature. Juveniles of 8-9 cm contain a few copepods but most stomachs were empty; above 12 cm they feed on penaeid prawns and small shoal fish such as *Gilchristella*, *Hepsetia* and juvenile mullets. *S. tala* is the commonest carangid in the Kosi system, juveniles occurring throughout the year and even adults are present in summer (Blaber, 1978c). Recorded salinity range 3-35 ‰.

Siganus rivulatus (Försk.); Mottled Rabbitfish. A tropical Indo-Pacific species which is abundant in Delagoa Bay and extends in decreasing numbers to Knysna Lagoon. Common in the weedbeds in Morrumbene estuary. Attains 35 cm but only juveniles less than 15 cm long have been taken in estuaries. They feed very largely on *Zostera*, *Halodule* and other sea grasses and algae.

Sillago sihama (Försk.); Silver Sillago. A tropical Indo-Pacific species abundant in Delagoa Bay. Recorded as common from Morrumbene estuary to Durban Bay but only rare specimens reach Knysna Lagoon. Attains 30 cm but is seldom over 12,5 cm in estuaries. Breeds in the sea but size at maturity and breeding season unknown. Juveniles in estuaries frequent sandy areas and feed on zooplankton up to a length of 7 cm; larger fish take shrimps, *Upogebia*, polychaetes and the siphons of bivalves.

Solea bleekeri Boulenger; Blackhand Sole. A subtropical species extending from Morrumbene estuary to Langebaan Lagoon. Attains 17,5 cm but seldom more than 10 cm. Size at maturity 10 cm. It breeds both in estuaries and the sea and one ripe-running female was taken in Lake St Lucia in a salinity of 47 ‰. The breeding season in Natal at least, is from June to August. This distinctive little sole is common on muddy sand and feeds mainly on amphipods, with smaller numbers of polychaetes, isopods, harpacticoid copepods, bivalves and *Assiminea* and is heavily parasitised by an acanthocephalan which lives in the rectum. It is preyed on by *Platycephalus indicus* and *Muraenesox bagio* in Natal and by the sandshark *Rhinobatos* in the Cape. Recorded salinity range 7,2-47 ‰.

Syngnathus acus Linn.; Longnose Pipefish. Cosmopolitan in warm seas and common in estuaries from Mngazana to the Olifants River. It attains 38 cm but is commonly 10-20 cm long and matures at 12-14 cm. It breeds both in estuaries and the sea and the male retains the developing embryos in a broodpouch in February and March. It is always found in weedbeds particularly *Zostera* in estuaries and lagoons (Day, 1959). It feeds on zooplankton particularly copepods which are sucked in individually and it also takes amphipods from the weeds.

Syngnathus djarong Blkr.; Belly Pipefish. A tropical Indo-Pacific species extending to Natal. Recorded as common in lagoons and estuaries from Morrumbene to Durban Bay. Attains 20 cm but commonly 10 cm long. Length at maturity unknown. Breeds in estuaries, the male retaining the developing embryos in a brood pouch. Shelters in *Zostera* and *Halodule* beds and feeds on zooplankton particularly copepods and amphipods which are taken individually. Recorded salinity range 8,3-37,6 ‰ (Millard & Broekhuysen, 1970).

Tachysurus feliceps Val; Sea Barbel. Reported as an endemic South African species 'right around our coasts' by Smith (1965) but also reported from India by Pillay (1967b). Common on Cape coasts and recorded in estuaries from the Buffalo River at East London to Hermanus Lagoon particularly on muddy bottoms. Attains 50 cm but usually less than 20 cm in estuaries. Dorsal and pectoral spines serrated and poisonous. Females may contain large eggs at a length of 13 cm but the testes are not developed in fish less than 22 cm long. Breeding occurs both in the sea and in estuaries and developing eggs are reported to be carried in the mouth of the male. Juveniles of 4,4-6 cm feed largely on zooplankton but take chironomid larvae, amphipods, isopods and bivalves as well. Fish longer than 7 cm cease feeding on zooplankton and concentrate on benthic crustacea; larger fish feed mainly on *Upogebia* and *Hymenosoma* but also take amphipods, gobies and small crabs including *Cleistostoma*. Salinity range not known.

Terapon jarbua (Försk.); Thornfish. A tropical Indo-Pacific species extending as far south as the Transkei and occasionally reaching the Cape. Common in estuaries from Morrumbene to Mngazana but only a few recorded from there on to the Swartkops estuary. Attains 30 cm but usually 5-15 cm in estuaries. Matures at about 16 cm and breeds in the sea in summer. Juveniles enter estuaries at a length of 1-2 cm and up to 5 cm they feed on copepods and amphipods. Larger juveniles prey on a wide range of invertebrates, fish fry and a few small fishes but the main food is fish scales (Whitfield & Blaber 1978a, Whitfield 1979). In an average *T. jarbua*, 66 % of the stomach contents are scales removed from living fish which are hunted by small groups of 2-7 *T. jarbua*. This is the first record of a marine fish feeding on scales. The scales belong to several species including mullet and kob many of which were larger than the *T. jarbua*. Apart

217

from scales, *T. jarbua* feeds mainly on amphipods, the crab *Hymenosoma,* insects (including winged ants) and the parasitic isopod *Anilocra* presumably removed from other fish. In fact, Whitfield & Blaber suggest that the scale-eating habit evolved from cleaning activities. It takes any bait and is regarded as a pest by anglers. It lives mainly in weedy shallows and the recorded salinity range is 7,2-70 ‰.

Thryssa vitrirostris (Bloch); Orangemouth Glassnose. A tropical Indo-Pacific species abundant in Mocambique and Natal with occasional specimens reaching Port Alfred. Recorded in the sea to a depth of 30-40 m and also in estuaries from Morrumbene to the Keiskamma River. Attains 20 cm and matures at 10 cm. Heeg & Blaber (1979) report that it enters St Lucia in late winter and spawns there during summer but few are present there in early winter. The juveniles are plankton feeders and change their diet in relation to the plankton cycle. *T. vitrirostris* swims in shoals and adults prey on *Gilchristella* and are themselves important in the diet of *Elops, Argyrosomus* and several other piscivorous fishes, as well as cormorants and terns. Recorded salinity range 7,2-65 ‰.

APPENDIX: COMMON ESTUARINE FISH IN SOUTHERN AFRICA

Species		1 Morrumbene Estuary	2 Inhaca Island	3 Kosi Bay	4 St Lucia	5 Richards Bay	6 Mlalazi	7 Durban Bay	8 Estuaries near Umkomaas	9 Estuaries near Port Shepstone	10 Estuaries near Port St Johns	11 Mngazana	12 Estuaries near Bashee River	13 Keiskamma River	14 Bushmans River	15 Sundays River	16 Swartkops River	17 Knysna River	18 Great Brak River	19 Breede River	20 Hermanus Lagoon	21 Milnerton Lagoon	22 Langebaan Lagoon	23 Berg River	24 Olifants River
Acanthopagrus berda (Försk.)	T	●	●	●	●	●	●	●	●	●	●	●	●				●								
Ambassis commersoni C. in C. & V.	T	●	●	●	●	●	●	●	●		●	●													
Ambassis natalensis G. & T.	ST	●	●	●	●	●	●	●	●		●	●	●												
Amblyrhynchotes honckenii (Bloch.)	T	●	●	●	●	●	●	●			●	●					●	●		●	●				
Argyrosomus hololepidotus (Lacep.)	E			●	●	●	●	●	●		●	●		●	●		●			●	●				
Arothron immaculatus (Bloch)	T	●	●	●	●	●	●	●	●		●	●					●	●							
Caranx ignobilis (Försk.)	T	●	●	●	●	●	●	●	●	●	●														
Caranx sexfasciatus Q. & G.	T	●	●	●	●	●	●	●			●		●					●							
Carcharinus leucas (M. & H.)	T			●	●	●	●			●	●	●		●											
Clinus superciliosus (Linn.)	E													●	●	●	●	●		●	●	●	●		
Caffrogobius caffer (Gunther)	E			●	●			●			●	●		●	●		●			●	●	●	●		
Oligolepis acutipennis (Cuv.)	T	●			●	●		●	●		●	●													
Diplodus sargus Linn.	ST							●	●		●	●		●	●	●	●	●	●	●	●	●	●	●	
Eleotris fusca (Bloch)	T	●	●	●	●	●		●	●	●		●													
Elops machnata (Försk.)	T	●	●	●	●	●	●	●	●		●	●					●	●	●						
Gerres acinaces Blkr.	T	●	●	●	●	●	●	●	●		●	●													
Gilchristella aestuarius (Gilchrist)	ST		●	●	●	●	●		●	●	●	●		●	●	●	●			●	●			●	●
Glossogobius giuris (Hamilton)	T	●	●	●	●	●		●	●			●	●				●								
Hepsetia breviceps (Cuv.)	E			●	●	●		●	●					●	●					●	●	●	●		
Hilsa kelee (Cuv.)	T	●	●		●	●		●	●		●	●													
Hyporhamphus knysnaensis (Smith)	E			●	●	●											●	●							
Johnius belengerii (Cuv.)	T			●	●	●		●																	
Leiognathus equulus (Försk.)	T	●	●	●	●	●	●	●																	
Lethrinus nebulosus (Försk.)	T	●	●		●			●																	
Lichia amia (Linn.)	E				●	●								●	●		●	●	●	●			●	●	
Lithognathus lithognathus (Cuv.)	E												●	●	●	●	●	●	●	●			●	●	
Lutjanus argentimaculatus (Försk.)	T	●	●	●	●			●	●	●	●	●													
Lutjanus fulviflamma (Försk.)	T	●	●	●	●	●		●			●	●		●											
Monodactylus argenteus (Linn.)	T	●	●	●	●	●	●	●									●	●		●					
Monodactylus falciformis Lacep.	T				●	●		●	●					●	●		●	●					●		●
Mugil cephalus Linn.	C		●	●	●	●		●	●	●	●	●		●	●		●	●					●	●	●
Mugil (Liza) macrolepis Smith	T	●	●	●	●	●		●	●	●	●	●		●	●										

Species		Morrumbene Estuary	Inhaca Island	Kosi Bay	St Lucia	Richards Bay	Mlalazi	Durban Bay	Estuaries near Umkomaas	Estuaries near Port Shepstone	Estuaries near Port St Johns	Mugazana	Estuaries near Bashee River	Keiskamma River	Bushmans River	Sundays River	Swartkops River	Knysna River	Great Brak River	Breede River	Hermanus Lagoon	Milnerton Lagoon	Langebaan Lagoon	Berg River	Olifants River
Mugil (Liza) richardsoni A.Smith	E											●	●	●	●	●	●	●	●		●	●	●	●	●
Mugil (Liza) dumerili Steindach.	ST	●	●	●	●	●	●	●	●	●	●	●	●	●	●	●	●	●	●		●				
Mugil (Valamugil) buchanani Blkr.	T	●		●	●		●	●																	
Mugil (Valamugil) cunnesius (Velanc.)	T	●		●	●		●	●	●	●															
Mugil (Valamugil) seheli Försk.	T	●			●		●	●	●																
Myxus capensis (Val.)	E			●	●			●	●	●							●	●		●					
Pelates quadrilineatus (Bloch)	T	●	●	●	●	●	●	●																	
Periophthalmus sobrinus Eggert	T	●	●	●	●	●	●	●																	
Platycephalus indicus (Linn.)	T	●		●	●		●	●								●	●	●							
Pomadasys hasta Bloch	T	●		●	●		●	●																	
Pomadasys commersonni (Lacep.)	T			●	●		●	●	●	●	●	●		●	●	●					●	●			
Pomadasys olivaceum Day	T			●	●			●													●	●			
Pomatomus saltatrix (Linn.)	C			●			●	●				●								●	●	●	●		
Psammogobius knysnaensis Smith	E											●		●			●	●		●	●	●	●	●	
Rhabdosargus globiceps (Cuv.)	E													●			●			●	●	●	●	●	
Rhabdosargus holubi (Steind.)	ST			●	●	●	●	●	●	●	●	●		●	●	●	●	●		●	●				
Rhabdosargus sarba (Försk.)	T	●		●	●		●	●	●	●				●	●		●								
Sarotherodon mossambicus (Peters)	ST	●		●	●		●	●	●	●		●		●											
Scomberoides tala (C. & V.)	T	●		●	●		●	●																	
Siganus rivulatus (Försk.)	T	●		●	●		●					●				●									
Sillago sihama (Försk.)	T	●		●	●		●	●										●							
Solea bleekeri Boul.	ST	●		●	●	●	●	●						●	●		●	●		●			●		
Syngathus acus Linn.	C							●						●	●	●	●	●		●			●	●	●
Syngnathus djarong Blkr.	T	●	●	●	●	●	●																		
Tachysurus feliceps Val.	E													●		●	●	●		●	●				
Terapon jarbua (Försk.)	T	●		●	●		●	●	●	●	●			●				●	●						
Thryssa vitrirostris (G..& T.)	T			●	●	●	●	●																	

REFERENCES

BARNARD, K.H. 1925. A monograph on the marine fishes of South Africa. Part 1. *Ann. S.Afr. Mus.* 21(1): 1-418.

BARNARD, K.H. 1927. A monograph on the marine fishes of South Africa. Part 2. *Ann. S.Afr. Mus.* 21(2): 419-1065.

BASS, A.J. 1976. Sharks in the St Lucia system. *In:* A.E.F. Heydorn (ed), *St Lucia Scientific Advisory Council Workshop – Charters Creek, February 1976.* Natal Parks Bd, Pietermaritzburg.

BASS, A.J., d'AUBREY, D.J. & N. KISTNASAMY 1973. Sharks of the east coast of southern Africa. I: The genus *Carcharinus* (Carcharinidae). *Invest. Rep. Oceanog. Res. Inst. Durban* 33: 1-168.

BIDEN, L. 1930, 1954. *Sea angling fishes of the Cape (South Africa).* 1st ed. 1930, Oxford Univ. Press; 3rd ed. Juta & Co, Cape Town.

BLABER, S.J.M. 1973a. Population size and mortality of juvenile *Rhabdosargus holubi* (Steindachner) (Pisces: Sparidae) in a closed estuary. *Mar. Biol.* 21: 219-225.

BLABER, S.J.M. 1973b. Temperature and salinity tolerance of juvenile *Rhabdosargus holubi* (Steindachner) (Teleostei: Sparidae) in a closed estuary. *J. Fish. Biol.* 6: 455-460.

BLABER, S.J.M. 1974a. Field studies of the diet of *Rhabdosargus holubi* (Pisces: Sparidae). *J. Zool. Lond.* 173: 407-417.

BLABER, S.J.M. 1974b. The population structure and growth of juvenile *Rhabdosargus holubi* (Steindachner) (Teleostei: Spardiae) in a closed estuary. *J. Fish. Biol.* 6: 455-460.

BLABER, S.J.M. 1974c. Osmoregulation in juvenile *Rhabdosargus holubi* (Steindachner) (Pices: Sparidae). *J. Fish. Biol.* 6: 797-800.

BLABER, S.J.M. 1976a. The food and feeding ecology of mullet in the St Lucia Lake system. *In:* A.E.F. Heydorn (ed), *St Lucia Scientific Advisory Council Workshop – Charters Creek, February 1976.* Natal Parks Bd, Pietermaritzburg.

BLABER, S.J.M. 1976b. Fishes of the Kosi system. *Lammergeyer*, March 1978: 28-41.

BLABER, S.J.M. 1979. The biology of filter-feeding teleosts in Lake St Lucia, Zululand. *J. Fish. Biol.* 15: 37-59.

BROEKHUYSEN, G.J. & H. TAYLOR 1959. The ecology of South African estuaries. Part 8: Kosi Bay estuary system. *Ann. S.Afr. Mus.* 44: 279-296.

BROWN, A.C. 1959. The ecology of South African estuaries. Part 10: Notes on the estuary of the Orange River. *Trans. roy. Soc. S.Afr.* 41(3): 283-294.

COETZEE, D.J. 1979. Maaginhoudontledings van vis in die Wildernismere, met spesiale verwysing na *Hyporhamphus knysnaensis* (Smith), *Research Rep.* Sept. 1979, Dept. Nature Environ. Conserv. Cape of Good Hope.

CRASS, R.S. 1964. *Freshwater fishes of Natal.* Shuter & Shooter, Pietermaritzburg.

DAY, J.H. 1951. The ecology of South African estuaries, Part 1: A review of estuarine conditions in general. *Trans. roy. Soc. S.Afr.* **33**(1): 53-91.

DAY, J.H. 1959. The biology of Langebaan Lagoon: a study of the effect of shelter from wave action. *Trans. roy. Soc. S.Afr.* **35**(5): 475-547.

DAY, J.H. 1974. The ecology of Morrumbene estuary, Mocambique. *Trans. roy. Soc. S.Afr.* **41**(1): 43-97.

DAY, J.H. 1975. The biology of planktonic Polychaeta near Nosy-Bé, Madagascar. *Cah. ORSTOM (ser. oceanog.)* **13**(3): 197-216.

DAY, J.H., MILLARD, N.A.H. & G.J. BROEKHUYSEN 1954. The ecology of South African estuaries, Part 4: The St Lucia system. *Trans. roy. Soc. S.Afr.* **34**(1): 129-156.

DAY, J.H. & J.F.C. MORGANS 1956. The ecology of South African estuaries. Part 7: The biology of Durban Bay. *Ann. Natal Mus.* **13**: 259-312.

DE SYLVA, D.P. 1969. Trends in marine sport fisheries research. *Trans. Am. Fish. Soc.* **98**(1): 151-169.

GREEN, J. 1968. *The biology of estuarine animals.* Sidgwick & Jackson, London.

GUNTER, G. 1961. Some relations of estuarine organisms to salinity. *Limnol. Oceanog.* **6**(2): 182-190.

GUNTER, G. 1967. Some relations of estuaries to the fisheries of the Gulf of Mexico. *In:* G. Lauff (ed), *Estuaries.* Am. Ass. Adv. Sci., Washington.

HARTLEY, P.H.T. 1940. The Saltash tuck-net fishery and the ecology of some estuarine fishes. *J. mar. biol. Ass. UK* **24**(1): 1-68.

HEDGPETH, J.W. 1967. Ecological aspects of the Laguna Madre, a hypersaline estuary. *In:* G. Lauff (ed), *Estuaries.* Am. Ass. Adv. Sci., Washington.

HEEG, J. & S.J.M. BLABER 1979. The biology of filter-feeding teleosts in Lake St Lucia, Zululand. *Abstract 4th (S.Afr.) Oceanog. Symp., Cape Town, July 1979.* SANCOR, CSIR, Pretoria.

HODGKIN, E.P. 1978. *An environmental study of the Blackwood River estuary, Western Australia, 1974-1975.* Govt. Printer, Perth.

KINNE, O. 1970. Temperature – invertebrates. *In:* O. Kinne (ed), *Marine ecology.* **1**(1): 407-514. *Environmental factors* Wiley-Interscience, London.

KINNE, O. 1971. Salinity – invertebrates. *In:* O. Kinne (ed), *Marine ecology.* **1**(2): 821-995. *Environmental factors.* Wiley-Interscience, London.

KORRINGA, P. 1967. Estuarine fisheries in Europe as affected by man's multiple activities. *In:* G. Lauff (ed), *Estuaries.* Am. Ass. Adv. Sci., Washington.

LONGHURST, A.R. 1957. The food of the demersal fish of a West African estuary. *J. anim. Ecol.* **26**: 369-387.

MACNAE, W. 1968. A general account of the fauna and flora of mangrove swamps and forests in the Indo-West-Pacific region. *In:* R.S. Russel & M. Yonge (eds), *Advances in marine biology.* Vol 6. Academic Press, London.

MACNAE, W. & M. KALK 1958. *A natural history of Inhaca Island, Mocambique.* Witwatersrand Univ. Press, Johannesburg.

MᶜHUGH, J.L. 1967. Estuarine nekton. *In:* G. Lauff (ed), *Estuaries.* Am. Ass. Adv. Sci., Washington.

MARAIS, J.F.K. & D. BAIRD 1979. Aspects of feed intake, feed selection and alimentary canal morphology of *Mugil cephalus, Liza tricuspidens, Liza richardsoni* and *Lisa dumerili.* Abstract *4th (S.Afr.) Natl. Oceanog. Symp., Cape Town, July 1979.* SANCOR, CSIR, Pretoria.

MEHL, J.A.P. 1973. Ecology, osmoregulation and reproductive biology of the White Steenbras *Lithognathus lithognathus* (Teleostel: Sparidae). *Zool. Afr.* **8**(2): 157-230.

MELLVILLE-SMITH, R. & D. BAIRD 1979. Abundance, distribution and species composition of fish larvae in the Swartkops estuary, Port Elizabeth. *Abstract 4th (S.Afr.) natl. oceanog. Symp. Cape Town, July 1979.* SANCOR, CSIR, Pretoria. (Also) *S.Afr. J. Sci.* **75**(12): 564.

MILLARD, N.A.H. & G.J. BROEKHUYSEN 1970. The ecology of South African estuaries. Part 10: St Lucia, a second report. *Zool. Afr.* **5**(2): 227-307.

MILLARD, N.A.H. & A.D. HARRISON 1954. The ecology of South African estuaries. Part 5: Richards Bay. *Trans. roy. Soc. S.Afr.* **34**(1): 157-179.

MILLARD, N.A.H. & K.M.F. SCOTT 1954. The ecology of South African estuaries. Part 4: Milnerton estuary and the Diep River, Cape Town. *Trans. roy. Soc. S.Afr.* **34**(2): 279-324.

MILNE, H. & G.M. DUNNET 1972. Standing crop, productivity and trophic relations of the fauna of the Ythan estuary. *In:* R.S.K. Barnes & J. Green (eds), *The estuarine environment.* Applied Science, London.

PARRY, G. 1966. Osmotic adaptation in fishes. *Biol. Rev.* **41**: 392-444.

PEARCY, W.C. 1962. Ecology of an estuarine population of winter flounder, *Pseudopleuronectes americanus* (Walbaum), III. Distribution, abundance, growth and production of juveniles; survival of larvae and juveniles. *Bull. Bingham Oceanog. Coll.* **18**: 39-64.

PERKINS, E.J. 1974. *The biology of estuaries and coastal waters.* Academic Press, London.

PILLAY, T.V.R. 1967a. Estuarine fisheries of West Africa. *In:* G. Lauff (ed), *Estuaries.* Am. Ass. Adv. Sci., Washington.

PILLAY, T.V.R. 1967b. Estuarine fisheries of the Indian Ocean coastal zone. *In:* G. Lauff (ed), *Estuaries.* Am. Ass. Adv. Sci., Washington.

SCOTT, K.M.F., HARRISON, A.D. & W. MACNAE 1952. The ecology of South African estuaries. Part 2: The Klein River estuary, Hermanus, Cape. *Trans. roy. Soc. S.Afr.* **33**(3): 283-331.

SMITH, J.L.B. 1949, 1965. *The sea fishes of southern Africa.* 1st ed. 1949, 5th ed. 1965. Central News Agency, Cape Town.

STEPHENSON, T.A. & A. 1947. The constitution of the intertidal fauna and flora of South Africa. Part III. *Ann. Natal. Mus.* **11**(2): 207-324.

TALBOT, F.H. 1955. Notes on the biology of the White Stumpnose *Rhabdosargus globiceps* (Cuvier) and on the fish fauna of the Klein River estuary. *Trans. roy. Soc. S.Afr.* **34**(3): 387-407.

THOMSON, J.M. 1959. Some aspects of the ecology of Lake Macquarie, NSW with regard to an alleged depletion of fish. IX: The fishes and their food. *Aust. J. mar. freshw. Res.* **10**: 365-374.

TODD, R.A. 1915. Report on the food of the plaice. *Fish. Invest., Lond.* **2**: 1-31.

VAN DER ELST, R.P. 1974. *Biology of the elf (Pomatomus saltatrix (Linn.)) in Natal.* Unpublished MSc thesis, Univ. of Natal, Durban.

VAN DER ELST, R.P. 1979. Changes in populations of estuarine-dependent sport fish in Natal. *Abst. 4th (S.Afr.) natl. oceanog. Symp., Cape Town, July 1979.* SANCOR, CSIR, Pretoria. (Also) *S.Afr. J. Sci.* **75**(12): 566.

VAN DER ELST, R.P., BLABER, S.J.M., WALLACE, J.H. & A.K. WHITFIELD 1976. The fish fauna of Lake St Lucia under different salinity regimes. *In:* Heydorn (ed), *St Lucia Scientific Advisory Council Workshop – Charters Creek, February 1976.* Natal Parks Bd, Pietermaritzburg.

WALLACE, J.H. 1975a. The estuarine fishes of the east coast of South Africa. I: Species, composition and length distribution in the estuarine and marine environments. II: Seasonal abundance and migrations. *Invest. Rep. Oceanog. Res. Inst., Durban.* **40**: 1-72.

WALLACE, J.H. 1975b. The estuarine fishes of the east coast of South Africa. III: Reproduction. *Invest. Rep. Oceanog. Res. Inst., Durban.* **41**: 1-51.

WALLACE, J.H. 1976. Biology of teleost fish with particular reference to St Lucia. *In:* A.E.F. Heydorn (ed), *St Lucia Scientific Advisory Council Workshop — Charters Creek, February 1976.* Natal Parks Bd, Pietermaritzburg.

WALLACE, J.H. & R.P. VAN DER ELST 1975. The estuarine fishes of the east coast of South Africa. IV: Occurrence of juveniles in estuaries. V: Ecology, estuarine dependence and status. *Invest. Rep. Oceanog. Rest. Inst., Durban* 42: 1-63.

WHITFIELD, A.K. 1976. An investigation of the family Mugilidae in Lake St Lucia, Zululand, with special reference to predation and population densities. *In:* A.E.F. Heydorn (ed), *St Lucia Scientific Advisory Council Workshop — Charters Creek, February 1976.* Natal Parks Bd, Pietermaritzburg.

WHITFIELD, A.K. 1979a. Field observations on the lepidophagous teleost *Terapon jarbua* (Forskal). *Env. Biol. Fish.* 4(2): 171-172.

WHITFIELD, A.K. 1979b. Quantitative study of the trophic relationships within the fish community of the Mhlanga estuary. *S.Afr. J. Sci.* 75(12): 565.

WHITFIELD, A.K. & S.J.M. BLABER 1978a. Scale-eating habits of the marine teleost *Terapon jarbua* (Forskal). *J. Fish Biol.* 12: 61-70.

WHITFIELD, A.K. & S.J.M. BLABER 1978b. Food and feeding ecology of piscivorous fishes at Lake St Lucia, Zululand. *J. Fish. Biol.* 13: 675-691.

WHITFIELD, A.K. & S.J.M. BLABER 1978c. Feeding ecology of piscivorous birds at Lake St Lucia. Part 1: Diving birds. *Ostrich* 49: 185-198.

WHITFIELD, A.K. & S.J.M. BLABER 1978d. Feeding ecology of piscivorous birds at Lake St Lucia. Part 2: Wading birds; Part 3: Swimming birds. *Ostrich* 50: 1-9; 10-20.

WHITFIELD, A.K. & S.J.M. BLABER 1978e. The distribution of the freshwater chiclid *Sarotherodon mossambicus* in estuarine systems. *Env. Biol. Fish.* 4(1): 77-81.

Chapter 13
The estuarine avifauna of southern Africa
W.R. Siegfried

Percy FitzPatrick Institute of African Ornithology, University of Cape Town

INTRODUCTION

In the estuarine environment, the characteristically high productivity of invertebrates dependent on the 'nutrient trap' effect permits high densities of vertebrate predators among which birds and fish predominate. While this is a valid generalization, it must be emphasized that very little is known about the details of the interactions between estuarine birds and their prey populations. Even less is known about the fish, because they are more difficult to study than the birds. However, an understanding of the role of fish in the predator-prey subsystem is crucial for a proper appreciation of avian species diversity and biomass in estuaries. Comparisons among congeneric animals have been made by estuarine biologists working in the temperate zones, but virtually no studies have been made of what may often be significant competition between taxonomically very different species at the same trophic level (Morse, 1975). This especially applies to tropical-zone estuaries whose fish and bird faunas have received even less study than those in the temperate zone.

It is only during the last decade that attempts have been made to investigate the role of avian predators within the estuarine ecosystem. These investigations are of both theoretical and practical interest, since they contribute to an understanding of estuarine species diversity and interdependence in relation to the stability of the estuarine community as a whole

(Milne & Dunnet, 1972). The present paper reviews in very general terms some of the features and principles attending the organization and functioning of bird populations in estuaries, and then goes on to describe southern African estuarine avifaunas in greater detail. In addition to summarizing the little that is known about the bird populations, I have attempted to develop, through partial syntheses, some of the opportunities for further research and some of the problems relating to the conservation of estuarine birds in southern Africa.

GENERAL CONSIDERATIONS

Temperate-zone estuaries characteristically support migrant populations of birds drawn chiefly from two taxonomic groups, Anatidae and Charadriiformes. The ducks, geese and swans are predominantly either primary consumers, feeding on submerged plants (eg *Zostera*) and plants growing above the tideline, or secondary consumers of bivalves (eg *Mytilus*) and other invertebrates. Recent studies indicate that the grazing pressure exerted by birds may be considerable in northern temperate-zone estuaries (Packham & Liddle 1970, Ebbing *et al* 1975). The charadriiforms include mainly plovers and sandpipers, collectively called waders or shorebirds, which feed primarily on invertebrates taken from the substrate at or near the water's edge. The invertebrates embrace mainly primary consumers and detritus feeders: mussels, snails, cockles, small crustaceans, small insects and polychaete worms (Barnes, 1974). Most of these prey animals have a restricted period of growth and reproduction in summer (Milne & Dunnet, 1972). Studies at the Ythan estuary in Scotland show that although up to 60 bird species may feed there during the year, only 20 species are present throughout the year or visit the estuary in large numbers (Milne, 1974).

Birds are exceptionally mobile animals capable of travelling long distances in short periods. This allows them to exploit temporally and spatially fluctuating resources outside the reach of many other animals. Birds can find a suite of resources and exploit it relatively quickly for as long as the opportunity remains economical. Just as important, birds can leave an area and move to another equally quickly. This must have repercussions on ecosystems in respect to import and export of minerals and energy. For instance, great numbers of migrants move northward to breed in areas which permit the birds to maximize their reproductive output during a relatively short and highly productive boreal summer. At the end of summer, the populations now greatly enlarged through recruitment of young, leave their breeding grounds. Since many birds in the combined populations will not return to breed having suffered mortality elsewhere, this represents a pathway for export of materials from the breeding areas. In ecosystems in the climatically more stable, humid tropics bird populations are mainly resident (Moreau, 1973), and they contribute more efficiently to recycling of materials.

Tropical estuaries characteristically support resident populations of birds mainly belonging to the Ciconiiformes and Pelecaniformes (Morse, 1975). The herons, storks, ibises, cormorants, anhingas and pelicans are predominantly at least tertiary consumers, taking mainly crabs and other large crustaceans, large gastropods, large insects, frogs and fish. The vegetation below, above and within the intertidal zone is not extensively exploited directly by birds. Conceivably, fish and other animals (eg turtles and sirenians in tropical estuaries) graze the submerged plants, and the vegetation above the tideline is cropped by insects and mammals such as rodents, hippopotami and ungulates.

Morse (1975) has argued persuasively in favour of competitive interactions over food resources between animals of widely unrelated taxa. This subject has received little attention from ecologists. However, one might postulate that certain avian groups (eg geese) are precluded by mammals and other animals from exploiting vegetation in tropical estuaries. In a similar context many fish species in tropical estuaries are primary, secondary or tertiary consumers of prey whose ecological counterparts form a major food resource for birds in temperate-zone estuaries. The greater species diversity of the fish fauna of tropical estuaries supports a relatively great diversity of fish-eating birds, many of which have no avian counterparts in the temperate zones. In this connection it is worth noting that whereas grey seals *(Halichoerus grypus)* and common seals *(Phoca vitulina)* occur in European estuaries, there are no seals in tropical estuaries. Tropical and subtropical estuaries have crocodiles and sharks which feed on fish and crabs.

In the temperate zones the major part of the fish production is lost from the estuarine ecosystem mainly to the sea (Barnes, 1974). Much of the invertebrate production, however, goes to the birds. For example, in the Ythan estuary the net production of *Mytilus* is about 700 kcal per m² per year of which about 275 kcal goes to eider ducks *(Somateria mollissima),* 100 kcal to oystercatchers *(Haematopus ostralegus),* 110 kcal to gulls *(Laridae)* and 210 kcal to man (Milne, 1974). It has been suggested that the Ythan mussel beds are being cropped to a maximum, so that the standing crop at the beginning of each year remains fairly constant (Milne & Dunnet, 1972). For other prey populations the effect of bird predation may be small (Barnes, 1974, Wolff *et al,* 1975). However, all predator-prey populations need to be studied to reach an understanding of the way in which the predators (fish, birds and others) partition the prey production. This knowledge is important if one is to predict changes in species diversity arising through human perturbations of the natural dynamic processes of the estuarine ecosystem. In this connection, it is not known whether food supplies in the estuarine wintering areas limit the migratory bird populations (Goss-Custard & Charman, 1976), although Baker & Baker (1973) have suggested that migratory populations of plovers and sandpipers are indeed regulated through competitive processes occurring in their north-temperate wintering areas.

THE ESTUARINE AVIFAUNA OF SOUTHERN AFRICA

This account deals only with zoogeographical South Africa or the South African Subregion of Wallace (1876), which is that part of the African continent south of the Kunene and Zambezi rivers. However, lack of information precludes mention of the relatively large estuaries of Mozambique. No previous review of the region's estuarine avifauna has been attempted, although Winterbottom (1972) includes tidal mudflats, salt marshes, estuaries and lagoons, and mangroves as four saltwater habitats in his discussion of the ecological distribution of birds in southern Africa. He does not, however, deal separately with estuaries and his remarks lumped together under lagoons and tidal mudflats are very brief: 'as the following list of dominant species in the south-west Cape shows, the non-breeding, migrant waders from the Palaearctic bulk large in the fauna: *Phoenicopterus*

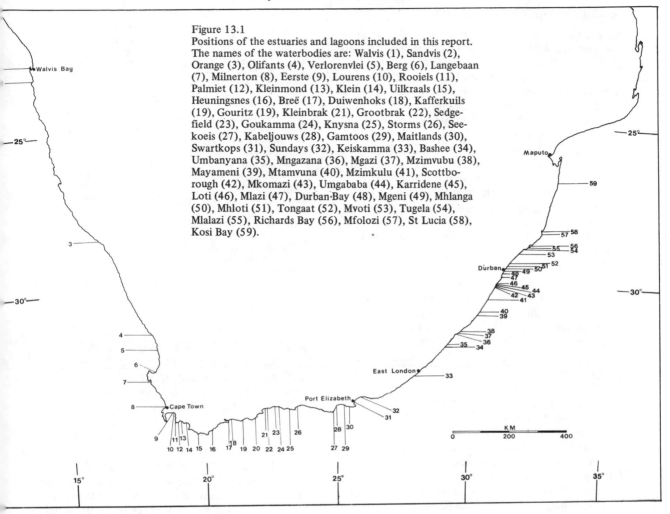

Figure 13.1
Positions of the estuaries and lagoons included in this report.
The names of the waterbodies are: Walvis (1), Sandvis (2),
Orange (3), Olifants (4), Verlorenvlei (5), Berg (6), Langebaan
(7), Milnerton (8), Eerste (9), Lourens (10), Rooiels (11),
Palmiet (12), Kleinmond (13), Klein (14), Uilkraals (15),
Heuningsnes (16), Breë (17), Duiwenhoks (18), Kafferkuils
(19), Gouritz (19), Kleinbrak (21), Grootbrak (22), Sedge-
field (23), Goukamma (24), Knysna (25), Storms (26), See-
koeis (27), Kabeljouws (28), Gamtoos (29), Maitlands (30),
Swartkops (31), Sundays (32), Keiskamma (33), Bashee (34),
Umbanyana (35), Mngazana (36), Mgazi (37), Mzimvubu (38),
Mayameni (39), Mtamvuna (40), Mzimkulu (41), Scottbo-
rough (42), Mkomazi (43), Umgababa (44), Karridene (45),
Loti (46), Mlazi (47), Durban·Bay (48), Mgeni (49), Mhlanga
(50), Mhloti (51), Tongaat (52), Mvoti (53), Tugela (54),
Mlalazi (55), Richards Bay (56), Mfolozi (57), St Lucia (58),
Kosi Bay (59).

*ruber, Arenaria inte:pres, Charadrius hiaticula, C. mar-
ginatus, C. pecuarius, Pluvialis squatarola, Calidris
minuta, C. ferruginea, Tringa nebularia, Numenius
arquata, N. phaeopus, Larus dominicanus, L. novae-
hollandiae, Sterna hirundo* and *Motacilla capensis.*
Almost as frequent are *Threskiornis aethiopica, Calidris alba, Tringa stagnatilis* and *Ceryle rudis.* The list
is not likely to be materially different from Walvis
Bay to Beira ... Tidal mudflats are usually found in
Lagoons, but in addition to the birds listed for the
mudflats, the following are dominant lagoon species
in the south-west Cape: *Phalacrocorax lucidus, P.
capensis, Egretta garzetta* and *Sterna sandvicensis'.*
From other 'faunistic' papers published by him (eg
Winterbottom 1967a), I assume that the species classed
as 'dominant' by Winterbottom (1972) are those which
feature in 40 % or more of his 'daily lists' for any single
habitat. Winterbottom, however, does not cite the
numbers of daily lists for the tidal mudflats, lagoons
and estuaries which he surveyed, nor does he name
the areas which were surveyed.

For purposes of this paper an estuary has been

taken as that area partially enclosed by land in which
freshwater of terrestrial origin and seawater meet.
Figure 13.1 shows the distribution of the estuaries
whose avifaunas are reported here. The avifaunas of
four sea-water lagoons, namely Walvis Bay, Sandvishawe, Langebaan Lagoon and Durban Bay were
included in some of the analyses, but they are identified clearly in the results. It appears, however, that
there is little, if any, difference between the avifaunas
of estuaries and coastal lagoons.

The analyses deal only with water birds *(sensu lato)*
which are listed in Appendix 13.1. In the main,
nomenclature follows McLachlan & Liversidge (1970).
The analyses relied on computer-assisted multivariate
cluster and regression techniques. Records for the
presence or absence of species and numbers of birds
at various estuaries were taken mainly from unpublished lists in the files of the Percy FitzPatrick Institute
of African Ornithology. Appendix 13.2 contains a
summary of all sources of published and unpublished
material.

225

Species composition and abundance

Very little information on the species composition of the southern African estuarine avifauna was gathered prior to 1970. Even straightforward lists of species present or absent were relatively scarce, either as published or unpublished documents. At about 1975 two small groups of birdwatchers independently started to count estuarine birds, particularly Palaearctic waders, in Natal and the south-western Cape province. Encouraged and guided by professional biologists, the amateur ornithologists subsequently extended their surveys to other areas and they broadened their censuses to include all estuarine birds. However, coverage of some regions, for example the eastern Cape province and Transkei, is still very incomplete, and relatively few censuses have been made during the austral winter.

One hundred and twenty-seven species of birds have been recorded for the estuaries included in Figure 13.1. These birds fall into four categories: residents throughout the year, seasonal migrants from the Palaearctic region, regular visitors which breed in southern Africa, and occasional visitors. The Palaearctic migrants comprise a little less than one-third of the total number of species (Table 13.1). More than half the species recorded belong to the orders Charadriiformes and Ciconiiformes. Hence, plovers, sandpipers, gulls, terns, herons, flamingos and storks predominate. Most of the Charadriiformes are Palaearctic migrants which visit southern Africa during the austral summer. In contrast, the representatives of the Ciconiiformes are essentially resident. The resident African fish eagle *(Haliaetus vocifer)* and the migratory osprey *(Pandion haliaetus)* are the two falconiforms.

During the austral summer, waders, which breed mainly in the northern tundra, feature prominently in the avifaunas of southern African estuaries. The waders and the terns (Sternini), and particularly the common tern *(Sterna hirundo),* are the two main groups of seasonal visitors from the Palaearctic. The common terns are drawn mainly from a population which breeds in Finland (Elliott, 1971).

Palaearctic waterfowl (Anatidae) normally do not reach southern Africa. All 16 anseriforms indigenous to southern Africa occur in estuaries, and all 16 belong to the family Anatidae. However, unlike estuaries in Europe, North America and North and West Africa, southern African estuaries generally do not support large populations of waterfowl. Swans, true geese and marine ducks are not indigenous, but a small feral population of mute swans *(Cygnus olor)* has been established at estuaries in the southern Cape province (Siegfried, 1970). A number of the southern Cape estuaries are important refuges for the Cape shelduck *(Tadorna cana)* during the dry season in the interior. After breeding inland, populations of the Cape shelduck move seasonally to estuaries where considerable

Table 13.1
Avian taxa at estuaries in southern Africa during the austral summer. Included under 'migrants' are those visiting species which do not breed in southern Africa; 'residents' are species which do breed. See Appendix 13.1 for particulars of individual species

	Number of species		
	Migrants	Residents	Totals
Podicipediformes	0	3	3
Pelecaniformes	0	7	7
Ciconiiformes			
Ardeidae (herons)	0	13	13
Ciconiidae (storks)	1	6	7
Others	0	7	7
Anseriformes	0	16	16
Falconiformes	1	1	2
Gruiformes	0	6	6
Charadriiformes			
Charadriidae (plovers)	6	7	13
Scolopacidae (sandpipers)	16	1	17
Laridae (gulls and terns)	8	8	16
Others	5	8	13
Coraciiformes	0	7	7
	37	90	127

numbers of flightless birds congregate while undergoing the annual moult. In this respect the Cape shelduck behaves in the same way as its congeners in Europe and Australia (Coombes, 1950; Frith, 1967).

There is a strong tendency for some Palaearctic waders to winter farther south in Africa than others. This is exemplified best by the curlew sandpiper *(Calidris ferruginea)* whose breeding grounds in the tundra extend from the Taimyr Peninsula east to Alaska (Glutz *et al*, 1975). Curlew sandpipers occur commonly and abundantly in southern Africa, but they are relatively uncommon at European estuaries in midwinter. There are also populations which winter regularly in New Zealand and Tasmania (Thomas, 1970). In contrast, the dunlin *(Calidris alpina),* which is very closely related to the curlew sandpiper and also breeds in the USSR, is the most abundant wader wintering in Europe or northwest Africa, but very few birds migrate farther south than Mauritania (Prater, 1976). Very little is known about the factors responsible for differences in wintering range between migratory wader species, although interspecific competition may be the mechanism promoting ecological segregation between certain species (Baker & Baker, 1973).

Many races and discrete populations of Palaearctic waders have been recognized, usually by plumage and/ or biometric criteria (Glutz *et al*, 1975, Prater, 1976). The more common species at southern African estuaries are apparently represented mainly by populations drawn from Greenland, Scandinavia and the USSR (Table 13.2) Evidence is accumulating in favour of the occurrence of *Ortstreue,* or fidelity to the same area, in a number of the wader populations which migrate to southern Africa (Middlemiss, 1961, 1962, Elliott *et al*, 1976). Indeed, Elliott *et al*, 1976) have shown that the same

Table 13.2
Origins of some of the main populations of Palaearctic waders which winter at estuaries in southern Africa. Data mainly from Glutz *et al* (1975) and Prater (1976)

	Breeding range of species	Origin of main population in southern Africa	Taxon of main population in southern Africa
Ringed plover*	Holarctic; tundra; boreal and temperate regions	USSR	*Charadrius hiaticula tundrae*
Sanderling**	Holarctic; tundra	Greenland, USSR	*Crocethia alba*
Curlew sandpiper**	Holarctic; tundra	USSR	*Calidris ferruginea*
Knot*	Holarctic; tundra	USSR	*Calidris c. canutus*
Little stint**	Palaearctic; USSR and Scandinavian tundra	USSR	*Calidris minuta*

* two or more races recognized   ** no races recognized

Table 13.3
Relative abundance of avian taxa at estuaries in southern Africa during the austral summer. Based on 264 870 birds counted at estuaries listed in Appendix 13.2. 'Migrants' and 'residents' defined in Table 13.1 and Appendix 13.1

	Percent no. of birds per estuary	Ratio residents : migrants
Podicipediformes	0,7	1 : 0
Pelecaniformes	1,8	1 : 0
Ciconiiformes	23,1	1 : 0
Anseriformes	6,7	1 : 0
Falconiformes	<0,1	1 : 0,03
Gruiformes	0,1	1 : 0
Charadriiformes		
Charadriidae	7,5	1 : 1,5
Scolopacidae	35,5	1 : 1 394,0
Laridae	21,7	1 : 8,5
Others	2,3	1 : 0,4
Coraciiformes	<0,1	1 : 0

curlew sandpipers return to Langebaan Lagoon each year, and they present data which suggest that some birds may travel together and stay in the same flock in southern Africa during one and subsequent migrations. One adult curlew sandpiper which was ringed in the south-western Cape was retrapped at the same locality 11 years later (Elliott *et al*, 1976).

In some years, Palaearctic waders are common at certain estuaries in winter as well as in summer. The curlew sandpipers which overwinter at Langebaan Lagoon in certain years are nearly all juveniles (Elliott *et al*, 1976), and it is likely, but not yet proven, that most overwintering waders of other species are also juveniles. However, some second-year curlew sandpipers do not travel northwards to the breeding grounds, but remain for 32 months in southern Africa. Thus, annual variation in the species composition and abundance of both wintering and overwintering waders in southern Africa may be explained partially by annual variation in breeding success and mortality in the northern hemisphere (Pringle & Cooper, 1975).

The estuarine populations of waders which are resident and breed in southern Africa apparently tend to be relatively stable from one year to the next. However, there are indications that the estuaries receive influxes of birds from seasonal wetlands when these are dry. For instance, Summers *et al* (1976) report some 3 500 Kittlitz's plovers *(Charadrius pecuarius)* at the Berg River estuary in summer. It is likely that these birds breed on the flood plain of the Berg River and other seasonal wetlands and spend the summer in the estuary when the breeding grounds are dry. Kittlitz's plover is common and abundant in the interior of southern Africa where large flocks are often observed for short periods, indicating some sort of local movement (McLachlan & Liversidge, 1970). The species has been observed 'wintering' in large numbers on the coast at East London (Skead, 1967). It is likely that the estuaries act as refuges for inland-breeding populations of this and other indigenous waders (eg the chestnut-banded plover *(Charadrius pallidus)* (Berry & Berry, 1975) during the dry season and during years of extended drought. Relatively few counts have been made of waders at southern African estuaries during the austral winter.

The total number of migratory individuals exceeded the total number of residents by a factor of 1,49 for those estuaries at which all birds were counted (Appendix 13.2). Depending on local water levels, there are more migrants than residents at both Lake St Lucia and Walvis Bay Lagoon, two extremes separated along an ecological gradient (see below).

The Cape cormorant *(Phalacrocorax capensis)* is probably the most abundant water bird along the southern African coastline, extending its range seasonally as far eastwards as Durban and occasionally to Maputo. The total population is in the region of one million birds (Rand 1960, Berry 1976), and large flocks occur at estuaries and lagoons. For instance, Berry & Berry (1975) report influxes of 200 000-300 000 Cape cormorants at Sandvishawe, and Frost & Johnson (1977) observed 15 000 birds nesting at the mouth of the Orange River in December 1976. However, such large flocks visit only sporadically at the few large estuaries and lagoons along the west coast, and typically Cape cormorants frequent the exposed marine coast (Siegfried *et al*, 1975, Berry, 1976). Although a super-abundant species, the relatively infrequent occurrence of large flocks of Cape cormorants at estuaries explains in part the low average abundance rating achieved by the Pelecaniformes (Table 13.3).

Table 13.3 shows that sandpipers (Scolopacidae)

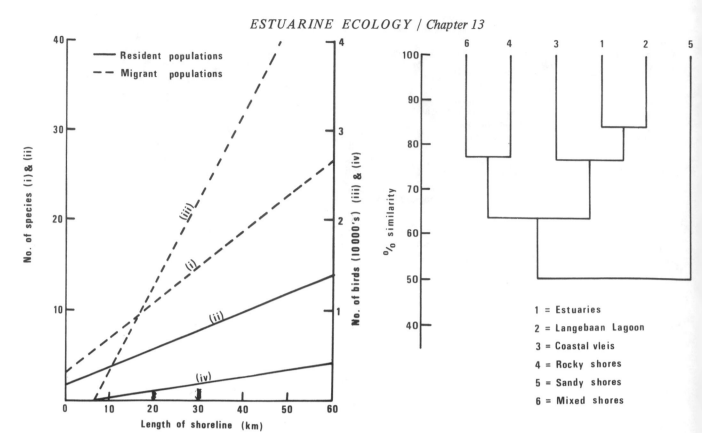

Figure 13.2
Species richness and abundance of waders in relation to length of shoreline of estuary in the south-western Cape during the austral summer. Based on census data taken from Summers *et al* (1976). The linear regression lines are:
  (i) $y = 2,64 + 0,46 \ x$ $r^2 = 0,64$ (sig. at 0,001 level);
  (ii) $y = 1,69 + 0,15 \ x$ $r^2 = 0,50$ (sig. at 0,01 level);
  (iii) $y = 0,59 + 0,09 \ x$ $r^2 = 0,74$ (sig. at 0,001 level);
  (iv) $y = -0,03 + 0,01 \ x$ $r^2 = 0,30$ (sig. at 0,05 level).
The waterbodies included are: Olifants, Berg, Milnerton, Eerste, Lurens, Rooi-els, Palmiet, Kleinmond, Uilkraals, Heuningnes, Breë, Duiwenhoks, Kafferkuils, Gouritz and Kleinbrak estuaries and Langebaan Lagoon.

Figure 13.3
A comparison between wader communities of different wetland types in the south-western Cape during the austral summer. The dendrogram was formed by the Bray-Curtis similarity coefficient and group-average sorting (Field 1970, 1971) using numbers of both migratory and resident wader species, based on census data taken from Summers *et al* (1976). The wetlands included are: estuaries as listed under Figure 13.2. Also: Langebaan Lagoon, Verlorenvlei, Rocherpan, Rietvlei, Paardeneiland-vlei, Wildevoëlvlei, Princess vlei, Rondevlei, Zeekoevlei, Paarde-vlei, Botriviervlei, Kleinriviersvlei, Voëlvlei (Elim), De Hoop vlei, Rietvlei (Albertinia). See Summers *et al* (1976) for definitions of shore types.

are the most abundant birds at estuaries in summer. With the exception of the Ethiopian snipe *(Gallinago nigripennis),* these birds are migrants from the Palaearctic. Next in order of abundance are members of the Ciconiiformes, which are residents. This taxon's relatively superior numerical status is mainly due to the two flamingos, *Phoenicopterus ruber* and *Phoeniconaias minor,* which together accounted for 98 % of all ciconiiforms counted at southern African estuaries. Berry (1975) records peaks of 40 000 and 15 000 flamingos at Walvis Bay and the Sandvishawe lagoon respectively. Smaller populations occur regularly at estuaries of the southern Cape province, and in Natal Lake St Lucia regularly supports some 10 000 flamingos. As far as is known, Lake St Lucia is the only southern African estuary at which *Phoenicopterus ruber* has bred successfully during the last four decades (Porter & Forrest, 1974). However, many decades may

separate breeding attempts there (Broekhuysen, 1975).
  Terns are the most abundant members of the Laridae at southern African estuaries. The 'comic' terns *(Sterna hirundo* and *S. paradisaea)* together accounted for 62 % of all the larids counted; both species are migrants which breed in the northern hemisphere. The ratio of residents to migrants is closer to parity in the Charadriidae (Table 13.3), with Kittlitz's plovers (41 %), chestnut-banded plovers (31 %) and white-fronted plovers *(Charadrius marginatus)* (25 %) making up the bulk of the resident species, and grey plovers *(Squatarola squatarola)* (53 %) and ringed plovers *(C. hiaticula)* (46 %) forming most of the migratory species. The chestnut-banded plover occurs along the south and west coasts, reaching maximum abundance at wetlands along the coastline of Namibia/South West Africa (McLachlan & Liversidge, 1970, Berry & Berry, 1975, unpublished data). The other four species, consi-

Table 13.4
Relative abundance of species of waders at coastal wetlands and on the marine coastline in the south-western Cape. All figures are percentages, based on census data taken from Summers *et al* (1976)

	Coastal wetlands			Marine coastlines		
	Langebaan Lagoon	Estuaries	Coastal vleis	Rocky	Sandy	Mixed
Painted snipe			0,001			
Black oystercatcher	0,09	0,08	0,05	9,48	3,25	7,26
European oystercatcher						0,01
Turnstone	2,63	0,55	0,18	44,16	1,03	18,41
Ringed plover	0,66	2,54	1,11	0,51	1,19	1,00
White-fronted plover	0,78	1,24	0,50	4,94	28,42	13,14
Chestnut-banded plover	0,11	0,31	1,34			
Kittlitz's plover	0,33	15,63	1,42	0,66	1,27	0,89
Three-banded plover		0,05	4,55	0,40		0,04
Golden plover						0,01
Grey plover	9,83	2,65	1,11	0,99	0,71	0,87
Blacksmith plover	0,04	0,57	0,88	0,18		0,12
Ethiopian snipe		0,02	0,05			
Curlew sandpiper	68,96	54,31	51,05	12,01	0,79	15,73
Little Stint	0,94	6,09	9,43	1,61	0,08	0,04
Knot	7,30	4,54	0,24	0,07		0,01
Sanderling	4,85	2,90	0,98	21,24	61,76	37,57
Ruff	0,22	2,49	15,74	0,04		0,46
Broad-billed sandpiper		0,000 4				
Terek sandpiper	0,21	0,32	0,001			
Common sandpiper	0,10	0,16	0,24	1,90	0,08	0,39
Marsh sandpiper	0,09	0,14	0,34			
Greenshank	0,84	1,84	0,83	0,14		1,83
Wood sandpiper		0,000 4	0,05			
Bar-tailed Godwit	0,09	0,07				0,01
Curlew	0,84	1,84				0,86
Whimbrel	1,26	1,95	0,03	1,65	0,08	1,27
Avocet	0,09	2,01	0,87		1,35	0,08
Stilt	0,02	0,45	0,09			0,01
Red-necked phalarope		0,000 4				
Number of birds counted	36 769	23 778	18 838	2 731	1 263	8 394
Number of wetlands surveyed*	1	15	14			
Length (km) of shoreline surveyed	34	128	112	80	100	120

* For names of wetlands, see captions for Figures 13.2 and 13.3.

dered above, are distributed along the entire coastline of southern Africa.

The curlew sandpiper is the most common and abundant Palaearctic wader at southern African estuaries. Leaving aside the common tern and the Cape cormorant, it may be that the curlew sandpiper is the most abundant of all birds encountered regularly in summer at these estuaries which probably support at least 100 000 individuals annually. This might represent about 75 % of the total curlew sandpiper population wintering in the sub-continent. However, to put this into perspective, about two million dunlins winter at estuaries in Europe and northwest Africa (Prater 1976).

During the austral summer, species richness is always greater for migratory than for resident waders at the estuaries of the south-western Cape. It is interesting to compare both the number of species and numbers of individuals of migratory and resident waders. With increasing length of estuarine shoreline (Figure 13.2), the number of species of migrants increases at a higher rate than the number of resident species. With the exception of relatively small estuaries (here defined as those having shorelines of less than about 3,5 km),

resident waders are also less abundant than migratory species whose abundance increases at a higher rate with increase in size of estuary. Figure 13.2 also shows that the rate of increase of migrant abundance is higher than that for species richness. For resident species, however, the slopes plotted (ii and iv in Figure 13.2) for species richness and species abundance, are similar.

Do estuarine waterbird communities differ, in respect to species richness and abundance, from the communities at other wetlands in southern Africa? An attempt to answer this question can be made at present only for waders at coastal wetlands in the south-western Cape region where comprehensive censuses have been undertaken recently (Summers *et al*, 1976). In this region, the estuarine community is most similar to that of the Langebaan Lagoon and least similar to the community of sandy shores along the exposed marine coast (Table 13.4, Figure 13.3).

Table 13.4 shows that the curlew sandpiper predominates in the wader communities of all three types of coastal wetlands in the south-western Cape (see Summers *et al* 1976 for classification of wetlands). The turnstone *(Arenaria interpres)* and the sanderling

(Crocethia alba) feature prominently amongst the waders found on rocky shores of the exposed coast, and the sanderling and the white-fronted plover together account for about 90 % of the wader population on sandy shores. These three species together make up only 4-5 % of the wader population at estuaries.

Figure 13.3 includes all wader species, and the pattern formed is very similar to a dendrogram (not reproduced herein) for the Palaearctic migrants only. A cluster analysis based on resident wader species (Figure 13.4) presents a different picture. A clear difference exists between the communities of the exposed marine coast on the one hand and those of estuaries, coastal vleis and Langebaan Lagoon on the other. The estuarine community is most similar to that of the coastal vleis.

Zoogeography

There have been differences of opinion as to the best biogeographical classification for the southern African marine coastal flora and fauna (reviewed by Brown & Jarman, 1978). However, following Day (1969), who built on the intertidal studies of Stephenson (summarized in T. & A. Stephenson, 1972), three South African provinces may be recognized: an East Coast biota extending on the east coast southwards to about the mouth of the Bashee River; a South Coast biota from the Transkei to Cape Agulhas; and a West Coast biota from Cape Agulhas to Walvis Bay. Day (1969) regarded the coastline between Cape Agulhas and Cape Town as an area of overlap between the South and West Coast provinces. The East Coast biota contains many tropical species, and the West Coast biota contains many temperate endemics. South Coast biota is characterized by a warm-temperate fauna. Brown & Jarman, (1978) have drawn attention to the controversy, some arguments merely semantical and absurd, over whether the West Coast biota should be characterized as either cold or warm temperate species. In this connection, the marine coastal avifauna may be considered to be warm-temperate rather than typically cold-temperate in designation. The evidence for this view includes the fact that marine waterfowl (Anatidae) do not occur in southern Africa (Siegfried, 1970) and also absent are the resident representatives of other groups of marine birds typical of cold-temperate coastal waters elsewhere in the southern hemisphere.

Although the physiographic factors affecting conditions in South African estuaries are varied and complex, according to Day (1974) the distribution of the estuarine biota accords in broad with the three faunistic provinces of the sea shore. Mangroves occur on the east coast southwards to Transkei, and thus fall within the area of the East Coast biota. On a global basis, the birds associated with mangals show no specializations to this environment, and only a few, such as the 'mangrove kingfishers', are restricted to mangals (Macnae

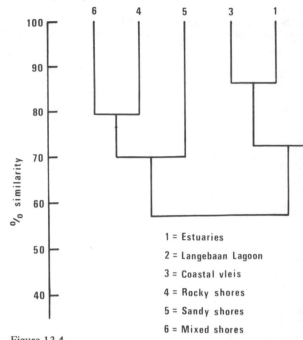

Figure 13.4
A comparison between resident wader communities of different wetland types in the south-western Cape during the austral summer. The dendrogram was formed by the Bray-Curtis similarity coefficient and group-average sorting (Field 1970, 1971) using numbers of resident wader species only, based on census data taken from Summers *et al* (1976). The wetlands included are listed under Figure 13.3.

1968). The sole southern African representative is the mangrove kingfisher *(Halcyon senegaloides)* which, however, is confined to mangrove swamps only during its non-breeding season, moving inland in October for three to four months (McLachlan & Liversidge, 1970).

The distribution of the estuarine avifauna may be considered on a wider basis. In the northern hemisphere the trend is from a migratory avifauna, embracing predominantly ducks, geese and waders in cool temperate-zone estuaries, to a largely resident, mainly fish-eating community of birds in tropical estuaries. On this basis one might predict that the species composition of the South African estuarine avifauna changes along a temperature gradient. In short, the bird communities of the estuaries along the northeastern coasts should conform more closely to the tropical condition than to those on the west coast. The validity of this premise is examined below.

Day's (1974) comparison of the biotas of three groups of South African estuaries (viz Natal and Transkei, the southern Cape, and the Atlantic coast) relies essentially on presence-absence records of species of benthic invertebrates and fishes. Comparable data on the presence-absence of species of waterbirds are available for only 10 of the 25 estuaries for which Day *(op cit)* compiled faunal lists. Nevertheless, it is possible to consider the composite avifaunas of a total of 26 estuaries (Table 13.5), in the three regions distinguished

Table 13.5

Composite totals of numbers of species in major groups recorded at estuaries in three regions of the South African coast. Number of migratory bird species in each group of estuaries shown in brackets. Data on plants, invertebrates and fish are from Day (1974) for the following estuaries: Umgazana, Bashee, Kosi Bay, St Lucia, Richards Bay, Mlalazi, Karridene, Umgababa, Mkomazi, Amanzimtoti, Mzimkulu, Mtamvuna, Mzimvubu, Mgazi and Umbanyana (Natal and Transkei); Zwartkops, Knysna, Klein River, Keiskamma, Great Brak and Breede (Southern Cape); Milnerton, Orange, Berg and Olifants (Atlantic coast). Data on the birds are for: Mhloti, Tongaat, Tugela, Mlazi, Mgeni, Mayameni, Mhlanga, Mvoti, St Lucia, Richards Bay, Mlalazi, Amanzimtoti, Mzimkulu, Mfolozi and Scottborough (Natal and Transkei); Zwartkops, Klein River, Gamtoos, Maitlands, Sundays, Seekoeis and Kabeljouws (Southern Cape); Milnerton, Berg, Olifants and Verlorenvlei (Atlantic coast)

	Natal & Transkei	Southern Cape	Atlantic coast
Total plants	25	35	30
Total benthic invertebrates	237	357	59
Total fish	159	53	18
Total birds	104(32)	97(30)	81(25)
Podicipediformes	1	3	3
Pelecaniformes	6	6	6
Ciconiiformes	24	21(1)	14
Anseriformes	13	9	8
Falconiformes	2(1)	2(1)	1
Gruiformes	3	5	4
Charadriiformes	48(31)	48(28)	42(25)
Coraciiformes	7	3	3

Table 13.6

Incidence of species belonging to the East African tropical avifauna (see Winterbottom 1967, 1972) in total number of bird species recorded at estuaries in three regions of the South African coast. All figures are percentages except those in parentheses which are absolute numbers of species recorded for the three groups of estuaries listed in Table 13.5

Natal & Transkei		Southern Cape		Atlantic Coast	
Res./All spp.		Res./All spp.		Res./All spp.	
67	48	63	44	58	41
(109)		(100)		(87)	

by Day. The most striking fact is the approximately 25 % increase in the number of bird species in a shift from the Atlantic coast estuaries to those in Transkei and Natal. However, this (25 %) is much smaller than the increases found for benthic invertebrates or fishes (Table 13.5).

Amongst the major taxonomic groups of resident birds, the number of ciconiiform species differed most (38 %) between Atlantic and Natal-Transkei estuaries. The storks, herons and their allies accounted for 66 % of the difference in the overall richness of resident species in the two regional groups of estuaries. The number of resident charadriiforms was the same (17) in both groups of estuaries. It may also be noted that there are very similar percentages (30-31 %) of migrants in all three regional groups of estuaries. The Atlantic coast group has only one migratory species,

the black tern *(Chlidonias niger)* which it apparently does not share with either of the two other regions. The southern Cape has two species which it does not share with the north-eastern group and seven not present in the Atlantic group. The north-eastern group has five and eight species which it does not share with southern Cape and Atlantic groups respectively. However, most of these species are normally rare in southern Africa, and it is likely that in future they will be recorded at estuaries in all three regions.

An apparently distinct fresh-water avifauna extends from Uganda southwards into central Africa and the north-eastern part of southern Africa. Winterbottom (1967b, 1972) listed 52 species as characteristic of this avifauna which he called the East African Tropical Avifauna. It should be noted that none of the Palaearctic migrants which winter in Africa is included in the list. Of the 52 species, 47 feature in the list of 127 species recorded at estuaries for the present survey. The remaining five include two normally very cryptic species, the water rail *(Rallus caerulescens)* and little bittern *(Ixobrychus minutus)*, the crowned crane *(Balaearica regulorum)*, the white-winged plover *(Hemiparra crassirostris)* and the red-winged pratincole *(Glareola pratincola)*. However, all five species have been recorded at estuaries in South Africa; the red-winged pratincole breeds regularly at the Mvoti estuary in Natal. Table 13.6 summarizes the incidence of the 52 species of the East African Tropical Avifauna in the three regional groups of South African estuaries. There is a decline from the full complement of 52 species at estuaries along the north-eastern coast to 36 species at Atlantic coast estuaries. Moreover, at least six of these species in both the southern Cape and Atlantic coast groups are rare, irregular stragglers. Examples include the dwarf goose *(Nettapus auritus)*, comb duck *(Sarkidiornis melanotus)*, the whistling ducks *(Dendrocygna* spp) and African jacana *(Actophilornis africanus)*.

The main conclusions emerging from the foregoing zoogeographical account are: estuaries along the north-eastern coast of South Africa tend to support more 'tropical' bird species than do Atlantic coast estuaries; these 'tropical' species form a higher percentage of the total species found at estuaries along the north-eastern coast than along the Atlantic coast; Atlantic coast estuaries tend to be poorer in bird species than estuaries along the north-eastern coast, but both regional groups of estuaries support about equal proportions of migratory species. Clearly, these differences provide some evidence in favour of a subtraction southwards of 'tropical' species and although the difference *(ca* 25 %) in species richness between the Atlantic and the north-eastern coast estuarine avifaunas is not great, it appears sufficient to gainsay Winterbottom's (1972) statement that a list of estuarine bird species 'is not likely to be materially different from Walfish Bay to Beira'. This does not, how-

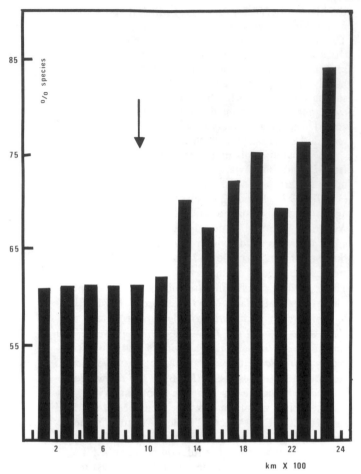

Figure 13.5. Frequency of occurrence of 90 resident bird species (recorded at estuaries included in this survey) in 13 200 kilometre sectors of the South African coastline from the Orange River (sector 1) around to the Natal-Mozambique border (sector 13). Approximate position of Cape Agulhas is indicated by an arrow.

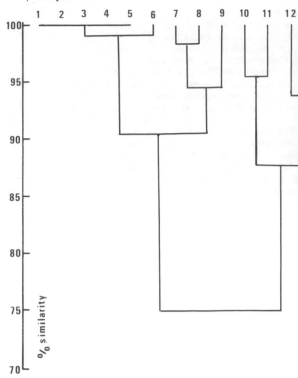

Figure 13.6
Dendrogram formed by the Bray-Curtis similarity coefficient and group-average sorting (Field 1970, 1971) using distributional ranges of 90 resident bird species (recorded at estuaries included in this survey) in thirteen 200 kilometre sectors of the South African coastline from the Orange River (sector 1) around to the Natal-Mozambique border (sector 13).

ever, help in resolving whether the distribution of the South African estuarine avifauna conforms with Day's (1969) three faunistic provinces. In other words, are there really three distinct estuarine avifaunal provinces along the South African coast and does an avifaunal ordination conform with either a benthic invertebrate or fish ordination, these two groups having been studied with a greater degree of thoroughness than the estuarine plants?

The voluminous literature dealing with the distribution of mainly intertidal animals along the South African coast (reviewed by Brown & Jarman, 1978) does not guide the reader clearly to definitions of the discreteness and attendant geographical boundaries of biotic provinces. Some of the interpretations offered for the different distributions of organisms are almost entirely subjective, and virtually all the zoogeographical accounts rely entirely on lists of the presence or absence of selected groups of invertebrates. It appears that there is a need for a critical testing and evaluation,

using several modern techniques, of the analyses and the biogeographical conclusions attending the distribution of the flora and fauna along the southern African coast.

Figure 13.5 is based on a partitioning of the South African coast, from the Orange River to the Natal-Mozambique border, into thirteen 200 km sectors. Computation of the percentage of the resident bird species (recorded at estuaries included in this survey) in each of the 13 sectors reveals again the relative poverty of the west coast estuaries, ending at about Cape Agulhas. Thereafter, a gradual increase in species richness occurs eastward. Forty (44 %) species occurred in all 13 sectors, and the breeding range of only one population (the roseate tern *Sterna dougallii*) does not extend beyond South Africa. However, breeding populations of this species do occur elsewhere in the world. Amongst the 90 resident species considered here six can be regarded as more or less southern African endemics, namely Cape gannet *(Sula capensis),* Cape cormorant, Cape shelduck, black oystercatcher *(Haematopus moquini),* Hartlaub's gull *(Larus hartlaubi)* and Damara tern *(Sterna balaenarum).* In addition, three ducks, the Cape shoveller *(Anas smithii),* yellowbill *(A. undulata)* and Maccoa *(Oxyura maccoa),* occur

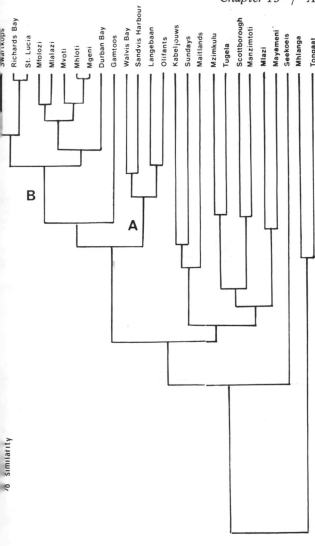

Figure 13.7
A comparison between the waterbird communities of selected southern African estuaries and lagoons. The dendogram was formed by the Bray-Curtis similarity coefficient and group-average sorting (Field 1970, 1971) using presence/absence of 127 species of birds (Appendix 13.1).

the resident avifauna of the Atlantic coast estuaries (contained in section 1-5). Thus, it seems fair to say that, only along the Atlantic coast of southern Africa is there a distinct indigenous estuarine avifauna, and that while some elements of this fauna extend around Cape Agulhas (the southern tip of the continent) fewer penetrate farther eastward than Transkei.

In summary, then, the geographical distribution of resident species of estuarine birds in southern Africa provides support for the notion of a distinct west-coast biotic province extending southwards from southern Angola. The area roughly between Cape Point and Transkei is best regarded as a zone of over-lap for elements of the west-coast province and a dif-fusion southwards of a tropical avifauna found along the east coast of Africa.

Turning to the individual estuaries and their res-pective lists of bird species, Figure 13.7 tends to re-affirm the relative discreteness of the Atlantic coast avifauna (A). The estuaries of the north-eastern coast tend to group together, but the cluster (B) also includes some estuaries of the southern Cape. This configuration may be due partly to enhanced species richness as an effect of the relatively large size of the estuaries concerned (see below). The balance of the Natal estuaries presents a confused picture and this may be due to their depauperate avifauna, and impo-verished invertebrate fauna (*vide* Day, 1974), as a con-sequence of their small size, paucity of intertidal banks of sandy mud, and artificially disturbed state.

'Area effect' as well as geographical location appears important in influencing species richness and abundance of avian communities at southern African estuaries. In North America Oviatt *et al* (1977) found a two-fold variation in species richness and a ten-fold variation in abundance of birds at intertidal salt marshes along the Rhode Island coast, and showed that these variations were more dependent on the size of the marsh than on any other factor. The 'area effect' may also be im-portant in relation to the fish population, both the pelagic filter feeders and the large swift predators but it would be difficult to find an objective method for separating groups of estuaries to test this assumption.

In recognizing that tropical regions generally sup-port a more diverse fauna than do colder regions, San-ders (1968) reviewed and criticized indices used to determine 'species diversity' (ie the number of species present and their numerical composition), since such indices are dependent on sample size, and he deve-loped a rarefaction method which allows samples of different sizes to be compared by interpolation. Figure 13.8 presents rarefaction curves calculated from the numbers of bird species and individuals recorded during the austral summer for selected estuaries and lagoons. The plots for estuaries in the north-eastern region of southern Africa tend to be high, indicating that these estuaries have greater avifaunal complexity than do the lagoons along the Atlantic coast. It must be stressed,

at estuaries only in southern Africa, although inland the species extend northward well beyond the borders of the subcontinent. All nine of these species are essen-tially confined to the temperate zone and occur most abundantly at estuaries along the west coast and to a lesser extent the south coast of southern Africa. There are no estuarine birds endemic to the east coast of the subcontinent. Thus, on the basis of endemic species alone, the avifaunas of the west and north-eastern coasts are distinct. Apparently there is no intermediate, distinct southern Cape coastal group.

This conclusion is supported by cluster analysis (Figure 13.6), showing the homogeneous character of

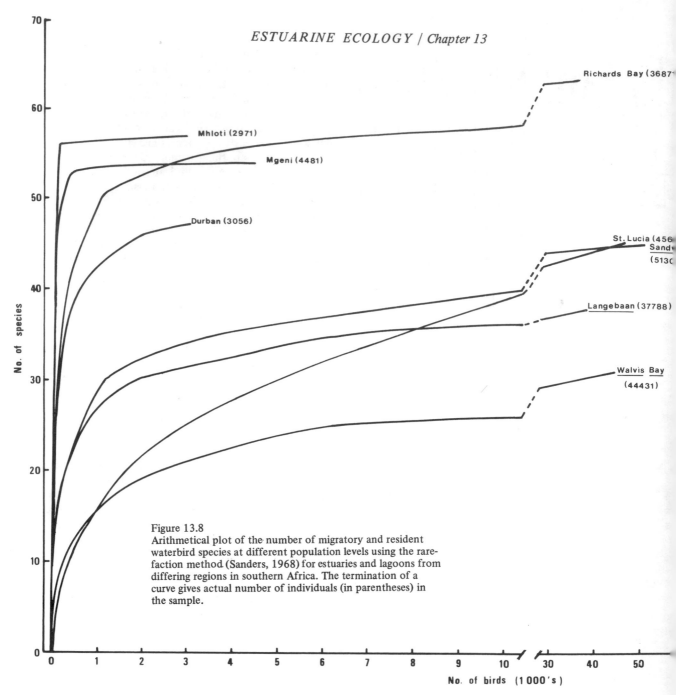

Figure 13.8
Arithmetical plot of the number of migratory and resident waterbird species at different population levels using the rarefaction method (Sanders, 1968) for estuaries and lagoons from differing regions in southern Africa. The termination of a curve gives actual number of individuals (in parentheses) in the sample.

however, that the analysis was based on a limited number of samples and that no Atlantic estuaries *(sensu stricto)* were included. The relatively low rank of St Lucia can be explained perhaps by the fact that the censuses of birds there were carried out during a year of exceptionally high water levels. Clearly there is a need for additional census data for birds at a greater number of estuaries in southern Africa than are available at present.

Trophic relations

The regular seasonal abundance of migratory birds at southern African wetlands is a major ecological phe-

nomenon. However, in spite of their numerical abundance, the migrants accounted for only about 5 % of the avian biomass for those estuaries at which all birds were counted in summer. The resident pelicans, cormorants and flamingos together made up 85 % of the biomass (Table 13.7).

All the migrants which visit southern African estuaries are carnivores, primarily eaters of invertebrates, while resident invertebrate feeders are also very common (Table 13.8). Taken together, the migratory and resident carnivorous species make up 89 % of the estuarine avifauna and 75 % of its biomass. Within the resident community 41 % of the species feed predominantly on fish and/or frogs, where 21 % of the migrants

Table 13.7
Relative biomass of avian taxa at estuaries in southern Africa during the austral summer. Based on 264 870 birds counted at estuaries listed in Appendix 13.2. 'Migrants' and 'residents' defined in Table 13.1 and Appendix 13.1

	Percent biomass of birds per estuary	Ratio residents : migrants
Podicipediformes	<1,0	1 : 0
Pelecaniformes	16,6	1 : 0
Ciconiiformes	68,4	1 : 0
Anseriformes	6,6	1 : 0
Falconiformes	<1,0	1 : 0,15
Gruiformes	<1,0	1 : 0
Charadriiformes		
Charadriidae	<1,0	1 : 5,05
Scolopacidae	2,7	1 : 38 622,0
Laridae	3,7	1 : 3,11
Others	<1,0	1 : 0,19
Coraciiformes	<1,0	1 : 0

Table 13.8
Classification of avian species by diet at estuaries in southern Africa during the austral summer. See Appendix 13.1 for particulars

Diet	Number of species		
	Migrants	Residents	Totals
Mainly fish and frogs	8	37	45
Mainly invertebrates	29	39	68
Mainly vegetable matter	0	14	14
Totals	37	90	127

Table 13.9
Classification of composite totals of numbers of avian species by diet at estuaries in three regions of the South African coast. All figures are percentages based on total numbers of species given in Table 13.5 which also lists the names of the estuaries in each region

Diet	Natal & Transkei		Southern Cape		Atlantic Coast	
	Migr.	Res.	Migr.	Res.	Migr.	Res.
Mainly fish and frogs	6	28	5	30	6	25
Mainly invertebrates	25	28	26	30	25	34
Mainly vegetable matter	0	13	0	9	0	8

Table 13.10
Classification of 127 avian species according to body mass. All figures are percentages. See Appendix 13.1 for particulars

Mass (g)	Migrants	Residents
1 - 100	44	14
101 - 300	38	22
301 - 1 500	15	41
1 501 - 4 500	3	19
4 501 - 13 500	0	4

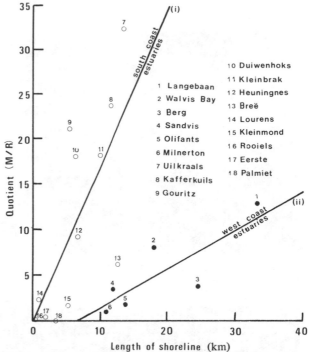

Figure 13.9
Biomass of migratory waders divided by biomass of resident waders in relation to length of shoreline of estuaries in South West Africa, and west coast (solid circles) and south coast (open circles) estuaries in the south-western Cape during the austral summer. Based on census data taken from Summers *et al* (1976) and Underhill & Whitelaw (1977). The linear regression lines are:

(i) $y = 0,61 + 1,79 \ x \ r^2 = 0,52$ (sig. at 0,01 level);
(ii) $y = -2,91 + 0,43 \ x \ r^2 = 0,71$ (sig. at 0,05 level).

The waterbodies include those listed under Figure 13.2, and also Sandvishawe and Walvis Bay lagoons.

prey species in the southern African estuarine systems, is likely to be considerably less than that exerted by the resident avian community.

Apparently the ratio of migratory to resident avian carnivores is similar in different geographical groupings of estuaries in South Africa (Table 13.9). However, there are almost twice as many fish and/or frog-eating resident species in Natal-Transkei than in the Atlantic coast region. Since the proportions of resident and migratory species taking live prey are similar, to what extent is there competition between these two groups for food resources? A partial answer, in very broad terms, is indicated by the fact that the migrant community contains a relatively greater proportion of small species which weigh less than 300 g (Table 13.10). This suggests that residents generally feed on larger prey than the migrants. In summary, the carnivorous resident avifauna contains more species but fewer individuals than the carnivorous migratory community which consists mainly of relatively small-bodied birds which take mainly small-bodied prey. There are no herbivorous migratory species in the South African

regularly do so (Table 13.8). These migrants comprise 11 % of the total biomass of all species which take vertebrate prey and are almost exclusively fish-eating terns (Sternini). Amongst the species which feed on invertebrates, the greater flamingo alone accounts for 90 % of the biomass. Thus, in spite of their superior abundance, the carnivorous migrants' impact, on the

Continued on page 240.

Curlew sandpiper at Langebaan (Photo: J. Cooper)

Sanderling feeding (Photo: J. Cooper)

Hottentot teal (Photo: J.C. Sinclair)

Waders, Duck and terns in Mvoti (Photo: J.C. Sinclair)

Flamingos in Lake St Lucia (Photo: A. Berruti)

Common tern (Photo: J.C. Sinclair).

Fish eagle (Photo: Peter Steyn)

Grey heron feeding (Photo: J.C. Sinclair)

White pelican (Photo: J. Cooper)

Pelican colony breeding at St Lucia
(Photo: J. Cooper)

Fishermen — hammerkop, grey heron,
goliath heron (Photo: George Begg)

estuarine avifauna, and the resident species which feed on algae and aquatic macrophytes comprise 25 % of the total avian biomass; the lesser flamingo and 11 anatids are the major elements of this trophic assemblage.

Inland waterbodies support a different trophic apportionment. Based on census data reported by Winterbottom (1960) for waterbirds at vleis in the south-western Cape during summer (December-February), I calculated that herbivores made up 54 % and carnivores 46 % of the avian biomass; coots *(Fulica cristata)* accounted for about 53 % of the herbivores, whereas waders (Charadriidae and Scolopacidae) represented 4 % of the carnivores. Curlew sandpipers contributed heavily (40 %) to the wader biomass, and two other important contributors were the ruff *(Philomachus pugnax)* (30 %) and the little stint *(Calidris minuta)* (13 %). Populations of these latter two species form a much smaller proportion of the wader biomass at estuaries, and they, together with the marsh sandpiper *(Tringa stagnatilis)* and wood sandpiper *(T. glareola),* belong typically to that part of the Palaearctic wader community which winters at freshwater wetlands.

Figure 13.9, based on quotients obtained by dividing the biomass of resident waders (R) into the biomass of migrants (M), demonstrates that the biomass of migrants is greater than that of residents except at those west coast estuaries which have shorelines of less than 5 km. The quotients tend to be lower for west coast estuaries which clearly are different from those of the south coast of the Cape. One might speculate that seasonal variations in productivity are greater in south-coast than in west-coast estuaries. The latter might be more productive throughout the year and, hence, capable of supporting relatively greater populations of waders in winter. A corollary of this is that migratory waders capitalize on the presumed higher productivity of south-coast estuaries in summer. In either event, relatively large-bodied species, of both migrants and residents, feature more abundantly at large than at small estuaries.

In the curlew sandpiper, males and females take slightly different food types at Langebaan Lagoon (Puttick, 1978). The females are significantly larger and have longer beaks than males, and a sex ratio favouring females was found at Langebaan Lagoon (Elliott *et al*, 1976). These workers suggested that the excess of females may be related to their relatively large size which may give them an advantage over males in times of food shortage.

The general principle that greater segregation of niches occurs when food is relatively scarce has been applied to waders by Baker & Baker (1973), who concluded that resource partitioning is greater in winter when density-dependent mortality may limit population size. This view may not apply generally to the migratory wader community of southern African estuaries, because southern Africa is right at the end of the species' wintering range and consequently several populations may normally be exploiting only a part of the suite of food resources potentially available. If this is true then competition between the sexes and the disparate sex ratio of the curlew sandpiper could be caused by conditions outside southern Africa.

Segregation between species and between sexes of Palaearctic waders with respect to use of habitat and food at Langebaan Lagoon has been investigated by Summers (1977) and Puttick (1978, 1979), but much remains to be studied for a clear understanding of the habitat requirements of each species. For instance, about 4 000 knots *(Calidris canutus)* winter at Langebaan Lagoon (Pringle & Cooper, 1975), which is at least 50 % of the estimated total population of 5 000-10 000 birds between Walvis Bay and Durban (Dick *et al*, 1976). What is special about Langebaan for the knot? Could other sites support more knots? Is Langebaan supporting a maximum number of knots? These questions are not only of academic interest; they bear strongly on conservation-management planning as well.

The Terek sandpiper *(Xenus cinereus)* is another species of interest to nature conservationists as well as academic scientists. It is known that the species prefers estuarine areas with soft mud (Summers *et al*, 1976), and it is relatively common in tropical mangrove estuaries. Substrata consisting of soft mud (ie high percentage silt and clay) occur in estuaries with Bokkeveld shales, such as the Gouritz, Breede or Olifants rivers where mud flats occur frequently and so do Terek sandpipers (personal observation). Is the species' distribution and abundance in southern Africa regulated by the availability of its preferred substrate? The question cannot be answered at present, but perhaps the Terek sandpiper and some of the other relatively scarce Palaearctic waders in southern Africa are 'under-represented'. If this is so, it implies at least in theory that the existing estuaries can accommodate more birds displaced from elsewhere. This raises the ecologically interesting point concerning the capacity of southern African estuaries to carry more predatory birds capable of exploiting a potential surplus of food. However, more needs to be known about the role of carnivorous fishes, as competitors for food, before the topic can be taken further.

Most of the Palaearctic waders which regularly visit southern African estuaries spend about eight months (September-April) in the south (Pringle & Cooper, 1975). Potential food supplies for Palaearctic waders are relatively high during the austral winter at Langebaan Lagoon (Puttick, 1977). The fact that first-year curlew sandpipers normally overwinter is apparently not related to any difficulty the birds might have in finding food but in feeding efficiently for premigratory fattening (Puttick, 1979). A point worth mentioning here is that the trophic apparatus (mainly beak morphology) of migratory Palaearctic waders is believed to be adapted primarily to dealing with prey found in the birds' wintering quarters rather than in their breeding areas (Baker & Baker, 1973).

Apparently, the waders do not depart for the north to escape adverse conditions in the south, and the advantages gained through breeding in the north must be considerable to offset the presumed hazards of a 30 000 km migration each year. Ringing studies have suggested that mortality is relatively high in the migratory populations of many Palaearctic Charadrii. For instance, in the European turnstone, as many as one-third of the adults may die each year, and about half of the fledglings survive the first year (Boyd, 1962). In answering why many wader species do not breed in southern Africa, it may be significant that these birds do not feed their young which are dependent on an abundance of protein-rich insects as food. The young hatch with relatively short beaks, and unlike their parents they are inefficient at probing the ground for food items. Very little is known about insects in southern African estuaries, particularly in salt marshes, but Puttick's (1977) studies at Langebaan Lagoon indicate that the above-ground insect biomass is relatively low. Apparently the greatest part of the available invertebrate biomass is buried in the top 10 cm of the sand and mud flats (Boltt 1975, Puttick 1977). Baker & Baker (1973) concluded that 'the great abundance of suitable food that becomes available on the breeding grounds for precocial shorebird wader young probably is the ultimate explanation for the evolution of migratory behaviour in these birds'.

The breeding performance of waders resident at southern African estuaries apparently differs markedly from the 'boom or bust' system of many of the Palaearctic migrants which use the highly productive, but seasonally short-lived, tundra. For instance, in the white-fronted plover, at Langebaan Lagoon, mated birds occur in pairs which retain exclusive territories of about 1,5 ha throughout the year. They breed during an eight-month season extending from June to January. In spite of the long season and repeated breeding attempts, reproductive output is very low. However, this is offset by a relatively long life expectancy (unpublished data).

A number of Palaearctic waders undergo post and pre-nuptial moults during their stay in southern Africa (Elliott *et al*, 1976, Summers, 1977, Summers & Waltner, 1979). Some first-year birds begin moulting their primary flight feathers within three months of their arrival, whereas others wait for nine months. A second moult starts immediately after the completion of the first moult, although it may start while the first one is still in progress. Therefore, the second moult involves the replacement of virtually new feathers. Although there may be other explanations, the significance of this unusual moult can be viewed as an adaptation for ensuring travelling with a set of new flight feathers when first returning north as a potential breeding bird, and ensuring that the wing does not become too worn

during the first long stay in southern Africa.

Migrating curlew sandpipers generally travel directly to the breeding grounds regardless of whether their route passes over areas like the Sahara, the Congo forests or the Arabian desert (Elliott *et al*, 1976). Apparently a similar direct route is followed by migrating knot (Dick *et al*, 1976). This direct route (on a great circle) is the shortest distance between Siberia and South Africa, and involves a flight of 13 700 km compared with, for instance, 15 700 km for a route via the west African coastline.

Probably most of the Palaearctic waders visiting southern African estuaries must make long, non-stop flights between resting and/or feeding stations while on migration. In southern Africa adult birds weigh least in mid-summer, and thereafter they accumulate fat reserves amounting to 30-50 % of their whole body mass in April, prior to northward migration. Such fat deposits allow the birds to fly non-stop for distances varying between 2 500 and 6 000 km, depending on the species concerned (Elliott *et al*, 1976, Summers & Waltner, 1979). Long, non-stop flights are normally made at night. Immature waders at southern African estuaries tend to weigh more in winter than during the summer. This may be an adaptive response to the cold weather during the austral winter.

The average body mass of premigratory curlew sandpipers at Langebaan Lagoon increases at a rate of 0,10 g.day^{-1} from mid-February to mid-March followed by 0,57 g.day^{-1} for the next two weeks and 0,73 g.day^{-1} for the first two weeks in April (Elliott *et al*, 1976). During this time the birds' feeding rate increases, concomitantly with an increase in food biomass potentially available to the birds (Puttick 1977, 1979). At Langebaan, the curlew sandpiper feeds mainly on the snail *Assiminea globulus*, and on polychaetes and small crustaceans, although insect larvae, mainly Stratiomyiidae, are taken in the salt marshes at high tide especially during the birds' premigratory, fattening period (Puttick, 1978). This extension in foraging habitat probably reflects the birds' need to feed for longer than the usual period of daylight exposure of the intertidal areas.

Very little is known about actual densities of birds at southern African estuaries, and the community of waders at Langebaan Lagoon is the only 'estuarine' avifauna which has been studied with any degree of thoroughness. At Langebaan, the average overall density of waders was 1 564 birds per km^2 yielding a biomass of 145 kg per km^2 (Summers, 1977). These figures compare favourably with those obtained for waders at the coastal wetland of Banc d'Arguin in Mauritania and at the Wash in England (Dick 1975). Summers (1977) has calculated that the waders at Langebaan deposited a total of 257,5 kg of feathers per year, or 0,06 kcal per m^2.yr^{-1}, in the lagoon system. Additionally, the birds returned annually 2,1-6,2 kcal per m^2 as faeces. The birds' contribution of minerals to the lagoon system has not yet been investigated. The estimated energy removed from the system in the form of food, by the wader community, amounted to 20,8 kcal per m².yr⁻¹ or 4,2 g ash-free dry mass of animal food per m² per year (Summers, 1977). These figures are similar to those for the Wadden Zee where 18,5 kcal per m².yr⁻¹ were removed by all carnivorous birds (Swennen, 1975), and 17,1 kcal per m².yr⁻¹ for the Grevelingen estuary in the Netherlands (Wolff *et al*, 1975).

The figures given above are slightly higher than those obtained for the energy consumption of birds in other ecosystems. Holmes & Sturges (1975) report 7,4 kcal per m² (*ca* 1,5 ash-free dry mass of food per m²) as the annual intake of birds in deciduous forest in New Hampshire (USA), representing 0,17 % of the above-ground primary production. Wiens & Nussbaum (1975) give 10-21 kcal per m² (*ca* 2,0-4,2 g ash-free dry mass per m²) per season (April-October) as the intake of birds in coniferous forest in Oregon. Estimates for the energy consumption of birds in other terrestrial ecosystems range between 0,2 and 3,3 g per m² (ash-free dry mass) per season (Wiens & Nussbaum, 1975).

The carnivorous birds studied by Swennen (1975) made up 96 % of the total avian biomass at the Wadden Zee and their energy intake represented about 0,5 % of the total amount of food (700 g dry weight per m² per year) available for consumers. In a comparative study of the trophic role of birds in the estuary and lake at Grevelingen, Wolff *et al* (1975) found that birds feeding on zoobenthos took about 6 and 3 % of the total zoobenthos production in the estuary and lake respectively. Apparently tides contributed to the greater exploitation of the zoobenthos by waders in the estuary. Wolff *et al (op cit)* also noted a relatively heavy level of predation by piscivorous birds in the lake, and they assumed that greater visibility in the water of the lake favoured better capture of prey. Wolff *et al (op cit)* concluded that the presence or absence of tides greatly influences the way birds may exploit a marine environment, and that a shift from secondary and tertiary consumers to primary consumers is predominant. It would be interesting to know whether this change occurs in open and closed estuaries in southern Africa. It should also be noted that increasing turbidity reduces the impact of avian piscivores and carnivorous fishes which seek their prey by sight, and favours fishes which hunt by other senses (smell and lateral-line system). Thus the clear waters of Langebaan Lagoon favour piscivorous birds and there are few piscivorous fish there. Similarly, seasonal changes in the incidence of piscivorous birds can be expected to occur in estuaries such as St Lucia in response to wind-induced seasonal changes in turbidity.

The full impact of birds on the ecosystem of any southern African wetland is not known, because of the paucity of information on the amount of food available for consumers. However, at Langebaan Lagoon the curlew sandpiper population alone consumed 12,9 % (91 kJ.m⁻².yr⁻¹) of the estimated gross annual production of the intertidal zoobenthos living on the surface or in the upper 10 cm of the substrate (Puttick, 1979). The effect of estuarine waders on their prey populations apparently varies considerably, but little is known about the behaviour of the prey species and the effects which this has on the feeding behaviour of the predators (Drinnan, 1957, Goss-Custard, 1969, Prater, 1972).

Studies on the fish-eating birds at southern African estuaries have only just begun, but there are hints that at times these predators can exert a considerable effect on fish populations. For instance, during the closure of the mouth of the West Kleinemond estuary, in the eastern Cape province, the population of juvenile bream *(Rhabdosargus holubi)* decreased in five months by 80 % (Blaber, 1973). This relatively high mortality rate was density dependent and apparently was mainly due to predation by piscivorous birds, the numbers of which were related to the density of fish. The grey heron *(Ardea cinerea)* and the two cormorants *Phalacrocorax lucidus* and *P. africanus* were the most numerous piscivorous birds. Whitfield (1978) reports a highly significant correlation between the relative density of fish and piscivorous bird numbers at Lake St Lucia where Whitfield & Cyrus (1978) found that the avian community associated with newly-created backwaters was dominated initially by piscivorous birds but with the decrease in water levels, invertebrate-eating wading birds became abundant.

The white pelican *(Pelecanus onocrotalus)* occurs mainly at relatively large wetlands, including estuaries in southern Africa, and tends to forage in areas of open water which are least affected by tides. Provisional, unpublished estimates for the fresh mass of fish taken by pelicans at Verlorenvlei (1 070 ha), St Lucia (30 050 ha) and Berg river estuaries (500 ha) are 6,8, 2,6 and 5,8 g per m².yr⁻¹ respectively. Annual average density and biomass of white pelicans at these three wetlands, taken together, are 5,2 birds per km² and 55 kg per km² respectively. This biomass is at least four times greater than that given by Swennen (1976) for all piscivorous birds in the Dutch Wadden Zee, suggesting that the ecological impact of fish-eating birds is greater in southern Africa than in north European estuaries. It is worth noting, too, that the white pelican's biomass (55 kg.km⁻²) falls within the combined biomass range (40-80 kg.km⁻²) for top mammalian predators (lion, leopard, cheetah, Cape hunting dog and hyaenas) in the African savanna biome (Bigalke, 1978).

The density of birds increases progressively from the open ocean in the North Atlantic across the North Sea and Dutch coast, culminating in the Wadden Zee (Swennen, 1976). This trend, expressed as 1:34:110:2 000, approximates a comparable trend, 1:1,8:2,4:7,0, for primary production (Swennen, *op cit*). However, differences in the number of trophic levels as well as differences in primary production influence the abundance and form of top predators in an ecosystem (Ryther, 1969). At an average ecological efficiency of 10 % (Slobodkin, 1960), progressively less energy is available for successive links in a food web. According to Swennen (*op cit*), 'in the Wadden Zee the top predators are much better off than those in the North Sea and the Atlantic Ocean', because they can operate early in a relatively richer food chain. The potential prey of avian carnivores in the Wadden Zee consists mainly of items, such as small and/or young fish, with which the birds can easily deal. By contrast, in the open sea, with its longer food chains, the biomass of potential prey consists mainly of units too small or too big to be caught or dealt with by birds (Swennen, *op cit*). Moreover, the potential prey biomass is spread over a much deeper water column than in estuaries, making prey-capture relatively difficult for birds. In short, differences in the number of trophic levels and in primary production between estuaries and the open sea can apparently be predictors of differences in avian species richness and abundance. The biomass of predatory birds is considerably greater in estuaries generally than in other marine biotopes, but with increasing depth and volume of water fish become more important than birds as the final predators in the food web (Macan, 1977).

Conservation

I will not outline here why birds are an important component of estuarine ecosystems, and why the estuarine avifauna deserves to be protected. Nor do I intend to assess what changes are likely to take place in the avifauna in the future. Indeed, man-induced changes in the southern African estuarine avifauna cannot at present be predicted in detail, because of the dearth of information on the ecological requirements of most individual species (Siegfried, 1978). However, some broad correlations between particular estuarine environments and communities of birds are evident.

Any reduction in tidal exchange, due to closure of the mouth of the river or a persistently strong flow of freshwater, will result in a reduction of waders dependent on regular daily inundation and exposure of tidal mud and sandflats. Conversion of a shallow estuary with extensive tidal mud and sandflats to a narrow, relatively permanent channel, as is promoted by the construction of constricting road and rail bridges, will have the same effect. These conditions, provided that the water is clear, will, however, in general favour fish-eating birds especially those species which actively pursue their prey underwater. This is perhaps best exemplified by Blaber's (1973) account of the bonanza exploited by certain fish-eating birds after closure of the mouths of estuaries in the eastern Cape Province.

The foregoing ecological correlations, promoted by man-induced changes to the estuarine environment, are already commonplace along the South African southern and eastern seaboards, but nowhere so obvious as in Natal. Extensive artificial alterations of estuarine ecosystems have taken place there, and concomitantly, conservation interests have not always been adequately represented. In considering the estuarine avifauna alone, present knowledge is inadequate for developing a fully rational argument in a national plan for the type, location, size and management of a series of estuaries as nature reserves. Surveys aimed at determining the species composition and abundance of avifaunal communities at South African estuaries have only just begun (this review), and basic inventories of the kind performed by Cooper *et al* (1976) in the south-western Cape reuire to be extended to the rest of the country's coastal regions. From a nature conservation point of view, the importance of the survey conducted by Cooper *et al* (1976) can be judged by the fact that the Berg River estuary and Langebaan Lagoon were found to qualify for international registration as wetlands of special importance to migratory waders. In 1975 South Africa became a contracting party to the 'Ramsar Convention on Wetlands of International Importance' but at the time the country only designated two of its wetlands (De Hoop & Barberspan nature reserves) to be of international importance. More than anything else, this is a reflection of the ignorance which largely still exists about the status of waterbirds and their habitats in southern Africa. Clearly, there is need for a survey to be made of the species composition and abundance of waterbirds at estuaries along the entire coastline of South Africa.

Estuarine avifaunas are relatively easy to census, most of the birds being diurnally active and conspicuous. It might be possible to classify South African estuaries along environmental gradients, based on an index to be derived from data on avian species richness and abundance. The data for such an index could be gathered rapidly, and the equipment and manpower required would be relatively inexpensive. Should such a scheme be found to be practicable, it would provide one 'short-cut' method for conservationists to judge the comprehensiveness of areas, as representative samples of estuarine ecosystems. The data on avian species richness and abundance could also provide base-lines for detecting changes caused by perturbations of the system. However, mere monitoring does not necessarily identify the source of the perturbation, and thus the exercise, while remaining useful, does not meet fully the requirements for management of a conserved area.

Birds and some other mobile animals are capable of altering the shapes of their niches (*sensu* Hutchinson 1957, 1965), in response to local habitat change. For instance, change in the spatial distribution of substrate type, caused by an abnormal erosion or deposition of silt in an estuary, might be reflected by certain species' populations altering their usual feeding behaviour and use of foraging areas. This may occur without any alteration in overall species composition or biomass, and one has to go beyond monitoring these parameters in order to detect such adaptations. Changes in status of niches, for birds at least, could be detected by monitoring the foraging behaviour and space utilization of selected species' populations.

Diamond (1975) has demonstrated the relevance of ecological principles to the design of nature reserves, in which one goal is to preserve certain species. Consideration of the concept of the niche is paramount in seeking a rational answer to the question: how large and of what shape should a nature reserve be if it is to provide a viable unit for the species in need of preservation? The African fish eagle, for instance, is an integral component of estuarine ecosystems in Africa. This species is at the top end of one of the estuarine food chains, and a pair of breeding birds occupies its own foraging territory usually encompassing several hectares. How many breeding pairs need to be conserved in order to ensure the continued viability of the species? The question cannot be answered with any surety, but in South Africa St Lucia is probably the only estuary that alone is capable of supporting the requisite number of birds. In any event, careful consideration of the niches of relatively large-bodied animals which are close to or at the top ends of food chains, can be very useful in the process of designating size and shape of estuarine nature reserves.

Although lists of bird species are still lacking for most South African estuaries, it appears that there are only two avifaunal provinces in the region (this review). Consequently, on this basis, at least one estuary on the west coast and one in the north-eastern sector (Zululand) of the country are required as nature reserves for birds. If maximum species richness is to be preserved, then, these two areas should be as large as possible and, once acquired as nature reserves, they should be managed so as to maintain maximum species richness. St Lucia, already proclaimed for many years as a nature reserve, and, on the west coast, the estuary of the Olifants River are leading candidates for selection. However, consideration will have to be given to additional sites containing special features such as endangered populations and unique areas for breeding, feeding, moulting or roosting. This is exemplified by the apparently unique phenomenon of about 200 crowned cranes *(Balearica regulorum)* which traditionally roost regularly on an island at the Nxaxo estuary in Transkei (Skead, 1972). The crowned crane is included in the South African Red Data Book: Aves (Siegfried *et al,* 1976). Doubtless there are other special sites, and surveys are needed to discover these and to investigate their distinctiveness. On balance, a few large estuaries rather than many small ones, will best serve the cause of nature conservation.

SUMMARY AND CONCLUSIONS

This paper describes the organization and ecology of waterbird populations at southern African estuaries. One hundred and twenty-seven avian species have been recorded for 45 estuaries and lagoons. More than one-third of the species belong to the orders Charadriiformes and Ciconiiformes. Hence, plovers, sandpipers, gulls, terns, herons, flamingos and storks predominate. Most of the charadriiforms are Palaearctic migrants which visit southern Africa during the austral summer. The representatives of the Ciconiiformes are southern African residents. The Palaearctic migrants comprise a little less than one-third of the total number of species; the waders (Charadrii) and the terns (Sternini) are the two main groups of seasonal visitors from the Palaearctic.

The total number of migratory birds exceeded the total number of resident birds by a factor of 1,49. Palaearctic sandpipers (Scolopacidae) are the most abundant birds at southern African estuaries during the austral summer. Next in order of abundance are the resident ciconiiforms. The two flamingos *(Phoenicopterus ruber* and *Phoeniconaias minor)* together accounted for 98 % of all ciconiiforms counted at southern African estuaries.

The estuaries along the north-eastern coast of South Africa tend to support more tropical bird species than Atlantic coast estuaries. The tropical species account for a slightly greater proportion of all the species found at estuaries along the north-eastern coast than along the Atlantic coast. Both regional groups of estuaries support about equal proportions of migratory species. The geographical distribution of resident bird species provides support for the notion of a distinct west-coast biotic province extending southwards from Angola. The area roughly between Cape Point and Transkei is best regarded as a zone of overlap for elements of the west-coast avifaunal province and a diffusion southwards of a tropical avifauna found along the east coast, including the equatorial belt of Africa.

During the austral summer, at estuaries in the south-western Cape, the number of migratory wader species increases at a higher rate than the number of resident waders in the larger estuaries. For migrants, the rate of increase for species abundance is higher than that for species richness. The biomass of migratory waders generally is greater than that of resident waders. Relatively large bodied species, of both migratory and resident waders, feature more abundantly at large than at small estuaries. The species richness and abundance of the southern African estuarine avifauna as a whole are influenced by both the geographical position and size of an estuary. The Atlantic coast estuaries apparently tend to be poorer in species, at least for residents than comparable estuaries in the north-eastern region of South Africa.

Palaearctic migrants accounted for about 5 % of the total avian biomass at estuaries during the austral summer. The resident pelicans, cormorants and flamingos together made up 85 % of the biomass. All the migrants which visit southern African estuaries are carnivores, primarily feeding on invertebrates. Taken together, the migratory and resident carnivores make up 89 % of the estuarine avifauna and 75 % of its biomass. Forty-one percent of the resident species feed on vertebrates, whereas only 21 % of the migrants do so. These migrants comprise 11 % of the total biomass of all species which feed on vertebrates and are almost exclusively fish-eating terns. Amongst the species which feed on invertebrates, the greater flamingo alone accounts for 90 % of the biomass.

The ratio of migratory to resident avian carnivores is similar in different geographical groupings of estuaries in South Africa. However, there are almost twice as many fish and/or frog-eating resident species in the north-eastern region as in the Atlantic coast region. In general, the carnivorous resident avifauna contains more species, but fewer individuals, than the carnivorous migratory community which consists mainly of relatively small-bodied birds which take mainly small-bodied prey. There are no herbivorous migratory species and the resident species which feed on algae and aquatic macrophytes make up 25 % of the total avian biomass; the lesser flamigo and 11 anatids are the major elements. Inland waterbodies support a different trophic apportionment of avian biomass.

Detailed examples are given of the food requirements of selected estuarine bird communities. The biomass of local waders can reach a fairly high level in summer, and food intake is large because of the birds' relatively high metabolic rate. There are indications that the ecological impact of fish-eating birds is greater in southern Africa than in north European estuaries.

The paper closes with a discussion of some of the problems relating to the conservation of estuarine birds in southern Africa.

Acknowledgements

This paper owes much to a few amateur ornithologists. Special thanks are due to Jean Spearpoint, Gerry Nicholls, Ian Sinclair and members of the Western Cape Wader Study Group who provided the bulk of the census data. I am grateful to Timothy Crowe and John Field for advice on computerization of the data. My debt to Anna Crowe as an enthusiastic assistant is gratefully acknowledged. Gillian Puttick, George Begg, Aldo Berruti, John Cooper and Ronald Summers contributed unpublished data. Richard Brooke, Alec Brown, John Cooper and John Field commented on sections of an earlier draft of the manuscript and offered helpful suggestions. Financial assistance was provided by the South African Council of Scientific and Industrial Research, and the then Department of Planning and the Environment.

APPENDIXES

Appendix 13.1

Waterbirds recorded at estuaries and lagoons (see Appendix 13.2) in southern Africa.

R = resident (species which breed in southern Africa), M = migrant (visiting species which do not normally breed in southern Africa). 1 = species which eat mainly fish and/or frogs, 2 = species which eat mainly invertebrates, 3 = species which eat mainly vegetable matter

Common name	Order, family and species		Status and diet	Body mass (g)
	O. Podicipediformes			
Great crested grebe	Fam. Podicipedidae	*Podiceps cristatus*	R1	1 030
Black-necked grebe	Fam. Podicipedidae	*Podiceps nigricollis*	R1	302
Dabchick	Fam. Podicipedidae	*Podiceps ruficollis*	R2	187
	O. Pelecaniformes			
Pink-backed pelican	Fam. Pelecanidae	*Pelecanus rufescens*	R1	10 000
White pelican	Fam. Pelecanidae	*Pelecanus onocrotalus*	R1	10 500
Cape gannet	Fam. Sulidae	*Sula capensis*	R1	2 644
White-breasted cormorant	Fam. Phalacrocoracidae	*Phalacrocorax carbo*	R1	2 506
Cape cormorant	Fam. Phalacrocoracidae	*Phalacrocorax capensis*	R1	1 280
Reed cormorant	Fam. Phalacrocoracidae	*Phalacrocorax africanus*	R1	692
Darter	Fam. Anhingidae	*Anhinga rufa*	R1	1 925

O. Ciconiiformes

Grey heron	Fam. Ardeidae	*Ardea cinerea*	R1	1 440
Black-headed heron	Fam. Ardeidae	*Ardea melanocephala*	R1	1 390
Goliath heron	Fam. Ardeidae	*Ardea goliath*	R1	4 690
Purple heron	Fam. Ardeidae	*Ardea purpurea*	R1	870
Great white egret	Fam. Ardeidae	*Casmerodius albus*	R1	1 100
Little egret	Fam. Ardeidae	*Egretta garzetta*	R1	500
Yellow-billed egret	Fam. Ardeidae	*Mesophoyx intermedius*	R2	500
Cattle egret	Fam. Ardeidae	*Bubulcus ibis*	R2	383
Squacco heron	Fam. Ardeidae	*Ardeola ralloides*	R2	225
Green-backed heron	Fam. Ardeidae	*Butorides striatus*	R1	196
Black heron	Fam. Ardeidae	*Melanophoyx ardesiaca*	R1	300
Rufous-bellied heron	Fam. Ardeidae	*Erythrocnus rufiventris*	R1	300
Night heron	Fam. Ardeidae	*Nycticorax nycticorax*	R1	635
Hamerkop	Fam. Scopidae	*Scopus umbretta*	R1	390
Marabou	Fam. Ciconiidae	*Leptoptilos crumeniferus*	R1	6 000
Openbill	Fam. Ciconiidae	*Anastomus lamelligerus*	R2	1 140
Saddlebill	Fam. Ciconiidae	*Ephippiorhynchus senegalensis*	R1	3 600
Wood stork	Fam. Ciconiidae	*Ibis ibis*	R1	1 757
Woolly-necked stork	Fam. Ciconiidae	*Dissoura episcopus*	R2	3 000
Black stork	Fam. Ciconiidae	*Ciconia nigra*	R1	3 000
White stork	Fam. Ciconiidae	*Ciconia ciconia*	M2	3 605
Sacred ibis	Fam. Threskiornithidae	*Threskiornis aethiopicus*	R2	1 586
Glossy ibis	Fam. Threskiornithidae	*Plegadis falcinellus*	R2	2 000
Hadeda	Fam. Threskiornithidae	*Hagedashia hagedash*	R2	3 000
Spoonbill	Fam. Threskiornithidae	*Platalea alba*	R2	1 790
Greater flamingo	Fam. Phoenicopteridae	*Phoenicopterus ruber*	R2	3 400
Lesser flamingo	Fam. Phoenicopteridae	*Phoeniconaias minor*	R3	1 900

O. Anseriformes

Spurwing goose	Fam. Anatidae	*Plectropterus gambensis*	R3	2 725
Egyptian goose	Fam. Anatidae	*Alopochen aegyptiacus*	R3	2 130
Cape shelduck	Fam. Anatidae	*Tadorna cana*	R2	1 322
Comb duck	Fam. Anatidae	*Sarkidiornis melanotos*	R3	1 547
Dwarf goose	Fam. Anatidae	*Nettapus auritus*	R3	275
Cape shoveller	Fam. Anatidae	*Anas smithii*	R2	661
Black duck	Fam. Anatidae	*Anas sparsa*	R2	1 015
Yellow-billed duck	Fam. Anatidae	*Anas undulata*	R3	992
Red-billed teal	Fam. Anatidae	*Anas erythrorhyncha*	R3	473
Cape teal	Fam. Anatidae	*Anas capensis*	R2	447
Hottentot teal	Fam. Anatidae	*Anas hottentota*	R3	251
White-faced duck	Fam. Anatidae	*Dendrocygna viduata*	R3	880
Fulvous duck	Fam. Anatidae	*Dendrocygna bicolor*	R3	800
Red-eyed pochard	Fam. Anatidae	*Netta erythrophthalma*	R3	772
Maccoa duck	Fam. Anatidae	*Oxyura maccoa*	R2	544
White-backed duck	Fam. Anatidae	*Thalassornis leuconotus*	R3	735

O. Falconiformes

Fish eagle	Fam. Accipitridae	*Haliaeetus vocifer*	R1	2 730
Osprey	Fam. Accipitridae	*Pandion haliaetus*	M1	1 481

O. Gruiformes

Baillon's crake	Fam. Rallidae	*Porzana pusilla*	R2	44
Black crake	Fam. Rallidae	*Limnocorax flavirostris*	R2	94
Red-chested flufftail	Fam. Rallidae	*Sarothrura rufa*	R2	120
Purple gallinule	Fam. Rallidae	*Porphyrio porphyrio*	R3	782
Moorhen	Fam. Rallidae	*Gallinula chloropus*	R2	295
Red-knobbed coot	Fam. Rallidae	*Fulica cristata*	R3	844

O. Charadriiformes

African jacana	Fam. Jacanidae	*Actophilornis africanus*	R2	150
*Painted snipe	Fam. Rostratulidae	*Rostratula benghalensis*	R2	104
*Black oystercatcher	Fam. Haematopodidae	*Haematopus moquini*	R2	685
*Turnstone	Fam. Arenariidae	*Arenaria interpres*	M2	117
* Ringed plover	Fam. Charadriidae	*Charadrius hiaticula*	M2	47
*Mongolian plover	Fam. Charadriidae	*Charadrius mongolus*	M2	62
*White-fronted sandplover	Fam. Charadriidae	*Charadrius marginatus*	R2	45
*Chestnut-banded sandplover	Fam. Charadriidae	*Charadrius pallidus*	R2	36
*Kittlitz's sandplover	Fam. Charadriidae	*Charadrius pecuarius*	R2	36
*Three-banded sandplover	Fam. Charadriidae	*Charadrius tricollaris*	R2	32
*Great sandplover	Fam. Charadriidae	*Charadrius leschenaultii*	M2	87

*Caspian plover	Fam. Charadriidae	*Charadrius asiaticus*	M2	75
*Golden plover	Fam. Charadriidae	*Pluvialis dominica*	M2	100
*Grey plover	Fam. Charadriidae	*Squatarola squatarola*	M2	217
*Crowned plover	Fam. Charadriidae	*Stephanibyx coronatus*	R2	197
*Blacksmith plover	Fam. Charadriidae	*Hoplopterus armatus*	R2	171
*Wattled plover	Fam. Charadriidae	*Afribyx senegallus*	R2	244
*Great snipe	Fam. Scolopacidae	*Gallinago media*	M2	159
*Ethiopian snipe	Fam. Scolopacidae	*Gallinago nigripennis*	R2	128
*Curlew sandpiper	Fam. Scolopacidae	*Calidris ferruginea*	M2	56
*Little stint	Fam. Scolopacidae	*Calidris minuta*	M2	24
*Knot	Fam. Scolopacidae	*Calidris canutus*	M2	143
*Sanderling	Fam. Scolopacidae	*Crocethia alba*	M2	55
*Ruff	Fam. Scolopacidae	*Philomachus pugnax*	M2**	87
*Terek sandpiper	Fam. Scolopacidae	*Xenus cinereus*	M2	67
*Common sandpiper	Fam. Scolopacidae	*Actitis hypoleucos*	M2	57
*Redshank	Fam. Scolopacidae	*Tringa totanus*	M2	70
*Marsh sandpiper	Fam. Scolopacidae	*Tringa stagnatilis*	M2	70
*Greenshank	Fam. Scolopacidae	*Tringa nebularia*	M2	200
*Wood sandpiper	Fam. Scolopacidae	*Tringa glareola*	M2	65
*Bar-tailed godwit	Fam. Scolopacidae	*Limosa lapponica*	M2	291
*Curlew	Fam. Scolopacidae	*Numenius arquata*	M2	769
*Whimbrel	Fam. Scolopacidae	*Numenius phaeopus*	M2	409
*Broad-billed sandpiper	Fam. Scolopacidae	*Limicola falcinellus*	M2	32
*Avocet	Fam. Recurvirostridae	*Recurvirostra avosetta*	R2	324
*Stilt	Fam. Recurvirostridae	*Himantopus himantopus*	R2	167
Red-necked phalarope	Fam. Phalaropodidae	*Phalaropus lobatus*	M2	58
*Crab plover	Fam. Dromadidae	*Dromas ardeola*	M2	300
Water dikkop	Fam. Burhinidae	*Burhinus vermiculatus*	R2	301
Cape dikkop	Fam. Burhinidae	*Burhinus capensis*	R2	200
Arctic skua	Fam. Stercorariidae	*Stercorarius parasiticus*	M1	500
Pomatorhine skua	Fam. Stercorariidae	*Stercorarius pomarinus*	M1	500
Kelp gull	Fam. Laridae	*Larus dominicanus*	R1	900
Lesser black-backed gull	Fam. Laridae	*Larus fuscus*	M2	800
Grey-headed gull	Fam. Laridae	*Larus cirrocephalus*	R2	300
Hartlaub's gull	Fam. Laridae	*Larus hartlaubi*	R2	296
Caspian tern	Fam. Sternidae	*Hydroprogne caspia*	R1	570
Common tern	Fam. Sternidae	*Sterna hirundo*	M1	138
Roseate tern	Fam. Sternidae	*Sterna dougallii*	R1	200
Arctic tern	Fam. Sternidae	*Sterna paradisaea*	M1	138
Sandwich tern	Fam. Sternidae	*Sterna sandvicensis*	M1	251
Lesser crested tern	Fam. Sternidae	*Sterna bengalensis*	M1	160
Swift tern	Fam. Sternidae	*Sterna bergii*	R1	200
Little tern	Fam. Sternidae	*Sterna albifrons*	M1	100
Damara tern	Fam. Sternidae	*Sterna balaenarum*	R1	40
White-winged black tern	Fam. Sternidae	*Chlidonias leucopterus*	M2	65
Wiskered tern	Fam. Sternidae	*Chlidonias hybridus*	R1	94
Black tern	Fam. Sternidae	*Chlidonias niger*	M2	150
Skimmer	Fam. Rynchopidae	*Rynchops flavirostris*	R1	153
	O. *Coraciiformes*			
Pied kingfisher	Fam. Alcedinidae	*Ceryle rudis*	R1	83
Giant kingfisher	Fam. Alcedinidae	*Megaceryle maxima*	R1	374
Half-collared kingfisher	Fam. Alcedinidae	*Alcedo semitorquata*	R1	43
Malachite kingfisher	Fam. Alcedinidae	*Corythornis cristata*	R2	17
Natal kingfisher	Fam. Alcedinidae	*Ispidina picta*	R1	20
Mangrove kingfisher	Fam. Alcedinidae	*Halcyon senegaloides*	R1	83
Brown-hooded kingfisher	Fam. Alcedinidae	*Halcyon albiventris*	R2	61

* Wader
** Included here amongst carnivorous species, *Philomachus pugnax* is granivorous in West Africa (Moreau, 1972).

Appendix 13.2 on next page

Appendix 13.2

Number of separate censuses of waterbirds at estuaries and lagoons in southern Africa. P/A = listing of presence/absence of species. C = count of number of individual birds.

	No. separate censuses						Source
	Waders		All species		Summer	Winter	
	P/A	C	P/A	C			
Walvis Bay lagoon	1	1	1	1	1		Underhill & Whitelaw (1977)
Sandvishawe lagoon	25	25	25	25	13	12	Berry & Berry (1975), Underhill & Whitelaw (1977)
Olifants	2	2	1		2		Summers *et al* (1976), Siegfried (*in litt* 1977)
Verlorenvlei	3	3	2	2	3		Summers *et al* (1976), Grindley (*in litt* 1977)
Berg	1	1	1		1		Summers *et al* (1976), Siegfried (*in litt* 1977)
Langebaan lagoon	1	1	1	1	1		Summers *et al* (1976), Pringle (*in litt* 1977)
Milnerton	1	1			1		Summers *et al* (1976)
Eerste	1	1			1		Summers *et al* (1976)
Lourens	1	1			1		Summers *et al* (1976)
Rooiels	1	1			1		Summers *et al* (1976)
Palmiet	1	1			1		Summers *et al* (1976)
Kleinmond	1	1			1		Summers *et al* (1976)
Klein	2	1	1		2		Scott *et al* (1952), Summers *et al* (1976)
Uilkraals	1	1			1		Summers *et al* (1976)
Heuningnes	1	1			1		Summers *et al* (1976)
Breë	1	1			1		Summers *et al* (1976)
Duiwenhoks	1	1			1		Summers *et al* (1976)
Kafferkuils	1	1			1		Summers *et al* (1976)
Gouritz	1	1			1		Summers *et al* (1976)
Kleinbrak	1	1			1		Summers *et al* (1976)
Sedgefield	1	1			1		Summers *et al* (1976)
Goukamma	1	1			1		Summers *et al* (1976)
Storms	1	1			1		Summers *et al* (1976)
Seekoeis	1		1		1		Spearpoint (*in litt* 1977)
Kabeljouw	1		1		1		Spearpoint (*in litt* 1977)
Gamtoos	53	53	52	32	22	31	Shewell (1950), Spearpoint (*in litt* 1977)
Maitlands	1	1	1	1	1		Spearpoint (*in litt* 1977)
Swartkops	3	1	3	1	3		Grindley (1974), Spearpoint (*in litt* 1977)
Sundays	1	1	1	1	1		Spearpoint (*in litt* 1977)
Mayameni	1	1	1	1	1		Cooper (*in litt* 1977)
Mzimkulu	1	1	1	1	1		Sinclair (*in litt* 1977)
Scottborough	1	1	1	1	1		Sinclair (*in litt* 1977)
Amanzimtoti	1	1	1	1	1		Sinclair (*in litt* 1977)
Mlazi	1	1	1	1	1		Sinclair (*in litt* 1977)
Durban Bay	11	11	11	11	7	4	Sinclair (*in litt* 1977), Nicholls (*in litt* 1977)
Mgeni	28	28	28	28	20	8	Sinclair (*in litt* 1977), Nicholls (*in litt* 1977)
Mhlanga	1	1	1	1	1		Nicholls (*in litt* 1977)
Mhloti	15	15	15	15	13	2	Sinclair (*in litt* 1977), Nicholls (*in litt* 1977)
Mvoti	15	15	15	15	12	3	Sinclair (*in litt* 1977), Nicholls (*in litt* 1977)
Tugela	3	3	3	3	3		Nicholls (*in litt* 1977)
Mlalazi	6	6	6	6	3	3	Sinclair (*in litt* 1977)
Tongaat	1	1	1	1	1		Nicholls (*in litt* 1977)
Richards Bay	3	3	3	3	3		Nicholls (*in litt* 1977)
Mfolozi	3	3	3	3	2	1	Sinclair (*in litt* 1977)
St Lucia	5	3	5	3	5		Day *et al* (1954), Berruti (*in litt* 1977)

REFERENCES

BAKER, M.C. & A.E.M. BAKER 1973. Niche relationships among six species of shorebirds on their wintering and breeding ranges. *Ecol. Monog.* **43**: 193-212.

BARNES, R.S.R. 1974. *Estuarine Biology.* Arnold, London.

BERRY, H.H. 1975. South West African flamingos. *In:* J. Kear & N. Duplaix-Hall (eds), *Flamingos.* pp 53-60. T. & A.D. Toyser, Berkhamsted, England.

BERRY, H.H. 1976. Physiological and behavioural ecology of the Cape Cormorant *Phalacrocorax capensis. Madoqua* **9** (4): 5-55.

BERRY, H.H. & C.U. BERRY 1975. A check list and notes on the birds of Sandvis, South West Africa. *Madoqua* **9**(2): 5-18.

BIGALKE, R.C. 1978. Mammals. *In:* M.J. Werger (ed), *Biogeography and ecology of southern Africa.* pp 981-1048. Junk, The Hague.

BLABER, S.J.M. 1973. Population size and mortality of juveniles of the marine teleost *Rhabdosargus holubi* (Pisces: Sparidae) in a closed estuary. *Mar. Biol.* **21**: 219-225.

BOLTT, R.E. 1975. The benthos of some southern African lakes. 5. The recovery of the benthic fauna of St Lucia Lake following a period of excessively high salinity. *Trans. roy. Soc. S.Afr.* **41**: 295-323.

BOYD, H. 1962. Mortality and fertility of European Charadrii. *Ibis* **104**: 368-387.

BROEKHUYSEN, G.J. 1975. South African flamingos. *In:* J. Kear & N. Duplaix-Hall (eds), *Flamingos* pp 61-64. T. & A.D. Poyser, Berkhamsted, England.

BROWN, A.C. & N. JARMAN 1978. Coastal marine habitats. *In:* M.J. Werger (ed), *Biogeography and Ecology of southern Africa.* pp 1239-1277. Junk, The Hague.

COOMBES, R.A.H. 1950. The moult migration of the Shelduck. *Ibis* **92**: 405-418.

COOPER, J., SUMMERS, R.W. & J.S. PRINGLE 1976. Conservation of coastal habitats of waders in the south-western Cape, South Africa. *Biol. Conserv.* **10**: 239-247.

DAY, J.H. 1969. *A guide to marine life on South African shores.* Balkema, Cape Town. 300pp.

DAY, J.H. 1974. The ecology of Morrumbene estuary, Mozambique. *Trans. roy. Soc. S.Afr.* **41**: 43-97.

DAY, J.H., MILLARD, N.A.H. & G.J. BROEKHUYSEN 1954. The ecology of South African estuaries. Part 4. The St Lucia system. *Trans. roy. Soc. S.Afr.* **34**: 129-156.

DIAMOND, J.M. 1975. The island dilemma: Lessons of modern biogeographic studies for the design of natural reserves. *Biol. Conserv.* **7**: 129-146.

DICK, W.J.A. 1975. *Oxford and Cambridge Mauritanian Expedition, 1973 Report.* Cambridge.

DICK, W.J.A., PIENKOWSI, N.W., WALTNER, M. & C.D.T. MINTON 1976. Distribution and geographical origins of Knot *Calidris canutus* in Europe and Africa. *Ardea* **64**: 22-47.

DRINNAN, R.E. 1957. The winter feeding of the Oystercatcher *(Haematopus ostralegus)* on the cockle *(Cardium edule). J. anim. Ecol.* **26**: 441-469.

EBBINGE, B., CANTERS, K. & R. DRENT 1975. Foraging routines and estimated daily food intake in Barnacle Geese wintering in the northern Netherlands. *Wildfowl* **26**: 5-19.

ELLIOTT, C.C.H. 1971. Analysis of the ringing and recoveries of three migrant terns. *Ostrich* (Suppl.) **9**: 71-82.

ELLIOTT, C.C.H., WALTNER, M., UNDERHILL, L.G., PRINGLE, J.S. & W.J.A. DICK 1976. The migration system of the Curlew Sandpiper *Calidris ferruginea* in Africa. *Ostrich* **47**: 191-213.

FIELD, J.G. 1970. The use of numerical methods to determine benthic distribution patterns from dredgings in False Bay. *Trans. roy. Soc. S.Afr.* **39**: 183-200.

FIELD, J.G. 1971. A numerical analysis of changes in the soft bottom fauna along a transect across False Bay, South Africa. *J. exp. mar. Biol. Ecol.* **7**: 215-253.

FRITH, H.J. 1967. *Waterfowl in Australia.* Angus & Robertson, Sydney.

FROST, P.G.H. & P. JOHNSON 1977. Seabirds on the Diamond Coast, South West Africa, December 1976. *Cormorant* **2**: 3-4.

GLUTZ VON BLOTZHEIM, U.N., BAUER, K.M. & E. BEZZEL 1975. *Handbuch der Vögel Mitteleuropas.* Bd. 6. Akademische Verlagsgesellschaft, Wiesbaden.

GOSS-CUSTARD, J.D. 1969. The winter feeding ecology of the Redshank *(Tringa totanus). Ibis* **111**: 338-356.

GOSS-CUSTARD, J.D. & K. CHARMAN 1976. Predicting how many wintering waterfowl an area can support. *Wildfowl* **27**: 157-156.

GRINDLEY, J.R. 1974. Estuarine Ecology. *In: Technical data report of environmental study of the Swartkops River basin.* Hill, Kaplan, Scott & Partners, Port Elizabeth.

HOLMES, R.T. & F.W. STURGES 1975. Bird community dynamics and energetics in a northern hardwoods ecosystem. *J. anim. Ecol.* **44**: 175-200.

HUTCHINSON, G.E. 1957. Concluding remarks. *Cold Spring Harbor Symp. Quant. Biol.* **22**: 415-427.

HUTCHINSON, G.E. 1965. *The ecological theatre and the evolutionary play.* Yale Univ. Press, New Haven, Conn.

MACAN, T.T. 1977. The influence of predation on the composition of fresh-water animal communities. *Biol. Rev.* **52**: 45-70.

MACNAE, W. 1968. A general account of the fauna and flora of mangrove swamps and forests in the Indo-west-Pacific region. *In:* F.S. Russel & M. Yonge (eds), *Advances in Marine Biology* **6**: 74-270. Academic Press, London.

MIDDLEMISS, E. 1961. Biological aspects of *Calidris minuta* while wintering in south-west Cape. *Ostrich* **32**: 107-121.

MIDDLEMISS, E. 1962. Return of Palaearctic waders to the same wintering grounds. *Ostrich* **33**: 53-54.

MILNE, H. 1974. Birds in the estuarine food chain. *Proc. Challenger Soc.* **4**(6).

MILNE, H. & G.M. DUNNET 1972. Standing crop, productivity and trophic relations of the fauna of the Ythan estuary. *In:* R.S.K. Barnes & J. Green (eds), *The Estuarine Environment* pp 86-106. Applied Science, London.

MOREAU, R.E. 1973. *The Palaearctic-African Bird Migration Systems.* Academic Press, London.

MORSE, D.H. 1975. Ecological aspects of adaptive radiation in birds. *Biol. Rev.* **50**: 167-214.

OVIATT, C.A., NIXON, S.W. & J. GARBER 1977. Variation and evaluation of coastal salt marshes. *Environmental Management* **1**: 201-211.

PACKHAM, J.R. & M.J. LIDDLE 1970. The Cefni salt-marsh, Anglesey, and its recent development. *Field Studies* **3**: 311-356.

PORTER, R.N. & G.W. FORREST 1974. First successful breeding of Greater Flamingo in Natal, South Africa. *Lammergeyer* **21**: 26-33.

PRATER, A.J. 1972. Food of Turnstones in Morecambe Bay. *Bird Study* **19**: 51-52.

PRATER, A.J. 1976. The distribution of coastal waders in Europe and North Africa. *Proc. IWRB Conservation of Wetlands and Waterfowl Conference, Heiligenhafen 1974:* 255-271.

PRINGLE, J.S. & J. COOPER 1975. The palaearctic wader population of Langebaan Lagoon. *Ostrich* **46**: 213-218.

PUTTICK, G. 1977. Spatial and temporal variations in intertidal animal distributions at Langebaan Lagoon, South Africa. *Trans. roy. Soc. S.Afr.* **42**: 403-440.

PUTTICK, G.M. 1978. The diet of the Curlew Sandpiper at Langebaan Lagoon, South Africa. *Ostrich* **49**: 158-167.

PUTTICK, G.M. 1979. The feeding ecology and energetics of the Curlew Sandpiper at Langebaan Lagoon. *Abstract 4th (S.Afr.) natl. oceanog. Symp., Cape Town, July 1979.* CSIR, Pretoria.

RAND, R.W. 1960. The biology of guano-producing sea-birds. 3. The distribution, abundance and feeding habits of the cormorants Phalacrocoracidae off the south-west coast of the Cape Province. *Invest. Rep. S.Afr. Div. Fish.* **39**: 1-26.

RYTHER, J.H. 1969. Photosynthesis and fish production in the sea. *Science* **166**: 72-76.

SANDERS, H.L. 1968. Marine benthic diversity: a comparative study. *Am. Nat.* **102**: 243-262.

SCOTT, K.M.F., HARRISON, A.D. & W. MACNAE 1952. The ecology of South African estuaries. Part 2. The Klein River estuary, Hermanus, Cape. *Trans. roy. Soc. S.Afr.* **33**: 283-331.

SHEWELL, E.L. 1950. Birds of the Gamtoos estuary. *Ostrich* **21**: 97-102.

SIEGFRIED, W.R. 1970. Wildfowl distribution, conservation and research in southern Africa. *Wildfowl* **21**: 89-98.

SIEGFRIED, W.R. 1978. Estuaries, ecology and conservation in South Africa. *S.Afr. J. Sci.* **74**: 406-407.

SIEGFRIED, W.R., WILLIAMS, A.J., FROST, P.G.H. & J.B. KINAHAN 1975. Plumage and ecology of cormorants. *Zool. Afr.* **10**: 183-192.

SIEGFRIED, W.R., FROST, P.G.H., COOPER, J. & A.C. KEMP 1976. *South African Red Data Book – Aves.* CSIR, Pretoria.

SKEAD, C.J. 1967. Ecology of birds in the eastern Cape Province. *Ostrich* Suppl. **7**: 1-103.

SLOBODKIN, L.B. 1960. Ecological energy relationships at the population level. *Am. Nat.* **94**: 213-236.

SUMMERS, R.W. 1977. Distribution, abundance and energy relationships of waders (Aves: Charadrii) at Langebaan Lagoon. *Trans. roy. Soc. S.Afr.* **42**: 483-495.

SUMMERS, R.W., PRINGLE, J.S. & J. COOPER 1976. *The status of coastal waders in the South-western Cape, South Africa.* Western Cape Wader Study Group, Cape Town.

SUMMERS, R.W. & M. WALTNER 1979. Seasonal variations in the mass of waders in southern Africa, with special reference to migration. *Ostrich* **50**: 21-37.

STEPHENSON, T.A. & A. STEPHENSON 1972. *Life between tide-marks on rocky shores.* Freeman & Co, San Francisco.

SWENNEN, C. 1975. Aspecten van voedselproduktie in Waddenzee en aangrenzende zeegebieten in relatie met de vogelrijkdom. *Het Vogeljaar* **23**: 141-156.

SWENNEN, C. 1976. Wadden Seas are rare, hospitable and productive. *Proc. IWRB Conservation of Wetlands and Waterfowl Conferences, Heiligenhafen, 1974:* 184-198.

THOMAS, D.G. 1970. Fluctuations of numbers of waders in south-eastern Tasmania. *Emu* **70**: 79-85.

UNDERHILL, L.G. & D. WHITELAW 1977. *Ornithological expedition to the Namib Coast.* Western Cape Wader Study Group, Cape Town.

WALLACE, A.R. 1876. *The geographical distribution of animals.* London.

WHITFIELD, A.K. 1978. Relationship between fish and piscivorous bird densities at Lake St Lucia. *S.Afr. J. Sci.* **74**: 478.

WHITFIELD, A.K. & D.P. CYRUS 1978. Feeding succession and zonation of aquatic birds at False Bay, Lake St Lucia. *Ostrich* **49**: 8-15.

WIENS, J.A. & R.A. NUSSBAUM 1975. Model estimation of energy flow in northwestern coniferous forest bird communities. *Ecology* **56**: 547-561.

WINTERBOTTOM, J.M. 1960. Report on the Cape Bird Club vlei counts, 1952-58. *Ostrich* **31**: 135-168.

WINTERBOTTOM, J.M. 1967a. The birds of three salt-water habitats in the south west Cape. *Ostrich* **38**: 148-154.

WINTERBOTTOM, J.M. 1967b. The relationships of some African aquatic avifaunas. *Rev. Zool. Bot. Afr.* **75**: 149-155.

WINTERBOTTOM, J.M. 1972. *The ecological distribution of birds in southern Africa. Monograph I.* Percy FitzPatrick Institute of African Ornithology, Cape Town.

WOLFF, W.J., VAN HAPEREN, A., SANDEE, A.J.J., BAPTIST, H.J.M. & H.L.F. SAEIJS 1975. The trophic role of birds in the Grevelingen estuary, the Netherlands, compared to their role in the saline Lake Grevelingen. *Comm. Delta Inst. Hydrobiol. Res.* Holland.

Chapter 14
Summaries of current knowledge of 43 estuaries in southern Africa

J.H. Day

Department of Zoology, University of Cape Town

INTRODUCTION

This chapter is an attempt to summarise what is known of each of the estuaries that have been studied with brief notes on a few others. The latter are poorly known but from their size or obvious characteristics they appear important and what *is* known should be recorded. The summaries are intended as a basis for further research and the more detailed accounts contained in publications and internal reports are quoted. Many of the original observations were made by expeditions from the University of Cape Town, the NIWR and NRIO, the Universities of Rhodes and Port Elizabeth, the Department of Planning and the Environment and the Cape Department of Nature Conservation. In addition, individual workers have freely supplied notes and photographs of their own unpublished work. I am grateful to all these institutions and individuals particularly Mr George Begg, whose inventory report on *The estuaries of Natal* and personally conducted tour of estuaries near Durban have been invaluable. The assistance of the many other workers is acknowledged in the text.

REGIONAL CHANGES ALONG THE COAST

The coastline of southern Africa from Inhambane in Mocambique, around the Cape to Walvis Bay in Namibia, is over 3 600 km long. It would not be expected that the estuarine environment or the plant and animal populations over this distance would be uniform. Indeed, the changes are obvious and it is useful to group the estuaries in some logical way so that broad changes may be readily appreciated. Ecological regions and provinces in the sea are based on the temperature of the various water masses, with an understanding that local differences will be related to depth, the degree of shelter from wave action, the nature of the substrate, the increased fertility due to upwelling and other factors. On the land, the regions and provinces are again related to temperature but in this case, precipitation and humidity become more important in warm climates and local differences are due to soil type, shelter from harsh winds and the nature of the vegetation. Since estuaries are formed where rivers meet the sea, both marine and terrestrial factors must be considered. It is usual to focus attention on salinity differences but in many estuaries these are not the most important environmental factors. With these points in mind, the factors which affect the estuaries in southern Africa may be discussed.

THE EFFECT OF TEMPERATURE

Estuarine temperatures are determined both by the temperature of the sea and the land, the coastal sea temperatures having a greater influence at the mouth and land temperatures causing a greater variation in the upper reaches. Along the coasts of southern Mocambique, Natal and the Transkei, the Mocambique and Agulhas currents have core temperatures of 25 ° or more. However, the Agulhas current tends to wander and the inshore surface temperatures which affect estuaries are cooler than this in winter and the seasonal range is about 18-25 °. The terrestrial temperatures along the coastal belt are more variable and the water temperatures in the upper reaches of estuaries range from 30 to 18 ° in Mocambique, 28 to 16 ° in Natal and 27 to 14 ° in the Transkei. In these warm estuaries, it is the minimum temperature which is the limiting factor and tropical species of plants and animals tend to decrease to the south.

Along the Eastern and Southern Cape coasts, the Agulhas current usually flows along the edge of the continental shelf and inshore counter currents have seasonal temperatures of 14-20 °. Very occasionally, upwelling in summer causes a sudden drop to 11 °. In the upper reaches of estuaries, the temperatures are slightly lower than in the Transkei with a seasonal range of about 26 to 12 °.

Along the West or Atlantic coast, upwelling often occurs in summer and coastal temperatures then vary from 10 to 16 °. In the winter the temperatures are more uniform at about 14 °. Land temperatures are similar to those on the coasts facing the Indian Ocean although the diurnal fluctuations are greater. Temperatures in the upper reaches of estuaries range from 24 to 10 ° in the Cape and 30 to 14 ° in Namibia.

Since an estuarine fauna is largely marine in origin, its distribution is related more closely to coastal sea temperatures than to river temperatures. As noted in chapter 9, there is a subtropical fauna extending from southern Mocambique to the Transkei and a warm temperate fauna extending from the Eastern Province of the Cape to Cape Agulhas. Whether the west coast estuaries should be regarded as forming a separate faunistic province is uncertain. As noted in Day (1974), all the species there are those found in estuaries of the southern Cape and there are no species peculiar to Atlantic coast estuaries. Nevertheless, these Atlantic coast estuaries have a fauna which includes only about one tenth of the species present in south coast estuaries. Too little is known of the species of fresh water origin to use their distribution as a basis for discussion. The distribution of macrophytic plants is also poorly known. As noted in chapter 6, several tropical sea grasses are limited to Mocambique and the species of mangroves decrease also but three, *Avicennia, Bruguiera* and *Rhizophora,* extend to southern Transkei. Little can as yet be said about the distribution of the low-growing salt marsh plants. The most obvious point is that, whereas many extend around southern Africa and indeed have a much wider distribution, *Limonium* spp and *Spartina maritima* first appear in the estuaries of the Eastern Cape. *Limonium* is present in Atlantic coast estuaries but *Spartina* in

southern Africa is restricted to the Eastern and Southern Cape except for an outlying record in Langebaan Lagoon which harbours many warm-temperate species.

On the basis of the available evidence, there seems to be no doubt that the fauna and flora of estuaries between southern Mocambique and southern Transkei are subtropical. The biota of the Eastern and Southern Cape is certainly warm temperate. The fauna and flora of the few west or Atlantic coast estuaries that have been sampled, appear to belong to a separate province which, for lack of a more appropriate name is called a Namaqua province. This agrees with the findings of Stephenson (1947) regarding the intertidal coastal biota.

TIDAL RANGE AND LITTORAL DRIFT

Several questions arise immediately. First, is the grouping based on estuarine temperatures and the distribution of the biota in line with changes in other environmental factors? In fact, do the estuaries, regardless of their fauna and flora, form natural groups along the coast? In view of the fact that there are several types of estuaries, do the differences between them make nonsense of the proposed grouping or are they variations within a group? Finally, how may the differences between individual estuaries be explained?

Two marine factors (apart from temperature), obviously affect estuaries. First the tidal range affects the strength of tidal currents, the volume of the tidal prism and the residence time of the estuarine water. A large tidal range exposes a larger intertidal area than a small range and the intertidal banks appear to be the most productive parts of an estuary. As shown in chapter 2, the spring tide range between Walvis Bay and Hermanus is of the order of 1,45 m. Between Mossel Bay and East London in the Eastern Cape it increases to 1,65 m and at Durban it increases to 1,8 m. Further north on the Mocambique coast it increases to 2,4 m at Inhaca Island, 3,7 m at Bazaruto Island and 5,6 m at Beira. In brief, the spring tide range is fairly low and uniform along the Atlantic coast but increases continually along the Indian Ocean shores of southern Africa.

The second important factor is wave action and the littoral drift of sand. These two are difficult to separate in regard to their effect on estuaries. High energy waves erode the shore and where they approach the coast obliquely under the influence of the wind, they generate longshore currents and transport the sand. Part of this is carried into the mouths of estuaries by the flood tide and part forms sandspits and shoals. If these are not eroded by the ebb tide as fast as they form, the estuary closes and the whole nature of the estuary changes.

Schulze (1965) states that the prevailing winds all

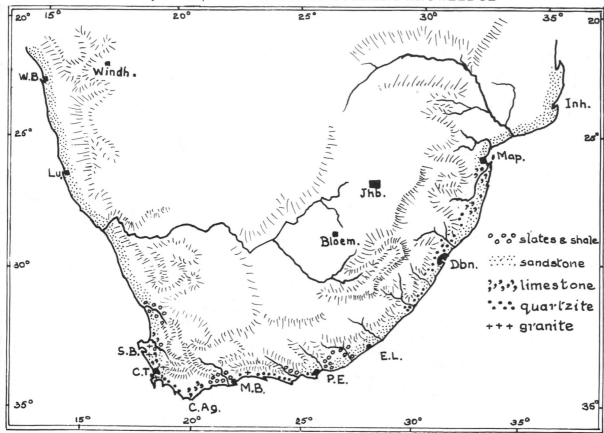

Figure 14.1
Geological map of the coastal belt of Namibia and South
Africa (based on a map of the geology of the continental
margin by D.V. Dingle & W.G. Siesser).

around South Africa blow parallel with the coastline.
On the west coast they blow from the north-west or
from the south or south-east. On the southern or eastern Cape coasts the predominant winds are from the
south-west or south-east, while on the Natal coast
they blow from the south-west or north-east. On the
coast of Mocambique the winds are more gentle
although there are squalls associated with the south-
westerly monsoon.

As noted in chapter 2, wave energy is very great all
along the coast of South Africa. It is strongest on the
shores of the south-western Cape where wave heights
of 12 m or more are recorded annually. Wave heights
decrease somewhat towards the coast of Namibia
where wave heights seldom exceed 8 m annually and
the same is true on Eastern Cape and Natal coasts. No
records are available for southern Mocambique but
personal observations suggest that apart from occasional typhoons, wave heights are low. Moreover the
submarine contours shelve gently and high waves
break well offshore.

The littoral drift of sand under the action of the
winds and waves is very evident all along the coast

from Namibia to Zululand. The contours of exposed
beaches may change as much as 2 m during a single
storm and the mouths of small estuaries may close
overnight. Usually littoral drift is from south to north
or south-west to north-east as is evident from the sandspits of many estuaries. The estuaries at the Matigulu
in Natal and the Gamtoos near Port Elizabeth are
good examples. Both show how floods break through
the sandspits and form new mouths nearer the axis of
the river while the old mouth silts up. Rocky promontories divert the littoral drift of sand and many estuaries
remain open in the lee of such promontories while
those on open sandy coasts close. On steep rocky
coasts, the sandy bed of the sea is below the reach of
the waves and sandspits and bars if present at all, are
insignificant. This may be seen on the rocky Tsitsikama
coast where the Knysna, the Caaimans and the Storms
River estuaries have deep and permanently open
mouths. To a lesser degree this is also true of estuaries
on the Pondoland coast on the Transkei-Natal border.
The nature of the estuaries on these sections of the
coastline are basically determined by the geomorphology of the coastal belt but elsewhere the effects of
winds, waves and littoral drift is predominant.

The influence of geomorphology on the nature of
rivers may be considered next.

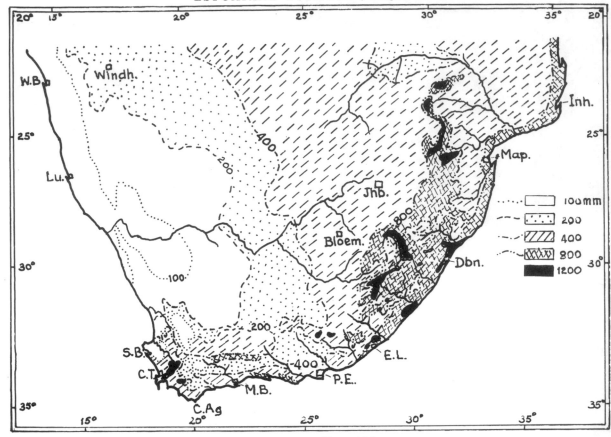

Figure 14.2
Map of the rainfall normals of the coastal belt of southern Africa (modified from Noble & Hemens, 1978).

GEOMORPHOLOGY OF THE COASTAL BELT

The width of the coastal belt and its elevation have obvious effects on the gradient of a river, its rate of flow and the length of an estuary. The geological nature of the coastal belt of southern Africa is indicated very crudely in Figure 14.1. Most of the coastal belt is narrow and this is particularly true of the Tsitsikama coast where the estuaries are seldom more than 10 km long. In contrast to this, the Breede in the broad low-lands of the southern Cape may be 40 km long during dry summers. The estuary of the Limpopo on the very broad coastal belt of Mocambique has not been inves- tigated; it is presumably long although it is obviously not comparable with the 180 km-long estuary of the Scheldt in Belgium.

Figure 14.1 also outlines the nature of the rocks and the soils derived from them. Since rivers with large drainage basins transport a variety of sediments from several geological formations, only sharply con- trasting types are worth discussing. The estuaries of the Knysna and Breede rivers provide such a contrast. Knysna River drains sandstone and quartzite forma- tions and its silt load is light and the estuary water is clear. The Breede River flows over slates and shales which erode to form silt and clay so that the banks of the estuary, where not covered by marine sand, are extremely muddy and the water is turbid. On the Natal coast the contrast is different. Here there are short rivers which drain the sandy coastal belt and long rivers which drain the Karoo formations inland. During the summer rains the estuaries of the coastal rivers remain fairly clear while those of the long rivers are extremely turbid and carry heavy sediment loads. Such estuaries lack the broad intertidal salt marshes and sand flats which elsewhere harbour a rich fauna of burrowing bivalves, prawns and polychaete worms. Unfortunately the mouths of the short rivers close rapidly and they tend to be impoverished for a diffe- rent reason.

RAINFALL AND RIVER FLOW

The most important factor which determines the nature of an estuary will be considered last. This is the volume of river flow and its seasonal variability. This obviously depends on the annual and seasonal precipitation and the rate of evapotranspiration. The rainfall normals in southern Africa are shown in out- line in Figure 14.2. Schulze (1965) gives both the mean annual rainfall and the annual evaporation for

254

Table 14.1
Rainfall and evaporation in seven stretches of the coastal belt of southern Africa (based on data from Schulze, 1965)

Coastal stretch	Rainfall (mm.yr^{-1})	Evaporation (mm.yr^{-1})	Rainy season	Remarks
(1) Namibia - St Helena Bay	<100 - >100 (unreliable)	2 473	autumn	desert to semi-desert
(2) S.W.Cape (Saldanha Bay - Hermanus)	500 - 700	1 830	winter	heavy rains in mountains
(3) S.Cape (Cp.Agulhas - Mossel Bay)	300 - 500	1 650	through the year	dry lowlands
(4) Tsitsikama (George - Humansdorp)	700 - 1 000	1 280	through the year	forest and mountain streams
(5) E.Cape (Port Elizabeth - East London)	500	1 830	mainly summer	wooded coasts, dry inland
(6) Transkei (Kei River - Mtamvuna River)	800 - 1 250	1 700	summer	wooded coasts, grassy inland
(7) Natal - S.Mocambique (Mtamvuna River - Inhambane)	1 000 - 1 250	—	summer	well-watered lowlands

South Africa and divides the country (including what is now to be termed Namibia) into 23 rainfall districts. While it is appreciated that the Orange River, the Great Berg, the Breede and other large rivers are mainly fed from inland areas, we are primarily concerned with conditions in the coastal belt. These are summarised in Table 14.1.

1. In the Namib desert and along the Namaqualand coast, the rain may fail entirely for a year or more and when it does come, most of it evaporates and the rest sinks rapidly into the sandy soil. The rivers flow briefly over the surface but mainly seep through sub-surface sands as in the case of the Quiseb and the Groenrivier. Only the Orange and the Olifants flow perennially. As described later in this chapter, the Orange River has no estuary for conditions are too variable; only the Olifants estuary is normal but it has a restricted fauna.

2. In the Southwestern Cape the rainfall is heavy on the mountains and poor on the low-lying plains, so that rivers with mountain sources such as the Great Berg are fresh to the sea in the winter rains but saline water penetrates well inland in the dry summer. Small rivers with a catchment on the plains cease flowing in summer and the estuary may or may not become hypersaline. Langebaan Lagoon has no significant fresh water supply and has the same salinity as the sea. It is not an estuary.

3. In the southern Cape, the rainfall is even lower than on the lowlands of the Western Cape and the estuaries are essentially similar. Only the Breede, which has a catchment in the mountain 'sponges' flows strongly through the year and has a well-developed estuary. The Gouritz has an even larger catchment but its source is in the dry Karroo and the estuary is not as well developed.

4. The mountain streams on the Tsitsikama coast which drain the wooded mountain slopes, have a good rainfall throughout the year and the estuaries, though short, are well-developed. The most easterly one is the Kromme where much of the good quality water is caught by the Churchill dam to supply Port Elizabeth and as a result its estuary may become hypersaline.

The Gamtoos nearby drains part of the Karroo where the rainfall is low and seasonal; moreover the Karroo shales form silts and clays so that the estuary is turbid, heavily silted and may close during droughts.

5. The Eastern Cape is dry inland and droughts are common, although the word 'drought' has a meaning which varies with the normal rainfall. The estuaries are very variable; many of the shorter ones close in the dry months while others are muddy or well-developed. The West Kleinmond, the Keiskamma and the Swartkops illustrate three extremes.

6. The rivers of the Transkei are basically similar although the rainfall is much higher and more definitely limited to summer. Again, the estuaries of the larger rivers such as the Kei and the Bashee which drain Karroo deposits inland carry heavy silt loads while short coastal rivers such as the Mngazana which drain the well-wooded coastal belt are well-developed. Conditions in all these estuaries improve in the dry winter months.

7. Natal, Zululand and southern Mocambique are all well-watered and floods are common in summer and droughts are less common to the north. The nature of the estuaries changes progressively from those in the Transkei. As the coastal belt broadens to the north, the soil also becomes more sandy and the bush-covered coastal dunes (exemplified by the Bluff at Durban) become more continuous. The estuaries broaden out behind these dunes and form lagoons, some of which have been cut off from the sea and become hyposaline or even fresh as described by Professor Brian Allanson in chapter 15. Silting is enhanced by the growth of sugar cane to the water's edge as described and illustrated by Begg (1978) and others. Several of the estuaries (such as the Lovu) are now mere films of water over a sandy bed. In contrast to this, the Kosi Bay system in an undeveloped part of Zululand, remains as a chain of deep saline lakes with a rich fauna. The richest estuary of all is the Morrumbene at Inhambane where there is a rich fauna in the lagoon and flourishing mangrove swamps along the muddy upper reaches.

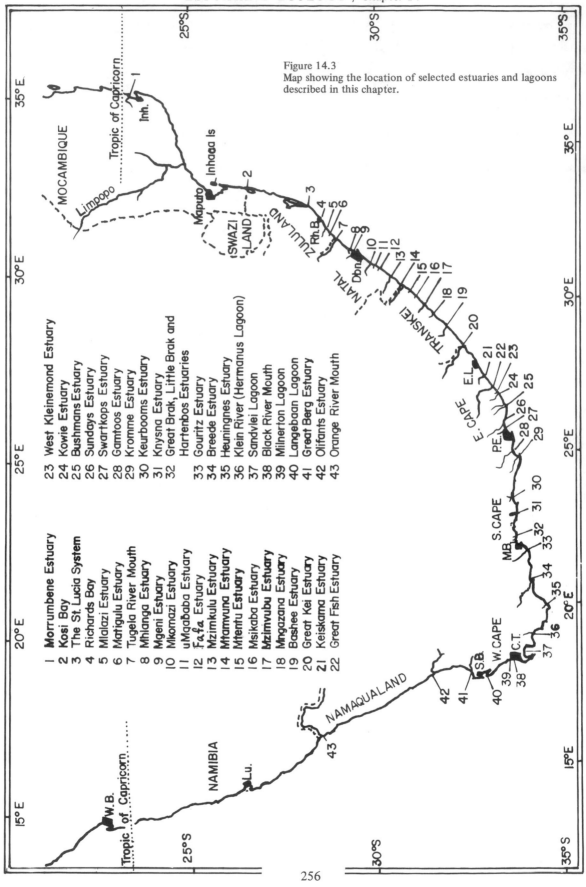

Figure 14.3
Map showing the location of selected estuaries and lagoons
described in this chapter.

1 Morrumbene Estuary
2 Kosi Bay
3 The St. Lucia System
4 Richards Bay
5 Mlalazi Estuary
6 Matigulu Estuary
7 Tugela River Mouth
8 Mhlanga Estuary
9 Mgeni Estuary
10 Mkomazi Estuary
11 uMgababa Estuary
12 Fafa Estuary
13 Mzimkulu Estuary
14 Mtamvuna Estuary
15 Mtentu Estuary
16 Msikaba Estuary
17 Mzimvubu Estuary
18 Mngazana Estuary
19 Bashee Estuary
20 Great Kei Estuary
21 Keiskama Estuary
22 Great Fish Estuary

23 West Kleinemond Estuary
24 Kowie Estuary
25 Bushmans Estuary
26 Sundays Estuary
27 Swartkops Estuary
28 Gamtoos Estuary
29 Kromme Estuary
30 Keurbooms Estuary
31 Knysna Estuary
32 Great Brak, Little Brak and
   Hartenbos Estuaries
33 Gouritz Estuary
34 Breede Estuary
35 Heuningnes Estuary
36 Klein River (Hermanus Lagoon)
37 Sandvlei Lagoon
38 Black River Mouth
39 Milnerton Lagoon
40 Langebaan Lagoon
41 Great Berg Estuary
42 Olifants Estuary
43 Orange River Mouth

THE NATURAL GROUPING OF ESTUARIES

Having presented the evidence regarding the interplay of marine and terrestrial factors on the development of estuaries along the 3 600 km of coastline, we now return to the original question. Is it possible to group the estuaries along successive stretches of coastline in a logical manner? In my opinion it is, although within each group the estuaries vary from one to the next in spite of their underlying similarity. As in the sea, the broad divisions are based on water temperature but in estuaries this is modified by the rainfall and river flow. There appear to be three main biotic provinces:

A. The subtropical estuaries of southern Mocambique to the Transkei. These have warm waters with minimum temperatures above 16 °, and a good summer rainfall and river discharge. Individual variations are due to the size and location of the watershed, the steepness of the coastal belt and the degree of silting and turbidity, due in many cases, to human activities.

B. The warm-temperate estuaries of the Eastern and Southern Cape. These have minimum winter temperatures of 12-14 °, a variable rainfall and they lie on a coastal belt which flattens in the south. The estuaries of the Tsitsikama coast may form a special subgroup for the rainfall is higher and more even through the

year. The coastal belt is high, the seaward edge is rocky and the rivers carry little silt. The variations among the estuaries south-west of Mossel Bay are due to the position of the main catchment and the permanence of the mouth which varies with the size of the river.

C. The cold-temperate estuaries of the Atlantic coast. As noted, there are few of these as the rainfall is so low and evaporation so high. Small coastal rivers are usually dry sand on the surface and their estuaries if present are due to seepage through the sand. It would be interesting to investigate their meiofauna. The only permanent estuaries are the Berg and the Olifants and these run fresh to the sea during winter floods. Verloerenvlei, of which a preliminary account is given by Grindley (1979) is basically similar to the coastal lakes discussed by Professor Brian Allanson in chapter 15. The great Orange River has no estuary at all. It has been classed as a river mouth.

What follows is a description of individual estuaries around the coasts of southern Africa. They are grouped into provinces A, B and C, as discussed above but further comparisons are omitted. The descriptions are primarily for reference as a basis for further research and for planning. The location of the various estuaries described is shown in Figure 14.3.

A. SUBTROPICAL ESTUARIES OF SOUTHERN MOCAMBIQUE, NATAL AND THE TRANSKEI

Morrumbene estuary
(23°44'S/35°24'E)

The following account summarises the only two reports. Day (1974) deals with the general ecology of the estuary and Day (1975) focusses attention on the mangrove fauna.

Morrumbene estuary has a normal salinity gradient throughout the year for it receives the run-off from 655 km² of hilly country with an annual rainfall of 919 mm. The heaviest rains fall in summer and the lightest during August, September and October. The mornings are bright and clear and the sun becomes intensely hot towards midday but in the afternoon clouds build up rapidly and a thunderstorm develops. Sometimes it is accompanied by fierce squalls which sweep across the broad lagoon and generate surprisingly large waves on lee shores. As a result, fine silt and organic matter only accumulates in sheltered areas.

The largest river flowing into the estuary is the Rio Inhanombe which drains open savannah. The sandy soil appears to be rather infertile for there is little cultivation apart from cocoanut plantations near the coast. The only modern town is the Porto do Inhambane and there are few peasant farmers. The alluvial soil along the banks of the streams is still uncultivated and is overgrown by reeds and swamp vegetation which acts

as a filter so that the water entering the estuary is clear but stained brown with peat.

Morrumbene estuary (Figure 14.4) is 20 km long with the upper reaches covered by an extensive mangrove swamp and the lower reaches forming a sandy lagoon opening into Inhambane Bay. The whole lagoon is very beautiful with golden sandflats and graceful dhows seen against a background of the cocoanut palms on Linga-Linga Point. The sandflats revealed at low tide are really extensive for the spring tide range is 2,8 m at Linga-Linga and still over 2 m at Mongué Point, halfway up the lagoon. The range decreases in the channels through the mangroves to a minimum of 1,2 m at the head of the estuary. Although the mudflats on which the mangroves grow are inundated at high tide, the channels are only about 1 m deep at low tide and 50 to 100 m wide. In the lagoon, however, the channels gradually deepen from little more than a metre between the maze of shoals opposite Mongué ferry to 5 and then 10 m at the mouth. As the tide falls, swift turbulent currents develop in the channels so that there is no stratification of salinity or temperature. The salinity varies with the tide from 0,1 to 10,0 ‰ at the head of the estuary and 31,2 to 35,0 ‰ at the mouth. Water temperatures change with the seasons; at the head of the range is 18,8 ° in winter to 28,0 ° in summer, while at the mouth the range is 21,0 to 25,5 °.

Although the river water that enters the estuary is normally clear, colloidal silt and organic matter flocculate where the water becomes saline and the turbidity in the channels increases. Much of the suspended matter including the silt brought down by the rivers in flood is trapped by the mangrove roots at high tide. As a result the percentage of silt in the mangrove mud is far higher than that in the sand of the open lagoon. Moreover, the water in the middle of the lagoon is clearer and the Secchi depth varies from 1,1 m to 3,0 m with the tides.

The character of the sediments has been tabulated by Day (1974). In the mangrove swamps the fine sand contains about 25 % of silt and between 3,6 and 7,1 % of organic matter. This decreases the permeability so that the permanent water table is at or close to the surface and the black deoxidised layer is less than a centimetre below even at the level of HWN. In the lower part of the lagoon, the percentage of silt and organic matter is much less and the sediment at the upper tidal levels is well drained and oxygenated.

Although three expeditions visited the estuary, there was too little time to study the plankton; all that is known is that the zooplankton is rich in species but low in biomass. Again, only the dominant estuarine plants have been identified and no studies of biomass or primary production have been made. Distribution lists of identified species are available in Day (1974) and it may be noted that six species of sea-grasses are present in the lagoon but only one, *(Halodule uninervis)* is present in the channels through the mangrove swamp. Of the six species of mangroves identified, only *Sonneratia alba* has not been recorded further south; *Avicennia, Ceriops, Bruguiera, Lumnitzera* and *Rhizophora* reach Kosi Bay on the Zululand coast.

Several intertidal transects were surveyed along the estuary and on these and at netting and dredging stations, 404 species of invertebrates and 114 species of fish were identified. Thus the estuary is very rich in species including 72 polychaete worms, 51 bivalves, 63 gastropods and 67 crabs. The transects showed that as usual, species richness increased towards low tide although the upper limits of individual species are more closely correlated with the depth of the permanent water table and the nature of the sediment than with the tidal level. This is well-illustrated by the distribution of *Alpheus crassimanus* and *Upogebia africana* which burrow in mud near low neap tide on sunlit shores but extend above mid-tide in waterlogged mud in the mangrove swamps.

It is well-known that although the mangrove fauna is characteristic, the species present are not restricted to mangrove habitats. *Littorina scabra, Balanus amphitrite* and *Saccostrea (= Crassostrea) cucullata* which are common on the mangrove trunks and pneumatophores, also occur on rocky shores in the estuary; species of *Uca* are more common on sunlit shores and the species of *Sesarma* which dominate the mangrove swamps also

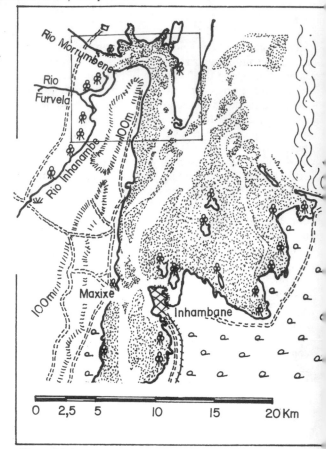

Figure 14.4
MORRUMBENE ESTUARY

harbour in dense salt marsh vegetation. Possibly the most characteristic species in mangroves is the gastropod *Cerithidea decollata* but this too has been found occasionally in salt marsh vegetation. Actually the mangrove fauna of Morrumbene is not rich in species possibly due to the monotony of the mangrove habitat and its restriction to upper tidal levels. Most of the species are detritus feeders although species of *Sesarma* in particular are reported to feed on mangrove leaves and seedlings and even climb among the branches at night. Apart from the oysters growing on the mangroves at the edges of the channels, bivalves are absent and polychaetes are poorly represented. Day (1974) illustrates the density of the dominant species on several transects. The general impression is that apart from the myriads of crabs, the biomass of the benthic fauna is low. The mudflats near low tide are richer; African women dig the larger bivalves and worms and collect the gastropods on the surface and the average catch per woman on a low spring tide is 1-2 kg.

The fish and prawn population is not rich although there is a wealth of small species. There are no large predatory barracuda or carangids and luckily for the netters, there were no sharks or crocodiles. Angling

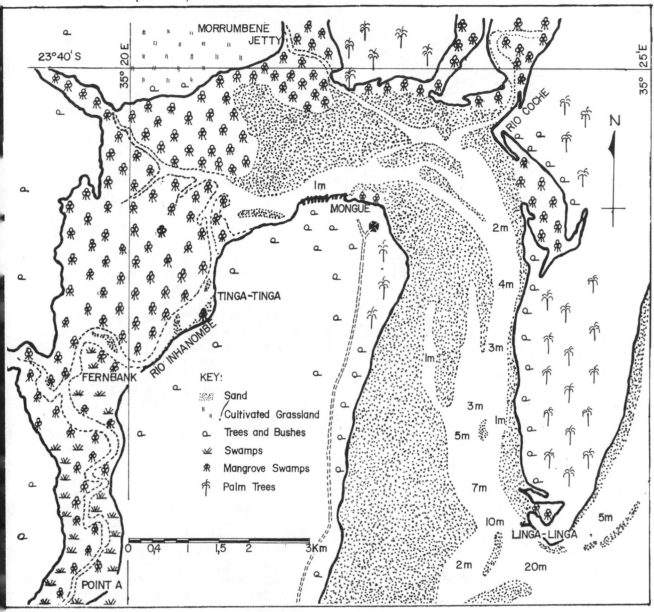

KEY:

Sand	Sand
Cultivated Grassland	Cultivated Grassland
Trees and Bushes	Trees and Bushes
Swamps	Swamps
Mangrove Swamps	Mangrove Swamps
Palm Trees	Palm Trees

was poor and although every suitable bank is beset with African fishtraps their catches were seldom more than a few swimming crabs and two or three kilograms of penaeid prawns and fish. Possibly the Africans hope for shoals moving into the estuary during floods. Nonetheless, there is evidence that the organic detritus derived from the mangroves supports a large population, for trawlers working at the mouth of Inhambane Bay make good catches.

The Kosi Bay system
(26°47′S/32°47′E)

Sixteen papers deal with the ecology of Kosi Bay. They are grouped below according to the main topics discussed:

Ecology and faunistic lists: Broekhuysen & Taylor (1959), Tinley (1976)

Bathymetry and origin: Hill (1969), Orme (1974)

Hydrology and chemistry: Allanson & van Wyk (1969), Oliff (1976), Cloete & Oliff (1976)

Phytoplankton productivity: Oliff (1976)

Mangroves: Macnae (1963), Breen & Hill (1969), Tinley (1976)

Benthos of Lake Nhlange: Boltt & Allanson (1975)

Fishes: Campbell & Allanson (1952), Pike (1967-71), Blaber (1978)

General summary: Begg (1978)

The parts of the Kosi Bay system (Figure 14.5) have received various names in official maps and publications. Begg (1978) has attempted to stabilize the nomenclature and his system of names is adopted here. The Kosi

Bay system lies in undeveloped country on the northern border of Zululand. The rainfall averages 1 250 mm per year, most of which falls in summer and as the 500 km² drainage basin is mainly flat and marshy, the several small streams which enter the system are slow flowing. The whole system consists of a series of five lakes linked by narrow channels. The first lake, Amanzimnyama is 2 m deep, while the next three, Nhlange, Mpungwini and Makhawulani are much deeper. All four are clear and are floored with sand overlain by a film of rotting vegetation while the fifth, variously referred to as Kosi Bay or Enkovugeni, is shallow with a bed of clean sand and a narrow exit to the sea with a reef on the south bank.

Lake Amanzimnyama, the first in the series is fresh, and Lake Nhlange is oligohaline with no sign of stratification in spite of its depth. The narrow channels between the lakes minimise salinity exchange and Lakes Mpungwini and Makhawulani show a stepwise increase to higher salinities. Both have marked haloclines over the deeper eastern depressions. Enkovungeni has a salinity of 10-35 ‰. Tidal range is first evident in Lake Makhawulani and increases to 0,6 m in Enkovugeni. Water temperatures in the deep lakes are 20-23 ° through the year.

The water throughout the system is very clear and both Nhlange and Mpungwini have a high NH_3 concentration so that phytoplankton production was relatively high (60 mg.Cm^{-3}.day^{-1}) in August 1971. Further details are discussed by Oliff (1976). The bottom water of Nhlange is surprisingly well-oxygenated even below 20 m but anoxic conditions may develop below the halocline in the deeper parts of Mpungwini.

The submerged vegetation in the shallows of Nhlange and Mpungwini includes *Chara* and *Potamogeton pectinatus*. *Enteromorpha* replaces *Chara* in Makhawulani and *Potamogeton* decreases. *Zostera* was searched for, but appears to be absent from the Kosi system and the bed of Enkovugeni is bare sand. The marginal vegetation changes from extensive reed beds (*Phragmites* sp) on the shores and channels between Nhlange and Mpungwini, to mangroves along sheltered shores of Enkovugeni. Five species are present and Breen & Hill (1969) report that after the mouth closed in 1965/66 and torrential rains increased the lake level, many of the partially submerged mangroves died through lack of tidal action.

Broekhuysen & Taylor (1959) list 106 species of macrobenthic invertebrates most of which are restricted to the reef in the entrance of Enkovugeni, only 44 extend into the estuarine channels and lakes. There is a normal mangrove fauna dominated by crabs and gastropods, of which *Sesarma* spp and *Assiminea bifasciata* extend into the reed beds. In the sands, *Callianassa* is abundant reaching densities of about 50-70 m^{-2}, but *Upogebia* has not been reported. Amphipods, *Hymenosoma* and *Lamya capensis* are widely distributed. Penaeid prawns are not common but reach Lake Mpungwini and

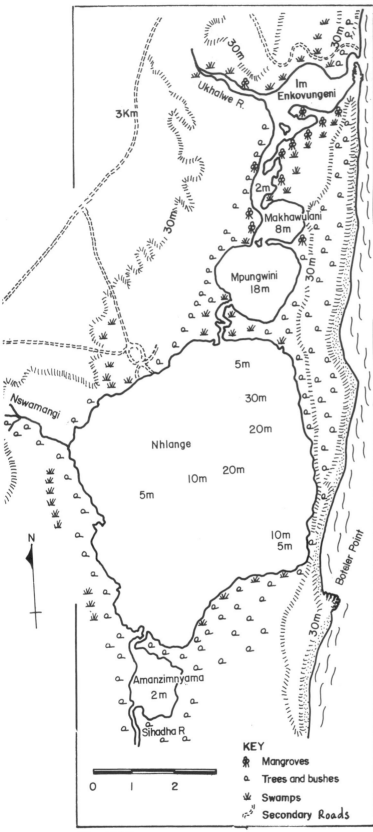

Figure 14.5
KOSI BAY SYSTEM

KEY
🜊 Mangroves
ª Trees and bushes
🜋 Swamps
⌇ Secondary Roads

are replaced by *Caridina nilotica* in the shallows of Lake Nhlange. Boltt & Allanson (1975) give a detailed and quantitative account of the benthos of this lake.

Blaber (1978) lists 124 marine and estuarine fish in the estuary of which 85 are restricted to Enkovugeni and the reef at its entrance. He discusses the relative abundance of the remaining 39 species in the different lakes and notes that almost all were found along the shallow margins. It is interesting that large adults of *Chanos chanos, Scomberoides tala* and *Pomadasys commersonni,* as well as a few others, reach the oligo-haline waters of Nhlange. The fish population is trapped extensively by Thanga fishermen; in fact there are so many fish traps in Enkovugeni that navigation is diffi-cult.

The Kosi system is completely unspoilt and the clear lakes are most attractive. It is important that con-servation measures be maintained.

The St Lucia system
(28° 23'S/32° 26'E)

Begg (1978) states that over 600 reports have been written about this system; of these, 60 are referred to here. The Natal Parks Board have established a Docu-mentation Centre at St Lucia as a repository for information about the system. Public outcry led to the appointment of a State Commission of Enquiry into the alleged threat to plant and animal life in St Lucia in 1963 and its findings, edited by Kriel (1967) was laid before Parliament. It contains several appen-dices by different authors and these are indicated by an asterisk in the list below. In 1976, the St Lucia Scientific Advisory Council held a workshop at Char-ter's Creek at which some 23 papers were presented. These have been edited by Heydorn (1976) and the ones referred to below are marked thus (†). Other papers have appeared in scientific journals or unpub-lished departmental reports. The papers are grouped below subject by subject for the convenience of later workers. The complete references are listed at the end of this chapter.

General ecology: Day *et al* (1954), Millard & Broekhuysen (1970), Begg (1978).

Commission of Enquiry: Kriel (1967).

Geology and sedimentation: Orme (1973), Hobday (1975), van Heerden (1976)†, Phleger (1976)†.

Climate and hydrology: Kriel (1967)*, Tinley (1969), Hut-chison & Pitman (1973), Hutchison (1974a, 1974b, 1975, 1976a†, 1976b), Alexander (1976)†.

Pesticides and trace metals: Oliff (1979).

Marginal and aquatic vegetation: Macnae (1963), Bayer & Tinley (1967)*, Ward (1976)†, Breen *et al* (1976)†.

Plankton: Cholnoky (1968), Grindley & Heydorn (1970), Grindley (1976)†, Johnson (1976)†, Blaber & Rayner (1979).

Benthic fauna: Boltt (1975), Hill (1976†, 1978).

Penaeid prawns: Joubert & Davis (1966), Champion (1976)†.

Fish: Wallace (1969, 1975a, 1975b, 1976†), Blaber & Whit-field (1972), Wallace & van der Elst (1975), Bass (1976)†,

Blaber (1976)†, van der Elst *et al* (1976)†, Whitfield (1976†, 1977), Whitfield & Blaber (1977, 1978a, 1978b, 1978c), Heeg & Blaber (1979), van der Elst (1979).

Crocodiles: Pooley (1976)†.

Hippopotami: Taylor (1976)†.

Birds: Forrest (1969), Whitfield (1977), Whitfield & Blaber (1978d, 1978e, 1978f), Whitfield & Cyrus (1978), Berruti (1979)

Reclamation proposals: Blok (1976)†, Hutchison (1976b, 1976c†), Hutchison & Pitman (1977), Mercer (1976)†.

St Lucia is the largest estuarine system in southern Africa. The water in the lakes and channels and a 1 km strip of land along the banks have been declared a nature reserve under the control of the Natal Parks, Game and Fish Preservation Board. Angling is permitted in certain areas but the fauna and flora is otherwise preserved. The St Lucia system (Figure 14.6) has a total area of 30 520 ha (reduced when the lake level falls) and includes three interconnected saline lakes leading into a long channel known as the Narrows. This passes under a bridge at St Lucia township and then broadens to form what is known as the Estuary Basin which finally opens 1,5 km from the mouth of the Mfolozi.

The St Lucia lakes are at the same level as the sea and receive the inflow of four main rivers with a total catchment of 9 000 km². Most of the rain falls in the coastal part of the catchment in summer and the inland areas are dry in winter. Moreover, part of the runoff which reaches the rivers is used for irrigation. The remaining inflow into the lakes, as estimated by Hut-chison (1976a)† in terms of 10^6 m^{-3}.yr^{-1}, is: Mzinene 22, Hluhluwe 28, Nyalazi 24 and the Mkuze 164. The Mpate which flows into the Narrows contributes 11 x 10^6 m^{-3}.yr^{-1}, while seepage from the eastern or coastal side of the system is 46 x 10^6 m^{-3}.yr^{-1}. Thus the St Lucia system receives a total inflow of 295 x 10^6 m^{-3} in a normal year. The lakes are large but the average depth is only 1-2 m and the continual winds which blow over the surface enhance evaporation. This is estimated at 397 x 10^6 m^{-3}.yr^{-1} and thus greater than the rainfall on the lakes which, in a normal year, is 268 x 10^6 m^3. The whole fresh water budget shows that in years of good rainfall, the lake level rises above sea level and the water flows down and out through the channel. In such years, a normal salinity gradient is established and in 1964 Millard & Broekhuysen recorded a salinity of 5,5 ‰ in False Bay and a gradual increase through the system to 24,8 ‰ in the Estuary Basin. Oliff (1979) records an essentially similar salinity range in May 1978. In years of drought the lake level falls due to evaporation and a reversed salinity graident is established with maximum values in False Bay. Sali-nities in excess of 50 ‰ have been recorded several times and in 1970 the salinity in the northern part of False Bay was over 100 ‰. South lake is seldom hyper-saline and the water at Brodie's Crossing and in the northern Narrows has the most stable salinity of the whole system.

The mouth of the estuary and the Mfolozi River

originally flowed into what was known as St Lucia Bay. In drought years, the estuary mouth closed but the Mfolozi, which is a very large river, remained open. Its water, which was filtered through extensive papyrus swamps was clear. Eventually the swamps were canalised by Warner's Drain in 1918 in order to plant sugar cane on the alluvial soil. Thereafter the Mfolozi floods filled St Lucia Bay with mud and part of the Mfolosi flowed into the St Lucia channel. This helped to keep the mouth open but as the tide rose, the muddy Mfolozi water flowed up the St Lucia channel and filled it with porridgey mud which smothered the aquatic plants and animals. Eventually a levee was built between the Mfolozi and the St Lucia channel and the latter was dredged. In years of drought the mouth of the St Lucia system is now kept open artificially.

Van Heerden (1976)† summarises the geological investigations of earlier workers and himself gives a fascinating account of the history of the lakes based on the examination of a series of cores. Details cannot be given here but it may be noted that there are few rock exposures. Cretaceous rocks containing numerous fossils may be seen at Lister Point and Hell's Gates. Elsewhere the estuary bed is floored with very soft mud with patches of sand particularly towards the eastern shores and at the mouth of the Estuary Basin.

The hydrology is fairly well-known. The spring tide range at the mouth is 1,9 m but decreases to 1,0 m at the bridge and then fades to 0,1 m in the Narrows. Thus there is no tidal range in the lakes but the persistent winds may raise or lower the water level on the shores by as much as 0,3 m. This seiche action and the wind-generated waves eliminate stratification and maintain the bottom water well-oxygenated but the waves also stir up the soft silt from the bottom. Normally a Secchi disc is visible at about 0,4 m but on a lee shore the Secchi depth may decrease to 0,1 m. Water temperatures in the lakes normally vary from a winter minimum of 18 ° to a summer maximum of 28 ° but extremes of 14,3 and 30,2 ° have been recorded.

Nutrient concentrations require further investigation. Johnson (1976)† reported conditions in False Bay during 1975. PO_4-P was 10 $\mu g.\ell^{-1}$ in July rising to 41 $\mu g.\ell^{-1}$ in December. 'NH_3 and NO_3-N' is shown in Johnson's figure 5 as averaging 170 $\mu g.\ell^{-1}$ in autumn, falling to 70 $\mu g.\ell^{-1}$ in winter and rising to a maximum of 175 $\mu g.\ell^{-1}$ in October. These figures do not indicate any nutrient deficiency but concentrations in the remainder of the St Lucia system have not been reported.

The phytoplankton also requires further study. Cholnoky (1968) has identified many of the diatoms and Johnson (1976)† reports that the phytoplankton is mainly autochthonous. The blue-green nanoplankter *Synechococcus* was abundant in the northern lakes during periods of low salinity but is contribution to phytoplankton biomass and productivity is unknown. On a basis of cell counts and cell volumes, her estimate

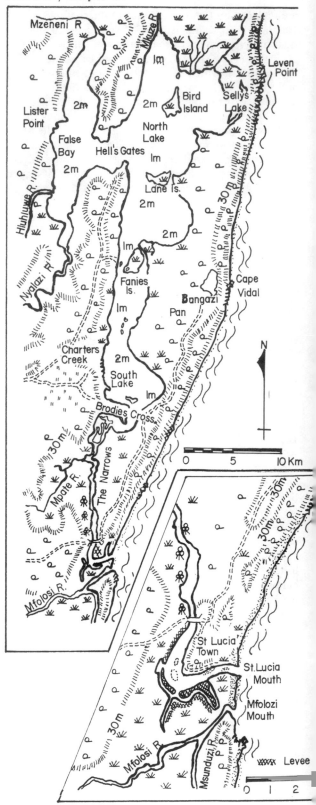

Figure 14.6
ST LUCIA LAKES AND CHANNEL

of phytoplankton volume was higher in False Bay and North Lake than elsewhere in the system. Over 50 % of this volume was due to the large diatom *Coscinodiscus granii.* Red water also appeared in the northern lakes during the winter of 1969 following hypersaline conditions earlier in the year. Grindley & Heydorn (1970) state that this was due to the proliferation of *Noctiluca* and speculate that this might have caused the death of fish that were washed up. Although *Noctiluca* is not known to be toxic, it has been associated with fish kills elsewhere. Zooplankton samples over 29 years have been studied by Professor J. Grindley. A final report is in preparation, but a preliminary account was given by Grindley (1976)†. Neritic marine species drift in and out of the estuary basin with the tides but the lakes are dominated by an autochthonous estuarine population which is surprisingly stable in composition over the years in spite of major changes in salinity. The two dominant copepods and the abundant mysids are surprisingly euryhaline. *Pseudodiaptomus stuhlmanni* and *Acartiella natalensis* tolerate salinities of 70 ‰ and breed in salinities exceeding 60 ‰ while *Mesopodopsis africana* tolerates 60 ‰ and breeds in salinities over 50 ‰. On windy days it congregates in the muddy water on lee shores in such numbers that a cupful can be caught by a few sweeps of a hand net, and it is often found in fish stomachs. The biomass of zooplankton catches in the South Lake in 1967 averaged 19 mg.m^{-3} and in 1969 it averaged 37 mg.m^{-3}. *P. stuhlmanni* and *Mesopodopsis* are both important in the diet of planktivorous fish.

The estuarine vegetation is best described by Ward (1976)†. *Enteromorpha* sp is the only macrophytic alga of importance and may grow anywhere in sheltered water even under hypersaline conditions. *Zostera* is absent from the Estuary Basin but appears in the Narrows and, with *Ruppia spiralis* and *Potamogeton pectinatus,* it dominates the shallows at Brodie's Crossing and extends along sheltered shores of South Lake. *Zostera* even extends into North Lake when the salinity is low and the water clear, but it disappears under hypersaline conditions.

Mangroves are restricted to tidal banks. They fringe sheltered banks in the Estuary Basin and extend for some distance along the Narrows. *Avicennia* and *Bruguiera* are common and the fern *Acrostichum aureum* appears with them but *Rhizophora* has disappeared, possibly when the mouth closed in the 1950's. Apart from the mangrove fringes, the marginal vegetation is dominated by tall species of *Phragmites, Scirpus, Cyperus* and *Typha* when the salinity is low but these die down when the salinity rises. At such times the dominant plants are grasses such as *Sporobolus* and *Stenotaphrum* at the water's edge and *Juncus* or *Sarcocornia* spp at higher levels.

The benthic fauna is not rich in species. Day *et al* (1954) and Millard & Broekhuysen (1970) list 84 species excluding the six penaeid prawns which will be discussed separately. About 17 of the 84 are restricted to the Estuary Basin and the number of species decreases further towards the northern lakes where only 29 have been identified.

Neither *Upogebia* nor *Callianassa* have been recorded from the system which is surprising for one or other dominates the benthic fauna of almost every South African estuary. Again, oysters *(Saccostrea cucullata)* which were once common on the reefs in the Estuary Basin are now absent. The dominant benthic forms that remain are amphipods, crabs and gastropods.

The shore fauna is only abundant in the mangrove fringes in the Estuary Basin and the Narrows and is well-illustrated by Millard & Broekhuysen (1970). Marsh crabs *(Sesarma* spp) and amphipods *(Orchestia* sp) extend into the marginal vegetation and *Hymenosoma orbiculare, Assiminea bifasciata* and several amphipods are common in the submerged vegetation and extend to the bottom of the lakes. All of these are important fish foods. Dredgings in the lakes however, was disappointing possibly due to the lack of tidal action and the accumulation of soft silt.

Boltt (1975) dealt with the recovery of the fauna on the bed of the lakes after a period of high salinity. As is well-known, the whole fauna of the northern lakes is impoverished by hypersaline conditions. Boltt found that few benthic animals except chironomid larvae and ostracods survived salinities of 55 ‰. Recovery occurs when salinities in the northern lakes drop below 40 ‰. Species from South Lake with short life cycles and planktonic larvae are the first to appear. The biomass of small benthic animals throughout the lakes is soon dominated by *Assiminea bifasciata* but other invertebrates increase very rapidly in spring.

The six species of penaeid prawns in the estuary are all migratory and breed in the sea. Juveniles enter the estuary mouth in spring and spread rapidly through the system. They are omnivorous and grow rapidly through the summer and most return to the sea in autumn. According to Champion (1976)† the bait fishery in the St Lucia system is the largest in South Africa yielding 15 000 kg.yr^{-1}, with *Penaeus indicus* as the most important species. In years when the northern lakes become hypersaline, the catch per unit effort decreases. In salinities over 60 ‰ it becomes insignificant.

Millard & Broekhuysen (1970), record 86 species of fish from the whole St Lucia system and according to Wallace (1975a, 1975b, Wallace & van der Elst 1975), 82 species are present in the lakes. Of these, five are fresh water fishes which extend into the lakes in periods of low salinity. At least 12 species breed in the lakes and the rest are migrants from the sea. Most of them enter the estuary as juveniles for food and shelter and return to the sea to breed on the approach of maturity so that the estuary is an important nursery area. A few species return as adults to feed on prawns and forage fish in summer and notable among them are sport fish such as the spotted grunter *Pomadasys commersonni* and the kob *Argyrosomus hololepidotus*. St Lucia is known as an angler's paradise during the grunter run and shark fishermen also congregate there. Several sharks enter the Estuary Basin to feed on mullet and *Carcharinus leucas* also spawns there and the young spread through the lakes.

Wallace (1975a, 1976†) has shown that 14 common species tolerate salinities of 60 ‰ or more but most species migrate out of the northern lakes during hypersaline periods and fish mortalities are rare. Apart from excessive salinities and closure of the mouth (now kept open artificially), there is little evidence that the increasing sport fishery has reduced the fish population since it is maintained by a yearly migration from the sea. Van der Elst (1979) has analysed the catch per unit effort of Natal angling clubs over the years and has shown that the catches of grunter and kob are stable; only the catch of river bream *(Acanthopagrus berda)* has declined.

In a long series of papers listed earlier, Blaber & Whitfield have discussed the feeding habits of three groups of fishes. The main plankton feeders are *Gilchristella, Thryssa* spp and *Hilsa kelee*. Mullets, milk fish and tilapia feed on 'detritus' (including diatoms and foraminifera), filamentous algae and *Assiminea*. Competition is reduced by differences in feeding areas and feeding times. Piscivorous fish feed predominantly on the small plankton feeders such as *Gilchristella* and *Thryssa vitrirostris* (the adults of which prey on larval *Gilchristella*) but many other prey species are taken as well. Feeding habits depend on the changing densities of the prey species and the larger predators are limited to deeper areas. All large predators take prawns and small ones feed on mysids, amphipods, shrimps, *Assiminea* and *Hymenosoma* in the weed beds.

St Lucia is famous for its bird life. Whitfield (1977b) lists 340 species of which about 90 are dependent on the estuary. Waders feed on the banks or in the shallows and flocks of flamingoes visit the lakes when the sandbanks are exposed. Piscivorous birds are particularly common although the turbid water provides the fish with some protection. Whitfield & Balber (1978d, e, f) discuss the feeding habits of the diving birds, the waders and the swimmers giving the list of prey species of each group. The divers which include the fish eagle, the terns and kingfishers take surface fish swimming in the upper 20 cm. The waders, including egrets and herons, feed on small fry and invertebrates in the shallows. The waders breed and feed most efficiently when lake levels are low and their prey is concentrated in the shallows (Berruti, 1979). The swimmers include cormorants and up to 15 000 white pelicans; the reed cormorant feeds mainly on *Sarotherodon*, the white-breasted cormorant feeds in deeper water on mullets, bream and clupeids while the pelican changes its habits. It is a wide-ranging bird and visits the Pongola pans 60 km to the north to catch chiclids when feeding its young but at other times it feeds mainly on the mullet shoals in the lakes.

St Lucia is a sanctuary for crocodiles and hippopotami as well as birds and fish. Both have increased. Over 400 crocodiles have been counted in winter. During periods of high salinity they move up the rivers but in the extreme salinities of 1970 about 40 died. In 1957, a census of hippopotami recorded 158 but in 1978 the numbers had risen to 522. They feed mainly on the short grasses near the banks but require fresh water for drinking and congregate near the seepage areas on the eastern shores. In drought years many congregate in the Mkuze swamps and others have to travel as far as Lake Bangazi.

It has long been realised that St Lucia is a very important national asset. Not only is it a nursery area for marine fish and penaeid prawns, it is also an important tourist area on account of its sport fish, its abundant birdlife, its wild game and its scenic attractions. It must be preserved. The key problem is the supply of sufficient fresh water to maintain salinities below 40 ‰ during drought years when the farmers require river water for irrigation. The obvious solution of diverting other large rivers into the system was proposed by Day *et al* (1954) and later workers. These schemes have since been examined by competent hydraulic engineers as shown by the list of references given earlier. It has been suggested that part of the Mfolozi be diverted into the Estuary Basin or the Narrows and a more ambitious and expensive scheme is to link the Mfolozi and the Nyalazi so that more water would flow into False Bay. Hutchison (1976c)† has compared the merits of these schemes as well as modifications of the early suggestion that water from the Pongola be diverted into the Mkuze. This would mean that the fresh water supply to North Lake could be augmented when required. Mr George Begg in a personal communication states: 'The situation today is that the Mfolozi link canal is being cut between the Mfolozi and the Mpate — a distance of 25 km. It is expected to divert 190 x 10^6 m^3 of water into the South Lake during the drought years, and thus retain a 'reservoir of life' in the south end of the system. The high salinities of the northern reaches will NEVER be controlled'. As noted earlier, the northern Narrows and the entrance to South Lake have a stable salinity even in times of drought and they always have acted as a 'reservoir of life'.

From an ecological viewpoint the diversion of more fresh water into the northern lakes (either False Bay or North Lake) would promote a normal salinity gradient through the system whereas the inflow of fresh water into the Estuary Basin or the Narrows would accentuate the reversed salinity gradient during droughts. The salt in the lakes would not be flushed out and evaporation would maintain higher salinities there than in the lower reaches.

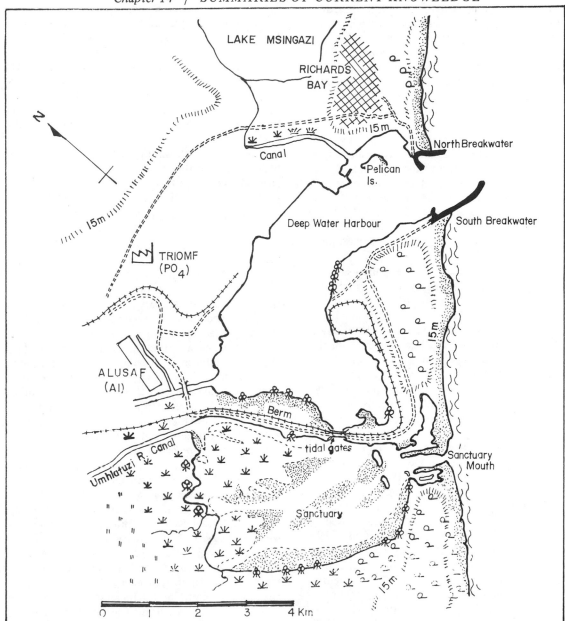

Figure 14.7
RICHARDS BAY

Richards Bay
(28° 49′S/32° 05′E)

Ten reports deal with various ecological aspects of the bay and with the Mhlatuzi River which supplies most of the fresh water. Begg (1978) cites over 50 references. Millard & Harrison (1954) describe the general ecology and more recent reports edited by Oliff (1976) and summarized by Cloete & Oliff (1976), bring together the detailed researches of teams of NIWR officers led by Dr J. Hemens and Dr A. Connell. These concern the sedimentology, hydrology, water chemistry and productivity of the Bay. McClurg (1979) and Connell & Airly (1979) assess the toxicity of fluoride, mercury and cadmium. Turvey (1960) describes the Mhlatuzi, Macnae (1963) deals with the mangroves, Cholnoky (1968) with the diatoms and Grindley & Wooldridge (1974) give an account of the zooplankton. Rossouw (1976) discusses the dynamics of the mouth.

In 1970 the modification of the Bay to form a deep-water harbour was commenced, and in 1974 the southern half of the Bay was separated off by a berm so as to form a wild life sanctuary. The berm was so positioned that the Mhlatuzi flowed into the sanctuary and a new mouth was cut through the sand dunes to give the sanctuary a direct opening to the sea. Tidal gates were also built into the berm in case the new mouth closed. It may be noted at this point that, in spite of solemn warnings by ecologists, the 6 km wide papyrus swamp which had acted as a filter at the mouth of the Mhlatuzi was canalized. The effects of the several modifications have been described in brief by Oliff (1976) and Emanuel (1977). A fuller account is given by Begg (1978). I wish to thank George Begg, Rudy van der Elst and Dr Julian Hemens for further information.

Although the prime objective of the work was the construction of an industrial harbour, the sanctuary was designed to conserve part of the original ecosystem and, amongst other things, to act as a nursery for prawns and juvenile game fish. It has proved to be an interesting ecological experiment. Many changes have occurred and in order to put these into perspective, an outline of conditions in the Bay prior to its modification is necessary.

Richards Bay was originally a large shallow estuarine lagoon, 2 890 ha in area and fed by the Mhlatuzi which has a normal run-off of 616×10^6 m^3 per year and by the outflow from Lake Msingazi which supplies $3,7 \times 10^6$ m^3 per year. The Bay had a very shallow muddy bottom about 0,9 m deep and a narrow exit channel to the sea. Tidal action was obvious only in the channel, and in the main body of the bay the range was about 0,1 m and the level at a particular point varied more with the stress of the wind. The winds blew from the NE or SW and had very important effects on the ecology of the bay. They maintained vertical circulation so that the water was well-oxygenated to the bottom but they also stirred up the soft mud and the Secchi depth was seldom as much as 0,3 m. In consequence, phytoplankton production was usually less than 100 mg.C.m^{-3}.day^{-1} and *Zostera* could only grow in the clearer water of the channel. Salinities varied between 12,5 and 35 ‰ (Grindley & Wooldridge, 1974) but was normally above 25 ‰. In May 1971, a major flood flushed out the estuary with fresh water. The seasonal temperature range was 17 to 28 °. Nutrient concentrations were fairly high with nitrogen (either in the form of ammonia or nitrate) varying from 48,6 μg atN.ℓ^{-1} in summer to 69,2 μg atN.ℓ^{-1} in winter. Total phosphorus was measured by the NIWR but not soluble phosphate.

The high turbidity provided protection for penaeid prawns and mullet against bird predation so that they were abundant in the open bay as well as in the *Zostera* beds. The zooplankton was also rich due to the long residence time of the bay water, but the benthic fauna in the soupy mud of the open bay was very poor apart from the mud crab *Tylodiplax*. Only in the sandy mud of the channel and its *Zostera* beds was animal life abundant. Of 183 species recorded in the estuary, 153 were found in the *Zostera* beds and mangroves. Penaeid prawns have been mentioned and shrimps, burrowing prawns, crabs, worms and small fish were abundant too. Sport fish such as kob,

spotted grunter, river bream, yellow-fin bream and silver bream were plentiful. Richards Bay was thus a favourite holiday resort.

The changes in the sanctuary following the construction of the berm wall, the cutting of a separate mouth and the canalization of the Mhlatuzi are described by Begg (1978). Mr Rudy van der Elst, in a personal communication, has provided further valuable information. Figure 14.7 is a map incorporating these observations.

After the mouth was opened, tidal scour aided by the discharge of the Mhlatuzi, rapidly increased its cross-sectional area from 200 m^2 to 900 m^2. The spring tide range which, it may be remembered, was 0,1 m in the Bay, increased to 1,0 m in 1976 and 1,8 m in 1978. The rate of siltation, originally estimated as 0,7 mm per year, increased enormously after the Mhlatuzi was canalised. During 1976 and 1977, five serious floods with flows exceeding 400 m^3.sec^{-1} filled the sanctuary with $3,2 \times 10^6$ m^3 of sandy mud so that at low spring tide half the estuary bed was exposed. The sanctuary had become the delta of the Mhlatuzi with distributory channels winding over the mud flats. The flood waters even carried the mud over the tidal gates and deposited it in the harbour along the northern side of the berm wall. The gates did not allow for the increased tidal range and at high spring tides pollution from the harbour can enter the sanctuary. Fluoride from the ALUSAF effluent reaches the harbour, but not in toxic concentrations; mercury and cadmium concentrations are very low.

With the infilling of the sanctuary and the greatly increased tidal range, the residence time of the estuarine water has decreased from months to hours. It is estimated that 88 % of the sanctuary water is exchanged during each tidal cycle. It is not surprising that the whole ecology of the sanctuary has changed. The salinity of the Bay was always close to that of sea water, but the turbid water has been swept out and the sea water that enters with each tide is much clearer. Phytoplankton production increased from less than 100 mg.cm^{-3}.day^{-1} to 40 mg.Cm^{-3}.hr^{-1} or about 320 mg. Cm^{-3}.day^{-1} in February 1976. The mangroves *Avicennia* and *Bruguiera* have colonised the new intertidal banks and *Rhizophora* is being introduced. *Juncus* and *Phragmites* are spreading over the Mhlatuzi delta and *Enteromorpha* is abundant. *Zostera* has not appeared as yet but conditions appear suitable for its establishment.

Although the plant life in what was originally the southern part of the turbid Bay has increased, the area left for aquatic animals has seriously decreased. Moreover the species composition has changed and the fauna is now more marine in character. Whereas the zooplankton was estuarine and abundant, it is now largely neritic and sparse. In 1974 a 50 m tow collected a settled volume of 9 mℓ, in 1975 it collected 1,9 mℓ and in 1976 only 1,05 mℓ. The benthic fauna of the intertidal banks has not been properly surveyed

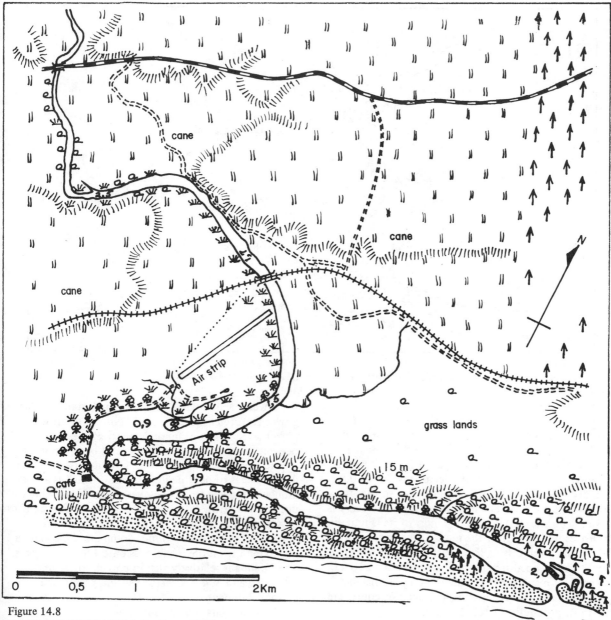

Figure 14.8
MLALAZI ESTUARY (after Begg, 1978).

since the sanctuary was formed and it would be interesting to do so for it originally harboured little except
penaeid prawns and the mud crab *Tylodiplax*. These
are still present and in addition, fiddler crabs *(Uca
urvillei* and *U. chlorophthalmus)* and marsh crabs
(Sesarma meinerti) have colonised the mud flats and
Scylla serrata is common in the shallows. Some 58
species of small marine fish have appeared (Begg,
personal communication) but fish are not as dense as
they were in the *Zostera* beds. The few large predatory
fish are limited to the remaining deep holes such as
the borrow pit near the mouth. On the other hand
game fish are plentiful in Richards Bay harbour.

The value of the sanctuary as a nursery area is

limited; on the other hand the shallows and intertidal
mud banks provide rich feeding areas for wading birds.
Begg (personal communication) reports 75 species of
birds associated with the estuary.

Mlalazi estuary
(28°58′S/31°48′E)

The few references include a general account of the
ecology by Hill (1966) and further observations made
by officers of the NIWR and reported by Oliff (1976).
Brief notes on the mangroves were made by Ward
(1960) and Macnae (1963). Summaries of these papers
will be found in Cloete & Oliff (1976) and Begg (1978).

267

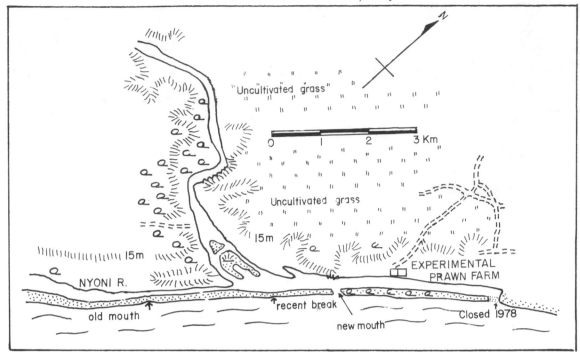

Figure 14.9
MATIGULU ESTUARY

This account is based on a draft by Dr Burke Hill written in 1976 and is amplified with observations contained in more recent papers.

The Mlalazi River rises near Eshowe and has a catchment of 415 km^2 with a summer rainfall of about 1 250 mm.yr^{-1}. About 72 % of the land is a Bantu reserve and the rest is under sugar cane. The estuary (Figure 14.8) varies in length, decreasing from 10,8 km in a dry season to 6 km after heavy rains so that the headwaters are fresh at times. Normally the estuary is 1-3 m deep and 100 m wide for most of its length, but increases to 200 m near the mouth. The bottom is sandy mud with clay in the upper reaches. The banks tend to be steep although there are more than 10 ha covered by mangroves. The area of the salt marshes and sand banks is very limited and these of course are the most productive parts of an estuary.

Although the mouth is narrow it is permanently open, the spring tidal range in the estuary is 1 m in the lower reaches decreasing to 0,6 m at the head. Water temperatures increase from a winter norm of 15-18 ° to a summer maximum of 23-28 °. The salinity has been discussed in detail in chapter 3 as an example of the normal regime in an open Natal estuary. In brief there are well-marked vertical salinity gradients in the upper reaches with the isohalines becoming steeper near the mouth where the salinity is 25-35 ‰. During the dry season, clear, blue sea water may extend to the surface 3 km from the mouth.

Nothing is known of the plankton although Oliff (1976) has shown that the concentration of NO$_3$ is high. The aquatic vegetation also requires further investigation. A few patches of *Zostera* may be present in the lower reaches, while there are small stands of mangroves and *Juncus, Scirpus* and *Sporobolus* grow at higher levels. Further up the estuary *Phragmites* sp become common but *Ruppia* has not been reported.

Hill (1966) and Oliff (1976) have identified a total of 84 species of benthic invertebrates which is a small number for a sub-tropical estuary. This is probably due to the narrow intertidal banks and the few patches of *Zostera*. Crabs, including *Sesarma meinerti, S. eulimine* and *Uca lactea annulipes* dominate the epifauna and extend from the mangroves onto the open shores. The infauna is relatively rich in polychaete worms, *Upogebia* and *Solen corneus*. Penaeid prawns are seldom abundant.

The fish fauna is diverse but the Mlalazi is not known as a rich angling resort. The usual game fish such as kob, grunter and bream are present and the presence of the man-eating shark *Carcharias leucas* is to be noted. The main shoals, however, are small fish such as *Ambassis* spp and juvenile mullet.

The Matigulu estuary
(29° 5'S/31° 37'E)

The observations on this estuary are due to officers of the NIWR; they are reported by Oliff (1976) and summarised by Cloete & Oliff (1976); Begg (1978) has summarised data from 12 references, and Emanuel (1977) has discussed conservation.

The Matigulu River has a catchment of 900 km^2 with a summer rainfall of 1 016-1 270 mm. Most of

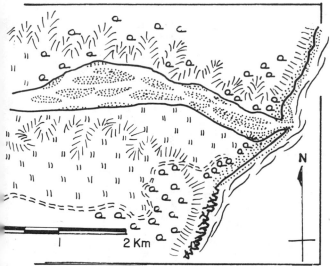

Figure 14.10
TUGELA RIVER MOUTH in May 1979 (location and shapes of shoals approximate)

the soil is sandy; over half is in a Bantu reserve and the rest is planted to sugar cane. The river, which rises west of Eshowe, passes close to the village of Amatikulu and may be polluted with sewage at times so that bacterial counts may be fairly high, particularly after heavy rains and consequent runoff.

As the river winds towards the sea it passes between grassy hills on the north bank and an indigenous forest on the south bank which was once part of the Amatikulu Leper Institute. The last 7,5 km (Figure 14.9) are estuarine and run north-east parallel to the seashore. Although the mouth closes during droughts the Matigulu captured the smaller Nyoni estuary during the floods of 1971 so that in 1977 there was a stretch of 14 km of low salinity water behind the coastal sand dunes; since then a new mouth has opened nearer the axis of the river.

Normally the estuary is 100-200 m wide between sandy banks and 1-2 m deep. Prior to 1977 there was a 3,8 m depression some 2 km from the mouth, but this has now silted up. Near the depression a small experimental prawn farm has been built by excavating ponds on the bank. Its success is doubtful, for the Matigulu estuary is not suitable for such work.

Normally the estuary is oligohaline with salinities below 1 ‰ although a bottom salinity of 21,3 ‰ has been recorded in the depression and salinities up to 32 ‰ have occasionally been recorded near the prawn farm. Temperatures varied between 25,2 and 28,8 ° in March 1972 but would be lower in winter. The water is well-oxygenated and the concentrations of both nitrogen and phosphorus compounds are of the same order as in other unpolluted estuaries in Natal.

Neither the plankton nor the estuarine vegetation has been investigated in detail. While the reed beds in the river extend into head waters of the estuary, there are no salt marshes or true mangroves. Small patches of *Zostera* are occasionally seen but neither *Ruppia* nor *Potamogeton* have been reported. There is little organic detritus in consequence.

The benthic fauna is very poor and there are few penaeid prawns. Some 14 species of fish are listed by Oliff (1976) including jvenile *Carcharhinus leucas,* surprisingly large kob and grunter and a number of mullet. George Begg (personal communication) reports 12 more species. Considering the low salinity, poor benthic fauna and few prawns, this is a good catch.

The Matigulu is an interesting example of an oligohaline estuary diverted for many kilometres along the coast by high energy waves and littoral drift.

The Tugela River mouth
(29°13'S/31°30'E)

Brand *et al* (1967) give a detailed account of the Tugela River from the sources of the several tributaries to Mandini some 8 km from the sea. A summary of conditions in the lower reaches and the mouth is given by Begg (1978) who quotes reports by nine earlier workers. The present account is based on Begg's work and observations during a brief visit to the mouth made by George Begg and myself at low spring tide on 27 May 1979.

The Tugela is the largest river in Natal and second only to the Orange River in the whole of southern Africa. Its tributaries arise on the slopes of the Drakensberg between Bergville and Wakkerstroom where the mean annual rainfall is 1 270 mm. The tributaries then flow over the uplands of Natal where the annual rainfall is much lower, averaging 442 mm. The river flow increases from a minimum of $73,6 \times 10^6$ m³ in winter to almost seven times this volume or 481×10^6 m³ over January, February and March. The summer floods are sudden and terrifying in their volume and velocity and it is from this that the Tugela got its Zulu name. The total run-off from the catchment is $5\ 071 \times 10^6$ m³ which is said to be 9,1 % of the total run-off in South Africa, with the Orange River being 13,5 %. The gradient of the river is so steep that the river rapidly erodes the softer soils of the uplands at the rate of 375 tons per km² per year, which may be compared with 154-230 tons per km² eroded by the Mississippi and 420 tons per km² eroded by the Indus. The sediment load of the Tugela is $10,5 \times 10^6$ tons yr⁻¹ and it is not surprising that the lower reaches are heavily silted. During summer the river spreads over its floodplain which is more than a kilometre wide in many places, and carries a brown flood far out to sea. Begg (1978) states that sediments from the Tugela supply sands to most of the beaches in northern Natal. The Tugela bank off the coast is a very large area of sandy mud fished by many Natal trawlers. During periods of low flow in winter the river changes markedly. The

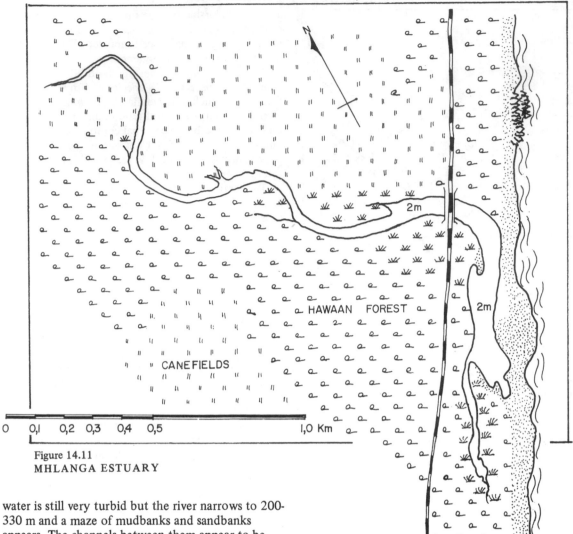

Figure 14.11
MHLANGA ESTUARY

0   0,1   0,2 0,3   0,4   0,5   1,0 Km

water is still very turbid but the river narrows to 200-330 m and a maze of mudbanks and sandbanks appears. The channels between them appear to be about 2 m deep but exact depths are not available.

As far down as Mandini about 10 km from the mouth, Brand *et al* (1967) report that the water is of good quality with no significant faecal pollution in the main river although stretches of some tributaries are polluted. Nutrient levels are low; the NH_3 concentration is 0,2 ppm, NO_3 is 0,8 ppm but PO_4 concentrations have not been reported. Dissolved oxygen concentrations are high. At Mandini there is the very large South African Pulp and Paper Industries (SAPPI) factory and the industrial effluent which contains much fibrous material as well as the effluent from the sewage works of the town pollute the river. Faecal *coli* counts and the organic content of the water are both high.

During summer the rate of flow is so high that the river is fresh to the sea and the mixing of salt and fresh water occurs offshore. Under such conditions there is no estuary. It is not known how long this persists for there are almost no records of salinity in the river mouth. In May 1979, George Begg and I recorded a salinity of 12 ‰ at the surface during low spring tide and probably a similar salinity exists at the bottom as the numerous shoals would promote turbulent mixing. The high tide salinity was not determined, but Begg (1978) states that in winter saline water extends about 800 m from the sea; certainly the water was fresh at a point 3 km from the mouth at low tide in May 1979.

As shown by Figure 14.10, the mouth of the river during the dry months is constricted by a sandspit which extends northward from the south bank, and Begg (1978) states that the mouth is about 50 m wide. Obviously this must vary with the flow of the river and the state of the tide. Within the area sheltered by the sandspits, the banks and shoals revealed at low tide are covered by a layer of fine silt and clay below which there is a layer of coarser sand. This is devoid of macroscopic vegetation but some areas are golden brown indicating benthic diatoms. *Phragmites australis* grows higher up the river but at the mouth only rotting stumps were found in clay at HWN.

The benthic fauna was very scarce and from a brief inspection, appeared to be limited to a few juvenile *Dotilla fenestrata* (about 1 m^{-2}) on the mudflats, a few *Sesarma eulimene* on the edges of the reeds and the holes of larger crabs. One of these contained juvenile *Palaemon pacificus.* The brief inspection suggested that the few benthic animals that colonised the banks in winter were either smothered by mud or washed out to sea in the summer.

In short, the estuarine phase of the Tugela is very brief and for most of the year the river is fresh to the sea. Somewhat similar conditions are found at the Orange River mouth and the mouths of other large, swift-flowing rivers in other parts of the world.

Mhlanga lagoon
(29°42'S/31°06'E)

This summary is based on an account by Begg (1978). Reference is made to an article on the effect of erosion by Heydorn (1977) and estimates of the area of the drainage basin and the volume of river flow have been must vary from Noble & Hemens (1978), which in turn is the mean of the values given by Begg (1978). Whitfield (1979) discusses the food resources available to the fish community.

Although Mhlanga lagoon is a small estuary and usually closed by a sandbar (Figure 14.11), it is very beautiful and of particular ecological interest. Not only is the mouth closed but the water is also hyposaline and this is unusual for evaporation in most estuaries maintains the salinity at a fairly high or very high level. In this case, river flow and precipitation must exceed evaporation, and such a hyposaline blind estuary has never been studied in detail.

Before the flood plain was drained, the lagoon was much larger. As Heydorn (1977) has pointed out, sugar cane is planted right to the water's edge (photo p380), and as a result, the banks are being seriously eroded. There were grandiose but unrealistic plans for marina development and the estuary deserves careful protection.

The Mhlanga has a catchment of 135 ha and the river discharge has been estimated as 19,7-29,5 x 10^6 m^3 per year. The lagoon has an area of 11,5 ha with the main channel normally 10-15 m wide and 1,5-2,0 m deep. During heavy rains, the level may rise 2,4 m flooding over the banks to increase the width to 100 m before the estuary bursts through the sandbar to the sea. Farmers whose lands are liable to be flooded also cut an exit through the bar.

The estuary basin empties rapidly and although sea water enters with the rising tide the bar closes before the salinity rises to polyhaline values. In the dry summer of 1975, the water near the blind mouth was stratified with a salinity of 7 ‰ at the surface and 32 ‰ near the bottom. In most years, however, the

salinity is much lower and in April 1977 the surface salinity was 1 ‰. At such a low salinity, the suspended silt brought down by the river tends to flocculate but the floccules are not precipitated unless the salinity rises further; as a result the estuary is fairly turbid. The Mhlanga sewage reaches the river 2 km above the lagoon. Nitrate values of 0,71-0,84 ppm and phosphate values of 0,23-0,30 ppm were recorded in the winter of 1964. However, there is no sign of bacterial pollution.

Little is known of the aquatic macrophytes. The phytoplankton was sampled by Professor Pienaar in May 1977 but the marginal and submerged vegetation have not been recorded. Rather surprisingly a few *Bruguiera* occur in one area and there are the usual freshwater mangroves *(Hibiscus tiliaceus* and *Barringtonia racemosa)* and large beds of the reed *Phragmites mauritianus.* These provide abundant detritus, which, with associated micro-organisms, has a biomass of 9,46 g (dry mass) m^{-2}, equal to 161 kJ.m^{-2} (Whitfield, 1979).

Few invertebrates have been recorded apart from *Callianassa* in the sand near the blind mouth. Crabs and chironomid larvae are said to be abundant but only *Sesarma meinerti* and *Varuna litterata* have been identified. Thirty-two species of fishes including mullet and juvenile game fish have been recorded; more could hardly be expected in such low salinities. More than 90 % of the fish population utilize the organic detritus as an energy source (Whitfield, 1979).

A thorough study of a hyposaline blind estuary such as Mhlanga is urgently needed as a basis for comparison. The Fafa in southern Natal has been studied intensively but the conditions there are artificial for the mouth is closed by a weir to maintain a depth suitable for aquatic sports. Ecologically it is therefore suspect, whereas the Mhlanga is quite natural.

The beautiful Hawaan forest with many rare species of plants and animals extends along the south bank of Mhlanga lagoon and the whole area is a nature resort. As Heydorn has so cogently advocated, the banks of the river urgently require protection from erosion. The existing legislation which prohibits cultivation within 10 m of the water's edge should be strictly enforced. In fact, the whole lagoon with the indigenous forest and the marshes should be declared a nature reserve.

Mgeni estuary
(29°48'S/31°05,5'E)

This is of particular interest as being one of the few large estuaries in southern Africa which receives a good deal of industrial and domestic sewage effluent.

References include Oliff (1976) and Cloete & Oliff (1976) which respectively give the details and a general summary of the hydrological, chemical and bacteriological investigations made by the NIWR in 1969 and 1972.

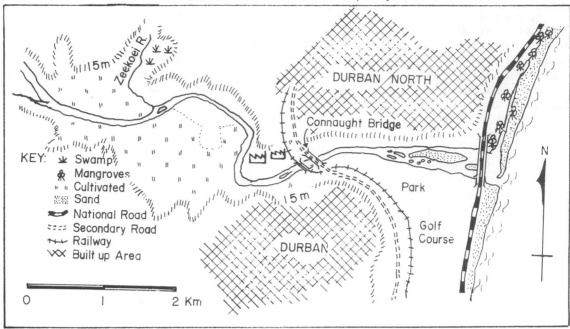

Figure 14.12
MGENI ESTUARY

Berjak *et al* (1977) published a general account of the mangrove swamps in the Beachwood tributary and Brown (1971) describes its gastropod fauna. Begg (1978) reviewed earlier studies and added further observations of his own (also tabulated in Nobel & Hemens, 1978).

The Mgeni River rises in the foothills of the Drakensberg and flows for 235 km through hilly but well-populated districts of Natal. The drainage basin is 4 871 km² in area and the run-off averages 707 x 10⁶ m³.yr⁻¹. The current is swift and the water is of good quality although it carries a fair load of silt during the summer rains. As it enters the outskirts of Durban, it receives the outflow from Zeekoei Lake which acts as a maturation pond for industrial and domestic sewage effluents so that the river is polluted.

The estuary (Figure 14.12) is 2,5 km long and runs between a park and the northern suburbs of Durban; it is crossed by a rainway and two road bridges before receiving the outflow of the Beachwood mangrove swamps and entering the sea. The mouth is narrow but seldom closes. The estuary is only 70 m wide in the upper reaches at Connaught Bridge and then broadens to form a lagoon 300 m wide with several intertidal shoals before it reaches the national road bridge. Over most of its length, the estuary is about 2 m deep at low tide but at high tide of springs the depth increases by more than a metre. The banks and bed are sand or even gravel in places, with relatively little fine silt although a black deoxidised layer is present 2-3 cm below the surface. Rather surprisingly, the polluted outflow from Zeekoei Lake causes little reduction in the dissolved oxygen concentration of the estuarine water, the minimum percentage saturation being 80 %.

Sea water of 35 ‰ salinity floods into the mouth of the estuary with the rising spring tide and extends as a bottom current with a salinity of 31 ‰ for 2 km along the estuary. At the head of the estuary, about 2,5 km from the sea, salinities of 2 ‰ extend from surface to bottom but to seaward the low salinity water tends to form a surface layer gradually mixing with the high salinity layer below, until the surface salinity is 23 ‰ at 0,5 km from the sea. Water temperatures are normal with a seasonal range from 17 ° in winter to 26 ° in summer.

The outflow from Zeekoei Lake increases the nutrient content of the river water before it reaches the estuary. A series of samples taken by the NIWR at several stations in the estuary during August 1972, showed consistently higher values of phosphorus and nitrogen compounds than those taken in May but the mean values for the two months show no obvious trends. Total phosphorus varied between 105 and 121 μg.Pℓ^{-1}, NH_3-N varied from 764 to 802 μg.ℓ^{-1} and NO_3-N varied from 1 390 to 1 470 μg.ℓ^{-1}. All these values are about ten times higher than those in unpolluted estuaries along the Natal coast. The relative concentrations of nitrate and ammonia suggest that oxidation is well advanced in Zeekoei Lake and the high dissolved oxygen concentrations in the estuary support this. Nutrient conditions in the sediments are given in Oliff (1976) but are not repeated here.

Bacteriological samples confirmed sewage pollution in the outflow from Zeekoei Lake but *E. coli* counts decreased in the saline water along the estuary. It was concluded that the estuary is moderately polluted with faecal coliforms but the samples taken in 1972

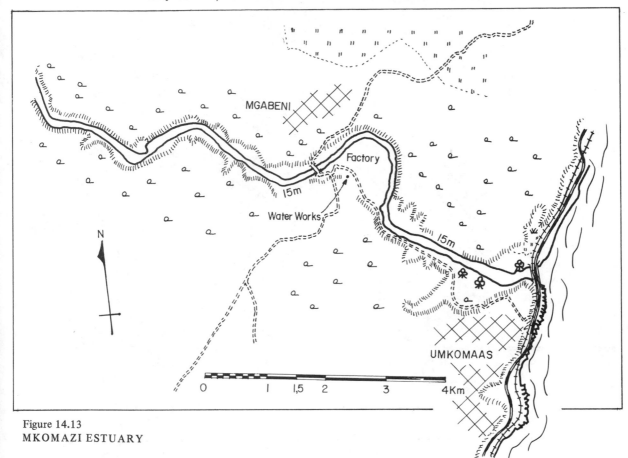

Figure 14.13
MKOMAZI ESTUARY

showed an improvement over conditions in 1968.

The plankton has not been investigated but as the estuary is short and the flow fairly rapid, the plankton is unlikely to be rich. The marginal and submerged vegetation might be expected to be high considering the nutrient concentration but in fact it is poor and there is no *Zostera* in the main estuary. The reason for this requires investigation.

There is a mangrove swamp in the Beachwood tributary which lies between the coastal sand dunes and the national road. Berjak *et al* (1977), describe how the mangroves were practically destroyed by the construction of a solid causeway across the swamps but now that the structure has been removed, tidal range has improved and the mangroves have recovered. The mangrove fauna requires further investigation but an interesting study of the gastropod fauna was made by Brown (1971). It is known that there are four species of *Uca*, at least three species of *Sesarma*, a few shrimps and prawns and 30 species of fish (Begg, 1978).

Since the Mgeni is the only estuary of any size that is at present polluted to a significant degree by sewage effluents, a more detailed study of the aquatic flora and fauna would be particularly valuable. It would be very interesting to know why the aquatic vegetation is not better developed in the presence of high nutrient concentrations and how the migration of juvenile fish and penaeid prawns is affected by existing conditions. The Mgeni must discharge valuable nutrient supplies to nearby coastal seas; is there any evidence of the results?

Mkomazi estuary
(30°12,5′S/30°47′E)

Little is known about the estuary of this large river. Unpublished notes and a catalogue of the fauna collected in January 1950 are available in the Zoology Department of the University of Cape Town. Begg (1978) provides further information from his own notes and a series of 18 references.

The Mkomazi rises in the foothills of the Drakensberg and has a catchment of 4 315 km². The run-off is 1 072 x 10⁶ m³ and the river flow during the summer rains is 34 000 $\ell.s^{-1}$; it carries a sediment load of 900 000 tons per year. Kemp *et al* (1976) report on the physico-chemical conditions in the river but not in the estuary.

The estuary (Figure 14.13) is about 5 km long, 100-150 m wide and the mouth is tidal and seldom closes. The estuary runs through a fairly steep valley with rocky cliffs and a few swampy areas here and

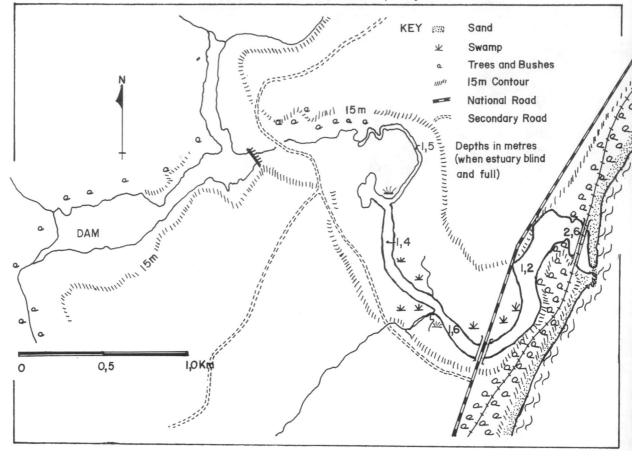

Figure 14.14
UMGABABA ESTUARY

there so that the intertidal banks are mainly narrow. The town of Umkomaas has been built on a hill on the south bank and slopes down to a rocky coast. To the north, the coast is sandy and a sandspit stabilised by embankments leading to the road and rail bridges, constricts the mouth of the estuary. In 1950 it protected a muddy cove on the north bank, overgrown by reeds and coarse grasses with a small stand of mangroves including *Avicennia, Bruguiera* and the fresh water mangrove *Hibiscus tiliaceus.* Since then, the cove has silted up and mangroves now occur on the south bank about 700 m from the sea.

Little hydrological information on the estuary is available. In January 1950 the river was flowing strongly at all states of the tide and the surface water was almost fresh to the mouth. It was so muddy that a Secchi disc disappeared below 5 cm. The temperature varied between 23 and 24,2 °.

Apart from the sandspit at the mouth the intertidal banks were all soft mud and there was no sign of *Zostera* or other submerged vegetation. No phytoplankton samples were taken.

Although little time was spent collecting, the fauna appeared to be poor along the whole estuary. Above the old road bridge 5 km from the sea, there was a typical fresh water fauna of insect larvae, tadpoles

and fresh water crabs *(Potamon)*. A few species which tolerate low salinities such as *Palaemon concinnus, Caridina nilotica, Macrobrachium scabriculum* and *M. equidens* were also present and these extended to within 2 km from the sea. At the same collecting station the crabs *Sesarma meinerti, S. eulimene, S. catenata* and *Uca lactea annulipes* were common. In the mud the tanaid *Apseudes digitalis* was abundant and the amphipod *Grandidierella bonnieroides* was common. The same estuarine species were found in the muddy cove near the mouth and with them was *Upogebia africana* but polychaetes and burrowing bivalves were not found. *Cerithidea decollata* was present on the mangrove trunks but there was no sign of barnacles or oysters.

Netting at three stations along the estuary revealed only nine species of small fish, none of them abundant.

The general impression of the estuary in summer was very poor due to the high rate of siltation. It is possible that the diversity and biomass of the fauna increases during the dry months when the inflow of sea water increases.

Umgababa estuary
(30°09'S/30°50'E)

The Umgababa estuary is a typical example of the many Natal estuaries that are closed by sandbars for part of the year. Three references are available. The ecology of the Umgababa was investigated by officers of the NIWR in June 1976 and February 1977 and their reports will be found in Oliff (1976, 1978). Notes on the fauna recorded in January 1950 are available in the Zoology Department of the University of Cape Town. a more recent account is that of Begg (1978) who summarised earlier reports and added addifurther notes made during his own visits.

The Umgababa river, some 35 km south of Durban, is about 15 km long and drains 37 km² of Zulu farmlands in the coastal belt. The rainfall is normally 1 100 mm per year and the run-off is 14 x 10⁶ m³. Much of this is retained in the Umnini dam 4 km from the mouth so that the overflow into the estuary is probably more even than the flow into most Natal estuaries.

The estuary (Figure 14.14) is 3 km long and gradually widens from a small stream 20 m wide to form an S-shaped lagoon with a maximum width of 150 m. It is about 1,5 m deep for most of its length but there is a depression 2,6 m deep near the mouth which is often closed by a sandbar. After heavy rains the water level rises until the sandbar is breached and then falls by about a metre and the estuary becomes tidal until the bar builds up again. The few records available suggest that the mouth is open fairly often during the summer rains and may also open briefly after a spate in winter.

Salinities at the head of the estuary, in the middle of the lagoon and at the blind mouth were recorded by officers of the NIWR and are quoted below as surface/bottom values.

	Station 1 (head)	Station 3 (middle)	Station 5 (blind mouth)
June 1976	1,8/12,6	9,6/11,0	11,4/31,6
Feb. 1977	0,8/0,8	5,0/12,0	16,2/32,7

These records illustrate four interesting points. First, in spite of the sandbar blocking the mouth and preventing tidal exchange, a normal salinity gradient is maintained along the length of the estuary. Records in other blind estuaries indicate that this is usually the case. The second point is that the salinity in the middle of the estuary is about 10 % in both winter and summer and the records from the head and blind mouth confirm that there is no marked seasonal change. This suggests that river discharge plus precipitation on the estuary surface is fairly well-balanced by outflow to the sea (when the mouth opens) plus evaporation through the year. The third point is that vertical salinity differences are maintained. The more detailed records in Oliff (1976) show there is no sharp stratification but partial mixing. This may indicate a persist-

ence of the mixing caused by tidal fluctuations when the mouth was open, but as the fresh water discharged by the river would flow over the surface of the saline water, it is more likely that the vertical salinity gradient is maintained by evaporation. The fourth point is that the bottom salinity at the blind mouth is over 30 ‰, ie higher than elsewhere in the estuary in both June and February. It has been noted that the depth at this point is 2,6 m whereas the rest of the estuary is only 1,2 to 1,5 m deep. The depth of 2,6 m is well below mean sea level and the higher salinity at this depth is an indication of seepage of sea water through the sandbar.

The temperature records range from 17,0 ° in winter to 29,4 ° in summer which corresponds fairly well with the seasonal range in an open estuary. The range of the bottom temperature at the blind mouth is 21,4 ° to 25,2 ° which is closer to the seasonal range in the sea. Secchi disc readings in January 1950 averaged 60 cm so that the water is fairly clear. Chemical analyses by the NIWR provide further information. Oxygen concentrations in the bottom samples were 30-60 % lower than in the surface samples but there was no evidence of a significant oxygen lack or of pollution. DDT and DDE residues were low in fish flesh samples. Again, trace metal concentrations in the sediments including Cd, Cu, Zn, Pb or Cr were low compared with samples from other estuaries but the Hg concentration in water samples averaged 0,822 $\mu g.\ell^{-1}$. This is 2,7 times higher than normal, possibly due to a geological cause. Kjeldahl nitrogen estimates of 636-995 $\mu g.\ell^{-1}$ are regarded as quite high but are similar to the concentrations in the dam above the estuary.

Brief notes were made of the estuarine vegetation in January 1950 but much additional information is required before the Umgababa can be used as a yardstock for other blind estuaries in Natal. The phytoplankton has not been investigated. The marginal vegetation in the upper reaches is dominated by *Juncus kraussi* and *Phragmites* sp. The salt marshes on the west bank of the lagoon include *Juncus kraussi, Sarcocornia* spp. and in 1950 there was a small but healthy submerged *Zostera* bed. The east bank of the lagoon from the national road bridge onward is overgrown by the grass *Stenotaphrum* in which there are one or two small mangroves *(Bruguiera)*.

Zooplankton samples collected in February 1976 showed a fairly rich fauna very similar to that in the blind Fafa estuary on the Natal south coast and comparable to the summer samples in the Swartkops estuary. The winter samples were richer, and Connell *et al* (in Oliff, 1978) suggest that this was due to the long period of stable conditions prior to sampling.

The macrobenthic fauna is poorly known. The few samples collected in January 1950 include 37 species of invertebrates. Among them are the usual species of polychaete worms, barnacles, amphipods, isopods, shrimps, penaeid prawns, *Callianassa, Upogebia,* crabs,

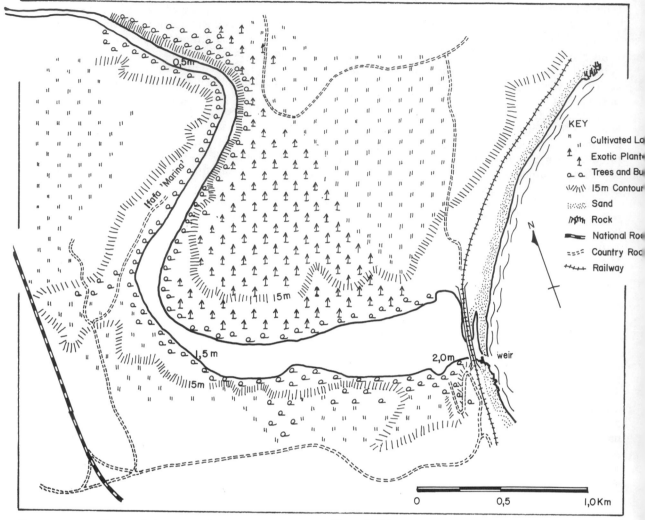

KEY

" "	Cultivated La
↑ ↑	Exotic Plant
ᴑ ᴑ	Trees and Bu
ᴎᴎ/ᴎᴎ	15m Contour
⣿⣿	Sand
ᴎᴝ	Rock
▬▬	National Roa
⁼⁼⁼⁼	Country Roc
++++	Railway

0   0,5   1,0 Km

Figure 14.15
FAFA ESTUARY

Saccostrea cucullata and *Assiminea bifasciata*. The presence of *Upogebia* is surprising as this burrowing prawn is seldom found in blind estuaries. Oliff (1978) reports that the meiofauna samples collected in June 1977 were rich.

The fish fauna is also incompletely known. Seine netting in January 1950 revealed 15 species but gill netting would certainly reveal additional ones. The fish fauna is, in fact, surprisingly rich and this is thought to be due to the stable salinity regime and the fairly frequent opening of the bar.

The Fafa estuary
(30° 27′S/30° 39′E)

Oliff (1976) records that officers of the NIWR made ten series of observations of hydrological, chemical, bacteriological and planktonic conditions in this estuary from 1970 onwards. A summary of this evidence and conclusions regarding the effect of a 'marina' are given by Cloete & Oliff (1976). Begg (1978) summarises these reports and adds further notes of his own.

The Fafa is a small river some 90 km south of Durban. It drains an area of 252 km² in the coastal belt most of which is planted to sugar cane. There are also fairly large eucalyptus plantations and a belt of natural bush along the river banks. The annual rainfall is about 1 000 mm and the run-off is estimated to be 30 x 10⁶ m³ per year. The river water is clear and of good quality.

Originally the mouth of the estuary was closed by a natural sandbar which burst after heavy rains so that the water level in the estuary fell and the estuary became tidal until the sandbar closed again. In order to maintain the water level at a minimum depth of 2 m for aquatic sports, a small weir was built into the lower part of the sandbar by local residents. As a result, the salinity of the estuary is very low and tidal effects are limited to brief periods after floods have breached the sandbar, or waves at high spring tide surge over the weir. Thus the estuary has been coverted into an artificial low-salinity lagoon. When a row of cabins was built in a caravan park and advertised as a 'marina' there was a danger that sewage effluents might cause pollution or eutrophication of the lagoon. It was for this reason that the NIWR made such careful observation of conditions in the river above the 'marina' and in the hyposaline lagoon below.

The blind estuary (or lagoon) is 2,8 km long, about 50 m wide and deepens from 0,5 m at the head to about 2,0 m near the weir. The banks are rather steep and covered with reeds and the fresh water mangroves *Barringtonia racemosa* and *Hibiscus tiliaceus;* there are no true mangroves.

The water is well-oxygenated and the Secchi depth is 40 to 100 cm. Summer and winter temperatures range from 25 to 14,2 °. For much of the year, the salinity is below 1,0 ‰ but when the bar is breached, saline water extends along the bed of the lagoon. Maximum salinities recorded in June 1971 were:

Distance from sandbar (km)	0,5	1,25	1,95	2,5
Surface salinity (‰)	14,2	13,0	2,4	0,1
Bottom salinity (‰)	32,9	23,5	12,8	0,1

During the next three months, the salinity decreased with continued discharge of the river and in October the whole lagoon was fresh, apart from the bottom water near the weir where the salinity was 0,5 ‰, possibly due to seepage of sea water through the sandbar.

Nutrient concentrations were always low. Nitrate rarely exceeded 1,0 $mg.NO_3.\ell^{-1}$, ammonia varied between 0,2 and 0,8 $mg.NH_3.\ell^{-1}$ and total phosphorus varied from 0,01 and 0,14 $mg.P.\ell^{-1}$. Other observations showed that the wastes of the developing marina up to July 1975 were absorbed in soakpits and septic tanks and did not increase nutrient concentrations at the head of the estuary. Nor did the estuarine water show signs of bacterial pollution at that time.

Both the river and the lagoon below the 'marina' were oligotrophic with an average phytoplankton cell count of 726 cells per millilitre. Primary production increased during November to a maximum of 29,93 mg. $Cm^{-3}.hr^{-1}$ as measured by oxygen production or 6,55 $mg.Cm^{-3}.hr^{-1}$ as measured by the ^{14}C method. *Potamogeton* has been recorded in the upper reaches. It is significant that in the Amanzimtoti estuary nearby, which is similar to the Fafa except that it receives the effluent from a sewage works, the nutrient concentrations are much higher. Below the effluent outfall the phytoplankton densities are four to six times higher than in the river above. Cloete & Oliff (1976) conclude that: 'unless the Fafa system is capable of tolerating a sewage effluent at the concentration that is planned by the marina complex, eutrophication must occur'. This seems doubtful.

Little is known of the fauna of the estuary. The records include 24 species of invertebrates of which nine are of marine origin; similarly netting revealed ten fishes of which nine are typical estuarine species. Thus the fauna is very poor but it is surprising how many species of marine origin manage to colonise and survive in this estuary which is almost fresh for most of the year.

The Mzimkulu estuary
(30°44'S/30°27'E)

Brief notes and a catalogue of the fauna of the estuary, made by a party in January 1950, are available in the Zoology Department of the University of Cape Town. Environmental conditions in the estuary were investigated in more detail by officers of the NIWR in August 1974 (Oliff 1976, Cloete & Oliff 1976). More recently Begg (1978) has summarised some 17 earlier reports and added observations of his own.

The Mzimkulu is one of the largest rivers in Natal after the Tugela. It rises on the slopes of the Drakensberg and with its tributaries it drains an area of 6 694 km^2. The average rainfall is about 1 000 mm per year, most of which falls in summer. The run-off averages 1 455 x 10^6 m^3. Floods often occur and as the upland part of the catchment is overgrazed and easily eroded, the river carries a very heavy sediment load.

The estuary (Figure 14.16), which lies in a narrow wooded valley, is 4-5 km long and the upper reaches are 130-180 m wide and 1-3 m deep. The main channel winds between muddy shallows and exposed sandbanks. Between the coastal road bridge and the sea the estuary narrows to 30 m and the depth increases to 4 m. The mouth is sheltered from littoral drift by a projecting reef and a concrete training wall on the west bank.

During January 1950 the surface current flowed out swiftly at all states of the tide and reached a velocity if 1 $m.sec^{-1}$. As often occurs, the golf course had been flooded and the water was fresh almost to the mouth and so muddy that a Secchi disc disappeared 2 cm below the surface. The more detailed observations made by the NIWR in August 1974, showed surface salinities of 4,2 to 7,8 ‰ along the last 2 km of the estuary and bottom salinities at high tide decreasing from 32,6 ‰ at the coastal road bridge to 29,0 ‰ some 2 km from the sea. No observations were made higher up the estuary. However, the salt wedge had obviously extended further from the sea with the decrease in river flow and the Secchi depth increased to 1,4 m. The surface temperature was 16,3 ° in August and 24,5 ° in January. These records are a little lower than the records in slower flowing estuaries in southern Natal.

Tests made by the NIWR indicate that the river water flowing into the estuary is unpolluted and there is only a small pesticide residue. Among trace metals only the zinc concentration was high possibly due to

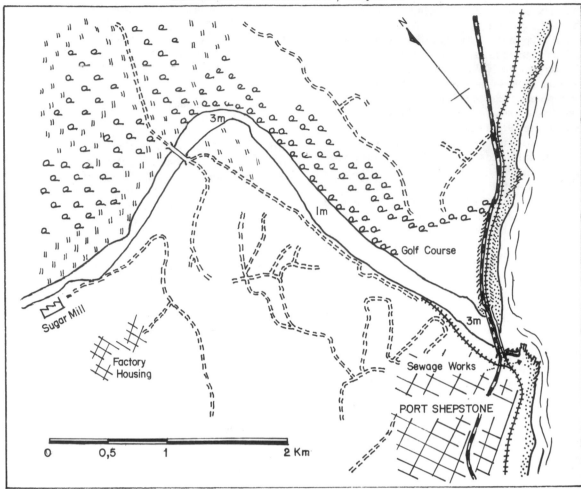

Figure 14.16
MZIMKULU ESTUARY (modified from Begg, 1979)

some geological feature. Samples taken at the mouth of the estuary near the sewage outfall, showed normal NO_3 and PO_4 concentrations but the NH_3 value was high and so was the concentration of Pb.

Bottom sediments appear to vary seasonally as does the turbidity of the water. During January 1950 soft silt was deposited all along the estuary to such an extent that the reed beds along the north bank were partially buried. In August 1974 not only was the water clear but much of the fine silt had been eroded leaving sandy shoals here and there.

The marginal vegetation consists of *Sporobolus* and *Juncus* at the high tide mark and *Phragmites* extending to low tide. *Hibiscus tiliaceus* is common but true mangroves are absent and neither *Zostera* nor other submerged vegetation was recorded. It is suggested that any seedlings that start to drow during the dry months are smothered by silt during the summer floods.

The benthic fauna is very poor and only ten macrobenthic species were recorded in 1950. Among them were three species of *Sesarma* and the small polychaete *Desdemona ornata* which is abundant in the muddy bed of the estuary. *Upogebia* has not been recorded

but a small colony of *Callianassa* is noted by Begg (1979). There are no burrowing bivalves and oysters and barnacles are restricted to the rocks at the mouth. The fresh water shrimps *Caridina nilotica* and *Macrobrachium equidens* were netted in the upper reaches in 1950 as well as ten species of fish. Residents report that fishing in the estuary is poor but good catches of kob, grunter and other game fish are made beyond the bar.

It is concluded that heavy silting, swift currents and low salinities combine to give this estuary a poor flora and fauna.

The Mtamvuna estuary
(31°04,5′S/30°12′E)

There are few references to the ecology of this estuary. The earliest consists of unpublished notes and catalogues of the fauna which are available in the Zoology Department of the University of Cape Town. Officers of the NIWR studied the hydrogy during September

278

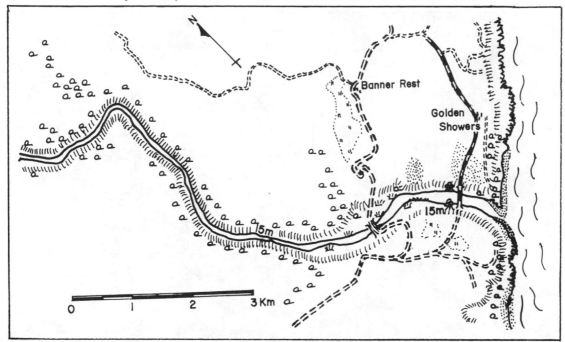

Figure 14.17
MTAMVUNA ESTUARY

1972 and their full report has been edited by Oliff (1976) while a summary is given by Cloete & Oliff (1976). More recently Begg (1978) has summarised the available data and added further notes of his own.

The Mtamvuna forms part of the southern border of Natal. The river rises in the highlands east of Kokstad and has a catchment of 1 589 km² with a summer rainfall of about 800 mm per annum. The run-off is 281 x 10⁶ m³. The lower parts of the river and the estuary have cut through the tablelands of the coastal belt and now run through a narrow wooded valley which in places becomes a gorge with steep rocky walls. The estuary (Figure 14.17) starts at a series of rapids and runs for 5,25 km to the sea. It is only 30-80 m wide for most of its length but broadens to 200 m near the mouth. The depth averages 4 m but there are two depressions 7 and 10 m deep. The mouth is filled with marine sand so that at times the estuary is closed or reduced to a shallow stream over the sandbar. The deeper parts of the estuary are floored with silt brought down by the river but the rate of siltation must be slow since the estuary is so deep and the beaches so narrow.

The vertical section of the estuary illustrated in Figure 3.6 (chapter 3), shows that there is a marked stratification of salinity, temperature and sometimes of dissolved oxygen when the mouth is closed. There is a surface layer of low salinity with well-oxygenated water, below which the salinity rapidly increases to more than 30 ‰ while the dissolved oxygen at times falls to less than 20 % saturation. When the mouth

opens, much of the low salinity surface water flows out and a bottom current of well-oxygenated sea water enters to refresh the system. Surface samples taken two weeks after a flood in January 1950 gave salinities of less than 1,5 ‰, temperatures of 28-29,8 ° and Secchi disc readings of 18-23 cm.

The banks are steep and rocky with muddy sediments in the coves. The bottom sediments contain over 75 % of subsieve particles (<0,063 mm) and are anoxic. Little is known of the aquatic vegetation. There is no *Zostera* or other submerged macrophytes but the muddy coves are overgrown with *Phragmites,* coarse grass and *Juncus* and there is a small stand of the mangrove *Bruguiera* about a kilometre from the sea.

No report on the plankton is available. The benthic macrofauna, recorded during a brief survey in January 1950, included only 15 intertidal species. The crab *Sesarma catenata* was common and *Uca* sp was found in a few places. No *Upogebia* or burrowing bivalves were found in the mudbanks and no oysters or barnacles were seen on the rocks. Grab samples taken by the NIWR in 1972 added 4-5 small invertebrates to the total list of species, including fair numbers of the polychaete *Prionospio sexoculata.* The general impression is a poor benthic fauna.

Netting in 1950 revealed 20 species of fish and a few shrimps and penaeid prawns. A list of the characteristic species in this and other estuaries will be found in the appendices to chapters 9 and 12.

Continued on page 284.

Tugela mouth at low tide (Photo: George Begg)

Fish traps in Kosi Bay (Photo: K.H. Cooper)

Mzimvubu in winter (Photo: E.J. Moll)

Mngazana, a mangrove estuary (Photo: A.E.F. Heydorn)

West and East Kleinemond estuaries (Photo: A.E.F. Heydorn)

Knysna lagoon (Photo: A.E.F. Heydorn)

Breede estuary (Photo: A.E.F. Heydorn)

Klein River (Hermanus lagoon) at low tide (Photo: A.E.F. Heydorn)

Great Berg River mouth (Photo: A.E.F. Heydorn)

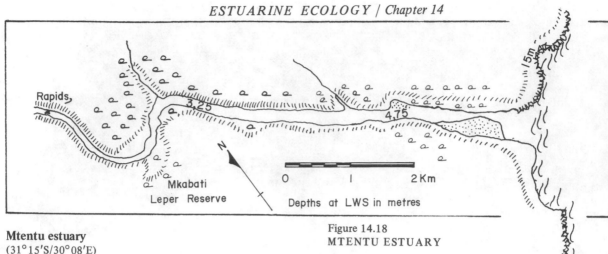

Figure 14.18
MTENTU ESTUARY

Mtentu estuary
(31°15'S/30°08'E)

A brief account of the ecology of this estuary is given by Connell (1974) as an introduction to his study of the planktonic Mysidacea.

The Mtentu is one of several rivers near the northern border of the Transkei that have cut through the elevated coastal plain so that it flows through a very steep valley. The drainage basin is 1 595 km² in area and the rainfall averages 1 100 mm per year and the run-off is 421 x 10⁶ m³. The main rains fall in summer. Thunderstorms and heavy spates may occur in any month but they pass rapidly through the estuary and have no persistent effect.

The estuary itself (Figure 14.18) is 3 km long, 50-150 m wide and mainly 4-5 m deep at low tide. The intertidal banks are narrow and either rocky or muddy. The mouth, which is constricted by a large bank of marine sand, is 30 m wide and 1,3 to 2,0 m deep but remained open all through 1971.

The spring tide range is 1,3 to about 2,0 m which is higher than on the coast. The water, as is usual in such deep narrow estuaries, is well-stratified. Bottom salinities at 34,8 to 35,5 ‰ are only slightly below sea water values. Surface salinities vary with the rains; salinities between 2 and 14 ‰ were recorded in the upper reaches and were 5 to 30 ‰ a kilometre from the sea. In the upper reaches, surface temperatures ranged from 17 or 17,5 ° in winter to 28 ° in summer, while bottom temperatures were more stable, ranging from 19,5 to 24,5 °. In spite of the vertical gradients of salinity and temperature, the bottom water was well-oxygenated with a saturation of 83 % or more, apart from an isolated record of 49 % saturation in November 1971.

Connell (1974) was not concerned with the vegetation, the benthic fauna or the flora. He merely notes that there is little marginal vegetation apart from one small *Zostera* bed, a few true mangroves *(Brugiera)* as well as fresh water mangroves *(Hibiscus tiliaceus)*. *Upogebia* is abundant and both oysters and barnacles extend along the estuary. He makes no remarks about the fishes.

His main account concerns the zooplankton, particularly the four species of mysids. Several tables and graphs provide evidence that the four mysids and two estuarine copepods contribute significantly to the biomass of the zooplankton. Details have been discussed earlier in chapter 8.

The Msikaba estuary
(31°29'S/29°59'E)

References to the ecology of this estuary will be found in Blaber *et al* (1974) and Wooldridge (1976). The following account is based on an article by Dr Burke Hill, now an officer of the Queensland Fisheries Department, Brisbane. It has been amended by the editor to conform with the format of other accounts.

The Msikaba River drains 1 629 km² of the Transkei. The inland part of the catchment has a mean annual rainfall of 864 mm and the coastal part 1 445 mm. Most of the rain falls in summer (October to April) and the run-off is 251 x 10⁶ m³. The sediment load must be light for the Msikaba is the deepest estuary in southern Africa, having a maximum depth of 35 m.

Part of the estuary lies in a deep canyon formed by an east-west fault in Table Mountain sandstone. The whole estuary (Figure 14.19) is 3,3 km long with an area of 0,18 km² and may be divided into three reaches. The upper reach which is 1,5 km long and 50 m wide, lies in a deep valley the bottom of which consists mainly of boulders. The maximum depth is 8 m, the tidal range is negligible and the water is well-stratified with a surface salinity below 10 ‰ and a bottom salinity of 15-35 ‰.

The middle reach is a fjord-like stretch 1 km long and 40 m wide between vertical rocky banks. The south bank is about 3 m high while the north bank is a precipice 20 to 30 m high. The maximum depth of 35 m is at the landward end but thereafter the bottom shelves to 10 m at the seaward end. The walls below

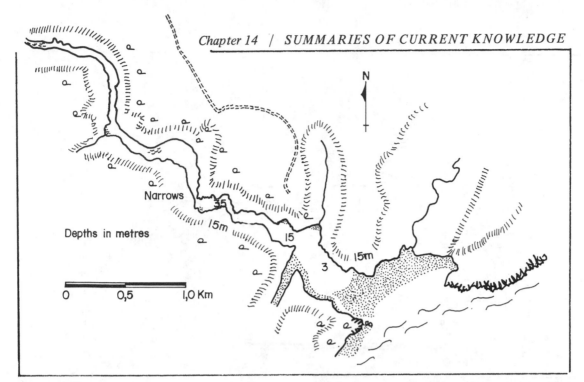

Figure 14.19
MSIKABA ESTUARY

the water surface are sheer except for a few isolated narrow ledges covered with silt while the estuary bed is soft mud. The tidal range is 0,5 m. The water is sharply stratified at 7 m, the surface salinity varying from 20 ‰ in the dry season to 2 ‰ in the wet season. Below 7 m the salinity remained at 35 ‰ through the whole year. In summer the dissolved oxygen concentration decreased with depth to 5 % saturation at 12 m and the bottom mud was anaerobic. In winter the oxygen concentration of the lower layer increased to about 50 % saturation due to the inflow of fresh seawater. The temperature below 12 m remained between 19,6 and 21,5 ° through the whole year. The surface layer is turbid and light was absent below 16 m.

The lower reach broadens to form a lagoon 500 m long and 250 m wide with a maximum depth of 10 m. It is permanently connected to the sea by a narrow channel through a large sandbank which almost occludes the mouth. The tidal range is 0,7 m. According to Wooldridge (1976), it is the deepening of the exit channel to 2,6 m after a flood which allows a bottom current of sea water to refresh the oxygen in the deepest part of the estuary. The salinity of the surface current increases from 5 to 30 ‰ and that of the inflowing bottom current increases from 30 to 35 ‰, indicating considerable mixing between the surface and bottom currents. The substrate is sandy with boulders on the south bank and solid rock on the north bank.

The aquatic vegetation is sparse. There are small stands of *Phragmites* near the head of the estuary and a small, isolated *Zostera* bed is sometimes present in the lagoon. The intertidal rocks bear small quantities of algae with a biomass of about 1 g.m⁻².

As might be expected in this steep walled estuary, the rock dwelling invertebrates display a clear vertical zonation. The barnacle *Balanus amphitrite* and the bivalve *Musculus virgiliae* are most numerous between tide marks and have a shell-free dry mass of 30 and 20 g.m⁻² respectively but they are absent below 9 m. *Saccostrea (= Crassostrea) cucullata* is common to a depth of 3 m. The sponge *Grantessa ramosa* extends from 3 to 7 m and reaches a maximum of 4 g.m⁻² at 5 m. The serpulid polychaete *Ficopomatus (= Mercierella) enigmatica* is present down to 32 m but is most numerous between 8 and 15 m where the biomass was 15 g.m⁻². It is interesting that it extends down in water which has such a low oxygen tension.

The non-sessile invertebrates, which are dominated by isopods including *Cirolana fluviatilis, Lanocira gardineri* and *Cyathura estuaria,* were most abundant between tide marks where they reached a biomass of 6,8 g.m⁻². However, they decreased rapidly with depth and none were recorded below 7 m. The silt on the rock ledges was occupied by *Upogebia.*

Neritic zooplankton extends far up the estuary in the high salinity bottom water while the estuarine species maintain station in the lagoon and exit channel by migrating vertically between the outflowing surface current and the inflowing bottom current.

There is a normal fauna of estuarine fish in the lagoon but very few were found further up the estuary. It is suggested that the paucity is due to the limited feeding areas and the anaerobic bottom sediments.

This is the only estuary in southern Africa in which the vertical distribution of the rock fauna has been investigated.

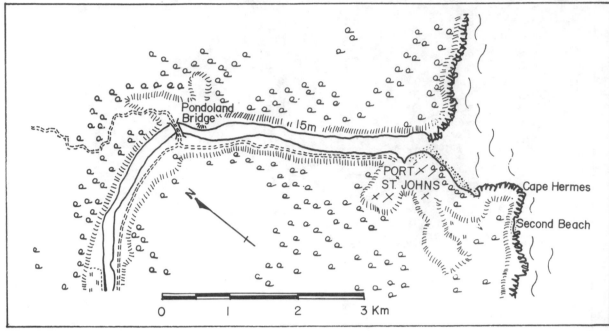

Figure 14.20
MZIMVUBU ESTUARY

The Mzimvubu and minor estuaries nearby
(31°26′S/29°34′E)

Ecological notes and faunistic catalogues of the estuaries near Port St Johns were made during a five-day visit in January 1950 and are available in the Zoology Department of the University of Cape Town. Oliff (1978) reports conditions in August 1977.

The Mzimvubu rises in the foothills of the Drakensberg and drains the northern Transkei between Umtata and Kokstad. The catchment covers 19 925 km² and has a summer rainfall of 800-1 000 mm per year. The run-off is 3 417 x 16⁶ m³ and, as much of the catchment is hilly and overgrazed, the summer floods carry heavy loads of silt.

The Mzinvubu has cut a deep valley through the tablelands of the coastal belt and the estuary runs in a deep gorge of Table Mountain sandstone covered with indigenous forest. The town of Port St Johns has been built on a Pleistocene flood plain on the south bank and is dominated by rocky cliffs known as 'The Gates', Mount Sullivan (340 m) on the north side and Mount Thesiger (370 m) on the south side. The estuary (Figure 14.20) is about 6-9 km long, 200 m wide and 2 m deep over most of its length but widens to 400 m at Port St Johns and the main channel between the shoals is up to 3 m deep. Surf breaks on the sandbar at the mouth.

The whole estuary is heavily silted but residents report that prior to 1940 small coasters entered the mouth and could sail 18 km upstream. Personal observations in 1930 confirmed that trading vessels tied up at the wharf below the Needles Hotel, and that the water was comparatively clear and the fishing good. In January 1950 the water was so turbid that a Secchi

disc was invisible 4-5 cm below the surface, the submerged vegetation had died and the estuary basin was covered with sand and soft silt. Even at the level of mid-tide the mud-flats at Port St Johns were bare and *Triglochin striata* and *Sarcocornia* sp only appeared at higher levels. There were no mangroves at St Johns but a few stunted bushes of *Bruguiera* grew at the bridge and there were dense stands of both *Scirpus* sp and *Phragmites australis* at the head of the estuary.

Few hydrological observations were made. The surface salinity at low tide in January 1950 was 1,1 ‰ at Port St Johns and the surface temperature was 24,4°. As noted above, the Secchi disc reading was 4-5 cm but the water was much clearer below the surface layer and the estuary was obviously stratified. Unfortunately, bottom conditions were not investigated. At the head of the estuary the surface temperature increased to 27°. In August 1977 on a falling tide, the surface salinity at the bridge (3,5 km from the mouth) was 16-18 ‰ and the bottom salinity was 23-33 ‰. The tip of the salt wedge with a salinity of 1,0 ‰ was found 8 km from the mouth.

The benthic fauna was very sparse and only 20 species of invertebrates were recorded. No barnacles or living oysters were found within the estuary. Although *Dotilla fenestrata* and *Ocypode kuhlii* were present on the sands at the mouth there were no bivalves or *Callianassa*. The absence of the latter was surprising as it is eubyhaline and it may even burrow in muddy sand. The mud banks further up the estuary harboured many amphipods (particularly *Grandidierella bonnieroides* and *G. lignorum*) and *Upogebia* holes

286

were common but all the specimens were small. The salt marsh vegetation near high tide harboured a normal subtropical fauna including *Assiminea bifasciata, Orchestia ancheidos,* four species of *Sesarma* and one or two juvenile *Uca annulipes.* The impression gained was that the estuary is colonised during the dry months of the year when the river carries little silt. During summer the plants and animals growing at below low tide are smothered by layer upon layer of silt and only a few juveniles survive to the following winter.

Pelagic animals were better represented that the benthic fauna. There was the usual variety of adult mullet, several juvenile fishes, penaeid prawns and young *Scylla serrata* but no *Palaemon* or gobies. In all, 17 pelagic species were obtained. Oliff (1978) reports over 38 species of zooplankton with a settled volume of 138 mℓ per tow at 7 km from the mouth. It was suggested that this impressive biomass is correlated with the poverty of the benthic fauna, but this seems doubtful as few macrobenthic species feed on zooplankton.

The environmental conditions and the fauna in small estuaries near Port St Johns provided sharp contrasts to the poverty of the muddy Mzimvubu. Two minor estuaries at Second Beach were investigated first. The larger western stream drains a few square kilometres of the nearby hillsides which probably have a rainfall of 1 000-1 250 mm per year. The estuary is about 550 m long, 25 m wide and 1 m deep. The mouth is narrow but permanently open and the tidal range is less than a metre. Salinities ranged from 21,7 to 35,7 ‰ with the tide and the temperature varied from 27,5 to 30,5 °. The important difference from the Mzimvubu was that the water was sufficiently clear to see the bottom at a depth of more than a metre. There was no *Zostera* but the marginal vegetation was well-developed with a few well-grown *Bruguiera* reeds, *Sarcocornia* sp and *Sporobolus.* The macrobenthic fauna included 26 species found in a single visit. Both *Callianassa* and *Dotilla* were common in the sands at the mouth and *Upogebia, Solen* and several species of polychaetes were found in the mud further up the estuary. Amphipods, isopods and several crabs were found in the marginal vegetation. Rocks and logs were covered with *Balanus elizabethae, Ficopomatus enigmatica, Musculus virgiliae* and *Siphonaria capensis.* Rather surprisingly no *Littorina scabra* were obtained nor any of the species characteristic of mangrove swamps although they are well-represented in the dense mangroves of the Mngazana estuary further south. Two hauls of a seine produced 16 species of small fish as well as *Palaemon pacificus, Penaeus indicus* and numerous *Gobius giuris.*

The Mngazi estuary some 8 km south-west of Port St Johns was investigated very briefly during the course of a day. The Mngazi is a much larger river than that at Second Beach having a catchment of 922 km² in the coastal belt and a run-off of 127 x 10⁶ m³. The estuary is about 4,5 km long, 50-170 m wide and 2-3 m deep. The upper reaches are muddy with flat, marshy banks covered with reeds and sedges. Lower down at the anglers' hotel some 800 m from the sea, the banks are steeper and the bed of the estuary is firm sandy mud without submerged vegetation. From the hotel to the sea the estuary broadens to form a shallow lagoon between the coastal sand dunes. The exit channel is 25 m wide and 0,7 m deep at low tide and runs over a stony foreshore to the sea. Sea water only enters the lagoon towards high tide and the tidal range is only about 0,5 m.

In January the salinity of the lagoon was 15,2 ‰, the temperature 26,5 ° and the water was sufficiently clear for the bottom to be visible at a depth of 0,8 m.

Although there was no submerged vegetation, the invertebrate fauna appeared rich considering the length of time spent collecting. Some 32 species were recorded most of which were the same as those in the minor estuaries at Second Beach. Netting revealed that mysids *(Mesopodopsis africanus)* were abundant and *Penaeus indicus* was common but rather small. Twelve species of fish were also obtained and anglers reported that large kob, leervis and spotted grunter are taken in the lagoon. A full scale biological survey would doubtless reveal many more.

Mngazana estuary
(31°42'S/29°25'E)

Six reports deal with the ecology of the estuary. Macnae (1963) studied the mangrove swamps, Glyphis (1976) gave a preliminary account of the ecology of the area, Wooldridge (1977) reported on the zooplankton, Oliff (1978) reported on the zooplankton, meiofauna, trace metals and pesticide residues, Dye (1979) gives a detailed account of the meiofauna and Branch & Grindley (1979) give a detailed account of the hydrology and the flora and fauna of the estuary throughout a year's observations.

The Mngazana River rises about 25 km from the sea. It is 150 km long and drains about 350 to 630 km² of the wooded coastal belt of the Transkei. The annual rainfall is 850-1 000 mm and falls mainly in summer although heavy floods occur irregularly. The run-off is 110 x 10⁶ m³. In spite of the floods the river carries a small sediment load and the estuary is fairly clear with a Secchi depth of 0,4-1,9 m.

The estuary (Figure 14.21) is 6 km long and winds through a Pleistocene flood plain between small cultivated fields and extensive marshes which include the best developed mangrove swamp in South Africa. At The bridge which marks the head of the estuary, the main channel is 40 m wide and 0,1-0,3 m deep but gradually widens to 100 m and deepens to 2,0-2,5 m before the exit channel narrows to 50 m between

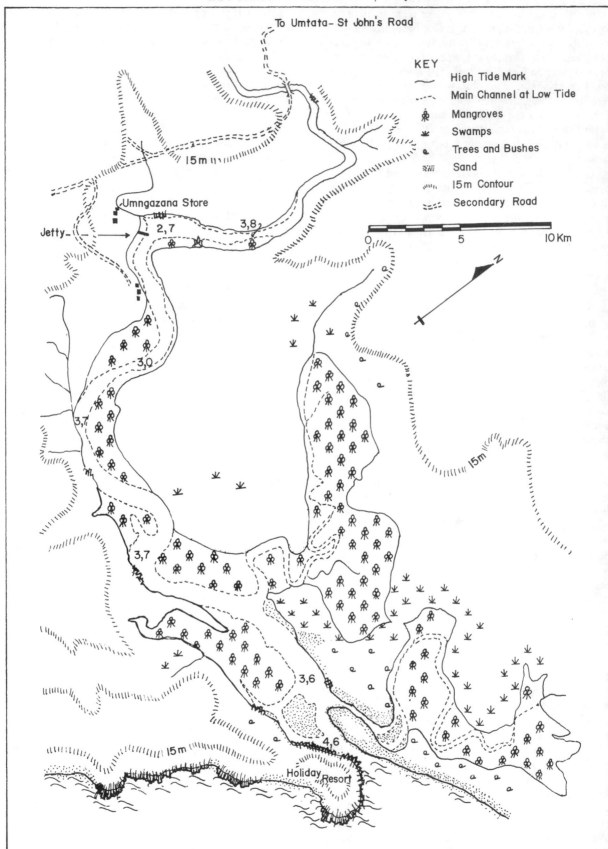

Figure 14.21

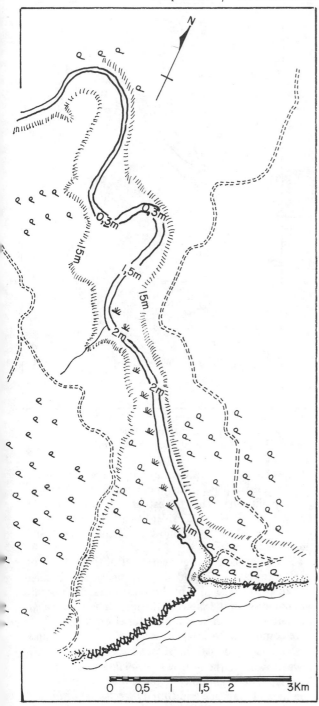

Figure 14.22
BASHEE ESTUARY

temperature range is 16-28 ° at the bridge and 18 to 24,5 ° at the mouth. Trace metal concentrations and pesticide residues are low. The estuary is well-oxygenated and unpolluted. The flora and fauna may be summarised as follows:

Vegetation: Phytoplankton poor; *Zostera* on mud-flats near mouth, extensive and well-developed mangrove swamps including *Avicennia, Bruguiera* and *Rhizophora;* scanty salt marshes on steep banks, wide marshes on lowlands.

Zooplankton: rich with a maximum of 1 200 mg (dry mass) m^{-2} in February (due largely to mysids), but poor in winter.

Benthic fauna: 209 species of invertebrates identified. Biomass and production of meiofauna high (Dye, 1979). Biomass of macrofauna 5-15 g (ash free) m^{-2} in *Zostera* beds due to *Upogebia,* polychaete worms and a few bivalves, and 15-35 g (ash free) m^{-2} in mangroves due to crabs, *Upogebia, Assiminea* and worms. Upper reaches of estuary poorer.

Fish: 62 species, including many juveniles of tropical species in summer and more endemics and mullet in winter. The *Zostera* beds are particularly rich. Sport fish include river bream, kob, dassies, wildeperd, leervis, red snapper, spotted grunter, elf, Natal stumpnose, small barracuda and ten-pounder.

In brief, this is one of the few well-developed estuaries in the Transkei with extensive mangrove swamps and *Zostera* beds, a rich fauna and good fishing. Although there are reports that it may be developed as a harbour, it is worthy of preservation as the finest coastal nature reserve in the Transkei.

The estuary of the Bashee River
and the minor Mbanyana nearby
(32°16,5'S/28°54,8'E)

Little information is available on the ecology of these estuaries. Records and collections made by a party of biologists in January 1950 are available in the Zoology Department of the University of Cape Town. Officers of the NIWR made observations on the hydrology and nutrient concentrations in the Bashee in June and December 1976 (Oliff, 1976). Chlorinated hydrocarbon residues in fish were reported by Oliff (1978). Brief references to the Bashee appear in an unpublished paper on the Dwessa by Siegfried (1977) and statistical data on the river are given by Noble & Hemens (1978).

The Bashee River rises in the foothills of the Drakensberg and drains 6 358 km^2 of the central part of the Transkei. The summer rainfall is very variable but the annual fall is of the order of 1 100 m; the run-off is estimated as 942 x 10^6 $m^3.yr^{-1}$ and carries a heavy sediment load.

The estuary (Figure 14.22) is about 8 km long and runs through a deep wooded valley with rocky cliffs

Brazen Head and a sand bank on the eastern shore. The mouth is permanently open with a spring tide range of 1,5 m decreasing to 1,35 m at the jetty and 0,7 m at the bridge. The banks are sandy at the mouth but thereafter muddy with occasional rocky outcrops. The salinity is usually 30-35 ‰ as far up as the jetty, but above this the water is well-stratified. The seasonal

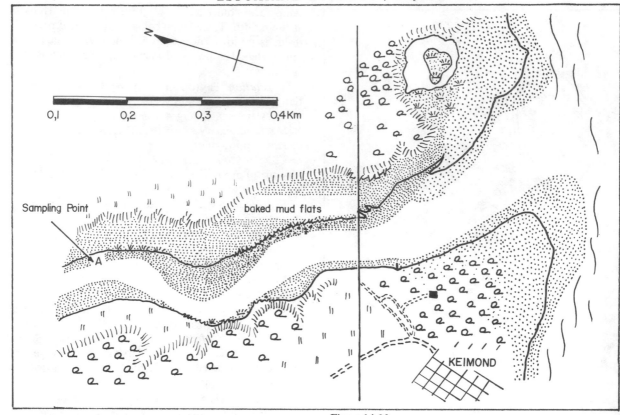

Figure 14.23
GREAT KEI ESTUARY (mouth based on an aerial photo May 1979, left of line based on a sketch made May, 1958).

here and there. It is about 100 m wide and 2 m deep for most of its length. The mouth is anchored by a rocky promontory on the west bank and is permanently open. A large sandspit tends to develop on the east bank and in some years (eg 1950) the mouth was narrowed against the rocks on the west bank being only 50 m wide and 1,5 m deep at low spring tide. When the sandspit develops, it shelters a cove with extensive muddy shallows near the east bank. In other years much of the sandspit is washed away and the cove disappears.

Apart from the sandbanks at the mouth, the intertidal banks are narrow and overgrown with reeds while the bed of the estuary is soft soupy silt. The cove near the mouth is overgrown with saltmarsh vegetation including *Phragmites, Sarcocornia* sp, dense *Triglochin striatum* and there is a small stand of mangroves including both *Avicennia* and *Bruguiera.* No *Zostera* or *Spartina marítima* was seen.

Records taken in summer of 1950 showed that the surface water was almost fresh within 3 km of the mouth. Records made by the NIWR in June 1976 showed that the spring tide range of 1,55 m extends halfway up the estuary. Salinities varied with the tide and the vertical salinity gradient was well marked. A summary of maximum and minimum values is shown in Table 14.2.

Nutrient values in June 1976 were low with 2-3 µg PO_4-P.ℓ^{-1}, 4-13 µg NO_3-N.ℓ^{-1} and 5-136 µg NH_3-N.ℓ^{-1}.

There were no obvious trends along the estuary. Bacteriological samples in June indicated that moderate faecal pollution reaches the estuary probably from African villages along the river. As might be expected, the concentrations of organo-chloride insecticides and trace metals were low.

The fauna as seen over neap tides in the summer of 1950 was poor but included those species which might be expected at the edge of the subtropics. *Callianassa* was common in the sands at the mouth and *Upogebia* was fairly common in muddy areas elsewhere. Crabs were common with *Sesarma meinerti, S. catenata* and *S. eulimene* in the salt marshes, *Cleistostoma edwardsi* lower down and the burrows of *Scylla serrata* near low tide. The amphipod *Orchestia rectipalma* was common at high tide and *Grandidierella lignorum* was extremely abundant in sloppy mud at and below low tide. No bivalves were found at all although some polychaetes were dug from the sandbank. The gastropod *Assiminea* sp was common in the salt marshes and *Littorina scabra* was found on the reeds but there were no *Cerithidea* on the rather scattered mangroves; apparently this species requires dense shade.

Netting in the soft mud was impossible and only a few species were taken in the seine on firmer bottoms. *Rhabdosargus holubi, Gilchristella,* juvenile mullets

Table 14.2
Hydrological records in Bashee River estuary

Position	1,1 km from sea	4,7 km from sea	7,9 km from sea
Surface salinity (‰)	16,2-30,0	1,5-17,0	0,0-6,0
Bottom salinity (‰)	23,6-30,0	17,7-23,0	0,0-8,0
Winter temperature	18,0 °		14,0 °
Summer temperature	22,0 °		26,0 °
Summer turbidity (1950)			
at low tide	5-8 cm (Secchi disc)		
at high tide	35 cm (Secchi disc)		

and *Gobius giuris* were the only common fishes but with them were a few prawns *(Penaeus indicus, P. monodon* and *Metapenaeus monoceros)* and a few shrimps *(Palaemon pacificus).*

Two small blind estuaries north of the Bashee were briefly examined. A small nameless estuary near the lighthouse was surprisingly rich for its size. Salinities, temperatures and lists of plants and animals are available in the University of Cape Town. Further north, about 3 km from the Bashee is the Mbanyana lagoon. The Mbanyana River is at least 10 km long and flows between steep wooded hills. It is saline for about a kilometre and forms a blind lagoon, 50-100 m wide and 2-3 m deep. At LWN it trickles out alongside a reef while at high tide waves slop over the sandbar and into the lagoon; it must burst open during heavy rains.

The mouth, of course, is sandy but inland the narrow banks are either rocky or muddy or both. The water was sufficiently clear for one to see the bottom at a depth of a metre and the temperature, at 29,5 was high (cf Bashee). Unfortunately, the salinity was not recorded but it is probably in the range of 10-20 ‰.

The narrow fringe of marginal vegetation included eight species of plants dominated by *Juncus kraussi* in mixed swards of *Sporobolus* and *Triglochin striatum.* There was also a field of *Paspalum vaginatum,* clumps of *Phragmites* and a few saplings of *Bruguiera.* It was interesting to find a patch of *Zostera* growing submerged on the sandy mud.

The fauna collected in a day's visit included 30 benthic species and 12 fishes, indicating that this small clear lagoon, although usually closed by a sandbar, has a much richer macrofauna than the large, open, but very muddy Bashee estuary. Dr Connell of the NIWR in a personal discussion, stated that the zooplankton of the Bashee is extraordinarily rich. He

suggested that this was in some way related to the poverty of the benthic fauna. As noted in the case of the Mzimvubu, this seems unlikely but it would be interesting to sample the zooplankton of the Mbanyana.

Great Kei Estuary
(32°40′S/28°23′E)

The only available information on the ecology of the Great Kei estuary consists of notes and an annotated sketch made by Mr van Wyk of the Cape Department of Nature and Environmental Conservation in May 1958. The mouth of the Kei changes and Mr van Wyk's sketch has been modified by reference to an aerial photograph taken in May 1979. The result is reproduced in Figure 14.23.

The Great Kei River is the southern boundary of the Transkei. Its tributaries drain 20 559 km² of the Eastern Cape from Aliwal North to Queenstown and the western parts of the Transkei from Cala to Tsomo. Most of the catchment has a rainfall exceeding 400 mm per year and the mean annual run-off is 1 001 x 10⁶ m³. The last 50 km of the river runs over the slates and shales of Karoo deposits so that silt and clay is carried down to the estuary and the flood plain is covered with dried mud. Marine sand is deposited in the mouth and as there are no reefs on the coast to anchor the exit channel, the latter varies from year to year. An isolated lagoon on the east bank probably represents the remains of an old channel.

Mr van Wyk's water samples taken during low tide at point A (about 0,8 km from the sea), showed that the salinity was less than 1,0 ‰ suggesting that the estuary was about 1 km long in May 1958. It probably varies with the rains. The estuary is about 80 m wide and is less than 1 m deep. The water temperature in May 1958 was 15,6 ° and the Secchi depth was 15 cm. With this turbidity, no submerged vegetation could be expected and the only report of marginal vegetation is a fringe of reeds *(Phragmites)* in the upper reaches.

The only report on the fauna is that *Upogebia* is common on the muddy banks and that flocks of waders feed on the broad mud flats 100-200 m from the head of the estuary.

B. THE WARM-TEMPERATE ESTUARIES OF THE EASTERN AND SOUTHERN CAPE

Keiskamma estuary
(33°17′S/27°29′E)

The earliest report was that made by a party from the University of Cape Town during January 9-10 in 1950. The notes on environmental conditions and the cata-

logues of the plants and animals collected are available in the Zoology Department. Later observations were made by Farrel & van der Walt (1976) and an account of the sediments, the marginal vegetation and the benthic invertebrates is given by Cowling *et al* (1979).

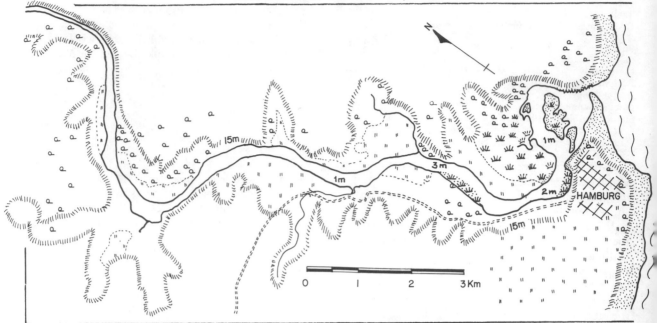

Figure 14.24
KEISKAMMA ESTUARY

The Keiskamma River rises near Stutterheim and winds through the rather dry hilly country of the Ciskei to the village of Hamburg. The 2 530 km² catchment has an erratic rainfall of about 500 mm per year with maximal falls in March and September. Periodic droughts and floods erode the clayey soil derived from the Beaufort beds so that the estuary is badly silted and the water turbid.

The head of the estuary is fixed by a causeway 29 km from the sea and from there on, the upper reaches run for 15 km through a narrow valley. This section of the estuary is about 100 m wide during floods with steep and often rocky banks. Thereafter the valley broadens and the muddy estuary (Figure 14.24) now 150-500 m wide, meanders through a 2-3 km broad floodplain. Most of this is under cultivation with salt marshes here and there. Below the village of Hamburg, some 2 km from the sea, the estuary makes an S-bend and in 1950 the oxbow formed an island at high tide but since then the eastern channel has silted up and the higher land on the west is now cultivated. The mouth is permanently open between rocks on the north bank and a broad sandspit on the south bank but the surf is too heavy for boats to enter.

The spring tide range is reduced to 1,3 m at Hamburg and most of the estuary is less than 2 m deep, with a sandy bottom at the mouth which rapidly changes through sandy mud to very soft mud along the rest of the estuary. The intertidal banks change in the same way. At Hamburg the upper tidal levels are very soft but firmer towards low tide; the main currents sweep along the eastern, convex side of the channel and the mudbanks are steeply eroded below the level of the *Spartina* marshes, the edges of which are further despoiled by bait diggers. Cowling *et al*

(1979) have analysed the particle size distribution of the bare mudflats on the west bank and although their ϕ cumulative curves are incorrect, there is no doubt that there is well over 50 % of subsieve (<0,063 mm) particles.

In 1950 there were a few patches of *Zostera* at low tide, but in 1979 these were rarer. At and above mid-tide, there are large stands of *Spartina maritima* in the lower reaches. These give way to patches of *Triglochin bulbosum* in extensive carpets of *Sarcocornia perennis* at slightly higher levels, and both are largely replaced by *Cotula coronopifolia* further up the estuary. *Phragmites australis* dominates the banks in the upper reaches. In the lower reaches, the upper fringe of *Sarcocornia* is replaced by dense growths of *Limonium linifolium* and *Chenolea diffusa* towards the level of HWS. *Sporobolus virginicus* or *Stenotaphrum secundatum* extends down to HWS from above and growing in this grassy sward are scattered clumps of *Sarcocornia pillansii*, *Disphyma crassifolium*, *Suaeda maritima* and other halophytes. Mangroves are entirely absent.

The benthic fauna includes most of the species present in the Eastern Cape (see chapter 9) but the fauna is by no means rich; in particular the fauna in the soft glutinous mud at low tide is very poor. *Assiminea* sp and/or *Hydrobia* sp and the crab *Sesarma catenata* are dominant in the vegetation at upper tidal levels while *Cleistostoma edwardsii* and *Nassarius kraussianus* occur on the flats below midtide and *Solen corneus* was dug from the mud near low tide. *Callianassa* is abundant in the sands near the mouth and *Dotilla fenestrata* was fairly common there. *Upogebia africana* dominates the infauna in the firmer mudbanks and

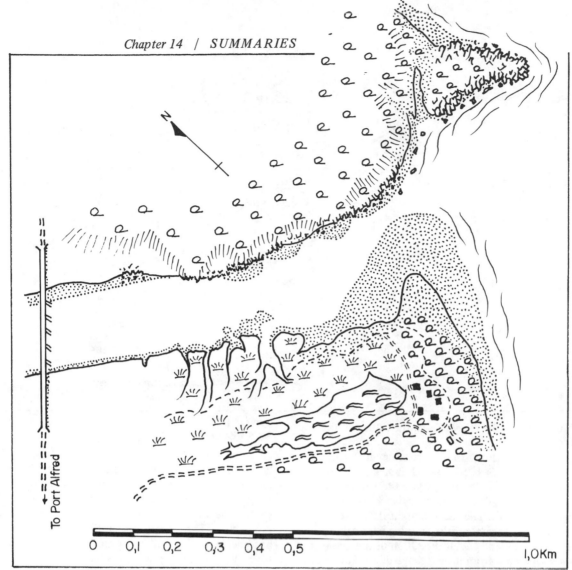

Figure 14.25
GREAT FISH ESTUARY

burrows deeply into the *Spartina* beds. No plankton samples have been collected but the mysid *Mesopodopsis slabberi* is abundant in turbid shallows, at least in January. *Palaemon pacificus* is common and both *Penaeus indicus* and *Metapenaeus monoceros* are present in January.

A few hauls with a small seine in January revealed 15 species of small fish; tropical species included *Terapon jarbua, Siganus rivulatus* and *Lutianus fulviflamma.* Anglers also caught several kob *(Argyrosomus hololepidotus)* near Hamburg; probably other grame fish are present.

As shown, the estuary is reasonably rich in species of marsh plants and intertidal animals. Nonetheless, the biomass of the fauna is low, particularly below midtide and the submerged vegetation is very poor. There can be no doubt that this is due to heavy siltation and the turbidity of the water.

The estuary of the Great Fish River
(33°30′S/27°08′E)

The only ecological information available regarding the estuary of this large river is an annotated sketch made in May 1960 by Mr van Wyk of the Cape Department of Nature and Environmental Conservation. The sketch has been amended from an aerial photograph taken during April 1979 and is reproduced below as Figure 14.25.

The great Fish River drains a large area of the Eastern Cape Province lying between Cradock and Queenstown. Although droughts occur fairly often, the mean annual rainfall is more than 400 mm. The catchment of the river and its tributories is 30 427 km^2 and the run-off is 479 x 10^6 m^3. The flood waters are known to rise over 10 m and carry heavy silt loads for the soft soil is derived from Karoo formations including slates and shales. During droughts, the river is reduced to a series of pools linked by stony runs.

Mr van Wyk recorded a salinity of about 0,4 ‰ some 20 km from the mouth which suggests that this point marks the head of the estuary at the end of

293

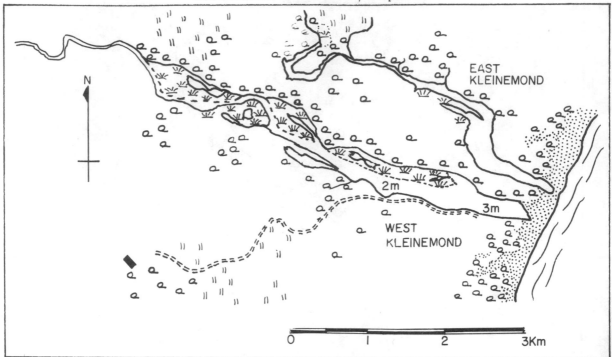

Figure 14.26
W. KLEINEMOND ESTUARY

autumn. During the rainy season the estuary is certainly shorter.

Figure 14.25 shows that there are many rocky outcrops along the north bank which is steep right to the sea with the reefs extending beyond to form Great Fish River Point. This promontory anchors the position of the mouth. The south bank is flat, muddy and low lying with several creeks draining wide salt marshes. The marshes extend to within 200 m of the seashore but are separated from it by a range of sand dunes, which project across the mouth as a bare sandspit, constricting the main channel against the rocky northern bank. The tidal currents in the mouth are thus strong and the sandspit at times breaks up to form shoals and minor channels.

Mr van Wyk records that the estuary was extremely muddy when he visited it in May 1960 and the surface temperature was 16,7 °. *Upogebia* was abundant in several places along the muddy and stony banks and crab holes were very common. The fauna of the sandspit was not recorded. Much more information is obviously needed.

The West Kleinemond estuary
(33°33′S/27°03′E)

There are two small blind estuaries some 16 km east of Port Alfred. The larger, or West Kleinemond estuary has been studied by Brown (1953) and Blaber (1973a, 1974). The following account is based on Dr Burke Hill's observations and summary.

The West Kleinemond drains 52 km² of dry farmland and scrub in the Bathurst district where the rainfall is erratic but averages 525 mm.yr⁻¹. The rains fall in spring and autumn and there are occasional floods so that the level in the estuary rises until the sandbar is breached. The estuary is then flushed out and becomes tidal for a month or more, but is often closed for a year and sometimes two years.

The estuary was originally 10 km long but a barrage was built in 1960 so that the upper part could be used for fresh water storage and the saline estuary (Figure 14.26) is now 5 km long. It may be divided for convenience into three reaches. The upper reach extends 2,5 km from the barrage and is 20-30 m wide, 2-3 m deep with a salinity of 5-10 ‰. It has steep muddy banks covered with *Phragmites australis* and there are dense growths of *Ruppia spiralis* in the shallows. The middle reach extends on for a further 2 km to the national road bridge. It widens from 50 to 200 m, is up to 2 m deep with a bottom of sandy mud, and the banks are covered with *Sarcocornia* spp to the water's edge. The salinity is 10-30 ‰. The final reach forms a sandy lagoon 500 m long, 200 m wide and up to 3 m deep after rains. Normally the salinity is above 25 ‰. In the dry months, bare sand flats are exposed for there is little marginal vegetation and the only submerged plants are clumps of *Codium tenue* and scattered growths of *Ulva* and *Enteromorpha*. *Zostera* is entirely absent. The water is well-mixed by the wind so that stratification does not persist long after the rain and the water is clear and well-oxygenated to the bottom. The

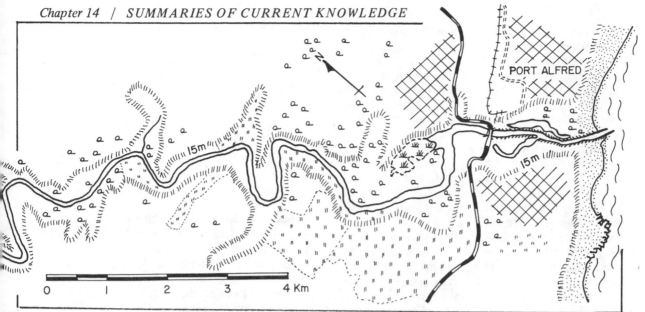

Figure 14.27
KOWIE ESTUARY

Kowie estuary
(33°36′S/26°54′E)

area is sparsely populated and there is no sign of pollution although there is a smell of H_2S when the muddy areas are disturbed. Water temperatures rise from 22 to 29° in summer and fall to 15 or even 12° in winter. The phytoplankton has not been investigated but the clarity of the water suggests that it is poor.

Although Brown (1953) recorded 12 species of insects and 34 invertebrates of marine origin, very few of them are common. Only *Callianassa krausii* is abundant and occurs even in mud which is unusual. The estimate of 600 holes per square metre suggests that the density may reach 300 m^{-2}. The serpulid polychaete *Ficopomatus enigmatica,* the isopod *Sphaeroma annandalei,* the decapods *Hymenosoma orbiculare* and *Palaemon pacificus* and the molluscs *Solen corneus, Lamya capensis, Nassarius kraussianus* and *Assiminea* sp are common. The reported absence of *Sesarma catenata* is surprising but *Upogebia africana* is seldom present in blind estuaries.

Some 30 species of fishes have been recorded but the number and composition varies with the opening of the sandbar since most are juveniles of species which breed in the sea. As detailed in chapter 12, Blaber (1973b, 1974b) has shown that the abundance of *Rhabdosargus holubi* varies from year to year and the final density is largely controlled by bird predation. A few small fishes breed in the estuary and *Sarotherodon mossambicus* is of fresh water origin.

The ecology of Kleinemond estuary is of particular interest as an example of the many small Cape estuaries which are closed and atidal during dry periods. In this case, evaporation is more or less balanced by river discharge, and the salinity regime is normal.

Although ecological notes are available in papers on estuarine crabs by both Alexander & Ewer (1969) and Hill (1975, 1976), no formal account of the ecology of the Kowie estuary has been published. Dr Hill's draft account has been amended by the editor.

The Kowie river rises in hilly country near Grahamstown and drains 576 km^2 with an average rainfall of 638 mm.yr^{-1} most of which falls in spring and autumn. Droughts and floods are not infrequent. The river winds through narrow wooded valleys and the estuary (Figure 14.27) begins below a stony run 19 km from the mouth. The upper reaches are 13 km long, 50-90 m wide and 2-6 m deep with steep, and occasional rocky banks 1-3 m high so that the intertidal zone is only 3-10 m wide. The tidal range at springs is 1,1 m and currents vary from 12 to 20 cm.sec^{-1}. The salinity is usually above 30 ‰ and may increase to 40 ‰ in dry years or decrease to nil at the surface after heavy rains. The middle reaches wind through a broad valley 3 km long; and the estuary here is 100-150 m wide and 3 m deep with troughs up to 8 m deep on bends. The bottom is sandy and the intertidal mud banks are up to 100 m wide. The maximum tidal range is 1,5 m, the currents up to 12 cm.sec^{-1} and the salinity usually above 30 ‰. The final 3 km to the mouth at Port Alfred was dredged, straightened and stabilized by stony embankments prior to 1890, so that the shallow salt marshes which lie on the east and west banks have been reduced to small areas which are only flooded at high tide. The mouth is guarded by breakwaters 75 m apart. The sandy bottom is about 3 m deep, the spring tide range is 1,7 m and the currents are up to 25 cm.sec^{-1}. The seasonal temperature range in the upper reaches is 14-27° (minimum 11°) and at the mouth it is 14-22°. The turbidity is high and during prolonged floods

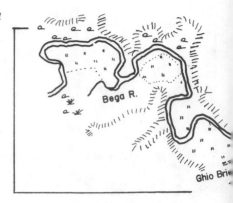

which may last for two to four weeks, the surface water of the whole estuary is almost fresh although it is normally 33-35 ‰ at the mouth.

The phytoplankton has not been investigated. The submerged and marginal vegetation is scanty although there are beds of *Ruppia spiralis* and dense fringes of *Phragmites australis* along the upper reaches. Patches of *Zostera* occur on the mud banks in the middle reaches and there are patches of *Spartina maritima* and *Sarcocornia* sp at higher levels; further investigations of the salt marshes are needed.

Neither the zooplankton nor the benthic fauna has been surveyed but appears to be poor apart from a few species. The rocky retaining walls are colonised by several marine forms including *Parechinus angulosus* which is devastated by periodic floods. At higher tidal levels, gastropods such as *Oxystele tigrina, O. variegata, Littorina knysnaensis* and the crab *Cyclograpsus punctatus* are common. The mudbanks in the middle reaches have a richer fauna; *Upogebia africana* may reach 600 m^{-2} and with it are *Nassarius kraussianus, Cleistostoma elwardsii* (and *C. algoense* in softer mud). *Sesarma catenata* is common towards high tide. The main invertebrates in the upper reaches are the crab *Sesarma eulimene,* and, at lower levels, the bivalves *Solen corneus* (up to 150 m^{-2}) and *Musculus virgiliae* both of which are preyed on by the swimming crab *Scylla serrata.* Smaller forms such as amphipods, isopods, hydrobiid gastropods and polychaete worms have not been investigated.

The fish fauna includes mullets, juvenile *Rhabdosargus holubi* and shoals of *Lithognathus lithognathus. Pomadasys commersonni* appears in late summer. The resident species include gobies, *Gilchristella aestuarius* and *Hepsetia breviceps.* A survey would doubtless reveal many other species as well as shrimps and a few penaeid prawns.

The poverty of the whole flora and fauna may be correlated with the narrow intertidal banks and rapid tidal exchange. This in turn is partly due to the canalisation of the mouth and the destruction of the main salt marshes and *Zostera* beds which are the most productive areas in an estuary. The widening of the connection between the main channel and the salt marshes on the east bank is being considered.

Bushmans River estuary
(33°41,6′S/26°42′E)

There are no formal reports on the ecology of this estuary. Extensive collections of the macrofauna and fish were made by Dr P. Jackson and Dr R. Liversidge over the spring tides of September 1950 and catalogues of the collection are available in the Zoology Department of the University of Cape Town. Unfortunately, no notes of environmental conditions remain apart from brief descriptions of the collecting stations. To

remedy this, Mrs Carolyn Palmer of the Zoology Department of Rhodes University with the aid of Mr A. Palmer and Mr A. Bok very kindly surveyed the estuary during the high spring tide of 20 May 1979. They noted the depth, nature of the substrate, salinity, temperature and Secchi depth from the coastal road bridge to a point 22,5 km up the estuary. Time did not permit observations at low tide and the estuary was too shallow and rocky to reach 'Ebb and Flow' which Dr Jackson had reported as the head of the estuary, some 40 km from the mouth. From the nature of the fauna recorded at Ebb and Flow the water is certainly saline at times.

In addition to these observations, Hill *et al* (1973) reported on the depth, substrate and current speeds at the coastal road bridge in 1969 and 1973. They show conclusively that the rock spoil and broken concrete beams deposited in the river while the bridge was being constructed, had reduced the depth by about 6 m. As a result, the surface current velocity under the bridge increased from 70-75 cm.sec^{-1} to more than 120 cm.sec^{-1}. Such a current is dangerous to bathers and at low tide broken concrete beams were exposed under the eastern span and were dangerous to small boats. The rubble under the western arch of the bridge was removed in 1976. Weaver (1979) reports that by 1977 the grain size of the bottom sediments seaward of the dredged channel had increased. Conversely the mean grain size of sediments adjacent to the undredged channel had decreased, showing that sediment had accumulated there.

The Bushmans River rises near Webster, some 60 km south of Somerset East. Its catchment is 2 678 km^2 but the area is rather dry and the mean run-off is only 38 x 10^6 m^3.yr^{-1}. The silt load has not been reported but a comparison of depths in the estuary between 1950 and 1979 suggests that siltation has increased.

As noted earlier, the estuary (Figure 14.28) extends from Ebb and Flow for some 40 km in the dry months. The countryside is hilly and the estuary runs in a wooded valley so that the flood plain narrows towards the sea and the banks are rocky in many places. The upper reaches above Ghio Bridge increase to a depth of 1,5 m with narrow intertidal banks and a bed of

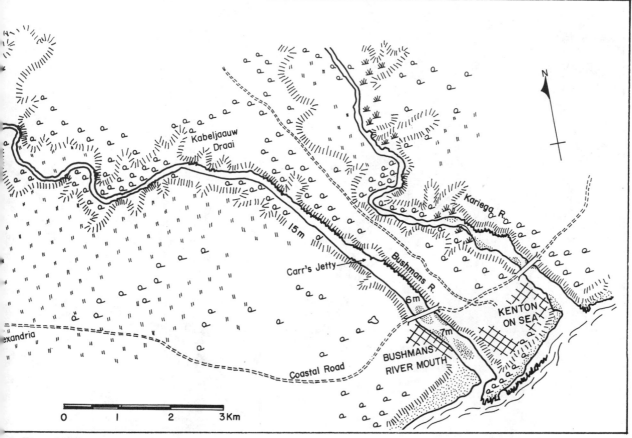

Figure 14.28
BUSHMAN ESTUARY

sandy mud. From Ghio Bridge to Kabeljaauw Draai the depth increases to about 2,5 m at high tide and the banks are steep and often rocky and the bed is muddy. From Kabeljaauw Draai to Carr's Jetty the estuary widens from 120 to 300 m, the depth increases from 2,5 to 5 m at high tide, the bed changes from mud to sandy mud, but the banks are unchanged. Along the 3 km lower reaches from Carr's Jetty to the mouth, the estuary widens to about 450 m, many shoals are exposed at low tide and the channels between them average 6-8 m in depth with a sandy bed (apart from the rocks under the bridge noted earlier). The mouth is anchored by a rocky promotory on the east bank and is constricted by a sandspit on the west bank.

Little is known about the hydrology of the estuary since observations were limited to high tide. The tidal range at springs which is 1,8 m on the coast must be reduced by the constriction at the mouth and the shallow waters under the coastal road bridge and is probably of the order of 1,2 m. However this range apparently persists far up the narrowing estuary and tidal effects are obviously responsible for the name Ebb and Flow. Incidentally, the neighbouring Kariega estuary is also tidal as far up as the causeway some 30 km from the sea. Tidal currents in the lower reaches of the Bushmans are of the order of 75 cm.sec^{-1}.

Surface salinities at spring tide are 30-34 ‰ as far up as Ghio Bridge and then decrease to 20 ‰ at 22,5 km from the mouth, but the salt wedge must reach Ebb and Flow in the dry months of the year. Thus the estuary is marine dominated as might be expected from the rather low run-off from the catchment. The water is fairly clear with high tide Secchi readings of about 0,5 m over the upper and middle reaches, increasing to 0,85 m at Carr's Jetty at 2,2 m at the coastal road bridge. Surface temperatures in May 1979 varied between 17 and 19 ° over the whole estuary which is normal in Eastern Province estuaries. In general, the environmental conditions reported appear to be very stable and the estuarine flora and fauna is rich in consequence.

The phytoplankton and zooplankton have not been investigated, and nutrient concentrations are unknown. Since the banks of the estuary are thinly populated pollution is unlikely.

The macrophytic vegetation is limited by the width of the intertidal banks. At Ebb and Flow, reeds *(Phragmites australis)* were reported, changing to *Scirpus littoralis* and other salt marsh plants at a station 22,5 km from the mouth. *Sarcocornia* sp is predominant at the

297

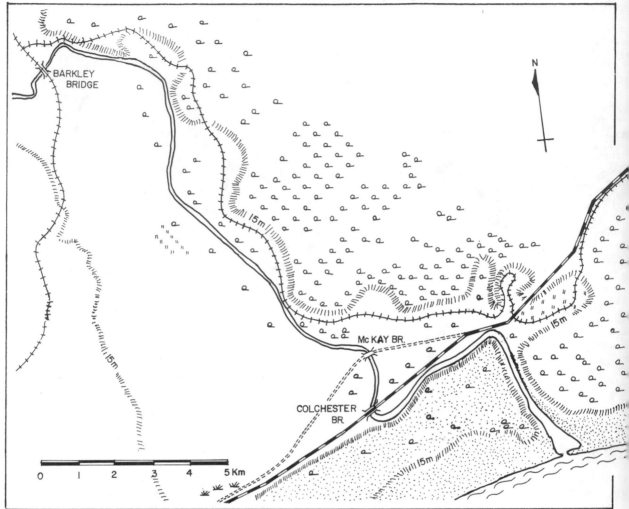

Figure 14.29
SUNDAYS ESTUARY

confluence of the Bega River tributary. From Ghio Bridge onwards, *Limonium linifolium* and *Sarcocornia* at high tide changes to *Spartina maritima* and *Triglochin* sp at lower levels; at Retreat, a salt marsh meadow develops with *Triglochin* dominant over *Spartina*. Scanty *Zostera* is present at low tide levels. At Kabeljaauw Draai, there is a narrow *Spartina* bed on the west bank and rocks on the east bank. In May 1950 it was noted that *Zostera capensis* was common at low tide and that *Caulerpa filiformis* colonised the sandy bed of the estuary. This is unusual in estuaries and confirms the statement that the estuary is marine dominated. Normal estuarine vegetation with *Limonium, Chenolea* and *Sarcocornia* above, *Spartina* at midtide and *Zostera* at, and below, low tide, extend seawards from Carr's Jetty to the coastal road bridge and it was noted that the *Zostera* becomes more luxuriant and *Codium tenue* is fairly common on submerged rocks. The final kilometre to the mouth is very sandy and the many shoals have a central patch of *Spartina* surrounded by luxuriant *Zostera* beds on muddy sands.

The meiofauna is unknown but the benthic macro-fauna is rich, 101 species being reported in 1950. Locality records are available in the Zoology Department of the University of Cape Town. Estuarine species including *Balanus elizabethae* and *Ficopomatus enigmatica* colonise the rocks at Ebb and Flow, *Lamya capensis* is present on the muddy bottom and *Palaemon pacificus* harbours amongst the reeds. A similar fauna was reported at Ghio Bridge and the fish *Gobius giuris* was common. At Retreat the number of species increased. *Sesarma catenata* and *Orchestia ancheidos* was common in the *Sarcocornia* beds, *Upogebia* and *Ceratonereis erythraeensis*, were present in the *Zostera* and *Teredo* sp bored rotting posts. Netting revealed several species of fish including abundant *Rhabdosargus holubi*, also *Sepia officinalis, Scylla serrata* and rhizostomid jellyfish. At Kabeljaauw Draai the crab *Cleistostoma algoense* and the pistol shrimp *Alpheus crassimanus* appeared with two stenohaline marine species, *Parechinus angulosus* and *Virgularia schulzei*.

Carr's Jetty was possibly the richest collecting

298

ground. Several marine species appeared on the rocks including *Oxystele variegata*, *Thais dubia*, *Chlamys tinctus* and *Patella longiscosta*. In the salt marshes *Assiminea bifasciata* and *Assiminea* sp were abundant and so were crabs and talitrid amphipods. *Callianassa*, *Upogebia* and *Solen capensis* burrowed at low tide and *Pinna squamifera* and burrowing holothurians were found with them. Several other species, including *Bullia laevissima* and *Parechinus* again confirm marine dominance. Many species of fish and shrimps were netted.

Much the same fauna persisted as far as the coastal road bridge but as the percentage of silt in the sand decreased below the bridge the burrowing fauna changed. *Callianassa* became abundant, *Nassarius kraussianus* reached densities of 200 m^{-2} and *Arenicola loveni* appeared; several more bivalves, gastropods, barnacles, sea anemones, sponges, ascidians and hydroids appeared on the rocks and piles.

It should be noted that these species of animals were all recorded in 1950 before the coastal road bridge was built. There is no doubt that the sand banks have grown since then and the depth of the remaining channels has decreased. It would be interesting to make a new biological survey of the estuary to see whether the fauna has changed. Nutrient concentrations should also be determined in the different seasons and both the phytoplankton and zooplankton biomass should be determined in what was, and still may be, a highly productive estuary.

Sundays River estuary
(33°42,4'S/25°51'E)

There are no published reports on the ecology of this estuary and this summary is based on a draft by Dr D. Baird supplemented by observations made by the editor in 1950.

The Sundays river rises in the Karoo near Nieu Bethesda and has a catchment of 16 826 km². The rainfall is only 550 m.yr^{-1} and the flow is regulated at the Metz dam near Tansenville. The estuary (Figure 14.29) is 21 km long and from its head below Barkley Bridge it winds over a clayey Pleistocene flood plain and has cut deeply into the soft soil so that the intertidal banks are steep and narrow with little marginal vegetation. At Barkley Bridge the river is about 20 m wide and over most of its length the estuary is only 50 m wide and 2,5 m deep with a muddy bottom. From Colchester onwards, the estuary skirts the edges of steep coastal sand dunes and 4 km from the sea the estuary bends abruptly to form a 5 m deep channel between the sand dunes and the rocky edge of the escarpment. Thereafter it broadens into a 200 m wide lagoon with a mouth that narrows to 30 m at low tide. Most of the estuary water is turbid with a Secchi depth of 20-40 cm but the lagoon is clearer at high tide and

Table 14.3
Hydrological conditions in Sundays River estuary

Reach		Head	Upper reaches	Lower reaches	Mouth area
Distance from sea (km)		18-21	4-18	1-4	0-1
Salinity (%oo)	Summer	2,6	20,0	28,0	28-35
	Winter	1,3	3,2	10,5	19-?
Temp. (°C)	Summer	25,4	25,4	23,5	23,5
	Winter	14,4	13,4	13,1	14,0

the bottom is visible at 1-2 m. There is no evidence of pollution. River flow maintains fairly steady gradients along the estuary and the tidal range at McKay bridge is about 1,2 m.

Hydrological conditions are tabulated above :

As noted, there is little marginal vegetation but small patches of *Zostera capensis* appear in the lower reaches. The fauna of the estuary is poor although many of the normal Cape species have been recorded. In the upper reaches, *Ceratonereis erythraeensis* is fairly common and tubes of *Ficopomatus enigmatica* form coralliform masses on the pylons of McKay Bridge with *Melita zeylanica* and *Cirolana fluviatilis* in the crevices. Below Colchester, *Callianassa kraussi* is common in the sandbanks reaching a maximum density of 90-180 m^{-2}; *Upogebia africana* is patchy but may reach similar densities in the mud banks. The crabs *Cleistostoma edwardsii*, *Hymenosoma orbiculare* and *Rhynchoplax bovis* are present and both *Cyclogopsus punctatus* and *Sesarma catenata* are fairly common at higher levels. Bivalves are not common but both *Solen corneus* and *S. capensis* have been recorded. Mysids are abundant in muddy shallows and both *Palaemon pacificus* and *Alpheus crassimanus* are present in *Zostera* beds.

Forty-one species of fish were recorded by Baird *et al* (1979) mainly from the lower reaches but the Sundays is regarded as a poor estuary for angling. Only a few aquatic birds are present.

The poverty of the estuary is probably due to its steep banks, poor aquatic vegetation and turbid water.

Swartkops estuary
(33°51'S/25°38'E)

The following summary is based on a more detailed draft by Dr D. Baird of the University of Port Elizabeth.

Macnae (1957) described the ecology of the macrobenthic fauna and flora. Since then 25 papers have dealt with the following aspects:

Water chemistry, sediments and pollution: Oliff (1976)
Phytoplankton, benthic algae and vascular plants: Pocock (1955), Macnae (1957), Henrici & Pienaar (1975), Dye (1978c), Baird (unpublished), Pierce (1979)
Zooplankton: Grindley (1976), Oliff (1976)
Meiofauna: Dye (1976, 1978a, 1978b), Dye & Furstenburg (1978), Dye *et al* (1978)

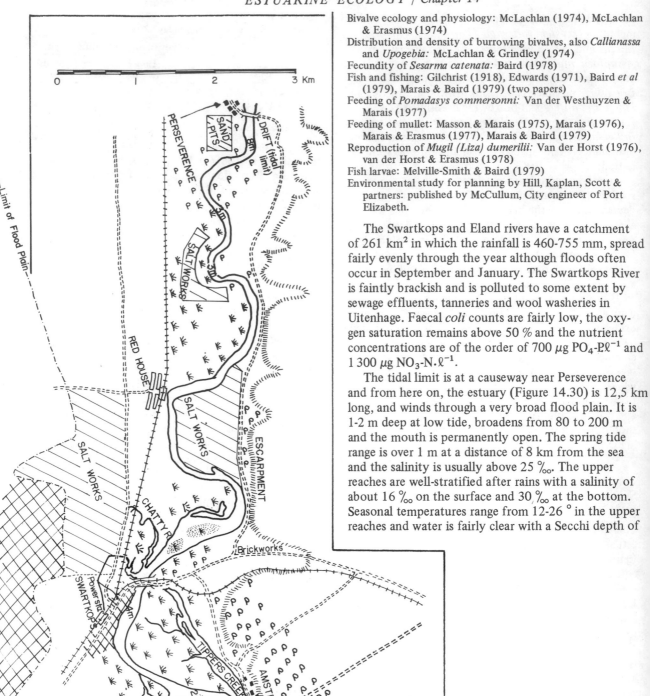

Bivalve ecology and physiology: McLachlan (1974), McLachlan & Erasmus (1974)

Distribution and density of burrowing bivalves, also *Callianassa* and *Upogebia:* McLachlan & Grindley (1974)

Fecundity of *Sesarma catenata:* Baird (1978)

Fish and fishing: Gilchrist (1918), Edwards (1971), Baird *et al* (1979), Marais & Baird (1979) (two papers)

Feeding of *Pomadasys commersonni:* Van der Westhuyzen & Marais (1977)

Feeding of mullet: Masson & Marais (1975), Marais (1976), Marais & Erasmus (1977), Marais & Baird (1979)

Reproduction of *Mugil (Liza) dumerilii:* Van der Horst (1976), van der Horst & Erasmus (1978)

Fish larvae: Melville-Smith & Baird (1979)

Environmental study for planning by Hill, Kaplan, Scott & partners: published by McCullum, City engineer of Port Elizabeth.

The Swartkops and Eland rivers have a catchment of 261 km^2 in which the rainfall is 460-755 mm, spread fairly evenly through the year although floods often occur in September and January. The Swartkops River is faintly brackish and is polluted to some extent by sewage effluents, tanneries and wool washeries in Uitenhage. Faecal *coli* counts are fairly low, the oxygen saturation remains above 50 % and the nutrient concentrations are of the order of 700 μg PO$_4$-Pℓ^{-1} and 1 300 μg NO$_3$-N$\cdot\ell^{-1}$.

The tidal limit is at a causeway near Perseverance and from here on, the estuary (Figure 14.30) is 12,5 km long, and winds through a very broad flood plain. It is 1-2 m deep at low tide, broadens from 80 to 200 m and the mouth is permanently open. The spring tide range is over 1 m at a distance of 8 km from the sea and the salinity is usually above 25 ‰. The upper reaches are well-stratified after rains with a salinity of about 16 ‰ on the surface and 30 ‰ at the bottom. Seasonal temperatures range from 12-26 ° in the upper reaches and water is fairly clear with a Secchi depth of

Figure 14.30
SWARTKOPS ESTUARY

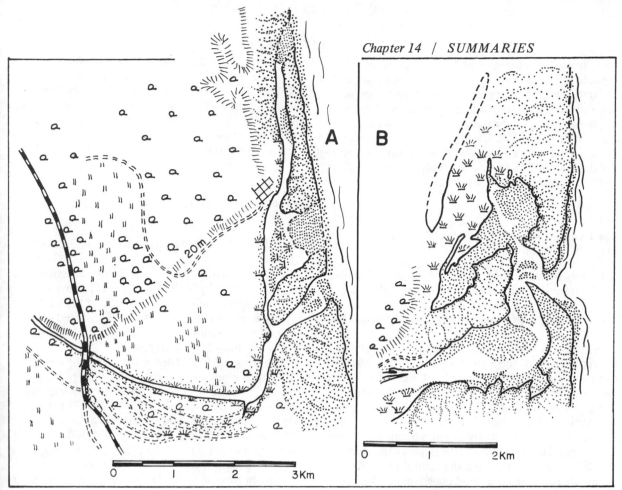

Figure 14.31
GAMTOOS ESTUARY : A in 1956; B mouth in 1979.

about 1 m. The substrate changes from boulders and coarse sand near the causeway, through soft mud with more than 20 % silt and clay in the upper reaches, to sandy mud and then clean sand at the mouth. Amsterdamhoek is guarded by a rocky embankment but the sediments of Tippers Creek are very soft mud with about 80 % silt and clay.

Phytoplankton production has not been recorded. The mean benthic microalgal production is 53 g.Cm.$^{-2}$ yr^{-1} in the sands opposite Amsterdamhoek and 116,5 g.Cm^{-2}.yr^{-1} on the mud near Redhouse. The macrophytic vegetation includes the normal Cape species (see appendix to chapter 6). Baird (unpublished) estimates the total area of the saltmarsh as 263,2 ha of which 22,7 % is covered by *Spartina maritima* with a net annual production of 509 g (dry mass).m^{-2}; he also records the biomass of other dominants.

Zooplankton is rich in the slow-flowing upper reaches. The meiofauna is rich in the muddy saltmarshes due mainly to nematodes; it is also well-developed in sandy *Callianassa* beds. In spite of this, the annual production of the meiofauna is only 0,24-1,72 g.C.m^{-2}.yr^{-1}. The common species of the benthic macrofauna are shown in the appendix to chapter 9. At upper tidal levels in the salt marshes, the hydrobiid snail *Assiminea* sp reaches densities of 5 620 m^{-2} while

the crab *Sesarma catenata* reaches 76-216 g.m^{-2} and its mean production is 9,33 g (ash free).m^{-2}.yr^{-1}. Towards midtide, another crab, *Cleistostoma edwardsii* occurs at densities of 76-164 m^{-2}. Holes of *Upogebia africana* appear at the same level in sandy mud but *Callianassa kraussi* prefers sandy sediments and extends below LWS. Burrowing bivalves dominate mud flats in which there is over 30 % silt. Maximum recorded densities approximate to 500 *Upogebia* m^{-2}, 400 *Callianassa* m^{-2}, 120 *Solen corneus* m^{-2} and 80 *Dosinia hepatica* m^{-2}. Heavy mortalities have occurred during prolonged floods and in the past the banks were devastated by anglers digging for bait. Digging is now illegal in this and other estuaries. A total of 42 species was recorded by Baird *et al* (1979). This is low.

The main estuarine fishes are listed in chapter 10. The sport fish include grunter, 83 % of catches by mass), leervis, kob, white steenbras, gurnard, elf and springer *(Elops)* (in that order), while juvenile *Rhabdosargus* spp are abundant and four species of mullet are common. The reproduction of the mullet *Liza dumerilii* (a summer breeder here) has been determined and the food of the grunter *(Pomadasys commersonni)* and the mullets has been investigated. Melville-Smith & Baird (1979) recorded 19 species of larval fish of which gobies and *Gilchristella* comprised over 90 %.

Swartkops is a rich estuary and the best-known in the Eastern Cape. There is danger that it will be spoilt by industrial development and there are already salt-pans, brickworks and factories on the flood plain and the outflow from the power station increases the concentration of trace metals to a small extent. The City Engineer has received a detailed environmental statement from Messrs Hill, Kaplan, Scott and Partners as an aid to planning and the Swartkops Trust is keeping a watching brief. Both the University of Port Elizabeth and the Museum are active in research.

The Gamtoos estuary
(33° 59'S/26° 57'E)

Relatively little is known about this estuary. Shelwell (1950) reported on the avifauna; Mr van Wyk of the Cape Department of Nature Conservation made brief notes and a sketch of the mouth in June 1956 and May 1959.

With its two tributaries, the Gamtoos River drains the central part of the Karoo. The 34 491 km² catchment is in an area of low rainfall and both droughts and floods occur at intervals. The run-off is of the order of 485 x 10⁶ m³ per year and floods erode the clayey soil so that the river is usually turbid.

The length of the estuary is very variable depending on the rains but the river is often saline at the coastal road bridge some 7 km from the mouth. From above the bridge to the sea the flood plain is about 3 km wide and so low-lying that whole width may be flooded and the course of the estuary is liable to change. Silt is deposited over wide areas. There are no rocks on the coast to anchor the position of the mouth and littoral drift of sand towards the east is considerable. Thus a sandspit grows on the west bank and the mouth moves eastward. During floods the estuary cuts through the sandspit to form a new mouth more in line with the river and parts of the old channel silt up to form a series of disconnected pools and lagoons behind the coastal dunes. Figure 14.31 based on aerial photographs taken in April 1979 illustrate the position of the mouth and the lagoons at that time.

Mr van Wyk's sketch showed that in 1959 the lower reaches of the estuary were between 130 and 200 m wide at high tide. The banks inland of the coastal sand dunes were muddy with salt marshes on the west bank and an almost isolated lagoon on the east bank. Where the estuary cuts through the bare coastal dunes, the wind-blown sand continually encroaches on the channel so that the banks are clean sand. At low tide, sandspits constrict the mouth to a width of 30 m and sandy shoals are revealed. During periods of low river flow the sandspits and shoals unite to close the mouth completely.

Mr van Wyk records that on 15 June 1956, water samples taken 730 m from the mouth showed a salinity of 36,7 ‰ and a surface temperature of 15°. The inter-

tidal banks are muddy near the same point and *Upogebia* was plentiful but *Solen* sp was scarce. *Callianassa* was abundant on sheltered intertidal sands nearer the sea and in the shallows of the isolated lagoon. On the other hand, the sifting sands along the edges of the dunes on the west bank were very barren. Much more information about this unusual estuary is required.

Kromme River estuary
(34° 08'S/24° 51'E)

D. Baird, Zoology Department, University of Port Elizabeth.

The first brief account of the estuary is contained in an MSc thesis by Mr T. Hecht (1973). Since then more extensive studies of the ecology of the whole estuary have been made by Dr Dan Baird, and an account of conditions in the canals of the developing marina has been published (Baird, 1979).

The Kromme River rises in the forests of the Outeniqua mountains and its main tributary is the Dwars. The catchment has an area of 1 085 km² and the mean annual run-off is 105 x 10⁶ m³. The water, which is clear and of good quality, flows into the Churchill dam which supplies Port Elizabeth. The overflow supplies the estuary which commences below a series of rapids and is later joined by the Geelhoutboom River and other minor streams. It flows for 14 km through a narrow valley with steep rocky banks so that in this stretch the estuary bed is coarse sand and stones. About 6 km from the sea the valley widens, the currents decrease and the banks and bed of the estuary become muddy and a wide saltmarsh develops on the north bank (Figure 14.32). The last 1-2 km are sandy and a marina has been developed on the south

Table 14.4
The density of the main macrobenthic invertebrates in the Kromme estuary

	Density m⁻²	Distance from mouth (km)
Polychaeta		
Arenicola loveni	1	1-5
Bivalvia		
Dosinia hepatica	20	8
Eumarcia paupercula	10	7-9
Loripes clausus	136	1,6
Macoma litoralis	10-45	5-8
Psammotellina capensis	1-5	0,5-5
Solen capensis	20	0,5
Solen corneus	5-15	4,7
Decapod Crustacea		
Alpheus crassimanus	10	5
Callianassa kraussi	About 10	Near mouth
	30-100	8-14
Cleistostoma edwardsi	5-20	5
Sesarma catenata	10-20	5
Upogebia africana	10-110	2-6

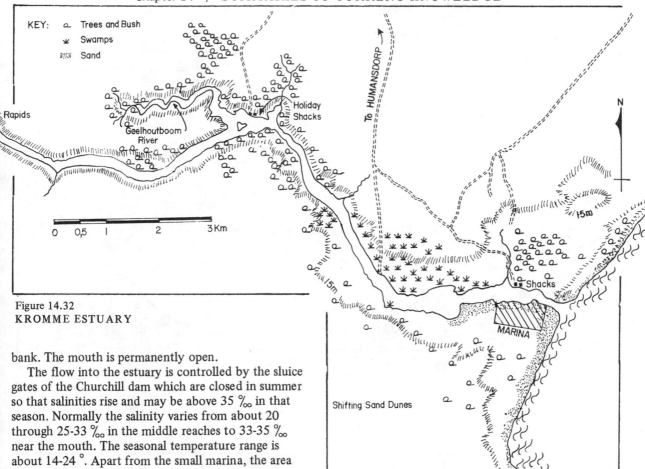

KEY:
- ℓ Trees and Bush
- ⚊ Swamps
- ⊞ Sand

Figure 14.32
KROMME ESTUARY

bank. The mouth is permanently open.

The flow into the estuary is controlled by the sluice gates of the Churchill dam which are closed in summer so that salinities rise and may be above 35 ‰ in that season. Normally the salinity varies from about 20 through 25-33 ‰ in the middle reaches to 33-35 ‰ near the mouth. The seasonal temperature range is about 14-24 °. Apart from the small marina, the area is very sparsely populated so that there is no pollution and the water is clear and well-oxygenated. An investigation of environmental conditions in the marina canals during summer and winter, showed that the tidal circulation was good and there was efficient flushing. Temperature, salinity and dissolved oxygen determinations of surface and bottom waters indicated well-mixed and air-saturated water. As a result, the plankton, benthic fauna and fish population is similar to that in the estuary nearby. Indeed, bait organisms *(Arenicola, Solen capensis* and *Callianassa)* which are better protected from bait collectors in the canals than in the open estuary, are more abundant in certain areas (Baird, 1979).

No submerged vegetation was found in the upper reaches. *Zostera capensis* first appears about 8 km from the mouth and the marginal vegetation includes *Spartina maritima, Triglochin* sp and *Sarcocornia* spp while *Chenolea* and *Limonium* grow at higher tidal levels. Neither the flora nor the fauna has been investigated in detail as yet, but estimates have been made of the maximum density of the dominant large invertebrates. These are shown in Table 14.4.

The main surprise is that *Callianassa,* which is abundant in the upper reaches, is scarce in the sands at the mouth where it is replaced by a higher density of

Loripes clausus than has been recorded elsewhere. As usual many species are more abundant in *Zostera* beds than on bare flats. Thus 12 species of invertebrates with a total wet mass of 419,3 g.m^{-2} were recorded in one *Zostera* bed compared with eight species with a wet mass of 44,4 g.m^{-2} on a bare bank.

This is a marine-dominated estuary with a rich fauna. Further ecological work is in progress.

Keurbooms estuary
(34° 03'S/23° 23'E)

There are few references. Day (1973) gave a brief account of ecological conditions in the summer of 1973 and Grindley (1973) discussed the danger of the proposed new bridges. Further observations made by a party from the University of Cape Town during July 1974 are available in the Zoology Department.

The estuary (Figure 14.33) is formed by the confluence of the smaller Bitou River and the larger Keurbooms. The combined catchment is 1 085 km^2 on the well-wooded Tsitsikama plateau where the annual rain-

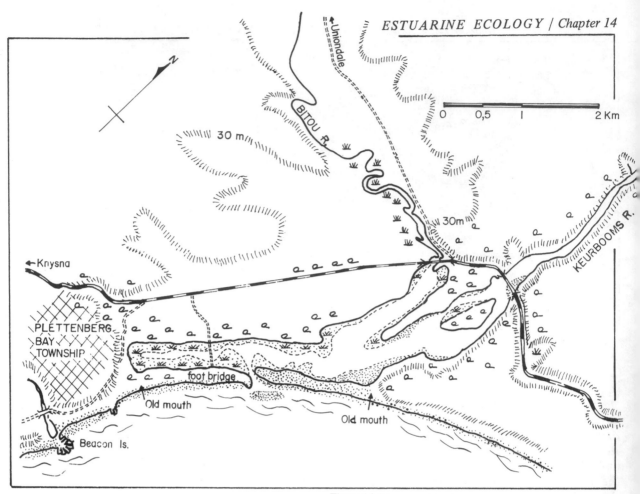

Figure 14.33
KEURBOOMS ESTUARY

fall is about 850 mm. The mean annual run-off is 160 x 10^6 m³. The lower reaches of the Bitou flow through a 1,5 km wide flood plain and become tidal and saline about 2 km above the national road bridge. The estuarine section is about 30 m wide and 1-2 m deep with a bed of sandy mud. The lower Keurbooms flows through a rocky gorge with no flood plain and the river becomes tidal and saline about 4 km above the national road bridge. The estuarine section increases in width from 40 to about 150 m and deepens to about 4 m with sand, mud and boulders on the bottom.

The bridges across both the Bitou and the Keurbooms and the remains of the old causeways severely restrict tidal flow. Above the bridges the tidal range is about 0,8 m at springs and below it is 1,1 m or more. Similarly the area of salt marsh above the Bitou bridge has been reduced. The estuary is best developed at the confluence below the bridges, where the channel is 3 m deep with the rich *Zostera* beds in the shallows. From here on, the estuary forms a sandy lagoon separated from the shore of Plettenberg Bay by a ridge of dunes which is breached by a narrow mouth. The mouth shifts over the years but remains permanently open.

Between the bridges and the mouth the salinity varies from about 13 to 30 ‰ with the rains; the temperature varies from 12 to 28 ° with the seasons and stratification is evident. The water is fairly clear with a Secchi depth of more than a metre.

The phytoplankton has not been studied. According to Professor J. Grindley the zooplankton is rich with a dry biomass of 2,9 to 108 mg.m⁻³. Catches with a 36 cm plankton net show *Pseudodiaptomus hessei* as the dominant copepod and it is particularly abundant in the reduced salinity above Keurbooms bridge. A variety of other copepods, amphipods, isopods, shrimps, mysids as well as the larvae of resident fishes and invertebrates are common in night plankton samples. The mysid *Gastrosaccus brevifissura* is common and, as mysids are known to avoid small nets, it may form the bulk of the planktonic biomass as in some other estuaries.

Below the bridges there are fairly rich *Zostera* beds and the marginal vegetation includes the usual Cape species; *Spartina maritima* is common but there are no extensive fields. Above Keurbooms bridge no submerged vegetation was found and marginal vegetation is scanty until *Phragmites* appears near the nead of the estuary. Above the Bitou bridge there are still

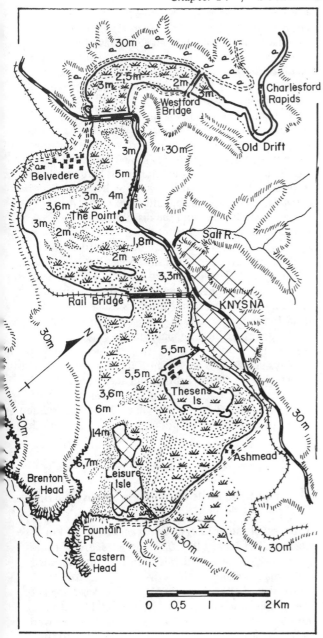

Figure 14.34
KNYSNA ESTUARY

extensive salt marshes dominated by *Sarcocornia* spp, *Juncus* and *Sporobolus*.

The benthic fauna is well-developed from the mouth to the bridges with *Callianassa* in sandy areas, *Arenicola* and *Solen capensis* in muddy sand, and *Upogebia* abundant in sandy mud. Many species harbour in the salt marsh vegetation and in the *Zostera* beds. *Assininea globulus, Sesarma catenata, Cleistostoma edwardsii* and *Nassarius kraussianus* are abundant and *Scylla serrata* is common. Above the Bitou bridge only *Callianassa* and *Sesarma* are common;

above Keurbooms bridge the benthic fauna is very poor.

The fish fauna is poorly known. Anglers report that the Keurbooms estuary is a fair but not rich angling resort. Large fish are rare but kob, grunter, white steenbras, elf, leervis, Cape stumpnose, dassies and wildeperd are taken. Mullet are common but not abundant. The angling above the bridges is disappointing.

The avifauna was not studied but waders are common near the confluence and duck, geese and coots frequent the Bitou marshes.

During 1978 most of the old causeway next to the bridge over the Keurbooms was excavated and residents report that conditions have improved although a flood is needed to scour out the channel. Improvements to the Bitou bridge are awaited. The National Road Board has promised that the bridges for the new national road will be supported on tall pillars and will not restrict tidal flow. It is hoped that the old bridges will be removed.

Knysna estuary
(34°04′S/23°03′E)

Thirteen of the reports listed at the end of the chapter deal with Knysna estuary; five are published and the rest are in typescript. They are grouped below by subject:

General ecology: Day et al (1952), Day (1964, 1967)
Sedimentation: Chunnet (1965)
Hydrology: Anderson (1976)
Trace metals: Watling & Watling (1975, 1976)
Oyster culture: Korringa (1956), Genade (1973)
Proposal for a marina: Anon. (1974)
Effect of proposed marina on ecology: Grindley (1976)
Method of evaluating financial effect: Chmelik et al (1974)
Effect of municipal sewage on ecology: Grindley (1978)
Methods of combatting oil pollution: Retief et al (1979)

Knysna River rises in the Outeniqua mountains and has a catchment of 526 km[2] with a rainfall of 922 mm spread fairly evenly through the year. The mean annual run-off is 110 x 10[6] m[3] and the river runs through one of the many gorges which dissect the Tsitsikama forest. Although the gradient is steep, erosion on wooded slopes is minimal so that the river water is clear though peat-stained and the flow is fairly uniform.

The estuary (Figure 14.34) is S-shaped, about 19 km long and gradually broadens and deepens to form a lagoon over 3 km wide and up to 6 m deep. The estuary is crossed by two road bridges and a rail bridge. All three have solid embankments which restrict tidal flow and encourage the deposition of sediments. The mouth of the estuary opens between two impressive rocky headlands which prevent longshore drift of sand so that marine sediments do not enter the lagoon. In fact Chunnett (1965) concludes that there has been little or no influx of sediment over the last 100 years. This means that the recent changes in the contours of the sand banks in the lower reaches and the mudbanks higher up are the result of engineering activities alone. At the head of the estuary, the channel has cut deeply into the flood plain but thereafter the banks become flatter and there are many intertidal salt marshes between Westford Bridge and Leisure Isle near the heads.

The spring tide range on the coast is 1,8 m and although there is some reduction in the broad lagoon, the maximum of 1,8 m persists at Belvedere and even increases slightly at the Charlesford Rapids which mark the head of the estuary (see chapter 3). At this point the water is well-stratified with salinities of about 0,4 ‰ on the surface and 5,1 ‰ on the bottom. However, the river discharge is minute compared with the tidal exchange at the mouth so that the isohalines rapidly become steeper (chapter 3). Salinities of 30-35 ‰ are recorded at the rail bridge and seaward of this point there is little indication of vertical gradients.

At the head of the estuary water temperatures range from a winter minimum of 13 ° to a summer maximum of 27,7 °. The stabilising effect of the sea becomes evident in the lagoon and at the mouth, the annual range is normally 15,3 to 22,4 °. Occasionally summer upwelling along the coast causes a sudden drop to 11,5 ° at the mouth and marine fish are either numbed or take refuge in the estuary.

Although the river water is peat-stained and turbidity increases in the upper reaches of the estuary, the water in the lagoon is sufficiently clear for the bottom to be visible at a depth of 2-3 m. Oxygen concentrations are high except near the outfall of the sewage effluent in the narrow channel north of Thesen's Island. Trace metal concentrations are low. In the same channel, nutrient concentrations are high and the marginal vegetation is lush as might be expected. Elsewhere in the estuary, nitrate nitrogen decreases from about 2,5 μg at NO_3-$N\ell^{-1}$ at the mouth to about 0,9 μg at ℓ^{-1} in the upper reaches during the winter. In the autumn the reverse is true and the nitrate concentration increases through the same range from the mouth to the upper reaches. The reason for these seasonal changes requires further investigation. Dissolved phosphate values remain more or less constant at about 1,2 μg at PO_4-$P\ell^{-1}$ throughout the estuary. Presumably this is due to the 'buffering' action of the bottom sediment (chapter 5).

The phytoplankton biomass has not been investigated but the clarity of the water suggests that it is low. The attached vegetation however is rich. The rocky banks at the heads which are washed by low swells are colonised by an unusually wide variety of algal macrophytes. Within the estuary there are exten-sive *Zostera* beds at and below low tide, and broad salt marshes at higher levels. *Spartina maritima* is abundant and all the other Cape species of vascular plants as well. Details are discussed in chapter 6. Estimates of biomass made by Grindley (1978) and discussed in chapter 16, are high but primary production has not been measured.

The benthic macrofauna includes 310 species with rich beds of *Upogebia, Arenicola,* also *Atrina squamifera, Solen corneus,* and other bivalves. The gastropods *Assiminea globulus, Hydrobia* sp and *Nassarius kraussianus* are abundant. There are numerous amphipods and isopods and the salt marsh vegetation harbours *Sesarma catenata, Cleistostoma* spp and other crabs. A list of characteristic species and the distribution of the benthic fauna is given in chapter 9.

Over 50 species of fish have been identified; most of them are juveniles of marine species but anglers also take large kob, leervis, spotted grunter, white steenbras, elf and even stenohaline marine fish such as *Sparodon durbanensis* on occasion. Forage fish such as *Gilchristella, Hepsetia breviceps* and three or four species of mullet are plentiful. The main species are listed in chapter 12.

Knysna lagoon is very beautiful and a popular tourist resort with excellent yachting and angling facilities. It is a national asset and planners should ensure that it remains so. Retief *et al* (1979) have already considered means of combatting oil pollution drifting in from the sea.

The Great Brak, Little Brak and Hartenbos estuaries (34°03,5'S/22°46,6'E)

Three small estuaries open between 10 and 30 km north-east of Mossel Bay. There is no record of ecological work on these estuaries apart from unpublished notes made during a visit by six biologists from the University of Cape Town in May 1950. A catalogue of the species of animals collected is available in the Zoology Department. In 1977 all three estuaries were polluted with oil drifting in from the sea due to the Venpet-Venoil collision. It is known that there was conserious destruction of the fauna and that the pollution was monitored (Moldan *et al,* 1979).

The environmental conditions in the three estuaries are set out in Table 14.5.

The Great Brak (Figure 14.35A) meanders through a broad flood plain, passes under the national road bridge and the rail bridge and then broadens to form a lagoon about 1 km wide. The whole estuary is tidal and the mouth is permanently open but narrow at low tide where it passes between a reef on the east bank and a sandspit on the west. Collections were made between the mouth and the road bridge and again at the head of the estuary.

The estuarine vegetation was not studied although

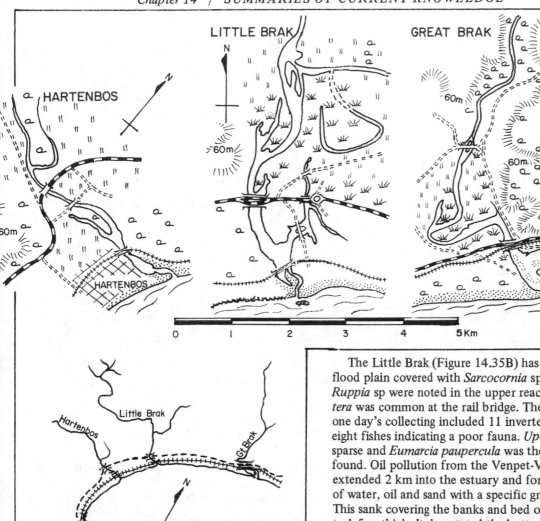

Figure 14.35
GREAT BRAK, LITTLE BRAK AND HARTENBOS
ESTUARIES

it was noted that the flood plain was covered with salt marsh (mainly *Sarcocornia* sp) and that there were well-developed *Zostera* beds. The aquatic fauna was rich and 90 species of invertebrates and 15 fishes were recorded. The usual Cape species were present. Both *Callianassa* and *Upogebia* were abundant, *Loripes clausus* was surprisingly common and, at the head of the estuary, *Palaemon capensis* the fresh water shrimp was recorded in an estuary for the first time. The fish *Abudefduf sordidus* was a long way south of its normal range.

When oil pollution entered the estuary from the sea most of the crabs and prawns were smothered in their burrows by chocolate mousse.

The Little Brak (Figure 14.35B) has an even wider flood plain covered with *Sarcocornia* sp. Reeds and *Ruppia* sp were noted in the upper reaches and *Zostera* was common at the rail bridge. The records of one day's collecting included 11 invertebrates and eight fishes indicating a poor fauna. *Upogebia* was sparse and *Eumarcia paupercula* was the only bivalve found. Oil pollution from the Venpet-Venoil collision extended 2 km into the estuary and formed a mixture of water, oil and sand with a specific gravity of 1,067. This sank covering the banks and bed of the estuary up to 1,5 m thick. It devastated the bottom fauna but did not affect the salt marsh vegetation (Moldan *et al*, 1979).

The Hartenbos is reported to close during dry months and as noted above, the salinity was very high in May. The whole estuary (Figure 14.35C) is shallow apart from a deep pool below the rail bridge. Like the other two estuaries, the broad flood plain is covered with salt marsh vegetation but no submerged vegetation was found. Collecting revealed only one shrimp, one water beetle and the mullet *Mugil cephalus*.

Table 14.5
Environmental conditions in the Great Brak, Little Brak and Hartenbos estuaries

	Great Brak	Little Brak	Harten-bos
Catchment (km²)	192	552	207
Annual run-off (x10⁶ m³)	29	45	5
Length of estuary (km)	6,5	7	8
Width of channel (m)	200-1 000	50	80-90
Depth of lower reaches (m)	2-4	2	1-3
Conditions in May 1950:			
surface temperature (°C)	18	18	17,5-18
salinity near mouth (‰)	33,7(LW)	34,1	38
salinity near head (‰)		6,0	41,8

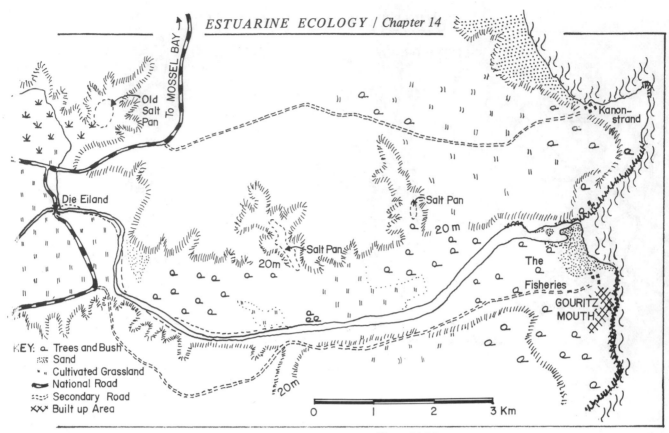

Figure 14.36
GOURITZ ESTUARY

The Gouritz estuary
(34°21'S/21°48'E)

Very little ecological information is available on the estuary of this large river and what follows is based on brief notes and an annotated sketch of the mouth made by Mr van Wyk of the Cape Department of Nature Conservation in January 1957. Further observations are urgently required.

The Gouritz river and its tributaries drains much of the Karoo. The catchment is 45 742 km² in area much of which has an annual rainfall of 200 mm and a potential evaporation rate of 1 500-2 000 mm per year so that the mean annual run-off is only 744 x 10⁶ m³. Moreover, much of the drainage basin is covered with silt and clay derived from the Karoo series of slates and shales so that the run-off carries muddy waters into the river. This has cut a deep ravine through the coastal plain but the valley widens to more than a kilometre as it approaches the sea.

The length of the saline intrusion and the tidal reach of the river has not been determined, but is probably of the order of 10 km. The estuary (Figure 14.36) gradually widens as it approaches the sea and is about 180 m wide for the last kilometre. The main channel runs close to the rocky north bank while the south bank shelves gradually to form sandflats. The mouth itself is constricted by a sandspit.

In January 1957, notes were made at a station about 0,8 km from the mouth. Near the time of low tide the salinity was 14,7 ‰, the temperature 24,5° and the Secchi reading was 38 cm. *Upogebia* was reported to be abundant in the muddy sandflats near the mouth, but there is no further biological information.

Breede River estuary
(34°24'S/20°50'E)

The only published account of this estuary is a report by the Department of Planning and the Environment in 1976. The present account is based on unpublished notes in the Department of Zoology of the University of Cape Town supplemented by further hydrological data recorded by the Cape Department of Nature and Environmental Conservation.

The Breede has been visited by parties of biologists from the University of Cape Town on several occasions between 1951 and 1980; four times in summer and three times in winter. Reference to conditions in the Breede estuary will be found in chapters 2 and 3 and the following account summarises the main observations.

The Breede River drains part of the mountainous winter rainfall area between Ceres, Worcester and Swellendam and then winds for 80 km over the dry

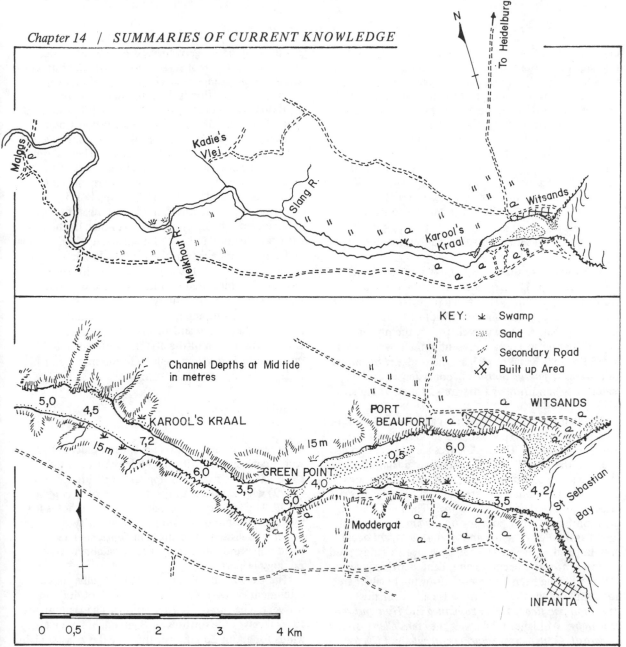

Figure 14.37
BREEDE ESTUARY

coastal plain to the village of Witsands. The catchment is 12 587 km² in area but only the mountainous part has a high rainfall. The river flows strongly in winter but even in summer pulses of brown berg water are felt in the estuary. The gradient over the coastal plain is so gentle, and the banks of the river are so steep that the river is tidal for about 50 km (Figure 14.37). Also it is more than 3 m deep and was once used by small coasters carrying farm produce to Cape Town. Since then the sandbar at the mouth has grown and, as noted by the Department of Planning and the Environment, it is now so dangerous that even small fishing boats must wait for a calm day and high tide to enter the estuary.

The saline intrusion varies in length. Residents report that in dry summers salt water and marine fish extend up to Malgas, 40 km from the mouth but when visited by ecologists only low salinities and a few brack water species were found. Normally, however, salinities are only appreciable within 16 km of the mouth in summer and during the heavy winter floods the whole estuary may be fresh from surface to bottom.

Environmental conditions in the estuary are briefly as follows. The banks are steep and largely composed of shales which erode to form clay and soft mud; the intertidal zone is narrow with soft mud or rocky outcrops except for the last 2 km where marine sands form shoals and extensive flats on the south bank. The spring tide range is about 1,7 m at the mouth, decreasing slowly to 1,2 m at Karool's Kraal where the low tide lag is 1,7 hours. Hydrological conditions change markedly with the seasons and the rains in the mountain catchment. In summer, the means of all salinity records from surface to bottom at high and low tide are: Port Beaufort 35 ‰, Karool's Kraal 25 ‰, Malgas 1,0 ‰; winter values are: Port Beaufort 17,7 ‰, Karool's Kraal 5 ‰, Malgas 0,0 ‰. Water temperatures for summer/winter are: Mouth 22/13 °, Port Beaufort 22,0/15,2 °, Karool's Kraal 22,6/13,1 °, Malgas 24/12,5 °. On a rising tide the clean sea water flows along the south bank at Port Beaufort and the contrast with turbid estuary water towards the north bank is sharply defined; vertical salinity gradients are also well-developed.

The aquatic vegetation is not extensive since the banks are steep. The largest salt marsh is at Green Point; elsewhere there is only a narrow fringe. *Zostera* beds are well-grown from Port Beaufort to above Karool's Kraal. As shown in chapter 6, the normal Cape estuarine species are present in the lower reaches. From Karool's Kraal to Malgas, the reeds *(Phragmites)* and rushes *(Cyperus textilis)* become denser at the water's edge and *Juncus* tufts are scattered at higher levels.

The benthic fauna is fairly rich and about 80 invertebrates have been recorded. The main species are listed in chapter 9. The rock ledges are not well-populated for the shale is always covered by a film of mud. Barnacles and *Siphonaria* spp are abundant and so is *Ostrea algoensis* but *Saccostrea (Crassostrea) margaritacea* is collected as fast as it grows. The sand flats at the mouth are rather poor for no *Callianassa* has been found and *Solen capensis* and *Arenicola loveni* occur only in patches. Where the banks change to sandy mud, *Upogebia africana* becomes abundant, densities of 400 m^{-2} having been recorded; burrowing bivalves are not common but the *Zostera* beds harbour many *Palaemon pacificus*, *Melita zeylanica* and *Hymenosoma orbiculare*. At higher tidal elvels the crabs *Cleistostoma edwardsii*, *C. algoense*, *Sesarma catenata* and *Cyclograpsus punctotus* are abundant under cover of salt marsh vegetation and talitrid amphipods harbour under drift weed. Small *Dotilla fenestrata* (a tropical crab) are sometimes common on sandbanks.

The Breede estuary is famous as an angling resort and many record catches of kob, white steenbras, grunter and leervis have been made there. To date, 34 species of fish have been recorded including such unusual estuarine catches as *Sparodon durbanensis* and *Seriola lalandii*. Large sharks occasionally take hooked fish and one or two up to 3 m long have been landed but have not been identified. The weight records at the Lucky Strike Botel are: kob 61 kg, leervis 27 kg, white steenbras 7 kg, grunter 4 kg. Kob have been caught all the way from the sea to Malgas but the best stretch for other sport fish is between Moddergat and Karool's Kraal for the extension of marine sand has largely covered the *Upogebia* beds lower down. The best season is late summer but white steenbras are common all through the year, except when the winter floods drive the fish seawards.

Extensive records of the vertical distribution of the macrofauna on the banks and the size and stomach contents of small fish are kept in the Zoology Department of the University of Cape Town. These should form the basis of a more complete account of the estuary. Witsands and Port Beaufort are unspoilt tourist resorts and would develop rapidly if fresh water supplies were not limited to rainwater tanks. A large dam is being built to divert much of the run-off from the mountain catchment to supply municipal and agricultural needs in the Western Province. There is good evidence that existing dams and irrigation works have already increased salinities and the penetration of marine sand into the estuary.

Further developments along the same lines will have both beneficial and adverse effects. A more detailed discussion of the effects of human activities on estuarine ecosystems will be found in chapter 17.

Heuningnes estuary
(34° 43′S/20° 07′E)

The only published reports are those of Mehl (1973) which deals mainly with the biology of the white steenbras *Lithognathus lithognathus* and Barham (1979) which deals with the spawning of *Arenicola loveni*. The following summary is based on notes made by a party of biologists from the University of Cape Town who visited the estuary in September 1973. Field notes and faunistic lists are available in the Zoology Department.

The Heuningnes River near Cape Agulhas has a catchment of 1 401 km^2 with a rainfall of 400 mm per year. The run-off is 78 x 10^6 m^3. The Heuningnes River receives the outflow of Soetendahl's Vlei and winds for about 20 km over the very flat coastal plain. The soil is sandy with limestone outcrops; part is under cultivation and part is covered with fynbos or coastal scrub. The estuary (Figure 14.38) is of interest in being the southernmost in Africa and in being closed for months at a time.

At the Struisbaai road bridge, some 12 km from the mouth, the estuary flows between steep banks and is about 20 m wide and up to 2 m deep. From there on to within a kilometre from the sea the estuary slowly widens to about 100 m but the depth remains much the same. Near the mouth there is a stony ford about 0,5 m deep below which the estuary broadens between salt marshes to form a shallow lagoon called Die Mond which is up to 300 m wide before reaching the coastal sand dunes. Monthly records made by Dr Mehl in the lagoon during 1971 give the following seasonal means:

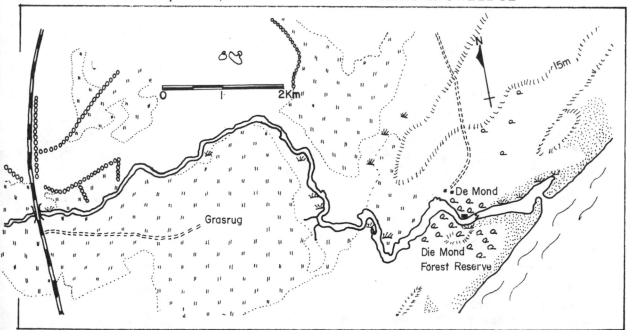

Figure 14.38
HEUNINGNES ESTUARY

	Summer	Autumn	Winter	Spring
Salinity (‰)	37,6	36,4	24,1	33,6
Surface temp. (°C)	19,8	17,5	13,3	17,3

The mouth is usually closed but opens after good rains. Weekly records made by the Cape Department of Nature and Environmental Conservation in 1973 showed that the mouth was closed from January and the salinity rose to about 50 ‰ in June. University of Cape Town observations in September showed that the salinity in the lagoon was 38,7 ‰ and at the Struisbaai bridge it was 37,8 ‰. Surface temperatures at four points between these limits varied between 13,9 and 14,8 °. The estuary obviously extends upstream of the bridge.

Neither the plankton nor the estuarine vegetation was investigated but it was noted that neither *Zostera* nor *Spartina* was present. This agrees with observations in most blind estuaries.

A total of 18 species of macro-invertebrates and 14 fishes were recorded, most of them from the lower reaches. Neither *Upogebia* nor burrowing bivalves were found but *Arenicola loveni* is locally common and *Callianassa* is abundant in the lagoon. Barham (1979) has shown that the gonads of *Arenicola* start growing in June-August and spawning reaches a peak in February. One surprising record was the tropical gastropod *Nerita albicilla* which had not been recorded south of Knysna estuary. The Department of Nature Conservation also had single records of *Scylla serrata* and *Penaeus japonicus.* Fish were quite plentiful and included the usual sport fishes such as kob, white steenbras, grunter, stumpnose *(Rhabdosargus globiceps),* dassies and leervis. Mullet *Mugil (Liza) richard-*

soni and *Mugil cephalus* were common and so too were *Gilchristella* and *Hepsetia breviceps.*

Klein River Lagoon, Hermanus
(34° 25'S/19° 18'E)

The following summary is based on an account of the general ecology of the lagoon by Scott, Harrison & Macnae (1952) supplemented by notes on the fishes by Talbot (1955). A few observations made by myself in January 1979 have been incorporated.

The Klein River drains 740-870 km[2] in the Caledon district which has an average rainfall of 510 mm per year. The mean annual run-off is 6×10^6 m³. December, January and February are the driest months. The estuary (Figure 14.39) is 12,8 km long and forms a beautiful lagoon 1-1,5 km wide and up to 3 m deep. It is used for yachting and angling by residents of Hermanus and Cape Town.

The north bank is formed by the slopes of the Hermanus mountains and the south bank is formed by a low-lying plain. There are outcrops of shale or sandy limestone on both banks and the estuarine sediments grade from sand or gravel with a film of silt in the lower reaches, to mud with occasional sandy beaches near the top of the lagoon. When the mouth is closed and tidal action is inhibited, the mud becomes anoxic and smells of H_2S when dug.

Normally the mouth is closed by a sandbank but it is open in winter and if the rains are good the mouth may remain open for months; it was still open in February 1980. Sometimes it is opened by farmers when the rise in level threatens to flood their lands, and when opened, the impounded waters rapidly erode a

311

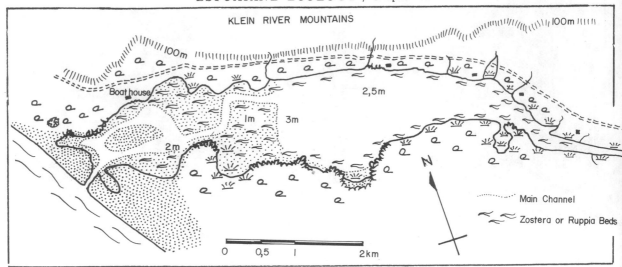

KLEIN RIVER MOUNTAINS

100m

Boat house

2,5m

1m 3m

2m

N

······· Main Channel

Zostera or Ruppia Beds

0 0,5 1 2km

Figure 14.39

KLEIN RIVER ESTUARY (HERMANUS LAGOON)

broad channel into Walker Bay. The level drops about 1 m within a day and the estuary becomes tidal. When closed, circulation is maintained by the wind so that the bottom water is well-oxygenated. The seasonal temperature range is 12 to 28 ° and as the discharge of the river and precipitation on the surface is approximately balanced by evaporation, salinities remain within the normal estuarine range. Only in very dry summers do salinities in the lower reaches reach 40 ‰. Usually there is a normal salinity gradient down the estuary and vertical stratification occurs in the upper reaches.

The phytoplankton has not been investigated. Of the macrophytes, *Zostera* is abundant over most of the lagoon and is accompanied and eventually replaced by *Ruppia* in the upper reaches. Both plants are heavily epiphytised by diatoms and filamentous algae. *Enteromorpha* grows densely in shallow coves and when dislodged by the wind, the drifting fronds are cast ashore and provide food and refuge for thousands of small crustacea.

As in other blind estuaries, *Spartina* is absent. Species of *Cotula* and *Triglochin* are common at the water's edge, *Sarcocornia* spp extends to higher levels and the grass *Sporobolus* forms a sward in which *Juncus* and several other halophytes occur. Reeds and sedges *(Phragmites* and *Cyperus textilus)* fringe the banks of the upper reaches and extend along the banks of the Klein Rivier.

In comparison with a permanently open estuary, the benthic fauna is impoverished. From the account by Scott *et al*, it is evident that many species including stenohaline invertebrates such as *Parechinus angulosus*, *Marthasterias glacialis* and dorid nudibranchs colonise the lower reaches when the mouth is open but die within a month after the mouth closes. A few euryhaline species such as *Arenicola loveni* and *Echinocardium cordatum* survive buried in the sandbar at the mouth where seepage from the sea must occur. Over

the rest of the lagoon the main forms at the water's edge are *Assiminea globulus*, talitrid amphipods, *Deto echinata* and *Ligia dilatata*. *Cyclograpsus* is present but not *Sesarma catenata* since it does not extend west of Cape Agulhas. In the shallows, the main epifauna is *Melita zeylanica*, *Exosphaeroma hylecoetes* and *Hymenosoma orbiculare*. *Callianassa kraussi* is the main burrowing form. All of these are common in fish stomachs and Talbot (1955) who studied the feeding habits of juvenile *Rhabdosargus globiceps*, found that shoals tend to concentrate on one food organism and then another or pack their stomach with the fronds of *Zostera* or *Ruppia*. In most estuaries in South Africa, the prawn *Upogebia africana* is one of the most important food organisms in the diet of demersal fishes. In Hermanus lagoon it is entirely absent and is replaced by *Callianassa* which then burrows in muddy as well as sandy areas.

Many species of aquatic insects have been reported, far more than in other South African estuaries, Possibly this was due to the enthusiasm of Dr Marjorie Scott and Dr Arthur Harrison who collected extensively in the upper reaches of the estuary.

Twenty-six species of fish have been reported and one specimen of *Pomadasys commersonni* was sent to me by Mr I. Williams. The common sport fish include kob, white steenbras, leervis and white stumpnose. The abundant forage fishes are *Gilchristella*, *Hepetia breviceps*, *Mugil cephalus* and *Liza richardsoni*. However, Hermanus Lagoon is better known for its yachting than its angling.

Sandvlei Lagoon

The ecology of Sandvlei Lagoon at Muizenberg has been studied by three zoology students: Miss M. Bourgeois (1948), Mr David Muir (1974) and Mr P. Skelton (1975). Their notes are being elaborated and presented as a thesis to the University of Cape Town. An account of the fish fauna has been published by Begg (1976). The following summary is based on these reports, on personal communications from officers of the City Engineer of Cape Town and on personal observations.

Three small rivers, the Sand, the Tokai and the Keysers flow into Sandvlei Lagoon. Their combined catchments cover about km² of mountain side, lowland and growing suburbs. The run-off is estimated as There is slight pollution from rural factories and stormwater drains but coliform counts decrease in salinities above 8-12 ‰.

The estuary (Figure 14.40) is 2,5 km long, 600 m at its widest point and 1-3 m deep. It tapers to a channel which winds through Muizenberg where it is blocked by a stony weir and a sandbar so that the level is stabilized. The estuary had long since been modified by dredging and canalisation of the channel and since 1974 a marina has been developed by cutting a series of key-shaped channels whose steep banks have been hardened with a cement mixture. As the water level rises during the winter rains, the sandbar bursts and the water level falls to some extent but a minimum depth is maintained by the weir. Some sea water enters at high tide but the tidal range is limited to about 7 cm. When the water level in the marina falls during summer it was planned that sea water should be pumped into the lagoon but this has not occurred for the marina has not been financially rewarding.

As will be obvious, environmental conditions in the lagoon are artificial. The banks are steep and mainly hardened, and the bottom sediments are sandy with very little silt but there is a fair amount of organic matter, increasing from 0,8 % to over 6 % at the head of the estuary which is marshy at the mouths of the rivers. The salinity is fairly constant for most of the year ranging from 10 ‰ at the head to 15 ‰ at the weir except in spring when it rises to between 25 and 32 ‰. Temperatures vary from 12-18 ° in winter to 15-21 ° in summer and the pH is between 8 and 9. The oxygen concentration is normally high but at the bottom it falls to about 50 % saturation in summer. Nutrient concentrations are high with 240 μg $NH_3.\ell^{-1}$, 620 μg $NO_3.\ell^{-1}$ and 50 μg $PO_4.\ell^{-1}$, due to seepage from sewage ponds nearby.

Under these eutrophic conditions it is not surprising that some aquatic plants are abundant. *Zostera* is entirely absent but occasionally there are dense blooms of nanoplankton and one toxic bloom of *Prymnesium parvum* is reported to have poisoned fish and many invertebrates (Begg, 1976). In summer, dense growths of *Potamogeton pectinatus* hamper boating in the upper

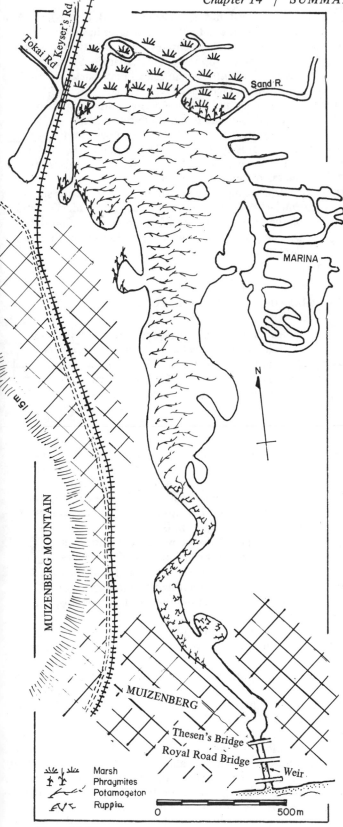

Figure 14.40
SANDVLEI LAGOON

Tokai Rd
Keyser's Rd
Sand R.
MARINA
N
15m
MUIZENBERG MOUNTAIN
MUIZENBERG
Thesen's Bridge
Royal Road Bridge
Weir

Marsh
Phragmites
Potamogeton
Ruppia
0   500m

half of the lagoon and *Ruppia maritima* extends down to the channel. Both aquatic grasses are 'mowed' when they get too dense and in any case the shoots die naturally in the winter. The blue-green *Lyngyba* carpets the shallows and *Enteromorpha intestinalis* grows luxuriantly in sheltered coves. All loose vegetation so formed drifts with the wind and rots on the banks. The stabilized water level, however, restricts the growth of emergent vegetation. *Spartina* is absent and *Sarcocornia, Cotula, Chenolea, Juncus* and *Limonium* are rare. On the other hand, *Phragmites australis* fringes the coves and backwaters and swards of *Sporobolus* grow to the water's edge where the banks are not cemented. *Triglochin* is abundant at the head of the estuary where the banks are muddy and low-lying.

The fauna is very poor. The only abundant large invertebrate is the burrowing prawn *Callianassa kraussi*

and up to 576 m^{-2} have been recorded. *Upogebia africana* is absent and burrowing bivalves have not been reported. Most invertebrates are small arthropods such as amphipods, isopods and insect larvae, the snail *Tommichia ventricosa* and small polychaete worms. In all, 28 species have been recorded compared with 129 in Hermanus and 310 in Knysna, although some of the species present are in dense concentrations.

Only eight species of fish are known and the angling is poor. This appears to be largely due to the sandbar and the weir which seldom allows post-larval fish from the sea to enter the lagoon.

Aquatic birds are not abundant. Fish-eating species find richer feeding areas elsewhere and waders depend on the fall of the tides to feed on exposed sand flats. Dredging of the shallows and stabilization of the water level inhibits this.

C. THE COLD-TEMPERATE ESTUARIES OF THE ATLANTIC COAST

The mouth of the Black (or Salt) River
(33°54'S/18°28'E)

A brief account of the Black River based on internal reports by NIWR officers edited by Oliff (1976) has been published by Cloete & Oliff (1976). In addition, the annual report of the City Engineer of Cape Town contains useful data on water chemistry. In 1971, three civil engineering students of the University of Cape Town, Messrs D.G. Banks, R.A. Ferrugia and R. McNamara recorded physical and chemical parameters in the river and its tributaries during October-November, while Mr B. Campbell (1971) reported on the fresh water fauna and Scarfe (1972) described the ecology of the river mouth and adjacent sandy beaches in an unpublished Zoology Honours project.

The four tributaries of the Black River have a total catchment of 113 km^2 in the densely populated northern suburbs of Cape Town and the Cape Flats. The area has a good winter rainfall of 400-800 mm per year and the mean annual run-off is 38 x 10^6 m^3. The Black River receives the effluents of the maturation ponds of the Athlone Sewage works (which may be overloaded at times) and the wastes from the industrial areas of Ndabeni and Paarden Island. Much of the river is canalised as shown in Figure 14.41. Thus the river is banks is sprayed with herbicides, algicides and insecticides when necessary. Traces of these biocides as well as hydrocarbons, oils, and other organic compounds and high levels of ammonia (13,5 mg.ℓ^{-1}), nitrate (2,8 mg. ℓ^{-1}) and SRP (2,31 mg.ℓ^{-1}) have been found in the water. Cloete & Oliff (1976) record OA levels of 0,3 mg.g^{-1} (30 times normal) at the river mouth. DO levels are fairly high and Hg levels are normal. The bacterial load of the river is high but *E. coli* dies rapidly in the sea; nonetheless an MPN count of 200 per 100 mℓ

indicated substantial sewage pollution had occurred.

Scarfe (1972) deals with the salinity, temperature, DO and COD values at the mouth and in adjacent tidal waters during June-July 1972 and discusses the density and species richness of the benthic fauna. His sampling stations in the very shallow canalised river extended 300 m from the sea and within this range the salinity had a mean value of 8 ‰ and a tidal variation of 1-28,5 ‰. In the sea, the decreased salinities showed that the plume of river water could be detected 120 m along the shore to the north. The length and direction taken by the plume must vary with the discharge of the river, the tide and currents in Table Bay. Cloete & Oliff report a length of 500 m.

Scarfe (1972) found no benthic animals in the final 300 m of the river canal but found 12 species of psammophilic animals on the sandy shores of the Bay. The number of species increased to a maximum of 12 away from the river mouth and the number of individuals increased as well. Scarfe concludes that the river water is toxic, and suggests the very variable salinity as an important factor. It is likely that the chemical composition of the water is much more important. Cloete & Oliff sampled the benthic fauna in the sea at a depth of 6 m at a distance of 150 m from the mouth. The northerly drift of the river plume was again shown by the density and nature of the fauna.

The Black River is obviously polluted. Its water is toxic but not anoxic for the percentage saturation has not been reported as falling below 47 %. Presumably the maturation ponds of the sewage works decrease the BOD load of the domestic sewage effluent. Nonetheless, the nutrient level of the effluent is high. Eutrophic conditions could not develop in the river in the presence of the biocides used. The nutrients are discharged to the sea but the Benguela current water

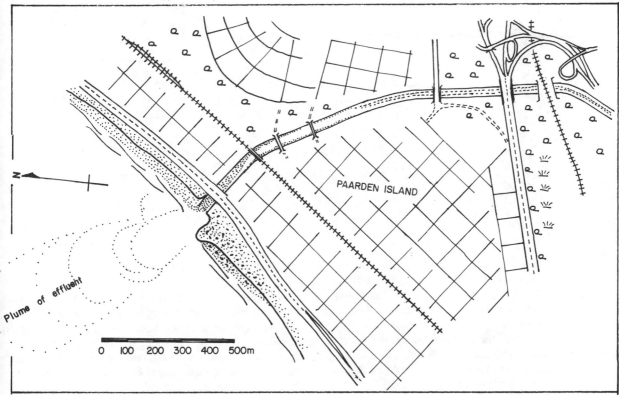

Figure 14.41
BLACK RIVER ESTUARY

in Table Bay is already rich in nutrients and it is unlikely that the effluent has any significant effect on primary production.

Seasonal changes in pollution have not been correlated with river flow which must obviously be important. As yet, there is no report on the concentration of toxic metals apart from mercury. Much remains to be done.

Milnerton Lagoon
(33°56'S/18°28'E)
The ecology of the estuary and Rietvlei which drains into it, was studied by Millard & Scott (1954); Scott (1954) also gave an account of the avifauna. More recently, when a marina was planned in part of Rietvlei, the history of Diep River, the vlei and the estuary was studied (Beaumont & Heydenrych, 1979).

Milnerton Lagoon (Figure 14.42) is formed where the Diep River flows into Table Bay. Dieprivier was so named because the mouth was permanently open up to 1888 and the river was deep enough for sailing barges to carry produce from farms 13 km up the river to Table Bay. Bad farming caused erosion of the sandy soil and progressive siltation of the river, and the weir built across the mouth in 1928 no doubt reduced the scouring action of the tides and winter floods. The

building of Blaauwberg bridge between Rietvlei and Milnerton Lagoon with its solid embankments up to the main channel, caused further siltation. The river now runs dry during the summer and much of Rietvlei becomes a saltpan.

The area of the Diep River catchment is 246 km², the annual rainfall is 593 mm and the mean annual run-off is 43 x 10⁶ m³. During the winter when the river is in flood, it inundates Rietvlei and the saline marshes south of Blaauwberg bridge. It then fills Milterton Lagoon until it breaches the sandbank blocking the mouth and flushes the salt water out of the estuary. Tidal conditions are then established with a rise and fall of about 20 cm at springs and saline water extends up the channel again.

The estuary is about 3,5 km long and extends beyond King George Port to the saline marshes. At the Old bridge the width is 200 m and the depth is about 1 m. The banks and bed of the estuary are soft mud, changing to sand at the mouth.

Hydrological conditions are very variable. During winter the mouth remains open and at the Old bridge halfway along the estuary, the salinity is about 4 ‰, the temperature is 11,0 ° and the Secchi depth is about 50 cm. In spring when the river stops flowing, evaporation proceeds rapidly in the broad saline marshes at the head of the estuary, sea water flows into the mouth at high tide, is carried up the channel and evaporates as it goes. Eventually the saline marshes become hypersaline. A reversed salinity gradient is thus established

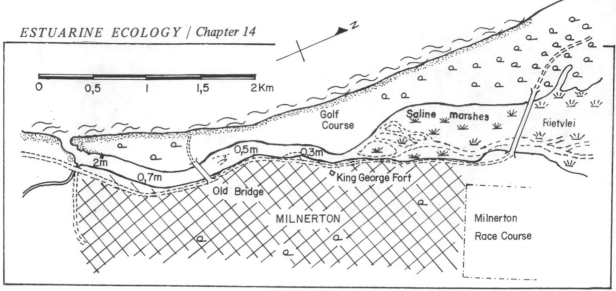

Figure 14.42
MILNERTON LAGOON

in summer with salinities rising to more than 350 ‰ before the marsh dries up and salt crystallizes out on the surface of the mud. Lower down the estuary, salinities never reach these extreme values; at the Old bridge they rise to 18 ‰ in December and 50 ‰ in March, and then, with the first good rains, the water becomes stratified with surface salinities of about 8 ‰. Summer temperatures reach a maximum of 24,3 ° and winter temperatures fall to a mean of 11,0 °.

As might be expected in an estuary that is both cut off from the sea and hypersaline in summer, both the flora and the fauna are restricted. Both *Zostera* and *Spartina* are absent. The submerged vegetation includes *Enteromorpha, Cladophora* and *Lyngyba* in the lower reaches, changing to *Ruppia maritima* and *Zannichellia Aschersoniana* in the saline marshes. Both of the latter die down when the salinity exceeds 65 ‰. The marginal vegetation includes *Triglochin bulbosum, Cotula coronopifolia* and *Scirpus maritimus* which form a lower belt and *Sporobolus virginicus, Sarcocornia* spp and *Juncus Kraussii* which form an upper belt. This grades into the surrounding scrub.

The whole fauna with its seasonal changes is listed by Millard & Scott (1954). When Rietvlei and the saline marshes are inundated by the winter rains, halophytic fresh water forms appear rapidly, particularly insects, entomostraca and snails. These become very abundant in spring and provide food for small flocks of waders in the saline marshes and enormous flocks in Rietvlei. All disappear when the water dries up in summer; the insects migrate while the snails and crabs burrow and aestivate in the damp mud and below the salt-encrusted surface layer.

In the channel between King George Fort and the mouth, water is present throughout the year although the salinity increases to more than 55 ‰ in autumn. The invertebrates include about 18 widespread estuarine species and a dozen or more insects. *Ficopomatus enigmatica* is the most conspicuous serpulid, forming coralliform masses on pylons and even on compacted mud. *Callianassa* is abundant in the sands at the mouth but *Upogebia* and burrowing bivalves are absent. The fish are all small and only 12 species have been reported. These include two fresh water species *(Galaxias punctifer* and *Sandelia capensis),* two gobies *(Caffrogobius nudiceps* and *Psammogobius knysnaensis)* and eight species of juvenile marine fish

The poverty of the fauna is a reflection of the extreme variation of salinity, the periodic closure of the mouth and the low summer temperature of the sea from which estuarine larvae are normally recruited.

Langebaan Lagoon
(33°06'S/18°01'E)

In an early paper, Day (1959) outlined the ecology of the Lagoon stressing that it is not an estuary but a sheltered arm of the sea without significant fresh water inflow. Nonetheless, the list of species includes many estuarine forms. Liversidge *et al* (1958) and later Pringle & Cooper (1975) described the rich avifauna and the importance of the Lagoon for palearctic migrants. Mostert (1972) reported that while there are good possibilities for mussel culture in Saldanha Bay, this does not hold for Langebaan Lagoon where the density of phytoplankton is greatly reduced.

At this stage it was announced that a deep water harbour was to be built in Saldanha Bay with facilities for loading iron ore. This would mean a large scale industrial development in the area and the danger that trace metals contained in the iron ore might reach Langebaan Lagoon. This possibility encouraged the Department of Planning and the Environment to finance a wide range of researches so that base line conditions would be known. Further developments could also be planned so as to minimise pollution in Saldanha Bay and Langebaan Lagoon.

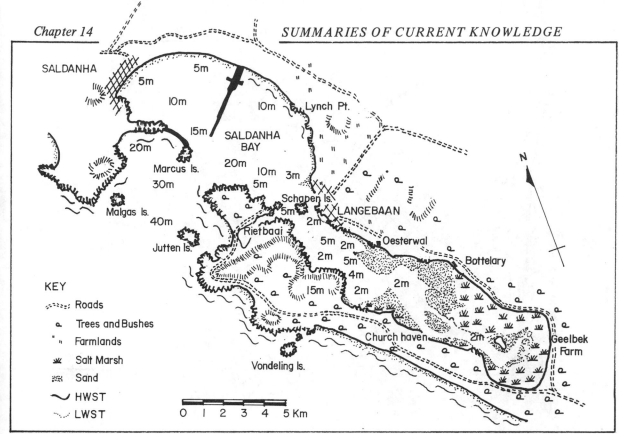

Figure 14.43
SALDANHA BAY AND LANGEBAAN LAGOON

Watling and Watling (1974) and Fourie (1975, 1976) determined the concentrations of trace metals and showed there was no evidence of pollution prior to development. Subsequently 17 papers were presented at a symposium held at Saldanha Bay. These were published in 1977 as volume 42 of the Transactions of the Royal Society of South Africa. They include most aspects of the natural history of the Bay and Lagoon and may be grouped as follows:

4 papers on geology, geophysics and geochemistry – Birch, Flemming, Du Plessis and de la Cruz, Willis *et al*

1 paper on hydrology – Shannon & Stander.

2 papers on phytoplankton and zooplankton – Henry *et al*, Grindley.

2 papers on macrophytes – Simons, Boucher & Jarman.

2 papers on macrobenthos – Christie & Moldan, Puttick.

1 paper on aquatic birds – Summers.

1 paper on conservation – Hey.

In an important paper, Flemming (1977) has since given a detailed account of the hydraulic processes which determine the nature and distribution of the sediments. More recently Ansell (1979) has constructed mathematical models of the Langebaan Lagoon ecosystem, Lucas (1979) has estimated the growth, metabolism and energetics of the bivalve *Mactra* and Puttick

(1979) has determined the feeding ecology and energetics of the curlew sandpiper.

Langebaan Lagoon (Figure 14.43) is a bottle-shaped arm of Saldanha Bay, some 16 km long, 2-3 km wide and up to 6 m deep. It is completely sheltered from wave action. The spring tide range is reduced from 1,7 m at the mouth to 1,4 m at the southern end and the strong tidal currents, which may exceed 100 cm. sec^{-1} in the entrance channel, decrease to less than half this velocity in the southern part of the lagoon. These currents and the wind-generated waves control the distribution of sediments. The banks of the channel are mainly rocky and there is a subsurface layer of calcrete and oyster shells over the whole lagoon. Above this the shoals and intertidal flats grade from medium sand in the channel to fine sand in the middle of the lagoon while at the southern end there is an increasing percentage of organic matter.

Although the salinity remains between 34 and 36 ‰ throughout, other hydrological conditions change along the length of the lagoon. The mathematical models reported by Ansell (1979) as dealing with the whole ecosystem of the lagoon in fact deal only with the plankton. As the density of phytoplankton decreases in the entrance channel, the water becomes very clear and the bottom is visible at 3 m. Mean summer temperatures increase from 16,5° at the entrance to 25,0° at the southern end; nitrate concentrations decrease markedly but phosphates increase and so do silicates. The biomass of the zooplankton is highest in the middle of

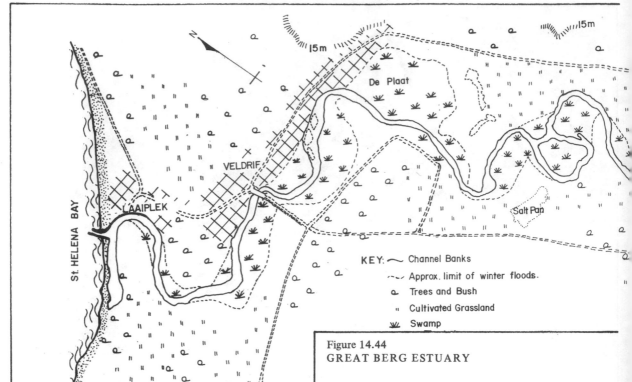

KEY: ～ Channel Banks
--- Approx. limit of winter floods.
a Trees and Bush
" Cultivated Grassland
wL Swamp

Figure 14.44
GREAT BERG ESTUARY

the lagoon but the predominant species change from neritic at the entrance to estuarine at the southern end.

Macrophytic algae, with the exception of the submerged *Gracilaria verrucosa,* decrease abruptly at the entrance while *Zostera* beds and salt marshes become extensive towards the southern end. Dr Nigel Christie has given estimates of biomass and primary production in chapter 7, and the species composition is shown in the appendix to chapter 6.

The benthic fauna also changes along the lagoon. At the northern entrance the intertidal sands have a poor fauna but the subtidal banks are very rich; concentrations of the bivalve *Mactra lilacea* with a biomass of 800 g (dry mass) m^{-2} have been recorded and Lucas (1979) estimated filtration rates, energetics and gamete production. The intertidal fauna increases on the sand flats in the middle of the lagoon and includes many estuarine species. *Upogebia* is common in muddy areas and *Callianassa* is abundant in the sand flats. Both make deep burrows and small polychaetes, crustacea and molluscs are found at or near the surface. Puttick (1977, 1978) while investigating the food available to the curlew sandpiper, recorded a biomass of 14,37-23,9 g (dry mass) m^{-2} in the upper 0-6 cm. The higher biomass on certain banks is largely due to *Assiminea globulus* which reaches a value of 11,28 g (dry mass) m^{-2} at HWN.

Fish are not abundant in the lagoon, possibly because the clear water provides little protection from fish-eating birds. Actually cormorants and terns are not abundant unless a shoal of fish is driven from Saldanha Bay into the lagoon. The main aquatic birds are waders of which there are 24 species including 15 palearctic migrants. Summers (1977) shows that the wader population increases from an average of 1 671 in winter to 36 759 in summer. They feed mainly on surface or shallow-burrowing invertebrates and consume 4,32 g (dry mass) m^{-2}.yr^{-1}. Since the mean annual biomass of the invertebrates is 18 g (dry mass) m^{-2}, the waders consume less than a quarter of the available food. Puttick (1979) discusses the energetics of the curlew-sandpiper and its seasonal variations. These birds take only 12,9 % of the available food representing 91 kJ.m^{-2}.yr^{-1} and Puttick argues that the carrying capacity of the Lagoon for these birds is much higher.

Langebaan Lagoon is very beautiful with flocks of flamingos and other birds feeding on the rich aquatic fauna. It is a favourite venue for yachtsmen and has rightly been declared a nature reserve. Nonetheless, pollution from the iron-ore berth and the industrial developments which will inevitably follow, pose a threat to its future.

The estuary of the Great Berg River
(32°46′S/18°09′E)

There are only a few reports on the ecology of this large estuary. Harrison & Elsworth (1958) give a detailed account of the ecology of the Berg River but

318

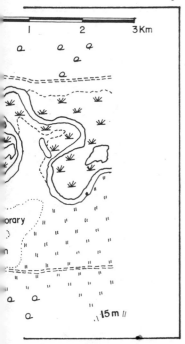

do not deal with the estuary. A few unpublished records of the flora and fauna are available in the Zoology Department of the University of Cape Town. The Fisheries Development Corporation (FISCOR) report F22-1 of June 1973, discussed the alledged effect on the salinity of the estuary when a new mouth was cut in 1966 to facilitate the entry of fishing boats to the harbour at Laaiplek. A preliminary note on water chemistry in a report edited by Oliff (1976) was later summarised by Cloete & Oliff (1976). A more complete account is at present being prepared for publication by Eagle, Gledhill & Greenwood. Gaigher (1979) reports on the bait organisms. I am grateful for permission to quote these data in the following account.

The Great Berg, which is the largest river in the Western Cape, rises in the Franschhoek Mountains and has a catchment of 4 012 km² in the winter rainfall area. The mean annual run-off is 234 x 10⁶ m³ and at Wellington the winter flow of the river may reach 500 m³.sec⁻¹. The river and its estuary (Figure 14.44) flow in a north-westerly direction past Veldrif, the marina now being constructed and on to the fishing harbour of Laaiplek on the coast of St Helena Bay. At Veldrif, the main channel is 100-200 m wide, 3 m deep and on a falling tide the flow velocity may reach 45 cm. sec⁻¹. The lower reaches of the river meander over very flat country so that the bed of the river falls only 1 m in the last 50 km. As a result, the heavy rains between May and August flood the extensive marshes along the banks and the river is almost fresh to the sea. Records at the mouth during winter show salinities of 2,0 to 4,8 ‰ and a temperature of 11,5-12 °. The turbidity in winter has not been recorded but the water is known to be very muddy.

As the river flow decreases in spring, the marshes start to dry and sea water enters the estuary with the flood tide. Thus at Veldrif, the salinity varied from 4,8 ‰ at low tide to 11,04 ‰ at high tide in October 1975 and from 29 to 35 ‰ in January 1976. Detailed records in the FISCOR report show that as evaporation continues through the summer, saline water is drawn further and further inland and the river is tidal for more than 70 km. By autumn, salinities of about 1,5 ‰ reach 38 km from the sea and the farmers complain. To limit the salt intrusion, the discharge from the Voelvlei dam was increased from 1,4 to 2,1 m³.sec⁻¹ in 1967. Even so, full strength sea water extended to 12,8 km from the sea in April 1968 and 9 ‰ salinity was recorded at Kersfontein 45 km along the length of the estuary, in February 1979. The presence of saline water so far up the estuary is partly due to the run-off from the Malmesbury shales in the drainage basin of the Soutrivier which joins the Berg 10 km above Kersfontein bridge. It will be evident that the length of the estuary increases very greatly from winter to summer.

Records of nutrient concentrations at Laaiplek show that nitrate concentrations range from a minimum of 5,42 to 14,7 μg.Nℓ^{-1} and total phosphorus from 1,68 to 2,05 μg.mol.ℓ^{-1}. The discharge of fish factory effluents greatly increases these values at times but the currents are too swift to allow anoxic or eutrophic conditions to develop. The seasonal temperature range is from 11,5 ° in winter to 27 ° in summer and the estuary is usually turbid with Secchi depths of about 20 cm in spring. In summer, however, the water becomes much clearer and in February 1979 the bottom at Veldrif was visible at 1,5 m. The sediment there is silty sand, Mdϕ being 3,40 with 41,7 % subsieve particles (<0,075 mm) and a negligible organic content.

The whole ecology of the estuary is determined by seasonal changes in river discharge and the consequent changes in salinity and turbidity. In the winter, when the estuary is flooded by muddy fresh water, most of the marine species disappear. As the floods recede in spring, the salinity increases and the marshes on the flood plain become saline as is evident from the film of salt on the dry mudflats in summer and the salinity of the pools that remain. The whole floodplain at Veldrif is covered with saltmarsh vegetation with *Sarcocornia natalensis* as the dominant. The intertidal mudbanks along the edges of the channel from Veldrif to Laaiplek are covered with *Sporobolus pungens, Chenolea diffusa, Sarcocornia perennis, Scirpus triqueter, Triglochin bulbosum, T. striatum* and several other species. Patches of *Zostera* grow at low tide and *Enteromorpha* is abundant. Further up the estuary, the marginal vegetation is largely replaced by *Phragmites australis* and *Zostera* is replaced by *Ruppia* sp. The phytoplankton has not been investigated.

The zooplankton changes with the seasons. In winter there are fresh water species including water beetles, cladocerans and copepods of the genera *Cyclops* and

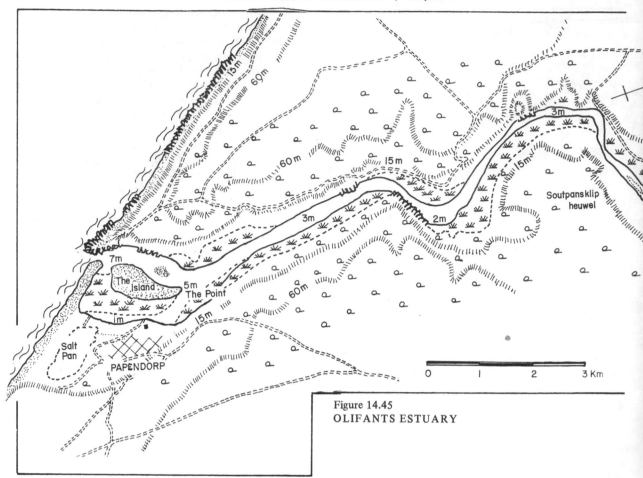

Figure 14.45
OLIFANTS ESTUARY

Diaptomus. In summer, marine plankton spreads up
the estuary including the mysid *Mesopodopsis slabberi,*
neritic copepods such as *Calanoides carinatus, Centropages brachiatus* and several other species as well as
ctenophores and crab zoeae. At times, the low salinity
surface water contains fresh water plankton and the high
salinity deep water contains marine forms but estuarine
plankton, if present, is poorly represented, the main form
being the copepod *Pseudodiaptomus hessei.*

The benthic macrofauna is scanty and restricted to
a few species, the records of which are available in the
University of Cape Town. The common species at
higher tidal levels include *Assiminea globulus, Deto
echinata, Anurida maritima* and *Orchestia rectipalma*
in weed beds and *Cyclograpsus* in eroding mudbanks.
Near low tide, *Hymenosoma orbiculare, Palaemon
pacificus* and *Exosphaeroma hylecoetes* are common
in *Zostera* beds. The common polychaete is *Ceratonereis erythraeensis. Upogebia africana* is not common
but *Callianassa kraussi* is plentiful in sand and is found
in muddy areas as well. Gaighter (1979) reports densities of 102-190 m^{-2}. No burrowing bivalves have been
recorded, but may be present in the blind channel
leading to the old mouth.

The few hauls that were made with a seine in September suggest that the fish fauna is poor. Eight
species were recorded. The common ones were *Gilchristella aestuarius, Hepsetia breviceps* and *Liza
richardsoni.* Stumpnose and kob are reported to enter
the estuary in summer.

As the shallow pools on the floodplain start to dry
up in spring, there is a marked increase in the avifauna.
Details will be found in Summers *et al* (1976, 1977).

Olifants River estuary
(31°42′S/18°11,5′E)

No report on the ecology of the Olifants River estuary
has been published. The following summary is thus
based on notes made by the staff of the Zoology
Department of the University of Cape Town and data
kindly supplied by officers of the National Research
Institute of Oceanology (NRIO). The Zoology Department notes and collections were made in January 1955
and in September 1973 while the hydrological and
chemical records were made by NRIO in February
1976 and July 1977. The NRIO records are being prepared for publication by Eagle, Gledhill & Greenwood.

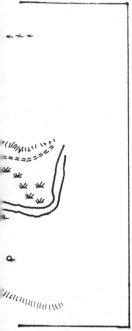

Professor John Grindley made further observations in January 1979 and Gaigher (1979) reports on the bait organisms. I am grateful for permission to incorporate some of these records in this summary.

The Olifants River rises in the mountainous winter rainfall area near Ceres and has a catchment of 46 084 km² much of which is in arid country. The river flows northward to Clanwilliam dam which has a capacity of 121,4 x 10⁶ m³. A canal takes the overflow to Bulshoek Barrage which has a capacity of 5,3 x 10⁶ m³ and supplies irrigation schemes along the river valley. The run-off from the valley and the overflow from the barrage eventually reaches the estuary. From December to March or April, there is little or no flow but from May to August the river is in spate.

Between Klawer and the sea, the river crosses the arid sandveld which rests on the slates and shales of the Malmesbury formation. These have been eroded by the river and the silts and clays that result are carried by the winter floods into the estuary basin. The flood plain is narrow and the flood waters flush the estuary rapidly so that the river runs fresh to the sea. In some years the flood waters fill the pans opposite Ebenhaeser and these may hold water for many years so that they form an important habitat for flamingos and other waders (Grindley, personal communication). During spring the river flow decreases rapidly and salt water extends up the channel; by summer it extends for 15 km to Olifantsdrif and probably further. Thus the length of the estuary increases greatly from winter to summer.

At Olifantsdrif the estuary (Figure 14.45) is 100-200 m wide and about 1 m deep but it gradually broadens to about 400 m and deepens from 3 m to about 5 m near Papendorp. The last 3 km and the posi-

tion of the mouth have changed. In 1925 a violent storm blocked the southern mouth with marine sand and the northern mouth deepened. The old southern channel has silted up and is now known as Die Dam and between it and the present channel there is a low island, the inland end of which is here referred to as the Point. The position of the present mouth is anchored by a reef and the bed is very uneven with depressions up to 7 m deep as well as rocky outcrops and sandy shoals. Further up the channel, the intertidal banks are mainly soft mud overgrown with salt marsh vegetation.

The sand at the mouth is fine with a negligible percentage of clay but the Point is muddy. A surveyed transect showed that at the level of midtide, Mdϕ is 2,4 with 15 % of silt and clay particles and 0,46 % of organic matter. It is very noticeable that the eroded channel side of the island has well-defined layers of coarser and finer sediments presumably deposited in summer and winter.

Hydrological conditions vary with the season. During the winter flood of 1977 the salinity was less than 2,85 ‰ all the way to the mouth and the water temperature was between 10,2 and 11,3 °. The water was muddy but the Secchi depth was not recorded. In September 1973, both the salinity and the temperature were higher; thus at Rhebokbaai, 4 km from the mouth, the surface salinity at high spring tide was 33 ‰ and the temperature was 14 °. In January and February, high salinity water extended about the same distance from the mouth but there was a vertical salinity gradient and considerable changes with the tide. At the Point, the surface salinity was 33 ‰ at high tide and 12,8 ‰ at low tide. The front where the clear sea water with a Secchi depth of 2,4 m flowed under the turbid estuary water with a Secchi depth of 0,4 m was easily seen. The upper reaches of the estuary are not very saline even in summer. In February 1976, salinities of 1,87 to 2,35 ‰ were recorded at Olifantsdrif 15 km from the mouth.

Nutrient concentrations were determined by officers of NRIO in the summer of 1976 and during a winter flood in 1977. During the flood there was little change along the length of the channel. Total phosphorus averaged 0,9 μmol.ℓ^{-1} and nitrate 10 μmol.ℓ^{-1}. During the neap tides of February 1976, total phosphorus increased from a mean of 1,4 μmol.ℓ^{-1} at Olifantsdrif to 2,85 μmol.ℓ^{-1} at the mouth. Similarly, nitrate concentrations increased from a mean of 12,6 μmol.ℓ^{-1} to 26,2 μmol.ℓ^{-1}. These increases in the higher salinities near the mouth show that most of the nutrients are derived from rich Benguela current water entering the estuary.

The phytoplankton has not been investigated. The zooplankton samples have been analysed by Professor John Grindley who notes that the lower reaches are dominated by neritic marine species in summer while estuarine species such as *Pseudodiaptomus hessei* dominate the upper reaches.

The macrophytes include a number of algae. There are fair growths of *Porphyra capensis, Chaetangium erinaceum, C. ovale* and *Chaetomorpha* sp on the rocks at the mouth. *Ulva* sp grows luxuriantly in Die Dam and *Enteromorpha* sp grows both here and in the marshes higher up the estuary. The vascular halophytes include rich beds of *Zostera capensis* at the Point and there are patches as far up as Rhebokbaai. The salt marsh vegetation at higher tidal levels is usually dense. *Sarcocornia perennis* forms carpets at midtide and both *Cotula coronopifolia* and *Triglochin* sp grow with it in waterlogged areas. *Spartina maritima* is absent. Near the high tide mark, *Chenolea diffusa* is dominant and *Limonium scabrum* is present. *Sarcocornia natalensis* extends to supratidal levels. In the upper reaches of the estuary, species of *Scirpus* become more obvious on the banks but reeds *(Phragmites australis)* are not common below Olifantsdrif.

The recorded distribution of 43 invertebrates is available in the Zoology Department of the University of Cape Town. The benthic fauna is not rich in species, but those that do occur, may be abundant in suitable habitats. On the rocks at the mouth there are fairly rich aggregations of *Choromytilus meridionalis* and among these, amphipods and isopods are common. Barnacles *(Balanus algicola* and *Tetraclita serrata)* are not abundant and oysters are absent. Browsing gastropods such as limpets and periwinkles are rather sparse at higher tidal levels but the shore crab *Cyclograpsus punctatus* is very common under stones.

The clean sands at the mouth are dominated by *Callianassa kraussi* but other species are poorly represented.

The highest density of benthic invertebrates was found at the Point. The following are mean values. In the salt marsh vegetation towards high tide the isopods *Deto echinata* and *Ligia glabrata* occurred in densities of 500 m^{-2} and the collembolan *Anurida maritima* reached 1 000 m^{-2}. *Assiminea globulus* tends to be patchy and averaged 380 m^{-2}. A surveyed transect showed that the infauna at low spring tide was dominated by *Upogebia africana* with a density of 500 m^{-2} while *Ceratonereis erythraeensis* varied from 125 to 1 500 m^{-2}. *Solen capensis* was present, and it is worth noting that this is the only living record from an Atlantic coast estuary although dead (?subfossil) shells have been found in the Berg estuary and in Milnerton Lagoon.

The epifauna at low tide included *Hymenosoma orbiculare, Exosphaeroma hylecoetes* and several species of amphipods, all of which were common in *Zostera* beds. The amphipods were even more common in the sheltered waters of Die Dam and with them were *Palaemon pacificus, Mesopodopsis slabberi* and *Nassarius kraussianus.* Densities were not recorded.

The fauna becomes impoverished about 4 km from the mouth although there are beds of *Upogebia* and *Callianassa* here and there and *Cyclograpsus* remains common. It should be noted, however, that the fauna

in the upper reaches of the estuary was poorly sampled and practically nothing is known of the aquatic insect population.

The fish fauna is also poorly known. Only a few hauls were made with a tuck seine at the Point and two hauls with a 1 mm mesh D-net in Die Dam. Only nine species were recorded, the commonest of which were *Liza richardsoni, Hepsetia breviceps* and *Gilchristella aestaurius.* Only a few small *Lithognathus lithognathus* were taken. We were informed that *Liza richardsoni* is caught commercially with a gill net and that *Lithognathus* up to 6 kg is occasionally taken by anglers at the mouth. Seining at Olifantsdrif revealed a few *Barbus serra* which normally lives in fresh water although it obviously tolerates low salinities.

On the whole, the fish fauna appears to be poor which is surprising seeing that plenty of invertebrate food is available in the lower reaches.

Brief notes were made of the estuarine birds in September 1973. The party noted a large flock of pelicans, about 20 little stint, some 20 spur-winged geese, about the same number of yellow-bill duck, a number of white-fronted cormorants, 3-4 white heron and a few whimbrel. In January 1979 Professor John Grindley made more detailed observations of the birds at the ten stations between the mouth and Ebenhaeser. Between Papendorp and the sea approximately 2 500 terns including the Common, the Swift, Caspian and Damara terns were roosting on the sand banks. More than 680 waders were present in the lower reaches including Avocet, Blacksmith Plover, Common Sandpiper, Curlew, Curlew Sandpiper, Greenshank, Sanderling, Turnstone, Whimbrel and White-fronted Plover. Other records include 13 Lesser Flamingo, 17 Spoonbill, 49 Pelican and one Fish Eagle. The pans at Ebenhaeser were occupied by 800 Greater and Lesser Flamingos and other wading birds including 45 Avocet. It is clear that the Olifants estuary and the adjoining pans provide important habitats for aquatic birds.

The ecology of the estuary seems to be determined by the seasonal changes in salinity and temperature. The major change in temperature between the cold sea water and the warm estuarine water in summer may inhibit the survival of marine larvae in the estuary, but extreme salinity changes are definitely important.

The Orange River mouth
(28°38,5'S/16°28,5'E)

The only references are papers on the avifauna by Plowes (1943) and Grindley (1959) and an account of the ecology of the mouth area by Brown (1959). The latter was based on a five-day visit by biologists from the University of Cape Town during spring tides in the winter of 1956. In January 1979, scientists of NRIO studied summer conditions (Orren *et al,* 1979).

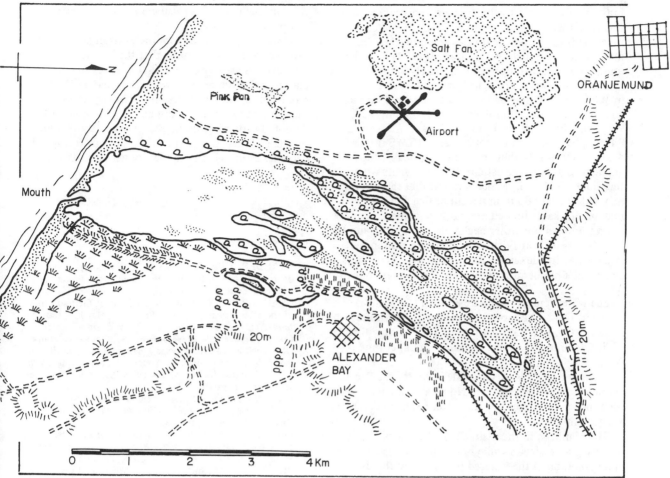

Figure 14.46
ORANGE RIVER MOUTH

Rogers (1979) gives an account of the distribution of the Orange River sediments in the sea.

The Orange River is the largest river in southern Africa. With its tributaries it drains most of the western part of southern Africa including parts of Namibia, Botswana, Transvaal, the Orange Free State, Lesotho and the northern Cape. Much of its catchment lies in the summer rainfall area and the water flow into the Verwoerd and le Roux dams which have capacities of 5 952,4 and 3 236,6 x 10^6 m^3 respectively. The middle reaches drain the southern part of Namibia but the run-off is very limited. Rogers (1979) states that the annual discharges of sediment and water average 60,4 x 10^6 tons and 9,3 x 10^9 m^2 respectively. The lower reaches flow through very arid country too but within 1-2 km of the sea, coastal fogs keep the air moist until mid-morning.

Seasonal changes in flow vary enormously. During summer, when the Orange River is in flood, the banks within the last 10 km from the mouth are inundated so that the river is about 2,5 km wide and the channel is up to 14 m deep. The flow is so strong that fresh water extends several kilometres out to sea. During the period of low flow in winter on the other hand, the river retreats to its channel which is then 1 500 m wide and about 2 m deep with many islands (Figure 14.46). The mouth itself may be constricted to 30 m by sandspits or may close completely in years of drought. Marshes enclosed by the coastal sanddunes persist on the south bank.

The following environmental conditions were recorded in July 1956. There are no rocky outcrops at the mouth and the steep sandspits vary in length. The sand becomes coarser towards the low tide mark. At this level the seaward slopes are composed of sand and gravel, the value of Mdϕ being 0,54 with a negligible percentage of fines and organic matter. The fine sediments from the river of course have been carried far out to sea and are deposited in a belt 500 km long and 40 km wide in depths of 70-120 m (Rogers, 1979). The waves along the coast continually erode and re-deposit the littoral sands so that the contours even on the inner sides of the sandspits may change by about half a metre during a single tide. In sharp contrast to this, the sediments on the river banks within the shelter of the sandspits change abruptly to very fine sand (Mdϕ = 3,98) with 74 % of silt and clay particles. This mud cakes as it dries and is black a centimetre below the surface. Orren *et al* (1979) have determined the concentrations of Cd, Cu, Pb and Zn.

The following hydrological observations were made in July. The low tide lag inside the bar is 300 minutes and the maximum ebb tide velocity is 0,78 m.sec^{-1}. While the mouth is open the salinity at HWS is over 34,7 ‰ from surface to bottom and at LWS it falls to 4,51 ‰. Obviously there must be a vertical salinity gradient higher up the estuary. In those years when the mouth is closed, residents report that saline water may extend over 8 km upstream or that the river water may be drinkable immediately inside the bar. Salinities during summer are either negligible or very low and the river flows so strongly that the pumping station that takes water to Alexander Bay township has been washed away and a plume of muddy water extends far out to sea.

Sea water temperatures are about 14 ° for most of the year. Inside the mouth the winter temperature increases from 15 ° at high tide to 16,5 ° at low tide, while in summer the mean maximum temperature is 24 °.

The river water is very turbid. In winter the Secchi depth at low tide was 25 cm. No measurements have been made in summer but the water has been described as liquid mud.

The plankton in the main channel consists largely of fresh water species while brackish water species are more common in the isolated pools left by floods along the south bank. In the extensive salt marshes on the south bank, only a few species of halophytes were recorded. No *Zostera* was seen; *Sarcocornia perennis*, *S. africana* and *S. asiatica* were abundant and *Salicornia meyeriana* was common.

As in the Berg and Olifants estuaries, meiofaunal counts were lower than in the sandy beaches of the west coast. There appears to be no distinct estuarine fauna. A few marine crustaceans including *Tylos granulatus*, *Eurydice longicornis* and *Talorchestia quadrispinosa* were found on the sandspit in winter. The kelp fly *Lamproscatella dischaeta* was common and both it and *Eurydice* extended to the salt marshes. The only other animal recorded there was the staphilinid beetle *Philonthus*. Presumably all these animals are swept out by the summer floods.

Seine netting was difficult but there seemed to be little to catch except a few small *Liza richardsoni* and *Mugil cephalus*. Residents reported they caught kob, white steenbras and skates but it is uncertain whether these were taken in the sea or the river. The fresh water yellow fish *Barbus holubi* is common and in summer it may even be caught outside the mouth (Grindley, personal communication). Observations of the avifauna were made by Grindley (1959) and the following were noted: two species of flamingo, avocet, African shellduck, Egyptian goose, *Anas capensis* and *A. erythrorhynchus*.

It seems clear, both from the environmental conditions and the aquatic fauna that there is no real estuary. The changes are so extreme that at times the mouth is marine and at times fresh water; Orren *et al* (1979) report no salt water incursion in January 1979; it is thus classed as a river mouth and not as an estuary. As noted earlier in this chapter, the swiftly flowing Tugela is similar for most of the year, but the mouth becomes estuarine during the dry winter months.

REFERENCES

ALEXANDER, J.J. & D.W. EWER 1969. A comparative study of some aspects of the biology and ecology of *Sesarma meinerti*. *Zool. Afr.* 4: 1-35.

ALEXANDER, W.J.R. 1976. Some aspects of the hydrology of the Mkuze swamp system. *In:* A.E.F. Heydorn (ed), *St Lucia Scientific Advisory Council Workshop – Charters Creek, February 1976*. Natal Parks Bd, Pietermaritzburg.

ANDERSON, F.P. 1976. Knysna Lagoon model investigation, Part I: Main report; Part II: Appendix. *Coastal Engineering and hydraulics division. NRIO, CSIR,* Stellenbosch (typescript, confidential).

ANON. 1974. Proposed Braamekraal Marina, Knysna model studies. *Coastal Engineering Division, CSIR Rep. C/Sea/ 74/6,* Stellenbosch. pp 1-23.

ANSELL, S.V. 1979. Mathematical models of Langebaan Lagoon ecosystem. *Abstract 4th (S.Afr.) natl. oceanog. Symp., Cape Town, July 1979.* CSIR, Pretoria.

BAIRD, D. 1979. The influence of a marina canal system on the ecology of the Kromme estuary. *Abstract 4th (S.Afr.) natl. oceanog. Symp., Cape Town, July 1979.* CSIR, Pretoria.

BAIRD, D. & P.E.D. WINTER 1979. Aspects of energy flow in the salt marshes of the Swartkops estuary, Port Elizabeth. *Abstracts 4th (S.Afr.) natl. oceanog. Symp., Cape Town, July 1979.* CSIR, Pretoria.

BAIRD, D., MARAIS, J.F.K. & P.E.D. WINTER 1979. Seasonal abundance, distribution and diversity in Eastern Cape estuarine fish populations. *Abstract 4th (S.Afr.) natl. oceanog. Symp., Cape Town, July 1979.* CSIR, Pretoria.

BARHAM, W.T. 1979. Spawning of *Arcnicola loveni* Kinberg in the Heuningnes River estuary, Bredasdorp. *S.Afr. J. Sci.,* 75: 262-264.

BASS, A.J. 1976. Sharks in the St Lucia Lake System. *In:* A.E.F. Heydorn (ed), *St Lucia Scientific Advisory Council Workshop – Charters Creek, February 1976.* Natal Parks Bd, Pietermaritzburg.

BAYER, A.W. & K.L. TINLEY 1967. The vegetation of the St Lucia Lake area. *In:* J.P. Kriel (ed), *Report of the Commission of Inquiry into the alleged threat to animal and plant life in St Lucia Lake.* Govt. Printer, Pretoria.

BEAUMONT, R.D. & P.J. HEYDENRYCH 1979. The effects of development on the Diep River estuarine system near Cape Town. *Abstract 4th (S.Afr.) natl. oceanog. Symp., Cape Town, July 1979.* CSIR, Pretoria. (Also) *S.Afr. J. Sci.* 75(12): 562.

BEGG, G.W. 1976. Some notes on the Sandvlei fish fauna, Muizenberg, Cape. *Piscator* 96: 4-14.

BEGG, G.W. 1978. *The estuaries of Natal.* Natal Town and Regional Planning Commission, Pietermaritzburg. Report Vol 41, 657pp.

BEGG, G.W. 1979. Water control and its relationship to resource management. *Abstract 4th (S.Afr.) natl. oceanog. Symp., Cape Town, July 1979.* CSIR, Pretoria.

BERJAK, P., CAMPBELL, G.K., HUCKETT, B.J. & N.W. PAMMENTER 1977. *In the mangroves of Southern Africa.* Natal Branch, Wildlife Soc. S.Afr., Durban. 72pp.

BERRUTI, A. 1979. Water levels, waterbirds and management at Lake St Lucia. *Abstract 4th (S.Afr.) natl. oceanog. Symp., Cape Town, July 1979.* CSIR, Pretoria.

BERRUTI, A. 1980. Birds of Lake St Lucia. *Southern Birds.* **8.**

BLABER, S.J.M. 1973. Temperature and salinity tolerance of juvenile *Rhabdosargus holubi* (Steindachner) (Teleostei: Sparidae). *J. Fish. Biol.* **5**: 593-598.

BLABER, S.J.M. 1974. Field studies of the diet of *Rhabdosargus holubi* (Steindachner) (Pisces: Sparidae) in a closed estuary. *J. Zool. Lond.* **173**: 404-417.

BLABER, S.J.M. 1976. The food and feeding ecology of mullet in the St Lucia Lake system. *In:* A.E.F. Heydorn (ed), *St Lucia Scientific Advisory Council Workshop – Charters Creek, February 1976.* Natal Parks Bd, Pietermaritzburg. (Also) 1976. *Biol. J. Linn. Soc.* **8**: 267-277.

BLABER, S.J.M. 1978. Fishes of the Kosi system. *The Lammergeyer* **24**: 28-40.

BLABER, S.J.M., HILL, B.J. & A.T. FORBES 1974. Infratidal zonation in a deep South African estuary. *Mar. Biol.* **28**: 333-337.

BLABER, S.J.M. & N. RAYNER 1979. Occurrence of *Daphnia pulex* in Lake St Lucia. *S.Afr. J. Sci.* **75**: 87.

BLABER, S.J.M. & A.K. WHITFIELD 1977. The feeding ecology of juvenile mullet (Mugilidae) in South-East African estuaries. *Biol. J. Linn. Soc. Lond.* **9**(2): 277-284.

BLOK, T.J. 1976. Lake St Lucia reclamation schemes: engineering progress and proposals. *In:* A.E.F. Heydorn (ed), *St Lucia Scientific Advisory Council Workshop – Charters Creek, February 1976.* Natal Parks Bd, Pietermaritzburg.

BOLTT, R.E. & B.R. ALLANSON 1975. The benthos of some southern African lakes. Part III: The benthic fauna of Lake Nhlange, KwaZulu, South Africa. *Trans. roy. Soc. S.Afr.* **41**: 241-262.

BOURGEOIS, M. 1948. *A study of two estuaries in the Cape Peninsula.* MSc thesis (Zoology), Univ. Cape Town.

BRAND, P.A.J., KEMP, P.H., PRETORIUS, S.J. & H.J. SCHOONBEE 1967. *Water quality and abatement of pollution in Natal rivers. Part III: The Tugela River and its tributaries.* Natal Town and Regional Planning Commission, Pietermaritzburg. 68pp.

BREEN, C.M., EVERSON, C. & K. ROGERS 1976. Ecological studies on *Sporobolus virginicus* (L) Kunth. with special reference to Lake St Lucia. *In:* A.E.F. Heydorn (ed), *St Lucia Scientific Advisory Council Workshop – Charters Creek, February 1976.* Natal Parks Bd, Pietermaritzburg.

BREEN, C.M. & B.J. HILL 1969. A mass mortality of mangroves in the Kosi estuary. *Trans. roy. Soc. S.Afr.* **28**: 285-303.

BROEKHUYSEN, G.J. & H. TAYLOR 1959. The ecology of South African estuaries. Part 8: Kosi Bay estuary system. *Ann. S.Afr. Mus.* **44**: 279-296.

BROWN, A.C. 1953. *A preliminary investigation of the ecology of the larger Kleinmond River estuary, Bathurst District.* MSc thesis, Rhodes University, Grahamstown.

BROWN, A.C. 1959. The ecology of South African estuaries. Part 9: Notes on the estuary of the Orange River. *Trans. roy. Soc. S.Afr.* **35**(5): 463-473.

BROWN, D.S. 1971. The ecology of gastropoda in a South African mangrove swamp. *Proc. malac. Soc. Lond.* **39** (263): 263-279.

CAMPBELL, B. 1971. *An ecological survey of the Black and Vygieskraal river complex with special reference to the effects of pollution.* Zool. Hons. project, Univ. Cape Town.

CAMPBELL, G.D. & B.R. ALLANSON 1952. The fishes of the 1947, 1948 and 1949 scientific investigations of the Kosi area, organised by the Natal Society for the Preservation of Wild Life and Natural Resorts. *Natal Soc. Pres. Wild Life Mag.* **April 1952:** 1-8.

CHAMPION, H.F.B. 1976. Recent prawn research at St Lucia with notes on the bait fishery. *In:* A.E.F. Heydorn (ed), *St Lucia Scientific Advisory Council Workshop – Charters Creek, February 1976.* Natal Parks Bd, Pietermaritzburg.

CHMELIK, F.G., VAN LOGGERENBERG, B.J., GRINDLEY, J.R. & A. DARRACOTT 1975. Economic model for estuarine valuation. *Proc. 10th Ann. Rep. mar. technol. Soc., Washington, DC:* 233-275.

CHOLNOKY, B.J. 1968. Die Diatomeenassociationen der Santa Lucia Lagune in Natal (Südafrika). *Bot. Mar.* **11**: (Suppl.).

CHUNNETT, E.P. 1965. Siltation problems in the Knysna Lagoon. *CSIR Rep.* **MEG 353.** pp 1-25.

CLOETE, C.E. & W.D. OLIFF 1976. South African marine pollution survey report 1974-1975. *S.Afr. Natl. Sci. Program. Rep. 8, Sept. 1976,* CSIR, Pretoria.

CLOUGH, B.F. & P.M. ATTIWELL 1975. Nutrient cycling in a community of *Avicennia marina* in a temperate region of Australia. *Proc. Internatl. Symp. Biol. Manage. Mangroves, Honolulu, October 1974.* pp 137-146 (Univ. Florida publ).

CONNELL, A.D. 1974. Mysidacea of the Mtentu River estuary, Transkei, South Africa. *Zool. Afr.* **9**(2): 147-159.

CONNELL, A.D. & D.D. AIREY 1979. The chronic effects of fluoride on the estuarine amphipods *Grandidierella lutosa* and *G. lignorum. Abstract 4th (S.Afr.) natl. oceanog. Symp., Cape Town, July 1979.* CSIR, Pretoria. (Also) *S.Afr. J. Sci.* **75**(12): 566.

DAY, J.H. 1959. The biology of Langebaan Lagoon: a study of the effects of shelter from wave action. *Trans. roy. Soc. S.Afr.* **35**: 475-547.

DAY, J.H. 1964. The origin and distribution of estuarine animals in South Africa. *In:* D.H.S. Davis (ed), *Ecological studies in southern Africa.* pp 159-173. W. Junk, The Hague.

DAY, J.H. 1967. The biology of Knysna estuary, South Africa. *In:* G. Lauff (ed), *Estuaries.* Am. Ass. Adv. Sci., Washington.

DAY, J.H. 1973. Notes on the Keurbooms estuary. *In:* P. Andrew, J. Moyle & Van Zyl Slabbert (eds), *Save the Garden Route.* Society for the protection of the environment, Stellenbosch (private publication).

DAY, J.H. 1974. The ecology of Morrumbene estuary, Mocambique. *Trans. roy. Soc. S.Afr.* **41**: 43-97.

DAY, J.H. 1975. The mangrove fauna of Morrumbene estuary, Mozambique. *In:* G. Walsh, S. Snedaker, H. Teas (eds), *Proc. Internatl. Symp. Biol. Management Mangroves, Honolulu, 1974.* Univ. Florida publication.

DAY, J.H., MILLARD, N.A.H. & G.J. BROEKHUYSEN 1954. The ecology of South African estuaries. Part 4: The St Lucia System. *Trans. roy. Soc. S.Afr.* **34**(1): 129-156.

DAY, J.H., MILLARD, N.A.H. & A.D. HARRISON 1952. The ecology of South African estuaries. Part 3: Knysna, a clear open estuary. *Trans. roy. Soc. S.Afr.* **33**: 367-413.

DEPARTMENT OF PLANNING AND ENVIRONMENT (Departement van Beplanning) 1970. *Riviermonde, strandmere en vleie. Vol 4: Knysna-Wildernes merekompleks.* pp 1-97. Dept. Environ. Planning, Pretoria.

DYE, A.H. 1976. Studies on the ecology of the estuarine meiofauna in South Africa. *Proc. 1st interdisciplinary Conf. far. freshw. Res. S.Afr., Port Elizabeth, July 1976.*

DYE, A.H. 1978a. An ecophysiological study of the meiofauna of the Swartkops Estuary. 1. The sampling sites: physical and chemical features. *Zool. Afr.* **13**(1): 1-18.

DYE, A.H. 1978b. Epibenthic algal production in the Swart-kops estuary. *Zool. Afr.* **13**(1): 157-161.

DYE, A.H. 1978c. Diurnal vertical migrations of the meiofauna in an estuarine sand flat. *Zool. Afr.* **13**(2): 201-206.

DYE, A.H. 1978d. Seasonal fluctuations in the vertical distribution of meiofauna in estuarine sediments. *Zool. Afr.* **13**(2): 207-212.

DYE, A.H., ERASMUS, T. & J.P. FURSTENBERG 1978. An ecophysiological study of the meiofauna of the Swartkops estuary. 3. Partition of benthic oxygen consumption and relative importance of the meiofauna. *Zool. Afr.* **13**(2): 187-200.

DYE, A.H. & J.P. FURSTENBERG 1978. An ecophysiological study of the meiofauna of the Swartkops estuary. 2. The meiofauna: composition, distribution, seasonal fluctuations and biomass. *Zool. Afr.* **13**(1): 19-32.

EDWARDS, M.S. 1971. *Fish and fishing: The Swartkops estuary. An ecological survey.* pp 24-32. Swartkops Trust, Port Elizabeth.

EMANUEL, A. 1977. Conservation at Richards Bay. *Omgewing. Environment RSA* **4**(12), December 1977: 5-7. Dept. Planning and Environment, Pretoria.

FISCOR 1973. Berg River salinity. *Report* **F22-1,** *Fish. deveop. Corp. S.Afr., Cape Town* (34pp, typescript).

FORREST, G.W. 1969. Some bird notes from St Lucia. *The Lammergeyer* **3**(10): 89-91.

FOURIE, H.O. 1976. Metals in organisms from Saldanha Bay and Langebaan Lagoon prior to industrialization. *S.Afr. J. Sci.* **72**: 110-113.

GAIGHER, C.M. 1979. A survey of the status and distribution of bait organisms in Cape estuaries. *Research Rep.: Estuaries* pp 1-18. Dept. of Nature and Environmental Conservation, Cape of Good Hope, Cape Town.

GENADE, A.B. 1973. A general account of certain aspects of oyster culture in Knysna estuary. *Abstract S.Afr. natl. oceanog. Symp., Cape Town.* CSIR, Pretoria.

GLYPHIS, J. (ed) 1976. *A report on the ecology of the Mngazana estuary (Transkei) and recommendation for a nature reserve.* Habitat Working Group, Univ. Cape Town (typescript).

GRINDLEY, J.R. 1959. Birds of the Orange River Estuary. *Ostrich* **30**: 127-129.

GRINDLEY, J.R. 1973. Estuaries of the Garden Route threatened by proposed highway. *East. Cape Naturalist* **50**: 9-12.

GRINDLEY, J.R. 1976. *Report on ecology of Knysna estuary and proposed Braamekraal marina.* School Environm. Studies, Univ. Cape Town (123pp typescript).

GRINDLEY, J.R. 1978. *Environmental effects of the discharge of sewage effluent into Knysna estuary.* School Environ. Studies, Univ. Cape Town (62pp typescript).

GRINDLEY, J.R. 1979. Plankton of west coast estuaries. *Abstract 4th (S.Afr.) natl. oceanog. Symp., Cape Town, July 1979.* CSIR, Pretoria.

GRINDLEY, J.R. & T.H. WOOLDRIDGE 1974. The plankton of Richards Bay. *Hydrobiol. Bull.* **8**: 201-212.

HARRISON, A.D. & J.F.E. ELSWORTH 1958. Hydrobiological studies of the Great Berg River, Western Cape Province. Part 1: General description, chemical studies and main features of the fauna and flora. *Trans. roy. Soc. S.Afr.* **35**(3): 125-329.

HECHT, T. 1973. *The ecology of the Kromme estuary with special reference to Sesarma catenata.* MSc thesis, Univ. Port Elizabeth.

HEEG, J. & S.J.M. BLABER 1979. The biology of filter-feeding teleosts in Lake St Lucia. *Abstract 4th (S.Afr.) natl. oceanog. Symp., Cape Town, July 1979.* CSIR, Pretoria.

HENRICI, R. & R.N. PIENAAR 1975. A preliminary study of the nanoplankton from Swartkops estuary, Port Elizabeth. *Proc. Electron Microsc. Soc. S.Afr.* **5**: 51-52.

HEYDORN, A.E.F. 1976. Research at St Lucia with reference to the situation in other estuarine systems. *In:* A.E.F. Heydorn (ed), *St Lucia Scientific Advisory Council Workshop – Charters Creek, February 1976.* Natal Parks Bd, Pietermaritzburg.

HEYDORN, A.E.F. 1977. Agriculture and earthworks – death knell of Natal' estuaries. *African Wildlife* **31**(6): 27-30.

HILL, B.J. 1966. A contribution to the ecology of Umlalazi estuary. *Zool. Afr.* **2**: 1-24.

HILL, B.J. 1969. The bathymetry and possible origin of Lakes Sibayi, Nhlange and Sifungwe in Zululand (Natal). *Trans. roy. Soc. S.Afr.* **38**(3): 205-216.

HILL, B.J. 1975. Abundance, breeding and growth of the crab *Scylla serrata* in two South African estuaries. *Mar. Biol.* **32**: 119-126.

HILL, B.J. 1976a. Studies on the benthos of Lake St Lucia by the late Dr R.E.Boltt. *In:* A.E.F. Heydorn (ed), *St Lucia Scientific Advisory Council Workshop – Charters Creek, February 1976.* Natal Parks Bd, Pietermaritzburg.

HILL, B.J. 1979. Biology of the crab *Scylla serrata* in the St Lucia System. *Trans. roy. Soc. S.Afr.* **44**: 55-62.

HILL, B.J., BOLTT, R.E. & S.J.M. BLABER 1973. *Report on conditions below the Bushman's River bridge.* Unpublished Report, Zoology Dept., Rhodes Univ.

HOBDAY, D.K. 1975. Quaternary sedimentation and development of the lagoonal complex, Lake St Lucia, Zululand. *S.Afr. Soc. quaternary Res.* (Also) 1976 *Ann. S.Afr. Mus.* **71**: 93-115.

HOWARD-WILLIAMS, C. 1979. The influence of bridges and artificial opening of the mouth on the biology of Swartvlei estuary. *Abstract 4th (S.Afr.) natl. oceanog. Symp., Cape Town, July 1979.* CSIR, Pretoria.

HUTCHISON, I.P.G. 1974, 1975. St Lucia lake research report. 1974. Vol 2. Hydrographic data. 1974. Vol 3. Mathematical models for simulating one-dimensional tidal propagation and dispersion in the St Lucia estuary. Vol 4. Mathematical models for simulating monthly water levels and salinities in Lake St Lucia. *Natal Prov. Admin. Hydrol. Res. Unit, Pietermaritzburg.*

HUTCHISON, I.P.G. 1976a. The hydrology of the St Lucia system. *In:* A.E.F. Heydorn (ed), *St Lucia Scientific Advisory Council Workshop – Charters Creek, February 1976.* Natal Parks Bd, Pietermaritzburg.

HUTCHISON, I.P.G. 1976b. St Lucia lake – development of mathematical models and evaluation of ameliorative measures. *St Lucia lake research report, Vol 5 (final).* Natal Prov. Admin. Hydrol. Res. Unit, Pietermaritzburg.

HUTCHISON, I.P.G. 1976c. Lake St Lucia: an assessment of possible remedial measures. *In:* A.E.F. Heydorn (ed), *St Lucia Scientific Advisory Council Workshop – Charters Creek, February 1976.* Natal Parks Bd, Pietermaritzburg.

HUTCHISON, I.P.G. & W.V. PITMAN 1973. Climatology and hydrology of the St Lucia Lake system. *St Lucia lake research report. Vol 1.* Natal Prov. Admin. Hydrol. Res. Unit, Pietermaritzburg.

HUTCHISON, I.P.G. & W.V. PITMAN 1977. Lake St Lucia Mathematical modelling and evaluation of ameliorative measures. *Civ. Engr. S.Afr.* **19**: 75-82.

JOHNSON, I.M. 1976. Studies of the phytoplankton of the St Lucia system. *In:* A.E.F. Heydorn (ed), *St Lucia Scientific Advisory Council Workshop – Charters Creek, February 1976.* Natal Parks Bd, Pietermaritzburg.

JOUBERT, L.S. & D.H. DAVIES 1966. The penaeid prawns of the St Lucia Lake System. *Invest. Rep. Oceanog. Res. Inst., Durban* **13**: 1-40.

KEMP, P.H., CHUTTER, F.M. & D.J. COETZEE 1976. Water quality and abatement of pollution in Natal rivers. Part V: The rivers of southern Natal. *Natal Town and Regional Planning Report. Vol.* **13**. 100pp, Pietermaritzburg.

KOK, H.M. 1979. The juvenile fish fauna of two Cape estuaries: Knysna and Swartvlei. *Abstract 4th (S.Afr.) natl. oceanog. Symp., Cape Town, July 1979.* CSIR, Pretoria.

KORRINGA, P. 1956. Oesterteelt in Suid-Afrika. Hidrographiese, biologiese en oestrologiese waarnemings in die Knysnastrandmeer, met aantekenings oor toestande in ander Suid-Afrikaanse waters. *Invest. Rep.* **20** *Dept. Industries (sect. Fish.) S.Afr.* 1-94.

KRIEL, J.P. 1967. Report on the hydrology of the St Lucia Lake system. Appendix 5. *In:* J.P. Kriel (ed), 1967, *Report of the Commission of Inquiry into the alleged threat to animal and plant life in St Lucia Lake.* Govt. Printer, Pretoria. 371pp.

LIVERSIDGE, R., BROEKHUYSEN, G.J. & A.R. THESEN 1958. The birds of Langebaan Lagoon. *Ostrich* **29**: 95-106.

LUCAS, M.J. 1979. Growth and metabolism of *Mactra lilacea. Abstract 4th (S.Afr.) natl. oceanog. Symp., Cape Town, July 1979.* CSIR, Pretoria.

MACNAE, W. 1957. The ecology of the plants and animals in the intertidal regions of the Swartkops estuary near Port Elizabeth, South Africa — Parts I and II. *J. Ecol.* **45**: 113-131 and 361-387.

MACNAE, W. 1963. Mangrove swamps in South Africa. *J. Ecol.* **51**: 1-25.

MARAIS, J.F.K. 1976. *Comparative studies on the nutritional ecology of mullet in the Swartkops estuary.* PhD thesis, Univ. Port Elizabeth, Port Elizabeth.

MARAIS, J.F.K. & D. BAIRD 1979a. Analysis of anglers' catch data from the Swartkops estuary, April 1972 to March 1978. *Abstract 4th (S.Afr.) natl. oceanog. Symp., Cape Town, July 1979.* CSIR, Pretoria.

MARAIS, J.F.K. & D. BAIRD 1979b. Aspects of feed intake, feed selection and alimentary canal morphology of *Mugil cephalus, Liza tricuspidens, Liza richardsoni* and *Liza dumerili. Abstract 4th (S.Afr.) natl. oceanog. Symp., Cape Town, July 1979.* CSIR, Pretoria.

MARAIS, J.F.K. & D. BAIRD 1979c. Seasonal abundance, distribution and catch per unit effort of gill net catches in the Swartkops estuary. *Abstract 4th (S.Afr.) natl. oceanog. Symp., Cape Town, July 1979.* CSIR, Pretoria.

MARAIS, J.F.K. & T. ERASMUS 1977. Chemical composition of alimentary canal contents of mullet (Teleosti: Mugilidae) caught in the Swartkops estuary near Port Elizabeth. *Aquaculture* **10**: 10pp.

MASSON, H. & J.F.K. MARAIS 1975. Stomach content analysis of mullet from the Swartkops estuary. *Zool. Afr.* **10**(2): 193-207.

M^CCLURG, T.P. 1979. The effects of industrial development on *Penaeus indicus* at Richards Bay. *Abstract 4th (S.Afr.) natl. oceanog. Symp., Cape Town, July 1979.* CSIR, Pretoria.

M^CLACHLAN, A. 1974. Notes on the biology of some estuarine bivalves. *Zool. Afr.* **9**: 15-34.

M^CLACHMAN, A. & T. ERASMUS 1974. Temperature tolerance and osmoregulation in some estuarine bivalves. *Zool. Afr.* **9**(1): 1-13.

M^CLACHLAN, A. & J.R. GRINDLEY 1974. Distribution of macrobenthic fauna of soft substrata in the Swartkops estuary, with observations on the effects of floods. *Zool. Afr.* **9**(2): 211-233.

MEHL, J.A.P. 1973. Ecology, osmoregulation and reproductive biology of the white steenbras, *Lithognathus lithognathus. Zool. Afr.* **8**: 157-230.

MELVILLE-SMITH, R. & D. BAIRD 1979. Abundance, distribution and species composition of fish larvae in the Swartkops estuary, Port Elizabeth. *Abstract 4th (S.Afr.) natl. oceanog. Symp., Cape Town, July 1979.* CSIR, Pretoria. (Also) *S.Afr. J. Sci.* **75**(12): 564.

MILLARD, N.A.H. & G.J. BROEKHUYSEN 1970. The ecology of South African estuaries. Part 10. St Lucia: a second report. *Zool. Afr.* **5**(2): 277-307.

MILLARD, N.A.H. & A.D. HARRISON 1954. The ecology of South African estuaries. Part 5: Richards Bay. *Trans. roy. Soc. S.Afr.* **34**(1): 157-179.

MILLARD, N.A.H. & K.M.F. SCOTT 1954. The ecology of South African estuaries. Part 6: Milnerton estuary and the Diep River, Cape. *Trans. roy. Soc. S.Afr.* **34**: 279-324.

MOLDAN, A., CHAPMAN, P. & H.O. FOURIE 1979. Some ecological effects of the *Venpet-Venoil* collision. *Mar. Pollution Bull.* **10**: 60-63.

MOSTERT, S.A. 1972. Preliminary report on black mussel culture in the Langebaan Lagoon. *S.Afr. Ship. News and Fish, Indust. Rev.* **27**: 59-63.

MUHR, D. 1974. *The ecology of Sandvlei.* Zool. Hons. project. Univ. Cape Town.

NOBLE, R.G. & J.R. HEMENS 1978. Inland water ecosystems in South Africa: A review of research needs. *S.Afr. Natl. Sci. Programmes. Rep.* **34**. CSIR, Pretoria. 148pp.

OLIFF, W.D. (ed) 1976 (unpublished). (South African) *National Marine Pollution Monitoring Program.* First annual report (509pp typescript), second annual report (172pp typescript), NIWR, Durban. (Includes many progress reports by the National Institute for Water Research first under the heading *Natal Rivers Research Fellowships,* second as reports on *Marine Disposal of Effluents* to the CSIR and from 1974 as progress reports on the *National Marine Pollution Surveys* (East Coast Section).

OLIFF, W.D. (ed) 1978. *National Marine pollution surveys East Coast Section. Third annual report.* NIWR, Durban (typescript 64pp). East London p 1-12; Buffalo estuary p 13-26; Bashee estuary p 27-28; Umgababa estuary p 29-34; Mzimvubu p 35-52; Mngazana p 53-64; ocean reference transect p 65-124.

OLIFF, W.D. (ed) 1979. *National marine pollution surveys: East Coast Section. Fourth annual report.* Natal Inst. Water Res., Durban (typescript). Chlorinated hydrocarbon levels in selected animals pp 1-4; Durban Bay complex pp 5-18; St Lucia estuary pp 19-33; oceanic reference transects pp 34-57. *Research papers:* 1. Connell & Avery (2 papers); 3. Connell (taxonomy of copepod spp).

ORME, A.R. 1973. Barrier and lagoon systems along the Zululand Coast, South Africa. *In:* D.R. Coates (ed), *Coastal geomorphology.* Publ. Geomorph. State Univ. NY, Binhampton NY: 181-216.

ORME, A.R. 1974. Estuarine sedimentation along the Natal coast, South Africa. *Tech. Rep.* **5**, *Office of Naval Research, USA:* 1-53.

ORREN, M.J., EAGLE, G.A., HENNIG, H.F. & A. GREEN 1979. Preliminary results of an environmental survey of the Orange River estuary. *Abstract 4th (S.Afr.) natl. oceanog. Symp., Cape Town, July 1979.* CSIR, Pretoria.

PHLEGER, F.B. 1976. Holocene ecology in St Lucia lagoon, Zululand, based on foraminifera. *In:* A.E.F. Heydorn (ed), *St Lucia Scientific Advisory Council Workshop — Charters Creek, February 1976.* Natal Parks Bd, Pietermaritzburg.

PIERCE, S.M. 1979. *The contribution of Spartina maritima (Curtes) Fernald to the primary production of the Swartkops estuary.* MSc thesis, Rhodes Univ., Grahamstown.

PIKE, T. 1967-1971. Fish Survey of Kosi Bay Lake system; 1968, Kosi Lake system fish survey; 1969, Continuation of the fish survey in Kosi Lake system; 1971, Kosi Lake system fish survey. *Internal reports Natal Parks Bd, Pietermaritzburg.*

PLOWES, D.C.N. 1943. Bird life at the Orange River mouth. *Ostrich* **14**: 123-138.

POCOCK, M.A. 1955. Seaweeds of Swartkops estuary. *S.Afr. J. Sci.,* **52**: 73-75.

POOLEY, A.C. 1976. Observations on the Lake St Lucia crocodile population. *In:* A.E.F. Heydorn (ed), *St Lucia Scientific Advisory Council Workshop — Charters Creek, February 1976.* Natal Parks Bd, Pietermaritzburg.

PRINGLE, J.S. & J. COOPER 1975. The palaearctic wader population of Langebaan Lagoon. *Ostrich* **46**: 213-218.

PUTTICK, G.M. 1977. Spatial and temporal variations in intertidal animal distribution at Langebaan Lagoon, South Africa. *Trans. roy. Soc. S.Afr.* **42**(3): 403-440.

PUTTICK, G.M. 1978. The diet of the curlew sandpiper at Langebaan Lagoon, South Africa. *Ostrich* **49**: 158-167.

PUTTICK, G.M. 1979. The feeding ecology and energetics of the Curlew Sandpiper at Langebaan Lagoon. *Abstract 4th (S.Afr.) natl. oceanog. Symp., Cape Town, July 1979.* CSIR, Pretoria.

RETIEF, G. de F., MULLIGAN, D. & A.P.M. VONK 1979. The prevention and combating of oil pollution at Knysna. *Abstract 4th (S.Afr.) natl. oceanog. Symp., Cape Town, July 1979.* CSIR, Pretoria. (Also) *S.Afr. J. Sci.* **75**(12): 563.

ROGERS, J. 1979. Dispersal of sediment from the Orange River along the Namib desert coast. *Abstract 4th (S.Afr.) natl. oceanog. Symp., Cape Town, July 1979.* CSIR, Pretoria. (Also) *S.Afr. J. Sci.* **75**(12): 567.

ROSSOUW, J. 1976. The dynamics of estuary mouths. *Proc. First Interdisc. Conf. Mar. Freshwater Res. South Africa, Port Elizabeth, July 1976.* CSIR, Pretoria.

SCARFE, A.D. 1972. *An ecological study of the psammofauna in the Black River mouth and along the sand beach of Table Bay, South Africa, with consideration of the effect of pollution.* Zool. Hons. project. Univ. Cape Town.

SCHULZE, B.R. 1965. *The climate of South Africa, Part 8: General Survey.* **WB28**, Govt. Printer, Pretoria.

SCOTT, K.M.F., HARRISON, A.D. & W. MACNAE 1952. The ecology of South African estuaries. Part 2: The Klein River estuary, Hermanus. *Trans. roy. Soc. S.Afr.* **33**: 283-331.

SHANNON, L.V. & G.H. STANDER 1977. Physical and chemical characteristics of water in Saldanha Bay and Langebaan Lagoon. *Trans. roy. Soc. S.Afr.* **42**: 441-459.

SHELTON, P. 1975. *The ecology of Sandvlei.* Zool. Hons. project, Univ. Cape Town.

SHEWELL, E.L. 1950. Birds of the Gamtoos estuary. *Ostrich* **21**: 97-102.

SIEGFRIED, W.R. (ed), 1977. *A report on preliminary surveys of selected communities of plants and animals at Dwesa Nature Reserve.* Univ. Cape Town. 36pp.

SUMMERS, R.W. 1977. Distribution, abundance and energy relationships of waders (Aves: Charadrii) at Langebaan Lagoon. *Trans. roy. Soc. S.Afr.* **42**: 483-494.

SUMMERS, R.W., PRINGLE, J.S. & J. COOPER 1976. *The status of coastal waders in the south-western Cape, South Africa.* Western Cape Wader Study Group, Cape Town. 162pp.

TALBOT, F.H. 1955. Notes on the biology of the white stumpnose, *Rhabdosargus globiceps* (Cuvier) and on the fish fauna of the Klein River estuary. *Trans. roy. Soc. S.Afr.* **34**: 387-407.

TAYLOR, R.H. 1976. Hippopotamuses at Lake St Lucia. *In:* A.E.F. Heydorn (ed), *St Lucia Scientific Advisory Council Workshop — Charters Creek, February 1976.* Natal Parks Bd, Pietermaritzburg.

TINLEY, K.L. 1958. *A preliminary report on the ecology of the Kosi Lake system. 18.10.1958 to 15.12.1958. Part I: Hippopotamus amphibius. Part II: General account of the environment.* (Typescript) Natal Parks Bd, Pietermaritzburg.

TINLEY, K.L. 1969. *The significance of hardpan horizon as the prime factor controlling the hydrology and life of sand country, with particular reference to the freshwater relation of the St Lucia lake system, Zululand* (unpublished report 15pp).

TINLEY, K.L. 1976. *The ecology of Tongaland: 3: Kosi Lake system.* Wildlife Society of Southern Africa (Natal Branch), Durban: 69-140.

TURVEY, L.G.S. 1960. Report on the lower Umhlatuzi River. *S.Afr. Dept. Water Affairs, Rep. HO File 3088* (typescript).

VAN DER ELST, R.P. 1979. Changes in populations of estuarine dependent sport fish in Natal. *Abstract 4th (S.Afr.) natl. oceanog. Symp., Cape Town, July 1979.* CSIR, Pretoria. (Also) *S.Afr. J. Sci.* **75**(12): 566.

VAN DER ELST, R.P., BLABER, S.J.M., WALLACE, J.H. & A.K. WHITFIELD 1976. The fish fauna of Lake St Lucia under different salinity regimes. *In:* A.E.F. Heydorn (ed), *St Lucia Scientific Advisory Council Workshop — Charters Creek, February 1976.* Natal Parks Bd, Pietermaritzburg.

VAN DER HORST, G. 1976. *Aspects of the reproductive biology of Liza dumerili (Steindachner 1869) (Teleostii, Mugilidae) with special reference to sperm.* PhD thesis, Univ. of Port Elizabeth.

VAN DER HORST, G. & T. ERASMUS 1978. The breeding cycle of male *Liza dumerili* (Teleostei: Mugilidae) in the mouth of the Swartkops estuary. *Zool. Afr.* **13**(2): 259-273.

VAN DER WESTHUIZEN, H.C. & J.F.K. MARAIS 1977. Stomach content analyses of *Pomadasys commersonni* from Swartkops estuary (Pisces: Pomadasyidae). *Zool. Afr.* **12**(2): 500-504.

VAN HEERDEN, I.L. 1976. The geology of the lake St Lucia and some aspects of its sedimentation. *In:* A.E.F. Heydorn (ed), *St Lucia Scientific Advisory Council Workshop — Charters Creek, February 1976.* Natal Parks Bd, Pietermaritzburg.

WALLACE, J.H. 1969. Some effects of recent high salinities on the fish and molluscan shell fish life in St Lucia Lake, Zululand (unpublished). *Rep. Oceanogr. Res. Inst., Durban,* 5pp.

WALLACE, J.H. 1975a. The estuarine fishes of the east coast of South Africa. 1: Species composition and length distribuion in the estuarine and marine environments. II: Seasonal abundance and migrations. *Invest. Rep. Oceanog. Res. Inst., Durban.* **40**: 1-72.

WALLACE, J.H. 1975b. The estuarine fishes of the east coast of South Africa. 3: Reproduction. *Invest. Rep. Oceanog. Res. Inst., Durban.* **41**: 1-51.

WALLACE, J.H. 1976. Biology of teleost fish, with particular reference to St Lucia. *In:* A.E.F. Heydorn (ed), *St Lucia Scientific Advisory Council Workshop — Charters Creek, February 1976.* Natal Parks Bd, Pietermaritzburg.

WALLACE, J.H. & R. VAN DER ELST 1975. The estuarine fishes of the east coast of South Africa. IV: Occurrence of juveniles in estuaries. V: Ecology, estuarine dependence and status. *Invest. Rep. Oceanog. Res. Inst., Durban.* **42**: 1-63.

WARD, C.J. 1962. *A report on weedgrowth in the upper reaches of St Lucia estuary.* Natal Parks Bd, Pietermaritzburg.

WARD, C.J. 1976. Aspects of the ecology and distribution of submerged macrophytes and shoreline vegetation of Lake St Lucia. *In:* A.E.F. Heydorn (ed), *St Lucia Scientific Advisory Council Workshop — Charters Creek, February 1976.* Natal Parks Bd, Pietermaritzburg.

WATLING, R.J. & H.R. WATLING 1974. Environmental studies in Saldanha Bay and Langebaan Lagoon. I: Trace metal concentrations in selected molluscs and algae. *FIS Special Rep.* **70**: CSIR, Pretoria. 77pp.

WALTING, R.J. & H.R. WATLING 1975. Trace metal studies in Knysna estuary. *Environment RSA* **2**(1): 5-7.

WATLING, R.J. & H.R. WATLING 1977. Metal concentrations in surface sediments from Knysna estuary. *Special FIS Rep.* **122**. CSIR, Pretoria.

WEAVER, A.v.B. 1979. The effects of flow restriction on selected grain size parameters of the sediments in the Bushman's River estuary. *Abstract 4th (S.Afr.) natl. oceanog. Symp., Cape Town, July 1979.* CSIR, Pretoria.

WHITFIELD, A.K. 1976. An investigation of the family Mugilidae in Lake St Lucia, Zululand, with special reference to predation and population densities. *In:* A.E.F. Heydorn (ed), *St Lucia Scientific Advisory Council Workshop — Charters Creek, February 1976.* Natal Parks Bd, Pietermaritzburg.

WHITFIELD, A.K. 1977. *Predation on fish in Lake St Lucia, Zululand.* MSc thesis, Univ. Natal, Pietermaritzburg.

WHITFIELD, A.K. 1979. A quantitative study of the trophic relationships within the fish community of the Mhlanga estuary. *Abstract 4th (S.Afr.) natl. oceanog. Symp., Cape Town, February 1979.* CSIR, Pretoria. (Also) *S.Afr. J. Sci.* 75(12): 565.

WHITFIELD, A.K. & S.J.M. BLABER 1978a. Distribution, movements and fecundity of Mugilidae at Lake St Lucia. *Lammergeyer* 26: 53-63.

WHITFIELD, A.K. & S.J.M. BLABER 1978b. Resource segregation among iliophagous fish in Lake St Lucia, Zululand. *Envir. Bull. Fish.* 3(3): 293-296.

WHITFIELD, A.K. & S.J.M. BLABER 1978c. Food and feeding ecology of piscivorous fishes at Lake St Lucia, Zululand. *J. Fish. Biol.* 13: 675-691.

WHITFIELD, A.K. & S.J.M. BLABER 1978d. Feeding ecology of piscivorous birds at Lake St Lucia. Part I: Diving Birds. *Ostrich* 49: 185-198.

WHITFIELD, A.K. & S.J.M. BLABER 1978e. Feeding ecology of piscivorous birds at Lake St Lucia. Part II: Wading Birds, Part III: Swimming Birds. *Ostrich* 50: 1-9; 10-20.

WHITFIELD, A.K. & S.J.M. BLABER 1978f. Predation on grey mullet (*Mugil cephalus* L) by *Crocodylus niloticus* at St Lucia, South Africa. *Copeia* (in preparation).

WHITFIELD, A.K. & D.P. CYRUS 1978. Feeding succession and zonation of aquatic birds at False Bay, Lake St Lucia. *Ostrich* 49: 8-15.

WOOLDRIDGE, T. 1976. The zooplankton of Msikaba estuary. *Zool. Afr.* 11: 23-41.

Chapter 15

The coastal lakes of southern Africa

B.R. Allanson

Institute for Freshwater Studies, Department of Zoology & Entomology, Rhodes University, Grahamstown

INTRODUCTION

The coastal lakes of southern Africa form an important component of the subcontinent's aquatic ecosystems and, by virtue of their proximity to the sea, their utilization by man has posed a considerable threat to their conservation. The International Biological Programme in South Africa provided the incentive for co-operative research in a number of the larger lakes. These research results have been published in a variety of media and in this chapter I have brought together some of the more significant works to show the limnological diversity within this particular ecosystem, the extent of tidal influence and the trophic status of the fauna and flora.

GEOMORPHOLOGY OF THE COASTAL RIMLAND AND THE ORIGIN OF THE COASTAL LAKES

The coastal regions of many of the Gondwanaland components have been subject to a variety of geological, marine and fluvial events since the time of their separation. Essentially all the important barrier systems in the southern hemisphere are the result of a reasonably well delineated system of quaternary sedimentation associated with eustatic marine transgressions and in Natal at least, with a tectonic movement of major importance — namely the Natal monocline (King, 1972).

A detailed analysis of these events has been described in three important papers by King (1970), Orme (1973) and Hobday (1976). Together they demonstrate unequivocally that during the early Pleistocene sea levels were higher than they are today and consequently inundated and contributed to the formation of the coastal plain, particularly in Natal. This was followed by an erratic retreat of the sea superimposed upon which were a series of glacio-eustatic sea-level oscillations (Hobday, 1979), resulting in north-south aligned dune ridges. Of prime importance to the building of the coastal lakes and lagoons as we know them today, was the effect of the marine regression associated with the last glaciation. Hobday (1979) has drawn attention to the rejuvenation of the lower courses of the coastal rivers consequent upon lowered sea level and the formation of steeply incised valleys. Sand eroded from the extensive sea shore assisted materially in the closure of the river mouths and heating of the land mass increased the strength of the landward breezes, and the construction of the barrier dunes began. The river mouths were blocked and the lagoons and lakes were formed. Some lagoon systems such as St Lucia with its associated Mfolozi River remained open to the sea although it is not always easy to establish whether or not the existing river mouths are, in fact, original. The position of existing submarine canyons across the narrow continental shelf of Natal suggests that during the Holocene, in the case of St Lucia, the mouth may have opened to the north of its present position, while those of Lake Nhlange and Lake Sibaya were more southerly in position.

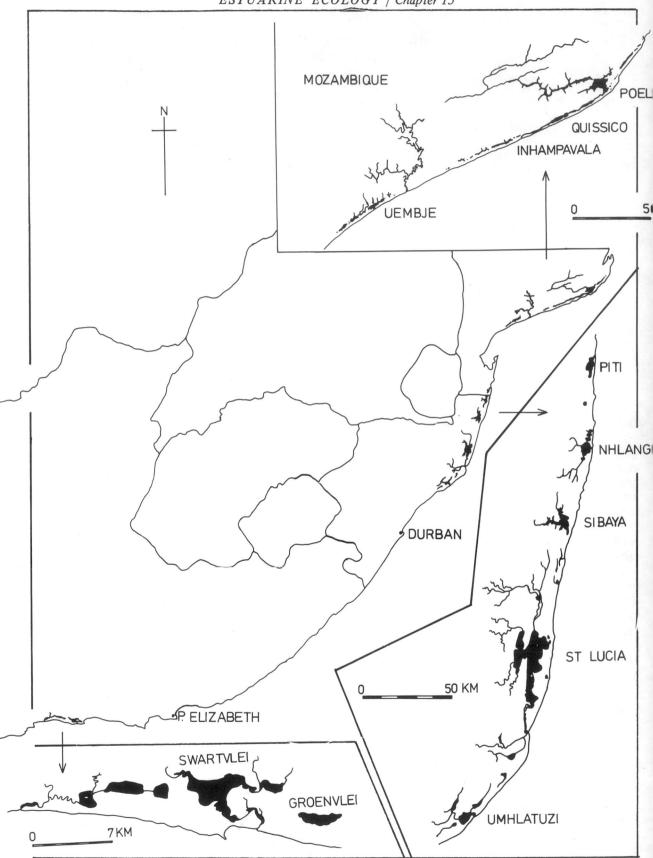

Figure 15.1
The main assemblages of coastal lakes in South and south-eastern Africa (modified from Hill, 1975).

The resulting lakes (Figure 15.1) were subject to considerable Holocene reduction in surface area. Orme (1973) has provided evidence that the Zululand lagoons have all experienced at least a 60 % reduction in area due to sedimentation, segmentation, and swamp encroachment. On the basis of Hill's (1975) discussion of the origin of southern African coastal lakes, a spectrum of sedimentary infilling, with Lake Sibaya being the deepest lake (>40 m) and St Lucia the shallowest (~2 m) is observed. Hill (1975) furthermore identifies three coastal lake types:

1. Drowned valleys associated with river systems. Examples are Sibaya, Nhlange, Poelela, Mhlatuzi and St Lucia in order of decreasing depth.

2. The narrow coastal lagoons of southern Moçambique, and Langvlei near Wilderness have been formed by marine inundation and are not necessarily associated with a river system.

3. Deflation basins such as Rondevlei at Wilderness.

An interesting example of a drowned valley which has not been subject to severe silting and remains open to the sea, is described by Blaber *et al* (1974) for the Msikaba estuary, some 30 km north of Port St Johns.

The description of the East Gippsland coastal lakes in the State of Victoria, Australia by Timms (1973) indicates yet another coastal lake system with origins similar to those in southern Africa. This in itself is not surprising, as many of the southern hemisphere seaboards would have been affected in the same way by the glacio-eustatic changes during the last glaciation. Bayly (1964) describes another mode of lake formation in the sand dunes on the coastal islands of Southern Queensland. Essentially, these coastal lakes are elevated well above sea level (120 m) to the extent that they have been described as 'perching', and appear to have been formed by a combination of wind action forming the deflation basin which then becomes impermeable to the passage of water by the accumulation of a layer of organic debris which together with the sand forms 'an organic cemented sand'. Their waters are acidic and in this respect parallel the Swartvlei system of our southern coastal belt.

METEOROLOGY OF THE COASTAL REGION

The coastal belt of southern Africa is subject in the south to south-westerly and south-easterly winds while on the eastern seaboard the prevailing winds are northerly and southerly. The modal wind speed at St Lucia for the period January 1969 to February 1973, was 4-6 m.sec^{-1} and occurred between January and July. From August to December the mode increased to 7-11 m.sec^{-1} (Hutchison & Pitman, 1973). These data accord very well with those reported by Allanson (1979) for Lake Sibaya some 80 km to the north and are characteristic of the wind regime of the eastern coastal belt.

Table 15.1
Variation in total annual rainfall (mm) at Lake Sibaya and Lake St Lucia

1964/65	1 059	834
1965/66	1 191	814
1966/67	1 222	842
1967/68	670	660
1968/69	891	891
1969/70	663	602
1970/71	876	1 078
1971/72	424	1 228
1972/73	996	–
1973/74	1 411	–
1974/75	1 314	–
1975/76	1 306	–

Rainfall in the east is largely restricted to the summer months of October to March, during which Lake St Lucia receives 67 % of the mean annual total of 890 mm calculated from 1918 to 1972. Wide variations in rainfall occur from year to year and from place to place on the eastern seaboard, as shown in Table 15.1.

These striking changes in rainfall and the probability of their past occurrence were examined for Lake Sibaya by Pitman & Hutchison (1975), who determined the water budget of the lake from the equation:

$$Q - L = S_e - S_b + A (E - P)$$

where the quantity $Q - L$ may be termed the 'net lake recharge' R, and is composed of Q, the inflow to the lake and L, loss due to seepage. E = evaporation loss from lake (m), P = depth of rainfall in lake (m), A = average area of lake, S_b = volume of water in lake at beginning of period (10^6 m^3), S_e = volume of water in lake at end of period (10^6 m^3). From this equation and the computer simulation derived from it, it was possible to simulate month-end lake levels for the period October 1914 to September 1974. Subsequent records of lake level have shown a rapid rise from 1975 to 1977 which exceeded in level (although rising at a similar rate to) that predicted by Pitman & Hutchison's model for 1943/44, when a lake level of 4,3 m above datum was recorded. The model indicates that since 1915 there have been three minimum levels of >1 m above datum and three peaks ranging separately from 3,3 in 1924, 4,3 in 1944, and 4,6 m in February 1977, this latter being an actually recorded value. Although the records are too few to be convincing, there is nevertheless a suggestion that these cycles occur with a periodicity of between 20 and 30 years. Such cycles emphasise the importance of long term recording schedules of essential hydrological and climatological parameters in assessing the effects of meteorological events upon the hydrology, physico-chemical and biological limnology of these lakes. Furthermore, while meteorological conditions may be fairly uniform over comparatively short distances, lake hydrological patterns are influenced by the characteristics of their respective catchments. In this regard, Hutchison &

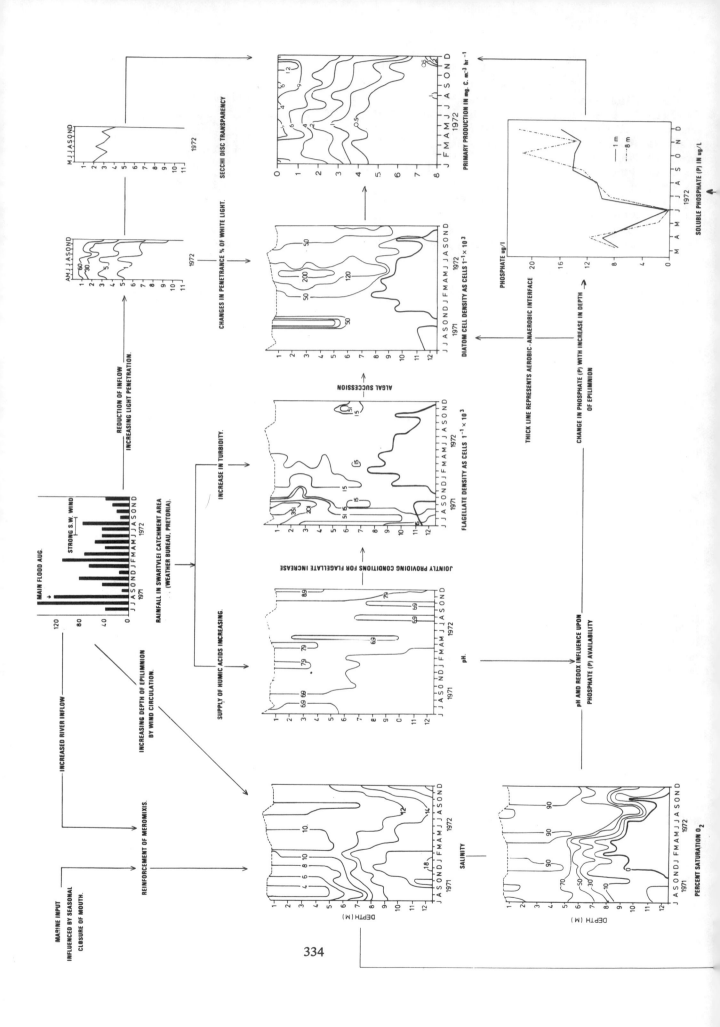

Pitman's (1973) report on Lake St Lucia is particularly relevant. Although it may be argued that there is some coincidence between lake level changes in Lake Sibaya and Lake St Lucia, obviously many more factors are influencing the hydrology of St Lucia and Lake Sibaya.

The two major regions of coastal lakes are separated by some 7 ° of south latitude such that the eastern lakes are subtropical while those of the southern Cape coast are warm temperate. The annual mean water temperature maxima and minima at Lake Sibaya are 27 °C and 17 °C, while at Swartvlei 26 °C and 6 °C have been recorded. Insolation in these latitudes is normally high with cloud cover increasing during summer. The mode of days during which the cloud cover is over 50 % was found to be during November and December at Lake Sibaya. The average number of hours of sunshine per day during 1972/73 was 6 (Allanson, 1979).

PHYSICO-CHEMICAL LIMNOLOGY

The influence of these meteorological features upon the limnology of the lakes is striking. The effect of wind speeds of between 18 and 32 km.hr^{-1} is sufficient to maintain homothermal conditions in all of the lakes with the exception of those which exhibit meromixis or salinity stratification. With these exceptions, the lakes exhibit only slight changes in dissolved oxygen concentration with depth during summer, although so far no summer records are available for L. Poelela (Hill *et al*, 1975) so that whether or not thermal stratification is accompanied by significant oxygen reduction is not known. The presence of *Najas* sp down to 12 m (Hill *et al*, 1975) and the wind-swept nature of the Mocambique coastal plain would seem to argue against this.

Lake Mpungiwini (part of the Kosi lakes system) and Swartvlei are essentially meromictic basins within estuarine complexes. They owe their meromixis to the inflow of sea during each spring tidal cycle. Whereas Lake Mpungwini possesses a stable meromixis, that of Swartvlei is developed every summer when the mouth is open, but is slowly destroyed during the following winter when the estuary mouth closes and freshwater inflow, precipitation and sensible heat loss destroys the density differential (Robarts & Allanson, 1977). It is re-established rapidly in early summer when the sandbar is breached and a tidal cycle is established. Details of the limnological consequences of meromixis in the Kosi lakes system are given by Allanson & van Wyk (1969) and for Swartvlei by Robarts & Allanson (1977). Both monimolimnia are anaerobic with typical reversed salinity and temperature profiles, and in Swartvlei Howard-Williams (1977) has shown that high concentrations of total phosphorus (103 µg.ℓ$^{-1}$) are to be expected below the halocline as compared with 5-10 µg. ℓ$^{-1}$ in the epilimnion.

This serves to introduce the phosphorus and nitrogen concentrations in these diverse water bodies. Allanson & van Wyk (1969) were the first to report the results of a series of soluble reactive phosphorus (SRP) analyses using Murphy & Riley's method with hexanol extraction (Golterman, 1969). The concentrations obtained in Lake Sibaya and Lake Nhlange varied from 19 to 55 µg.ℓ$^{-1}$. Hart & Hart (1977) reported 3 to 5 µg.ℓ$^{-1}$ of SRP for Lake Sibaya during 1970-71. The most detailed work on phosphorus dynamics in a coastal lake is that of Howard-Williams and his team in the field laboratory on the Swartvlei lagoon. In an introductory note, Howard-Williams (1977) reports values for (1) soluble reactive phosphorus, (2) total soluble phosphorus, (3) particulate phosphorus and these together are recorded as (4) total phosphorus. The effect of the meromixis is clearly seen in his Table 2 with epilimnetic total phosphorus varying between 19 and 27 µg.ℓ$^{-1}$ while that of the monimolimnion is 103 µg.ℓ$^{-1}$. All these data, including those of Robarts (1973), fall within close limits.

Soluble reactive phosphorus (µg.ℓ$^{-1}$)			
Lake Sibaya	Nhlange	Mpungwini	Swartvlei
55 ± 25	19*	100*	17 ± 1

* composite samples.

These nutrient concentrations indicate that the surface waters of these lakes are at the oligotrophic-mesotrophic boundary.

All the coastal lakes with the exception of Lake Sibaya are brackish, having normal salinity levels varying from:

	Nhlange	Mpungwini	Swartvlei	Poelela	Sibaya
S ‰	3	4/13,8*	4,3/31*	8	135 mg/ℓ.Cℓ$^{-1}$

* salinity (‰) above and within monimolimnion

The eastern coastal lakes tend to have alkaline, well-buffered water systems with the pH varying from 8,0 to 8,3 while the southern lakes, and in particular Swartvlei, receive large quantities of humate-stained water from catchments which are entirely confined within Table Mountain sandstone. In this respect, the southern lakes are reminiscent of the humic 'perched' lakes described by Bayly *(op cit)*, albeit their respective sources of humic materials differ somewhat. The ionic proportions in the various eastern lakes, although tending towards equivalence to sea water, demonstrate a remarkable elevation in sodium thus upsetting the ion ratios expected from salt derived from the sea. Allanson & van Wyk (1969) and Hill *et al* (1975) have drawn attention to this. The elevated Na$^+$ in the coastal lakes may in fact be influenced by the ionic concentrations of the waters contained within the sand aquifers of their respective catchments.

The biological significance of these ionic proportions is the maintenance of a euryhaline estuarine biota in the lakes which reaches its extreme in the fresh, but nonetheless Na$^+$ and Cl$^-$ enriched waters of Lake Sibaya.

e 15.2
nfluence of mouth closure, increased river flow and wind on some aspects of the physico-chemical limnology of vlei lagoon and planktonic primary productivity.

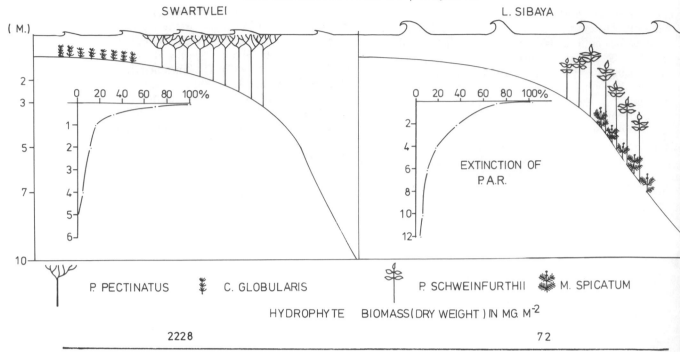

Figure 15.3
The vertical distribution of hydrophytes in Swartvlei and Lake Sibaya and their respective biomass in dry mass g.m^{-2}. The extinction (as %) of photosynthetically active radiation (PAR) in the water column is given for each lake.

BIOLOGICAL CONDITIONS

In view of the excellence of Tinley's unpublished report (1958) on the plants and animals of Lake Sibaya and the northern Lake Nhlange, it was possible to concentrate on specific problems which relate directly to the IBP programme and the application of the ecosystem concept in analytical limnology.

The plant and animal community in Lake Sibaya was described by Allanson *et al* (1974), and although this particular coastal lake has no surface connection with the sea, the checklist provided a basis for comparison with the other lake systems referred to earlier. None of the lakes so far studied is enriched to such a degree that the phytoplankton community dominates the primary producer component. Indeed, in all of the lakes, the macrophyte (hydrophyte) community is obvious to different degrees. While Swartvlei possesses the most striking littoral hydrophyte biomass, those of Lake Sibaya and Lake Poelela are, by comparison, greatly reduced. The hydrophyte community is largely dominated by two species of *Potamogeton. P. pectinatus* is common to all the lakes while *P. schweinfurthii* has so far only been found in Lake Sibaya. *Myriophyllum spictatum* is found over quite wide areas of Lake Sibaya down to a depth of 6-7 m, while in Swartvlei the algal species *Chara globularis* and *Lamprothamnion tepulosum* are abundant in the shallow littoral. A comparison of hydrophyte biomass and vertical distribution between Lake Sibaya and Swartvlei is illustrated in Figure 15.3.

Lake Poelela is certainly the most transparent of the series. Hill *et al* (1975) report 1 % of surface irradiance at 22 m as compared with 12 m in Lake Sibaya and 2-4 m in Swartvlei. So far, both *Najas* sp and *Chara* sp have been found at 10-12 m in Lake Poelela while Boltt *et al* (1969) demonstrated that in Lake Sibaya the compensation depth for *Myriophyllum spicatum* was 5 m. Both Lake Poelela and Lake Sibaya demonstrate the influence of wave action upon the upper limit of distribution of hydrophytes. In neither lake are they to be found above a depth of 3 m except in very sheltered areas where the community is dominated by *Ceratophyllum demersum*. In contrast to this, Swartvlei, being relatively protected from prevailing winds, has developed a luxuriant hydrophyte community. Clearly this is not the whole story: for example, the sediment phosphorus reserves differ greatly between Swartvlei and the more subtropical lakes. Howard-Williams (1979) has also stressed the importance of hydrophyte beds for the accumulation of allochthonous and autochthonous organic matter which, while significant in Swartvlei, is poorly represented in the macrophyte beds of Lake Sibaya. This is due largely to currents of 0,2-0,6 kts over the weed bed and the paucity of allochthonous material entering the lake.

An estimate of macrophyte production for Lake Sibaya and Swartvlei is given in Table 15.2.

This table demonstrates the remarkable difference which can exist between lakes as to where production

Table 15.2
Submerged macrophyte standing stock and productivity as compared with pelagic (phytoplankton productivity) in Lake Sibaya and Swartvlei

	Macrophytes	Phytoplankton
Lake Sibaya		
area occupied (km²)	5,00	65
mean biomass (g.m⁻²)	72	
calculated productivity (g.m⁻².yr⁻¹)	220 (1 100)	
observed productivity (g.m⁻².yr⁻¹)		294 (19 110)
Swartvlei		
area occupied (km²)	2,25	11
mean biomass (g.m⁻²)		
P.pectinatus	2 088	
Chara globularis	140	
observed productivity (g.m⁻².yr⁻¹)		8 (88)
P.pectinatus	1 841 x 10⁶	

The numbers in brackets are the productivity (in g.yr⁻¹) over the total area in km².

Table 15.3
A comparison of the mean primary productivity of a number of coastal lakes and the completely tidal Langebaan Lagoon, near Saldanha in mg.C.m⁻³.day⁻¹

		Langebaan Gran's technique	Lake Sibaya ¹⁴C	Swartvlei ¹⁴C
winter	Stn. 6	183	97	43
summer	Stn. 6	294	144	92

emphasis lies. Thus, in Lake Sibaya, phytoplankton production is about 20 times greater on an areal basis than macrophyte production in the main basin of the lake, while in Swartvlei, the opposite pertains. The reasons for these differences are not immediately obvious. Two major factors appear to be the humic staining of the water of Swartvlei with associated reduction in light penetration, and the low values of soluble reactive phosphate in the epilimnion. Lake Sibaya is a great deal more transparent and this, coupled with a higher SRP concentration, will explain the higher standing stock of phytoplankton and its production. It does not explain the comparatively low macrophyte production.

The structure and seasonal cycles of the phytoplankton of Lake Sibaya have been reported by Hart & Hart (1977), while Allanson & Hart (1975) have recorded the primary production of this community as estimated by ¹⁴C fixation for the period July 1973 to August 1974. The maximum recorded phytoplankton productivity was 1 847 mg.C.m⁻².d⁻¹ during November 1972 while the minimum, 227 mg.C.m⁻².d⁻¹, was obtained during February under conditions of 10/10 cloud. This demonstrates the sensitivity of subtropical lake systems to light intensity even during summer.

Comparable data for Swartvlei were published by Robarts (1976), and are co-ordinated in Figure 15.2. The low phosphate values and the transparency have been referred to earlier. These, together with the high

variable salinity regime in the euphotic zone, are responsible for the lowest phytoplankton productivity recorded so far in southern African coastal lakes, namely 22,9 mg C.m⁻².hr⁻¹. The rôle of the extensive beds of *Potamogeton pectinatus* and *Chara* spp in storing and cycling phosphorus is slowly emerging, largely as a result of Dr Howard-Williams' work. What remains enigmatic is the effect, if any, this macrophyte subsystem has upon the phosphorus dynamics of the open water and hence the growth of phytoplankton. The effect of this subsystem upon the phytoplankton community in Lake Sibaya appears to be minimal, judging from the remarks of Howard-Williams (1979). The phytoplankton biomass and its production is governed very largely by phosphorus and nitrogen levels and the optical density of the lake. These factors appear to remain fairly constant during the year, at levels insufficient to cause rapid growth of algal populations at specific periods, so that biomass, as represented by α-chlorophyll concentration, has no systematic variation, and remains between 2 and 4,5 μg.ℓ⁻¹ (Hart & Hart *op cit*).

A comparable, but by no means similar coastal environment is found in Langebaan lagoon which has been subject to a detailed investigation by a group from the University of Cape Town. Of particular interest was the measurement of primary production in the lagoon using the light and dark bottle technique. Henry *et al* (1977) have demonstrated a marked seasonal variation in gross primary productivity with lowest values during the southern winter, namely 183 mg C.m⁻³.day⁻¹ and the highest, 885 mg C.m⁻³.day⁻¹ during summer. These authors have reported a striking horizontal variation in the lagoon, with productivity decreasing towards the head of the lagoon where nitrate concentrations are low. Notwithstanding the differences in technique used, Table 15.3 records the present low productivity of the coastal lakes in which marine influence is minimal.

The important influence of the sea is underlined by these results and is confirmed very largely by Liptrot's (1978) work on community metabolism in the Swartvlei channel. His work has demonstrated the dependence of the system upon tidal introduction of nutrients into the channel. With the exception of those estuaries affected by known sources of nutrients either in the form of sewage effluents or agricultural run-off, the coastal lakes so far studied show the opposite condition from those in the northern hemisphere, where the river inflow is a substantial source of nutrients.

A direct comparison of the data from coastal lakes and Langebaan Lagoon, given in Table 15.3, is unrealistic as the data have been collected by two dissimilar methods. Nevertheless, there is evidence in the literature (Lewis, 1974) which suggests that the ¹⁴C measurement of productivity can be equivalent to gross production as measured by Gran's technique. Thus the data in Table 15.3 show that marked differences in energy fixation occur in these coastal systems.

While our understanding of the nanoplankton community of these lakes is poor, Robarts (1976) has demonstrated the existence of acetate uptake by flagellate populations which increase in density during flooding of Swartvlei. Allanson & Hart (1975) have pointed to the importance of the nanoplankton community in energy fixation in Lake Sibaya. Comparison of carbon fixation in water filtered through a 35 μm Nitex filter, with fixation in unfiltered water, showed no significant change. This emphasises the importance of the microphytoplankton in energy fixation whether by autotrophic or heterotrophic organisms in these lakes.

The fauna of the coastal lakes, as might be expected, is influenced markedly by the extent to which the sea influences the salinity regime. Thus the structure of their faunal community is governed almost entirely by the species' ability to withstand *either* an elevated ionic concentration *or* a lowered concentration. Subsidiary factors such as wind-induced wave action, sand abrasion and silt concentrations are largely responsible for distribution within the lakes.

The interaction between community structure and the lake environment is best illustrated in Lake Sibaya. Allanson *et al* (1966) recorded the existence of a normal freshwater fauna in the lake and discovered the occurrence of a fauna hitherto only known from estuaries in South Africa. Of particular note was the presence of Amphipoda, Tanaidacea, Isopoda, polychaetes and both zoeae and adults of the crab *Hymenosoma orbiculare*. In an earlier unpublished report, Tinley had recorded the presence of the euryhaline fish *Gilchristella aestuarius* and *Hepsetia breviceps*.

The presence of this essentially 'relict' fauna pointed to the estuarine origin of the lake and this has been adequately confirmed. It also pointed to the remarkable osmotic tolerance of groups hitherto believed to be unable to function physiologically at such low ionic strengths. As noted in chapter 11, Forbes & Hill (1969) demonstrated that hyperosmotic regulation is maintained in *Hymenosoma orbiculare* below 31 ‰ and that below 6,5 ‰, a further drop in blood osmotic pressure occurred. The individuals living in Lake Sibaya maintained the osmotic pressure of the haemolymph at 538 m osmoles. This second decrease in blood osmolarity was matched by individuals taken from sea water in the Kowie estuary and obviously represents a physiological mechanism to reduce the amount of energy that would be required to maintain steep osmotic gradients. These authors also showed that the survival of the zoeae depended upon whether or not the eggs were extruded in high or low salinity water. The zoeae from eggs extruded in low salinity showed maximal tolerance as compared with those from eggs extruded in middle range salinities and hatched in low salinity (3,5-1,7 ‰). The beneficial effect of temperature upon tolerance at low salinity

Table 15.4
The structure of the planktonic faunal community in three coastal lakes in ascending order of salinity. Only the common animal groups have been listed.
(1) Lake Sibaya (135 mg/ℓ.Cl⁻)
(2) Lake Nhlange (S ‰ = 3)
(3) Swartvlei (S ‰ = 3-12)

	(1)	(2)	(3)
Plankton			
Coelenterata medusae*		●	
Ostracoda	●	●	●
Cladocera			
Moina sp	●	●	−
Bosminia longirostris	●	●	−
Copepoda			
Mesocyclops aequatorialis	●		
Halicyclops sp		?	●
Thermocyclops emini	●		
Tropocyclops brevis	●		
Pseudodiaptomus hessei	●	●	●
Acartia natalensis		?	●
Amphipoda			
Grandidierella lignorum	●	●	●
Brachyura			
Hymenosoma orbiculare (zoeae)	●	●	●
Mollusca			
Veliger larvae *(Musculus virgiliae)*	−	●	●

* During January 1966 medusae, probably of *Oustromovia inkermanica* were particularly abundant in daytime hauls (>80 % frequency)

Table 15.5
Zooplankton settled volumes in millilitres from H_{10} hauls using a 158 μm mesh net with a mouth opening of 300 mm. Mean values in brackets

Lake Sibaya	Lake Nhlange	Swartvlei
5-9 (9)	3-19 (8)	0,5-2 (?)

was shown experimentally to be maximal at 22 °C and 0,8 ‰.

Essentially similar work has been done by Boltt (1969) on the amphipod fauna in both Lake Sibaya and the Kowie River estuary. Once again a remarkable osmoregulatory ability has been demonstrated. A similar inference can be drawn for the calanoid copepod *Pseudodiaptomus hessei* which is known from Lake Sibaya, and from fully tidal estuaries and salinities up to 60 ‰.

The structure of the zooplankton community of Lake Sibaya, Lake Nhlange and Swartvlei is given in Table 15.4.

The wide tolerance of salinity variation is stressed once again, with the cladocerans providing the exception. In fact, the Cladocera in Lake Sibaya are unimportant elements of the zooplankton. Likewise, some of the cyclopoid Copepoda found in Lake Sibaya do not occur in the more saline lakes of Nhlange and Swartvlei. In the latter, the genus *Halicyclops* is found. In all the lakes *Pseudodiaptomus hessei* is numerically dominant with frequency of 0,8, if collections are made at night. On two occasions in Lake Nhlange, spat of the bivalve *Musculus virgiliae* were dominant during winter.

The biomass of the zooplankton community is reflected in Table 15.5

The numerical importance of *Pseudodiaptomus hessei* was closely matched by its importance in the settled volume of the plankton and led to a detailed investigation of its productivity and turnover by Hart & Allanson (1975). Using Winberg's method for the estimation of mean daily production, these authors report a mean standing stock over the 27 month period of 5,7 mg (dry mass).m^{-3} with a mean productivity of 0,634 mg dry mass $m^{-3}.day^{-1}$. The daily P/B ratio was 0,11 which is high compared to other copepod values. The biomass and productivity remains low when compared with values in eutrophic lakes. In Lake George in Uganda, Burgis (1971) reports a mean daily production of *Thermocyclops hyalinus* 750 times as great as that for *P. hessei* on a unit volume basis. The low production in Lake Sibaya is paralleled by the low nitrogen and phosphorus values, and seasonal changes in biomass appear to be more strongly correlated with temperature than with algal standing stock. *P. hessei* ingests bacterioplankton in addition to nanoplankton as well as some net phytoplankton so that it is difficult to decide whether the poor correlation betweeen zooplankton production and chlorophyll content of phytoplankton reflects the calanoids' mixed diet or that temperature does indeed exert an over-riding influence on productivity.

Comparative data from the other coastal lakes are not available. We do not understand the importance if any of this zooplankter in the trophic dynamics of of such lakes and estuaries. The species has, however, provided information upon the factors which control its diel vertical migration. Hart & Allanson (1976) have provided evidence which indicate that *Pseudodiaptomus* is highly photosensitive so that even at a depth of 40 m in Lake Sibaya, light energy levels are very slightly above the threshold of the instars (C_{II}-adults) involved. Consequently any tendency to planktonic distribution is overridden. These authors argue that notwithstanding this high photosensitivity, *P. hessei* can remain on shallow illuminated substrates at light intensities several orders of magnitude higher than its preferred level of light intensity in deep water. Thus the photosensory mechanism tending to drive later instars downwards, is overridden by thigmokinesis on contact with a substrate. This is a logical shut-off mechanism which would prevent undue activity in shallow waters and which may have selective advantage in reducing predation. This argument is supported by the discovery of an endogenous rhythmicity in *P. hessei,* but what is not known at this stage is whether the copepod is responding directly to changes in light intensity, or whether these changes modulate the intrinsic rhythm which controls the migratory pattern. Furthermore, as Lake Sibaya is homothermal, this

migratory pattern would not provide the energetic advantages envisaged by McLaren's hypothesis. Hart & Allanson (1976) suggest that energy conservation, analogous to that arising from occupation of warm and cooler waters, may arise from the *P. hessei's* habit of spending the daylight hours resting on or in the bottom sediments. The possibility that such a diurnal behaviour pattern exists, in what is essentially a fine particle feeder, may support the view that diel migration may provide a mechanism whereby the primary producers in the euphotic zone are allowed a period of recovery, and consequently there is a more economic utilization of food resources.

There remains a wide diversity of biological problems associated with the zooplankton of the coastal lakes. No account of their productivity can be completed until this has been done. At present our understanding of energy flow within this component of the ecosystem is at best fragmentary.

The structure of the benthic community of the coastal lakes based upon common animals is reported in Table 15.6.

Wherever there is a band of littoral hydrophytes, as in Lake Sibaya and particularly in the Swartvlei lagoon, there is a marked increase in insect diversity. In this respect, tolerance of salinity variations is quite remarkable. For example Forbes & Allanson (1970) have shown the tolerance to increased salinity within the genus of Baetid mayflies, *Cloeon*. While *Cloeon crassi* possesses the ability to hyporegulate between 8 and 20 ‰ salinity, this ability is not shared by either *C. virgiliae* or *C. africanum* which osmoregulate up to 8 ‰. The presence of *Cloeon crassi* in Swartvlei lagoon is, therefore, not surprising as the salinity rarely rises above 13 ‰.

The tendency to report faunal abundance in the form of biomass is to be welcomed. Due very largely to the work of the late Dr R.E. Boltt, we are in a position to compare the biomass of many of the common benthic animals in the lakes. In Table 15.7 maximum values are given for three of the lakes.

Of particular interest in this table is the similarity between the standing stocks of a variety of benthic species in the lakes, notwithstanding the striking difference in the littoral sediments shown in Table 15.8.

It would seem that the autochthonous production of detritus does not correlate with the biomass of benthic animals. Although the surface provided by the stems of the dense palisade of *Potamogeton pectinatus* in Swartvlei is used extensively by *Musculus virgiliae* and the tubiculous polychaete *Ficopomatus enigmatica*, Boltt (1973) has recorded a dry biomass of 414,0 gm. m^{-2} for shell-free *M. virgiliae* and 2,9 gm.m^{-2} for *Ficopomatus enigmatica*. These are exceptionally high values and as both species are planktonic feeders, the food resource of this niche within the littoral needs examination.

It is appreciated that these biomass data only be-

Table 15.6
The common animals of the benthos of Lakes Poelela, Nhlange, Sibaya and Swartvlei lagoon (● indicates presence and gaps mean that the taxon has not been recorded).

(1) Lake Poelela
(2) Lake Nhlange
(3) Lake Sibaya
(4) Swartvlei

	(1)	(2)	(3)	(4)
NEMATODA	●	●	●	●
CRUSTACEA				
Amphipoda				
Grandidierella bonnieroides		●		
G. lignorum			●	
Corophium triaenonyx		●	●	●
Orchestia ancheidos		●		
Tanaidacea				
Paratanais sp	●			
Apseudes digitalis			●	●
Isopoda				
Cyathura carinata			●	●
Pontogeloides sp	●		●	
Pseudosphaeroma barnardi			●	●
Cirolana fluviatilis		●		●
Decapoda				
Hymenosoma orbiculare	●	●	●	●
Rhyncoplax bovis		●		
Caridina nilotica			●	
MOLLUSCA				
Musculus virgiliae	●	●	●	●
Bellamya capillatus			●	
Melanoides tuberculata			●	
Bulinus natalensis			●	
ANNELIDA				
Polychaeta				
Ceratonereis keiskama	●	●	●	
Oligochaeta		●	●	●
COELENTERATA				
Hydroidea				
Ostroumovia inkermanica	●	●		
Hydra sp			●	
INSECTA				
Ephemeroptera				
Povilla adusta	●		●	
Baetis ?bellus			●	
B. ?crassi			●	
Cloeon crassi				●
Austrocaenis capensis			●	●
ODONATA				
Anisoptera				
Aeshna minuscula			●	
Paragomphus hageni			●	
Enallagma spp			●	●
Pseudagrion spp			●	●
TRICHOPTERA				
Economus thomasetti			●	
Oecetis sibayiensis			●	
DIPTERA				
Chironomidae				
Chironomus sp	●	●	●	●
Polypedilum sp	●		●	
Tanypus sp			●	
Cryptochironomus sp	●		●	●

Table 15.7
Maximum values of the biomass (in dry grams.m^{-2}) of some common benthic animals in three coastal lakes

	L. Poelela	L. Sibaya	Swartvlei
Polychaeta			
unsorted	–	–	0,338
Ficopomatus enigmatica	–	–	2,9
Ceratonereis keiskama	0,009	0,020	
Amphipoda			
Grandidierella lignorum	–	0,366	0,368
G. bonnieroides	0,143	–	
Corophium triaenonyx	–	0,133	0,058
Melita zeylanica	0,009	–	0,065
Tanaidacea			
unsorted			1,056
Apseudes digitalis	–	0,713	
Isopoda			
Pontogeloides latipes	0,012	ND	ND
Sphaeromidae	ND	ND	
Decapoda			
Hymenosoma orbiculare	ND	0,036	0,134
Caridina nilotica	ND	3,520	absent
Mollusca			
Prosobranchia			
Bellamya capillatus and			
Melanoides tuberculata		4,9	absent
Lamellibranchia			
Musculus virgiliae	2,3	absent	414,0

Table 15.8
The physico-chemical structure of littoral sediments

	L. Sibaya	Swartvlei
Detritus (sediment surface)	1 g.m^{-2}	115 ± 13 g.m^{-2}
% total nitrogen	0,04	1,52
% total phosphorus	–	0,11

come significant when they are linked to turnover times. Dr R.E. Boltt's unpublished data on the productivity of *Grandidierella lignorum* show that with a P/B.d^{-1} of 0,078, under the environmental conditions of Lake Sibaya, *G. lignorum* produces a total biomass of 10,4 gm.m^{-2}.yr^{-1}. From these data and the low value of the standing stock (0,366 g.m^{-2}), it follows:

(a) that utilization of this production by a higher tropic level is high, *or*

(b) that the production is rapidly degraded by bacteria to CO_2 and detritus.

An exactly similar situation exists for *G. lignorum* in Swartvlei where an annual production of 5,7 g.m^{-2} has been reported.

The contribution and importance of these secondary producers in the energy transfer processes in these coastal lakes is thus unknown. As such it represents an area of fruitful research.

The role of herbivorous fish and predators is equally doubtful, although the work of Bowen (1976) and Caulton (1977) on the herbivorous cichlid *Sarotherodon mossambicus,* has provided some information on the energy turnover of this demersal feeding fish.

The data used in this argument, come from Bruton's (1973) work on the structure of *S. mossambicus* populations on the terraces of the south basin of Lake Sibaya

Table 15.9
Estimates of total metabolic energy requirements in kJ per day^{-1} for six juvenile fish weighing 13-14 g and feeding from an area of 1 m^{-2} of terrace sand in Lake Sibaya. The efficiency of utilization of the net production of benthic algae is also reported.
(1) Temperature (°C)
(2) Total metabolic energy (Caulton (1977) for *Tilapia rendalli)*
(3) Total assimilated energy (Bowen (1976) for *Sarotherodon mossambicus*
(4) Total energy requirements of kJ – six fish^{-1}.m^{-2}.d^{-1}
(5) Percentage efficiency utilization Allanson (1979) has reported a net production of 1 630 mg C.m^{-2}.d^{-1} by epissammic algae which is equivalent to 61,32 kJ.m^{-2}.d^{-1})

(1)	(2)	(3)	(4)	(5)
18	0,37	–	2,2	3,6
28	1,02	1,4	6,12-8,4	10-14
35	1,20	–	7,2	12

Table 15.10
The degree of isolation from the sea and the composition of the fish fauna of five African coastal lakes (after Blaber, 1978).

Lake	Degree of isolation	Fish fauna
Poelela	75 km tenuous connection with sea	Marine fish present but fauna dominated by freshwater Cichlidae
Nhlange	Connected to sea via Kosi lakes and estuary	Freshwater fish present but the fauna is dominated by marine species
Sibaya	Isolated from sea	Estuarine relict fauna present dominated by freshwater Cichlidae
St Lucia	Directly connected by an estuarine channel to sea	Dominated by marine fish
Swartvlei	Directly connected by an estuarine channel to sea	Dominated by marine fish

during summer when terrace water temperatures are about 29-31 °C. Bowen (1976) estimated the food resources of the terrace sands and their utilization by juveniles of *S. mossambicus*. Caulton (1977) also determined the total metabolic requirements of the macrophyte feeding cichlid *Tilapia rendalli.*

Bruton (1973) recorded that during summer, the terraces were utilized by juvenile *S. mossambicus* with a wet mass of 13-14 gm, and estimated their density as six individuals per square metre. Using these data, and the daily energy assimilation of 1,25 kJ per fish from Bowen (1976), it is possible to compare these data for *S. mossambicus* with the more extensive data for *T. rendalli* published by Caulton (1977). The influence of temperature upon total metabolic energy requirements may thus be estimated. In Table 15.9 these data are combined with data of Allanson (1979) on the net production of the benthic algal community of Lake Sibaya to give an estimate of the utilization of this production by juveniles of *S. mossambicus. sambicus.*

The discrepancy in Table 15.9 between Caulton's (1977) data for *T. rendalli* and Bowen's (1976) data for *S. mossambicus* at 28 °C is due to two factors. First, different herbivorous cichlid species are being compared and second, I have used Caulton's total metabolic energy equation without correction for the effect of size difference upon oxygen uptake.

This table shows the prime importance of temperature on the efficiency of utilization of a food resource. Bowen (1976) has also shown that on the littoral, wave action decreases feeding intensity. Thus, on days of heavy wave action estimates of food ingested are 240 mg per dry gm of fish, increasing to 271 mg per gm during periods of moderate wave action. This is equivalent to 450 and 500 Joules assimilated per day respectively.

Some idea of the complexity of the factors influencing this component of the coastal lake ecosystem is summarised below:

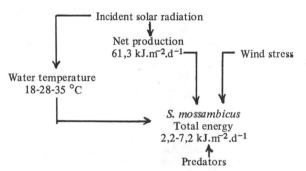

Wind stress is also shown to be important and these two factors may reinforce each other in controlling the use of the littoral habitat by *S. mossambicus.*

While this is essentially a summary of the data presented earlier, it does demonstrate the great importance of abiotic factors upon resource utilization. Considering the effort that has gone into investigating this part of the ecosystem model, we should be exceedingly careful about computer simulation of quantitatively poorly known systems.

UTILIZATION BY MAN

The east coast lakes are remote and are only marginally utilized by the Amatongans. In Mocambique, important fisheries have developed in Lake Poelela and to a lesser extent in Lake Piti, Lake Nhlange and its associated estuary is used only by the local Amatongan people and, as far as I know, there is no important export of processed fish from the area. Blaber (1978) records that the main fish species caught in the fishermen's traps are carnivorous species of *Gerres;* also the spotted grunter *Pomadasys commersonni* and the river bream *Acanthopagrus berda.* Lake Sibaya is hardly used as a source of fish, although Bruton (1973) has recorded some use of 'umono' traps placed in sheltered bays and flooded

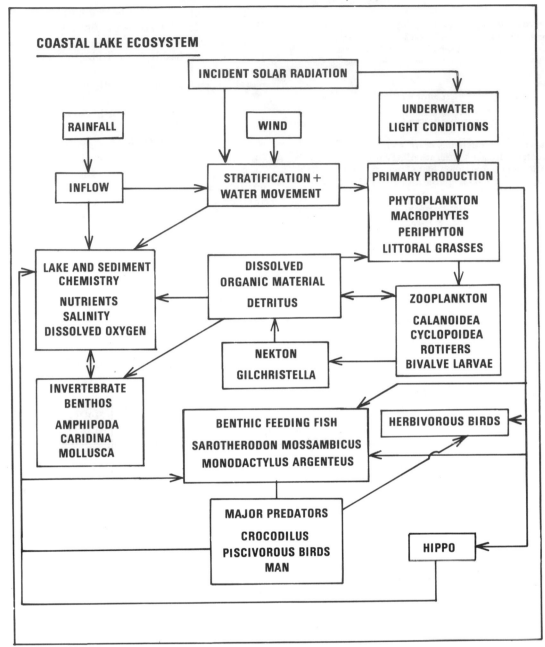

Figure 15.4
Coastal lake ecosystem

trenches between potato fields. The principal fishing areas in this lake are the narrow inlets and Etsheni Bay to the north.

The importance of the fish community to man may well depend upon the degree of isolation of the lakes from the sea. Blaber's (1978) table with the addition of data from Swartvlei is reproduced as Table 15.10.

All the lakes provide a rich food source for a wide diversity of birds. The eastern lakes and lagoons vary dramatically in light transparency. St Lucia is essentially a turbid system in which aquatic macrophytes are only weakly established, while in the lakes with a greater transparency, such as Sibaya and Poelela, wave action on the terraces prevents the establishment of rich beds of macrophytes. As a consequence, the avifauna in these systems is largely piscivorous and dominated by cormorants, pelicans and fish eagles. The southern coastal lakes and lagoons are less windswept and this, coupled with the more advanced succession, provides a very effective habitat for a variety of herbivorous birds, particularly yellow-billed duck and coot.

Figure 15.4 is an attempt to assemble these facts in a model of the coastal lake ecosystem. Various subcomponents of the model are currently being studied in detail, and they demonstrate the complexity of interactions which exist in both space and time. A detailed example of the effects of several changes in the hydroclimate of Swartvlei upon planktonic primary production is given in Figure 15.2. This figure has been constructed from data by Robarts (1976) and Robarts & Allanson (1977). There is no doubt that synthesis at this level helps us to understand the way these ecosystems function. I believe that it is only by this approach that we will be able to make the predictions required for the future management of these coastal lakes. Human pressures are increasing year by year and it is hoped that our work will prevent the *ad hoc* decisions of land developers, and provide a sound ecological framework upon which their rational utilization may be based.

REFERENCES

ALLANSON, B.R. 1979. Lake Sibaya. *Monographiae Biologicae.* **36.**

ALLANSON, B.R., BRUTON, M.N. & R.C. HART 1974. The plants and animals of Lake Sibaya, KwaZulu, South Africa. A checklist. *Rev. Zool. Afr.* 88(3):507-532.

ALLANSON, B.R. & R.C. HART 1975. The primary production of Lake Sibaya, KwaZulu, South Africa. *Verh. Int. Verin. Limnol.* 19: 1426-1433.

ALLANSON, B.R., HILL, B.J., BOLTT, R.E. & V. SCHULTZ 1966. An estuarine fauna in a freshwater lake in South Africa. *Nature*, Lond. 209(5022): 532-533.

ALLANSON, B.R. & J.D. VAN WYK 1969. An introduction to the physics and chemistry of some lakes in Northern Zululand. *Trans. roy. Soc. S.Afr.* 38(3): 217-240.

BAYLY, I.A.E. 1964. Chemical and biological studies on some acidic lakes of East Australian sandy coastal lowlands. *Aust. J. mar. freshwat. Res.* 15: 56-72.

BLABER, S.J.M. 1978. Fishes of the Kosi system. *Lammergeyer* 24: 28-41.

BLABER, S.J.M., HILL, B.J. & A.T. FORBES 1974. Infratidal zonation in a deep South African estuary. *Mar. Biol.* 28: 333-337.

BOLTT, R.E. 1969. *A contribution to the benthic biology of some southern African lakes.* PhD thesis, Rhodes Univ., Grahamstown.

BOLTT, R.E. 1973. Coastal lakes benthos. *Rep. Inst. freshw. Studies 1972/73.* Rhodes Univ., Grahamstown.

BOLTT, R.E. 1975. The benthos of some southern African lakes. Part IV: The benthos of Lagoa Poelela. *Trans. roy. Soc. S.Afr.* 41(3): 273-281.

BOLTT, R.E. 1975. The benthos of some southern African lakes. Part V: The recovery of the benthic fauna of St Lucia Lake following a period of excessively high salinity. *Trans. roy. Soc. S.Afr.* 41(4): 295-323.

BOLTT, R.E., HILL, B.J. & A.T. FORBES 1969. The benthos of some southern African lakes. Part 1: Distribution of aquatic macrophytes and fish in Lake Sibaya. *Trans. roy. Soc. S.Afr.* 39(3): 241-248.

BOWEN, S.H. 1976. *Feeding ecology of the cichlid fish Sarotherodon mossambicus in Lake Sibaya, KwaZulu.* PhD thesis, Rhodes Univ., Grahamstown.

BRUTON, M.N. 1973. *A contribution to the biology of Tilapia mossambica Peters in Lake Sibaya, South Africa.* MSc thesis, Rhodes Univ., Grahamstown.

BURGIS, M.J. 1971. The ecology of production of copepods, particularly *Thermocyclops hyalinus,* in the tropical Lake George, Uganda. *Freshwat. Biol.* **1**: 169-192.

CAULTON, M.S. 1977. The effect of temperature on routine metabolism in *Tilapia randalli* Boulenger. *J. Fish. Biol.* **11**: 549-553.

FORBES, A. T. & B.R. ALLANSON 1970. Ecology of the Sundays River. Part I: Water chemistry. *Hydrobiologia* 36(3-4): 479-488.

FORBES, A.T. & B.J. HILL 1969. The physiological ability of a marine crab *Hymenosoma orbiculare* Desm. to live in a subtropical freshwater lake. *Trans. roy. Soc. S.Afr.* 38(3): 271-283.

GOLTERMAN, H.L. 1969. *Methods for chemical analysis of fresh waters.* IMP Handbook No 8, Blackwell, Oxford.

HART, R.C. & B.R. ALLANSON 1975. Preliminary estimates of production by a calanoid copepod in subtropical Lake Sibaya. *Verh. Int. Verein. Limnol.* 19(2): 1434-1441.

HART, R.C. & B.R. ALLANSON 1976. The distribution and diel vertical migration of *Pseudodiaptomus hessei* (Mrazek) (Calanoida: Copepoda) in a subtropical lake in southern Africa. *Freshwat. Biol.* 6: 183-198.

HART, R.C. & R. HART 1977. The seasonal cycles of phytoplankton in subtropical Lake Sibaya: A preliminary investigation. *Arch. Hydrobiol.* 80(1): 85-107.

HENRY, J.L., MOSTERT, S.A. & N.D. CHRISTIE 1977. Phytoplankton primary production in Langebaan Lagoon and Saldanha Bay. *Trans. roy. Soc. S.Afr.* 42(3/4): 383-398.

HILL, B.J. 1975. The origin of southern African coastal lakes. *Trans. roy. Soc. S.Afr.* 41(4): 225-240.

HILL, B.J., BLABER, S.J.M. & R.E. BOLTT 1975. The limnology of Lagoa Poelela. *Trans. roy. Soc. S.Afr.* 41(3): 263-271.

HOBDAY, D.K. 1976. Quaternary sedimentation and development of the lagoonal complex, Lake St Lucia, Zululand. *Ann. S.Afr. Mus.* 71: 93-113.

HOBDAY, D.K. 1979. Lake Sibaya. *Monographicae Biologicae* **36.**

HOWARD-WILLIAMS, C. 1977. The distribution of nutrients in Swartvlei, a southern Cape coastal lake. *Water S.A.* 3(4): 213-217.

HOWARD-WILLIAMS, C. 1979. The growth and production of aquatic macrophytes in a south temperate saline lake. *Verh. Int. Verein. Limnol.* 20.

HOWARD-WILLIAMS, C. 1979. Lake Sibaya. *Monographiae Biologicae* 36.

HUTCHISON, I.P.G. & W.V. PITMAN 1973. *St Lucia Lake Research Report. Vol 1. Climatology and hydrology of the St Lucia Lake system.* Hydrological Research Unit, Univ. Witwatersrand, Johannesburg.

KING, L. 1970. Uloa revisited. *Trans. roy. Soc. S.Afr.* 73: 151-158.

KING, L. 1972. *The Natal monocline explaining the origin and scenery of Natal, South Africa.* Geol. Dept. Univ. Natal, Durban.

LEWIS, W.M. 1974. Primary production in the plankton community of a tropical lake. *Ecol. Monogr.* 44: 377-409.

LIPTROT, M.R.M. 1978. *Community metabolism and phosphorus dynamics in a seasonally closed South African estuary.* MSc thesis, Rhodes Univ., Grahamstown.

ORME, A.R. 1973. Barrier and lagoon systems along the Zululand coast, South Africa. *Tech. Rep., US Office of Naval Research* No 1. (Also) *In:* D.R. Coates (ed), *Coastal geomorphology:* 181-217. Geological State Univ., Binhampton, NY.

PITMAN, W.V. & L.P.G. HUTCHISON 1975. *A preliminary hydrological study of Lake Sibaya. Hydrological Research Unit Report* No 4/75. Univ. Witwatersrand, Johannesburg.

ROBARTS, R.D. 1973. *A contribution to the limnology of Swartvlei: The effect of physico-chemical factors upon primary and secondary production in the pelagic zone.* PhD thesis, Rhodes Univ., Grahamstown.

ROBARTS, R.D. 1976. Primary productivity of the upper reaches of a South African estuary (Swartvlei). *J. exp. mar. Biol. Ecol.* **24**: 93-102.

ROBARTS, R.D. 1976. A preliminary study of the uptake of dissolved organic compounds by the heterotrophic microbial populations of Swartvlei, South Africa. *Trans. Rhod. Sci. Ass.* **57**(5): 35-44.

ROBARTS, R.D. & B.R. ALLANSON 1977. Meromixis in the lake-like upper reaches of a South African estuary. *Arch. Hydròbiol.* **80**(4): 531-540.

TIMMS, B.V. 1973. A limnological survey of the freshwater coastal lakes of east Gippsland, Victoria. *Aust. J. mar. freshwat. Res.* **24**: 1-20.

TINLEY, K.L. 1958. *A preliminary report on the ecology of the Kosi Lake system, 18.10.1958 to 15.12.1958. Part I: Hippopotamus amphibius. Part 2: General account of the environment.* (Roneoed report). Natal Parks Bd, Pietermaritzburg.

Chapter 16

The estuarine ecosystem and environmental constraints
J.H. Day and J.R. Grindley

Department of Zoology   School of Environmental Studies
University of Cape Town

INTRODUCTION

In this chapter an attempt will be made to consider the estuarine ecosystem as a whole. In earlier chapters the environmental conditions in several types of estuaries were discussed and their effects on the various types of living organisms were described. It was made abundantly clear that estuarine communities are adapted to very variable conditions and individual species can tolerate wider ranges of salinity, temperature and other factors than normally occur. Here, attention is focussed on fundamental processes which link the members of the community under normal conditions and on the physical and chemical constraints imposed by unusual conditions. It is these extreme and infrequent changes such as floods, droughts, serious pollution or the closure of the mouth for long periods which affect the welfare and productivity of an estuarine community.

These primary changes often have a number of secondary effects. Thus floods not only increase water levels and current velocities but also affect salinity distribution, turbidity, and they completely change the distribution of sediments that have been sorted and resorted over many years. It would not be surprising if the effects of such disasters on the estuarine plants and animals were profound. On the contrary, when normal conditions return, the estuarine biota recovers surprisingly rapidly, and the underlying mechanism of this stability must be considered.

Obviously, estuarine ecosystems are not independent of external factors and biological events. The devastating floods are mediated by the geological, climatic and botanical conditions in the catchment which affect not only the volume of river flow but also the rate of siltation. Human activities such as agriculture, drainage, dam construction and bridge building have marked effects which finally reach the estuary. Similarly the size and permanence of the estuary mouth is affected not only by high energy waves and the littoral transport of sediments but also by harbour works. It is emphasised again that estuarine ecosystems are not independent. The proper management of an estuary must include the management of the catchment and due consideration of conditions at the estuary mouth. The problems will be discussed in the next chapter.

ORGANIC DETRITUS

Primary production by autotrophic plants provides the basis for all ecosystems and the biosphere as a whole. The products of photosynthesis may be used directly by herbivorous heterotrophs or indirectly by detritivores. In either case the energy content provided by the plant material drives the whole ecosystem. Primary production is thus the fundamental process in an estuary.

The net annual primary production of estuarine autotrophs has been discussed in chapter 6. There are large variations from one type of estuary to another but the order of magnitude may be summarised as: phytoplankton 50-200 g $C.m^{-2}.yr^{-1}$ in shallow turbid estuaries and 100-500 g $C.m^{-2}.yr^{-1}$ in clear deep estuaries; benthic microalgae 100 g $C.m^{-2}.yr^{-1}$; seagrasses about 500 g $C.m^{-2}.yr^{-1}$; salt marsh macrophytes (including mangroves) 500-1 000 g $C.m^{-2}.yr^{-1}$. In any individual estuary the proportions depend on the relative areas of the intertidal banks and the deep channels.

There are surprisingly few macrophagous herbivores. Sheep, rabbits, water rats, insects and even hippopotami graze on the marginal vegetation and a number of birds and crabs eat the seeds but the total effect is small. Geese and swans crop the fronds of the sea grasses and so do a number of fishes. From the work of Blaber (1973) it would appear that the fish do not digest the leaves and are only able to assimilate the epiphytes and epizoans growing on them. The cropped leaves pass out with the faeces and decay to form detritus. The remainder of the primary consumers are microphagous feeders. Phytoplankton is consumed by zooplankton and other suspension feeders such as barnacles, oysters, mytilids and cockles but, as much organic detritus is suspended in the currents, they must ingest such material as well. It would be interesting to know how much is assimilated and how much it contributes to their energy requirements. The feeding mechanisms of deposit feeders are non-selective and it would also be interesting to know how much the living micro-algae as distinct from micro-heterotrophs and dead organic matter contributes to their metabolism. Estimates by Perkins (1974), Barnes (1974) and Head (1976) suggest that between 80 and 95 % of an estuarine fauna are detritus feeders. The nature and origin of this material is thus of great importance.

Organic detritus includes dissolved organic carbon compounds (DOC) and particulate organic carbon (POC) retained by a 45 μm filter. DOC is formed mainly by living plants and Hellebust (1967) estimated that 10 % of the photosynthate of phytoplankton is excreted as DOC. Head (1976) quotes an estimate of 40 % for the alga *Fucus*. The exudates of living vascular plants are not known but probably much less. Moribund phytoplankton releases an increasing amount of DOC (Wood, 1965) and still more is released during autolysis and microbial decomposition. Hellebust (1967) states that 15 % is lost during zooplankton grazing. Considering the relative biomass of autotrophs and heterotrophs it is presumed that the bulk of DOC is due to the decay of the primary producers. In an analysis of the DOC produced by living phytoplankton, Hellebust found that the main compounds were trihydric alcohols such as glycerol and mannitol and that the nitrogen compounds were mainly polypeptides. Brown algae such as *Fucus* and *Laminaria* also produce complex heterocyclic compounds. These combine with other algal exudates to form humates which are very resistant to aerobic bacterial decomposition.

Similar materials are formed during the anaerobic decomposition of marsh plants.

Particulate organic detritus (POC) consists of dead organic matter and a community of living micro-organisms including not only bacteria and fungi but also micro-algae (many of which are facultative heterotrophs), protozoa and small metazoans (Wood, 1965). POC forms the bulk of the organic detritus; Head (1976) quotes concentrations of 0,5 to 5,0 mg.ℓ^{-1} in suspension and the analysis of various grades of sediment from sand to silty mud gives values of 0,1 to 10,0 % of dry organic matter per gram of dry sediment. The bacterial content also increases in the sediment since their extracellular enzymes and the digestion products are more concentrated (Wood, 1965). Barnes (1974) quotes values of 170 to 460 x 10^6 bacteria per gram of surface sediments. In contrast to this Kirby-Smith (1976) quotes the findings of Wiebe & Pomeroy (1972) that detrital particles in suspension contain few bacteria. He also notes that Thayer *et al* (1974) found that the biomass of suspended phytoplankton was 1,7 to 20 times the mass of heterotrophs in the seston while on the sediments the biomass of the heterotrophs was 11 to 49 times that of the autotrophs.

The origin of organic detritus. Organic detritus in an estuary is derived from three primary sources. From the land, the river carries in fragments of terrestrial plants and riverine vegetation including marginal plants such as reeds and fresh water algae. From the sea, detached algae, sea grasses, marine plankton and decomposing material of all three types is carried in by the tides. Finally, the banks and waters of the estuary itself provide both emergent plants which may range from mangroves to subterrestrial algae and fully aquatic forms including vascular plants and algae of all sizes.

Another method of dividing the sources of detritus would be a distinction between fully terrestrial organisms, marginal forms which are subject to both atmospheric and aquatic conditions and fully aquatic forms.

All the living material dies and decays to form organic detritus and it is both interesting and of great practical importance to determine the proportions of the material derived from the three primary sources. It is almost impossible to do this by conventional means but within the last ten years a promising new technique has been developed. In essence, this is the determination of the ratio of different carbon isotopes in the detritus.

Stable carbon isotopes. Carbon exists in three isotopic forms of which ^{12}C and ^{13}C are stable and ^{14}C is radioactive. ^{14}C is formed in the upper atmosphere through the action of cosmic rays and is very rare (10^{-12} natural carbon). ^{13}C constitutes about 1 % of natural carbon. The ratio of $^{13}C:^{12}C$ is normally related to a standard which is usually a marine belemnite limestone referred to as the 'Chicago PDB standard'. Values are recorded as a deviation (usually negative) from the standard in parts per mille. Thus atmospheric CO_2 has $\delta^{13}C = -7\ \permil$.

Stable carbon isotope ratios can be used with caution to interpret the sources of organic carbon in the environment. This is a method that has been used to examine the origin of organic detritus in the marine environment and promising results have been obtained. The isotopes of carbon are partitioned in a number of ways during the process of photosynthesis. Plants segregate into three groups with respect to their stable carbon isotope composition. Vascular plants divide into two classes. Those plants having the C-3 pathway of photosynthesis, have low $\delta^{13}C$ values of -24 to $-34\ \permil$. Second, those plants having the C-4 pathway of photosynthesis, have higher $\delta^{13}C$ values of -6 to $-19\ \permil$. Third, algae have intermediate $\delta^{13}C$ values of -12 to $-23\ \permil$.

Emery (1969), in his study of the 'Oyster Pond' at Woods Hole, found that organic carbon from eight sediment samples ranged from $-22,8$ to $-25,1\ \permil$ which suggests land plant, or perhaps phytoplankton, sources. The salt-marsh grass *Spartina* and marine macro-algae produce much higher values.

Haines (1976) in a study of the salt marsh near Sapelo Island, Georgia, sampled organic carbon from biota, soils and tidal water. *Spartina alterniflora* had $\delta^{13}C$ values of $-12,3$ to $-13,6\ \permil$ while other salt marsh plants had values between $-22,8$ and $-26,0\ \permil$. Benthic diatoms had intermediate values of $-16,2$ to $-17,9\ \permil$. Marsh soils, and to some extent the invertebrate fauna reflected the carbon isotope ratios of the major plants in the various zones. Rather surprisingly, particulate organic carbon (POC) in the water showed $\delta^{13}C$ values of $-19,8$ to $-22,8\ \permil$. These values fall within the range of $\delta^{13}C$ found for offshore POC, which is usually assumed to originate from phytoplankton photosynthesis.

Tan & Strain (1979) studied organic carbon isotope ratios in recent sediments in the St Lawrence Estuary and Gulf of St Lawrence. The results showed that the organic matter deposited in the Saguenay Fjord ($\delta^{13}C$ $-25,9 \pm 0,4\ \permil$) and the upper St Lawrence Estuary ($\delta^{13}C$ $-25,0 \pm 0,6\ \permil$) is of terrestrial origin. The variability of values in the lower estuary ($-25,6$ to $-21,8\ \permil$) suggests the presence of organic matter from both terrestrial and marine sources. In the open Gulf of St Lawrence remarkably uniform values ($-22,4 \pm 0,2\ \permil$) suggest that the organic matter in the sediments is derived from marine sources.

Although there are limitations to the reliability of the carbon isotope method, it would appear to be the best method available at the present time for determining the origin of organic detritus in estuaries.

Contributions to the detritus in different types of estuaries. The proportion from each primary source depends on the type of estuary. In the salt marshes of Georgia where the ratio of intertidal marsh to open channels is very large, Odum & de la Cruz (1967) found that 94 % of the suspended particulate detritus in the creeks originated from *Spartina,* 5 % from micro-algae and about 1 % from animals. The detritus formed about 90 % of the net seston and phytoplankton about 10 %. In the small Ythan estuary near Aberdeen where the tidal inflow is about 20 times the river flow and most of the estuary bed is exposed at low spring tide, Leach (1971) estimates that 'the marine contribution is at least an order of magnitude greater than that of the fresh water'. The major part of the input is organic detritus and the autochthonous estuarine contribution is small. Benthic micro-algae provide 31 g $C.m^{-2}$ as compared with about 100 g C in many other estuaries while the mean standing crop of *Enteromorpha* is 150 g (dry mass).m^{-2}. Phytoplankton drifts in and out of the estuary with the tides but the net estuarine production is not stated.

The broad, deep waters of Chesapeake Bay provide a sharp contrast. Head (1976) quotes a budget by Biggs & Flemer (1972) of the suspended POC in the upper and middle reaches. If the measurements in these two sections of the estuary are combined it appears that 55 % of the POC is derived from fresh water (mainly the Susquehana River) while 45 % is autochthonous and due to phytoplankton production. The contribution of salt marsh vegetation and benthic algae is not mentioned and is presumably not significant. Of the total POC, 20 % is permanently lost in the sediment, 73,5 % is taken up by hetrotrophs including bacteria and 65 % is transported further down the estuary.

There are few reports regarding the source of DOC. Head (1976) quotes a provisional report by Duursma that 87 % of the DOC in the Wadden See is derived from fresh or saline inflow. Presumably this is mainly refractory material. Much of the colloidal organic matter carried down by the river flocculates where it meets salt water as described in chapter 4 and is eventually deposited in the sediment where it is decomposed by bacteria and fungi. The humic acids are very resistant but the simpler organic compounds are rapidly absorbed. According to the CSIRO report for 1974-76 the turnover of labile DOC in Port Hacking estuary in New South Wales is 8,75 hours. During the phase of active growth, microheterotrophs such as bacteria absorb DOC to form new protoplasm so that it contributes to the food web but during the stationary phase they respire DOC to CO_2 and thus short circuit this fraction of primary production.

The rates of decomposition of different organic compounds. The rate at which the various compounds in POC are decomposed varies from days to months. Fibrous materials such as lignin are very resistant. Odum & de la Cruz (1967) found that after 300 days the percentage dry masses of the original 'dead stand-ing' plants were:

65 % of *Juncus*

47 % of the grass *Distichlis*

42 % of *Spartina*

6 % of *Salicornia*

The crab *Uca* had completely decomposed in 180 days.

During decomposition, the original material divided into smaller and smaller particles and part dissolved. An analysis of the smallest particles of 'nanodetritus' derived from *Spartina,* showed that while the percentage of crude plant fibre remained unchanged, the carbohydrate fraction decreased slightly from 50 to 40 %. The percentage of protein first decreased from 10 % in the living material to 6 % in the dry leaves and then increased to 24 % in the ash-free dry mass of the nano-detritus. This must have been due to the action of bacteria which use the carbon compounds as an energy source to build up bacterial protein. Somewhat similar results were obtained by Fell *et al* (1975) during the decomposition of mangrove leaves. They found that complete decomposition took four months and during the process the ratio of nitrogen to carbon increased.

The food value of organic detritus. It is obvious that organic detritus with its community of micro-organisms is a richer source of protein than the dead macrophytes and phytoplankton from which it is derived. It has long been known that the percentage of organic detritus increases in finer sediments since fine silt, clay and organic particles have similar settling velocities. Long-bottom (1968) summarised the results of earlier workers which showed that the percentage of organic carbon increased logarithmically with decreasing particle size. Newell (1970) reports a similar logarithmic relationship between particle size and organic nitrogen content. He states that his own results and those of other workers show that sediments with a median particle diameter of 0,15 mm may contain between 0,01 and 0,02 % organic nitrogen. The C/N ratio in the deposits is thus less than 10 suggesting that it is mainly heterotrophic material and Newell notes that the C/N ratio in bacteria is of the order of 7. He concludes that 'micro-organisms represent the dominant part of the organic carbon and organic nitrogen values'.

It is well-known that some invertebrates such as *Urechis* and *Mytilus* will grow on a diet of bacteria but Newell (1970) points out that there are others including oyster larvae, which will not. In a classical experiment Newell (1965) demonstrated that the detritus-feeding gastropod *Hydrobia ulvae* digests the micro-organisms which colonise its own faeces. *Hydrobia* were placed in filtered sea water and their faeces were collected. Part were analysed and the dried mass was found to contain 10 % of organic carbon and 0,25 % of organic nitrogen. The remaining faeces were cultured for three days at the end of which the organic carbon content had fallen to 8 % while the organic nitrogen

had increased to 1,75 %. It is presumed that these changes were due to micro-organisms which have used the undigested organic carbon as an energy source and had built up proteins. The enriched faeces were fed to starved *Hydrobia,* and when their own faeces were analysed it was found that the organic carbon content was similar to the material on which they had been feeding but the organic nitrogen content had fallen to ca 0,075 %. Subsequent culturing of the faeces again showed an increase in organic nitrogen. Newell (1970) concludes that dead organic matter which contains much indigestible matter 'is subjected to a cycle in which bacteria develop on its surface and are subsequently digested by deposit feeding organisms. The voided faecal material is then recolonised by bacteria and thus again becomes suitable for food. The whole cycle may be repeated many times until the organic carbon comprising the debris is used up (also Seki & Taga, 1963)'.

It is important to know how widely this system of feeding on bacteria and recycling may be applied. It has been noted that oyster larvae do not grow on a diet of bacteria. The faeces of some bivalves do not decay and are preserved in fossil deposits. Kirby-Smith (1976) has questioned the case of benthic suspension feeders. He reports that the total energy needs of the scallop *Argopecten irradians* in a North Carolina estuary 'could be supplied by phytoplankton even though the algae contribute only 20 % of the total particulate carbon in suspension'. In a later experiment he tested the growth of the scallops in four different media over a period of 18 days. The first was filtered sea water; the second contained a mixed phytoplankton culture with a slightly lower chlorophyll a content than the estuarine water. The third contained freshly ground *Spartina* powder and the fourth contained aged *Spartina* powder kept in suspension for six months.

The scallops in the phytoplankton culture showed the same rate of growth as those in the field while those fed on fresh *Spartina* powder grew at about one-third this rate. Those in filtered sea water and those fed on aged detritus did not grow after the first week. At the end of the experiment all scallops except those in filtered sea water had full stomachs and intestines. An analysis of the phytoplankton and aged detritus showed that the C/N ratio was 11/1 and 12/1 respectively which suggests a fairly high protein content while the fresh detritus (C/N = 34) had a low protein content. 'From these data . . . it would appear that the C/N ratios, and presumably the protein content, do not necessarily indicate the nutritive value of a food resource for scallops'. He further suggests that 'micro-autotrophs (phytoplankton) in suspension are the only major energy sources for suspension feeders'. Whether this suggestion has wide validity remains to be proved. It may be noted that the living carbon content of the aged detritus was low and that suspended detritus is known to be poorly colonised by bacteria.

Table 16.1
Summary of salt marsh energetics derived from studies at
Sapelo Island, Georgia, USA (after Teal, 1962)

Input as light	600 000 kcal.m^{-2}.yr^{-1}
Loss in photosynthesis	563 000 or 93,9 %
Gross production	36 380 or 6,1 % of light
Producer respiration	28 175 or 77 % of gross production
Net production	8 205 kcal.m^{-2}.yr^{-1}
Bacterial respiration	3 890 or 47 % of net production
1st consumer respiration	596 or 7 % of net production
2nd consumer respiration	48 or 0,6 % of net production
Total energy dissipation by consumers	4 534 or 55 % of net production
Export	3 671 or 45 % of net production

The existing techniques of ATP analysis and chlorophyll a determination make it possible to estimate
the proportion of living heterotrophs and autotrophs
in organic detritus. It is an important line of research
and an analysis of growth rates provides a good estimate
of assimilation.

Part of the organic detritus is permanently lost to
the sediments. Biggs (1967) estimated that there was
about 2,8 % of organic carbon at a depth of 25 cm in
the fine bottom sediments of Chesapeake Bay and a
higher percentage may be expected in the subsurface
sediments of mangrove swamps and salt marshes. There
are further losses of organic carbon during microbial
decomposition on the sediment surface. Teal (1962)
has summarised the energy losses at different trophic
levels in a Georgia *Spartina* marsh, as shown in Table
16.1.

BIOTIC RELATIONSHIPS

Any ecosystem will include autotrophs, dead organic
materials, herbivores, detritivores and carnivores. Each
of these groups in turn includes a number of competing
species, some abundant and some rare, some with a
rapid rate of turnover and some long-lived. Migrants
move in and out of the system and materials are gained
and lost. The ecosystem is so complex that some simplified model is obviously needed to appreciate the
material content of each component and the rate of
energy flow or of materials between them. The technique used is systems analysis, originally developed for
physical systems and known as systems ecology when
applied to ecosystems. A general account of the construction of mathematical models, their use in organis-
ing research and their value in predicting the effects of
human intervention is given by Walters in Odum (1971),
by Collier *et al* (1974 — 2nd edition) and in more
recent texts.

Models with varying degrees of complexity may be
constructed according to the information required. A
simple compartment model may be built to indicate
the route along which carbon from the environment is
transferred through the different trophic levels:

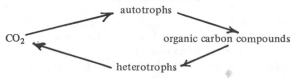

Such a model provides an outline but it is incomplete.
A more complex model is required to show the loss of
carbon by autotrophs during respiration and the direct
consumption of autotrophs by herbivores in addition
to the indirect transfer by detritivores. The predator-
prey relationship among heterotrophs needs to be
elaborated. Moreover quantitative data are required
concerning the carbon content of each compartment
and the quantity transferred from one compartment
to another. The mass of carbon would be a suitable
unit or currency in the above example and ^{14}C is a useful tracer in experiments. Alternatively, nitrogen or
phosphorus might be used as units as these are essential constituents of protoplasm; in such cases the inorganic input and end product would be nitrogen compounds or phosphate. Energy units (kilocalories or
kilojoules) are often used to measure the input or the
conversion of light energy to chemical bond energy in
the process of photosynthesis; next its transfer through
the food web as energy-rich organic compounds and the
loss as heat during metabolism. Finally all the energy
losses are summed as the output. When converting the
compartment model into a mathematical model, each
compartment is represented by a symbol; e.g. A (autotrophs), O (organic carbon), H (heterotrophs) and the
transfers become AO, OH, etc. The total energy loss or
output (z) is the sum of the losses AZ + OZ + HZ.
Since the compartments may include one or more elements, the state of any compartment at any time is
represented by one or more units termed system variables. Interactions between compartments are represented by equations called transfer functions. As there
are inputs and output the boundaries to the ecosystem
are defined. Inputs are represented by equations termed
forcing functions and the constants in the equations
are termed parameters. The mathematical model thus
becomes a set of equations representing the rate of
change of the system with time.

The mathematical model is valuable in many ways.
It provides an understanding of the quantitative
relationships throughout the ecosystem and indicates
those relationships which are most important and
must be measured most accurately. If the ratio of input
to output remains constant when environmental conditions vary within measured limits, it indicates that
within these limits the ecosystem is in a steady state.
The effect of human changes of the environment or
changes in the ratio of plants to herbivores to carnivores by importations or culling may thus be evaluated.

The analysis of a whole ecosystem is a long-term
project demanding many specialists. Environmental

conditions must be monitored so that they may be correlated with system variables and the fluxes between compartments must be calculated. The changing growth rates and the biomass changes of each major element in a compartment is obtained by field measurements while energy losses may be obtained by the difference between gross and net production or laboratory experiments under known conditions. Experiments also provide the parameters required for the equations representing transfer functions between compartments, such as plants to herbivores or herbivores to carnivores. Both approaches are used in studies of complete ecosystems.

A proposal for the systems analysis of an estuary may be quoted from the Australian CSIRO progress report for 1974-1976. In it, the Division of Fisheries and Oceanography has outlined a simple mathematical model designed to provide the information required for the rational management of an estuary and the effects of human manipulations of the environment or the culling of the population. Carbon has been chosen as the unit of measurement and the south-west arm of Port Hacking near Cronulla has been selected as the test estuary. Since it is a clear deep estuary, phytoplankton is the most important source of primary production. Salt marsh plants are disregarded but the sea-grasses *Zostera capricorni* and *Posidonia australis* are important sources of organic detritus. As noted in chapter 4, this includes dissolved and particulate organic matter and a complex community of living macro-organisms including autotrophs as well as heterotrophs such as bacteria, fungi, protozoans and small metazoans. For the carbon model it is important to evaluate the proportions of dead and living organic carbon. The total mass of living matter is determined by adenosine triphosphate (ATP) analysis and the percentage of autotrophs in the same sample is estimated by the chlorophyll a content less its degradation products in the dead material.

The compartments in the model are thus CO_2, autotrophs (AUT) heterotrophs (HET), dead particulate detritus (DET) and dissolved organic carbon (DOC). These may be arranged in a matrix of donor and recipient compartments as shown in Figure 16.1.

'The matrix displays *(inter alia)* that CO_2 is utilized by plants, that plants contribute CO_2 in their respiration, are eaten by bacteria and animals, become detritus when they die and release dissolved organic material as part of their metabolism.' (CSIRO report). The complete carbon model displays that these fluxes occur both in the water column and the sediments, and that there are inputs and outputs across the boundaries of the whole ecosystem. For example detritus is received from the land, part is permanently deposited in the sediments and there are both gains and losses of detritus and living organisms to the sea.

In practice the environmental variables are monitored and for ease of analysis, the biota is divided into

		Recipient			
	CO_2	AUT	HET	DET	DOC
CO2		+			
AUT	+		+	+	+
HET	+		+	+	+
DET			+		+
DOC	+	+	+	+	

Figure 16.1 Matrix of a simplified five compartment system

a number of habitat groups each subject to the same environmental conditions. Each may be investigated independently and particular attention is paid to important primary producers and consumers.

BIOMASS, NET PRODUCTION AND ENERGY FLOW

It may be noted that almost half the energy fixed as chemical bonds during net production by primary producers is lost during bacterial decomposition. Further the primary and secondary consumers that live in salt marshes are only part of the consumer population in a complete estuarine ecosystem so that the export of 45 % of net production reported by Teal (1962) is not necessarily the normal export to the sea. In Long Island Sound which is partly estuarine, Odum (1971) quotes net primary production as 2 500 kcals $m^{-2}.yr^{-1}$ which is balanced by heterotroph respiration so that the net community production is 'very little or none'. This assumes no export by migratory fishes or birds of the loss of organic detritus to the open sea which must occur.

No energy budgets of an entire estuarine ecosystem have yet been completed but many workers have contributed information on growth rates, biomass or net production of important species in different habitat groups. Estimates of the net production of primary producers have been given earlier. Estimates of secondary production or the rate of storage at consumer's levels (Odum, 1971) are usually determined for each of the size group; micro-heterotrophs, meiofauna and macrofauna. Estimates for meiofauna production in the past have often included the production of the micro-organisms. Dye *et al* (1978) quote estimates of the combined contribution as varying between 11 and 58 % of the total secondary production. In Dye *et al*'s study of meiofauna production in sandy and muddy areas of the Swartkops estuary, Port Elizabeth, the total secondary production was estimated from the oxygen consumption of the sediment and organisms

Table 16.2
Mean secondary production at four levels from HW to below LW at sandy and muddy stations in Swartkops estuary, Port Elizabeth (modified from Dye *et al*, 1978)

Total secondary production g C.m^{-2}.yr^{-1}	Percentage contributed by:		
	macrofauna	meiofauna	microfauna
Sandy station	5,75	2,10	92
Muddy station	24,40	0,03	75,5

Table 16.3
Mean biomass, production and P/B ratios of the six main species of infauna in soft mud below midtide in the Tamar estuary (data from Warwick & Price, 1975)
(1) Mean biomass (g dry mass.m^{-2})
(2) Production (g dry mass.m^{-2}.yr^{-1})

Rank and species	(1)	(2)	P/B ratio
(1) *Nepthys hombergi*	3,947	7,335	1,9
(2) *Mya arenaria*	5,537	2,659	0,5
(3) *Ampharete acutifrons*	0,426	2,322	5,5
(4) *Scrobicularia plana*	2,146	0,482	0,2
(5) *Macoma balthica*	0,337	0,308	0,9
(6) *Cerastoderma edule*	0,847	0,205	0,2
Total	13,204	13,311	

in it by the light and dark bottle technique. Due allowance was made for oxygen production of micro-algae. The biomass of *Callianassa kraussi, Psammotellina capensis* and *Diogenes brevirostris* representing the main elements of the macro-fauna was known and an RQ of 0,85 was assumed from which the annual production was calculated in terms of g C.m^{-2}.yr^{-1}. The biomass of the meiofauna was also known and in this case the annual production of the dominant nematodes based on the RQ method was checked by the P/B ratio assuming a turnover of eight generations of nematodes per year. The results agreed closely. Secondary production of the 'microfauna' (? = microheterotrophs) was estimated as the difference between the total secondary production and the combined production of the macro- and meiofauna. The total secondary production in sandy and muddy sediments and the percentage contribution of the macrofauna, meiofauna and 'microfauna' are shown in Table 16.2.

Dye *et al* conclude: 'that the meiofauna in these exposed sand and mud flat areas is not quantitatively important but the qualitative importance may be considerable. 'The annual production of the macrofauna in the bare sandy and muddy areas was 4,74 and 196,02 g.C.m^{-2} respectively'.

Warwick & Price (1975) have estimated the mean biomass and production of the macrofauna in the Tamar estuary near Plymouth at a station in soft silty mud where the organic content is 12,2 to 13,6 %. The macrofauna was extracted with a 0,5 mm mesh sieve. The six most productive species are shown in Table 16.3.

Nepthys and *Ampharete* are both polychaete worms, *N.hombergi* being a carnivore and *A. acutifrons* a detritivore. It is interesting that they have the highest P/B ratios. Warwick & Price suggest that *N. hombergi*

may feed on the meiofauna. The other four species are bivalves; *Mya* and *Cerastoderma* being suspension feeders and the other two deposit feeders.

It is interesting to compare the biomass of an estuarine mudflat with that of sandy shores in a sheltered lagoon without fresh water inflow. The sediments of Langebaan Lagoon are medium to fine sand with 2 % organic carbon in more sheltered areas but no silt or clay. In the mouth of the lagoon, where the currents are swift and phytoplankton abundant, the intertidal sands are relatively barren and there are few species (Day, 1959) but below low tide at a depth of 1-5 m there are about 20 species with a biomass of 47 to 119 g (ash free).m^{-2} (Christie & Moldan, 1977). The main contributor was the suspension feeding bivalve *Mactra lilacea* which occurred in concentrations up to 96 g.m^{-2}. Polychaetes such as *Orbinia angrapenquensis* and crustacea such as *Callianassa kraussi* and *Ampelisca* spp dominated sandy areas and *Upogebia africana* was abundant in muddy sand. Towards the southern end of the lagoon where the currents were weak, phytoplankton sparse but salt marsh plants abundant, the sediments at 2-3 m level contained up to 15 % organic matter and the biomass was dominated by *Callianassa kraussi* and amphipods (60-66,6 %) with polychaetes contributing 20 to 28,8 %. The total biomass was 12-24 ash free g.m^{-2}. It is interesting that the fauna in this part of the lagoon is dominated by those species which are common in estuaries.

Puttick (1977) has estimated the biomass of the epifauna and shallow-burrowing infauna on the intertidal banks in the same southern end of the lagoon. Her aim was to determine the amount of food available to the curlew sandpiper *Calidris ferruginea* the numbers of which often exceed 20 000. She sampled the fauna on transects at Bottelary, Geelbek and Scrywers Hoek to a depth of 6 cm and extracted with a 1 mm mesh sieve. This technique misses the large, deep-burrowing species such as *Callianassa kraussi, Arenicola loveni*, and only a few juveniles of *Upogebia africana* were recorded. In her appendix, she lists about 30-40 species most of which contribute little to the total biomass. From the lists the ten top-ranking species have been extracted and their biomass is shown in Table 16.4.

These ten top-ranking species differ on the three transects and reference to Puttick's appendix shows that there are obvious seasonal variations, while her Figure 2 shows changes of density at different levels. These correlate with the location of weed beds and probably with seepage from higher levels too (Day, 1959). The Bottelary transect lay on a bank of fine sand (Mdϕ = 2,5) with 0,81 % of organic matter. It has the highest biomass of 23,9 g (dry mass).m^{-2} half of which is due to *Assiminea* living at the edge of the salt marsh and about one-quarter due to *Scoloplos* at low tide. The Geelbek transect lay over muddy sand (Mdϕ = 2,8) with a high organic content of 4,0 %. It is well-

Table 16.4

Mean annual biomass (g dry mass.m^{-2}) of the ten top ranking species on intertidal transects at Bottelary, Geelbek and Scrywers Hoek in Langebaan Lagoon (data recalculated from Puttick, 1977)

Species	Bottelary	Geelbek	Scrywers Hoek	Mean
Polychaeta				
Ceratonereis erythraeensis	1,056*	1,677*	0,582*	0,105
Cirriformia tentaculata	0,863*	5,352*	2,650*	2,955*
Lumbrineris tetraura	0,030	0,433*	0,520*	0,328*
Notomastus latericeus	0,362*	0,823*	0,813*	0,666*
Orbinia angrapequensis	2,520*	2,770*	2,323*	2,538*
Scoloplos johnstonei	5,190*	0,072	3,502*	2,921*
Telothelepus capensis	0,749*	0,870*	0,005	0,541*
Crustacea				
Ampelisca palmata	0,005	0,757*	0,072	0,278
Urothoe grimaldii	0,498*	0,027	0,092	0,206
Exosphaeroma hylecoetes	0,173	0,478*	0,260*	0,304*
Cleistostoma edwardsii	0,313	1,870*	0,767*	0,983*
Hymenosoma orbiculare	0,372*	0,308	0,138	0,273
Upogebia africana (juv.)	0,564*	0,005	0,080	0,216
Mollusca				
Assiminea globulus	11,280*	4,992*	0	5,424*
Nucula sp	0,003	0,315	0,347*	0,222
Enteropneusta				
Balanoglossus capensis	0	0	2,603*	0,868*
Biomass of ten top-ranking species marked*	23,904	20,02	14,37	17,528

covered with salt marsh vegetation and *Zostera*. The biomass is slightly above average with *Assiminea, Cirriformia, Orbinia* and *Cleistostoma* as the main contributors. *Cirriformia* is particularly abundant in organic mud while *Scoloplos* which prefers sandier sediments, is poorly represented. The Scrywers Hoek transect runs over medium to coarse, well-sorted sand (Mdϕ = 1,56) with 1,2 % organic matter, very little salt marsh vegetation and no *Zostera*. The biomass is well below average due to the virtual absence of *Assiminea* on the dry sand above mid-tide. *Cirriformia, Scoloplos* and *Balanoglossus* however are well represented near low spring tide.

It may be noted that the main contributors to the biomass are all deposit feeders. The gastropod *Assiminea* is the only member of the epifauna in this group, all the rest being shallow-burrowing worms. The amphipods, isopods, crabs and the small bivalve *Nucula* which are all members of the epifauna, contribute little to the total biomass although Puttick mentions that the amphipod *Urothoe* is particularly abundant with an annual (gross) biomass of 0,51 g.m^{-2} at Bottelary and a P/B ratio of 1,36. The corresponding figures for *Assiminea ?globulus* were 19,21 g.m^{-2} and a P/B ratio of 4,96.

When the biomass of 14,37 to 23,9 g (dry mass).m^{-2} is compared with the value of 12 to 24 dry g.m^{-2} recorded by Christie & Moldan (1977) at depths of 2-3 m in the same part of the lagoon the results are very similar. Since the latter workers used a suction sampler penetrating to 60 cm in the sediment they obtained *Callianassa* and other deep burrowers. Sampling to the same depth on intertidal banks would undoubtedly increase the biomass there very considerably.

Considering the high biomass of the benthic fauna

there are surprisingly few predatory fish in the southern half of the lagoon. Day (1959) records four species as common. The sandgoby *Psammogobius knysnaensis* feeds on amphipods and polychaetes; small sharks *(Mustellus nigropunctatus)* and sandsharks *(Rhinobatus blochii)* move onto the banks with the rising tide and feed mainly on *Callianassa, Upogebia* and *Hymenosoma*. Migratory wading birds are the main predators. Summers (1977) estimated that 30 000 waders roost in the salt marshes with 20 000 curlew sandpipers forming 42 % of the total biomass. The 24 species of waders which include 15 palaearctic migrants during the austral summer, consume 4,320 g (dry mass).m^{-2}.yr^{-1} with an energy value of 20,8 kcals. Summers estimates that they return between 2,1 and 6,2 kcals in the form of faeces and 0,06 kcals in the form of feathers. Since the mean annual biomass of the benthic macrofauna on the intertidal banks where the waders feed is 17,53 g (dry mass).m^{-2}, the waders consume less than a quarter of the food available. Puttick notes that the benthic biomass is not significantly different from winter to summer although Summers' Table 3 shows that the wader population in the lagoon increases from 1 671 in winter to 36 759 in summer.

Earlier estimates of the intertidal macrofauna in the estuaries of southern Africa were presented in the form of kite diagrams showing the changes in the density of common species at different tidal levels. Day (1967) shows transects in the different reaches of the temperate Knysna estuary and Day (1974) shows similar transects in the subtropical Morrumbene estuary. Both indicate a higher density in the shelter provided by salt marsh vegetation or mangroves and an increase in both density and diversity from the upper reaches to the lower reaches. More recently McLachlan & Grindley

Table 16.5
Mean biomass of five dominant species in salt marsh vegetation and *Zostera* at Ashmead, Knysna (modified from Grindley, 1978)

Level and vegetation	Biomass g dry mass. m⁻²	Contribution of dominant macrofauna (g.m⁻²)
HW	*Limonium, Chenolea & Sarcocornia*	8,94	*Sesarma catenata* 8,1; *Assiminea ponsonbyi* 0,84
HWN	*Triglochin*	14,07	*Sesarma* 12,7; *Upogebia africana* 0,62; *Assiminea* 0,59
MTL	*Spartina*	15,42	*Sesarma* 13,5; *Upogebia* 1,7
LWN-LWS	*Zostera*	23,09	*Nassarius kraussiana* 4,4; *Diogenes brevirostris* 2,4; *Upogebia* 7,5; *Sesarma* 5,9

Mean biomass at all levels 15,38 (g dry mass.m⁻²)

(1974) estimated the biomass of burrowing prawns and bivalves at LWS in four reaches of the Swartkops estuary, Port Elizabeth. The dominant forms were *Callianassa kraussi* in the sands of the mouth area and at the top of the estuary, *Upogebia africana* and *Solen capensis* in the muddy sand of the lower reaches and the bivalves *Eumarcia paupercula, Solen corneus, Macoma litoralis* and *Dosinia hepatica* in the soft mud of the middle to upper reaches. They sampled to a depth of 60 cm and extracted with a 4 mm mesh sieve so that small crabs, amphipods, isopods and polychaetes which usually comprise half the biomass of the benthic fauna were omitted. Nonetheless, their kite diagrams indicate a mean of about 125 g (wet mass).m⁻² and a maximum of 450 g.m⁻² in the lower reaches, about 50 g.m⁻² in the middle to upper reaches and about 35 g.m⁻² at the top of the estuary. Since the salinity is normally 26-35 ‰ along the estuary, the changes in the biomass are due to the changes in the percentage of subsieve particles (<0,063 mm) in the different reaches. In the mouth area the percentage subsieves is ca 2 %, in the lower reaches it is less than 20 %, in the middle and upper reaches it is 20-40 % and at the top of the estuary it is about 5 %. McLachlan & Grindley suggest that the distribution of the bivalves *Eumarcia, Solen corneus, Macoma* and *Dosinia* is restricted by competition with *Upogebia* but as these bivalves prefer to burrow in the soft mud of the upper reaches both in Swartkops and Knysna (Day, 1967), changes in the nature of the sediments provide a simpler explanation.

One of us (Grindley, 1978) estimated the biomass of the larger macrofauna in salt marshes and *Zostera* beds in the lower reaches of Knysna estuary. He sampled along transects from HWS to LWS, digging to a depth of 25 cm and extracting with a 4 mm sieve so that very few polychaete worms and no amphipods were included in his Ashmead transect. The mean biomass at four different levels is shown in Table 16.5. *Assiminea* makes a surprisingly small contribution to

the total biomass compared with its importance at Langebaan and many may have escaped through the coarse sieve. *Sesarma* which contributes two-thirds of the biomass does not occur at Langebaan; it is mainly herbivorous but also scavenges in organic detritus. *Upogebia* filters detritus and plankton drifting over the mouth of its burrow while *Diogenes* is both a scavenger and filter feeder. Most species of *Nassarius* are carnivorous but *N. kraussiana* occurs in such high densities (up to 274 m⁻²) that it is suspected of being a deposit feeder as well. The absence of the crab *Cleistostoma edwardsii* from the Ashmead transect is surprising for it is common in all South African estuaries and Grindley records 440 m⁻² on bare mud east of Thesens' causeway.

Table 16.5 shows that the biomass of the larger macrofauna increases to a maximum at LWS. The mean of 15,38 g (dry mass).m⁻² for all intertidal levels is similar to the values obtained by Puttick (1977) in the Langebaan marshes although the different techniques sampled different parts of the benthic fauna. It is suggested that digging to spade depth (25-30 cm) and sieving through a 1 mm mesh sieve would reveal a biomass of the order of 30 g (dry mass).m⁻². Even this is far less than the estimate of Dye *et al* (1978) for the macrofauna in muddy areas of the Swartkops estuary.

In Langebaan Lagoon there are relatively few fish feeding on the benthic fauna and the main predators are the wading birds. In Knysna, the position seems to be reversed. Summers *et al* (1976) recorded 1 521 waders in December (summer) as compared with over 36 000 in Langebaan in the same season. Moreover, few of these birds feed in salt marshes and prefer thin *Zostera* beds or sand flats. It would appear that the birds are relatively unimportant as predators. Unlike Langebaan, Knysna is rich in demersal fishes. Mrs M. Smith (in Grindley, 1976b) lists numerous species of which about 50 feed on the benthic macrofauna on soft sediments. No estimates of numbers or biomass are available but Knysna lagoon is regarded as an excellent angling area. As shown in chapter 12, *Lithognathus lithognathus* up to 5 kg, as well as small species and juveniles are present through the year and large benthic feeders such as kob (*Argyrosomus hololepidotus* up to 50 kg), and spotted grunter (*Pomadasys commersonni* up to 4 kg) are present in summer. Large kob feed mainly on mullet and cephalopods but the smaller ones as well as the white steenbras and grunter feed on *Upogebia, Callianassa, Hymenosoma*, bivalves and polychaetes. Gobies, small soles and large numbers of juvenile *Rhabdosargus holubi, Diplodus sargus, Pomadasys olivaceum, Tachysurus felicips, Sarpa salpa* and *Spondylisoma emarginatum* feed on amphipods, isopods, shrimps, burrowing prawns, polychaetes, chironomid larvae and small molluscs. All the small fish in turn are taken by pelagic predators such as *Lichia amia* and *Pomatomus saltatrix;* fisheating birds are not common.

The whole fauna is remarkably diverse for a warm temperate estuary with 310 species of benthic invertebrates, more than 50 species of fish and 35 species of birds. Visits over 30 years indicates that the whole ecosystem is effectively stable.

The most complete account of the quantitative relationship between the elements of an estuarine fauna is that given by Milne & Dunnet (1972). Over the years they and many other workers have investigated the 8 km Ythan estuary near Aberdeen. Two-thirds of the estuary bed is exposed at low spring tide and for convenience 90 % may be regarded as muddy sand flats and the rest as mussel beds. The work is still in progress but in 1972 attention had been focussed on the benthic macrofauna and the fishes and birds that feed on them. Of the total of 63 invertebrates, about 40 live on the mud flats. The dominant species are detritus feeders including *Hydrobia ulvae, Corophium volutator, Nereis diversicolor, Arenicola marina, Macoma balthica* and *Cerastoderma (Cardium) edule.* Of the 24 fishes in the estuary, the main predators on the mud flats are gobies and flounders. The gobies breed in the summer and reach a biomass of 0,7 g (wet mass).m^{-2} in autumn but they are preyed on to such an extent during the winter that their biomass falls to less than 0,2 g (wet mass).m^{-2} in spring. The founders migrate into the estuary in spring and reach a biomass of 35 g (wet mass).m^{-2} in summer. Of the 55 species of aquatic birds, the main predators on the mudflats are gulls, shelduck, dunlin, knot, redshank and oyster-catchers. The food web has been worked out and in some cases the quantitative relationships between the predators and their prey have been determined. Thus *Hydrobia* reaches a peak biomass of 26 g (wet mass).m^{-2} in winter and *Corophium* reaches a peak of 13 g.m^{-2} in summer but as their predators feed at different states of the tide and reach maximum abundance in different seasons, competition is reduced and the prey are able to recover.

The energy relationships in the mussel beds have been investigated in detail. *Mytilus edulis* is the main prey but there are also numerous amphipods and shore crabs *(Carcinus maenas)* which are eaten by turnstones, blennies, gobies and butter fish. *Mytilus* increases from a winter minimum of about 130 g (wet mass).m^{-2} to 230 g.m^{-2} in spring when it suffers heavy predation and then increases to an autumn maximum of 400 g (wet mass) m^{-2}. Its gross annual production in 1969 was 268 g (dry mass).m^{-2} or 1 340,8 kcals.m^{-2}, half of which was due to the 1967 cohort. The partitioning of the annual production is shown in Figure 16.2 reproduced from Milne & Dunnet's Figure 10.

The three bird predators feed mainly in winter but their feeding is separated in space and time and they feed on different sizes of mussels. Eider ducks feed when the mussels are submerged and take mussels averaging 18 mm long. Oyster-catchers and gulls feed when the mussel beds are exposed, the oyster-catchers

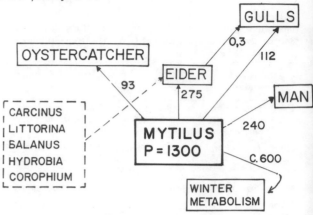

MYTILUS PRODUCTION TO PREDATORS (KCAL M^{-2} YR^{-1})

(gross production = 5400 g (wet wt) m^{-2} or 270 g (dry wt) m^{-2})

Figure 16.2 The partitioning of *Mytilus* production among the various predators of the mussel bed community in the Ythan estuary (after Milne & Dunnet, 1972).

taking mussels averaging 33 mm while the gulls take either very small mussels 2-10 mm long, which they swallow whole, or large ones (>35 mm) which they drop so that they break on the rocks. Man also takes the largest mussels.

Milne & Dunnet estimate that 'all the gross production is accounted for in terms of predation and over-wintering metabolic requirements. This suggests that the mussel bed is cropped to the maximum, so that the standing crop at the beginning of each year remains fairly constant and no net changes in the size of the mussel bed are occurring'.

STABILITY AND DIVERSITY

Howard Sanders (1968), in a classic paper on benthic diversity, discussed diversity (meaning species richness) in contrasting environments. He compared the fauna of boreal shores where environmental conditions are often severe and the number of species low, with the fauna of coral reefs where the environment is optimal and the diversity of the fauna is high. He also compared three estuarine faunas. The lowest diversity, ie the lowest number of species per unit number of individuals was in a cold boreal estuary, that of the Pocasset River, Massachusetts. In that estuary, diversity decreased up-estuary and was lowest at the top station where the salinity was low and very variable. Higher diversities occur in tropical estuaries. Sanders (1968) notes that the Vellar estuary at Porto Novo and the Godavari estuary at Karinda in India, harbour many species of polychaetes and bivalves although the salinity may be reduced to zero during the monsoon. It was suggested that the fauna was better able to tolerate reduced salinities at the high temperatures in tropical estuaries than was possible in the low temperatures of boreal estuaries.

Such comparisons led to Sander's 'stability-time hypothesis' which postulates that where physiological stresses have been low, biologically accommodated communities have evolved, while in estuaries where the gradient of physiological stress increases under harsh conditions, the communities become predominantly physically controlled. The number of species diminishes continuously along the stress gradient.

It is widely agreed that estuaries are highly productive systems. Since they act as nutrient traps, they are able to maintain high rates of primary production and they also benefit from the energy subsidies of tidal flow. However, the variability of environmental conditions imposes stresses which severely limit the diversity of animals and plants in such a rich environment.

Low diversity in a community is often associated with instability. This raises questions about the stability of estuarine ecosystems. Sanders' hypothesis suggests that high diversity is a property of stable natural systems which might lead one to believe that estuarine systems must be 'unstable'.

Orians (1975) has discussed the various concepts to which the term 'stability' has been applied. He refers to constancy, persistence, elasticity, amplitude, cyclical stability and trajectory stability, to illustrate the many different meanings that may be involved. Ecologists are mainly concerned with some measure of stability to a perturbation; some measure or resiliance. This is only meaningful if considered in relation to evolutionary responses to past perturbations, both physical and biotic. In general, it would seem that there is no simplistic relationship between diversity and stability which may be applied generally.

The concept of stability in natural systems such as estuaries, implies a stable adjustment brought about by self-regulating mechanisms, some type of ecological feed-back. The energy flow through the system should tend to return to steady rates. The self-regulating mechanisms should cause changes leading towards the 'stable' condition when outside influences which have disturbed the balance are relaxed.

In accordance with the second law of thermodynamics, systems tend towards a state of entropy with the conversion of energy to less available forms and a more dispersed state (Odum, 1971). However, this running down of natural systems is prevented by the influx of solar energy and the more highly organised structure of living forms. Living plants and animals maintain a state of low entropy by absorbing energy to build and maintain their complex organisations. In an estuarine ecosystem the ratio of total community respiration to total community biomass (R/B) may be considered as a thermodynamic order function or a 'maintenance' to 'structure' ratio. This turnover rate is referred to as the 'Schrödinger ratio'. If expressed in appropriate units, and divided by the absolute temperature, the rate becomes the ratio of entropy increase of maintenance to the entropy of ordered structure. It is not yet clear whether nature maximises the ratio of community structure to maintenance metabolism or whether it is the energy flow itself that is maximised.

In estuaries characterised by physically unstable conditions such as fluctuations of salinity, temperature, light penetration and other environmental variations, biological diversity is normally low. This implies that there are fewer possibilities for the establishment of stabilizing feedback mechanisms than in highly diverse communities. Where the 'antithermal maintenance costs' imposed by the physical environment are high, the R/B ratio is high and less of the community energy can go into diversity. Consequently, communities in physically stable environments such as tropical coral reefs have a higher species diversity than communities in cold-temperate estuaries. However, it is not yet clear whether an increase in community diversity in a particular environment such as an estuary can in itself increase the stability of that ecosystem in the face of physical changes. Certainly diversity is high in the older communities of undisturbed estuaries such as Knysna and low in newly established ones such as artificial marina canals. However, it is difficult to measure the stability or resiliance of such different systems and thus to compare them.

Productivity does not appear to be related to diversity in any simple way although both are related to total energy flow. High productivity can occur in coral reefs where the diversity is high as well as in estuaries where the diversity is relatively low. The correlation between the diversity of the biota and the stability of the environment seems to be clearer than any relationship between diversity and resiliance or productivity. The causal relationship between diversity and community stability, though generally accepted needs further investigation.

While physical stress tends to reduce diversity as if by reversing succession to a subclimax community, biological stress may increase diversity. Paine (1966) showed that intertidal organisms, for which space is the limiting factor, had higher diversities when first and second order predators were present. Experimental removal of the predators reduced the species diversity. Paine concluded that species diversity is directly related to the efficiency with which predators prevent monopolization of space or other important requisites by one or a few species.

The nature of the predator must also be considered. Paine (1966) dealt with the effects of predation by starfish which are selective predators. R.W. Day (1977), who dealt with the effect of several fishes grazing on the biota of coral reefs, carried the discussion further. He provided evidence that the effect of herbivorous fish grazing on the algae which grow on lighted surfaces, is to reduce species richness by 19 %. In contrast to this, the grazing by other species of fish which feed on sedentary invertebrates (particularly ectoprocts and ascidians growing on the walls of submarine caves) is to increase species richness by 20 %.

At both sites, space was the limiting factor. On lighted surfaces some algae covered or shaded part of the area and the scarid or acanthurid fish predators scraped such areas of dense algal growth irrespective of the species growing there. They were non-selective. As a result, the density and species richness was reduced, and opportunistic, rapidly-growing algae such as *Ectocarpus* replaced them. On cave walls, ascidians and ectoprocts cover large areas and pomacanthid and balistid fish graze on them. These fish are selective predators which reduce the area covered by the more-abundant prey species and the grazed areas are then available for other species. There is thus an increase in species richness. In brief, it would seem that the effect of predation depends on the habits of the predator. Non-selective predators decrease species richness while selective predators increase it.

The stability of estuarine communities may be more directly related to community structure than to species diversity. The activities of burrowing deposit feeders can maintain such a physically unstable environment so as to exclude tube-dwelling animals (Rhoads & Young, 1970). In this way the stability of the burrowing community is protected. Conversely dense assemblages of tube builders can structure a community by restricting the amount of surface available for settlement. By feeding on planktonic larvae, settlement can be limited while the accumulation of their faecal material creates an unsuitable environment for larval settlement (Woodin, 1976). This can result in a situation where the stability of the tube-building community is maintained to the virtual exclusion of suspension feeders and infaunal deposit feeders.

An interesting example of the stability of a mixed deposit feeding community of a mud flat in the Tees estuary has been described recently by Kendall (1979). During an 18 month study of the Seal Sands the numerical structure of the fauna proved to be highly stable. Apparently the same community structure existed in the area during surveys in 1971 and 1973 (Gray, 1976). This is remarkable in that the community in question consists of large numbers (up to $1,4 \times 10^6 \text{ m}^{-2}$) of the tube building polychaete *Manayunkia aestuarina* and infaunal deposit-feeding annelids including *Peloscolex benedeni*. Up to 18 species were present at the study sites but the number of species present each month followed a similar pattern to the changes in abundance of the two dominant species. Thus the degree of dominance of *Manayunkia* is a major factor in determining the calculated index of diversity. Therefore the index is of little value in describing changes in this community for there is a highly significant correlation between diversity and evenness. The combination of dense concentrations of tube builders and infaunal deposit feeders

in the Seal Sands of the Tees estuary would appear to be unusual and potentially unstable. However, this mixed community appears to have persisted for a long period. Several factors appear to have aided the maintenance of this stability. The year-round production of benthic larvae by *Manayunkia* may preclude catastrophic mortalities followed by sediment instability. Competition for food with infaunal deposit feeders is limited as *Manayunkia* takes small particles between 1 and 2 μm in size. Perhaps most significantly, *Manayunkia* has shown itself to be able to avoid any adverse effects of sediment reworking activities by the infaunal deposit feeders. As its larvae are brooded in the parental tube they can also avoid these effects. It may well be that it is this immunity from the effects of interactions with the deposit feeders which accounts for the unusual stability of this mixed community.

The effects of a severe river flood on the stability of benthic communities in the Swartkops estuary has been described by McLachlan & Grindley (1974). Before the flood, the middle reaches of the estuary were dominated by bivalves including *Dosinia hepatica* and *Macoma litoralis* and there was a distinct boundary between that community and the mud prawn *Upogebia africana* lower down the estuary. Nine months after the flood the boundary had shifted more than a kilometre upstream. The bivalves had been greatly reduced and *Upogebia* moved up the estuary. Fifteen months after the flood the situation was beginning to return to normal. Due to recruitment and growth of spat, bivalve numbers were beginning to recover in the middle reaches.

Severe river floods may flush estuarine plankton completely out of an estuary so that when normal stable conditions return a new population must develop. The rate of recovery of the estuarine plankton in Richards Bay after a major flood in May 1971 indicated high levels of secondary productivity (Grindley & Wooldridge, 1974). Such resilience is not always apparent and weekly sampling for 10 weeks after a flood in the Sundays river provided no evidence of recovery of the previously existing estuarine plankton community.

The closure by a sand-bar of a previously open estuary creates environmental changes that stress the stability of the enclosed communities. Some organisms like fish may maintain a stable if slowly declining population (Blaber, 1973). Others, dependant on the tidal rise and fall of water level such as mangroves, may succumb rapidly as described by Breen & Hill (1973) in Kosi Bay. Such differences in the stability of particular populations are however related to their individual requirements rather than to any feature of community structure or diversity.

In general it seems that some estuarine systems such as the long-lived salt marsh communities may be remarkably stable and resilient to natural perturbations despite a low diversity. Odum (1975) suggests that when sources of high utility energy coupled with inputs of growth promoting substances are available in excess of maintenance needs, low diversity has advantages. High energy, low diversity systems can be quite stable in time and in terms of resistance to perturbation.

ECOLOGICAL EFFICIENCY AND HABITAT GROUPS

The data presented in earlier chapters show clearly that not only does the fauna and flora differ from one estuary to another but there are also differences between the reaches of an individual estuary. The changes in the vegetation along an estuary were discussed in chapter 6 and the changes in the macrobenthic fauna in each of the four reaches were discussed in chapter 9. It is suggested that these changes both in the flora and fauna are basically due to specific differences in salinity tolerance, substrate preferences, variation in turbidity and other factors of the physical environment. Competition for requirements in short supply such as light in dense vegetation, food in densely populated areas or shelter from predators reinforce these differences in tolerance to abiotic factors.

The fact that plants and animals *do* change along an estuary is so well-known that a recapitulation of the evidence is unnecessary. However, within each reach of an estuary there are further subdivisions based on tidal level (or periods of exposure and immersion), more subtle changes in the nature of the substrate at the same tidal level, degrees of shelter from frost or strong sunlight or differences in the availability of nutrients for plants or food for animals. As a result, several habitat groups may be recognised in each reach of an estuary. Two examples may be quoted. The fauna at the drift line includes a number of small animals sheltering under dead leaves or algae cast up by the tide and partially embedded in sand or mud. The characteristic animals are talitrid amphipods and oniscid isopods capable of living in a humid atmosphere for long periods and with them are a number of terrestrial insects such as staphilinid beetles, the larvae of kelp flies and predatory forms including small centipedes and wolf spiders. A number of small wading birds such as sanderlings feed on the drift line from time to time. The fauna at the same level on rocky shores may be very different.

At the level of low tide in the same reach there may be well-developed sea grass beds growing on sandy mud. Life is much richer at this level. *Zostera* commonly bears many small epiplytes and epizoans such as diatoms, spirorbid polychaetes, foraminifera and herbivorous isopods. Sheltering between the leaves are shrimps, amphipods and many juvenile fishes. The epifauna on the mud includes the microbial and meiofaunal communities associated with organic detritus, larger animals such as gastropods and crabs while the infauna is dominated by burrowing polychaetes, bivalves or thalassinid prawns. At low tide, herons and waders feed in the *Zostera* beds and at high tide shoals of fish replace the birds. On banks of coarse sand at the same level, macrophytic plants are conspicuously absent; the composition of the fauna changes but a few species may be abundant.

Habitat groups such as these obviously include a number of species with trophic links but it is not suggested that each habitat group forms a single food web or indeed, that there must be trophic links between the majority of the species present. The habitat group is basically a statistical concept (Mills, 1969) based on the evidence of samples. The Czechanowski coefficient or the Bray & Curtis coefficient which measures the index of similarity of abundance of species pairs between samples drawn from the same habitat group is significantly higher than the index of similarity between samples drawn from different habitat groups. Although habitat groups are easily recognised by a skilled observer in the field, this statistical analysis is the only objective method of verifying their existence. The interesting and important question is, why the species composition does not change so gradually that habitat groups cannot be distinguished. To pose the question in a different way, why do ecologically equivalent species tend to replace one another in different habitat groups? Why do the limits of distribution of most species in a habitat group tend to coincide? A brief summary of the various habitat groups which commonly occur in the four reaches of an undisturbed and unpolluted estuary is set out below as a basis for the discussion which follows.

(i) *Mouth region.* If rocky banks are present, the biota is a poor reflection of that on a rocky seashore. The upper tidal levels are colonised by browsers feeding on lichens, microalgae and stranded plankton. Fishes are seldom found in the pools and although gulls and other birds may roost on the more inaccessible rocks, they feed at lower levels. Rocks at low tide or permanently submerged, usually have scanty growths of macrophytic algae unless the water is very clear (as in a fjord), so that rocky surfaces are dominated by suspension feeders such as barnacles, mussels and/or oysters. The resident fishes are mainly blennies and gobies harbouring in the crevices although shoals of larger predatory fish move in from the sea with the rising tide. Gulls, oyster-catchers and turnstones feed on the mussels, barnacles and amphipods when the tide is low. In contrast to this, mobile sandbanks at the estuary mouth are very barren and usually lack macrophytes. Only a few talitrid amphipods and isopods are found on temperate shores while ocypodid crabs may replace them on tropical shores. On more stable sandbanks at subtidal levels dense populations of burrowing suspension feeders such as the bivalve *Mactra* may be present.

(ii) *Lower reaches*. On sandy mud, important changes in the epifauna are related to differences in plant cover and tidal level. As noted earlier, there is a characteristic group at the drift line. The composition of the saltmarsh vegetation changes with the tidal level and as Ranwell has emphasised, the water potential is related to the soil type. Microbenthic algae are abundant and organic detritus is plentiful. The epifauna is dominated by hydrobiid and litorinid gastropods and amphipods. In the subtropics, mangroves become increasingly important and in well-developed forests even minor differences in tidal level may be correlated with changes in the dominant species of mangrove. The fauna hardly reflects this. The sesarmid crabs and gastropods which dominate the fauna show little or no relation to the species of mangrove trees and vary more with the density of shade. *Sesarma* spp is largely replaced by *Uca* spp in open glades and in the lighted lower fringes at about mid-tide. Suspension feeders on the mangrove trees including barnacles and oysters, are largely restricted to the lower fringes. The infauna of salt marsh vegetation and of mangrove swamps is poorly developed and only those species which are adapted to low oxygen levels in the soil are common. Predators are also poorly represented. Wading birds roost in salt marsh vegetation but feed mainly on the open sand flats at lower levels. Similarly birds are seldom seen in mangroves and grapsid crabs are the main carnivores. On open sunlit banks above mid-tide the epifauna of temperate shores consists of hydrobiid and litorinid gastropods and two groups of amphipods. Some feed when submerged and resist desiccation by burrowing into the moist sediment when the tide is low, while the other group emerges when the tide falls and feeds over the moist sand until forced to retreat to higher banks as the tide rises. Ocypodid crabs have the same habit in tropical and subtropical estuaries.

The characteristic flora and fauna of *Zostera* and other sea grass beds at, and a little below low tide has been described earlier. This community appears to be restricted to the same low level as the sea grasses and below the beds of *Zostera, Posidonia* or *Thallasia* there is a marked reduction of the benthic macrofauna. Some species such as heart-urchins, *Pecten* among bivalves and *Diopatra* among polychaetes do extend to the bed of the estuary. It is suggested that these may be representatives of a different habitat group although many of them also occur in the sea grass beds. Bottom feeding fish are numerous but as they move onto higher banks as the light falls and the tide rises and move in and out of the estuary with the seasons. They are not closely correlated with the submerged bottom fauna of the lower reaches.

(iii) *Upper reaches*. The obvious environmental changes are the lower and more variable salinity, reduced current velocities, higher turbidy and the increased percentage of silt in the sediment which often becomes porridgey mud. The drift line fauna, often found where masses of leaves have killed the high tide vegetation, at first appears much the same as that in the lower reaches but the two species of amphipods and isopods are usually different. Several species have disappeared and there are few replacements so that the diversity is reduced.

The salt marsh vegetation on the upper banks shows somewhat similar changes although the plants appear to be less affected by reduced salinity than the animals. Reeds *(Phragmites)* and sedges (*Cyperus* sp or *Scirpus* sp) which grow tall, shade out low-growing plants. The epifauna is poorer in species but those present are not very different from those in the lower reaches. The infauna in the soft black mud is very poor and those that require permanent burrows are limited to infrequent sandspits or eroding mudbanks along the main channel. Few species live on the muddy bottom below low tide. The fauna is often dominated by a few specialised amphipods and capitellid polychaetes or oligochaetes such as *Tubifex* which are specially adapted to hypoxic conditions. The fish fauna is also limited; plankton feeders such as atherinids, and mullets which feed on detritus and epiphytes are the most common forms. Wading birds are not abundant but herbivorous geese and duck tend to be more common here than in the lower reaches.

(iv) *Head of estuary*. The vegetation changes slowly to a fresh water flora with abundant reed beds along the margins and *Ruppia* or *Potamogeton* replacing *Zostera*. The fauna is very scanty and varies greatly with the nature of the substrate. Brack water mussels, barnacles and *Ficopomatus (= Mercierella)* may cover rocks below mid-tide and provide shelter for brack water amphipods and isopods. Salt-tolerant fresh water species of insects and pulmonate gastropods become relatively important in the vegetation.

Planktonic habitat groups. Changes in the plankton along an estuary are less marked than those of benthic forms and do not coincide with the four reaches of a typical estuary. Neritic phytoplankton and zooplankton including a wealth of species, drift from the sea into the lower reaches and may extend further depending on tidal exchange. The truly autochthonous estuarine plankton, as well as halophytic fresh water species, are centred on the upper reaches and extend to the head of the estuary for fresh water forms are not numerous in rivers.

Ecological efficiency and competition. We return now to the question as to why habitat groups such as those described above exist in an estuary and why ecologically equivalent species tend to replace one another. It is generally agreed that the bulk of an estuarine fauna is of marine origin and it has been repeated almost *ad nauseum* that estuaries are highly productive. Yet the

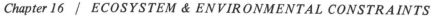

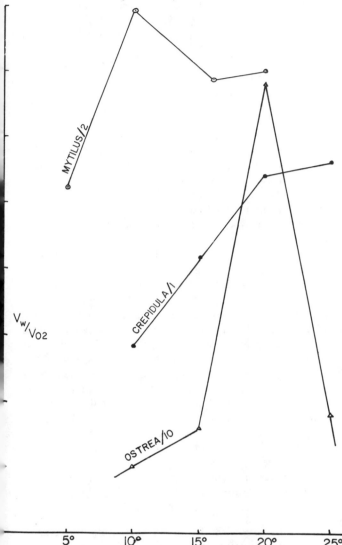

Figure 16.3 Comparison of the energetic cost of feeding of *Mytilus edulis, Ostrea edulis* and *Crepidula fornicata* when acclimated to temperatures between 5° and 25°. Data for *Mytilus* from Widdows & Bayne (1971) divided by 2, values for *Ostrea* and *Crepidula* from Newell (1979), that of *Ostrea* divided by 10, that of *Crepidula* as given.

fauna of an estuary is relatively poor in species in comparison to shallow seas. It has been argued that while estuaries contain rich supplies of food, the environment is harsh and changeable and it is only those few species that are adapted to such a stressed environment that can take advantage of the abundant food. However, this still does not explain the distinction of the habitat groups or the changes in ecological efficiency between replacing species.

Richard Newell (1979) has developed the concept of the cost of feeding. This is outlined below. Newell quotes the balanced energy equation of Winberg (1956):

$$C = P + (R + F + U)$$

where C = the energy content of the good absorbed
P = energy production as growth or gametes
R = energy loss through metabolism
F = energy loss through faeces
U = energy loss through dissolved organic matter (including urine)

Newell points out that this equation may be re-formulated as follows:

$$A = C - (F + U + R) = P_g + P_r \text{ (Newell 1979, p 494)}$$

or net energy = food consumed − losses
through faeces, dissolved organic matter and respiration = energy available for growth and reproduction.

For practical reasons it is easier to work with suspension feeders ingesting cell suspensions of known concentration and thus of known energy content. For such suspension feeders the energy gain may be expressed as the clearance rate (V_w litres of water per hour) and the oxygen consumption or energy loss by respiration may be expressed as $V_{O_2}.m\ell.hr^{-1}$. The ratio V_w/V_{O_2} gives 'the convection requirement' used to compare the energy gain between organisms under different conditions. Newell points out that the ratio V_w/V_{O_2} is greater in small specimens than in larger ones of the same species. Thus small specimens may channel more energy into growth than larger ones at the same feeding ration.

The ratio V_w/V_{O_2} may be adjusted by acclimiation: 'Adjustment of the ratio may thus be an important index of the ability of suspension feeding organisms to maintain optimal filtration efficiency in the face of a change in environmental conditions' (Newell, 1979). This is obviously relevant to the environmental changes in estuarine habitats.

Three possibilities exist for the adjustment of the V_w/V_{O_2} ratio. These are referred to as Types 1, 2 and 3 by Newell & Branch (1979).

(Type 1) The clearance rate may be adjusted to correspond with environmental temperature and energy losses may be relatively reduced. The essential point is the simultaneous compensation of two components of energy balance. One example is *Crepidula fornicata.*

(Type 2) In Type 2 the clearance rate increases with the environmental temperature but metabolic losses also increase. Thus the adjustment of the V_w/V_{O_2} ratio must compensate for the increased energy expenditure. An example is *Ostrea edulis.*

(Type 3) In this type the clearance rate remains constant but the energy expenditure is decreased following warm acclimation. An example is *Chlamys opercularis.*

Newell goes on to quote physiological examples and the subcellular mechanisms involved. These do not concern us in this ecological discussion. More relevant is the energetic cost of feeding at different temperatures as expressed by the V_w/V_{O_2} ratio. Newell (1979) quotes evidence from Widrow (1978) and Widrow & Bayne (1971) that *Mytilus edulis* has a maximum filtration rate at 10 °, the greatest energetic gain at 15 ° and is 'significantly depressed at 25 °C'. The mean values given by Widrow & Bayne for different temperatures are shown graphically in Figure 16.3. Newell's graphs for *Ostrea edulis,* based on data from Newell, Johnson & Kofoed (1977), and for *Crepidula fornicata,* based on Newell & Kofoed (1977) are also reproduced in Figure 16.3.

The absolute values are not comparable for those of *Mytilus* are twice those shown in Figure 16.3 while those of *Ostrea* are ten times as much. However, the trends are obviously different. *Mytilus* obtains the greatest energy gain from feeding at 10 ° (or 15 ° according to Newell), *Ostrea* has the greatest gain at 20 ° and is much less efficient at 25 °, while *Crepidula* is efficient at 20 ° and increases slightly in efficiency at 25 °

These three suspension feeders are competitors living near low tide on rocky or shelly bottoms at the mouths of estuaries in southern England. *Mytilus* and *Ostrea* are endemic and *Crepidula* was unfortunately introduced from the warmer shores of the USA and is now a pest on oyster beds.

Newell (1979, p 683) concludes that: 'The ability to maintain optimal values for the V_w/V_{O_2} ratio over the range of environmental temperatures prevailing in the habitat may thus be one factor which controls the replacement of one species by another when both are competing for a potentially limiting food resource, since the lower the energetic cost of filtration, the more energy is available to be channelled into growth and reproduction'.

There are, of course, many other endogenous and environmental factors which affect the relative abundance and competitive replacement of *Mytilus, Ostrea* and *Crepidula.* These do not concern us here. What is important is the type of argument that may be used when discussing ecological efficiency and competitive replacement of equivalent species.

We may now consider other environmental changes which affect competitive replacement from one estuarine habitat to another. Some at least are interrelated. Thus the availability of food may affect the cost of feeding at different temperatures.

Food availability. There are, of course, limits to temperature tolerance and there are even narrower limits to temperature acclimation. There are also limits to the maximum rate of food intake even under optimal temperatures and maximal food availability. Response to temperature change may ultimately depend on the availability of food both as regards quantity and predictability. Where food is plentiful Newell suggests that the reduction of energy losses may not be necessary *or even desirable.* Animals may adopt an 'exploita-

tive strategy' in which growth, reproduction and energy turnover are high. Conversely in species which suffer food shortage, metabolic losses may be critical in an energetic gain from the environment. Too much energy may be lost in attempting to feed in such a lean environment; it may cost too much to feed.

Organisms adopting such a conservative strategy may include most of the animals on the upper shore. Their energetic turnover can be adjusted to conform with local or seasonal variations in food supply. Newell (1979) quotes examples of barnacles living at different tidal levels from the work of Crisp and his colleagues and work on species of *Patella* at different tidal levels from Newell & Branch (1979).

It is interesting to discuss the reverse. One may speculate that where food is freely available, oxygen supplies may be low (as in hypoxic waters) or oxygen may be difficult to obtain due to clogging of the gills or drying of the respiratory surface. Under such conditions feeding activities must be restricted. Anaerobic glycolysis would provide energy at a slow rate but it is wasteful of the food digested and thus only effective in food-rich environments. Animals living in organically rich mud which is poor in oxygen would be subject to such limitations except for those species which build up large oxygen reserves in their blood (eg the polychaete *Capitella capitata*) or supplement their oxygen supplies from overlying water while feeding in the deoxygenated mud below (eg *Arenicola* spp). Talitrid amphipods and oniscid isopods which feed at the drift line have rich food supplies but the oxygen supplies in the sediments below the water table are low. The talitrids (eg *Orchestia* spp) and oniscids (*Ancinus* sp in Australia and *Tylos* sp in South Africa) have 'tracheate' pleopods and are able to refresh their gills in moist air.

Salinity variations. The effects of salinity on oxygen consumption are complicated by activity and other endogenous factors as well as the interactions between the effects of external factors. Newell (1979) draws attention to the increase in salinity tolerance at higher temperatures. However, he is cautious of accepting the evidence of increase in respiration rates under salinity stress since the higher oxygen consumption may reflect an increase in activity rather than the cost of osmoregulation.

Potts & Parry (1964) estimated that the work involved in ion transport is only about 1 % of total metabolism in those animals that they tested. However, such an estimate is controversial; it may not have wide application and Newell quotes four references which give contradictory estimates. The energetic cost of osmoregulation must vary according to the departure of ambient salinity from the optimism for each species. Further, the extent of acclimation in both salinity and temperature and the energetic cost of salinity change to osmoconformers such as *Mytilus edulis* and *Arenicola marina* requires further investigation.

Rao (1958) studied the effect of salinity on two populations of the penaeid prawn *Metapenaeus monoceros,* one adapted to marine and other to brackish water, taking into account the body size. The rate of respiration of the brack water population was lowest in 50 % sea water and increased in both hypo- and hypersaline conditions. In contrast to this, the respiratory rate of the marine population was lowest at 100 % sea water and maximal in 25 % sea water. Acclimation led to a decreased oxygen consumption. However, work was not measured so that Newell contends that the effect of salinity on the respiratory rate may be due to increased activity in conditions of osmotic stress. The metabolic oxygen requirement for osmoregulation thus remains controversial.

The changes from one habitat group to another may include changes in temperature and salinity (as at different tidal levels on an exposed sandflat) and a decrease in oxygen availability in organically rich sediments and on drying surfaces. Changes in the availability of food are obvious. The effect of all these factors on the energetic gain from feeding help to explain competitive replacement of species from one habitat to another. Quite apart from energy gain are the factors which affect the intensity of predation. One of them is the density of the vegetation. Local concentrations of prey species occur in dense sea grass beds, salt marsh vegetation or mangroves which reduce predation by aquatic birds and predatory fish.

The most obvious change in ecologically equivalent species which is evident to an experienced collector, is the change from one type of substrate to another. Such changes are also the most difficult to explain. As shown in chapter 9, macrobenthic larvae are attracted to settle on a substrate more by the nature of the microbial community than by the grade of the sediment. However, the microbial community, particularly the bacterial population is affected by salinity variations, the degree of desiccation and the percentage of particulate organic matter in the sediment. Thus the settlement of larvae is affected by more factors than would at first appear. The factors which affect the survival of these larvae, their metamorphosis and subsequent growth are not sufficiently well-known to permit useful generalisations.

In conclusion it is evident that we can appreciate some of the factors which are responsible for differences between one habitat group and the next but by no means all.

CONSTRAINTS ON THE WHOLE ESTUARINE BIOTA

Having discussed the factors which affect the various habitat groups within a single estuary we now turn to consider the differences between the estuaries within the same biotic province. Typical estuarine plants and animals are both eurythermic and euryhaline and the whole community has an underlying resiliance so that it remains stable in spite of 'normal' fluctuations of the environment. It is the extreme and unusual changes which affect the nature and productivity of an estuarine biota. Such changes include prolonged floods, droughts extending over several months, the protracted closure of the mouth or stabilisation of the water level and persistent pollution. The emphasis is on long term changes from normal conditions since all these events (apart from the inflow of toxic pollutants) occur naturally and the fauna and flora is adapted to survive brief changes. The human activities, which often intensify or prolong natural fluctuations in estuarine environments, will be discussed under the heading 'Management' in the next chapter.

Floods. Apart from raising the water level and increasing current velocities, floods have many obvious secondary effects. The increased discharge of the river at first reduces the salinity of the surface layers and then as turbulence increases, the bottom layers are affected and in extreme cases the whole estuary may run fresh to the sea. Rivers in flood usually carry an increased load of silt and when the turbulent waters reach the estuary the water becomes increasingly turbid. In such cases a Secchi disc may disappear within 5 cm of the surface. The erosion of sediment from one part of the estuary and its deposition elsewhere may completely change the bottom topography. As noted in chapter 4, sediments that have been sorted and resorted during ten years of normal flow may be swept away during a single flood. Sandspits which have constricted or even closed the mouth disappear so that the tidal range regains its normal amplitude.

The extremes of all these factors as well as others which are liable to change the nature and productivity of an estuary are discussed below.

Rise in water level and increased currents. The prolonged rise in water level due to any cause will of course inundate the banks above the normal spring tide level. As the depth increases the progression of the tidal wave upstream would be less inhibited by bottom friction and the tidal range should increase. These effects seem to depend on the shape of the estuary basin. When the Swartkops estuary was subjected to a severe flood in September 1971, it was expected that the spring tide range would increase markedly. Tidal heights along the length of the estuary were recorded by the consulting engineers, Messrs Hill, Kaplan, Scott & Partners and in their report to the City Engineer of Port Elizabeth they noted that although water levels had increased, there was no significant increase in tidal range. Possibly the flood waters had spread so widely over the estuarine flood plain that bottom friction had increased in the shallows. It is also probable that the water drained back into the channel during the ebb so as to counterbalance the fall in level at low tide of springs.

Such effects would not operate in an estuary with steep banks where the flood waters would merely increase the depth of the channel.

The flood waters carry salt and plant debris to supratidal levels and if the deposits are extensive and of long duration, the terrestrial vegetation dies. Poorly drained depressions become more saline as the floods recede and evaporation proceeds. Macnae (1957) reports that such areas in the flood plain of the Swartkops are colonised by *Sarcocornia (= Arthrocnemum) pillansi* and the opportunistic annual *Salicornia meyeriana.* Similar effects may be seen on the banks of the Berg River (chapter 14).

The swiftly flowing flood waters tend to sweep down the plankton so that neritic marine species are carried to the sea and estuarine species are dislodged from their normal station in the upper reaches and are replaced by fresh water forms. Thus the character of the plankton may change and plankton becomes scarce in swiftly flowing estuaries. Both Ketchum (1954) and Margalef (1967) have pointed out that in such situations there is a premium on adaptation to maintain station and on rapid multiplication to maintain population levels. It is not surprising that rapidly reproducing nanoplankton forms up to 80 % of the phytoplankton biomass and that estuarine copepods are smaller and grow faster than marine species (Grindley & Wooldridge, 1974).

Well-established estuarine plants, most of which are long-lived perennials with either matted rhizomes or extensive root systems, tend to survive floods. Sieman (in Ferguson Wood & Johannes, 1975) quotes a report by Scoffin (1970) which gives one of the few quantitative reports available on the effect of currents on such plants. He states: 'that in dense long-bladed beds of *Thalassia testudinum,* currents up to 40 cm/sec measured immediately above the blades are reduced to zero at the sediment-water interface'. Similar measurements in other types of aquatic vegetation would be most valuable.

Erosion, turbidity and the deposition of sediment. Although these three effects of floods and the similar effects of dredging and filling of estuaries by engineers are usually associated, it is convenient to discuss them separately.

The strongest currents during a flood occur at, or near the time of mid-tide during the ebb and often erode the channel margins at this tidal level where plant cover is minimal. Seedlings with small roots are severely affected (Ranwell, 1972) and one may speculate that floods during the early growing season are more destructive than those at other seasons. However, adult plants established at higher levels along the channel margins may be undercut and swept away. Even mangrove trees may be affected; *Avicennia* is usually absent on the margins of eroding banks and *Rhizophora* in spite of its strut-roots may fall into the

channel (Day, 1974). The benthic fauna is affected in the same way. Perkins (1974) reported that a flood in a tributary of the Solway in Scotland cut a new channel and thousands of bivalves were swept away or buried too deeply for survival. McLachlan & Grindley (1974) reported that the 1971 flood in the Swartkops had also devastated the bivalve fauna although in this case it was impossible to distinguish whether the damage was due to erosion, the deposition of heavy layers of sediment or the reduced salinity. The prolonged reduction of salinity certainly caused part of the damage for dead bivalves were found on the surface (McLachlan & Grindley, 1974).

Silt in suspension not only reduces light penetration, it also reduces dissolved oxygen concentrations due to the suspension of decaying organic matter. Ferguson Wood & Johannes (1975) discuss the damage caused to plants and animals by these effects of increased turbidity. They quote the work of Odum (1963) on decreased primary production and the observations of van Eepoel (1971) on the decrease in depth at which *Thalassia testudinum* grows in Lindberg Bay, Virgin Island. In 1967 *Thalassia* was visible at a depth of 10 m and covered most of the bottom of the bay; in 1971, after a period of high turbidity, *Thalassia* and *Syringodium* were sparse with 'only very few plants being found deeper than 2,5 m; in the most turbid portions of the bay they were limited to 1,5 m in depth'.

Similar effects have been noted in estuaries in southern African subject to heavy floods carrying silt-laden water. For example *Zostera* is scanty and limited to intertidal levels in the Sundays estuary, it is virtually absent in the Bashee and completely absent in the Mzimkulu and Mzimvubu where the Secchi depth is 0,1 m or less in summer. In contrast to this, it is present in small coastal estuaries nearby which are not subject to silt-laden floods. Emergent salt marsh plants such as *Spartina* spp are less affected by turbidity. Even so, the shoot length decreases at lower tidal levels (Ranwell, 1972). Differences in shoot length of *Spartina alterniflora* in North America and *S. capensis* in the Cape have often been reported but there are no concurrent data to show that the distribution of shoot length is related to turbidity.

Benthic micro-algae appear to be affected in a different way. *Euglena,* diatoms and dinoflagellates normally migrate up to, and project from the sediment surface when exposed by the receding tide during the day. Green (1968) states that this occurs on the banks of the turbid Avon estuary but quotes a report by Perkins (1960) that this does not occur in the estuary of the River Eden where the water is clear. In the Eden the diatoms *Pleurosigma aestuarii* and *Suriella gemma* do not burrow when covered by the rising tide but remain on the sand surface as long as the light is of sufficient intensity.

Suspended silt appears to have little effect on the

normal estuarine fauna although corals are absent from estuaries and encrusting bryozoans, sponges and hydroids are rare. Many of the common bivalves are adapted to exclude silt; oysters close their shells when conditions are unfavourable and grow well on mangrove roots and man-made racks raised above the mud surface. They filter the silt as well as the food particles and consolidate the unwanted material into pseudofaeces, thus helping to clarify the water. Hobbie (1976) estimates that an oyster filters between 70 and 300 litres of water per week and that its pseudofaeces are sufficient to cover it in 36 days.

Rather surprisingly there is some evidence which suggests that silt may be advantageous in limiting predation. It has been noted that in Langebaan Lagoon where the water is so clear that the bottom is often visible at a depth of 2 m, fish are not abundant, and when a shoal does enter the lagoon, it is attacked by hundreds of cormorants and flocks of terns. In the more turbid estuaries along the south and east coasts of Africa where fish are far more common, piscivorous birds are less numerous. Even in Lake St Lucia, where piscivorous birds *are* common they are mainly concentrated in the deeper parts of the lake where the water is much clearer. It has also been noted that mysids and penaeid prawns tend to congregate in turbid shallows presumably because they are sheltered from predation.

The deposition of silt in estuaries is of course a natural process to which the estuarine biota is adapted. It is only when sedimentation is abnormally rapid and prolonged that the plants and animals are seriously affected. Macnae (1957) noted that after the Swartkops estuary had been subjected to a heavy flood the freshly deposited layer of silt lacked the crumb-structure normally seen on the banks. Possibly this was an indication that the microbial community had been smothered by the heavy deposit but more detailed observations are needed.

The smothering of the microbiota is relatively unimportant compared with the burial of aquatic macrophytes and macrobenthic fauna in estuaries subject to severe floods. As noted in chapter 14, this impoverishes many of the large estuaries draining the inland areas of the Eastern Cape Province, the Transkei and southern Natal. To quote just two examples, layers of dead shells and rotting plants 10 cm below the surface of the Mkomazi and the Mzimkulu were found after summer floods. Summer floods and the heavy siltation occur annually and although the biota recovers to some extent in the dry months, it is either eroded or smothered the following summer. There is abundant evidence in historical records that the rate of siltation has accelerated with more intensive cultivation and the draining of swamps. Moreover siltation not only impoverishes the estuarine fauna, it also fills up the estuary basins. The Mzimkulu at 'Port' Shepstone and the Mzimvubu at 'Port' St Johns were trading ports

within living memory but are now only suitable for small craft. Begg (1978) quotes many examples of estuaries along the Natal coast that have been completely spoilt by heavy siltation. One example is the Lovu some 37 km south of Durban. Once a pleasant estuary 2-3 m deep, it is now so full of silt that it is only 0,4 m deep to within a kilometre from the sea. The bed is so shallow that flood waters rise 4-5 m above the normal level each year washing away one bridge after another. Saline penetration is very limited and the river is often fresh to the sea. Aquatic vegetation is practically non-existent and the macrofauna, apart from *Callianassa* is not worth recording.

Extreme salinities. During a flood, oligohaline water in the range of 0,5 to 5,0 ‰ flows down the estuary. Although the bottom water of deep estuaries may retain its salinity, in shallow estuaries the whole estuary basin may be filled with fresh water. If the flood is of brief duration the decreased salinity has little effect on the benthic fauna although the neritic plankton and the fish will temporarily return to the sea. The aquatic plants are more resistant to reduced salinities and are hardly affected. If, however, the oligohaline conditions persist for long periods (weeks to months) the whole character of the estuarine biota will change. Macrophytic algae of marine origin disappear although *Enteromorpha* which tolerates salinities between 0 and 140 ‰ remains and Green (1968) quotes a report that in Randersfjord, *Fucus vesiculosus* persists in salinities of 6 ‰. *Zostera* is replaced by species of *Ruppia* and *Potamogeton* in oligohaline waters and mangroves (in Natal at least) are replaced by *Barringtonia racemosa* and *Hibiscus tiliaceus;* similarly *Spartina maritima* is replaced by *Phragmites australis.*

Stenohaline animals die rapidly during floods. In chapter 10, Dr Burke Hill has recorded that colonies of *Parechinus angulosus* which normally colonises the rocks near the mouth of the Kowie estuary, die in thousands when floods reduce the salinity below 5 ‰ for a week while the normal euryhaline animals survive. However, if the flood persists even these die. McLachlan & Grindley (1974) report that after the 1971 flood in the Swartkops estuary, when the salinity was reduced to 2 ‰ for five days, the bivalves *Solen corneus* and *Macoma litoralis* were found dead on the surface while *Dosinia hepatica* which has a stout, tightly closed shell remained alive.

Possibly the best documented account of the effect of floods of long duration is provided by Hodgkin (1978) in his description of the ecology of the Blackwood estuary in south western Australia. This large estuary some 40 km long, includes two large shallow lakes connected by a channel up to 4 m deep as well as two partially isolated lagoons near the sea. The temperature range of 10-28 ° is normal in a temperate estuary. During the winter rains the river flow increases rapidly to 20×10^6 m^3 per day and floods may discharge up to 10^8 m³ per day. As a result the salinity falls rapidly to below 5 ‰ except in deep depressions in the lower channel and in the semi-isolated lagoons. Since the tidal range on the coast is only 0,7 m at springs, low salinities persist for about six months. As the salt wedge creeps slowly along the estuary bed in spring, the water becomes highly stratified. During summer, surface salinities of 20 ‰ extend only 5 km up the channel and even in autumn they reach only 12 km. In brief, salinities on the shallow banks and intertidal margins are below 5 ‰ for half the year.

The effect on the aquatic vegetation is to restrict *Zostera* to within 1-2 km from the sea and even there it is scanty. *Ruppia* grows well in the lagoons and *Potamogeton* colonises the lakes. The resident fauna is very poor. A total of 55 macrobenthic species has been reported where over 200 might be expected. No estimate of biomass is available but it is probably low for all the benthic animals are small except for two burrowing bivalves which are restricted to the higher salinities near the mouth and one euryhaline shrimp which is ubiquitous in the weedbeds. Of the 55 species, 13 die during the low salinity phase. These are replaced by new recruits every summer, either from the sea or from the lagoons which act as faunistic reservoirs during the winter floods. The small species multiply rapidly and although they become well-established in the lakes by autumn, they die during the next winter floods. The fish population is less impoverished than the benthic fauna, for of the 57 species reported, 50 are migrants from the sea whose juveniles enter the estuary during summer. The seven resident species include gobies, atherinids and one bream *(Mylio butcheri)* all of which breed in the estuary.

The Great Berg estuary on the Atlantic coast of the Cape appears to be similar to the Blackwood in several respects. It is also affected by heavy floods and low salinities during winter. Details are given in chapter 14. The fauna is not as well-known as that of the Blackwood but appears to be better developed. It is probably significant that in this case the tidal range on the coast at 1,5 m during springs is about double that in the Blackwood so that saline replacement of the fresh water after the floods is more rapid and the oligohaline phase is of shorter duration.

This brief summary of the ecology of the Blackwood together with the evidence from other estuaries shows clearly that it is not so much the absolute value of the minimum salinity which impoverishes the fauna as the duration of oligohaline conditions.

The effects of extremely high salinities have been investigated in Laguna Madre on the Texas coast and in the St Lucia system in Natal.

The Laguna Madre as described by Hedgpeth (1967) is a very long coastal lagoon separated from the Gulf of Mexico by a range of sand dunes broken at Port Aransas near Corpus Christi Bay in the north and by Brazos Santiago opposite Port Isabel in the south.

These rather shallow channels allow the exchange of surface water with the Gulf of Mexico but not the more saline deeper water in the Laguna. Thus the Laguna normally has a salinity over 40 ‰ and on occasion the salinity may rise to more than 100 ‰ in certain areas. Marginal vegetation is virtually absent and even the submerged weed beds of *Ruppia maritima* and *Halodule beaudetti (Olim Diplanthera wrightii)* are thin although there are thick mats of micro-algae. Although the fauna has been sampled several times Hedgpeth (1967) is doubtful of the records. He reports, however, that there are 34 species which persist in salinities of 75-80 ‰. Four of them are planktonic including the dominant copepod *Acartia tonsa;* 11 are macrobenthic invertebrates and 19 are fishes. The number of species increases when salinities fall to about 35 ‰ and there is a total record of 70 species of fish. Most of them as well as the decapod crustaceans migrate out of the Laguna when salinities rise above 60 ‰ although *Penaeus aztecus* tolerates 70 ‰ and the fish *Cyprinodon variegatus* has been found in a salinity of 142 ‰.

The ecology of the St Lucia system has been outlined in chapter 14. It consists of large saline lakes connected to the sea by a long narrow channel. The salinity of the lakes is extremely variable. During rainy years the salinity may fall to below 10 ‰ and during years of drought the salinity of the northern lakes may rise to 70 ‰ and in one area it has reached 102 ‰. The southern lake and the exit channel have a more stable salinity in the range of 30-35 ‰.

When the salinity of Lake St Lucia falls below 20 ‰ the marginal and aquatic vegetation is lush as noted in 1964 by Millard & Broekhuysen (1970). As salinities rise to hypersaline values, emergent plants such as *Cyperus* spp and *Phragmites* die back although the green shoots of the latter remain on the soil above the water level for some time. Submerged forms such as *Zostera* disappear when the salinity exceeds 40 ‰ and the shoots of all rooted vegetation die in salinities above 55 ‰. It is suspected that the rhizomes persist for a longer period for when the salinity falls to mesohaline values there is a very rapid growth of all the common species.

Grindley (1976a) who has examined zooplankton collected over a period of 29 years, finds that in years of severe drought the biomass is halved. In the 1969-71 period for example, when salinities in the northern lakes varied between 50 and 100 ‰, most of the plankton disappeared, while in the southern lake where the salinity was 35 ‰ in 1969, the biomass was 37 mg (ash free).m⁻³. In spite of this the species composition remained remarkably uniform over the whole 29 year period. All the species in the lakes are typically estuarine and although many disappear during hypersaline periods, those that remain continue to breed. For example *Pseudodiaptomus stuhlmanni* which is dominant in many subtropical estuaries in Natal, tolerates salinities of <1 to 75 ‰ and breeds in 60 ‰. Several other species of copepods, mysids, amphipods and insect larvae have similar salinity ranges. This suggests that factors other than salinity determine their distribution and abundance.

The benthic fauna is certainly reduced when the salinity rises above 50 ‰. Day *et al* (1954) reported that when the salinity in the northern lakes reached 52 ‰, the macrobenthic fauna was very reduced although it was still fairly rich in the southern lake where the salinity was about 35-40 ‰. Boltt (1975) concluded that the main reduction occurs at salinities above 55 ‰ and that chironomid larvae and ostracods then form the bulk of the remaining population. When salinities fall to between 20 and 25 ‰ there is rapid recolonisation from the south lake. The first species to appear in the northern lakes are those with planktonic larvae and short life cycles. Thus the gastropod *Assiminea bifasciata* increased ten-fold in six months and spread from three to 17 stations in the Lake St Lucia North. Champion (1976) notes that catches of penaeid prawns were high all over the lakes in 1967 when the salinity was 35-40 ‰. If the 1967 catch per unit of effort is regarded as 100 %, then the catch in 1968 fell as the salinity increased. In salinities of 45-55 ‰ it fell to 50 % and in salinities of 50-80 ‰ it fell to 20 %. In salinities above 60 ‰ only one large shoal was netted at the mouth of the Mzeneni river.

Fish are remarkably euryhaline; elvers and salmon swim rapidly from the sea through estuaries to fresh water. *Mugil cephalus* can live in fresh water, the sea or in salinities above 70 ‰ and Wallace (1976) has reported that 17 of the 19 common fishes in Lake St Lucia tolerate salinities above 60 ‰. Nonetheless, the fish do migrate out of hypersaline areas possibly due to the reduction of food supplies.

When the observations and records in the Laguna Madre are considered in conjunction with those from St Lucia, there can be no doubt that excessively high salinity is a constraint on the development of the flora and fauna. Although the species in the two estuaries differ, both show a marked reduction in species richness above 55 ‰ salinity and an increase when salinities fall to 35 ‰ or lower. Estimates of changes in biomass and productivity are not so well documented but appear to be equally marked.

Temperature extremes. Death from very low temperatures is exceptional. The surface water of many fjords in Norway and Canada freezes in winter but the deep lower layers remain above 5 °. Thus Hardanger fjord has a deep water temperature of 6-8 ° through the year (Saelen, 1967). Surface freezing may also occur in Scottish and even English estuaries during severe winters such as 1962/63 (Perkins, 1974) but the fauna is adapted to withstand such conditions. The main damage is due to ice erosion during the spring thaw. Perkins also reports that forms such as cockles and

scallops on intertidal banks become comatose at very low temperatures and fall an easy prey to wading birds.

In the temperate estuaries of southern Africa minimal temperatures of 10-12 ° have no noticeable effect on the fauna although active forms such as crabs remain dormant in their burrows. On the south coast where the summer inshore temperature is normally 18-20 °, local upwelling may reduce the temperature to 11 ° within a day (Day *et al*, 1952, Korringa, 1956). This numbs some of the coastal reef fish which may be caught by hand; others take refuge in the estuaries and are driven further in as the cold sea water enters with the rising tide, but the estuarine fauna itself appears to be unaffected.

It is only in the subtropics that freezing temperatures are lethal, and such conditions are of course abnormal. In the Laguna Madre, Hedgpeth (1967) reports a seasonal temperature range of 10-30 °. Gunter (1967), however, reports that on occasion 'on the south Texas coast mush ice forms in large quantities along the shores of the bays. These extreme cold waves cause catastrophic mortalities of aquatic organisms in that area (Gunter 1952a) but at no time have I seen the temperature of the open bay lower than 4 °C'.

In the subtropical estuaries of Natal and Mocambique open water temperatures rise to a maximum of 30 °, ie a temperature similar to that in the Laguna Madre. Shallow pools on the mud flats may even reach 36 ° and the dry sand of course becomes much hotter. Edney (1961) reports that at Inhaca Island (26 °S latitude) sand banks may rise to 46 ° in areas where the fiddler crab *Uca lactea annulipes* makes its burrows. However, these burrows extend down some 20 cm to where the temperature is about 32 °. The crabs remain active on the surface through the day but return to their burrows at 15 minute intervals to refresh their gills and the blood temperature remains below 38,7 °. Edney determined that their median lethal temperature is 42 °. Similar examples are quoted by Newell (1970) which indicate that while tropical animals live close to their upper lethal temperature, heat death has not been recorded in nature.

Ferguson Wood & Johannes (1975) also stress the fact that subtropical and tropical marine and estuarine animals live closer to their upper lethal temperature than do temperate animals. They point out that for this reason the threat of heated effluents from power plants and other industrial installations is greatest in the tropics although the effluent is seldom more than 3 or 4 ° above the ambient water temperature.

Restriction of tidal action and stabilization of the water level. The propagation of tidal energy along an estuary and the factors which affect it were discussed in chapter 3. It may be recalled that in deep estuaries with tapering channels, the tides may be propagated for more than 100 km. Thus tidal effects extend

240 km along the Hudson and in the Scheldt they reach Antwerp 180 km from the sea. Dissipation of tidal energy and reduction of range may occur naturally due to the formation of shallow sand bars and constricting sand spits at the mouth. In many cases the tidal range is decreased as the flood tide enters a narrow mouth and then spreads out over a broad lagoon as occurs in bar-built estuaries along the Atlantic coasts of the United States, South Australia and Victoria. Long narrow channels beset with islands and shoals also reduce tidal range as may be seen in the Kosi Bay system and St Lucia (chapter 14). The saline lakes beyond such channels show no rise and fall with the tides.

Tidal energy may also be reduced or even eliminated by human activities. Solid road embankments leading to the narrow bridge spans across the Keurbooms and other South African estuaries, the dumping of rubble under the bridge over the Bushmans estuary, the construction of a causeway across the Kariega and weirs across the mouth of the Fafa estuary in Natal and the Sandvlei lagoon at Muizenberg are all obvious examples. Range action and tidal exchange with the sea are both reduced to varying degrees and in extreme cases, due to the growth of the bars and sandspits at the mouth, the estuary may close and a blind estuary results. Such a condition may persist for months or even years until the obstruction is washed away by a flood.

Before the direct biological effects of restriction of tidal action and the closure of the mouth are discussed, it may be noted that they cause several effects on other environmental factors. The estuary may become hypersaline or hyposaline. Decreased tidal currents lead to a longer residence time of the estuarine water or, from the opposite point of view, a decreased flushing time. Reduced currents also allow the suspended silt to settle and the clarity of the water increases. In deep estuaries stratification may develop and the lower layers become depleted of oxygen, as has been noted in two lakes in the Kosi system and in the Mtamvuna and Msikaba estuaries (see chapter 5 and 14 for details). In Lake St Lucia and Hermanus Lagoon, both of which are broad and shallow, wind action is sufficient to maintain vertical circulation and the bottom water remains well-oxygenated. The biological effects of all these secondary changes have been discussed earlier and need not be repeated here. We may concentrate on the direct biological effects of reduced tidal range, the closure of the mouth and the consequent stabilization of the water level.

With the reduction of tidal exchange it would be expected that the biomass of plankton in the estuary would increase. The maximal effect should occur when the estuary mouth closes. Unfortunately quantitative comparisons between closed and neighbouring open estuaries are lacking. Moreover, other environmental changes such as the increased clarity of the water and alterations of salinity cloud the issue. Begg (1976) reported that dense blooms of a toxic nanoflagellate caused heavy mortality to fish and other aquatic animals in Sandvlei. It should be noted however that such blooms are rare and the one observed might have been due to the inflow of sewage effluent.

Changes to benthic vegetation are more obvious. As the tidal range is reduced, the upper tidal banks dry out and become highly saline and the area covered by salt marsh vegetation decreases. Above the tidal obstruction caused by the bridge over the Bitou which flows into Keurbooms estuary, large areas of dead and drying salt marshes have been reported (Day, 1973a). When the water level in an estuary is completely stabilized the effects on the vegetation are more severe. As noted in chapter 6, mangroves whose pneumatophores are submerged die as has occurred in the Beachwood tributary of the Mgeni (Berjak *et al*, 1977). In fact mangroves are either absent or very scanty in blind estuaries. *Spartina maritima* which grows at the same mid-tidal levels as mangroves, is absent from most blind estuaries in the Cape, including West Kleinemond, Heuningnes, Hermanus Lagoon, Bot River, Sandvlei and Milnerton Lagoon. In these estuaries the effect on plants which normally grow at high tide of springs varies from one species to another. *Juncus kraussii, Sporobolus pungens* and *Sarcocornia pillansii* are not obviously affected while *Limonium linifolium* and *Chenolea diffusa* are rare or absent. Similarly among the submerged sea grasses, *Ruppia* spp and *Potamogeton pectinatus* seem to be unaffected while *Zostera capensis* is absent from almost all blind estuaries except Hermanus Lagoon. Possibly its healthy growth in this estuary is due to turbulence caused by wind-generated waves or the more frequent opening of the mouth; further research is obviously necessary.

The estuarine fauna is seriously impoverished when an estuary is cut off from the sea either by the formation of a sand bar or the building of a weir. Since the bulk of an estuarine fauna is of marine origin, recruitment of larvae from the sea stops. Most of the fish, the penaeid prawns and the large swimming crabs spend only a part of their life cycle in estuaries and breed in the sea; their populations are obviously reduced if the mouth is closed for many months. Wallace (1975) has suggested that the estuary-dependent fishes in South Africa have a prolonged breeding season to take advantage of the brief and irregular periods when many Natal estuaries are open. Since most of these fish have a wide distribution in the Indian Ocean this seems doubtful.

As the tidal range in an estuary is reduced and eventually eliminated by closure of the mouth, the feeding area of the intertidal fauna is reduced and sessile species such as barnacles which have settled above what becomes the permanent water level die as has been observed in Hermanus Lagoon. Permanently submerged species are affected in a different way. Those that depend on tidal currents for the supply of food are stunted and some disappear. In an open estuary the rise and fall of the tides over a sandflat allows the interstitial water in the sediment to drain away during the ebb and as the tide rises, oxygenated surface water replaces it. Moreover as the tide falls, the sandflat acts as a filter and at low tide the plankton and seston are caught as on a filter paper. This provides food both for the epifauna and those burrowing species which gather surface deposits. Burrowing deposit feeders including many polychaete worms, bivalves, amphipods and ocypodid crabs are thus poorly represented in blind estuaries. The burrowing prawn *Upogebia africana* obtains its food from organic detritus drifting over the mud flats and is one of the most important fish foods in most of the open South African estuaries. It is completely absent in blind estuaries such as Hermanus Lagoon (Scott *et al,* 1952), West Kleinemond (Brown, 1953), Umgababa (Day, unpublished) and Sandvlei (Begg, unpublished). On the other hand, *Callianassa kraussii* which feeds on buried detritus in sandbanks and creates its own feeding currents, persists in all these estuaries and, in the absence of *Upogebia* it has extended its burrows from sandy into muddy areas.

One further interesting point may be noted. The distribution of the fauna in Hermanus Lagoon (Scott *et al,* 1952) shows that burrowers such as *Arenicola loveni, Solen capensis* and *Echinocardium cordatum* are confined to the sandbar at the mouth. Their presence here and absence from sandbanks elsewhere in the estuary suggests that they are supplied with food and oxygen by changes in water level due to the tides and waves in the sea. As the water level in the sea rises and falls relative to the stable water level in the estuary, seepage must occur through the sandbar and it is upon this that the burrowers depend. Measurement of changes in the interstitial water table and dissolved oxygen concentrations in this and other sandbars in the estuary are needed to confirm this speculation.

Persistant pollution. This is an attempt to define the intensity and duration of pollution that will change the nature of an estuarine community. The chemical effects of organic pollution and trace metals in sewage effluents from industrial towns have been outlined in chapter 5.

In dealing with the long term effects of pollution, it is necessary to say something of light or short term pollution. However this is merely explanatory and is certainly not intended as a complete account.

The decomposition of organic matter is a natural process which occurs in all estuaries. The bacteria and other micro-organisms metabolise the more labile organic compounds, absorbing oxygen, multiplying and releasing much of the nutrient salts, particularly nitrogen compounds and phosphate. So long as sufficient dissolved oxygen is available, the estuarine community flourishes and indeed phytoplankton blooms may result and the attached vegetation grows lush. It is only when the organic matter is supplied at a rate which exceeds the oxygen supply that anaerobic bacteria flourish. This normally occurs in subsurface mud but when pollution becomes serious the overlying water becomes anoxic, first nitrates and finally sulphates are reduced and H_2S is liberated. This itself is toxic and both the submerged plants and animals die adding to the pollutant load.

It will be obvious that toxic chemicals such as cyanides, phenol and sulphides or sulphites often found in industrial wastes, increase the effect. Rooted plants along the margin of an estuary which obtain oxygen supplies from the air, persist longer than do submerged forms but as the concentration of herbicides, detergents and toxic metals such as copper and arsenic increase, they also die. Throughout the process BOD and COD levels increase so that measurement of these parameters may be used as a guide to the intensity of pollution. Finally H_2S is evolved from the water surface and the whole estuary is converted into a stinking sewer in which all forms of life except bacteria are absent. This has occurred in many estuaries flowing through large industrial cities such as the Thames, the Elbe, the Scheldt and the Hudson to name but a few.

There are of course less serious stages of pollution in many estuaries. The South African estuaries described in chapter 14 are seldom polluted to any extent. The small Black (or Salt) River in Cape Town is canalized and the last 300 m are effectively without life, but the sewage effluent is relatively clear and has a low BOD unless the sewage works are overloaded. Presumably it is the toxic chemicals from factories at Paarden Island which cause the serious pollution. The Mgeni estuary in Durban is moderately polluted as judged by bacterial tests. There is practically no marginal vegetation and sea grasses are absent but there is some animal life although it is scanty. The comprehensive account of the Natal estuaries given by Begg (1978) stresses that several of them are polluted by effluents from sugar cane mills. Begg states that the worst is the Sezela which receives not only the effluent from the mill but also the clinker and ash from the coal-fired power plant. In spite of the povery of the originally rich fauna, the swamps in the oligohaline water carry flourishing stands of reeds and other vegetation.

It is difficult to distinguish the biological effects of heavy metals from those due to organic pollution, although experimental evidence is available from laboratory cultures. Many of the heavy metals such as mercury, chromium, arsenic, cadmium and copper are toxic even in trace concentrations and are concentrated to varying degrees by plants and animals. Minimata disease due to mercury caused death to humans in

Japan; dangerous levels of cadmium occur in Scandinavian countries and fish in some Caribbean areas contain such a high copper concentration that they have been declared unfit for sale. However, metal pollutants carried by rivers are largely deposited in the muddy upper reaches of estuaries. Wollach & Peters (1975) give a budget of several metals reaching Antwerp and show how much is deposited in that area and the small proportion that is carried on to the sea.

The heavy metal concentrations in the fine sediments of several Dutch and German estuaries are tabulated by de Groot *et al* (1976). In the Rhine, concentrations have increased by a factor of ten or more since the 16th century; Korringa (1967) indicates how the fisheries have been devastated over the same period.

A review of oil pollution in the sea has been made by an International Working Group and published in an FAO report (Anon. 1977). Researches on the ecological effects of oil pollution in Milford Haven (which is the largest oil port in the United Kingdom) have been edited by Cowell (1971). The latter publication includes a valuable account of pollution by oil and surfactants on rocky shores and on salt marsh plants but there is surprisingly little information on the effects on the fauna of muddy shores.

Many statistics are given of the toxicity or sublethal effects of oil and surfactants on development, feeding and reproduction of individual species of those plants and animals which are common in British or North American estuaries. When it is remembered that the toxicity varies with the origin of the crude, the boiling point fraction used, the effect of weathering or age of the spill, the temperature of the water, the presence and nature of surfactant used and the species of plant or animal damaged, it seems useless to quote such statistics here. There are too many variables to appreciate the significance of a 48 hour LD 50 of 1 ppm or the other concentrations quoted. A few generalizations regarding the toxicity of different crudes and surfactants have been quoted in several publications which suggests that they are of wide validity.

In regard to the serious biological effects of oil pollution and surfactants, all observers agree that oil on the surface of the sea (or on fresh water) has caused the death of many thousands of aquatic birds and that the toll is increased by the use of surfactants. To concentrate attention on estuarine birds, Anon (1977) quotes an observation that an oil spill in an estuarine area in the Netherlands in 1971 killed 5 000 birds including ducks, geese, swans and coot. It is known that the water-repellent feathers lose their waterproofing so that they become waterlogged and lose their insulating effect. The birds cannot fly and die of exposure. Toxic light oils are ingested during preening and damage the alimentary and excretory organs. A small spill of number 2 diesel oil in Puget Sound in 1971 killed 30 000 Brand geese. The effects of oil on sea birds is better documented and oil pollution is

equally damaging particularly to species such as cormorants, razor-bills and penguins which spend much time in the water. Surfactants increase the mortality to such an extent that ornithologists suggest they should not be used even in the open sea. However, other effects of oil pollution apart from destruction of aquatic birds must be considered.

The skin of a fish is slimy and oil-repellent and adult fish are seldom killed even by serious oil spills although their flesh may be tainted and unpalatable for a few days. It is even reported that *Mugil* kept in tanks with oil on the surface swallowed the oil but when the mullets were transferred to clean water the oil passed out through the vent and the fish were apparently unharmed. Some fish eggs such as those of pilchards and anchovies float near the surface and when fertilized they become very susceptible to oil and surfactants. The toxicity increases again at the hatching stage and even fish fry are damaged. Pacific salmon fry suffered 100 % mortality in 96 hours when subjected to 500 ppm Venezuelan crude.

Pelagic invertebrates suffer some mortality but again, early developmental stages are more susceptible than adults. The evidence in regard to the effect of oil pollution on phytoplankton production is not convincing.

The main damage due to oil pollution is caused when an oil spill reaches the shore. Surfactants, even the less toxic ones, cause so much more damage than the oil that experts recommend that it should not be used in water less than 5 m deep. Most of the observations have been made on littoral communities living on rocky shores. The oil floating on the water tends to adhere to dry rocks, shells and algae at the upper tidal levels. Thus plants and animals that live below low tide are not seriously harmed by floating oil but may be seriously damaged by the oil-in-water emulsions promoted by surfactants. Similarly plants and animals that remain moist during low tide are less affected than those near high tide. This is supported by the illustrations made by Crapp (in Cowell, 1971) which show changes in abundance before and after oil spills. There are however many complications. The decrease of grazing gastropods allows algae to increase and the settlement of mussels and barnacles may be reduced; similarly the decrease of carnivorous whelks allows their barnacle prey to extend their range. A striking case is the decrease of echinoids which are particularly susceptible to oil pollution. Following the death of echinoids, there is a massive proliferation of algae.

When the evidence of many observers is considered, the most obvious conclusion is that species even within the same genus differ so much in their tolerance to oil pollution that it is dangerous to draw conclusions concerning one species from observations on another. As Day *et al* (1973) showed in the case of the *Wafra* oil spill near Cape Agulhas, *Patella cochlear* may be decimated while *P. longicosta* nearby survives and even extends onto sites previously occupied by *P. cochlear.* *P. oculus* is badly affected but *P. granularis* survives even when its shell is black with oil. Different species of barnacles and crabs provide similar examples.

The effect of oil pollution on sandy shores is less obvious than the effect on rocky shores. Again it is evident that oil adheres more readily to the dry sand at upper tidal levels than to the wet sand lower down. However, the oil, particularly the lighter fraction, seeps into the sand and extends to lower levels. Its effect on the subsurface microbial community and the meiofauna has not been reported but according to Sanders *et al* (1972), oil extends well below low tide and causes serious destruction of polychaetes, bivalves, and amphipods. Moreover, in anaerobic subsurface sediments, biodegradation is extremely slow and the oil may persist for years.

Little has been published on the effect of oil pollution on estuarine animals; in particular there is little evidence regarding the effect on the burrowing fauna of mud banks apart from recent observations on South African estuaries by Moldan *et al,* 1979. They report that after 14 days weathering the chocolate mousse drifted into the Great Brak and Little Brak estuaries where it covered the sediment surface and smothered the burrowing prawns and crabs. Their new and important observation is that the chocolate mousse mixed with sand to form a water/oil/sand mixture with a specific gravity of 1,067. This dense mixture sank covering the bed of the Little Brak estuary to a depth of 1,5 m in some places. The bottom fauna was completely destroyed in consequence. Eventually the oil had to be removed by suction pumps.

The experiments of Dr Jennifer Baker (in Cowell, (1971)) give a clear insight into the effects of oil and surfactants on salt marsh vegetation. It would appear that blue-green algae in the creeks where the oil tends to accumulate are resistant to oil and nigrogen-fixing bacteria in the microbial community may metabolise the oil. Filamentous algae such as *Enteromorpha* are susceptible to oil spills but regrowth is rapid. As oil spreads over the salt marsh, it sticks to the leaves and shoots of rooted plants so that these act as absorbants and prevent more widespread pollution. The oil also lies on the surface of sandy mud but penetrates no more than a centimetre, although emulsifiers aid penetration. The presence of this oil on the surface limits the penetration of oxygen into the burrows of the infauna and the reducing environment in the sulphide layer below becomes toxic.

Oil damages rooted plants in two ways. When an oily layer covers the leaves it blocks the stomata and cuts down the transmission of light thus reducing photosynthesis. This means a reduction of oxygen and food reserves. Oil also damages cell membranes particularly those at the growing apices of leaves and in the leaf axils. It enters the cells, inhibits RNA formation and passes on into the intercellular spaces. In this way it reduces translocation and the oxygen supply to the roots, most of which penetrate subsurface anoxic soils. Further, it upsets water intake from the soil and thus the rate of transpiration.

The survival of a marsh plant depends to some extent on its food reserves. A single oil spill kills growing shoots but does not cause long-term damage to the perennials which form the bulk of the vegetation. However, shallowly rooted plants such as *Suaeda maritima,* annuals such as *Salicornia* spp and seedlings usually die. So long as the food reserves of the perennials remain, new shoots may be formed. Thus the effect of a succession of oil spills varies from one species to another. Some perennials such as *Juncus maritimus* and *Spartina anglica* with food reserves in the rhizomes recover from four light spills. Other perennials such as *Spergularia media* and *Limonium* spp with large food reserves in their tap roots or *Triglochin maritima* with large food reserves in bulbs are very resistant and survive eight or more oil spills. The resistance of many other European marsh plants is compared by Baker; only those genera that are common in southern Africa are quoted here.

Odum & Johannes (in Ferguson Wood & Johannes, 1975) review the effects of oil pollution on mangroves. They suggest that the damage results from the clogging of lenticels and 'air holes' on prop-roots and pneumatophores. Presumably this implies that the trees badly damaged or killed are those growing at the level where floating oil accumulates thickly. This would explain the conflicting reports regarding the damage caused by different oil spills.

Zieman (in Ferguson Wood & Johannes, 1975) gives a brief review of oil pollution on sea grasses. As would be expected from their habitat at and below low tide, floating oil seldom contacts these plants and they are less liable to damage than intertidal salt marsh vegetation and mangroves. Nonetheless, there is a report of severe damage to *Thalassia testudinum* in Puerto Rico. After several months the *Thalassia* disappeared and was replaced by blue-green algae. Zieman also notes that sea grasses stabilize the sediments and when they are dead, severe erosion of sediments may occur.

In general one may conclude that single oil spills cause short term damage to intertidal plants. Obviously oil spills during the growing season are more destructive than those which occur in the winter resting period. Although plants regenerate as long as food reserves are sufficient, four recurrent oil spills kill many species and few resist eight or more.

The effect on adult fish is marginal and as most estuarine species breed in the sea, the greater susceptibility of embryonic and larval stages is not relevant to the estuarine population in the short term. The effect on plankton is not conclusive but probably it is not severe to holoplankton although larval stages of meroplankton would suffer from spills in spring. The effect of oil spills on macrobenthic invertebrates, particularly those that inhabit intertidal mudbanks is poorly documented and further research is necessary. However it seems evident that an estuary subject to repeated oil spills would be devastated. Moreover the evidence of many workers indicates that once the oil has penetrated into sand or mud it will remain there for many years.

There are few conclusions that can be drawn from a consideration of all the constraints that may limit the productivity and species richness of estuaries. The normal estuarine biota is very hardy and tolerates a wide range of conditions so that only severe changes of long duration produce significant effects. Even so the plants and animals are so resiliant that when the stress conditions are relaxed the biota reverts to the norm. Heavy pollution, such as that which converted the Thames into a stinking sewer, is widely regarded as the most serious threat to life in an estuary. It took over 40 years of research and improvement of the sewage works to improve conditions but many species of fish are now passing up and down the estuary. It *can* be done. Nonetheless geological evidence shows that estuaries have a limited life; they are all silting up. Possibly the most serious constraint to life in an estuary is the acceleration of silting brought about by man.

REFERENCES

ANON. 1977. Impact of oil on the marine environment. *Rep. Stud. GESAMP*, **6**. 250pp. FAO, Rome.

BARNES, R.S.K. 1974. *Estuarine biology*. Edward Arnold, London. 76pp.

BEGG, G.W. 1976. Some notes on the Sandvlei fish fauna, Muizenberg, Cape. *Piscator* **96**: 4-14.

BEGG, G.W. 1978. *The estuaries of Natal*. Natal Town and Regional Planning Commission, Pietermaritzburg. 657pp.

BIGGS, R.B. 1967. The sediments of Chesapeake Bay. *In:* G. Lauff (ed), *Estuaries*. Am. Ass. Adv. Sci., Washington.

BIGGS, R.B. & D.A. FLEMER 1972. The flux of particulate carbon in an estuary. *Mar. Biol.* **12**: 11-17.

BLABER, S.J.M. 1973. Population size and mortality of juveniles of the marine teleost *Rhabdosargus holubi* in a closed estuary. *Mar. Biol.* **21**: 219-225.

BOLTT, R.E. 1975. The benthos of some South African lakes. Part 5: The recovery of the benthic fauna of St Lucia Lake following a period of excessive high salinity. *Trans. roy. Soc. S.Afr.* **41**: 296-323.

BRANCH, G.M. & J.R. GRINDLEY 1979. The ecology of South African estuaries. Part XI: Mngazana, a mangrove estuary. *S.Afr. J. Zool.* **14**(3): 149-170.

BREEN, C.M. & B.J. HILL 1969. A mass mortality of mangroves in the Kosi estuary. *Trans. roy. Soc. S.Afr.* **38**: 285-303.

BROWN, A.C. 1953. *A preliminary investigation of the ecology of the larger Kleinemond River estuary, Bathurst District*. MSc thesis, Rhodes Univ., Grahamstown.

CHAMPION, H.F.B. 1976. Recent prawn research at St Lucia with notes on the bait fishery. *In:* A.E.F. Heydorn (ed), *St Lucia Scientific Advisory Council Workshop — Charters Creek, February 1976*. Natal Parks Bd, Pietermaritzburg.

CHOLNOKY, B.J. 1968. Die Diatomeenassociationen der Santa Lucia Lagune in Natal (Südafrika). *Bot. Mar.* **11** (suppl).

COLLIER, B.B., COX, G.W., JOHNSON, A.W. & P.C. MILLER 1974. *Dynamic ecology*. Prentice/Hall Internatl., London. 563pp.

COWELL, E.B. (ed) 1971. *The ecological effects of oil pollution on littoral communities*. Inst. of Petroleum, London. 250pp.

CHRISTIE, N.D. & A. MOLDAN 1977. Distribution of benthic macrofauna in Langebaan Lagoon. *Trans. roy. Soc. S.Afr.* **42**(3-4): 273-284.

CSIRO, Division of Fisheries and Octeanography. *Estuarine Project: Progress report for 1974-76*. Cronulla.

CSIRO, Division of Fisheries and Oceanography. *Research Report 1974-1977*. Cronulla.

DAY, J.H. 1959. The biology of Langebaan Lagoon: a study of the effect of shelter from wave action. *Trans. roy. Soc. S.Afr.* **35**: 475-547.

DAY, J.H. 1967. The biology of Knysna estuary, South Africa. *In:* G. Lauff (ed), *Estuaries*. Am. Ass. Adv. Sci., Washington. 757pp.

DAY, J.H. 1974. The ecology of Morrumbene estuary, Mozambique. *Trans. roy. Soc. S.Afr.* **41**: 43-97.

DAY, J.H., COOK, P.A., ZOUTENDYK, P. & R. SIMONS 1971. The effect of oil pollution from the tanker *Wafra* on the marine fauna of the Cape Agulhas area. *Zool. Afr.* **6**: 209-219.

DAY, J.H., MILLARD, N.A.H. & G.J. BROEKHUYZEN 1954. The ecology of South African estuaries. Part 4: The St Lucia System. *Trans. roy. Soc. S.Afr.* **34**(1): 129-156.

DAY, J.H., MILLARD, N.A.H. & A.D. HARRISON 1952. The ecology of South African estuaries. Part 3: Knysna, a clear open estuary. *Trans. roy. Soc. S.Afr.* **33**: 367-413.

DAY, R.W. 1977. Two contrasting effects of predation on species richness on the coral reef habitats. *Mar. Biol.* **44**: 1-5.

DE GROOT, A.J., SALOMONS, W. & E. ALLERSMA 1976. Heavy metals in sediments from the Rhine and Meuse estuaries. *In:* J.D. Burton & P.S. Liss (eds), *Estuarine chemistry*. Academic Press, London, New York, San Francisco. 229pp.

DYE, A.H., ERASMUS, T. & J.P. FURSTENBURG 1978. An ecophysiological study of the meiofauna of the Swartkops estuary. 3: Partition of benthic oxygen consumption and relative importance of the meiofauna. *Zool. Afr.* **13**(2): 187-220.

EDNEY, E.G. 1961. The water and heat relations of fiddler crabs (*Uca* spp). *Trans. roy. Soc. S.Afr.* **36**(2): 71-91.

EMERY, K.O. 1969. *A coastal pond studied by oceanographic methods*. Elsevier, New York. 80pp.

FELL, J.W., CEFALU, R.C., MASTER, L.M. & A.S. TALLMAN 1975. Microbial activities in the mangrove *(Rhizophora mangle)* leaf detrital system. *In:* G. Walsh, S. Snedaker & H. Teas (eds), *Proc. Inst. Symp. Biol. and Management of Mangroves, October 1974, Honolulu*. Vol II: 661-679.

GRAY, J.S. 1976. The fauna of the polluted River Tees estuary. *Estuar. coast. mar. Sci.* **4**: 653-676.

GREEN, J. 1968. *The biology of estuarine animals*. Sidgwick & Jackson, London. 401pp.

GRINDLEY, J.R. 1976a. Zooplankton in St Lucia. *In:* A.E.F. Heydorn (ed), *St Lucia Scientific Advisory Council Workshop — Charters Creek, February 1976*. Natal Parks Bd, Pietermaritzburg (23 contributions, typescript).

GRINDLEY, J.R. 1976b. *Report on the ecology of Knysna estuary and proposed Braamekraal marina*. School of Environmental Studies, Univ. Cape Town (123pp, typescript).

GRINDLEY, J.R. 1977. The zooplankton of Langebaan Lagoon and Saldanha Bay. *Trans. roy. Soc. S.Afr.* **42**(3-4): 341-370.

GRINDLEY, J.R. 1978. *Environmental effects of the discharge of sewage effluent into Knysna estuary.* School of Environmental Studies, Univ. Cape Town (62pp, typescript).

GRINDLEY, J.R. & A.E.F. HEYDORN 1970 Red water and associated phenomena in St Lucia. *S.Afr. J. Sci.* **66**: 210-213.

GRINDLEY, J.R. & T.H. WOOLDRIDGE 1974. The plankton of Richards Bay. *Hydrobiol. Bull.* 8: 201-212.

GUNTER, G. 1967. Some relationships of estuaries to the fisheries of the Gulf of Mexico. *In:* G. Lauff (ed), *Estuaries.* Am. Ass. Adv. Sci., Washington. 757pp.

HEAD, P.C. 1976. Organic processes in estuaries. *In:* J.D. Burton & P.S. Liss (eds), *Estuarine chemistry.* Academic Press, London. 229pp.

HEDGPETH, J.W. 1967. Ecological aspects of the Laguna Madre, a hypersaline estuary. *In:* G. Lauff (ed), *Estuaries.* Am. Ass. Adv. Sci., Washington. 757pp.

HELLEBUST, J.A. 1967. Excretion of organic compounds by cultured and natural populations of marine phytoplankton. *In:* G. Lauff (ed), *Estuaries.* Am. Ass. Adv. Sci., Washington. 757pp.

HOBBIE, J.E. 1976. Nutrients in estuaries. *Oceanus* **19**(5): 41-47.

HODGKIN, E.P. 1976. The history of two coastal lagoons at Augusta, Western Australia. *J. roy. Soc. Austr.* **59**: 39-45.

HODGKIN, E.P. 1978. *An environmental study of the Blackwood River estuary, Western Australia.* 78pp. Dept. of Conservation and Environment, Perth.

JENNINGS, J.N. & E.C.F. BIRD 1967. Regional geomorphological characteristics of some Australian estuaries. *In:* G. Lauff (ed), *Estuaries.* Am. Ass. Adv. Sci., Washington.

JEFFERIES, R.L. 1972. Aspects of salt marsh ecology with particular reference to inorganic plant nutrition. *In:* R.S.K. Barnes & J. Green (eds), *The estuarine environment.* Applied Science, London. 133pp.

JOHNSON, I.M. 1976. Studies of the phytoplankton of the St Lucia system. *In:* A.E.F. Heydorn (ed), *St Lucia Scientific Advisory Council Workshop – Charters Creek, February 1976.* Natal Parks Bd, Pietermaritzburg (23 contributions, typescript).

KENDALL, M.A. 1979. The stability of the deposit feeding community of a mud flat in the River Tees. *Estuar. coast. mar. Sci.* 8: 15-22.

KETCHUM, B.H. 1954. Relation between circulation and planktonic populations in estuaries. *Ecology.* **35**: 191-200.

KIRBY-SMITH, W.W. 1976. The detritus problem and the feeding and digestion of an estuarine organism. *In:* M. Wiley (ed), *Estuarine processes. Vol 1. Uses, stresses, and adaptations in the estuary.* Academic Press, New York, San Francisco, London. 541pp.

KORRINGA, P. 1967. Estuarine fisheries in Europe as affected by man's multiple activities. *In:* G. Lauff (ed), *Estuaries.* Am. Ass. Adv. Sci., Washington. 757pp.

LEACH, J.H. 1971. Hydrology of the Ythan estuary with reference to distribution of major nutrients and detritus. *J. mar. biol. Ass. UK* **51**(1): 137-158.

LONGBOTTOM, M.R. 1968. *Nutritional factors affecting the distribution of Arenicola marina L.* PhD thesis, Univ. London.

MACNAE, W. 1957. The ecology of the plants and animals in the intertidal regions of the Swartkops estuary near Port Elizabeth, South Africa – Parts I and II. *J. Ecol.* **45**: 113-131 and 361-387.

MARGALEF, R. 1967. Laboratory analogues of estuarine plankton systems. *In:* G. Lauff (ed), *Estuaries.* Am. Ass. Adv. Sci., Washington.

McLACHLAN, A. & J.R. GRINDLEY 1974. Distribution of macrobenthic fauna on soft substrata in the Swartkops estuary, with observations on the effects of floods. *Zool. Afr.* 9(2): 211-233.

MILLARD, N.A.H. & G.J. BROEKHUYSEN 1970. The ecology of South African estuaries. Part 10: St Lucia: a second report. *Zool. Afr.* **34**(1): 157-179.

MILLARD, N.A.H. & A.D. HARRISON 1954. The ecology of South African estuaries. Part 5. Richards Bay. *Trans. Roy. Soc. S.Afr.* **34**(1): 157-179.

MILLARD, N.A.H. & K.M.F. SCOTT 1954. The ecology of South African estuaries. Part 6: Milnerton estuary and the Diep River, Cape. *Trans. roy. Soc. S.Afr.* **34**: 279-324.

MILLS, E.L. 1969. The community concept in marine zoology with comments on continua and instability in some marine communities: a review. *J. Fish. Res. Bd. Can.* **26**: 1415-1428.

MILNE, H. & G.M. DUNNET 1972. Standing crop, productivity and trophic relations of the fauna of the Ythan estuary. *In:* R.S.K. Barnes & J. Green (eds), *The estuarine environment.* Applied Science, London. 133pp.

MOLDAN, A., CHAPMAN, P. & H.O. FOURIE 1979. Some ecological effects of the *Venpet-Venoil* collision. *Mar. Pollution Bull.* 10: 60-63.

NEWELL, R. 1965. The role of detritus in the nutrition of two marine deposit feeders, the prosobranch *Hydrobia ulvae* and the bivalve *Macoma balthica. Proc. zool. Soc. Lond.* **144**: 25-45.

NEWELL, R.C. 1979. *Biology of intertidal animals* (3rd ed). Marine Ecological Surveys Ltd., Faversham, UK. 781pp.

ODUM, E.P. 1971. *Fundamentals of ecology* (3rd ed). Saunders Co, Philadelphia, London, Toronto. 574pp.

ODUM, E.P. 1975. Diversity as a function of energy flow. *In: Unifying concepts in ecology.* pp 1-14. Junk, The Hague.

ODUM, E.P. & A.A. DE LA CRUZ 1967. Particulate organic detritus in a Georgia salt marsh – estuarine ecosystem. *In:* G. Lauff (ed), *Estuaries.* Am. Ass. Adv. Sci., Washington. 757pp.

OLIFF, W.D. (ed) 1976 (South African) *National marine pollution monitoring program.* CSIR National Institute for Water Research, Durban. 681pp typescript (includes many progress reports by officers of the NIWR; 1st annual report 1-509; 2nd annual report 510-581).

ORIANS, G.II. 1975. Diversity, stability and maturity in natural ecosystems. *In: Unifying concepts in ecology.* pp 139-150. Junk, The Hague.

PAINE, R.T. 1966. Food web diversity and species diversity. *Am. Naturalist.* **100**: 65-75.

PERKINS, E.J. 1960. The diurnal rhythm of the littoral diatoms of the River Eden estuary, Fife. *J. Ecol.* **48**: 725-728.

PERKINS, E.J. 1974. *The biology of estuaries and coastal waters.* Academic Press, London. 678pp.

PUTTICK, G.M. 1977. Spatial and temporal variations in intertidal animal distribution at Langebaan Lagoon, South Africa. *Trans. roy. Soc. S.Afr.* **42**(3-4): 403-440.

RANWELL, D.S. 1972. *Ecology of salt marshes and sand dunes.* Chapman & Hall, London. 258pp.

RHOADS, C.D. & D.K. YOUNG 1970. Influence of deposit feeding organisms on sediment stability and community trophic structure. *J. mar. Res.* **28**: 150-178.

RILEY, G.A. 1967. The plankton of estuaries. *In:* G. Lauff (ed), *Estuaries.* Am. Ass. Adv. Sci., Washington. 757pp.

SANDERS, H.L. 1968. Marine benthic diversity: a comparative study. *Am. Nat.* **102**: 243-282.

SANDERS, H.L., GRASSLE, J.F. & G.R. HAMPSON 1972. The West Falmouth oil spill. I. Biology. *Tech. Rep. Woods Hole Oceanogr. Inst.* **72-20**: 48pp.

SCOTT, K.M.F., HARRISON, A.H. & W. MACNAE 1952. The ecology of South African estuaries. Part 2: The Klein River estuary, Hermanus. *Trans. roy. Soc. S.Afr.* **33**: 283-331.

SUMMERS, R.W. 1977. Distribution, abundance and energy relationships of waders (Aves: Charadrii) at Langebaan Lagoon. *Trans. roy. Soc. S.Afr.* **42**(3-4): 483-496.

SUMMERS, R.W.. PRINGLE, J.S. & J. COOPER 1976. *The status of coastal waders in the south-western Cape, South Africa.* 162pp. Percy Fitzpatrick Institute of African Ornithology, Univ. Cape Town.

TAN, F.C. & P.M. STRAIN 1979. Organic isotope ratios in the St Lawrence Estuary and Gulf of St Lawrence. *Estuar. coast. mar. Sci.* 8: 213-226.

TEAL, J.M. 1962. Energy flow in the salt marsh ecosystem of Georgia. *Ecology.* 43: 614-624.

THAYER, G.W., FERGUSON, R.L. & M.A. KJELSON 1974. Pools of organic matter, carbon, nitrogen and energy in the Newport River estuary. *Ann. Rep. Atl. estuarine Res. Center, Beaufort, NC:* 245-252.

WALLACE, J.H. 1975. The estuarine fishes of the east coast of South Africa. 3: Reproduction. *Invest. Rep. Oceanog. Res. Inst., Durban* 41: 1-51.

WALLACE, J.H. 1976. Biology of teleost fish, with particular reference to St Lucia. *In:* A.E.F. Heydorn (ed), *St Lucia Scientific Advisory Council Workshop – Charters Creek, February 1976.* Natal Parks Bd, Pietermaritzburg (23 contributions, typescript).

WARD, C.J. 1976. Aspects of the ecology and distribution of submerged macrophytes and shoreline vegetation of Lake St Lucia. *In:* A.E.F. Heydorn (ed), *St Lucia Scientific Advisory Council Workshop – Charters Creek, February 1976.* Natal Parks Bd, Pietermaritzburg (typescript).

WARWICK, R.M. & R. PRICE 1975. Macrofauna production in an estuarine mud-flat. *J. mar. biol. Ass. UK* 55(1): 1-18.

WIEBE, W.J. & L.R. POMEROY 1972. Micro-organisms and their association with aggregates and detritus in the sea: A microscopic study. *Mem. 1st Ital. Idrobiol.* 29, suppl. 325-352.

WOLLACH, R. & J.J. PETERS 1978. Biogeochemical properties of an estuarine system: The River Scheldt. *In:* Ed. Goldberg (ed), *Biogeochemistry of estuarine sediments.* UNESCO, Paris.

WOOD, E.J.F. 1965. *Marine microbiol ecology.* Chapman & Hall, London. 243pp.

WOOD, E.J.F. & R.E. JOHANNES (eds) 1975. *Tropical marine pollution.* Elsevier, Amsterdam. 192pp.

WOODIN, S.A. 1976. Adult-larval interactions in dense natural assemblages: patterns of abundance. *J. mar. Res.* 34: 23-42.

Chapter 17

The management of estuaries
J.H. Day and J.R. Grindley

Department of Zoology   School of Environmental Studies

University of Cape Town

INTRODUCTION

This chapter has been written for decision makers particularly legislators who frame the laws and regulations and the planners and administrators who have the difficult task of holding the balance between the conflicting interests of developers and conservationists. The earlier chapters reviewed what is known about estuarine ecology in broad outline with a more detailed account of conditions in southern Africa. This review is intented as a basis for further research for there is much that we do not know and there are several areas of grey knowledge which are contentious. However, detailed studies in Europe, North America and southern Africa have provided a solid foundation of information as a basis for legislation and the management of estuaries.

The evidence regarding the interrelations within estuarine ecosystems has been presented in technical terms to avoid ambiguity. This evidence will not be repeated here. However the main conclusions are presented as an introduction to the problems of management. The chapter is thus complete in itself and is written in non-technical terms for busy decision-makers.

With the growth of human populations and the increase in urbanisation and industrial activities, the development of certain estuarine systems is both inevitable and necessary. Protection of the estuarine environment in the long term is also essential if the natural resources provided by estuaries and the quality of human life near them is to be maintained. Such protection is costly and many would claim that conservation hinders progress, inhibits industrialization and thus leads to unemployment and increased costs of manufactured products. Conservation and development thus appear to have rival aims. Planners must decide which estuaries must be developed and which must be preserved in their natural state or restored where they have become open sewers. Administrators must ensure that where development is necessary it takes place with the minimum disturbance of the environment and where development is not necessary they must maintain the beauty and productivity at its highest level.

It must be appreciated that estuaries are not independent ecosystems and should not be managed as such. Since an estuary is formed where a river meets the sea, both the river and the sea affect conditions in the estuary (see chapters 1-3). This applies not only to the dilution of sea water by river water but to many other factors as well. The river water carries fertilizing salts leached from the soil of the drainage basin, persistent pesticides used in agriculture (chapter 5) and the eroded soil transported down to the estuary (chapter 4). Much of the silt is deposited and mixes with marine sands brought in by the tide until a flood erodes the estuary bed and carries the sediment out to sea. Thus it is claimed that the sandy shores of the Lincolnshire coast are derived from the Humber estuary (King, 1962) and that the maintenance of sandy beaches north of the Tugela River in Natal is dependent on supplies from the drainage basin of the river (Begg,

1978). Due to the inflow of fertile water from the river, estuaries are highly productive but conditions of life are harsh and variable. In dry climates the salinity tends to rise rapidly and it is the inflow of the river which maintains the salinity within limits which the aquatic plants and animals can tolerate. The inflow of sea water with the rising tide is also important (chapter 16) and when tidal action is reduced or inhibited the estuary suffers. Tides carry in the young stages of fish and invertebrates and they maintain the circulation of the water so that a healthy amount of oxygen is carried down to the bed of the estuary. Both fish (chapter 12) and invertebrates (chapter 9) grow rapidly in the sheltered waters of estuaries so that the latter have been called 'the nurseries of the sea'. This rich estuarine life attracts many predators. The predatory fish eventually return to the sea to breed while the smaller fish and invertebrates attract enormous flocks of aquatic birds (chapter 13). Man also takes his toll of fish, prawns, crabs, mussels and oysters. Eventually the estuarine plants and animals die and the products of decay drift out to sea. Thus the coastal seas near estuaries are well fertilized and form richer fishing grounds than the waters of the open ocean.

This in brief is the reason why estuaries should not be managed as independent units. It is realized of course that it is difficult to integrate the various laws and local regulations that apply to farms in the drainage basin, with those that apply to the river itself or to the supply of fresh water to urban areas and the other laws deal with the saline estuary and its tidal banks and the coast on which the estuary opens. Moreover the estuaries are but one type of environment and the laws and regulations must apply to all. In his book, *South African environmental legislation,* Dr Andre Rabie (1976) states: 'It seems more satisfactory, at least at the present state of environmental law to deal separately with separate environmental problems as they are treated in legislation . . .' Thus there are innumerable difficulties to the co-ordinated management of estuarine ecosystems but at least the regulations that affect estuaries should be drawn up after consultation between the various administering authorities and representatives of the local community. The ideal of course is a River Authority for each major river system with co-ordinated regulations applicable from the source to the estuary mouth. The Thames Conservancy Board, and in South Africa the Swartkops Trust, the Kowie Trust, the Outeniqualand Trust and the Bushmans-Kariega Trust are steps in this direction.

THE VALUE OF ESTUARIES

The development of wetlands and estuarine areas for agricultural, residential or industrial purposes gives immediate financial rewards while conservation is an ongoing expense. Why then should wetlands and estuaries be conserved? Why should salt marshes be protected when they appear to be obvious areas to reclaim for agriculture, marina construction or dumping rubble and dredge spoil until the level rises sufficiently high for building purposes? What in general is the value of estuaries that makes them worthy of conservation?

Estuarine productivity

Chapters 6 and 16 provide evidence that estuarine marshes are as productive as the same area of good agricultural land and are far more productive than the open sea. Odum (1971) compares the number of grams of dry organic matter produced by different environments per m^2 per day. In the open sea the value is less than 1; in coastal waters it is 0,5-3; in wild grasslands 0,5-3; in ordinary farm lands 3-10; in intensively cultivated farmland 10-25; and in estuaries 10-25. Few estuarine plants are of direct commercial value. The marsh grasses in Europe are sometimes used for grazing or may be cropped as nutritious hay, while the farming of ducks and geese in salt marshes has proved financially rewarding. Mangrove trees which, under ideal conditions in the tropics, grow to a height of 20-40 m have been used for boat building and furniture since the timber is resistant to shipworm and termites. Traditionally the poorer trees have been used for charcoal production and the bark for tanin but more recently Japanese concessionaires have converted large stands of mangroves into wood pulp. When the trees are thinned and felled at 20-30 year intervals by experienced foresters, the logging causes little damage and production may be increased two-fold over that in unexploited stands.

Estuaries are unusual in that few aquatic animals feed directly on growing plants apart from microscopic forms. At the end of the annual growing season, much of the protein of the larger plants is withdrawn to the stems and roots and the leaves fall and decay. As they disintegrate into minute fragments they support a wealth of micro-organisms, some of which build up the protein content until the 'organic detritus' is richer than that in the original leaf. About 90 % of the invertebrates and fishes feed directly or indirectly on this organic detritus. Some of the detritus is buried in fresh layers of sediment and forms fertile soil but much is washed out to sea during floods and enriches the fishing banks of coastal seas.

It will be evident that this transport of basic plant food to the sea makes it difficult to define the extent of estuarine-dependent fisheries. There are further difficulties. Many fish, prawns and edible crabs which breed in the sea enter estuaries as juveniles to feed and shelter and later migrate back to the sea. Again fishery statistics seldom separate fish caught within estuaries from those caught in coastal seas nearby, and it is seldom possible to monitor the catches of sport fishermen.

Table 17.1
Estimates of the quantity and value of estuarine fishery products

Locality	Product	Total quantity (wet meat) and value		Quan.	Value per hec.	Reference
		Quantity	10^6 US \$ yr^{-1}	kg.ha^{-1} yr^{-1}	10^3 US \$ yr^{-1}	
SHELLFISH						
USA, Connecticut	All types		1,5			Valiela & Vince, 1976
USA, Mass. (Falmouth)	All types		0,151			Valiela & Vince, 1976
USA, Main	Soft clams		5,7		17,54	Lindsay & Savage, 1978
USA, Chesapeake Bay	Oysters		13,6	673		Walne, 1962; McHugh, 1967
USA, Eastern States	Oysters (cultivated)				2,2	Chapman, 1977
UK, England & Wales	All types		1,0			Walne, 1972
UK (2 localities)	Cockles			1 918		Walne, 1972
UK (4 localities)	Mussels			2 688		Walne, 1972
UK, Fal estuary	Oysters (cultivated)			319		Walne, 1972
Spain (NW estuaries)	Mussels (cultivated)	30 t per raft				Walne, 1972
Japan	Oysters (cultivated)	350 tons				Walne, 1972
New South Wales (1976)	Oysters (cultivated)	140 201 bags	10,936	1 794	30,76	NSW Fish. Rep., 1977
SHRIMPS, PRAWNS, CRABS						
USA, Florida mangroves	Crabs	6 805 100 kg	1,241			Walsh, 1977
USA, Florida mangroves	Prawns	14 550 100 kg	15,719			Walsh, 1977
Liberia (St Paul estuary)	Shrimp *(Macrobrachium)*	7 128 kg	0,007 5			Miller, 1971
W.Australia (Shark Bay)	Prawns	$9,24 \times 10^6$ kg	27,72			Meany, 1979
New South Wales (1976)	Prawns	$2,47 \times 10^6$ kg	6,303			NSW Fish. Rep., 1977
EDIBLE FISH						
USA, Florida mangroves	Mullet	605 100 kg	0,1			Walsh, 1977
USA, Florida mangroves	Spotted sea trout	1 679 700 kg	0,132			Walsh, 1977
USA, Florida mangroves	Red drum	396 500 kg	0,132			Walsh, 1977
USA, Chesapeake Bay	Striped bass			40		McHugh, 1967
USA, Chesapeake Bay	All fish			216		McHugh, 1967
USA	Salmon	14 345 t	56,66			McHugh, 1967
USA	Menhaden	10 672 472 t	25,87			McHugh, 1967
Canada (British Colombia)	Salmon	64 269 t	63,42			Frazer, 1979
USA	Sport fishing		Approx. 600			Begg, 1978
UK	Salmon		4,512			Walne, 1972
ALL ESTUARINE DEPENDENT FISHERY PRODUCTS						
Holland (Wadden Zee)			400			Chapman, 1977
USA (Gulf coasts)				322		McHugh, 1967
USA (Florida)					985	Walsh, 1977
USA (Georgia)					10 246	Valiela & Vince, 1976
South Africa (commercial fishing prohibited)						
South Africa (St Lucia)	Bait sales		0,063			Van der Elst, 1979
S.Mocambique (Morrumbene estuary)				±1 200		Day, 1974
ANNUAL PRODUCTION OF COMMERCIAL AND SPORT FISHERY PRODUCTS						
SE Asia (Fertilized ponds)				1 000-2 000		McHugh, 1967
Peru current (Anchovy fishery)				370		McHugh, 1967
USA (Chesapeake Bay)				140		McHugh, 1967
Japan Sea				28		McHugh, 1967
North Sea				28		McHugh, 1967
Banks of Nova Scotia				13,5		McHugh, 1967

The landed mass and/or the value of several estuarine-dependent fisheries are set out in Table 17.1. Unfortunately the figures are seldom comparable for several reasons. The areas covered by the various fisheries are unequal; some figures represent the wild catch alone while others include the enhanced production due to aquaculture. The landings and values per hectare are more informative.

The commercial catch of shellfish such as oysters, mussels, clams and cockles is easier to delimit and quantify but the value varies greatly from country to country and the distance from the market. Shrimps, prawns and large edible crabs are particularly valuable products. The largest landings are made in the Gulf of Mexico, South East Asia, Japan and Australia. Of the many types of fin fish caught in estuaries, the most valuable is undoubtedly the salmon although these are estuarine only in the sense that they are caught at the mouths of estuaries on their migrations from the sea to their spawning grounds in the rivers. In England and Wales the Department of the Environment report for 1973 (ISBN 11 750396 7) states that: 'In England and Wales the sale value of salmon fisheries based on average catches during the last five years, is from £500 to £700 per fish caught'. Eels are also highly appreciated in Europe and sturgeon were once so common in European estuaries that they were regarded as food for peasants. Nowadays of course, pollution has largely restricted them to Russia (Korringa, 1967). Other commercial fish such as plaice, flounders and herring which are commonly thought to be restricted to the sea, tend to spawn in sheltered estuarine waters. The Zuider Zee was an important spawning area until it was reclaimed but the Firth of Forth and the Baltic are still important nursery areas. Several other plankton feeders such as menhaden (*Brevoortia* spp) in the United States, the bonga *(Ethmalosa dorsalis)* in tropical west African estuaries and the Bombay duck *(Harpondon nehereus)* in Indian estuaries are also commercially important (Pillay 1967a, 1967b). Commercial netting is illegal in South African estuaries but similar filter feeders including white bait *(Gilchristella)* and atherinids (*Hepsetia* spp) are the main food of prized game fish such as leervis, kob, kingfish and ten-pounders.

The types of predaceous fish in estuaries are too numerous to be treated individually and in any case common names change from one country to another. Various types of kob *(Sciaenidae),* bream *(Sparidae),* grunters *(Pomadasidae),* kingfish and leervis *(Carangidae)* and ten-pounders (*Elops* spp) are common in all warm estuaries. These as well as local endemics such as the white steenbras *(Lithognathus)* of South Africa or the striped bass *(Roccus saxatilis)* in the United States are mainly caught by sport fishermen.

Aquaculture is practised in many estuaries. Oyster farming has been developed most successfully in Japan, South-East Asia, Australia and the United States. Unfortunately the pilot experiments in South Africa have been beset with many difficulties but the outlook is now more hopeful (Genade, 1973). Mussel culture is most productive in north-western Spain where the deep ravine-like estuaries shelter rafts from which long ropes are suspended. The mussels grow so thickly on the ropes that 30 tons of mussel flesh are produced by each raft. Possibly the sheltered waters of Saldanha Bay would provide similar facilities.

Prawn culture requires great expertise. It is most successful in Japan where shallow areas of the estuarine Inland Sea are enclosed by fences. Juvenile prawns bred by the government are sold to farmers and grow to marketable size in about eight months. In South-East Asia fertilized ponds or tambaks are flooded with estuarine water containing juvenile prawns, milk fish and edible crabs. Carefully regulated ponds are extremely productive but farmers who use primitive methods live at subsistence level. Estuarine dependent prawns are caught in southern Mocambique particularly near Inhambane and in Delagoa Bay and are sold for bait at St Lucia and Richards Bay by the Natal Parks Board. Prawn culture has not been economically successful in southern Africa.

The farming of estuarine fin fish has been practised for centuries in the far east. Milk fish were the main species used and these were harvested as juveniles. Attempts to farm plaice and other flat fish in Europe were initially unsuccessful for the fish migrated to the sea before they reached marketable size. The methods have recently been improved and it is claimed that the yield of the coastal fisheries has increased but this is difficult to prove. Mullet are successfully farmed in South-East Asia and in Israel even in fresh water ponds, but they are not highly regarded as a table fish elsewhere. Several species of fish are farmed in Japan but the techniques require further development. Production of fish protein for food is successful in South-East Asia where methods are traditional and labour is cheap. As marine fish are over-exploited almost everywhere, fish farming is the hope for the future.

The comparison of fish production in different waters shown at the end of Table 17.1 are very revealing. In the open sea the Peruvian banks are far more productive per hectare than the traditional fishing banks in the North Sea, the Japan Sea and those off Nova Scotia. The banks off the west coast of southern Africa are comparable to those off Peru. On the other hand, the estuarine fish ponds of South-East Asia are over three times more productive than the Peruvian banks. It should be emphasised that these valuable fish ponds do not decrease the normal productivity of the estuaries.

Water transport

Estuaries are areas of sheltered water suitable for harbours and connect ocean-going vessels with cheap riverine transport to inland areas. Fresh water is available from the river and the estuary carries sewage, factory wastes and heated effluents from power stations out to sea. It is not by chance that large metropolitan areas such as London, Glasgow, Hamburg, Antwerp, New York, Quebec, Buenas Aires, Cairo, Calcutta, Shanghai and Sydney have been built on the banks of estuaries. The estuary of the Scheldt for example carries ocean-going vessels for 180 km to Antwerp and the Nile provides water transport for 194 km from Cairo to Luxor. South African estuaries are now too badly silted for water transport but in the early days it was cheaper and quicker to ship farm produce the 450 km from Malgas on the Breede estuary to Port Beaufort and then around Cape Agulhas to Cape Town than it was to take it by the 200 km road.

RECREATION

With the growth of human populations, increasing

affluence and greater leisure, more and more people are attracted to water sports. In many parts of the world boating and other aquatic sports are among the fastest growing outdoor activities. Sailing, fishing, speed-boating, water skiing, sail-surfing, swimming and diving are steadily growing in popularity. The unprecedented growth in such activities has resulted in tremendous pressures for the development of holiday resorts adjoining estuaries. Residential sites with water frontage are at such a premium that in many parts of the world marinas have been constructed to increase such facilities. In many cases the disruption of environmental conditions by these developments have been severe (Barada & Partington, 1972). Many aquatic recreational activities are incompatible with one another. The disposal of sewage is always a problem. Yachtsmen and speed-boat enthusiasts do not like to share the same water while water-skiing and bird-watching are equally incompatible. Further, the desire for private waterfront properties conflicts directly with the concept of public access along shorelines. Thus the realization of the value of estuaries and lagoons for human recreation has brought a host of planning and management problems to the fore. The need for various kinds of controls including the zoning of areas for different activities has become apparent. In particular it has become clear that the natural balance of physical processes and biological interactions in estuarine ecosystems need to be jealously guarded if their value for recreation is to be maintained or enhanced.

HUMAN ACTIVITIES AND CHANGES IN THE ECOSYSTEM

Although the estuarine environment is harsh, the plant and animal populations grow rapidly and are hardy enough to survive normal fluctuations. Floods and hurricanes may decimate the population but it soon recovers. As detailed in chapter 16, it is only extreme changes of long duration which act as constraints and impoverish the flora and fauna. Human activities in the drainage basin of the river, in the estuary itself or in the coastal seas nearby tend to accentuate the effects of nature. They may increase the salinity and turbidity, cause sedimentation, reduce the oxygen content of the water by pollution, restrict tidal action and, by overfishing, reduce the population. 'Unnatural' effects include pollution by persistent pesticides which drain into the estuary from farmlands and the toxic metals and oily wastes from industry and oil spills which may drift in from the sea. As will be detailed below, not all human activities are detrimental; some improve conditions of life in an estuary and increase its productivity.

Sedimentation

The flow of a river and thus the amount of sediment it can carry, increases with the steepness of the river bed. In hilly country bad farming methods and overstocking lead to loss of ground cover and hence erosion. Much of the eroded soil is caught before entering the river by the vegetation along the banks. If the farmer ploughs to the water's edge there is nothing to prevent the good soil from being carried down the river.

Where the gradient is gentle the river meanders slowly over its flood plain. These meanders decrease the velocity of flow even more so that fine alluvial soil is deposited and this encourages the growth of reeds and other marsh plants. These act as a filter and as more sediment is deposited, the level of the river banks rises to form a levee higher than the general level of the flood plain. During floods the river overflows its banks and cuts through the levees to deposit its silt elsewhere until a broad swamp is formed with many distributaries which carry clear water on to the estuary.

When reclaiming swampland for agriculture, a straight canal is cut through the swamp and with the increased flow, floods carry most of their sediment to the estuary, so that it is rapidly filled with muddy water and eventually forms another marsh. This happened long ago in what was once St Lucia Bay in Zululand (Day *et al*, 1954). It has recently occurred again in what was intended as the 'sanctuary' of Richards Bay (Emanuel 1977, Heydorn 1978, Begg 1978). In both cases, the land drained by the canal was planted with sugar cane and in both cases the normal cycle of levee formation along the banks of the canal occurred. Eventually the floods breached the levees and inundated the low-lying cane fields beyond.

Dredging for harbour development has both short term and long term effects. While dredging is in progress, the suspended silt and organic detritus decreases light penetration and absorbs oxygen. When it is deposited, it may kill valuable algae such as *Gracilaria* as occurred in Langebaan Lagoon (Simons, 1977). In Lindbergh Bay, Virgin Island, it killed all the deeper *Thallasia* beds and their associated fauna (van Eepoel, 1971). The long term effect is the removal of the shallow banks where fish feed and the consequences of this are obvious.

The important sea grasses that stabilize the shallows and provide a refuge for young fish and prawns may be seriously damaged by mechanical action. Like most estuarine plants they are perennials but the root systems grow slowly. The Australian CSIRO Oceanographic Report for 1974-1976 states that continaul pulling of small boats up to the high tide mark over sea grass beds, the use of tracked vehicles in the shallows and practice with explosives during war training have left bare areas which had not regrown several years later.

Restriction of tidal action

Tidal range varies from 0,5 m to over 6,0 m in different seas. In wide-mouthed tapering estuaries the range and velocity of tidal flow may be enhanced until a tidal bore is formed. In most estuaries however, the range is reduced by sandspits at the mouth and obstructing rocks and constrictions along the channel. Tidal action may even be eliminated completely. On the coasts of southern Africa where wave action is high, littoral drift along sandy shores builds bars and sandspits across the mouths of slow flowing rivers until the estuary is cut off from the sea and does not open until a flood sweeps the sand away. Details are discussed in chapter 2. The effects of tidal action on life in an estuary are described in chapter 16. In brief, the turbulence resulting from tidal action increases the mixing of river water and sea water and, by breaking down the density layers, it facilitates the downward penetration of oxygen. The flood tide also carries in the larvae of invertebrates and juvenile fish and the ebb tide flushes pollutants out to sea. By the same token it decreases the residence time of estuarine water and estuarine plankton is carried out to sea before it can reach maximal abundance. A large tidal range also increases the area of the shoals and banks exposed at low tide and these intertidal flats are the most productive parts of an estuary. The microtidal estuaries of South Africa with a range of 2 m or less have far smaller intertidal flats than the macrotidal estuaries along the English channel where the range is 4-6 m (Hayes, 1977).

Many types of engineering works have the unfortunate effect of decreasing tidal flow and reducing the intertidal area. This is inevitable where a weir is built to retain a minimum depth of water for aquatic sports or to reclaim the upper part of the estuarine basin as a reservoir of fresh water.

The most detrimental work as far as life in an estuary is concerned, is to dredge sediment from the shallows on to marsh land or to deposit rubble there so as to raise the level above high tide. The reclaimed land is used for building and the deepened channel is faced with stone and used as a quay wall. Alternatively, the marsh may be incised with key-shaped canals and the dredge spoil is dumped in the intervening areas to form building lots for a marina. The financial gain from such work is obvious and immediate but the estuary is impoverished in the long term. The area of the highly productive marsh is decreased and the intertidal feeding grounds for wading birds disappear. In marinas the disposal of sewage and domestic waste becomes a problem. Even where there are strict regulations against the use of the canals as sewers, rotting algae accumulate so that the lower levels of the water lose oxygen, and in extreme cases they develop stinking sulphuretted hydrogen which is highly toxic. To prevent this the canals must be designed so that they are flushed by the tide and there are no 'dead end'

canals where wastes can accumulate.

Breakwaters provide shelter from storm waves and divert the littoral drift of sand so that the estuary mouth remains deep and tidal action is uninhibited. Moreover, the rocky crevices between the concrete blocks provide habitats for mussels and oysters and other forms of life which cannot live in the mudbanks so that diversity is increased. In some estuaries however, rocky training walls prolong the breakwaters inland along the edges of the estuary channel in an attempt to increase the scouring action of the ebb tide. These walls prevent the flood tide from spreading over much of the fertile marshes on the sides of the channel. The land may be reclaimed for playing fields or golf courses but again the estuary suffers.

Long bridge spans and the pylons which support them are expensive structures. To minimise costs the consulting engineer first determines the maximum river discharge during the heaviest flood which may be expected in 50 years. To this a safety factor may be added, for the bridge span or spans must be sufficiently long to allow such a flood to pass under the bridge without danger. It may be expected that heavy floods will erode the sediments under the bridge and ideally the supporting pylons extend down through the sediments to bed-rock. This adds greatly to the expense, for bed-rock may be 30 m below the bed of the estuary. Normally the width of the channel at mid-tide is sufficient for the passage of the flood and this is spanned by the bridge. The approaches to the bridge across the saltmarshes are then filled with cheap rubble and when this settles the roadway is built on top. The solid fill is of course an obstruction to tidal flow and at high tide there is an area of dead water on either side of the bridge span which acts as a silt trap. In time, sandbanks and muddy shoals grow which reduce the tidal prism or volume of tidal water moving in and out of the estuary mouth. The first effect is that the higher levels of the marsh above the bridge dry out and their production is lost to the estuary. The second effect becomes important during droughts; as river flow decreases, the tidal prism becomes more important in maintaining the scouring action of the ebb tide and when this flow is reduced the cross sectional area of the mouth is lessened and tidal action is decreased; eventually the mouth may close.

Extremes of salinity

The salinity in an estuary depends on the ratio of river discharge to sea water inflow and evaporation. It obviously varies with the tides and the seasonal rainfall and most estuarine plants and animals tolerate variations between 1 to 4 % of salt (usually expressed at 10-40 ‰ salinity). During droughts in arid countries evaporation in shallow saline lakes and lagoons may boost the salinity to above 50 ‰ or even above

100 ‰. Most resident species die and the migratory fish and prawns move back towards the sea while hippopotami and crocodiles move into the rivers and the birds desert the area.

It is appreciated that during such droughts farmers must irrigate their crops and where possible dams are constructed for the purpose. However, there is a tendency for farmers to cultivate high value crops such as sugar cane which require more water than is available in drought years. There is thus a conflict of interests between the irrigators and the conservators as to the amount of water released to the estuaries. Lake St Lucia is a case in point. The water budget calculated by Hutchison (1975) and summarised in chapter 14, shows that the abstraction of water from the rivers during droughts has increased the salinity of the lakes to values of 70-115 ‰, well beyond those which occurred before dams were built. Van der Elst (1979) shows that when the salinity in Lake St Luci is excessively high, the financial returns from tourism fall drastically.

In other estuaries by contrast, the prolonged presence of fresh water during the rainy season may impoverish the estuary. In large rivers with mountain catchments where the rainfall is high during one season, the river may run fresh to the sea and the estuarine fauna becomes very poor. This occurs in the Blackwood estuary in Western Australia as described by Hodgkin (1978) and summarised in chapter 16. A similar but less extreme condition occurs in the Great Berg estuary in the Western Cape and in the Tugela River in Natal. In the latter, a small estuarine fauna develops only during the dry winter months. The construction of dams on such rivers would not only conserve fresh water for urban and agricultural use but would regulate the discharge into the estuary. Alternatively, by the system of river regulation used in Britain and South Africa, the peak flow in one river may, by the use of canals and tunnels, be transferred to another river during periods of low flow.

Pollution

Pollution is always objectionable and if pathogens are present it may be dangerous to health. However, pollution is difficult to define in ecological terms since some organic matter is normally present in rivers and estuaries. The complex chemical changes during decomposition of organic wastes have been outlined in chapter 5. In essence, the organic compounds are broken down into simple carbon compounds such as carbon dioxide while phosphates and nitrogen compounds are released. During the process a large amount of oxygen is absorbed but normally the water and the surface of the sediments remain oxygenated and only the deeper layers of the sediment lack oxygen. To keep the water healthy, the organic waste must be added slowly enough for the water surface to absorb the oxygen required.

If the small quantity of sewage produced by farms and villages is comminuted and screened to break up the faeces the effluent when discharged into an estuary will not damage the aquatic fauna. The phosphates and nitrogen compounds fertilize the aquatic plants and may even cause entrophication or excess growth.

Sewage from large industrial towns is another matter. Industrial wastes contain many toxic materials including heavy metals such as mercury, arsenic, cadmium, chromium and zinc, poly-chlorinated biphenols (PCB) used in plastics, often sulphuric acid and occasionally cyanide. It is illegal for such trade wastes to be flushed into the sewage system, and industrial effluents must pass strict standards before they are allowed to flow into natural waters. This may demand a change in the manufacturing process but during purification valuable by-products such as sulphuric acid may be recovered. In spite of legal restrictions, toxic substances *do* reach the sewage systems where they may cause breakdowns by poisoning the sewage plant. Costly sewage plants must be employed to treat the sewage of large cities by decomposing the sewage with the aid of micro-organisms and precipitating the remaining sludge to minimize the organic content of the effluent. Even so, the estuaries which flow through large cities such as New York, Tokyo, Hamburg, Antwerp, Glasgow and many others are grossly polluted and only a few hardy aquatic animals survive. London was equally bad and the Thames was once called an open sewer. However, extensive research extending over 50 years and the construction of highly efficient sewage plants has reduced the organic load until in 1964, salmon and other fish could pass up the Thames to fresh water. It can be done!

High percentages of toxic metals are precipitated where the fresh water in which they are discharged mixes with salt water at the head of an estuary. Thousands of tons of several different metals are present in the mud of the Elbe, the Rhine, the Scheldt, and other European estuaries (de Groot *et al*, 1976). Aquatic plants and animals tend to concentrate these metals many thousands of times and the metals may be passed up the food chain through fish and man. Minimata disease in Japan was due to mercury poisoning in this way; cadmium has reached dangerous levels in flat fish in the Baltic and the level of copper in fish from some West Indian islands has made them unsaleable.

Intensive agriculture demands the use of large quantities of fertilizers and an increasing quantity of pesticides. These are leached from farm lands and flow into the rivers. Small quantities of fertilizers of course increase the growth of aquatic vegetation, but large quantities may cause dense blooms of phytoplankton and floating weeds such as *Salvinia* and water hyacinths. The weeds may be so dense as to obstruct boating and when the plants die and rot they remove the dissolved oxygen and cause pollution.

There are many types of pesticides used in agricul-

379

Continued on page 382

Cultivation of sugar cane to edge of Mhloti river (Photo: George Begg)

Silt plume from Mkomazi (clean sea water lower right) (Photo: E.J. Moll)

Dead mangroves in Kosi Bay (Photo: E.J. Moll)

Breakwaters prolonged
as stony embankments
along Kowie estuary
(Photo: A.E.F. Heydorn)

Ash dams, heated effluent and mud banks in Sezela estuary due to sugar mill (Photo: George Begg)

Oil pollution in Great Brak estuary (Photo: J.R. Grindley)

ture, forestry and mosquito control and as the insects become more resistant, there is a tendency to increase the dosage rate. Thousands of tons are now sold annually. Parathion and related substances which contain phosphorus are very dangerous to man and livestock but fortunately they are biodegradable forming phosphates which add to fertility. DDT, dieldrin and other chlorinated hydrocarbons resist bacterial decay and persist for years. Luckily they are almost insoluble in water but traces are absorbed by the small animals that are preyed on by prawns, fish and birds where they accumulate in the body fat until they reach toxic concentrations. Prawn are unsaleable in some Texas estuaries and fish-eating birds eggs and the population of certain species is declining in the United States. DDT and allied pesticides are so destructive that their sale is banned in many countries but they are still used in the tropics for mosquito control.

Oil pollution was discussed in chapter 16. Briefly oil loses much of its txicity in a few days while drifting on the surface but persists in subsurface mud for years. Its effect on plants and animals is very variable. Most marsh plants survive three or four spills but are killed by continual pollution. Fish are not seriously affected but crabs, sea urchins and mud-dwelling animals are very sensitive to oil. As yet no practical method of preventing oil spills from drifting into estuaries or onto sea shores has been developed and the surfactants that are used to disperse oil at sea are more harmful than the oil itself when used on the shore.

Power stations require a through-flow of coolant water which returns as a heated effluent to the river or estuary. Nuclear power stations require a flow of coolant water equal to that of a fair sized river. It is normal practice to regulate the flow so that the effluent is 8-10 °C above ambient. At the outlet, the heated effluent may cause some distress to the aquatic fauna but if mixing is adequate the heated plume of the estuary may be no more than 4 °C above ambient. This causes no damage to aquatic plants and animals in temperate estuaries for the temperature tolerance of the fauna exceeds this. Indeed, it may promote early spawning and allow warm-water animals to colonise the area. A subtropical tube worm *(Ficopomatus)* has been reported to have invaded warmed British estuaries in such numbers that its coral-like colonies interfere with the closing of sluice gates (Tebble, 1953). Elsewhere the culture of tropical prawns in warmed estuaries is being investigated.

The case is different in subtropical and tropical estuaries. During summer the aquatic plants and animals live close to their upper temperature limit, and even an increase of 2 °C may be fatal. Wood & Johannes (1975) quote several reports that: 'Thermal effluent from a power plant at Turkey Point, Biscayne Bay (Florida) has in recent years killed virtually all the plants and killed or greatly reduced animal populations in the bay in an area circumscribed closely by a +4 °C isotherm . . . Mortalities extend more than 1,5 km from the outfall'.

Wood & Johannes (1975) also state that the effluents from desalination plants are very destructive. Not only are the effluents heated but they are also highly saline and contain concentrations of toxic metals such as copper 30-40 times greater than normal. This is due to corrosion in the distillation plant. As the demand for fresh water increases so more and more desalination plants are being built in arid areas. As yet there is little research on how to reduce their damaging effects.

ECONOMIC EVALUATION PROCEDURES

The value of an unspoilt estuary has been summarised and the results of those human activities that affect estuaries have been stated. The value of an unspoilt estuary as a nursery for coastal fisheries or for aquaculture, sport fishing, scenic beauty and tourist attraction is obvious. However, it is difficult and in some respects impossible to translate all these advantages into monetary terms. Moreover, the financial returns will be spread over many years and, where access is difficult, they may be potential rather than actual. There are many alternatives. Some estuaries are best left as wildlife sanctuaries under the control of conservancy boards. Others may be developed as fishing villages and/or tourist resorts, while yet others are essential for industrial townships and harbours as the population expands.

In contrast to the long-term gains from conservation, there is an immediate financial gain from the sale of building lots and the construction of factories along the banks of an estuary. Both towns and factories require roads, bridges, power stations and facilities for the disposal of wastes but such requirements need not impoverish the estuary or spoil its beauty. The land used for housing and roads need not extend over the productive marshes; bridges may be built on pillars which do not restrict tidal flow and the effluents from factories and sewage works may be purified to the extent that they do not seriously pollute the estuary. The immediate costs are greater but the returns over the years should repay these. Decision makers require some yardstick to measure immediate costs against long term benefits.

To an administrator spending public funds for the benefit of the community at large, the most useful yardstick is money. Although he is aware that some benefits (such as scenic beauty) are difficult to evaluate even in terms of tourism, there are other benefits such as the rateable value of property that can easily be evaluated. There are many alternatives in regard to the siting and construction of bridges or the design of factories and sewage works and the administrator must ensure that the alternative he chooses is both economical and effective. It is thus desirable that the relative economic efficiency and the effectiveness of the various schemes be established. The concept of effec-

Table 17.2
Comparison of the characteristics of the three evaluation techniques

Method	Goal system	Units of measurement		Investigation criteria	Output (for decision-making)
		Costs	Benefits		
COBA = Cost-benefit analysis	Goal of one dimension Maximization of the economic efficiency	Monetary	Monetary	Difference or quotient from benefits and costs	Economic efficiency of an action and the order of alternatives
CEA = Cost-effectiveness analysis	A single goal up to a multi-dimensional system with benefit factors as goal criteria	Monetary	Physical units or cardinal scaling of quantitative factors	Comparison of effectiveness and costs 1. Comparison in pairs 2. Fixed least effectiveness 3. Fixed budget restriction	Order of alternative actions not only according to economic efficiency but according to an 'overall effectiveness' of satisfying the complex set of goals and objectives
UA = Utility analysis	A single goal up to a multi-dimensional goal system with benefit and cost factors as goal criteria	Weighted degrees of goal attainment (eg points) for both benefits and costs		Hierarchy of the utility	

iveness is related to the success of the scheme in attaining the goals and objectives desired. These goals may be economic, social, recreational or relate to the environmental quality. Different sectors of the community such as anglers or property developers are liable to be affected in different ways by a proposal and the maximum benefit to all is not easy to assess. The three assessment techniques that are widely used are cost-benefit, cost-effectiveness and utility analysis (Baxa, 1978). The major differences between these mthods are outlined in Table 17.2 extracted from Baxa, p 5.

Cost-benefit analysis

This is a method of determining the extent to which the benefits of a project differ from the costs. The method can deal only with those costs and benefits which can be expressed in monetary units since these are the only ones that are included in the analysis. The result may be expressed as a ratio of benefits to costs or as the difference between positive benefits and negative costs. Since aspects such as environmental disturbance, public utility and aesthetic values are difficult to express in monetary terms they are usually excluded from the analysis. However, these aspects may be so important that projects which appear from the analysis to be 'economically beneficial' may not be acceptable. When consulting engineers are planning a new development their basic criterion is the cost and durability of the construction. Cost-benefit analysis appears attractive but may be misleading.

Although it is difficult, there is an urgent need to assess the environmental resources (such as fertile salt marshes) and tourist attractions (such as wildlife) of an unspoilt estuary in monetary terms. If this can be done, any economic changes due to a proposed development can be included in the cost-benefit analysis. Chmelik *et al* (1975) have proposed a method for doing this.

They have illustrated their method by reference to a hypothetical causeway between Thesen's Island and Leisure Isle in Knysna estuary. The value of the natural resources of the whole of Knysna estuary were unknown when Chmelik *et al* constructed their model and the significance of the area enclosed by the hypothetical causeway was uncertain for much of the ecological information was not available at that date. Chmelik *et al* thus had to make estimates to illustrate their method and some of these were exaggerated so that the final monetary values they calculated are unreliable but this does not invalidate the method.

Economists could assess the existing rateable value of property and the income derived from trade. Reasonable financial projections can be made for the present environmental conditions over a period of several years. This is referred to as the Base Case.

Ecologists were required to assess what environmental changes would occur if the proposed causeway was constructed, including its effects on sport fishes, aquatic birds, water quality, marsh production and other estuarine features. The percentages by which these resources would be degraded by the hypothetical causeway were also assessed. Since such ecological assessments were limited by available knowledge, confidence limits were required. Within such limits economists could assess how the degradation of that part of the lagoon affected by the proposed causeway would affect the local economy. The impact of these changes in the lagoon were assessed as distinct from the other attractions of the area such as the coastal bathing beaches and Knysna forests which would not be affected. The economic assessment in terms of property values and business income after the construction of the causeway is referred to as the Expansion Case. In both the Base and Expansion Cases all relevant benefits and costs such as savings in transport less construction and maintenance costs of the causeway were included in the economic model.

As ecological changes may take place over a period of years and the full impact on the economy may be delayed, the time scale must be included in the analysis. For this purpose a discounted cash flow procedure is employed in a time series analysis. When all monetary values are presented in terms of discounted net present values, decisions can be made on a cost-benefit basis.

It will be evident that Chmelik's method of expanding cost-benefit analysis to include a monetary evaluation of environmental features contains several approximations. These limit the reliability of the method. As ecological knowledge increases, the significance of environmental features may be defined more exactly.

It must be appreciated that the Knysna model provides an economic assessment of the estuary in relation to its effect on property values and local trade. This is not the intrinsic value of the estuary but a derived value based on its effect on the economy of the community. A completely unspoilt estuary may have great intrinsic value as a nature reserve and as a nursery for sport fish and swimming prawns or it may have great potential value for aquaculture. However, if there is no local community to profit from these advantages, Chmelik's method would not indicate their value.

In Knysna, property values and the income from tourism are largely related to the lagoon and in all such areas where the economy is greatly affected by the attractions of the estuary such an evaluation is significant. Different developments in other estuarine areas such as the mining of titanium sands at Umgababa, or the construction of a deep water harbour at Richards Bay would hardly be affected by the quality of the environment. In terms of Chmelik's method of monetary evaluation, the environmental quality of the estuary would be insignificant.

Thus a derived evaluation of an environmental feature such as an estuary based on its importance to the local economy includes only certain aspects of its potential value. It is important to recognise that it is only a partial value. Alternative uses or developments may increase that value.

The aesthetic importance of quiet estuaries and wilderness areas is widely recognised and is part of the potential value which is difficult to express in monetary terms. Nevertheless in the stress of modern life, man's need for an unspoilt natural environment for his leisure and recreation becomes steadily more important.

Cost effectiveness analysis

This is an alternative to cost-benefit analysis. The main difference in establishing cost-effectiveness is that a required 'level of effectiveness' is established in advance. Alternative projects are then evaluated and compared on the basis of the cost of achieving this required level of effectiveness. Costs may be considered as the actual cash flows or they may be considered to include the equivalent monetary value of all impacts of the project. In the latter case the method becomes somewhat similar to the cost-benefit approach. The main advantage of cost-effectiveness analysis is that it is not necessary to assign monetary values to all the impacts concerned. Levels of effectiveness can be defined using physical dimensions such as tidal flow rates or harbour areas. One can compare the cost involved in maintaining a defined tidal exchange volume without attempting to assign a monetary value to the tidal exchange required. The major disadvantage of cost-effectiveness analysis is that the non-costable impacts are only considered if they are included in the objectives or levels of effectiveness in the analysis.

Utility analysis

In a number of ways *utility analysis* has advantages over both cost-benefit and cost-effectiveness analysis. An outline of the method is given by Baxa (1978) and further details are given by Zangemeister (1973). 'Utility analysis is in effect a semi-quantitative means of 'trading off' the effects of implementing any given scheme, that is, the relative desirability of achieving a given set of goals and objectives and the degree to which this target system is fulfilled, are combined to give a measure of how far each scheme will go in meeting all or any of the goals and objectives, and so provides the answer to the question of the effectiveness of the scheme' (Baxa, 1978). Utility analysis has the advantage that it can handle financial, quantitative and qualitative effects simultaneously. All of the elements of a proposal or a management scheme can be considered together in this form of analysis. The procedure is similar to that employed intuitively by any individual making a decision. Having a goal or a set of goals, a decision must be made regarding several alternative courses of action in order to determine the 'best'. In the case of a public decision this is the greatest utility or effectiveness for those concerned. In order to reach the goals, certain objectives must be met in relation to the effects or impacts of the actions. Each objective is further broken down into a number of criteria which are used to measure the degree to which alternative actions satisfy the objectives and finally reach the goals. Utility analysis provides solutions to complex questions where there are sets of goals and objectives and a variety of alternative courses of action with different effects.

Utility analysis does not necessitate expressing all effects in monetary terms, it can accommodate multidimensional goal systems and the implications of the actions considered are clarified; moreover it can be supplemented by calculation of the economic efficiency of schemes. The relative preference for a number of goals is indicated by an index value determined by distributing 100 points among the set of goals (eg marina development, aquaculture development, etc). The relative preference for objectives defining each goal (eg financial return, conservation of resources, etc), would be determined by distribution of the points allocated to the goal amongst its objectives. The criteria (eg construction costs, maintenance costs, indexes of environmental quality, etc) would then be weighted to reflect their relative importance by a panel of arbiters including experts. Performance graphs are then set up and each scheme is scored on the basis of the performance graphs. The resultant matrix shows the utility or effectiveness of each of the proposals considered. Although the actual mechanism of utility analysis is essentially simple, there may be problems in formulating goals, objectives and criteria and in carrying out the procedure in practice.

METHODS OF CONTROL

There is an age-old tradition that it is the duty of the state to protect the rights and welfare of the community. This includes the protection of public health and the care of the environment as well as many other important matters. The central government divides its responsibilities between various ministries and delegates its authority in local affairs to provincial councils, county councils and borough councils. The names and powers of these various authorities differ from one country to another but the essential point is that ultimately the state makes the laws of the land and its officers administer them. Where a proposed development approved by the government affects the property of an individual, he has the right to claim compensation and contest the decision in the courts. Where the development affects public land or waters controlled by the state, the matter is decided by the government or one of its subsidiary bodies. No reasons for the decision need be given. Where a river, an estuary and the lands along its banks are owned by the state any developments that are planned are official matters only. Private residents in the area have no legal right to contest the siting of a bridge of its method of construction, the building of a dam or the draining of wetlands. Even though such developments may have a deleterious effect on the local environment, private individuals have no *locus standi* in the courts. If there is a public protest the government may call for a commission of enquiry but it still decides the issue and still need not give any reasons for its final decision. With the recent growth of public interest in the quality of the environment and the protection of nature, this has led to dissatisfaction.

This, in outline is the method of environmental control in European countries and indeed, in most countries of the world. In 1970 the United States of America enacted a new environmental law which requires a developer, whether he represents an organ of government or a private business concern, to assess the impact of the proposed development on the environment. This assessment is open to the public and can be challenged in the courts. Thus the private individual gained a *locus standi* and only when the court had approved the proposal was a permit issued for the development to proceed.

Within their limits of jurisdiction, this method of procedure has been adopted in various forms by the several state legislatures of the United States. Various forms have also been adopted by Canada and Australia and there are reports that it is being seriously considered by China and the Republic of South Africa. Although it provides the private citizen with some say in the quality of his environment, some critics object that the procedure stultifies development while others say it does not go far enough. The systems in the United States and Australia will be discussed later.

As noted earlier, an estuary is not an independent ecosystem. Thus the effective management of an estuary depends on regulations controlling what may be done in the river as well as in the estuarine area itself.

Where the river and its estuary flow through land belonging to the state, management would appear to be simple. It is usual however to refer administrative duties to departments such as Forestry, Public Health, Sea Fisheries and the Navy or to local authorities. Thus rivers, estuaries and sea shores tend to be controlled by different authorities each of which has its own regulations and administrative officers. The legal definition of an estuary and the extent of its banks thus becomes important. The scientific definition of an estuary has been discussed in chapter 1 but legal definitions vary; they are seldom based on salinity variations and more often extend to the limits of tidal action so that stretches of fresh water where conditions of aquatic life are different, may be included in the legal definition of an estuary. There are other legal limits in some countries such as the limits of navigation. Moreover the banks of an estuary to which the regulations of one authority apply may be fixed at the mean level of high tide while the extreme upper limit of storm tides may be used by another authority.

In arid countries such as South Africa, South Australia and Western Australia tidal action in some estuaries is excluded for many months or even years by the formation of a sandbar across the mouth or by the building of a weir. In such estuaries legal definitions based on tidal action make nonsense of the law and lead to confusion among administrative officers. Wherever there are conflicting regulations that apply to rivers and estuaries and the lands along their banks, the management of the ecosystem will suffer. There is an urgent need for co-ordination of regulations between the various authorities. While this presents many difficulties, consultative panels could be formed to simplify existing regulations that apply to each estuary and the adjacent wetlands in the watershed.

Many of the difficulties of estuarine management arise from private ownership of land along the banks of an estuary or the river which flows into it. The private owner with riparian rights naturally wishes to use or develop his land for his own benefit. He may wish to use the river water for irrigation or the disposal of waste; he may wish to dredge the sand or gravel along its banks, also to reclaim swamps or construct marinas or factories. These operations are restricted in various ways in different countries for the benefit of the public. Thus the public benefits but the landowner feels that his land has decreased in value so that he is the loser. He may claim compensation or the removal of the restriction. If legal action ensues and the case is brought to court, the public benefit of the restriction is closely examined. The public benefit which is regarded as most important is usually the protection of public health. Thus the restrictions regarding the standard of effluent which may reach the river or estuary are strictly upheld by the courts. The public benefit of protecting the environment and wildlife has been regarded as less important but public concern is increasing and the authority concerned with the protection of wildlife is gaining strength.

The outright appropriation of private land by the State provides fewer difficulties. In many countries one of the conditions of land tenure is the right of the state to appropriate land required for public transport. Should the state appropriate such land for the building of a bridge, the landowner is recompensed.

Management in the United States

Conservationalists in many countries are particularly interested in the methods used to manage wetlands and estuaries in the United States. That country has had 10 years experience of legislation which requires a developer to provide an environmental impact statement showing the effect of the proposed development on the environment. This legislation provides all interested bodies including private citizens with the legal right to test the accuracy of the impact statement in the courts before the proposed development is approved.

The constitutional issues connected with estuarine management in the USA are discussed by Banta (1976). The advantages and disadvantages of requiring an impact statement for any development which may affect the quality of the air, the land or the water in the United States are discussed by Edmunds (1978). The following paragraphs are based largely on these two papers.

Landowners in the States base their rights on the Fifth Amendment of the United States Constitution which states *inter alia*, '. . . nor shall private property be taken for public use without just compensation'. This is referred to briefly as *The Taking Issue*. Just compensation is usually defined as the fair market value of the land prior to the improvement due to the proposed development (Banta, 1976).

For over a century the US regulations for the protection of wetlands and estuarine areas were based on 'police power' or the power of the police to protect public health, safety and general welfare. While the danger of pollution to public health has long been the most important issue, the rights of coastal authorities to make regulations to ensure adequate navigational facilities within the tidal waters of estuaries and rivers has been recognised by the courts. These rights include the protection of natural features that protect harbours. Thus a regulation prohibiting the removal of gravel from tidal beaches and sandbars was supported by the court of Massachusetts in the 1950's.

The importance of protecting aquatic birds and estuarine fisheries received slow recognition. In the 1960's proposals to restrict dredging or the dumping of the spoil and the discharge of effluents which would affect aquatic life were regarded with indifference. However, public outcry for the protection of the environment and wildlife soon led to the National Environmental Policy Act of 1970 and the Clean Air Act. The Federal Water Pollution Control Act Amendments of 1972 extended the requirement for a permit for developments on river banks beyond the traditional limit of navigable waters. As a result, local authorities gained the power to manage wetlands and 'flood hazard areas' including flood plains. Moreover the National Environmental Policy Act of 1970 required developers to issue an Environmental Impact Assessment (EIA) which could be challenged in the courts both by administrative authorities and private citizens. They had gained a *locus standi* in the courts and to those concerned with the protection of the environment this was a most important advance. There were, however, checks and balances. A zoning regulation restricting the use of private property in an estuarine or wetland area (or elsewhere) could be challenged in the courts as 'a taking'. According to Banta (1976) courts base their decisions on a statement by the Supreme Court in 1922: 'The general rule at least is, that while property may be regulated to a certain extent, if regulation goes too far it will be recognized as a taking'. Although useful, the statement leaves the decision open to doubt in many cases and as a result several factors are taken into consideration by the courts. Banta states that in estuarine areas the most important of these are: '1) the public purpose served by the regulation that is open to challenge; 2) competing public interests in the affected property (easements, public trust, navigational rights); and 3) loss of value'. These three issues are discussed below.

The 'public purpose' finds expression in local, state and federal regulations to protect natural resources. Some public purposes are regarded by the courts as particularly important; tough regulations to eliminate quarries and dirty industry are strongly supported and the taking argument is rejected. On the other hand, aesthetic considerations such as the restriction of unsightly advertisements have not been supported and between the two extremes there are many doubtful issues. Environmental protection in coastal habitats falls in this grey area for the effects of human activities upon coastal and estuarine ecosystems are complex. As knowledge increases damaging activities are more easily identified. Meanwhile the courts are influenced by earlier decisions in important cases. One of these concerned San Francisco Bay. The legislature had set out clear objectives for the San Francisco Bay Conservation and Development Commission and challenges to the Commission's strict dredge and fill regulations were rejected by the courts.

'Competing public interests' include the right of public access to tidal areas even where the adjoining land is privately owned. Public interests are held in trust by the state and 'the doctrine protects against shortsighted stewardship of the public resource'. In Wisconsin for example, the Act includes fresh water wetlands in its public trust definition and in Marinette County a Shoreland Protection ordinance had been adopted. 'The Justs owned lakefront lots and began filling the shorefront swamp contrary to the ordinance that required a special permit to fill'·(Banta, 1976). The ordinance was challenged and the Wisconsin Supreme Court was asked to review the Act. In its judgement in 1972, the court emphasised 'the interrelationship of the wetlands, the swamps and the shorelands to the purity of the water and to such natural resources as navigation, fishing and scenic beauty'. It found that: 'The changing of wetlands and swamps to the damage of the general public by upsetting the natural environment and the natural relationship is not a reasonable use of that land which is protected from police power regulation'.

Banta goes on to deal with the significance of 'loss of value' of private property in judicial decisions. Financial loss is of course an important source of controversy but restrictions to abate a public nuisance may be upheld even when they result in an 80 % reduction of the value of the land. On the other hand, if the regulation results in the inability of an owner to undertake an improvement of his land this may invalidate the regulation. Decisions regarding the validity of regulations thus differ according to circumstances and there are many doubtful cases. A successful court challenge may invalidate the relevant regulation but 'typically a property owner will not collect money as a result' (Banta, 1976). Some states have adopted laws requiring compensation but this often leads to further litigation. Legal experts have suggested different solutions to the problem. Banta concludes that: 'unless a legislative or judicial solution of this sort emerges, controversy is likely to continue over the constitutional limits on regulations for estuarine management'.

Edmunds (1978) is not particularly concerned with regulations concerning estuarine areas and deals with the broad effects of the environmental protection acts and the advantages and disadvantages of Environmental Impact Assessments. He discusses two classes of 'trade-offs' namely: '1) The techno-economic trade-offs of specific environmental programs and 2) the political trade-offs relating to forging a revised regulatory process and a workable decision mechanism'. Some of his main points are discussed below.

Techno-economic trade-offs

Since the environmental laws were enacted, methods of measuring and monitoring the quality of air and water and the concentration of toxic substances in wastes have been developed. Levels of toxicity to humans and some elements of wildlife have been determined although sublethal and ecological effects remain uncertain. Using these measurements, the effects of different industries, mines, power plants and other human influences on the environment have been assessed. Major sources of pollution have been identified and regulations have been drafted to set minimum standards of purity for the discharge of wastes into air and water. Some industries adopted alternative technical processes to meet these standards while others contested the regulations in the courts.

Before a new development is permitted an Environmental Impact Assessment (EIA) has to be prepared, which is available to public scrutiny for 30 (later 45) days, and can then be tested in the courts. The preparation of the EIA's is both costly and time consuming and before they are challenged the administrators and/ or private bodies have to make further investigations. The findings of the court may lead to prohibition of the proposed development, or permission to proceed, or to an amendment of the relevant regulation. According to Edmunds the whole regulatory process leads to increased costs to business and usually results in delayed decisions. The total cost of administrative work, business delays and litigation is eventually paid by the public in the form of higher prices for manufactured articles or higher taxes. He quotes a number of examples.

In 1977 the Dow Chemical Company estimated that the cost of complying with environmental regulations was $147 million, of which $50 million was regarded as excessive. Delayed decisions were also costly. The Dow Chemical Company spent $6 million in buying property near San Francisco for the site of a chemical plant and then $4 million in costs and finally abandoned the project after two years since it had obtained only one of the 65 necessary permits. It may be noted in passing that the environmental effects of the proposed chemical plant are not stated.

Edmunds also states that a consortium abandoned plans for the Kaparowitz power plant in Utah because of unending delays in obtaining permits. Edmunds estimates that environmental regulations have increased the time required for deciding on a site from five to ten years.

The total US expediture for all types of pollution control have been officially estimated as $32 x 10^9$ for 1975 rising to $64 x 10^9$ in 1984. About half of this cost would arise whether environmental regulations existed or not. Thus the increment due to environmental protection is about $15 x 10^9$ per year or 1 % of GNP. The increased inflation rate due to pollution control is 0,3-0,5 % per year compared with the general inflation rate of 6-12 %. Unemployment resulting from the closure of older industrial plants is more than offset by increased employment in pollution control devices such as sewage works.

On the credit site, Edmunds (1978) is concerned mainly with the effects of environmental regulations on human health. The quality of water for human consumption has improved markedly but by 1974 there was little improvement in the quality of water for aquatic life. The regulations have not been in force for long and improvements may be slow. Human intake of DDT has been halved since its production and use were restricted. However, the concentrations of dieldrin and aldrin in surface waters have continued to rise in spite of restrictions on their use as insecticides. Polychlorinated biphenols (PCB's) used for many purposes are somewhat similar substances and are also persistent in the environment as they are effectively non-biodegradable. They are concentrated up the food chain and reach high concentrations in the bodies of top predators such as fish-eating birds where they cause high mortality and reduced breeding success. The population of the brown pelican and the fish eagle in particular have declined. PCB's may also cause cancer and the dietary intake by humans in the USA ranges from 1 to 3 ppm. Mercury is used in agriculture and is also a waste product of several industries. Human intake has decreased by a third since the use of mercury as a fungicide was banned in 1972.

Regulatory and decision trade-offs

'The ultimate socio-political trade-off in the light of these health and cancer hazards, is how much environmental quality is enough and who shall make the environmental decisions' (Edmunds, 1978). He discusses whether business alone or government alone should make the decisions or whether there is some way of combining the technical knowledge of business and the social responsibility of government.

Predicting the effects of pollution requires very extensive knowledge of interactions within the ecosystem and the ultimate hazards to human life. Industrial pollution may be decreased by a number of alternative technical processes best known to the industry concerned. In the long run the decrease of pollution must be paid for by the public in one way or another.

How much is the public willing to pay for the knowledge necessary to improve human health and protect the environment? 'That is the information cost trade-off' (Edmunds, 1978).

Edmunds suggests that one method is for industry to balance costs and benefits within its own organisation and thus 'to internalize the decision process'. He quotes an example of the phosphate fertilizer industry in Florida. The fluoride in the factory smoke was settling on the surrounding grasslands and affected the grazing cattle so that they slowly became immobilized and starved to death. The cattle farmers brought the matter to court and the court injunction demanded that the emission of fluoride be decreased. The cost of increasing the purity of the smoke from 90 to 95 % would cost the industry $6,7 million and to remove further fluoride until the smoke was 99 % pure would cost a further $16 million.

Rather than increase the purity of the gases emitted beyond 95 %, the fertilizer companies bought 200 000 acres of the land at a cost of $25 million. They thus became large cattle owners themselves and then reduced pollution to a level which protected their livestock. They had avoided external information costs and internalized the decision as to how much pollution abatement was necessary (Edmunds, 1978). But was this all? What happened when the fluoride was leached from the land and reached the streams? Is it really possible to limit the effect of a pollutant?

Under the existing laws business is not allowed the final judgement. Legislators are too busy to deal with individual cases and administrators do not have the latitude for final decisions. Thus litigation ensues and the final decision is left to the overburdened courts. The whole process is costly and time consuming. The problem is to make administrative regulations more functional and to give administrators greater discretion. Edmunds suggests that the courts, having gained experience of many environmental cases, should accept important and clear-cut decisions as precedents and leave regulation to executive agencies. Only where it is shown that there is a reasonable cause for complaint should the courts intervene.

Another possibility is to allow business to decide how to meet the requirements of the regulations and if a complaint is made, business must appear before a government agency and explain how it has reached its decision. If the explanation is not reasonable and sufficient, the administration should indemnify the complainant and retroactive retribution would fall on the business. Edmunds suggests that this procedure would reduce delays in decision making. It would '1) provide for citizen access by filing complaints to be heard by administrative agencies, and 2) expose business to potential liabilities for poorly informed judgements in environmentally significant decisions' (Edmunds, 1978).

The merit of either of these alternative procedures of leaving decisions to administrative agencies or to business with penalities for wrong decisions may be debated. Edmunds concludes that: 'The costly delays in environmental procedures have caused government and business alike to question the workability of the regulatory process. Hence the environmental program has come full circle to the point from which it started, namely, to try to establish a workable decision making process'.

Edmunds article shows clearly that business has grounds for objecting to the procedures of environmental control in the United States that existed prior to 1978. He states that government also questions the procedures but provides no evidence.

Edmunds does not deal with the views of conservationists regarding environmental impact statements. Their view is that EIA's are highly important but they may be misleading. In a personal communication, Dr R.W. Day of the University of California at Santa Barbara states that the main objection is how the raw data are interpreted and the way the EIA is prepared.

The developer often employs a firm of consultants to gather the environmental and biological data as a basis for the impact statement. The consultant may then be required to do one of two things. He may be asked to interpret the data and prepare the whole EIA, or he may be asked to prepare an 'internal report' which the developer can use when drawing up his own impact statement. The accuracy of an EIA depends on the interpretation of the data, and incomplete biological data in particular are capable of many interpretations. The environmental consultant wishes to maintain his good reputation but he also has an interest in keeping his client happy. He thus interprets the data in the most favourable light. If he is merely required to prepare an internal report he is not responsible for the final EIA and the developer who produces the EIA is not responsible for the data. Thus the responsibility for producing a misleading EIA cannot be pinned on to either party. Moreover, some environmental consultants have found that their internal reports have been misrepresented in the EIA or that unfavourable parts have simply been omitted.

Management in Australia

Australia has adopted a modification of impact assessment procedure. The Environmental Protection (Impact of Proposals) Act was passed by the federal government in December 1974. Between 1974 and 1976 all the states except Western Australia had developed, or were developing, similar laws. The federal laws deal only with issues concerning federal territory and the whole of Australia (such as imports and exports and the noise of international aircraft) and the State laws deal with more local issues. Both have raised considerable public discussion. So much so, that the journal *Search* (Sydney) devoted a special number (volume 7, number 6 of June 1976) to a sym-

posium on impact assessment and environmental protection. The comments of the several contributors were introduced by Professor E. Linacre and summarised by Professor F. Talbot. The following account is based mainly on these two papers.

The Australian system is obviously inspired by the American legislation of 1970, in that an Environmental Impact Assessment (EIA) is required for all developments that may affect the environment. However, the system is modified by the British tradition that the final decision as to how the development should proceed is decided by an administrative authority. There is no reference to the courts so that litigation is avoided.

The sequence of stages leading to a decision in the federal parliament is summarised by Linacre. First the proponent of the development must send a Notice of Intent to the federal minister concerned. Next, he or his environmental consultant must draft an EIA which is open to public comment for four weeks or, alternatively, the minister may call for a public enquiry on the proposed development. A final Environmental Impact Statement (EIS) is then issued by the federal or state authority concerned and the government decides the issue. According to Linacre: 'The federal act does not give the environment minister any powers beyond gathering information, forcing developers to consider the environment, obtaining facts to help in decision-making, providing the public with a case which can be argued and forcing governments to justify their decision'.

Put in this way the whole system seems a sensible procedure. However, the Australians in their usual forthright manner have criticised the system without mincing words. Their EIA legislation has been described by several critics as a sham; a sop to the ecology lobby; a salve to the conscience of developers and politicians; an expensive exercise providing big returns to environmental consultants and useful delays to local authorities who need time to co-ordinate their plans.

It is said that developers may lose an advantage over their competitors if their provisional plans are made public too early; indeed local authorities in New South Wales are restricted from disclosing more than the bare outlines of their plans. Often the EIA's are not available to the public or to conservation groups. Nevertheless strikes by labour unions or 'green bans' to prevent inroads into the parklands around Sydney by highway engineers and the public outcry against the disposal of sewage into the Paramatta River have made administrative bodies aware of the strength of public feeling. Indeed Linacre quotes a bitter remark by Recher that EIA's have little effect unless there *is* a public outcry. One of the main criticisms is that the community cannot participate effectively in plans which affect their environment. It is argued that the 28 days allowed for public comment on an EIA is too short for the people to formulate their opinion. Conservationists feel that the EIA which comes *after* the initial planning comes too late and that the environmental impact of the project should be considered from the beginning together with its technical and financial feasibility. No developer, having spent time and money on making his plans, wishes to change them. A plan which incorporates environmental considerations from the beginning would certainly be more acceptable to the local community.

There is another objection. In states other than Western Australia, it is common practice for a developer to employ a firm of environmental consultants to prepare the EIA. The interactions within the environment are complex and so too are the effects of environmental factors on man. Thus the preparation of an EIA requires a broad spectrum of knowledge and an expensive training. There is a danger that the firm of consultants may be biased or consider only a few of the impacts of the proposed development so that the EIA is misleading. Indeed, it is suggested that the alternative proposals included in some EIA's are intentionally poor so that the plan recommended stands in a better light. Linacre suggests that there should be three EIA's, one dealing with the biophysical effects (presumably including human health), a second to cover social effects and a third to deal with economic aspects. Thus each aspect would receive due consideration. This is one man's opinion; international opinion as formulated by SCOPE is that all aspects should be combined since there are interactions between them. Talbot agrees and stresses the need for more ecological research and a more exact definition of the jurisdiction of state and local authorities so as to reduce conflicting regulations and simplify the task of drawing up an EIA.

As noted, the formal Environmental Impact Statement (EIS) is drawn up after public discussion of the EIA and comments by other government departments which are affected by the proposal. However, Talbot states that the government has, on occasion, approved a proposal before the EIS was completed. The intense indignation of conservationists can be appreciated! Moreover he states that there is no clear procedure for assessing whether the impact of the proposal on the environment is so severe that a proposal should be abandoned or an alternative approved.

The final decision regarding the proposal is made by the government which need give no reasons and there is no appeal to the courts. Talbot says: 'There is neither a clear legal enforcement of disclosure nor enforcement of review of the process leading to the decision and the citizen has no recognisable legal rights to environmental quality under law'.

Taken all in all, the articles in the symposium are a severe indictment of the system as it was in 1976. No doubt many counter arguments might be advanced by developers and administrative authorities but it is clear that the public at least is not satisfied. On the other hand there is no suggestion that EIA's should be abolished. There is also a feeling that runs through the symposium that since the system had only been in operation from 1974 to 1976 it was too soon to expect perfection; it was still suffering from teething troubles. The main need was an earlier and more effective method of involving the community in environmental protection.

The symposium dealt with the environment in broad outline and little was said about estuaries. Australia is not highly industrialised or densely populated and most estuaries are not polluted. Nonetheless there is obvious dissatisfaction about pollution in the Paramatta estuary that flows through the heart of Sydney, the Hunter estuary at Newcastle, parts of Port Phillip Bay at Melbourne, the accumulation of paper mill wastes in the Shoalhaven and particularly the pollution of Botany Bay. According to Linacre: 'there have been numerous EIS concerning various developments in Botany Bay but no overall study summing the total impact of extensive changes. Large issues like the dumping of only partially-treated sewage into the ocean appear to be the concern, officially, only of the instrumentality whose finances benefit by continuing in the same way. No EIS is likely for such customary practices. Little attention is paid to possible future claims on land affected by the project'.

Estuarine management in Britain

Great Britain follows the tradition that the government is responsible for the protection and welfare of its peoples. It makes the laws and administers them and although informative documents are issued from time to time there is no statutory requirement that the government must explain the reasons for its decisions. In cases where new developments or new restrictions which affect the environment have been decided, no impact statement is issued. The private citizen thus has no right to contest environmental laws unless his personal health is affected, his property is appropriated without fair compensation or his traditional use of waterways is infringed.

Prior to 1974 the control and administration of rivers and many estuaries was one of the duties of the River Authorities. The Thames estuary was an exception being controlled by private legislation of the Port of London Authority. This body included representation of the Essex and Kent River Authorities, the Lee Conservancy Catchment Board and the Greater London Council.

In such a densely populated, highly industrialised and low-lying country as England, the main problems are the supply of wholesome fresh water, the control of pollution so that the water returned to the rivers can be used again, and the prevention of flooding either by rivers or storm tides. The marine and fresh water fisheries (particularly salmon and trout fisheries) are very valuable and must be protected. Surveys up to 1972 showed that the 3 000 000 anglers spent on average £80 per head per annum on their sport. Aquatic recreation was also becoming very popular; about 6 000 000 people took part and sailing clubs had increased from 400 to 1 600 over 20 years. As explained in document SBN 11 750570 published by HM Stationary Office in 1973, these matters were controlled by several organisations, so that over-lapping was inevitable. A national plan was required to provide a strategy within which more detailed planning and subsequent executive action could proceed. Accordingly the Government in 1973 proposed that ten Regional Water Authorities be formed in England and Wales. 'Each Authority will take over the responsibilities of water undertakers and of local authorities in relation to sewerage and sewage disposal for its area. They will thus be responsible for water resources and supply; sewerage and sewage disposal; the prevention of pollution; land drainage and flood protection; fisheries, and the recreational and amenity use of their water space and in some cases for navigation'.

It had been estimated that the demand for wholesome fresh water would increase from the 13,9 million m^3 per day supplied in 1970 to 28 million m^3 in the year 2 000; direct abstraction of lower quality water from rivers by agriculture and industry would increase from 32 million m^3 per day to 34 million m^3 most of which would return to the rivers. On this basis a further 9 million m^3 of water from new sources was required for deployment from areas of high rainfall to areas where the demand exceeded the existing supply. To meet this, new reservoirs were needed and the enlargement of existing reservoirs was necessary; underground storage and storage in parts of large estuaries was planned and river regulation or the transfer of water from one river system to another had to be extended. Such undertakings would be expensive. About half of the proposals for the reorganisation was passed by Parliament in 1974 and in 1978 the remainder still awaited the necessary funds. Much of the legislation concerning pollution control and a review of legislative and administrative procedures was published by HMSO in 1976 and a second edition appeared in 1978 (ISBN 0 11 7513679).

The division of responsibilities is interesting. The Department of the Environment (DOE) and the Department of Transport in co-operation with the Scottish Office and the Welsh Office are responsible *inter alia* for the supply of fresh water and the control of pollution in inland waters but the responsibility for the control of pollution in tidal waters and estuaries is not clear. The Ministry of Agriculture, Fisheries and Food (MAFF) is responsible for the protection of fresh water and marine fisheries and the control of pesticides used in agriculture and also their effects on pollution. The Department of Trade is responsible for marine pollution and the Department of Health and Social Security is responsible for medical aspects of environmental pollution.

This division of responsibilities in regard to pollution suggests a lack of co-ordination. In fact co-ordination at ministerial level takes place within the Cabinet Office. Further, there is a Central Unit on Environmental Pollution on which the various departments are represented. Advice is provided by five research councils financed by the Department of Education and Science. The Royal Commission on Environmental Pollution provides advice on problems of national and international importance.

Central government retains control of certain pollution problems, for example exposure to radio-active materials and products which have more than a local effect since they are sold throughout the country. In general, however, the implementation of the law is delegated to regional or local authorities where the effects of pollution are first experienced. Such authorities include representatives of the local community as well as officers of the various departments mentioned earlier and they may exercise a considerable degree of discretion. Further, the Regional Water Authorities are required when planning within the limits of the national strategy, to take into account 'the views of other interested parties who will have the same statutory rights to individual schemes when they are put forward as they have under the present (pre-1974) law'. To this extent representatives of local communities are responsible for the quality of their environment.

There is no fixed standard for the quality of water. Water used for agriculture or industrial cooling may be of low quality. Although water supplied to urban areas for human consumption normally meets the standards of the World Health Organisation, the only specification is that it must be safe and 'wholesome'. This requirement allows the standard to be changed in accordance with local facilities for effluent treatment and increasing knowledge of water-borne diseases. Thus soft water has been correlated with cardiovascular disease; excessive nitrate is said to harm babies and a high ammonia content affects the amount of chlorine required to kill pathogens.

If the discharge of 63 million m^3 of coolant water and the one million m^3 of mine effluents are excluded, the flow of sewage and industrial effluents into canals and non-tidal rivers in 1970 was over 21 million m^3 per day. This comprised:

   10 million m^3 of treated sewage containing 18 % industrial effluent;
   1,2 million m^3 of crude sewage;
   11 million m^3 of industrial effluent.
Put in another way:
   12 % of crude sewage was discharged to the sea;
      6 % was disposed by 'other means' and 81 % was discharged into rivers;

16 % of industrial effluent was discharged to the sea; 3 % was ducted to sewage works and 81 % was discharged into rivers.

By 1972 the quality of river water had been considerably improved 'in that there are 240 miles less of heavily polluted non-tidal stretches of river'. However the condition of tidal waters and estuaries remained much the same.

The 1970 report of the Royal Commission on Environmental Pollution emphasised that 'estuaries remain more vulnerable to pollution than any other part of the British environment'. Thus in 1973 it was estimated that one-fifth of the surface water and one-fifth of the effluent of England drains through the Humber estuary. The estuaries of the Tyne, Tees and the Mersey are also badly polluted. Effluents first discharged after 1960 are controlled under the Clean Rivers (Estuaries and Tidal Waters) Act of 1960. 'Unaltered pre-1960 discharges can only at present be controlled if an order is made by the Secretary of State under the Rivers (Prevention of Pollution) Act of 1960. Very few such orders have been made' (SBN 11 750570 6). The Royal Commission on Environmental Pollution recommended in 1970 that: 'full control should be extended to all discharges to tidal waters and that, pending legislation, River Authorities should seek ministerial approval for this purpose *when they think an estuary is at risk*'. Although only part of the legislation for the formation of Regional Water Authorities had been passed by 1978, the government have proposed that these Authorities be given full control over all discharges to estuaries and the sea. The Authorities aim 'to achieve a massive clean up of the country's rivers and estuaries in the early 1980's'.

Obviously there is much to be done but the full implementation of the 1974 proposals must await the necessary funds. Meanwhile there have been relatively few criticisms. Doubts have been expressed regarding the proposed use of parts of the major estuaries for fresh water storage and the Morecambe Bay scheme appears to be impractical for technical reasons. Indignation has also arisen regarding the dumping and infilling of marshes south-east of London to form a new airport. Business is also dissatisfied about the powers delegated to local authorities to refuse permission for developments. Thus the ICI claim that after careful planning to meet environmental standards for a factory on the Tees, it was refused by the local authorities who, they say, do not have the necessary expertise to judge the plans.

PROBLEMS AND CONTROLS IN SOUTHERN AFRICA

It is obvious that environmental management must concentrate on the most important problems of a country. In the densely populated and highly industrialised countries of Europe, North America and Japan the most important problem is pollution. Southern Africa is a young country and is neither heavily industrialised nor overpopulated. Nonetheless there *are* pollution problems due to the shortage of water. Apart from the well-watered eastern sector stretching from Mocambique to Transkei, southern Africa is arid, grading into desert in the western part of Namibia (see Figure 14.2). Thus water must not only be conserved, it must be re-used as often as possible and this demands the control of pollution and the purification of effluents. This does not only apply to water for domestic consumption. It has been reported that water reaching the rivers from mine shafts and slime dams in the Transvaal was in some cases as saline as the sea and in other cases it was extremely acid. When used for irrigation it killed the crops. This discharge of polluted water has been stopped and the effluents from these sources as well as those from sewage works and factories are closely monitored by the South African National Institute for Water Research. However, it is an ongoing problem as new mines and factories are developed and old ones fall into disuse.

Diseases transmitted by water also cause serious problems. Bilharzia (or schistosomiasis) is endemic in the subtropical lowveld of Transvaal, Mocambique and Natal and cases have been reported from Transkei and even further south. In spite of continuous research on the ecology of the vector water snails in rivers and the parasitic worms in humans, control is not effective. Luckily the snails cannot live in estuaries but the main hope is the education of the Black population who are the main carriers of the disease.

Pollution in fresh waters and estuaries is controlled by the Department of Water Affairs in terms of the Water Act of 1956 and health aspects are controlled by the Department of Health in terms of Health Act No 63 of 1977.

The saline waters of estuaries are commonly used for aquatic sports so it is important to consider the danger to bathers of pathogenic bacteria and viruses. Estuaries become polluted from domestic sewage but unless the flow is continuous, the estuaries do not remain polluted for long for the pathogens die rapidly in salt water particularly if it is sufficiently clear for the transmission of sunlight. Most estuaries are reasonably clean. Nonetheless there is moderate pollution in the Mgeni at Durban and slight pollution in the Swartkops at Port Elizabeth. The Black River at Cape Town is grossly polluted when the sewage works are overloaded or break down but normally the faecal coli counts are low. The lower reaches of the river and the estuary which receive a heavy load of industrial pollutants are little better than an industrial canal. Undoubtedly there will be a tendency for pollution to increase as the towns grow and industry increases, but we have been warned by what has happened in Europe and the controls which already exist can be more strictly enforced as the need arises.

Two-thirds of the world's oil is shipped around the Cape. Serious oil spills have occurred and there is evidence that oil slicks have drifted ashore particularly along the southern Cape coasts and into estuaries near Mossel Bay causing considerable damage (see the report on the Little Brak in chapter 14). The prevention of oil pollution in territorial waters is covered by the South African Act No 67 of 1971. It is administered by the Department of Industries which is also responsible for cleaning-up operations. Unfortunately no satisfactory method of preventing oil slicks from drifting into estuaries has yet been developed.

The degradation of South African estuaries is mainly due to siltation. Many examples were given in chapter 14 and the seriousness of the situation has been stressed earlier in this chapter. There are two basic causes. One is the inward transport of marine sand due to heavy wave action and the inflow of the tide. During periods of low river flow the ebb current is too weak to scour the estuary and the mouth may be completely blocked. Nothing can be done about this unless the mouth is continually dredged and this is extremely expensive as has been found in St Lucia.

The other basic cause is erosion in the drainage basin of the river. This is no new phenomenon. There is geological evidence that it has been going on for thousands of years but it has been balanced by the flushing action of periodic floods. It is the acceleration of erosion by human activities that has upset the balance so that siltation has increased in living memory. Estuaries that were clear and sufficiently deep to form small harbours for coastal traffic have been filled with mud and the wharves are left derelict. In other cases the estuary basins have become so shallow that in times of normal river flow they are less than a metre deep. When the floods come the waters cover the flood plain to abnormal heights washing away bridges and inundating cultivated lands, golf courses and amusement parks that encroach on the flood plain. The parlous condition of many Natal estuaries has been stressed by Begg (1978), who estimates that 45 of the 73 estuaries are heavily silted. Further examples may be seen in the Transkei and the Eastern and Southern Cape.

Soil erosion is one of the worst environmental problems in South Africa. Rabie (1976) documents the many attempts at control. Since the Soil Conservation Act of 1946 was ineffective, a further Soil Conservation Act (No 76 of 1969) was promulgated. It attempted to improve methods of tilling and soil use and to protect the sources of the streams and the riverine vegetation. Attempts have been and are being made to educate the population in better methods of agriculture but, like all farmers the world over, the Black farmers are very conservative. Moreover the Black community evaluates a man's wealth by the number of cattle he keeps with the inevitable result of overstocking. Among other protective measures the Act of 1969 states: 'No land shall be plowed or cultivated and no vegetation (excepting proclaimed weeds and noxious plants) shall be destroyed within 10 m of the edges or banks of rivers, brooks, springs, vleis, marshes, etc . . .' The Act is administered by the Department of Agriculture. Nevertheless the land is often cultivated to the water's edge not only in the Black homelands but also in the White-owned sugar farms of Natal. Thus there is no marginal vegetation on the flood plains to hold the valuable top soil. Photographic evidence showing bank erosion may be seen in Heydorn (1978) and Begg (1978). According to Anderson (1979) this is the major cause of siltation in estuaries. Obviously the law should be more strictly enforced and the canalisation of the filtering swamps prohibited.

Other problems arise from the low and markedly seasonal rainfall. During the dry months of the year many human activities are affected. Farms, urban areas and factories all need more water than is usually available in the rivers. Small dams are reported to be inefficient on account of the high rate of evaporation but many large dams have been built and water from rivers with a large seasonal flow has been diverted to rivers with a poor flow. The Orange River Scheme is a well-known example and work is proceeding on the headwaters of the Tugela and on the sources of the Berg, Breede and Eerste as part of the Three Rivers Scheme.

By such river regulation the needs of irrigation areas and towns are supplied. In some cases, notably in the Tugela, the Great Berg and possibly the Breede, where the salinity of the estuary falls to such low levels during the rainy season that the fauna is impoverished, the estuary benefits as well. But the opposite is also true. There has been an impression in the Department of Water Affairs and among civil engineers that any fresh water that runs through an estuary to the sea is wasted. This was hotly debated in the meetings of the State Commission of Enquiry into the Alleged Threat to Aquatic Life in St Lucia. Although the high salinities in the St Lucia lakes during 1969 and 1970 were disastrous to aquatic life, they at least convinced the authorities of the danger of starving the lakes of river water. However, the idea that estuaries *do* require periodic floods to flush out highly saline water and the accumulated sediment is not generally recognised.

Under the Roman Dutch system of law in South Africa, the *res communes* including the air, running water, the sea and the seashore belong to no one but may be used and enjoyed by all. Estuaries are legally regarded as part of the sea. Thus estuaries and tidal waters up to the high tide mark of storm tides belong to the State President and under the Seashore Act of 1935 the beds and intertidal banks of estuaries are administered by the Department of Agricultural Credit and Land Tenure. In terms of our common law, parts of the foreshore were alienated prior to the Seashore Act of 1935 and further areas may be alienated and thus owned privately. Likewise the State may grant special rights to corporations which amount to alienation. For example parts of the seashore including estuarine areas have been delegated to the South African Railways and Harbours for harbour construction. Such alienation, however, is always limited by the right of the public to enjoy the use of the foreshore. Further, the Minister of Agriculture may let any portion of the seashore so long as the letting is in the interests of the public and does not affect the public's enjoyment of the seashore and the sea. Where the seashore: 'is situated within or adjoins the area of jurisdiction of a local authority, the Minister shall first consult that local authority'. The Seashore Act further proclaims that the land in the sea or on the seashore (including estuaries) that is reclaimed becomes State land. Thus the State may make a grant of such land for the construction of a marina after an investigation. Rigid controls for marinas have been established in consultation with the various government departments and the provincial authorities concerned.

For the reasons stated earlier in this chapter and discussed in more detail in chapter 16, the draining of marshy shores, the dredging of sand or gravel and the dumping of rubble, refuse or materials harmful to estuarine life should be prohibited. Regulations in this regard are now being drafted by the Department of Environmental Conservation and Energy. After consultation with the coastal provinces it is proposed that local authorities shall have the power to construct jetties, launching ramps and similar facilities rather than permit individuals to construct private jetties.

These several laws and regulations apply to the sea and seashores including the 'tidal waters' of estuaries. As noted earlier it is difficult to define the landward limit of the tides. Moreover many estuaries are not tidal when closed by a sandbar (eg Mhlanga in Natal and the Hermanus lagoon and Bot River lagoon in the Cape); tidal action may also be prevented by the construction of a weir (eg the Fafa estuary in Natal and part of the West Kleinemond and Kariega in the Cape). Even the upper reaches of some open, saline estuaries such as Lake St Lucia and the upper lakes of the Kosi Bay and Wilderness systems are non-tidal. It is obvious that to define the limits of an estuary in relation to tidal action rather than by the admixture of fresh water and sea water, makes the interpretation of the laws difficult and allows clever lawyers to distort the intention of the law. Begg (1978) has proposed that closed estuaries be distinguished from open estuaries as 'lagoons'. As explained in chapter 1, the word lagoon is used all over the world as an expanse of sheltered water such as that enclosed by a coral reef, a sheltered part of a bay (eg Langebaan lagoon opening into Saldanha Bay) or merely an expanded part of an open estuary (eg Knysna lagoon). It is a useful word without a restricted meaning and would cause many difficulties if adopted as a legal term.

Estuarine tourist resorts are thriving in South Africa as elsewhere and their popularity is due to aquatic sports such as sailing, water skiing, bathing, fishing and the quiet enjoyment of watching aquatic birds. Such recreation is controlled by the ordinances of the coastal provinces which define certain zones for power boats and water skiing and others for various other activities. The Department of Recreation has also drafted regulations for the zoning of water areas for different forms of recreation. There appears to be an overlapping of control.

Commercial fishing is illegal in South African estuaries with the exception of the Berg and the Olifants. The living resources of the seashores, estuaries and rivers of Natal are controlled by the Natal Provincial Council which delegates its authority to the Natal Parks, Game and Fish Preservation Board. The Cape Provincial Council has a more limited authority, for the living resources on the seashore and those estuaries where there are harbours (eg Knysna estuary, Langebaan Lagoon and the lower reaches of the Great Berg estuary) are controlled by the Sea Fisheries Institute of the Department of Agriculture under the Sea Fisheries Act of 1973. The life in other estuaries is controlled by the Cape Department of Nature and Environmental Conservation. In both provinces, regulations limit the minimum size of game fish that may be taken but as these fish breed in the sea and enter the estuaries every year there is little danger that the population will be depleted by estuarine anglers alone.

This, however, does not apply to bait organisms. Swimming prawns (penaeids) also breed in the sea but the populations in South African estuaries are limited and the catch in Natal is controlled by the Natal Parks Board. Sand prawns *(Callianassa)* and mud prawns *(Upogebia)* are abundant but the banks where both species occur are easily damaged if dug over with a spade. Moreover the important sea grass *(Zostera)* is easily damaged by mechanical action and takes years to recover. Blood worms *(Arenicola)* are in short supply and only occur in certain sandbanks. The use of a spade to collect all these baits is thus illegal and the number that may be taken by each angler is restricted. Legislation is adequate but the main problem is sufficient staff to enforce the regulations.

The whole question of maintaining adequate stocks of bait animals in estuaries is being studied by the Natal Parks Board and the Cape Department of Nature and Environmental Conservation in co-operation with the universities. The present findings are that if each bait area is rested for a year to allow stocks to recover, the supplies of bait can be maintained. Whether this will continue as the number of anglers increases remains to be seen. As in all cases of nature conservation the main threat is the deterioration of the environment.

Several environmental problems have been outlined above with a statement of the controlling authorities

Table 17.3
Summary of estuarine problems, controlling authorities and laws

Problems	Controlling authorities*	Laws
Definition of estuarine limits	Dept. of Agricultural Credit and Land Tenure	Seashore Act of 1935
Soil erosion and siltation	Dept. of Agriculture and Technical Services	Soil Conservation Act of 1969
Canalisation of marshes	Dept. of Agriculture and Technical Services	Soil Conservation Act of 1969
Reclamation of intertidal marshes	Dept. of Agric. Credit and Land Tenure *with* Local Authority	Seashore Act of 1935
Dredging of sand and gravel	Dept of Environ. Planning *with* Dept. of Mines *with* Provincial Authority	Environ. Planning Act of 1966 as amended 1977
Marina construction	Dept. of Environ. Planning *with* Provincial Authority *with* Local Authority	Seashore Act of 1935
Restriction of tidal flow by bridges	National and Provincial Roads Boards	
Water quality – pathogens	Dept. of Health	Public Health Act of 1977
Water quality – water chemistry	Dept. of Water Affairs *with* Provincial Authority	Water Act of 1956; Nature Conservation Ordinance of 1974
Shortage of river water (high salinity)	Dept. of Water Affairs	Water Act of 1956
Oil pollution	Dept. of Industries	Prevention and Combatting of Pollution of the Sea by Oil Act of 1971
Protection of fauna: Natal	Natal Parks Game & Fish Preservation Board;	Natal Conserv. Ordinance of 1974;
Protection of fauna: Cape	Cape Dept. of Nature and Environmental Conservation. *Also* Sea Fisheries Institute	Cape Nature and Environ. Ordinance of 1978; Sea Fisheries Act of 1973
Zoning for acquatic recreation: Natal	Dept of Sport and Recreation. *Also* Natal Parks Board	Under consideration; Natal Prov. Ordinance of 1978
Zoning for acquatic recreation: Cape	*or* Cape Dept. of Nature Conservation	Cape Prov. Ordinance of 1978
Estuarine Nature Reserves		Under consideration

* These authorities were correct in 1979 but some have changed in 1980.

and the laws under which they act. A summary is provided in Table 17.3. The government departments are highly independent and the multiplicity of laws and diversity of control in regard to environmental matters leads to overlapping and delayed decisions. Rabie (1976) discusses the question whether environmental contol should be placed under a single all-embracing Environmental Protection Agency as in the United States. He argues (p 3) that environmental law 'does not, however, constitute a separate part of the law in the sense that it contains separate legal principles: Legal provisions relating to the environment are encountered in many conventional fields of law, such as administrative law, constitutional law, criminal law, tax law, the law of derelict and jurisprudence'. And he goes on to say: 'It seems more satisfactory, at least at the present stage of development of environmental law, to deal separately with separate environmental problems as they are treated in legislation . . .' In regard to administration he concludes (p 8): 'It is simply not practical, nor necessary, to integrate all the separate environmental sections of the numerous government departments (even if restricted to the most important central government departments) that deal with environmental affairs into one gigantic department of the environment as has been done in the United States with the creation of the Environmental Protection Agency in 1970'.

The United States National Environmental Policy Act of 1970, discussed earlier in this chapter was, as Rabie (1976) comments, enacted with the purpose of compelling government agencies 'to consider environmental values in their decision-making processes and to make public, *inter alia,* the potential impact which their actions may have on the environment'. In South Africa, too, public indignation has been aroused by the decisions taken by government departments and semi-government agencies which seriously damage the environment or might do so. Details have been given in chapter 14 of the damage done to Lake St Lucia by the damming of rivers and the planting of pine forests on the eastern shores; of the damage to the 'sanctuary' at Richards Bay by the canalisation of the Mhlatuzi River; by the dumping of rubble under the national road bridge over Bushman estuary and by the solid approaches to the bridges over Keurbooms estuary. The National Transport Commission is now planning a six-lane highway through the Wilderness lakes, the remnants of the Tsitsikamma Forest and another bridge over the Keurbooms estuary at Plettenberg Bay. As Rabie remarks: 'the government itself has become a major contributory factor to pollution and the depletion of resources'. The advice of conservationalists if asked at all, often goes unheeded. No reasons for the decisions are made public. Although criticisms of Environmental Impact Assessment as it exists in the United States and Australia have been noted earlier, some form of Impact Assessment is urgently required in South Africa. Moreover the assessment should be made at an early stage before a department has spent time and money on detailed plans.

Co-ordination between government departments in regard to environmental matters takes place at cabinet level in the Planning Advisory Council to the Prime Minister. Guide plans for land use in rural and urban areas are prepared by the Department of Environmental

Planning and Energy in consultation with the relevant government and provincial departments, local authorities and private organisations such as the Council for the Habitat or the Soil Conservation Society. Such guide plans are available for inspection and are discussed by the Prime Minister's Planning Advisory Council but the Minister of the Environment has no execuive authority and cannot himself prevent environmentally detrimental actions by other departments.

Co-ordination with foreign governments in regard to lands, rivers and estuaries on the boundaries of South Africa, eg Mocambique, Transkei (and Namibia when its independence is established) take place through the Department of External Affairs. Up to the present, the coastal 'homelands', eg Kwazulu and the Ciskei at present adopt the relevant provincial ordinances but after independence they may at any time promulgate their own environmental laws.

In addition to co-ordination at government level, there is an urgent need for co-ordination at the local level. What co-ordination there is, is *ad hoc* as a problem arises and planning is piecemeal. As noted earlier, an estuarine ecosystem is not an independent unit and is affected by events such as erosion and pollution in the river, the condition of the sea mouth, and the migration of marine fish. There is need for a statutory body such as a River Authority to integrate plans and methods of control from the source of a river to its mouth. Local communities should be represented on such an Authority as well as government and provincial departments for it is their environment that is affected. The Outeniqua Land Trust and the Swartkops Trust are steps in this direction.

Southern Africa is famous for its terrestrial nature reserves and the abundance and variety of its wild life. Not only do they provide an international tourist attraction they also provide facilities for ecologists to determine the complex interactions within the ecosystem upon which control must be based. Moreover they afford an opportunity for our own citizens to enjoy the beauties of nature and rest from the stresses of city life. But we need marine and estuarine reserves as well. The richness and diversity of aquatic and bird life in Morrumbene estuary near Imhambane, Inhaca Island in Delagoa Bay, Kosi Bay in Zululand, Mngazana in the Transkei, Knysna on the Tsitsikama coast and Langebaan lagoon in Saldanha Bay are second to none. Studies of the ecology of these sheltered waters have provided us with a better appreciation of what an unspoilt estuary or lagoon can be, than is available anywhere in the world. We have a yardstock against which to measure the detrimental effects of pollution, siltation and urban development. Unless we act now and declare a representative series of estuaries as nature reserves in which no exploitation or development is permitted, our children will wonder what an unspoilt estuary was like.

REFERENCES

ALEXANDER, W.J.R. 1979. Sedimentation of estuaries: causes, effects and remedies. *Abstract 4th (S.Afr.) natl. oceanog. Symp., Cape Town, July 1979.* CSIR, Pretoria. (Also): *S.Afr. J. Sci.* 75: 569.

ANON. 1977. *Report on the fisheries in New South Wales for 1975-1976.* Govt. Printer, NSW.

BANTA, J.S. 1976. Constitutional issues and management. *Oceanus* 19(5): 64-70.

BARADA, W. & W. PARTINGTON 1972. *Report of investigation of the environmental effects of private waterfront canals.* Environmental Information Centre; Winter Park, Florida.

BAXA, J.V. 1978. *An evaluation procedure for short-term planning.* Transport Planning Research Report **TPRR9**; CSIR, Pretoria.

BEGG, G.W. 1978. *The estuaries of Natal,* Natal Town and Regional Planning Commission, Pietermaritzburg.

CHAPMAN, V.J. 1977. *Ecosystems of the world. I: Wet coastal ecosystems.* Elsevier Sci. Publ. Co, Amsterdam.

CHMELIK, F.B., VAN LOGGENBERG, B.J., GRINDLEY, J.R. & A. DARRACOTT 1974. Economic model for estuarine valuation. *Proc. 10th ann. Rep. mar. technol. Soc., Washington, DC* 233-275.

CLARK, J.R. 1977. *Coastal ecosystem management. A technical manual for the conservation of coastal zone resources.* Wiley & Sons, New York.

COMMONWEALTH SCIENTIFIC AND INDUSTRIAL RESEARCH ORGANISATION OF AUSTRALIA. *Fisheries and Oceanog. Rep. 1974-76.* Cronulla, NSW.

DAY, J.H. 1974. The ecology of Morrumbene estuary, Mocambique. *Trans. roy. Soc. S.Afr.* **41**(1): 43-97.

DAY, J.H., MILLARD, N.A.H. & G.J. BROEKHUYSEN 1954. The ecology of South African estuaries. Part IV: The St Lucia system. *Trans. roy. Soc. S.Afr.* **34**(1): 129-156.

DE GROOT, A.J., SALOMONS, W. & E. ALLERSMA 1976. Processes affecting heavy metals in estuarine sediments. *In:* J.D. Burton & P.S. Liss (eds), *Estuarine chemistry.* Academic Press, London.

DEPARTMENT OF THE ENVIRONMENT 1974. *Report of a river pollution survey of England and Wales, vol 3. Discharges of sewage and industrial effluents to estuaries and coastal waters excluded from volume 2 of the 1970 survey; and a summary of all effluents recorded in the survey and to other coastal waters.* HMSO, ISBN 0 11 750843 8.

DEPARTMENT OF THE ENVIRONMENT: CENTRAL UNIT ON ENVIRONMENTAL POLLUTION. 1978. *Pollution control in Great Britain: How it works. A review of legislative and administrative procedures.* HMSO, ISBN 0 11 7513679.

DEPARTMENT OF THE ENVIRONMENT AND THE WELSH OFFICE 1974. *A background to water reorganisation in England and Wales.* HMSO, ISBN 11 750570 6.

DE SYLVA, D.P. 1969. Trends in marine sport fisheries research. *Trans. Am. Fish. Soc.* **98**(1): 151-169.

EDMUNDS, S. 1978. Trade-offs in assessing environmental impacts. *Environmental Management* 2(5): 391-401.

EMANUEL, A. 1977. Conservation at Richards Bay. *Environment RSA* 4(12): 5-7.

FRAZER, G.A. 1979. Limited entry: experience of the British Columbia salmon fishery. *J. Fish. Res. Bd. Can.* **36**(7): 754-763.

GENADE, A.B. 1973. A general account of certain aspects of oyster culture in Knysna estuary. *Abstract 1st (S.Afr.) natl. oceanog. Symp., Cape Town, July 1973.*

HAEDRICH, R.L. & C.A.S. HALL 1976. Fish and estuaries. *Oceanus* 19(5): 55-63.